DNA Damage and Repair

Contemporary Cancer Research
Jac A. Nickoloff, Series Editor

Breast Cancer, edited by **Anne M. Bowcock,** 1998

DNA Damage and Repair, Volume 2: DNA Repair in Higher Eukaryotes, edited by **Jac A. Nickoloff** and **Merl F. Hoekstra,** 1998

DNA Damage and Repair, Volume 1: DNA Repair in Prokaryotes and Lower Eukaryotes, edited by **Jac A. Nickoloff** and **Merl F. Hoekstra,** 1998

DNA Damage and Repair

Volume 1: DNA Repair in Prokaryotes and Lower Eukaryotes

Edited by

Jac A. Nickoloff

University of New Mexico, Albuquerque, NM

and

Merl F. Hoekstra

ICOS Corporation, Bothell, WA

Humana Press Totowa, New Jersey

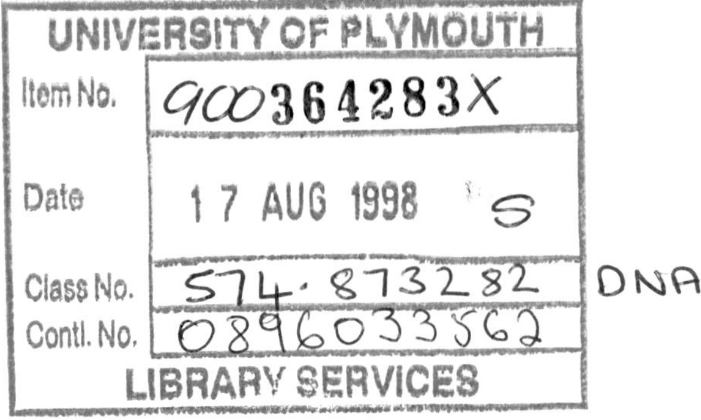

999 Riverview Drive, Suite 208
Totowa, New Jersey 07512

This publication is printed on acid-free paper. ∞
ANSI Z39.48-1984 (American Standards Institute) Permanence of Paper for Printed Library Materials.

Cover design by Patricia F. Cleary.

For additional copies, pricing for bulk purchases, and/or information about other Humana titles, contact Humana at the above address or at any of the following numbers: Tel: 973-256-1699; Fax: 973-256-8341; E-mail: humana@mindspring.com, or visit our Website: http://humanapress.com

Printed in the United States of America. 10 9 8 7 6 5 4 3 2 1

Library of Congress Cataloging in Publication Data

DNA damage and repair/edited by Jac A. Nickoloff and Merl F. Hoekstra
p. cm.—(Contemporary cancer research)
Includes index.
Contents: v. 1. DNA repair in prokaryotes and lower eukaryotes—v. 2 DNA repair in higher eukaryotes.
ISBN 0-89603-356-2 (v.1: alk. paper.—ISBN-89603-500-X (v.2: alk. paper)
1. DNA repair. I. Hoekstra, Merl F. II. Series.
[DNLM: 1. DNA Repair. 2. DNA Damage. 3. DNA—physiology. 4. Prokaryotic Cells. 5. Eukaryotic Cells. QH 467 D629 1998]
QH467.D15 1998
572.8'6—dc21
DNLM/DLC
for Library of Cengress

97-28562
CIP

Preface

DNA Damage and Repair grew from a conversation between the editors at the 1994 Radiation Research meeting in Nashville, TN. At that time, Errol Friedberg, Graham Walker, and Wolfram Seide were just releasing the second edition of the outstanding textbook *DNA Repair and Mutagenesis,* and many of the human genes involved in nucleotide excision repair were being described and their connection to xeroderma pigmentosum and trichothiodystrophy was being outlined. Many of these mammalian genes had homologs in model systems and investigators were making connections between NER and transcription. This was an exciting and confusing time. Nomenclature for radiation repair genes lacked consistency and new genes (and even pathways) were being described in many organisms.

The result of our conversation was the decision to ask experts working with various organisms and in subdisciplines of DNA repair whether a volume reviewing these major advances might be helpful. The resounding answer was yes, quickly followed by the question of why anyone would want to take on this task. We took on the challenge with the goal of collecting peer-reviewed chapters on a variety of repair topics, from a variety of organisms, written by experts in the field. It has been quite a learning experience about the challenges of editorship and the breadth of DNA repair.

We thank all of the authors for their fine submissions. We also owe a great debt to the many individuals who made suggestions about our initial list of chapter titles and/or reviewed and corrected chapters, including Brenda Andrews, David Ballie, James Cleaver, Pricilla Cooper, Elena Hildago, Lawrence Grossman, Philip Hanawalt, John Hays, Etta Kafer, Robert Lahue, Richard Kolodner, David Lilley, Michael Liskay, Lawrence Loeb, Kenneth Minton, Rodney Rothstein, Roy Rowley, Leona Samson, Alice Schroeder, Gerry Smith, Michel Sicard, Wolfram Siede, and Graham Walker. A comparison of this list with the table of contents shows that many of these individuals did double-duty, both as a chapter contributor and as reviewer. We owe special thanks to Phil Hanawalt for providing an excellent overview of the field and for his continuing encouragement. We are in debt to these individuals for their time and assistance, without which this volume would not have been possible.

During the formation of the book, a variety of exciting breakthroughs occurred and we are pleased that a number of authors were able to make last-minute changes to include these discoveries. In particular, the fields of mismatch repair, cell-cycle checkpoints, and DNA helicases have grown significantly by the discovery of disease genes, such as *MLH, MSH,* and *ATM,* and the genes involved in Bloom's and Werner's syndromes. Unfortunately we were not able to cover all topics and regret that we could not dedicate individual chapters for such organisms as bacteriophage and *Drosophila* and on such processes as X-ray crystallography/NMR and the impact this has on DNA repair. Nevertheless, we are grateful to all that have participated in this effort: It started modestly at 25–30 chapters, but rapidly grew to two volumes, a testament to the explosive growth that the field has enjoyed in recent years.

Finally, we thank our families for their patience and understanding: Denise, Jake, Ben, Courtney, Debra, Brad, Lauren and Brielle.

Jac A. Nickoloff
Merl F. Hoekstra

Contents

Contents for the Companion Volume DNA Damage and Repair, Volume 2: *DNA Repair in Higher Eukaryotes*

Contributors

YUNGCHAN AHN • *Department of Biochemistry, Johns Hopkins University, Baltimore, MD*

JEFFREY B. BACHANT • *Department of Biochemistry, Baylor College of Medicine, Houston, TX*

JOHN R. BATTISTA • *Department of Microbiology, Louisiana State University, Baton Rouge, LA*

RICHARD L. BENNETT • *Department of Microbiology, Cornell University Medical College, New York, NY*

RICHARD BOCKRATH • *Department of Microbiology and Immunology, Indiana University School of Medicine, Indianapolis, IN*

ANNE B. BRITT • *Section of Plant Biology, University of California, Davis, CA*

ANTONY M. CARR • *MRC Cell Mutation Unit, University of Sussex, Brighton, UK*

DANA CARROLL • *Department of Biochemistry, University of Utah, Salt Lake City, UT*

ALLYSON COLE-STRAUSS • *Department of Pharmacology, Thomas Jefferson University, Philadelphia, PA*

GRAY F. CROUSE • *Department of Biology, Emory University, Atlanta, GA*

STEPHEN J. ELLEDGE • *Department of Biochemistry, Baylor College of Medicine, Houston, TX*

BEVIN P. ENGELWARD • *Department of Molecular and Cellular Toxicology, Harvard School of Public Health, Boston, MA*

DAVID O. FERGUSON • *Department of Microbiology, Cornell University Medical College, New York, NY*

DOMINIC J. F. GRIFFITHS • *MRC Cell Mutation Unit, University of Sussex, Brighton, UK*

LAWRENCE GROSSMAN • *Department of Biochemistry, Johns Hopkins University, Baltimore, MD*

PHILIP C. HANAWALT • *Department of Biological Sciences, Stanford University, Stanford, CA*

PHIL S. HARTMAN • *Department of Biology, Texas Christian University, Fort Worth, TX*

ZAFER HATAHET • *Department of Microbiology and Molecular Genetics, University of Vermont, Burlington, VT*

MERL F. HOEKSTRA • *ICOS Corporation, Bothell, WA*

WILLIAM K. HOLLOMAN • *Department of Microbiology, Cornell University Medical College, New York, NY*

HIROKAZU INOUE • *Faculty of Science, Saitama University, Saitama, Japan*
ETTA KAFER • *Institute of Molecular Biology and Biochemistry, Simon Fraser University, Burnaby, British Columbia, Canada*
BÖRRIES KEMPER • *Institute of Genetics, University of Cologne, Germany*
ERIC B. KMIEC • *Department of Pharmacology, Thomas Jefferson University, Philadelphia, PA*
WALTER H. KOCH • *Molecular Biology Branch, Food and Drug Administration, Washington, DC*
SANFORD A. LACKS • *Biology Department, Brookhaven National Laboratory, Upton, NY*
CHIEN-LIANG LIN • *Department of Neurology, Johns Hopkins University, Baltimore, MD*
LAWRENCE A. LOEB • *Department of Pathology, University of Washington School of Medicine, Seattle, WA*
DAVID LYDALL • *Department of Molecular and Cellular Biology, University of Arizona, Tucson, AZ*
M. G. MARINUS • *Department of Pharmacology and Molecular Toxicology, University of Massachusetts Medical School, Worcester, MA*
GREGORY S. MAY • *Department of Cell Biology, Baylor College of Medicine, Houston, TX*
JEFFREY H. MILLER • *Department of Microbiology and Molecular Genetics, University of California, Los Angeles, CA*
GREG NELSON • *Loma Linda University Cancer Institute, Loma Linda, CA*
TERRY G. NEWCOMB • *Department of Pathology, University of Washington School of Medicine, Seattle, WA*
JAC A. NICKOLOFF • *Department of Molecular Genetics and Microbiology, University of New Mexico School of Medicine, Albuquerque, NM*
KENAN ONEL • *Department of Microbiology, Cornell University Medical College, New York, NY*
LENE JUEL RASMUSSEN • *Department of Chemistry and Life Sciences, Roskilde University, Roskilde, Denmark*
MARA H. RENDI • *Department of Microbiology, Cornell University Medical College, New York, NY*
MICHAEL L. RICE • *Department of Pharmacology, Thomas Jefferson University, Philadelphia, PA*
MATTHEW S. SACHS • *Department of Chemistry, Biochemistry, and Molecular Biology, Oregon Graduate Institute of Science and Technology, Portland, OR*
LEONA SAMSON • *Department of Molecular and Cellular Toxicology, Harvard School of Public Health, Boston, MA*
ALICE L. SCHROEDER • *Department of Genetics and Cell Biology, Washington State University, Pullman, WA*
WOLFRAM SIEDE • *Department of Radiation Oncology, Emory University School of Medicine, Atlanta, GA*

GERALD R. SMITH • *Fred Hutchinson Cancer Research Center, Seattle, WA*
MICHAEL P. THELEN • *Lawrence Livermore Laboratory, Livermore, CA*
SUSAN S. WALLACE • *Department of Microbiology and Molecular Genetics, University of Vermont, Burlington, VT*
TED WEINERT • *Department of Molecular and Cellular Biology, University of Arizona, Tucson, AZ*
BERNARD WEISS • *Department of Pathology, University of Michigan Medical School, Ann Arbor, MI*
DAVID M. WILSON III • *Department of Molecular and Cellular Toxicology, Harvard School of Public Health, Boston, MA*
ROGER WOODGATE • *Section on DNA Replication, Repair and Mutagenesis, National Institute of Child Health and Human Development, Bethesda, MD*

1
Overview

Philip C. Hanawalt

In recent years, the field of DNA repair has attained widespread recognition and interest appropriate to its fundamental importance in genomic maintenance. An essential set of repair pathways must be operative in all living systems to maintain genomic stability in the face of the natural endogenous threats to DNA as well as those resulting from cellular exposures to radiation and chemicals in the external environment. The intrinsic chemical lability of the DNA molecule poses a formidable threat to its persistence and even the polymerases that replicate DNA occasionally make mistakes that must be rectified. Fortunately, the inherent redundancy of the genetic message assured by the two complementary strands of the duplex DNA molecule facilitates the recovery of information through excision repair when one of the strands is damaged, or when incorrect base pairings or small loops of unpaired bases are present. The fundamental research that led to the discovery of excision repair in *Escherichia coli* in the early 1960s has now developed to the point that we realize that DNA repair is ubiquitous and essential for life. Furthermore, we have learned that DNA repair interfaces in some manner with each of the other cellular DNA transactions, including replication, transcription, recombination, and regulation of the cell cycle. We are also finding remarkable sequence and functional homologies among the essential repair enzymes as different species are compared. This has spurred the rate of discovery, most remarkably through the parallel analyses in yeast and mammalian systems in which the complex multicomponent systems for excision repair are virtually identical.

The collection of chapters in this remarkable two-volume treatise attests to the explosive rate at which the DNA repair field is evolving. Recently published texts in this field are in need of updating as soon as they hit the bookshelves. Fortunately, the cutting edge researchers who have contributed chapters in the immediate area of their expertise are in a good position to anticipate significant upcoming developments in their respective subfields. The unique perspectives of researchers attacking similar problems using diverse approaches and different biological systems constitute another important strength of the present collection of chapters. For general background and an introduction to the field of DNA repair and mutagenesis, the reader is referred to the comprehensive treatment by Friedberg et al. *(5)*. An excellent set of current reviews covering excision repair *(15,18)*, mismatch repair *(13)*, and relationships between repair and transcription *(4)* can be found in the 1996 volume of *Annual Reviews of Biochemistry*.

From: DNA Damage and Repair, Vol. 1: DNA Repair in Prokaryotes and Lower Eukaryotes
Edited by: J. A. Nickoloff and M. F. Hoekstra © Humana Press Inc., Totowa, NJ

In this short overview, I would like to step back and peruse the broad field of DNA repair to ask: What is left to learn? What are some of the important new directions in which the field is moving—or should be moving? What is the significance of this field for human health? I will attempt to address just a few of these questions to provoke thought and debate as the reader approaches the many facets of the subject covered in the following comprehensive chapters.

To set the stage, we should begin with a cursory review of the sorts of deleterious alterations that can occur in DNA and the various classes of mechanisms by which those alterations can be repaired or accommodated in the cell. Nearly twenty years ago, my colleagues and I published a comprehensive review on "DNA repair in bacteria and mammalian cells" for *Annual Reviews of Biochemistry (7)*. At that time most of the cellular pathways for repairing or tolerating DNA damage were already known but the detailed enzymology had not been elucidated. We now know essentially all of the proteins required for each of the DNA repair pathways and for most of them we can also assign them unique roles *(5,13,15,18)*. However, there is still much to be learned about how the damaged DNA substrate is recognized.

We define damage as "any modification of DNA that alters its coding properties or its normal function in replication or transcription." Thus, some minor base modifications or base replacements might simply alter the coding sequence, whereas other types of damage, such as bulky chemical adducts or UV-induced cyclobutane pyrimidine dimers, may distort the DNA structure and interfere with the translocation of polymerases. To be repairable, the damage must be recognized by a protein that can initiate a sequence of biochemical reactions leading to its elimination and the restoration of the intact DNA structure. It is a formidable challenge to the lesion-recognition enzymes to detect the damaged sites in the context of the variable and dynamic structure of the normal DNA molecule. The normal DNA structure includes nucleotide-sequence-dependent bends and kinks, as well as unique secondary structures, such as Z-DNA. The DNA in cells is also decorated with a variety of ligands, including tightly bound proteins that may alter its intrinsic structural features and that would be expected to encumber any simple damage recognition system. The repair machinery must rigorously avoid mistaking a normal DNA–protein complex for damaged DNA. The detection machinery must also be able to distinguish a bona fide lesion from a normal variant of the DNA structure. An additional yet related problem is raised by the requirement that the first step in the excision-repair sequence involves cutting the DNA. It could be potentially disastrous if there were many individual nucleases diffusing freely about the cell in search of generalized structural distortions in DNA. The nucleotide excision-repair mechanism employs multienzyme complexes rather than single proteins to recognize broad classes of structure-distorting lesions. Thus, in the *E. coli* system the individual polypeptides encoded by the *uvrA*, *uvrB*, and *uvrC* genes are not independently operating nucleases. Lesion recognition requires a complex interaction between UvrA and UvrB. UvrB is loaded onto the DNA at the lesion site, and only then does the UvrC protein (the limiting element with only a dozen copies per cell) join the UvrB-DNA complex to unleash cryptic nuclease activities to produce dual incisions bracketing the lesion to initiate the excision repair process.

In the case of base-excision repair, the initiation steps are simpler and evidently more specific. The general mechanisms appear to require the swinging out of the altered or incorrect base from the DNA helix so that its detailed dimensions and charge structure can be assessed in a "pocket" of the relevant enzyme. Such a recognition scheme also operates in the direct repair of cyclobutane pyrimidine dimers by photolyase. In that case the catalytic cofactor that splits the dimer is found in the pocket and is activated by light once the dimer has been bound. It is likely that the photon-catalyzed reversal of pyrimidine dimers *in situ* was one of the earliest repair mechanisms to emerge as life evolved on the primordial earth. The intense flux of ultraviolet light from the sun, unattenuated by an ozone layer, must have been the predominant threat to the survival of primitive life forms.

Some types of damage to DNA may not be recognized as such and may have no deleterious consequences. Thus, phosphotriesters in mammalian DNA do not appear to be subject to repair and do not appear to have any adverse effects for the cell. On the other hand, in bacteria the phosphotriesters are repaired and are also utilized to warn the cell of the presence of DNA alkylating agents that cause them so that an adaptive response to upregulate repair of alkylated DNA is triggered. A number of the enzymatic pathways for dealing with environmental threats to cellular DNA appear to be similarly inducible whereas, not surprisingly, a constitutive level is maintained of those enzymes needed to repair endogenous damage to DNA.

The principal endogenous or so-called spontaneous DNA lesions are caused by deamination of cytosine to uracil (or methyl cytosine to thymine), loss of purines to yield abasic sites, and reactive oxygen species that produce strand breaks directly but also at least 20 different sorts of base damage *(10,12)*. Notable are thymine glycols and 8-oxo-dG, which are repaired by the base excision repair pathway. However, it is important to remember that the other sorts of base damage as well as the strand breaks may also contribute to the biological consequences. Some potential problems are dealt with at the nucleotide level, even before incorporation into DNA might occur from precursor pools. Thus, the dUTP levels are modulated by dUTPases, and a major product of endogenous reactive oxygen species, 8-oxo-dGTP, is converted to the benign 8-oxo-dGMP by a dedicated phosphatase. The mere existence of the latter pathway attests to the need to control the level of damage inflicted by endogenous reactive oxygen species.

One of the most important excision-repair schemes is mismatch repair, which reduces the error rate in cellular DNA replication by three orders of magnitude. Destabilization of tracts of simple repetitive DNA occurs in bacteria defective in mismatch repair genes. There also may be important interactions between mismatch repair and other repair pathways, notably nucleotide excision repair. Deficiency in mismatch repair has been shown to attenuate transcription-coupled repair in human cells as well as in bacteria. The importance of these connections is highlighted by the implication of the human MutS and MutL mismatch-repair homologs in hereditary nonpolyposis colorectal cancer. The enhanced tumorigenesis is likely caused by the resultant microsatellite instability when mismatch repair is defective *(13)*.

Among the more impressive successes in recent years have been the development of cell-free systems from bacteria, yeast, and mammalian cells that carry out all of the essential steps in excision repair; indeed, such soluble DNA repair factories have now been reconstituted from purified proteins *(15,18)*. These successes have spawned models in which the requisite proteins are assembled into large, unwieldy complexes termed "repairasomes" that presumably diffuse through the cell in search of defective sites in the genomic DNA. This would be analogous to taking the lumber company into the woods in search of trees to cut. Models are usually drawn to show repair enzymes diffusing up to (or along) the DNA duplex. It is notable, however, that these soluble in vitro biochemically reconstituted repair reactions are remarkably inefficient, as are many other such reconstituted systems, including those for transcription. It is likely, indeed certain, that in the intact cell these transactions are carried out by spatially organized arrays of enzymes on surfaces (e.g., the nuclear matrix) and in subcellular compartments. These features could enhance the efficiency and certainty of repair by reducing a three-dimensional diffusion search for lesions to two dimensions, and/or by concentrating particular repair events to those genomic domains in which the repair process is essential to normal cellular functions. Recent studies have hinted that on recognition of lesions, the damaged DNA may be recruited to the nuclear matrix, where the repair factories may reside *(8)*. The damaged DNA once repaired would then be released from these sites (Fig. 1). An important emerging area for research is the elucidation of the intracellular localization of repair events.

We have learned a lot in recent years about the complex relationships between DNA damage and biological endpoints, such as mutagenesis, cell death, and tumorigenesis. We have learned that mutagenesis is under genetic control—thus, for example, ultraviolet (UV) light is not mutagenic in *E. coli* deficient in the *umuC* gene, evidently because almost no translesion DNA synthesis occurs. We have also been surprised to learn that cells carrying damaged genomes may choose to commit suicide, a process technically termed apoptosis, although severely damaged cells may be nonviable in any case. Apoptosis embodies a concept that is difficult to understand in the context of freely living cells and seemingly at odds with the idea of cellular competition for survival. Why should a single free-living cell exhibit such seemingly altruistic behavior? The answer is probably that apoptosis is a process that has evolved to promote development and maintenance of tissues in multicellular organisms; and that in fact it has no relevance to cells in culture. When a tissue is damaged, such as skin from intense solar exposure, the tissue must be remodeled. The dead and damaged cells must be removed so that a new epidermal layer can be assembled. The pathways to apoptosis are complex and under genetic control in damaged cells just as in the programmed cell death that occurs during normal embryonic development.

Susceptibility to cancer can be the consequence of a complex interplay between intrinsic hereditary factors and persisting damage to DNA. We need to learn the nature of these interactions as well as the genetic defects that confer enhanced risk. In xeroderma pigmentosum (XP) the increased risk of skin cancer correlates with a defect in nucleotide excision repair (NER). In Cockayne syndrome (CS) a specific defect in the subpathway of transcription-coupled DNA repair (TCR) does not predispose the patients to the sunlight-induced skin cancer characteristic of XP. The TCR pathway is targeted to lesions in the transcribed strand of expressed genes that arrest the translocation of RNA polymerase. The demonstration of TCR in rodent cells, which are generally deficient in the global genomic DNA-repair pathway, indicates that UV resistance correlates with repair of cyclobutane pyrimidine dimers in expressed genes rather than with global NER. The other major lesion produced by short wavelength UV light, the pyrimidine (6-4) pyrimidone photoproduct, is more efficiently recognized by excision

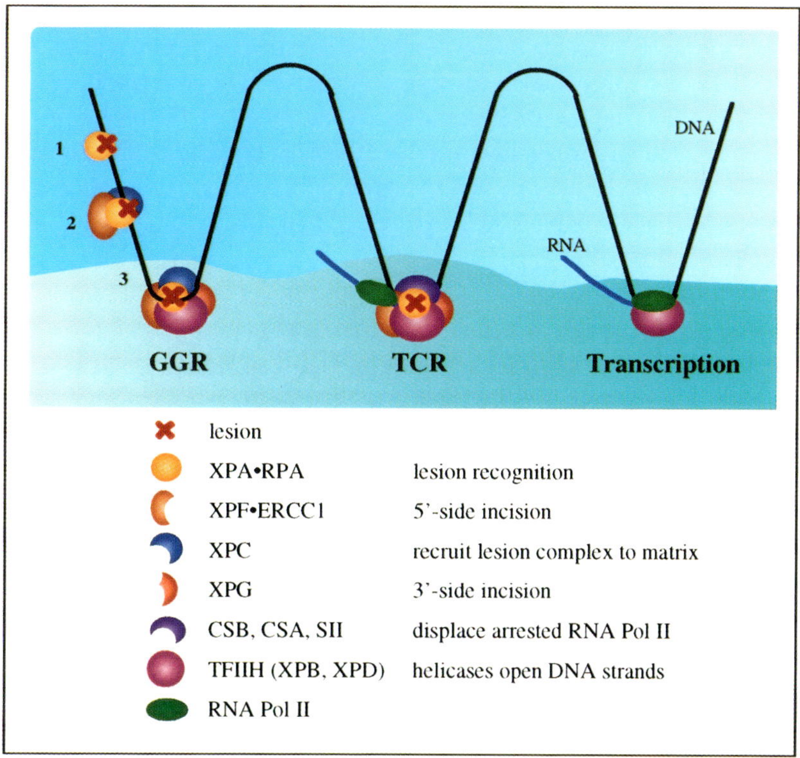

Fig. 1. A conceptual model for the localization of nucleotide excision repair in the nuclear matrix. Transcription occurs in association with the nuclear matrix, as shown, and requires a number of factors for its initiation, including TFIH, which is also essential for excision repair. In the model we assume that TFIIH is localized in the matrix to carry out its dual responsibilities in transcription and repair. When transcription is arrested at a lesion, various factors are required to transiently displace the polymerase and facilitate access of repair enzymes, including those that recognize the lesion and those that produce incisions in the damaged strand to initiate repair. As shown, the XPC protein is not required for transcription-coupled repair (TCR) but it is essential for global genomic repair (GGR). For GGR, we postulate that lesion recognition takes place in regions not associated with the matrix (Step 1), but that additional proteins are then employed (Step 2) to recruit the damaged DNA segments to "repair factory" sites in the nuclear matrix (Step 3) to complete the repair process. Once repaired, these genomic regions would then be released from the matrix. (Figure prepared by Graciela Spivak).

repair enzymes than is the cyclobutane pyrimidine dimer, such that the TCR of this lesion is usually masked by efficient global repair. It is important to consider genomic locations of DNA damage for a meaningful assessment of the biological importance of particular DNA lesions *(6)*.

Mutations in the p53 tumor suppressor genes occur in most human tumors. In the cancer-prone Li-Fraumeni syndrome, fibroblasts expressing only mutant p53 exhibit less apoptosis and are more UV resistant than are normal human cells. The p53-defective cells are deficient in global NER but have retained TCR. The loss of p53 function may lead to genomic instability by reducing the efficiency of global NER, whereas cellular survival may be assured through the operation of TCR and the elimination of apoptosis *(3)*. A specific role for the wild-type p53 gene product in the induction of global NER and apoptosis has been demonstrated in p53 homozygous mutant fibroblasts into which an exogenous wild-type p53 gene has been introduced under control of a tetracycline (Tet)-repressible promoter. Withdrawal of Tet results in the induction of wild-type p53 gene expression, transcriptional activation of p21, and cell-cycle checkpoint activation. Induction of wild-type p53 results in the recovery of normal levels of global NER and enhanced UV-induced apoptosis, compared to the NER-deficient p53 mutant cells *(3a)*.

The arrest of transcription at unrepaired lesions may be a signal for p53 stabilization and apoptosis *(11,19)*. In the case of CS, deficient in the removal of transcription blocking lesions, apoptosis is induced by very low UV doses, thus eliminating many cells that might otherwise undergo oncogenic transformation. Of course, dead cells do not form tumors. Those CS cells that survive are fully proficient in global genome repair. This may explain the lack of sunlight-induced tumors. In contrast, the cells from xeroderma pigmentosum, complementation group C, are proficient in TCR, so survival is high; unfortunately, these cells are defective in global genome repair so a high level of mutation may be induced. Cells from CS patients have also been shown to be deficient in TCR of certain lesions, including thymine glycol, produced by ionizing radiation *(9)*. Ionizing radiation produces lesions largely through oxidation by free radicals that attack DNA. The same types of free radicals are generated endogenously as byproducts of oxidative metabolism in nonirradiated cells. The characteristic developmental problems in CS may be the consequence of inappropriate apoptosis initiated by endogenous DNA damage in expressed genes *(1,6)*.

The importance of the TCR pathway to normal human development is highlighted by the clinical features of Cockayne syndrome patients *(14)*. On the other hand, the clinical consequences of xeroderma pigmentosum, except for the cases in which CS is also present, are not as severe; with the notable exception of the adverse response to sunlight. The lack of a major predisposition to internal cancers in XP patients assures us that there is not a substantial amount of carcinogenic DNA damage from endogenous sources that requires the intervention of nucleotide excision repair for its amelioration. However, the revelation that some rare lesions resulting from reactive oxygen species are subject to nucleotide excision repair has led to the suggestion that such lesions, if unrepaired, may cause the neurological degeneration that typically accompanies later development in the most severe XP patients *(16)*.

It is important that we continue to analyze DNA damage processing in so-called simple bacterial model systems. There have been some recent surprises, such as the discoveries that the genes defective in Bloom's syndrome and in Werner's syndrome in humans are both homologs of the *E. coli recQ* gene that was originally isolated as a gene that, when mutated, confers resistance to the lethality caused by thymidine starvation (i.e., thymineless death). RecQ has been shown to be a helicase and, in fact, there are three homologues of the *recQ* gene in humans. Adding to the intrigue is the fact that in *E. coli* RecQ is implicated in the very minor *recF* recombination pathway. Recent studies suggest that an important—if not the most important—role of *recF* is in enabling the resumption of DNA replication at arrested replication forks, such as at lesions and following their repair *(2)*. The situation of a blocked DNA polymerase complex at a lesion may be quite analogous to that of a blocked transcription complex. The polymerase encumbers access of repair enzymes to the lesion until it is displaced; furthermore, the DNA strands must be reannealed at the lesion site before excision repair can take place. It is likely that basic studies of the process of DNA polymerase displacement, repair, and replication restart in *E. coli* will help us to understand how the process might operate in mammalian cells.

New genes implicated in DNA repair continue to surface as genomes are sequenced and cancer susceptibility genes are characterized. Thus, the BRCA1 and BRCA2 genes that predispose to breast cancer, if either one is inherited in mutated form, have been shown to interact with RAD51, so they both may be involved in the same DNA-repair pathway *(17)*. Mouse embryos lacking BRCA2 are extremely sensitive to ionizing radiation. Once again, a possible implication is that endogenous DNA damage caused by reactive oxygen species may be the culprit in the mutation cascade from normal cells to tumors. In the realm of risk assessment it is becoming increasingly important to determine the relative contributions to DNA damage from endogenous vs environmental factors. Such understanding should lead to new approaches to cancer prevention.

REFERENCES

1. Cooper, P. K., T. Nouspikel, S. G. Clarkson, and S. A. Leadon. 1997. Defective transcription-coupled repair of oxidative base damage in Cockayne syndrome patients from XP group G. *Science* **275,** 990–993.
2. Courcelle, J., C. Carswell-Crumpton, and P. C. Hanawalt. 1997. *recF* and *recR* are required for the resumption of replication at DNA replication forks in *Escherichia coli*. *Proc. Natl. Acad. Sci. USA* **94,** 3714–3719.
3. Ford, J. M. and P. C. Hanawalt. 1995. Li-Fraumeni syndrome fibroblasts homozygous for p53 mutations are deficient in global DNA repair but exhibit normal transcription-coupled repair and enhanced UV-resistance. *Proc. Natl. Acad. Sci. USA* **92,** 8876–8880.

3a. Ford, J. M. and P. C. Hanawalt. 1997. Expression of wild-type p53 is required for efficent global genomic nucleotide excision repair in UV-irradiated human fibroblasts. *J. Biol. Chem.*, in press.

4. Friedberg, E. C. 1996. Relationships between DNA repair and transcription. *Annu. Rev. Biochem.* **65,** 13–42.
5. Friedberg, E. C., G. Walker, and W. Siede. 1995. DNA repair and mutagenesis. ASM, Washington, DC.
6. Hanawalt, P. C. 1994. Transcription-coupled repair and human disease. *Science* **266,** 1957–1958.
7. Hanawalt P. C., P. K. Cooper, A. K. Ganesan, and C. A. Smith. 1979. DNA repair in bacteria and mammalian cells. *Ann. Rev. Biochem.* **48,** 783–836.
8. Koehler, D. R. and P. C. Hanawalt. 1996. Recruitment of damaged DNA to the nuclear matrix in hamster cells following ultraviolet irradiation. *Nucleic Acids Res.* **24,** 2877–2884.
9. Leadon, S. A. and P. Cooper. 1993. Preferential repair of ionizing radiation-induced damage in the transcribed strand of an active human gene is defective in Cockayne syndrome. *Proc. Natl. Acad. Sci. USA* **90,** 10,499–10,503.
10. Lindahl, T. 1993. Instability and decay of the primary structure of DNA. Nature **362,** 709–715.
11. Ljungman, M., and F. Zhang. 1996. Blockage of RNA polymerase as a possible trigger for U. V. light-induced apoptosis. *Oncogene* **13,** 823–831.
12. Marnett, L. J. and P. C. Burcham. 1993. Endogenous DNA adducts: potential and paradox. *Chem. Res. Tox.* **6**:771–785.
13. Modrich, P. and R. Lahue. 1996. Mismatch repair in replication fidelity, genetic recombination, and cancer biology. *Ann. Rev. Biochem.* **65,** 101–33.

13a. Mullenders, L. H. F., A. C. V. van Leeuwen, A. A. van Zeeland, and A. T. Natarajan. 1988. Nuclear matrix associated DNA is preferentially repaired in normal human fibroblasts. *Nucleic Acids Res.* **16,** 10,607–10,622.
14. Nouspikel, T., P. Lalle, S. A. Leadon, P. K. Cooper, and S. G. Clarkson. 1997. A common mutational pattern in Cockayne syndrome patients from xeroderma pigmentosum group G: implications for a second XPG function. *Proc. Natl. Acad. Sci. USA* **94,** 3116–3121.
15. Sancar, A. 1996. DNA excision repair. *Ann. Rev. Biochem.* **65,** 43–81.
16. Satoh, M. S., C. J. Jones, R. D. Wood, and T. Lindahl. 1993. DNA excision-repair defect of xeroderma pigmentosum prevents removal of a class of oxygen free radical-induced base lesions. *Proc. Natl. Acad. Sci. USA* **90,** 6335–6339.
17. Sharon S. K., M. Morimatsu, U. Albrecht, D.-S. Lim, et al. 1997. Embryonic lethality and radiation hypersensitivity mediated by Rad51 in mice lacking *Brca2*. *Nature* **386,** 804–810.
18. Wood, R. D. 1996. DNA repair in eukaryotes. *Ann. Rev. Biochem.* **65,** 135–167.
19. Yamaizumi, M. and T. Sugano. 1994. UV-induced nuclear accumulation of p53 is evoked through DNA damage of actively transcribed genes independent of the cell cycle. *Oncogene* **9,** 2775–2784.

Part I

Prokaryotic Responses to DNA Damage

2

Nucleotide Excision Repair in *Escherichia coli*

Lawrence Grossman, Chien-liang Lin, and Yungchan Ahn

1. INTRODUCTION

Nucleotide excision repair (NER) consists of a series of enzymatic reactions required to remove virtually all DNA lesions *(61)*, including the majority if not all of those removed by base excision repair (BER) (Table 1). Damaged nucleotides are excised *(57)* as oligonucleotide fragments, whose size is generally fixed and is independent of the nature of the damage. This pathway consists of five general steps: damage recognition, incision, excision, repair synthesis, and ligation. The substance of the processes leading to incision will be emphasized, because it is the rate-limiting step in NER and because of the intrinsically interesting damage-recognition process. Other details about NER in bacteria are reviewed in ref. *13*.

In the past few years, the application of recombinant DNA techniques contributed to our progress in characterizing the proteins that mediate the individual steps in NER of *Escherichia coli*. The first three steps of this process are carried out by an ensemble of three proteins termed UvrABC endonuclease encoded by the *uvrA*, *uvrB*, and *uvrC* genes (Table 2) in an ATP hydrolysis-dependent series of reactions. The incision reaction generates both 5'- and 3'-incisions to an adducted nucleotide leading inevitably to a "dual incision" reaction *(44,79)*. UvrABC endonuclease possesses a broad-spectrum substrate specificity that acts on seemingly unrelated species of UV-mimetic DNA damage. It has been proposed that this almost indiscriminate specificity of the UvrABC complex is owing to its ability to recognize and conform to damage-induced conformational changes in the DNA and not necessarily to the chemically modified bases per se *(14,44,46,70,76)*.

2. EARLY GENETIC ANALYSIS OF UV REPAIR

The fate of UV-induced cyclobutane pyrimidine dimers (CPDs) in cellular DNA was studied in UV-resistant (*E. coli* B/r or K-12) and in sensitive strains (*E. coli* Bs-1 or AB 1886). It was concluded that CPD removal during dark reactivation involved repair processes in *uvr*$^+$ strains that were not functional in sensitive *uvr*$^-$ strains *(4,58)*. This dark reactivation process was designated excision repair. Twenty-three mutants sensitive to UV were isolated and mapped. All the uvr genes are unlinked and map at one of three loci designated *uvrA* (92 min), *uvrB* (17 min), and *uvrC* (41.5 min) as

From: DNA Damage and Repair, Vol. 1: DNA Repair in Prokaryotes and Lower Eukaryotes
Edited by: J. A. Nickoloff and M. F. Hoekstra © Humana Press Inc., Totowa, NJ

Table 1
Substrates of *E. coli* UvrABC Endonuclease[a]

DNA-damaging agent	Adduct(s)
N-acetoxy-2-acetylaminofluorene	C-8-Guanine
Aflatoxin B1[b]	N-7-Guanine
Anthramycin	N-2-Guanine
AP sites	Base loss
AP sites (reduced)[c]	Ring opened AP site
Alkoxamine-modified AP sites	AP site analog
Benzo[α] pyrenediolepoxide	N-2-Guanine
CC-1065	N-3-Adenine
Cisplatin and transplatin	N-7-Guanine
Cyclohexylcarbodiimide	Unpaired G and T residues
Ditercalanium	Noncovalent bis-intercalator
Doxorubicin and AD32	Intercalated compounds
N-hydroxyaminofluorene	C-8-Guanine
N,N'-bis(2-chloroethyl)-*N*-nitrosourea	Bifunctional alkylation
N-methyl-*N'*-nitro-*N*-nitrosoguanidine	O6-Methylguanine
Mitomycin	*N*-7-Guanine
Nitrogen mustard	Bifunctional alkylation
4-Nitroquinoline-1-oxide	C-8, N-2-Guanine N-6-Adenine
UV light	
(6-4)PD	C-6, C-4-Pyr
CPD	C-5, C-6-Pyrimidine
Psoralen + UV, crosslink + monoadduct	C-5, C-6-Thymine
Ionizing radiation, thymine glycol	C-5, C-6-Thymine

[a]Adapted from Van Houten et al. *(69)*.
[b]Oleykowski et al. *(36)*.
[c]Snowden and Van *(62)*.

determined from the time of entry of markers in crosses with various *Hfr* donor strains. None of the fully UV-sensitive *uvrA*, *uvrB* or *uvrC* mutants were able to excise detectable amounts of CPDs during incubation. These mutants were also sensitive to bifunctional alkylating agents, nitrous acid and mitomycin C *(13)*.

3. ISOLATION OF *UVR* GENES

3.1. uvrA

The *uvrA* structural gene was obtained from *E. coli* K-12 chromosomal DNA inserted into plasmids that complemented the *uvrA* mutant *(41,84)*. *uvrA* is one of a series of genes collectively referred to as SOS genes, which are induced to increased levels of transcription by agents that cause DNA damage (*77,78*; *see* Chapter 7). A binding site for the LexA repressor protein specific for the *uvrA* promoter (the so-called LexA box or SOS box) has been identified *(43)*. Sequence analysis of the 2.82-kbp *uvrA* gene indicates that it is translated into a protein consisting of 940 amino acids

Table 2
Properties of *E. coli* UvrA, UvrB, and UvrC

Property	UvrA	UvrB	UvrC
Mol. mass (Dalton)	103,874	76,118	66,038
No. of amino acids	940	763	610
No. of Trp residues/molecule	3	0	2
Molar extinction coefficient	46,680	27,699	36,200
pI	6.5	5.0	7.3
Stokes radius (A)	59	41	41
Sedimentation coefficient	7.4	4.2	4.9
Intrinsic metal	2 Zn^{2+}	None	None
DNA binding	Yes	No	Yes
Nucleotide binding motifs	2	1	None
ATPase activity	Yes	No	No
SOS regulation	Yes	Yes	No
No. mol/cell	20 (250)[b]	250 (1000)$_b$	10

[a]Adapted from Van Houten *(70)*.
[b]Values in parentheses denote levels induced under SOS conditions; UrrC is not under direct SOS control.

(M_r =103 kDa) and contains two ATP binding motifs, two zinc finger motifs and a helix-turn-helix (HTH) motif *(9,20)*. These motifs provided for the basis for site-directed mutagenesis of *uvrA* (*see* Section 5.1.).

3.2. uvrB

The *uvrB* structural gene was obtained from *E. coli* AB1157 chromosomal DNA *(77)* and plasmid p25-23 in the Carbon-Clarke collection *(40,67)*. It was selected by complementation of the *uvrB* mutant *E. coli* AB1885. *uvrB* is also a member of the SOS regulon and is inducible by DNA damage. The gene is transcribed from two overlapping promoters P1 and P2. A LexA binding site is present in the promoter region. Transcription from P2 is inhibited by LexA repressor protein, whereas that from P1 is unaffected. A third promoter, designated P3, is 320 bp upstream of P2. In vitro transcription from P3 is directed toward *uvrB* but terminates in a region of the LexA binding site. Interestingly, the putative stem-loop structure in P3 shares sequence homology with the binding site for DNA A protein in the *E. coli* replication origin *oriC*. *uvrB* is 2019 bp in length *(1)*, which translates into 673 amino acids (76.6 kDa), that includes a consensus Walker type A nucleotide binding motif.

3.3. uvrC

The *uvrC* structural gene was cloned from *E. coli* K-12 AB1885 *(83)* and the Clarke-Carbon plasmid pLC13-12 carrying *flaD* and *uvrC (42)*. *uvrC* is not inducible by DNA damage and is not a member of the SOS regulon. *uvrC* is 1,830 bp in length *(45)* encoding a polypeptide of 610 amino acids (66 kDa). The RNA polymerase binding site is 0.9 kbp upstream.

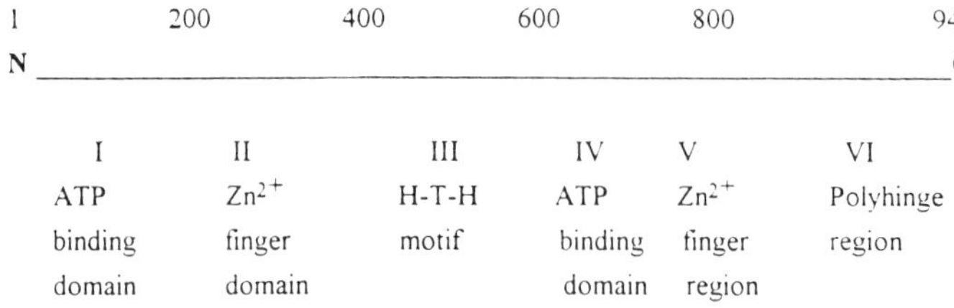

1	200	400	600	800	940
N					C

I	II	III	IV	V	VI
ATP binding domain	Zn^{2+} finger domain	H-T-H motif	ATP binding domain	Zn^{2+} finger region	Polyhinge region

I: 31-GLSGSGKSSL-40

II: 253-CX2CX20CX2C-280

III: 494-RIRLASQI GAG LVGVMYVLD-513

IV: 640-GVSGSGKSTL-649

V: 740-CX2CX19CX2C-766

VI: 900-WIVDLGPEGGSGGGEILVSGTPETVAECEASHTARFLKPML-940

Fig. 1. Schematic diagram of structural and functional domains of UvrA protein. Amino acid sequences for indicated domains are shown below. H-T-H = helix-turn-helix region.

4. PROTEIN COMPLEMENTATION

The first demonstration of in vitro ATP-dependent UV-endonuclease activity by Seeburg and his colleagues *(49,50)* revealed that damage-dependent activity required the complementary action of the *uvrA*, *uvrB*, and *uvrC* gene products. Neither UvrA nor UvrB showed any appreciable endonuclease activity on UV-irradiated DNA. ATP is clearly required for the uvr^+ encoded endonuclease activity. The UvrAB complex was shown to recognize the damaged site in an ATP hydrolysis-dependent reaction. Strand cleavage by UvrABC was not inhibited by ATP-γ-S, suggesting that damage recognition required the presence of ATP, whereas the endonucleolytic step was ATP-independent. This finding supports the observation that the UvrAB complex first binds to a nonspecific site on the DNA and thereafter is translocated to the site of the lesion by a process that requires the hydrolysis of ATP *(15)*.

5. PROPERTIES OF UvrA, UvrB, AND UvrC

5.1. UvrA

UvrA is a DNA-independent ATPase that binds DNA. Both of these functional attributes correlate with specific structural motifs in the translated nucleotide sequence (Fig. 1). UvrA has an ATPase whose catalytic activity is dependent on the lysine residues K37 and K646 *(30,51,63)*. UvrA was found to also hydrolyze GTP *(7)*. The ATPase of UvrA is not stimulated by DNA, but instead it is modulated by DNA *(15,34)*. The reactive species of UvrA is a dimer. It is the binding energy of ATP (or ATP-γ-S) that drives dimerization of the protein *(32)*, whereas the hydrolysis of ATP leads to monomerization of UvrA. The binding of $(UvrA)_2$ protein to DNA has been demonstrated using a variety of experimental techniques *(26,69,79)*. The binding of UvrA to undamaged DNA is 10^3–10^4-fold weaker than that for damage-specific binding *(26)*.

ATPase

```
MSKPFKLNSAFKPSGDQPEAIRRLEEGLEDGLAHQTLLGVTGSGKTFTI
ANVIADLQRPTMVLAPNKTLAAQLYGEMKEFFPENAVEYFVSYYDYYQP
EAYVPSSDTFIEKDASVNEHIEQMRLSATKAMLERRDVVVVASVSAIYG
LGDPDLYLKMMLHLTVGMIIDQRAILRRLAELQYARNDQAFQRGTFRVR
GEVIDIFPAESDDIALRVELFDEEVERLSLFDPLTGQIVSTIPRFTIYP
KTHYVTPRERIVQAMEEIKEELAARRKVLLENNKLLEEQRLTQRTQFDL
EMMNELGYCSGIENYSRFLSGRGPGEPPPTLFDYLPADGLLVVDESHVT
IPQIGGMYRGDRARKETLVEYGFRLPSALDNRPLKFEEFEALAPQTIYV
SATPGNYELEKSGGDVVDQVVRPTGLLDPIIEVRPVATQVDDLLSEIRQ
RAAINERVLVTTLTKRMAEDLTEYLEEHGERVRYLRSDIDTVERMEIIR
DLRLGDFDVLVGINLLREGLDMPEVSLVAILDADKEGFLRSERSLIQTI
GRAARNVNGKAILYGDKITPSMAKAIGETERRREKQQKYNEEHGITPQG
LNKKVVDILALGQNIAKTKAKGRGKSRPIVEPDNVPMDMSPKALQQKIH
ELEGLMMQHAQNLEFEEAAQIRDQLHQLRELFIAAS        ↑
```

ompT

Fig. 2. Functional domains of UvrB. The large "C" residue is a target for fluorescence derivativization.

The specificity for UV-irradiated DNA is completely abolished by ATP-γ-S while enhancing nonspecific binding *(67)*. The dissociation rate of UvrA from damaged sites is fast in the presence of ATP. Thus, it appears that ATP hydrolysis increases the specificity of binding to damaged DNA, but lowers the equilibrium binding constant by stimulating dissociation *(15)*. Each UvrA molecule has two zinc finger motifs. Extended X-ray absorption *(31)* and atomic absorption *(8,9)* reveal that the UvrA protein contains two Zn^{2+} atoms. When UvrA was renatured without Zn^{2+} after denaturation, it showed no DNA binding activity *(8)* until at least one Zn^{2+} atom was restored to the protein. The mutation in a zinc finger motif located in the C-terminal domain confers extreme UV sensitivity. A protein deficient in the C-terminal zinc binding site is unable to bind DNA, indicating that this region may be involved in nonspecific DNA binding of UvrA *(74)*. However, substitution of C 253 or C 256 with serine in the N-terminal zinc finger motif had little or no effect on DNA binding *(75)*. Monoclonal antibodies to this domain failed to inhibit the DNA binding activity of UvrA protein *(23)*.

Two mutations created in the HTH motif by degenerate oligonucleotide-directed mutagenesis (G502D and V508D) had no effect on UvrA binding to unirradiated plasmid DNA. However, both mutants failed to respond to UV damage in the filter binding assay, suggesting that the HTH motif may be involved in the recognition of UV damage *(73)*. The deletion of 40 amino acids in the C-terminus glycine-rich sequence resulted in loss of damage recognition, suggesting that this region of UvrA may stabilize the protein–DNA complex when damage is encountered *(9)*.

5.2. *UvrB*

Although the sequence of *uvrB* codes for 673 amino acids, the purified UvrB protein contains 672 amino acids, since it lacks the first amino acid (Met); it has a calculated mol wt of 76 kDa (Fig. 2). The purified protein has no detectable ATPase activity. ATPase activity is observed, however, when UvrB interacts with UvrA in the presence of DNA (double- and single-stranded DNA and UV-irradiated DNA) *(7)*. This dramatic increase is owing to the activation of the cryptic ATPase on forming a $UvrA_2B$–DNA complex. Mutation of lysine 45 to alanine in the ATP binding motif of UvrB leads to an acute defect in ATP hydrolysis and significant suppression of the activated ATPase in the presence of DNA *(51)*. During the isolation of UvrB protein, it was discovered that it was cleaved near its C-terminus, yielding a 70-kDa protein, UvrB*. A comparison of the UvrB sequence with the adaptive protein (Ada) suggested conserved regions near the C-terminus end of the molecule, which is a potential cleavage site for the Ada protease: MSPKALQQ. The "cryptic UvrB ATPase" exposed in UvrB* is activated by single-stranded DNA or chaotropic salts *(7)*. The presence of the UvrB ATPase in $UvrA_2B$–DNA complex was also observed using the formation of complexes of wild-type UvrB with the UvrA double mutant (K37A-K646A) *(63)*. UvrB can bind to DNA only in the presence of UvrA *(81)*. UvrB has a hydrophobic region that may be exposed in solution allowing binding to a hydrophobic column during purification *(65,80)*. The amino acid sequence of the UvrB protein shows homology with two limited regions of UvrC.

5.3. *UvrC*

Two different translational products of *uvrC* have been reported. The first begins at an ATG codon, yielding a protein containing 588 amino acids (M_r = 66 kDa) *(45)*. The second is initiated at a GTG codon, 66 bp upstream of the ATG. The C-terminal domain of UvrC has a high degree of homology with the C-terminal domain of human ERCC-1 (*68*; *see* Chapter 18, vol. 2). UvrC is a single-stranded DNA binding protein and binds to it with a relatively high affinity. Mutations in the C-terminal region cause extreme sensitivity to UV and defects in 5'-incision activity. UvrC can associate with a UvrB–DNA complex *(82)*.

6. MOLECULAR MECHANISM OF DUAL INCISION

The repair of UV damage is a multistep process that involves a variety of protein complexes with the Uvr proteins, DNA, and nucleotide cofactors as diagrammed in Fig. 3. Each of these steps is considered in turn.

6.1. *UvrA Dimer*

Nucleotide (ATP) binding to UvrA drives UvrA dimerization, whereas the hydrolysis of ATP or the presence of ADP drives monomerization *(34)*. In glycerol gradients, UvrA migrates as a monomer in the absence of ATP. In the presence of ATP, a single peak migrating between monomer and dimer species is observed. In the presence of ATP-γ-S, the labeled protein migrates as a dimer. The UvrA dimer is the active form that binds to undamaged or UV-irradiated DNA *(26)*.

6.2. *$UvrA_2B$ Complex in Solution*

Purified UvrA protein associates with UvrB protein to produce the $UvrA_2B$ complex. The modulated ATPase of UvrA in the presence of UvrB reflects the interaction of UvrA with UvrB *(32)*. From gel-filtration experiments, the Stokes radius of the complex is 6.01 ± 0.48 nm, which corresponds to a molecular weight of 201 kDa.

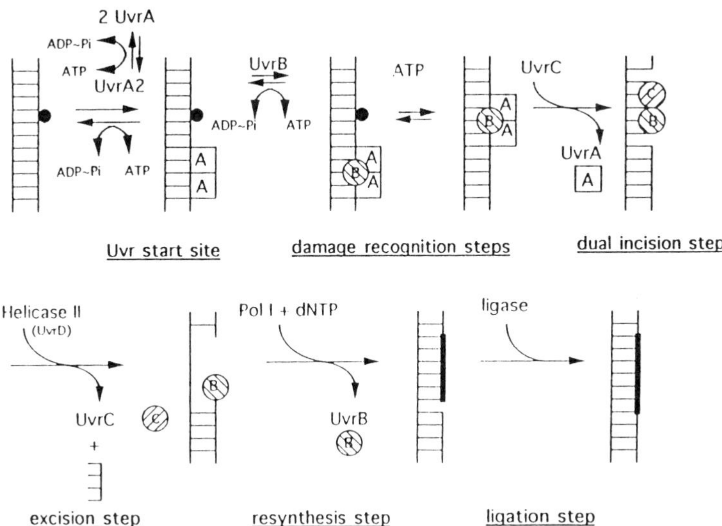

Fig. 3. Nucleotide excision repair pathway in *E. coli*. $UvrA_2$ dimer binds to damaged DNA. UvrB then binds and the $UvrA_2B$ complex translocates to the damaged site. UvrC displaces UvrA, and dual incision is catalyzed by UvrBC. UvrD displaces UvrC and an oligonucleotide containing the damage. DNA pol I fills the single-stranded gap, and DNA ligase seals the nick.

Velocity sedimentation experiments indicate that the UvrA and UvrB subunits interact to form a complex in solution with a UvrA to UvrB ratio of 2:1 *(34)*.

6.3. Nucleoprotein Complex Formation

UvrA binds to DNA containing psoralen-thymine monoadducts at defined positions with an apparent binding affinity (K_a) of 0.7–1.5 × $10^8 M^{-1}$ covering a 33-bp region that surrounds the modified thymine *(69)*. The equilibrium constant for nonspecific binding (K_{ns}) of UvrA is 0.7–2.9 × $10^5 M^{-1}$. ADP decreases the UvrA binding affinity two to threefold, and ATP is not required for the specific binding to a damaged site. ATP-γ-S quantitatively inhibits the specific binding while enhancing nonspecific binding. The dimer of UvrA binds to UV-damaged DNA with an affinity (K_{uv}) of 10^8–$10^9 M^{-1}$ in the presence or absence of ATP *(26,69)*. The binding constant for supercoiled DNA is 3 × $10^5 M^{-1}$ and for linear DNA is 2 × $10^6 M^{-1}$ which is lower than that for single-stranded DNA *(69)*. The addition of UvrA to unirradiated pBR322 in the presence of ATP-γ-S induces significant unwinding of the DNA, by about 100% per UvrA dimer. A small but detectable amount of UvrA unwinding was observed with ATP. It was also found that the addition of UvrA (in the presence of ATP) to UV-irradiated DNA induced unwinding of the DNA helix in proportion to the UV fluence *(32)*.

6.4. $UvrA_2B$-DNA Complex

The formation of the $UvrA_2B$ complex results in a DNA (single- and double-stranded, UV-damaged)-dependent ATPase through activation of its cryptic ATPase on binding to DNA. The mutant $UvrA_2B$(K45A) binds irreversibly to undamaged DNA and is unable to translocate because of its inability to hydrolyze ATP *(52)*. The UvrAB single-stranded DNA-dependent ATPase is associated with the helicase activity, requiring DNA binding. Less than 20 nucleotides are actively displaced from a duplex and a D-loop DNA *(33)* substrate when the UvrAB complex translocates along DNA in a 5'- to 3'-direction *(35)*. When the UvrAB protein is incubated with relaxed closed circular duplex DNA this helicase activity can generate negative and positive supercoils in the presence of ATP indicating translocation (or tracking) along DNA *(22)*. When ATP-γ-S is included in a reaction mixture instead of ATP, the addition of UvrB to UvrA results in locally unwound undamaged DNA between 180 and 220°.

It has been observed by gel mobility shift assays that a 49-bp duplex oligonucleotide, containing a benzoxyamine-modified abasic site, can interact with a $UvrA_2B$ complex generating three detectable intermediates, which are consistent with the presence of a $UvrA_2$–DNA complex, a $UvrA_2B$–DNA complex, and a UvrB–DNA complex. When damaged DNA was examined by DNase I protection, the footprint decreased to 19 bp in the presence of UvrA and UvrB *(69)*, reflecting the formation of a UvrB–DNA complex.

6.5. Tracking Along DNA

In examining the degree of association of UvrA or $UvrA_2B$ to DNA, it was found that the discrimination factor Df (K_{uv}/K_{ns}) is only 10^3 *(26)*. This level of damage recognition is too low for the Uvr system to recognize a single damaged site per genome by a passive diffusion-controlled reaction. Thus, initial Uvr protein–DNA recognition probably occurs at undamaged rather than at damaged sites in DNA. It is suggested that a more concerted mechanism, such as one involving tracking, could play a role in damage site recognition. It has been shown that the $UvrA_2B$ associated 5-' to 3'-helicase activity that displaces short regions in an unwinding, reannealing mode of translocation is inhibited by the presence of bulky base damage in UV-irradiated DNA. Consistent with this model, mutation of the lysine residue located in the Walker type A nucleotide binding motif of UvrB (K45A) or UvrA (K646A) leads to defective ATPase and DNA helicase activity in the $UvrA_2B$ complex and a failure to support the incision of UV-damaged DNA *(64)*. These findings indicate that the translocation of the $UvrA_2B$ complex is required for damage recognition. The transient supercoiling activity of the $UvrA_2B$ complex is a possible reflection of a translocational mechanism.

6.6. Sensing Mechanism

The ability of the $UvrA_2B$ helicase to cause supercoiling implies that the search for damage along DNA is a consequence of the ATP $\leftrightarrow$ ADP~Pi equilibrium that leads to microprocessive association–dissociation to DNA coupled to unidirectional translocation *(15)*. The UvrA component in a $UvrA_2B$ complex may sense the presence of DNA damage.

6.7. Recognition of Damaged Sites

The UvrABC endonuclease does not recognize a specific chemical group or structure in the damaged nucleotide, nor does it recognize a specific backbone deformity in the duplex induced by a wide variety of different genotoxic chemicals. The enzyme recognizes damage ranging from bulky lesions, such as CPDs, pyrimidine (6-4) pyrimidone photoproducts [(6-4)PDs], 2-aminofluorene DNA adducts, aflatoxin B1 adducts, psoralen monoadducts, and cross-links to smaller lesions including *O*6-methylG, thymine glycol, and apurinic/apyrimidinic (AP) sites (Table 1). Initial damage recognition may be a consequence of the combined conformational change of the HTH and polyhinge regions of UvrA protein. Even though molecular mechanisms by which the UvrB protein is loaded into a damaged site are not clear, the UvrB–DNA (damaged DNA) complex is apparent in footprinting experiments *(69)*, by electron microscopy *(60)*, and by gel-mobility shift (72), and the complex is stable in gel-filtration columns *(37,38)*. The formation of this complex requires UvrA and ATP, and occurs after the recognition of a damage site by the UvrAB complex. The DNA molecule in this complex is kinked by 130° *(59,60)*. The rate of UvrB loading to DNA is low ($K_{on} = 6 \times 10^{-4} M^{-1}$), suggesting that this step may be limiting for damage-specific incision of DNA. Effective loading of the UvrB protein occurs over a limited range of UvrA protein concentrations (5–50 n*M*) and is inhibited at the higher concentrations.

6.8. Dual Incision

The excision of CPDs is preceded by dual incision by the UvrABC endonuclease *(44,79)*. The binding of UvrC to the UvrB-DNA complex results in the incision of DNA on both sides of the lesion in the damaged strand *(38)*. The precise location of the 5'- and 3'-incision sites is affected by DNA sequence context. Studies with mutant UvrB and UvrC proteins have indicated that the incision process is not a concerted one. The 3' incision is apparently influenced by the UvrB protein, and this incision precedes 5' nicking by UvrC protein. The release of the damaged fragment from the UvrBC-incised DNA and turnover of the UvrBC subunits bound to the damage are dependent on the coordinated excision–resynthesis reaction by UvrD and DNA polymerase in the presence of dNTPs *(5)*. The damaged site is released as a 12–13 bp long oligonucleotide fragment. UvrD affects release of both the excised oligomer and UvrC, but not UvrB. UvrB is released only upon inclusion of DNA pol I and dNTPs *(39)*.

7. TRANSCRIPTION COUPLED REPAIR

Several discoveries have led to the concept that NER is coupled to transcription. Preferential repair occurs in the transcribed strand in actively expressed genes *(3,17,27,28)*. Components of the transcription factor, TFIIH, that are essential for transcription initiation are also required for NER (*11,12,47,48,74*; *see* Chapter 10, vol. 2). It had been reported that the efficiency of repair of UV-induced CPDs in fibroblasts from XP-C patients is 50% of normal. A biphasic time-course for the removal of DNA lesions was observed in cultured human cells *(16)*. The rapid removal of pyrimidine dimers in the first few hours after UV irradiation was followed by a gradually decreasing rate, approaching a plateau by 24 h *(21)*. The rapid recovery of RNA synthesis in UV-irradiated mammalian cells led to the suggestion of preferential repair in transcriptionally active chromatin.

7.1. Selective and Strand-Specific Repair in an Active Gene

Measurements of CPDs in the active *DHFR* gene (amplified ~50-fold) in UV-irradiated Chinese hamster ovary (CHO) cells indicated that 70% were removed. Little repair was detectable in a fragment 30 kbp upstream from *DHFR*, and no repair was detected in flanking sequences at the 3'-end or upstream of the promoter region. Only 15% of the dimers were lost from the genome overall in 24 h. These results indicated that there was intragenic repair heterogeneity.

To elucidate further the nature of preferential repair in an active gene, the repair in the transcribed and nontranscribed DNA strands in *DHFR* in CHO and human cells was measured. In the CHO cells, nearly 80% of the dimers had been removed from the transcribed strand within 4 h, whereas almost no repair had occurred in the nontranscribed strand. Similar but less dramatic differences were evident in the analysis of strand-specific repair in human cells (*see* Chapter 18, vol. 2).

7.2. *Strand-Specific Repair in* E. coli

Measurements of NER in the two DNA strands of the lactose operon in UV-irradiated *E. coli* indicated that the level of repair in the uninduced condition is about 50% after 20 min in both strands. As a consequence of a 436-fold induction of transcription, most of the dimers (70%) were removed from the transcribed strand of the induced operon within 5 min, whereas repair in the nontranscribed strand was similar to that of the uninduced state.

7.3. *Effect of CPDs and RNA Polymerase on Transcription and Repair*

CPDs in the template strand constitute an absolute block for transcription, whereas those in the complementary strand have no effect on transcription. In addition, UV irradiation of cells results in truncated transcripts. In vitro experiments showed that the nucleotide added opposite to the 3' T- moiety of CPDs was the 3'-terminal nucleotide in nearly all truncated RNA molecules; similar results were obtained with a psoralen monoadduct *(59)*. The effect of transcription on repair was monitored using a defined system consisting of two DNA duplexes containing a CPD in either the transcribed or nontranscribed strands, purified *E. coli* RNA polymerase (RNAP), UvrA, UvrB, and UvrC. When repair was measured in the absence of ribonucleoside triphosphates (rNTPs), it was found that the promoter-bound RNAP had no effect on repair of a CPDs downstream from the transcriptional initiation site whether the CPD was in the transcribed or in the nontranscribed strand (Fig. 4). However, in the presence of rNTPs, transcription specifically interfered with the repair of the transcribed strand *(53)*. This result demonstrates that a stalled complex on its own does not target UvrABC endonuclease to the lesion (Fig. 5).

7.4. *Transcription Repair Coupling Factor*

7.4.1. *In* E. coli *Repair* (57)

An *E. coli* cell-free system was used to explore the existence of a factor that couples transcription to repair. By measuring repair DNA synthesis by assaying incorporation of label into plasmid DNA, a factor was identified that can supplement preferential repair in a defined system containing UvrA, B, C, D, DNA pol I, DNA ligase, RNAP, rNTPs, and dNTPs. A large (130-kDa) protein was partially purified and called transcription repair coupling factor (TRCF) *(55)*. It was later shown that cell extracts from *mfd* mutant cells lacked strand-specific repair *(54)*. This defect was complemented with partially purified TRCF, and it was concluded that the *mfd* gene encoded TRCF, and that the mutation frequency decline phenotype in *mfd* mutants was owing to a lack of

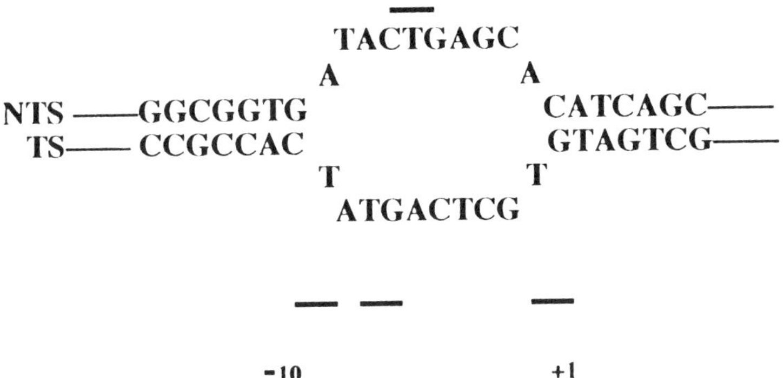

Fig. 4. $UvrA_2B$ landing site. Regions protected by the $UvrA_2B$ complex are over- and underlined. NTS and TS indicate transcribed and nontranscribed strands.

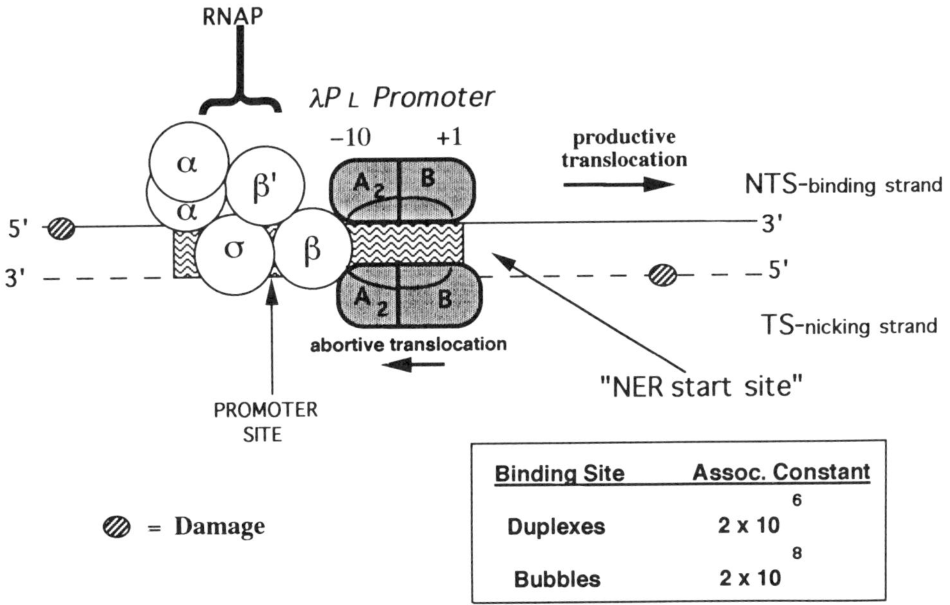

Fig. 5. Strand selectivity by $UvrA_2B$ helicase. *See text* for details.

transcription repair coupling. A genomic DNA fragment complementing *mfd* mutant WU3610-45 was isolated, and it encoded the 130-kDa Mfd protein (TRCF), which is a relatively abundant protein of about 500 copies/cell *(56)*. Mfd is a monomer, binds to DNA nonspecifically, and has a weak ATPase activity (K_{cat} = ~3/min), but lacks helicase activity. *E. coli* TRCF specifically interacts with RNAP stalled at a lesion, and dissociates the ternary complex and the nascent RNA. Although no interaction between TRCF and UvrA could be detected by standard hydrodynamic methods, UvrA does bind specifically to a TRCF affinity column, which indicates that the interaction is relatively weak. Thus, the action of TRCF overcomes the inhibitory effect of the stalled complex on repair by UvrABC endonuclease. However, the mechanism of stimulation is still not clear, because some of the key intermediates have not been experimentally demonstrated. In particular, intermediates involving DNA-TRCF, DNA-TRCF-UvrAB, or DNA-RNA-RNAP-TRCF-UvrAB have not been identified by either hydrodynamic or footprinting techniques *(52)*.

An examination of the *E. coli* genes controlling transcription-coupled repair by Mellon and Champe *(29)* revealed that mutation in the mismatch repair genes *mutS* or *mutL* abolished transcription-coupled repair. NER in the *lac* operon in *mutS* or *mutL* strains resembled repair found in *mfd* mutants. These results suggest an important link between NER and mismatch correction in transcription-coupled repair. For more details on TCRF, *see* Chapter 9.

7.4.2. In Mammalian Repair

Patients with Cockayne's syndrome (CS) are characterized by their photosensitivity, neurological abnormalities and moderate UV sensitivity at the cellular level (*see* Chapter 18, vol. 2). Genetically, three CS groups have been defined *(24)*. Biochemically, cells of groups A and B (CS-A and CS-B) carry out normal levels of nicking and repair synthesis following UV irradiation, but are defective in recovery of UV-induced inhibition of RNA synthesis. This led to the proposal that CS gene(s) might be involved in the repair of transcriptionally active genes *(25)*. Indeed when CS-A and CS-B cells were tested for gene-specific repair by measuring T4 endo-V-sensitive sites, CPD repair was defective *(71)*. The product of the *CSB* gene, *ERCC6*, has been implicated in the coupling process *(66)*. Additionally, it has been reported that CS-B cells fail to repair CPDs, but are able to repair other photoproducts in transcriptionally active genes *(2)*. As yet there have been no biochemical studies of transcription repair coupling in human systems; the mechanism is probably more complicated than that in bacteria. Mammalian genes are generally much longer and transcribed more slowly than genes in *E. coli*. Transcription of the 2.5-Mb human dystrophin gene, for example, requires over 8 h. It would seem inefficient to abort nearly completed transcripts of such genes every time RNAP encounters a lesion. The transcription elongation factor SII provides for an alternative mechanism. This factor catalyzes nascent transcript cleavage by RNAP II at natural pause sites, enabling the polymerase to back up and try again without aborting the incomplete transcript. A similar reaction has been demonstrated at CPDs in a model DNA template in vitro *(2)*, suggesting that this cleavage activity may be a key feature of transcription-coupled repair. Transcription factor SII may be the *CSA* gene product *(18)*.

7.5. Role of RNAP in Strand Selection by UvrA$_2$B

Transcription coupled to NER specifies the location in active genes where DNA repair is to take place. During transcription repair coupling, there is a physical association of the β-subunit of *E. coli* RNAP and the UvrA component of the repair apparatus. This molecular affinity is reflected in the synergistic ability of the RNAP to increase, in a promoter-dependent manner, DNA supercoiling by the UvrA$_2$B helicase complex

(1a). The RNAP-induced DNA structure (Fig. 4) around the transcription start site signals a landing locus for the $UvrA_2B$ complex. Because of the presence of the RNAP, the $UvrA_2B$ complex is only able to translocate along the nontranscribed strand because of the intrinsic 5' → 3' directionality of the associated helicase activity. As a consequence of this directionality, its binding to the transcribed strand would inevitably lead to a collision with RNAP bound to its promoter. Since incision occurs on the strand opposite to preferential binding to the nontranscribed strand, preferential incision of DNA damaged sites on the transcribed strand occurs as a consequence of the specific translocation of the $UvrA_2B$ complex along the nontranscribed strand (*see* Fig. 5).

It was found that a $UvrA_2B$–DNA complex formed with a synthetic DNA oligomer contained a bubble comparable to that seen in promoter regions, but in the absence of RNAP (Fig. 4). These sites appeared to be start sites for the $UvrA_2B$ complex with the same affinities for damaged sites in DNA. The preferential recognition and incision of damaged sites was observed downstream of bubble and loop regions in the strand complementary to the translocating strand. These results imply that the bubble region generated in duplex DNA by RNAP serves as a preferred start site for the translocation of the $UvrA_2B$ complex and predetermines where incision is to occur *(1b)*.

8. FUTURE DIRECTIONS

Current understanding of mechanisms of DNA repair is limited by the lack of structural information. The inability to prepare large or stable crystals of the Uvr proteins has severely limited structure–function studies of this biological process. In addition to a static structure, mechanistic insights will derive from time-resolved techniques as well as stoichiometric relationships between ATP binding hydrolysis levels coupled to rates of complex movement during translocation. In this manner, the fundamental thermodynamics of movement may be understood in terms of power stroke and protein conformational changes during translocation. This level of understanding has yet to be gained in other systems, such as muscle contraction. However, DNA repair mechanisms may be quite amenable to studies of the energetics of protein complex movement and the recognition of altered biomolecular structures.

ACKNOWLEDGMENTS

The work described in this chapter attributable to the authors was supported by a grant from the National Institutes of Health—GM 22846.

REFERENCES

1a. Ahn, B. and L. Grossman. 1996. RNA polymerase signals UvrAB landing sites. *J. Biol. Chem.* **271:** 21,453–21,461.

1b. Ahn, B. and L. Grossman. 1996. The role of DNA loops and bubbles as start sites in nucleotide excision repair. *J. Biol. Chem.* **271:** 21,462–21,470.

2. Barrett, S. F., J. H. Robbins, R. E. Tarone, and K. H. Kraemer. 1991. Evidence for defective repair of cyclobutane pyrimidine dimers with normal repair of other DNA photoproducts in a transcriptionally active gene transfected into Cockayne syndrome cells. *Mutat. Res.* **255:** 281–291.

3. Bohr, V. A., D. S. Okumoto, L. Ho, and P. C. Hanawalt. 1986. Characterization of a DNA repair domain containing the dihydrofolate reductase gene in Chinese hamster ovary cells. *J. Biol. Chem.* **261:** 16,666–16,672.

4. Boyce, R. P. and P. Howard-Flanders. 1964. Release of ultraviolet-light induced thymine dimers from DNA in *E. coli* K-12. *Proc. Natl. Acad. Sci. USA* **51:** 293–300.
5. Caron, P. R., S. R. Kushner, and L. Grossman. 1985. Involvement of helicase II uvrD gene product and DNA polymerase I in excision mediated by the uvrABC protein complex. *Proc. Natl. Acad. Sci. USA* **82:** 4925–4929.
6. Caron, P. R. and L. Grossman. 1988. Potential role of proteolysis in the control of UvrABC incision. *Nucleic Acids Res.* **16:** 9641–9650.
7. Caron, P. R. and L. Grossman. 1988. Involvement of a cryptic ATPase activity of UvrB and its proteolysis product, UvrB* in DNA repair. *Nucleic Acids Res.* **16:** 9651–9662.
8. Claassen, L. A., B. Ahn, H. S. Koo, and L. Grossman. 1991. Construction of deletion mutants of the *Escherichia coli* UvrA protein and their purification from inclusion bodies. *J. Biol. Chem.* **266:** 11,380–11,387.
9. Claassen, L. A. and L. Grossman. 1991. Deletion mutagenesis of the *Escherichia coli* UvrA protein localizes domains for DNA binding, damage recognition, and protein–protein interactions. *J. Biol. Chem.* **266:** 11,388–11,394.
10. Cleaver, J. E. 1968. Defective repair replication of DNA in xeroderma pigmentosum. *Nature* **218:** 652–656.
11. Drapkin, R., J. T. Reardon, A. Ansari, J. C. Huang, L. Zawel, K. Ahn, A. Sancar, and D. Reinberg. 1994. Dual role of TFIIH in DNA excision repair and in transcription by RNA polymerase II. *Nature* **368:** 769–772.
12. Feaver, W. J., J. Q. Svejstrup, L. Bardwell, A. J. Bardwell, S. Buratowski, K. D. Gulyas, T. F. Donahue, E. C. Friedberg, and R. D. Kornberg. 1993. Dual roles of a multiprotein complex from S. cerevisiae in transcription and DNA repair. *Cell* **75:** 1379–1387.
13. Friedberg, E. C., G. C. Walker, and W. Siede. 1995. *DNA Repair and Mutagenesis.* ASM, Washington, DC.
14. Grossman, L., P. R. Caron, S. J. Mazur, and E. Y. Oh. 1988. Repair of DNA-containing pyrimidine dimers. *FASEB J.* **2:** 2696–2701.
15. Grossman, L. and S. Thiagalingam. 1993. Nucleotide excision repair, a tracking mechanism in search of damage. *J. Biol. Chem.* **268:** 16,871–16,874.
16. Hanawalt, P. C., P. K. Cooper, A. K. Ganesan, and C. A. Smith. 1979. DNA repair in bacteria and mammalian cells. *Annu. Rev. Biochem.* **48:** 783–836.
17. Hanawalt, P. C. 1991. Heterogeneity of DNA repair at the gene level. *Mutat. Res.* **247:** 203–211.
18. Hanawalt, P. C. 1994. Transcription-coupled repair and human disease. *Science* **266:** 1957,1958.
19. Howard-Flanders, P., R. P. Boyce, and L. Theriot. 1966. Three loci in *Escherichia coli* K-12 that control the excision of pyrimidine dimers and certain other mutagen products from DNA. *Genetics* **53:** 1119–1136.
20. Husain, I., H. B. Van, D. C. Thomas, and A. Sancar. 1986. Sequences of *Escherichia coli* *uvr*A gene and protein reveal two potential ATP binding sites. *J. Biol. Chem.* **261:** 4895–4901.
21. Kantor, G. J. and R. B. Setlow. 1981. Rate and extent of DNA repair in non-dividing human diploid fibroblast. *Cancer Res.* **41:** 819–825.
22. Koo, H. S., L. Claassen, L. Grossman, and L. F. Liu. 1991. ATP-dependent partitioning of the DNA template into supercoiled domains by *Escherichia coli* UvrAB. *Proc. Natl. Acad. Sci. USA* **88:** 1212–1216.
23. Kovalsky O. I. and L. Grossman. 1994. The use of monoclonal antibodies for studying intermediates in DNA repair by the *Escherichia coli* UvrABC endonuclease. *J. Biol. Chem.* **269:** 27,421–27,426.
24. Lehmann, A. R. 1987. Cockayne's syndrome and trichothiodysrophy: defective repair without cancer. *Cancer Rev.* **1:** 82–103.
25. Mayne, L. V. and A. R. Lehmann. 1982. Failure of RNA synthesis to recover after UV irradiation: an early defect in cells from individuals with Cockayne's syndrome and xeroderma pigmentosum. *Cancer Res.* **42:** 1473–1478.

26. Mazur, S. J. and L. Grossman. 1991. Dimerization of *Escherichia coli* UvrA and its binding to undamaged and ultraviolet light damaged DNA. *Biochemistry* **30:** 4432–4443.
27. Mellon, I., G. Spivak, and P. C. Hanawalt. 1987. Selective removal of transcription-blocking DNA damage from the transcribed strand of the mammalian DHFR gene. *Cell* **51:** 241–249.
28. Mellon, I. and P. C. Hanawalt. 1989. Induction of the *Escherichia coli* lactose operon selectively increases repair of its transcribed DNA strand. *Nature* **342:** 95–98.
29. Mellon, I. and G. N. Champe. 1996. Products of DNA mismatch repair genes *mutS* and *mutL* are required for transcription-coupled nucleotide-excision repair of the lactose operon in *Escherichia coli. Proc. Natl. Acad. Sci. USA* **93:** 1292–1297.
30. Myles, G. M., J. E. Hearst, and A. Sancar. 1991. Site-specific mutagenesis of conserved residues within Walker A and B sequences of *Escherichia coli* UvrA protein. *Biochemistry* **30:** 3824–3834.
31. Navaratnam, S., G. M. Myles, R. W. Strange, and A. Sancar. 1989. Evidence from extended X-ray absorption fine structure and site-specific mutagenesis for zinc fingers in UvrA protein of *Escherichia coli. J. Biol. Chem.* **264:** 16,067–16,071.
32. Oh, E. Y. and L. Grossman. 1986. The effect of *Escherichia coli* Uvr protein binding on the topology of supercoiled DNA. *Nucleic Acids Res.* **14:** 8557–8571.
33. Oh, E. Y. and L. Grossman. 1987. Helicase properties of the *Escherichia coli* UvrAB protein complex. *Proc. Natl. Acad. Sci. USA* **84:** 3638–3642.
34. Oh, E. Y., L. Claassen, S. Thiagalingam, S. Mazur, and L. Grossman. 1989. ATPase activity of the UvrA and UvrB protein complexes of the *Escherichia coli* UvrABC endonuclease. *Nucleic Acids Res.* **17:** 4145–4149.
35. Oh, E. Y. and L. Grossman. 1989. Characterization of the helicase activity of the *Escherichia coli* UvrAB protein complex. *J. Biol. Chem.* **264:** 1336–1343.
36. Oleykowski, C. A., J. A. Mayernik, S. E. Lim, J. D. Groopman, L. Grossman, G. N. Wogan, and A. T. Yeung. 1993. Repair of aflatoxin B1 DNA adducts by the UvrABC endonuclease of *Escherichia coli. J. Biol. Chem.* **268:** 7990–8002.
37. Orren, D. K. and A. Sancar. 1989. The ABC excinuclease of *Escherichia coli* has only the UvrB and UvrC subunits in the incision complex. *Proc. Natl. Acad. Sci. USA* **86:** 5237–5241.
38. Orren, D. K. and A. Sancar. 1990. Formation and enzymatic properties of the UvrB–DNA complex. *J. Biol. Chem.* **265:** 15,796–15,803.
39. Orren, D. K., C. P. Selby, J. E. Hearst, and A. Sancar. 1992. Post-incision steps of nucleotide excision repair in *Escherichia coli*. Disassembly of the UvrBC–DNA complex by helicase II and DNA polymerase I. *J. Biol. Chem.* **267:** 780–788.
40. Sancar, A., N. D. Clarke, J. Griswold, W. J. Kennedy, and W. D. Rupp. 1981. Identification of the *uvr*B gene product. *J. Mol. Biol.* **148:** 63–76.
41. Sancar, A., R. P. Wharton, S. Seltzer, B. M. Kacinski, N. D. Clarke, and W. D. Rupp. 1981. Identification of the uvrA gene product. *J. Mol. Biol.* **148:** 45–62.
42. Sancar, A., B. M. Kacinski, L. D. Mott, and W. D. Rupp. 1981. Identification of the *uvr*C gene product. *Proc. Natl. Acad. Sci. USA* **78:** 5450–5454.
43. Sancar, A., G. B. Sancar, W. D. Rupp, J. W. Little, and, D. W. Mount. 1982. LexA protein inhibits transcription of the *E. coli* uvrA gene in vitro. *Nature* **298:** 96–98.
44. Sancar, A. and W. D. Rupp. 1983. A novel repair enzyme: UvrABC excision nuclease of *Escherichia coli* cuts a DNA strand on both sides of the damaged region. *Cell* **33:** 249–260.
45. Sancar, G. B., A. Sancar, and W. D. Rupp. 1984. Sequences of the *E. coli uvr*C gene and protein. *Nucleic Acids Res.* **12:** 4593–4608.
46. Sancar, A. and M. S. Tang. 1993. Nucleotide excision repair. *Photochem. Photobiol.* **57:** 905–921.
47. Schaeffer, L., R. Roy, S. Humbert, V. Moncollin, W. Vermeulen, J. H. Hoeijmakers, P. Chambon, and J. M. Egly. 1993. DNA repair helicase: a component of BTF2 TFIIH basic transcription factor. *Science* **260:** 58–63.

48. Schaeffer, L., V. Moncollin, R. Roy, A. Staub, M. Mezzina, A. Sarasin, G. Weeda, J. H. Hoeijmakers, and J. M. Egly. 1994. The ERCC2/DNA repair protein is associated with the class II BTF2/TFIIH transcription factor. *EMBO J.* **13:** 2388–2392.
49. Seeberg, E. 1976. Incision of ultraviolet-irradiated DNA by extract of E. coli requires three different gene products. *Nature* **263:** 524–526.
50. Seeberg, E. 1978. Reconstitution of an *Escherichia coli* repair endonuclease activity from the separated *uvr*A$^+$ and *uvr*B$^+$ *uvr*C$^+$ gene products. *Proc. Natl. Acad. Sci. USA* **75:** 2569–2573.
51. Seeley, T. W. and L. Grossman. 1989. Mutations in the *Escherichia coli* UvrB ATPase motif compromise excision repair capacity. *Proc. Natl. Acad. Sci. USA* **86:** 6577–6581.
52. Seeley, T. W. and L. Grossman. 1990. The role of *Escherichia coli* UvrB in nucleotide excision repair. *J. Biol. Chem.* **265:** 7158–7165.
53. Selby, C. P. and A. Sancar. 1990. Transcription preferentially inhibits nucleotide excision repair of the template DNA strand *in vitro*. *J. Biol. Chem.* **265:** 21,330–21,336.
54. Selby, C. P., E. M. Witkin, and A. Sancar. 1991. *Escherichia coli mfd* mutant deficient in mutation frequency decline lacks strand-specific repair: *in vitro* complementation with purified coupling factor. *Proc. Natl. Acad. Sci. USA* **88:** 11,574–11,578.
55. Selby, C. P. and A. Sancar. 1991. Gene- and strand-specific repair in vitro: partial purification of a transcription-repair coupling factor. *Proc. Natl. Acad. Sci. USA* **88:** 8232–8236.
56. Selby, C. P. and A. Sancar. 1993. Molecular mechanism of transcription-repair coupling. *Science* **260:** 53–58.
57. Selby, C. P. and A. Sancar. 1994. Mechanism of transcription-repair coupling and mutation frequency decline. *Microbiol. Rev.* **58:** 317–329.
58. Setlow, R. B. and W. L. Carrier. 1964. The disappearance of thymine dimers from DNA; an error correcting mechanism. *Proc. Natl. Acad. Sci. USA* **51:** 226–231.
59. Shi, Y. B., H. Gamper, and J. E. Hearst. 1987. The effects of covalent additions of a psoralen on transcription by *E. coli* RNA polymerase. *Nucleic Acids Res.* **15:** 6843–6854.
60. Shi, Q., R. Thresher, A. Sancar, and J. Griffith. 1992. Electron microscopic study of ABC excinuclease. DNA is sharply bent in the UvrB-DNA complex. *J. Mol. Biol.* **226:** 425–432.
61. Snowden, A., Y. W. Kow, and H. B. Van. 1990. Damage repertoire of the *Escherichia coli* UvrABC nuclease complex includes abasic sites, base-damage analogues, and lesions containing adjacent 5' or 3' nicks. *Biochemistry* **29:** 7251–7259.
62. Snowden, A. and H. B. Van. 1991. Initiation of the UvrABC nuclease cleavage reaction. Efficiency of incision is not correlated with UvrA binding affinity. *J. Mol. Biol.* **220:** 19–33.
63. Thiagalingam, S. and L. Grossman. 1991. Both ATPase sites of *Escherichia coli* UvrA have functional roles in nucleotide excision repair. *J. Biol. Chem.* 266: 11,395–11,403.
64. Thiagalingam, S. and L. Grossman. 1993. The multiple roles for ATP in the *Escherichia coli* UvrABC endonuclease-catalyzed incision reaction. *J. Biol. Chem.* **268:** 18,382–18,389.
65. Thomas, D. C., T. A. Kunkel, N. J. Casna, J. P. Ford, and A. Sancar. 1986. Activities and incision patterns of ABC excinuclease on modified DNA containing single-base mismatches and extra helical bases. *J. Biol. Chem.* **261:** 14,496–14,505.
66. Troelstra, C., A. J. van Gool, J. de Wet, W. Vermeulen, D. Bootsma, and J. H. Hoeijmakers. 1992. ERCC6, a member of a subfamily of putative helicases, is involved in Cockayne's syndrome and preferential repair of active genes. *Cell* **71:** 939–953.
67. van den Berg, E., R. H. Geerse, J. Memelink, R. A. Bovenberg, F. A. Magnee, and P. van de Putte. 1985. Analysis of regulatory sequences upstream of the *E. coli uvr*B gene; involvement of the DnaA protein. *Nucleic Acids Res.* **13:** 1829–1840.
68. van Duin, M., J. de Wit, H. Odijk, A. Westerveld, A. Yasui, H. M. Koken, J. H. Hoeijmakers, and D. Bootsma. 1986. Molecular characterization of the human excision repair gene ERCC-1: cDNA cloning and amino acid homology with the yeast DNA repair gene RAD10. *Cell* **44:** 913–923.

69. Van Houten, B., H. Gamper, A. Sancar, and J. E. Hearst. 1987. DNase I footprint of ABC excinuclease. *J. Biol. Chem.* **262:** 13,180–13,187.
70. Van Houten, B. 1990. Nucleotide excision repair in *Escherichia coli. Microbiol. Rev.* **54:** 18–51.
71. Venema, J., L. H. Mullenders, A. T. Natarajan, A. A. van Zeeland, and L. V. Mayne. 1990. The genetic defect in Cockayne syndrome is associated with a defect in repair of UV-induced DNA damage in transcriptionally active DNA. *Proc. Natl. Acad. Sci. USA* **87:** 4707–4711.
72. Visse, R., M. de Ruijter, G. F. Moolenaar, and P. van de Putte. 1992. Analysis of UvrABC endonuclease reaction intermediates on cisplatin-damaged DNA using mobility shift gel electrophoresis. *J. Biol. Chem.* **267:** 6736–6742.
73. Wang, J. and L. Grossman. 1993. Mutations in the helix-turn-helix motif of the *Escherichia coli* UvrA protein eliminate its specificity for UV-damaged DNA. *J. Biol. Chem.* **268:** 5323–5331.
74. Wang, Z., J. Q. Svejstrup, W. J. Feaver, X. Wu, R. D. Kornberg, and E. C. Friedberg. 1994. Transcription factor b TFIIH is required during nucleotide-excision repair in yeast. *Nature* **368:** 74–76.
75. Wang, J., K. L. Mueller, and L. Grossman. 1994. A mutational study of the C-terminal zinc-finger motif of the *Escherichia coli* UvrA protein. *J. Biol. Chem.* **269:** 10,771–10,775.
76. Weiss, B. and L. Grossman. 1987. Phosphodiesterases involved in DNA repair. *Adv. Enzymol. Biol.* **60:** 1–34.
77. Witkin, E. M. 1967. The radiation sensitivity of *Escherichia coli* B: a hypothesis relating filament formation and prophage induction. *Proc. Natl. Acad. Sci. USA* **57:** 1275–1279.
78. Witkin, E. M. 1969. Ultraviolet-induced mutation and DNA repair. *Annu. Rev. Microbiol.* **23:** 487–514.
79. Yeung, A. T., W. B. Mattes, E. Y. Oh, and L. Grossman. 1983. Enzymatic properties of purified *Escherichia coli* UvrABC proteins. *Proc. Natl. Acad. Sci. USA* **80:** 6157–6161.
80. Yeung, A. T., W. B. Mattes, E. Y. Oh, G. H. Yoakum, and L. Grossman. 1986. The purification of the *Escherichia coli* UvrABC incision system. *Nucleic Acids Res.* **14:** 8535–8556.
81. Yeung, A. T., W. B. Mattes, G. H. Yoakum, and L. Grossman. 1986. Protein complexes formed during the incision reaction catalyzed by the *Escherichia coli* UvrABC endonuclease. *Nucleic Acids Res.* **14:** 2567–2582.
82. Yoakum, G. H., S. R. Kushner, and L. Grossman. 1980. Isolation of plasmids carrying either the *uvr*C or *uvr*C *uvr*A and *ssb* genes of *Escherichia coli* K-12. *Gene* **12:** 243–248.
83. Yoakum, G. H. and L. Grossman. 1981. Identification of *E. coli uvr*C protein. *Nature* **292:** 171–173.
84. Yoakum, G. H., A. T. Yeung, W. B. Mattes, and L. Grossman. 1982. Amplification of the uvrA gene product of *Escherichia coli* to 7% of cellular protein by linkage to the pL promoter of pKC30. *Proc. Natl. Acad. Sci. USA* **79:** 1766–1770.

3

Prokaryotic Base Excision Repair

David M. Wilson III, Bevin P. Engelward, and Leona Samson

1. INTRODUCTION

Our genetic material is continually exposed to physical and chemical agents that induce a wide variety of DNA modifications. These agents are abundant in our environment, in our food, and some DNA-damaging agents are produced intracellularly as natural metabolites *(59)*. Alterations in DNA can also arise via spontaneous hydrolytic decay or by the incorporation of inappropriate nucleotides during chromosomal replication *(107,96)*. DNA modifications may decrease the fidelity of replication and transcription and, in some instances, may block these essential cellular processes. In order to prevent DNA-damaging agents from causing mutation or cell death, organisms are equipped with several mechanisms to prevent and repair DNA damage.

A wide variety of DNA lesions are repaired via the multistep base excision repair (BER) pathway that executes the following steps (Fig. 1):

1. Release of an inappropriate base by a specific DNA glycosylase;
2. Incision of the sugar phosphate backbone at the resulting abasic site by an endonuclease;
3. Removal of the nonconventional DNA terminus created by the endonuclease;
4. DNA repair synthesis; and
5. Ligation of the remaining single-stranded DNA break.

Although no human disease has yet been attributed to a defect in BER, the putative role of this repair process in protecting human cells from the accumulation of DNA damage cannot be overemphasized, particularly since unrepaired DNA damage may be involved in carcinogenesis, neurodegeneration, and aging. BER has been studied most extensively in *Escherichia coli*, and this chapter will focus on the molecular, biochemical, and genetic properties of the BER process in prokaryotic cells.

BER is the main system for the repair of modified bases, sites of base loss, single-strand breaks, and short gaps in DNA. The first step in BER involves either the spontaneous loss of a base or the removal of a particular improper base from DNA by a DNA glycosylase (Fig. 1). DNA glycosylases cleave the *N*-glycosylic bond to release an unwanted base from DNA and, as a result, produce an apurinic/apyrimidinic (AP) site. Some DNA glycosylases also display a class I AP lyase activity that incises the phosphodiester linkage immediately to the 3'-side of an AP lesion (Figs. 1 and 2). This incision activity generates a natural 5'-phosphate group and a 3'-terminus that requires removal by a class II AP endonuclease/3'-diesterase prior to repair synthesis and liga-

From: DNA Damage and Repair, Vol. 1: DNA Repair in Prokaryotes and Lower Eukaryotes
Edited by: J. A. Nickoloff and M. F. Hoekstra © Humana Press Inc., Totowa, NJ

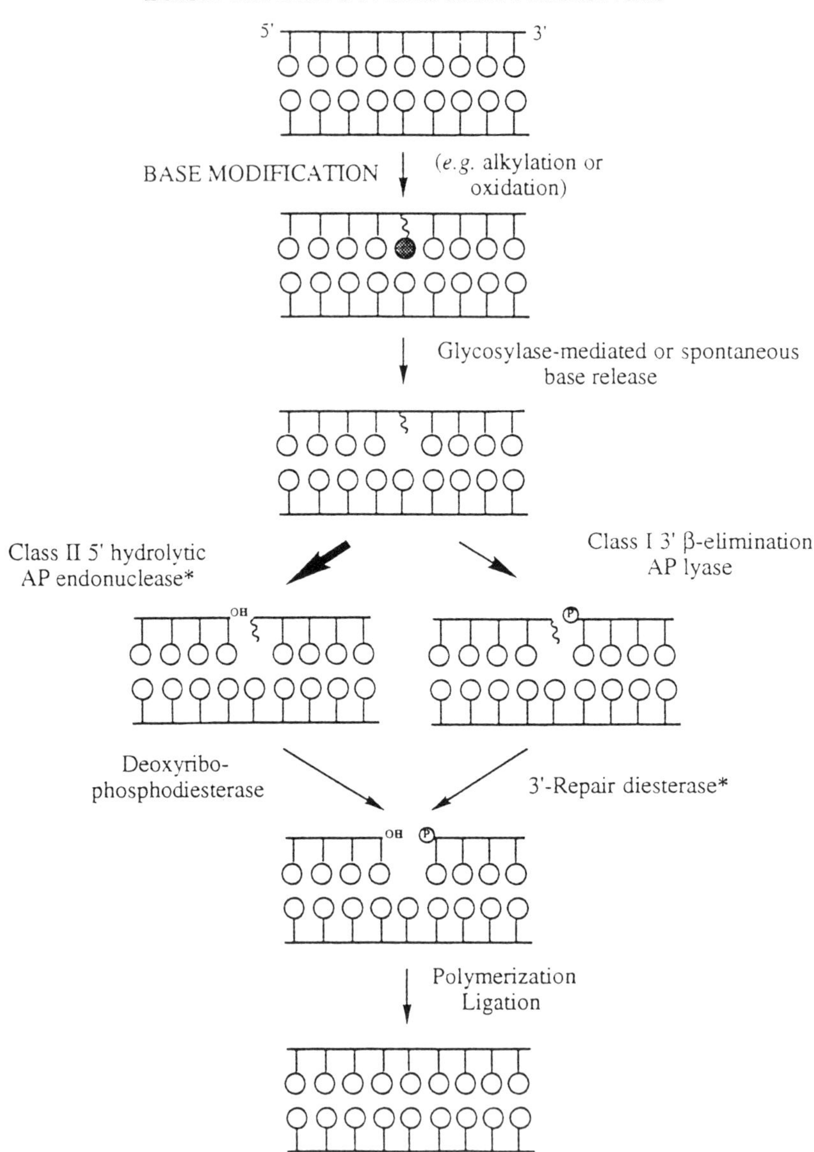

Fig. 1. Schematic of the various steps in the BER pathway. The darkened circle represents a modified or inappropriate base that can be removed by a specific DNA glycosylase, initiating BER. The class II AP endonuclease and 3'-diesterase functions are executed by the same repair protein, as indicated by the asterisk. Each step is detailed in the text.

tion (Fig. 1). However, cleavage of the phophodiester bond at AP sites is thought to be catalyzed primarily by class II AP endonucleases (Figs. 1 and 2). These enzymes incise to the immediate 5'-side of an abasic site, leaving a 3'-OH terminus (that can act as a primer for DNA synthesis) and a 5'-abasic residue that is removed by a deoxyribophosphodiesterase to produce a nucleotide gap and a conventional 5'-phosphate.

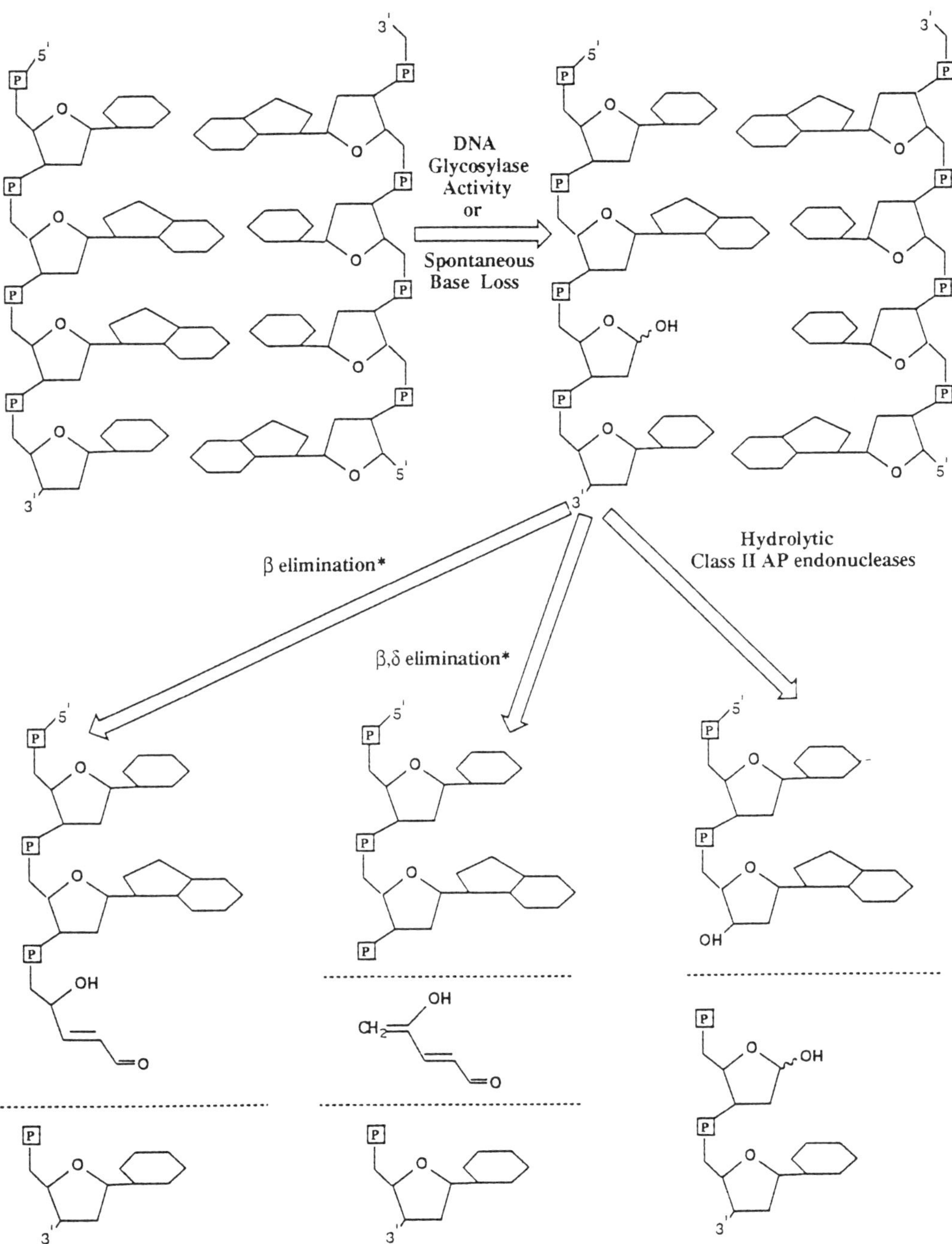

Fig. 2. The three forms of incision at abasic sites in DNA. Following spontaneous base loss or the enzymatic release of an inappropriate base by a DNA glycosylase, cleavage at the resulting abasic site occurs by one of three mechanisms: β elimination, β,δ elimination, or hydrolysis. The first two incision reactions (indicated by the asterisk) can be catalyzed by DNA glycosylases that possess an associated AP lyase activity. The cleavage products from the three AP incision events are depicted.

Thus, the stage is set for DNA polymerase to fill the gap and for DNA ligase to seal the single-strand nick, restoring DNA to its original state and completing BER (Fig. 1). The five steps of BER in prokaryotic cells will each be considered in turn.

2. DNA GLYCOSYLASES

The specificity of BER is dictated by the substrate specificity of the initiating DNA glycosylase. A summary of the substrates recognized by the known *E. coli* DNA glycosylases is shown in Table 1. The abasic sites created by the action of DNA glycosylases are mutagenic and cytotoxic, so the protective effects of DNA glycosylases depend on the subsequent action of the other enzymes involved in BER (Fig. 1). Taken together, the DNA glycosylases of *E. coli* release a stunning array of damaged or inappropriate DNA bases, the majority of which are induced by endogenous and exogenous agents.

As mentioned above, some DNA glycosylases have an associated AP lyase function. At first glance, it may seem efficient to couple base release and backbone cleavage, but the in vivo significance of the AP lyase activity remains unclear. As shown in Figs. 1 and 2, backbone cleavage by an AP lyase produces a 3'-terminus that requires further processing by a class II AP endonuclease/3'-diesterase. Thus, AP endonucleases are required for the completion of BER even when BER is initiated by a DNA glycosylase that executes AP lyase cleavage. A summary of the salient features of the known *E. coli* DNA glycosylases follows.

2.1. *Fpg/MutM DNA Glycosylase*

The *E. coli* Fpg/MutM DNA glycosylase, expressed from the *fpg/mutM* gene, was first characterized for its ability to release the ring opened form of 7-methylguanine, namely 2,6-diamino-4-hydroxy-5-*N*-methylformamidopyrimidine (FAPY; Fig. 3; *30*). This lesion has been shown to block DNA replication, and it is therefore considered to be potentially cytotoxic *(12)*. FAPY lesions were initially considered to be the most relevant Fpg/MutM substrates, since the FAPY precursor 7-methylguanine appears to arise spontaneously in cells and since ionizing radiation can convert purines into their ring-opened form *(140,174)*. However, *fpg* mutant *E. coli* do not have an increased sensitivity to agents that are thought to generate FAPY lesions *(10)*. Thus, FAPY is either a minor product of these treatment conditions, or their repair by the Fpg/MutM DNA glycosylase is not critical for cell survival.

Interestingly, this DNA glycosylase, which was purified based on the ability to repair ring opened 7-methylguanine, was later shown to be identical to a previously characterized 8-oxoguanine (8oxoG) endonuclease (*30,31,172*; *see* Chapter 6). 8oxoG (Fig. 3) is highly mutagenic, since it can pair with adenine during DNA replication and thus drives GC to TA transversions *(154,189)*. 8oxoG is formed not only from exposure of cells to oxidizing agents, but also from endogenous sources of free radicals. Thus, it is not surprising that *E. coli* mutants lacking the Fpg/MutM DNA glycosylase demonstrate an increased spontaneous G:C to T:A mutation rate *(24,122)*. Given the phenotype of the *fpg/mutM* mutants, it is now generally accepted that 8oxoG is the most biologically important substrate of Fpg/MutM DNA glycosylase.

Similar to most other DNA glycosylases, Fpg/MutM DNA glycosylase has only been shown to release damaged bases from duplex DNA (it does not release mismatched bases). However, the Fpg/MutM DNA glycosylase differs from most DNA glycosylases in that it has both an associated class I AP lyase activity and a deoxyribophosphodiesterase (dRpase) activity (*see* Section 4.) *(67,131)*. The β,δ lyase activity of Fpg/MutM DNA glycosylase produces a single base gap flanked by a 3'- and 5'-phosphate terminus, the former of which requires a 3'-diesterase for removal (*4*; Fig. 2). Thus, despite its many repair activities, Fpg/MutM DNA glycosylase cannot sequentially release a base, cleave the backbone, and process the ends ready for subsequent repair synthesis.

Fpg/MutM contains a single zinc atom that is essential for enzymatic activity, presumably because it is required as part of the DNA binding domain *(173)*. Recent analysis of Fpg/MutM has revealed that this protein forms a Schiff-base enzyme–substrate intermediate *(171)*, a mechanism that was previously described for PD DNA glycosylase (*see* Section 2.5.) and may be utilized by all combined DNA glycosylases/AP lyases (reviewed in *49*). The biological significance of the AP lyase and dRpase activities associated with Fpg/MutM DNA glycosylase is currently under investigation.

2.2. *MutY DNA Glycosylase*

E. coli MutY DNA glycosylase acts in concert with the Fpg/MutM DNA glycosylase to prevent the potentially mutagenic consequences of 8oxoG (Fig. 4). 8oxoG lesions that escape repair by Fpg/MutM DNA glycosylase frequently pair with A during DNA replication, producing an 8oxoGA: base pair. MutY releases the undamaged A from this base pair, initiating BER; presumably MutY will continue to initiate BER until C is incorporated opposite 8oxoG during DNA repair synthesis. Interestingly, the Fpg/MutM DNA glycosylase does not efficiently recognize 8oxoG when paired with A, thus preventing this DNA repair enzyme from driving GC to TA transversions *(112)*. Like Fpg/MutM, MutY possesses an AP lyase activity, but the biological role of this function remains to be determined.

In the absence of MutY, DNA replication past the 8oxoG:A mismatch results in thymine incorporation opposite adenine on one of the daughter strands, creating a fixed mutation. Therefore, *mutY E. coli* display a greatly elevated spontaneous G:C to T:A transversion rate *(129)*. In addition to releasing A residues incorporated opposite 8oxoG, the MutY DNA glycosylase releases A residues opposite C or G residues *(178)*. It should be noted, however, that *mutY* mutants only display increased G:C to T:A spontaneous mutations, presumably because the *mutHLS* methyl-directed DNA mismatch repair pathway (described in Chapter 11) eliminates C:A and possibly G:A mismatches, making this feature of MutY redundant.

The *mutY* gene encodes the only known *E. coli* DNA glycosylase that functions as a "mismatch" repair enzyme to remove normal unmodified bases from DNA. Unlike the *E. coli mutHLS* mismatch repair system that serves to correct a variety of mismatched bases that arise primarily from errors made during normal DNA replication, MutY serves to remove adenines that have been misincorporated opposite oxidatively damaged guanine. That *E. coli* have evolved multiple systems to prevent mutations arising from 8oxoG attests to the significance of this selective pressure during evolution.

2.3. *Thymine Glycol DNA Glycosylase (DNA Endonuclease III)*

Endonuclease III was first identified for its ability to incise DNA that had been damaged by ultraviolet (UV) or ionizing radiation *(60,139,163)*. Later, it was shown that endonuclease III is actually a DNA glycosylase with an associated AP lyase activ-

Table 1
***E. coli* DNA Glycosylases**

DNA glycosylase	Substrate (listed as base released)	Ref.
Fpg/MutM *(fpg/mutM)*[a]	2,6-Diamino-4-hydroxy-5-*N*-methyl)formamidopyrimidine	*30*
	4,6-Diamino-5-formamidopyrimidine	*17*
	8-oxoguanine	*31,172*
	5-Hydroxycytosine	*79*
	5-Hydroxyuracil	*79*
	Aflatoxin-bound imidazole-ring-opened guanine	*29a*
	Imidazole ring opened *N*-2-aminofluorene-C8-guanine	*9a*
MutY *(mutY)*[a]	A of A:8oxoG	*122*
	A of A:C or A:G	*1a,178*
TG DNA glycosylase *(nth)*[a]	Cis- and transthymine glycol	*45,91*
	5,6-Dihydrothymine	*45*
	5,6-Dihydroxydihydrothymine	*45*
	Urea	*19,91*
	5-Hydroxy-5-methylhydantion	*18*
	Methyltartronylurea	*18*
	6-Hydroxy-5,6-dihydro-pyrimdines	*16,20*
	5-Hydroxycytosine and 5-hydroxyuracil	*79*
	5-Hydroxy-6-hydrothymine	*48a*
	5,6-Dihydrouracil	*48a*
	Alloxan	*48a*
	Uracil glycol	*48a*
	5-Hydroxy-6-hydrouracil	*48a*
Endonuclease VIII *(nei)*[a]	Substrates tested are the same as that of TG DNA glycosylase	*180*
PD DNA glycosylase *(denV^{+})*[a]	Cyclobutane pyrimidine dimers	*45,138*

Enzyme	Substrate	Reference
Uracil DNA glycosylase *(ung)*	Uracil	*108*
	5-Hydroxy-2'-deoxyuridine	*79*
	5,6-Dihydroxyuracil	*192*
	5-Fluorouracil	*182a*
3MeA DNA glycosylase I *(tag)*	3-Methyladenine	*141,90*
	3-Methylguanine	*8*
	3-Ethyladenine	*141,175*
	3-Ethylthioethyladenine	*73*
3MeA DNA glycosylase II *(alkA)*	3-Methyladenine	*90*
	3-Methylguanine	*175*
	7-Methylpurines	175
	7- and 3-Ethylpurines	*175*
	1-Carboxyethyladenine	*175*
	7-Carboxyethylguanine	*175*
	O^2-Methylpyrimidines	*119*
	7(2-Ethoxyehtyl)guanine	*71*
	7(2-Hydroxyehtyl)guanine	*26,71*
	7(2-Chlorothyl)guanine	*26,71*
	1,2-Bis(7-guanyl)ethane	*26,71*
	3-Ethylthioethylpurines	*73*
	N2,3-Ethenoguanine and N2,3-ethanoguanine	*72,117,149a*
	Hypoxanthine	*150*
	5-Hydroxymethyluraci and 5-formyluracil	7

[a]Indicates AP lyase activity.

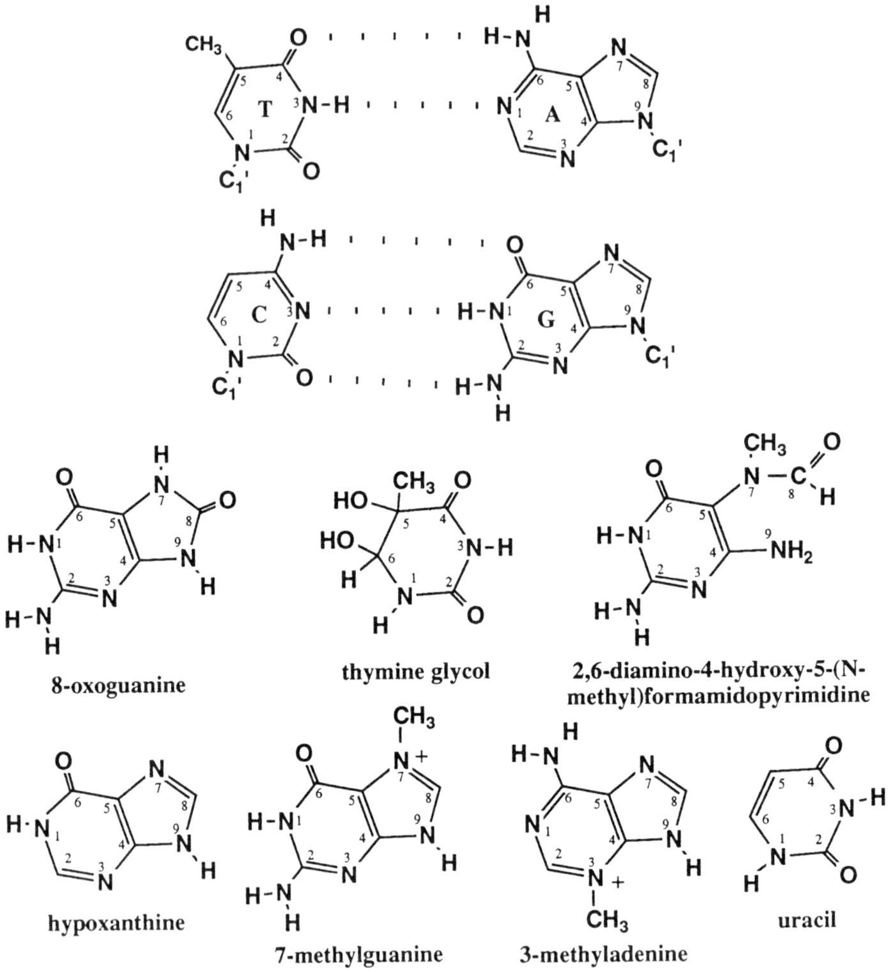

8-oxoguanine

thymine glycol

2,6-diamino-4-hydroxy-5-(N-methyl)formamidopyrimidine

hypoxanthine

7-methylguanine

3-methyladenine

uracil

Fig. 3. Structures of normal DNA bases and several modified DNA bases. The top portion depicts the chemical structures and the base pairing arrangements of A, G, T, and C. Several nonconventional, oxidized, and alkylated bases (labeled accordingly) are indicated below.

ity *(2,45,91)*. Since one of the first substrates identified for this enzyme was thymine glycol (Fig. 3), this enzyme is often referred to as the thymine glycol DNA glycosylase (Tg DNA glycosylase).

The fact that Tg DNA glycosylase can nick DNA that had been treated with either X-rays, osmium tetroxide, UV, or acid suggested that this enzyme may repair several types of DNA damages *(60)*. As it turns out, Tg DNA glycosylase can, in fact, repair a wide range of damaged pyrimidines, including ring saturation, fragmentation, contraction, and oxidation products caused by many DNA damaging agents (*see* Table 1, *97*, reviewed in *50* and *181*). Although most of the substrates shown to be repaired by Tg DNA glycosylase are modified thymines, Tg DNA glycosylase also cleaves UV treated DNA at cytosines *(185)*. It was later demonstrated that 6-hydroxy-5,6-dihydrocytosine and 5-hydroxy-2'-deoxycytidine are at least two types of damaged cytosines recog-

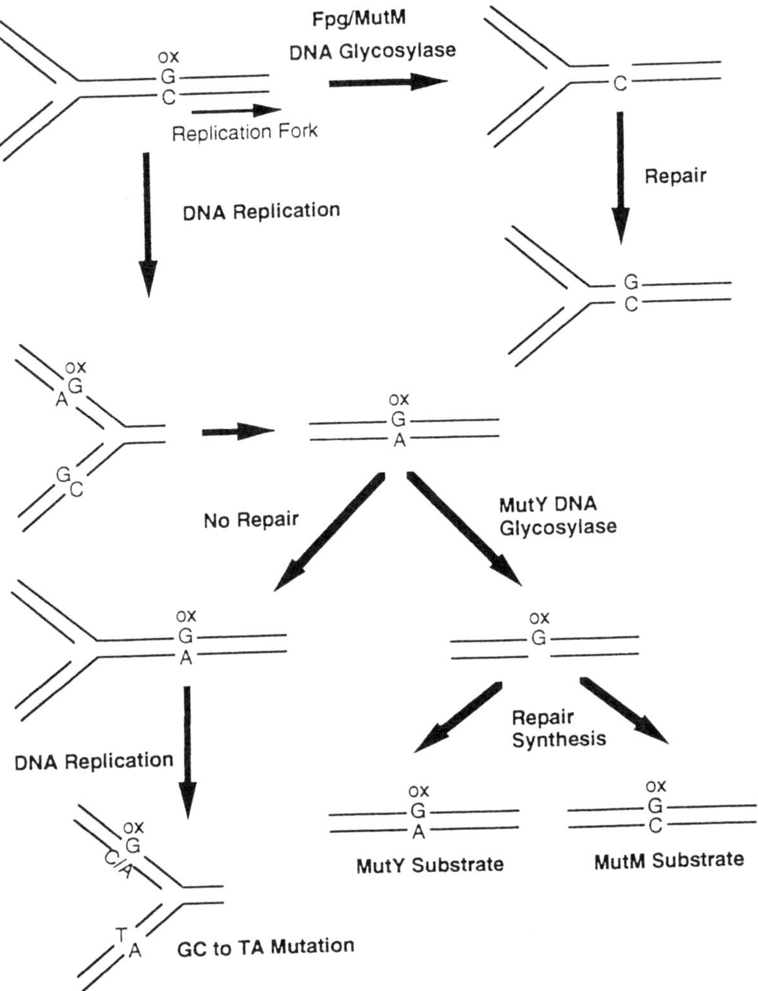

Fig. 4. Pathways for preventing mutation at 8oxoG in DNA. Schematic of the DNA repair machinery for 8oxoG. As indicated, Fpg/MutM DNA glycosylase initiates the BER process prior to DNA replication. However, replication may occur before proper repair by MutM and can result in incorporation of A opposite 8oxoG. (C may also be incorporated opposite Gox [not shown] and this replicated Gox:C would again be a substrate of MutM.) Gox:A is a substrate for MutY, which would remove the A. Repair synthesis may result in either recreation of Gox:A (a substrate for MutY again) or in a return to Gox:C (a substrate for MutM again). If repair is not successful (indicated as "No Repair"), a second round of DNA replication will result in a fixed mutation, since T will be incorporated opposite the aberrant A.

nized by Tg DNA glycosylase *(16,79)*. Tg glycosylase is a fairly small enzyme (23.4 kDa), so it is likely to have a single active site *(1)*. One must therefore ask: what common structural feature is shared by this diverse collection of substrates? The Tg DNA glycosylase may recognize the saturated 5-6 double bond, which causes the loss of the planar structure of these bases *(18,91)*. However, not all of the substrates of this enzyme share this characteristic; thus, the substrate recognition mechanism of the PD; glycosylase is still under investigation.

Isolation of the gene-encoding Tg DNA glycosylase *(nth)* permitted high-level expression and subsequent purification of the Tg DNA glycosylase protein *(1,38)*. Using elemental analysis, Mössbauer spectra, and electron paramagnetic resonance (EPR data), it was established that the Tg DNA glycosylase is an iron-sulfur containing DNA repair enzyme *(36)*. Crystals of Tg DNA glycosylase suggest that this iron–sulfur cluster is not involved in catalysis, but rather is important for maintaining structural integrity in an area of the protein thought to be important for nonspecific DNA binding *(100)*. This iron–sulfur DNA binding motif has been referred to as a "primitive zinc finger" and appears to be a novel interacting domain *(174b)*. Analysis of the MutY amino acid sequence suggests that it too may contain such a motif.

The biochemical mechanism of the DNA glycosylase and AP lyase activities of Tg DNA glycosylase has been studied in great detail. Kow and Wallace observed that Tg DNA glycosylase cleaves as a DNA lyase at abasic sites by β-elimination *(2,99)*. Altering the reaction conditions inhibited specifically AP lyase activity at preformed abasic sites, but did not inhibit the AP lyase activity that was directly linked to DNA glycosylase activity, indicating that the glycosylase and lyase activities of the Tg DNA glycosylase are inextricably associated *(99)*.

Although many of the damaged bases shown to be repaired by Tg DNA glycosylase in vitro are formed as a result of oxidation or radiation damage, *E. coli nth* mutants are no more sensitive than wild-type cells to killing by the oxidizing agent hydrogen peroxide (H_2O_2) or by ionizing radiation *(38)*. However, these repair-deficient *E. coli* do have an increased number of spontaneous mutations owing to the existence of unrepaired mutagenic damages produced by endogenous agents.

2.4. Endonuclease VIII

As described above, Tg DNA glycosylase repairs a wide range of free radical-induced base damages, some of which are thought to block DNA replication. Surprisingly, *E. coli* lacking Tg DNA glycosylase are not hypersensitive to X-rays *(38)*. This raised the question of whether a redundant pathway for the repair of X-ray-induced cytotoxic base damages exists in *E. coli*. Indeed, endonuclease VIII, was subsequently purified from endonuclease III-deficient *E. coli* cells *(120)* and found to exhibit a thymine glycol DNA glycosylase activity as well as an AP lyase activity. The gene for endonuclease VIII *(nei)* has been isolated *(87)*. Endonuclease VIII is similar in most repair capacities to Tg DNA glycosylase, except that it has β,δ lyase activity rather than β lyase activity *(87)*. All of the endonuclease III substrates thus far analyzed are also substrates of endonuclease VIII (Table 1; *87*, reviewed in *180*). When considering the total repair capacity of *E. coli* toward thymine glycol, endonuclease VIII contributes <10% (reviewed in *181*), and it is not yet clear why *E. coli* has this apparently redundant enzyme.

2.5. Pyrimidine-Dimer DNA Glycosylase (DenV$^+$)

UV irradiation of DNA causes saturation of the 6-4 bonds of adjacent cytosine or thymine residues, and formation of a four-membered cyclobutyl ring linking the pyrimidines (Fig. 5). These cyclobutane pyrimidine dimers (CPDs) can be lethal or mutagenic if unrepaired. In *E. coli*, CPDs are either monomerized by photolyase or repaired by nucleotide excision repair (*see* Chapter 2, this volume, and Chapter 2 in vol. 2).

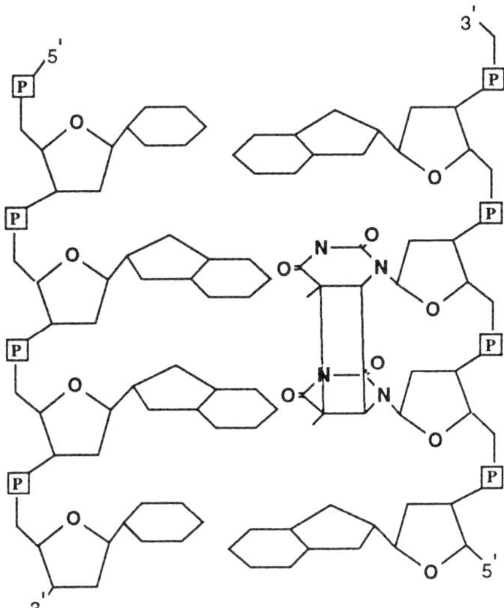

Fig. 5. A schematic representation of the structure of CPD in DNA drawn to indicate the position of the bonds connecting the pyrimidines. The bond lengths and distances between atoms are not proportional.

When *E. coli* viruses infect cells, damage on the phage DNA is generally repaired by the host DNA repair systems. However, *E. coli* T4 phage maintains its own DNA repair gene, namely the T4 *denV*+ gene (DNA endonuclease V) that encodes a pyrimidine-dimer DNA glycosylase (PD DNA glycosylase; *58*). Should *E. coli* become infected with T4 phage, the bacteria acquire an enhanced ability to repair CPDs, which in turn provides the phage with increased resistance to UV irradiation *(58,113,152)*. A similar PD DNA glycosylase has also been characterized from *Micrococcus luteus (89,155,161)*.

Like endonuclease III, PD DNA glycosylase was first identified as an endonuclease that cleaves damaged DNA. In this case, however, the activity was specific for UV-irradiated DNA. Moreover, the number of enzyme-generated DNA strand breaks corresponded to the number of CPDs in the substrate, so the protein was presumed to be a CPD-specific endonuclease.

Years later, experiments were done to investigate the possibility that neighboring pyrimidines may be differentially susceptible to dimerization depending on the identity of the surrounding nucleotides. To measure dimer formation in different nucleotide contexts, DNAs of known sequence were treated with UV and subsequently incubated with the *denV*$^+$ gene product. Since the *denV*$^+$ gene product was known to cleave at CPDs, the lengths of the enzyme-generated DNA fragments could be analyzed to reveal the location of CPDs. As predicted, dimer formation depended on the neighboring nucleotides. Unexpectedly, however, careful analysis of the DNA cleavage products revealed that incision did not occur 5' to the dimer as originally believed, but rather between the pyrimidines of the dimer (Fig. 6; *78*). What was most surprising was that the mobility of the DNA fragments indicated that the 5' incision product had lost its pyrimidine, suggesting that the *denV*$^+$ gene product had an associated glycosylase activity.

The cleavage mechanism of PD DNA glycosylase was ultimately clarified by adding photolyase into the reaction (Fig. 6). In these experiments, UV-damaged DNA was first incubated with the *denV*$^+$ gene product and subsequently incubated with photolyase. When photolyase monomerized the pyrimidine dimers, a free thymine residue was released, thus indicating that *denV*$^+$ codes for an enzyme with DNA glycosylase activity *(78)*. Similar studies were done using additional UV exposure (in place of photolyase) to monomerize the thymine dimers, and these results likewise showed that the *denV*+ encoded enzyme cleaves the *N*-glycosylic bond, permitting the release of the thymine base on monomerization *(45,138)*. Together these studies showed definitively that the PD protein acts as a DNA glycosylase to release the 5'-pyrimidine and then cleaves the backbone 3' to the newly formed abasic site (Fig. 6). The PD DNA glycosylase is quite different from other DNA glycosylases in that its substrate is not released from DNA as a free base, but rather remains attached to its neighboring pyrimidine. Removal of this damage is probably mediated by the combination of monomerization and 5'- to 3'-exonuclease degradation. Crystal structure of the enzyme complexed with a duplex DNA substrate revealed an unusual DNA conformation where the base complementary to the 5' side of the CPD is flipped out of the duplex DNA structure and into a cavity of the bound protein *(178a)*. Base flipping may be a common mechanism for substrate recognition by DNA glycosylases (*see* Section 2.6.).

2.6. *Uracil DNA Glycosylase*

Uracil is a normal constituent of RNA. If available, deoxyuridine is efficiently integrated into *E. coli* DNA opposite adenine without significantly affecting cell viability *(182)*. Since uracil pairs with adenine as effectively as thymine in DNA, why do cells have thymine at all? Under physiological conditions, cytosine frequently undergoes deamination to form uracil *(107)*. Clearly, it is critical that cells remove uracil from U:G base pairs, since subsequent DNA replication would result in CG to TA mutations. Thus, one major evolutionary advantage of thymine over uracil is that it facilitates specific recognition and repair of potentially mutagenic deaminated cytosine residues of U:G base pairs in DNA.

To reduce dUTP incorporation into DNA, deoxyuridine triphosphatase (dUTPase) converts dUTP to dUMP, thus limiting the intracellular pool of available dUTP (an intermediate in the biosynthesis of thymidine). In wild-type *E. coli*, some uracil still arises in DNA from low-level incorporation of dUTP, from spontaneous deamination of cytosine, or from exposure to DNA-damaging agents, such as nitrous acid or ionizing radiation *(151,181)*. Uracil DNA glycosylase is highly efficient in releasing uracil from a UG mispair, and will also release uracil from UA pairs, but less effectively *(179)*. Removal of U from UA pairs is not critical, since it is not mutagenic; certainly this enzyme has evolved for efficient release of U from UG mispairs. The biological role of this enzyme is underscored by the observation that *E. coli* lacking uracil DNA glycosylase have increased susceptibility to spontaneous CG to TA mutations *(51)*.

In the absence of both dUTPase and uracil DNA glycosylase, as much as 20% of the thymine in the DNA is substituted with uracil without ill effect *(182)*. However, cells in which nearly all thymine has been replaced by uracil have a reduced viability, presumably because DNA that contains a high number of uracil residues has an altered secondary structure that interferes with critical DNA–protein interactions *(52)*. When

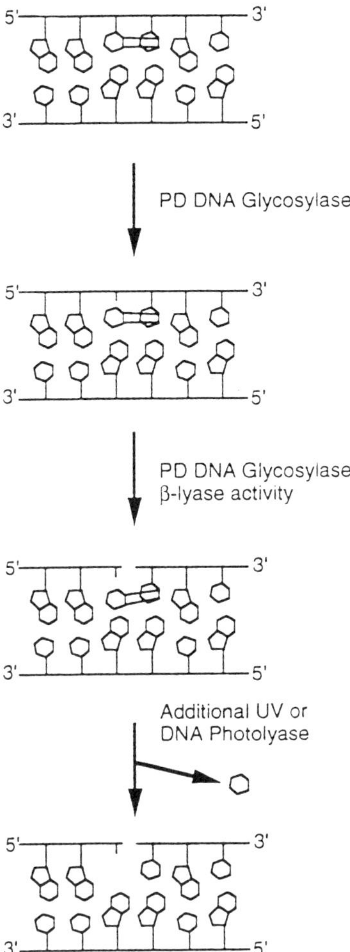

Fig. 6. The mechanism of CPD repair by the PD DNA glycosylase, as described in the text. Note that the 5' pyrimidine of the CPD is released from the backbone and the β-lyase activity cleaves the backbone 3' to this released pyrimidine.

E. coli M13 phage are propagated in cells lacking both dUTPase and uracil DNA glycosylase, their single-stranded DNA accumulates high levels of uracil. These phage tolerate uracil without ill effects, but lose their ability to infect wild-type *E. coli* cells. Apparantly, the infecting phage DNA is subject to "repair" by uracil DNA glycosylase and AP endonucleases, which together fragment the single-stranded genome.

Recent structural studies has shown that uracil DNA glycosylases employ a common enzymatic mechanism, termed "base flipping," to catalyze specific removal of uracil from DNA (reviewed in ref. *144*). In this reaction, the substrate uracil base is "flipped" out of the DNA and into the active site pocket of the protein. Special constraints and specific interactions between the flipped base and the amino acids lining the binding pocket restrict the substrate range of the enzyme.

Unlike most other DNA glycosylases that catalyze base release from only double-stranded DNA, uracil DNA glycosylase releases uracil from either double- or single-stranded DNA. Thus, there appears to be an unknown mechanistic difference between uracil DNA glycosylase and other DNA glycosylases. This distinction may be related to the necessity for uracil DNA glycosylase to act exclusively on DNA. Clearly, efficient removal of uracil from RNA would be detrimental.

2.7. 3-Methyladenine DNA Glycosylases (Tag and AlkA)

Although the 3-methyladenine (3MeA) lesion (Fig. 3) is a small minor groove lesion that does not appear to cause a dramatic distortion in DNA secondary structure, it blocks chromosome replication in *E. coli* and is therefore potentially lethal *(11,101)*. Since organisms produce endogenous methylating agents *(114,140)*, it is logical to speculate that DNA repair systems would have evolved to cope with 3MeA lesions. *E. coli* have two 3MeA DNA glycosylases, Tag and AlkA. Tag, a 21 kDa protein, is constitutively expressed, whereas the 30-kDa AlkA protein is induced as part of the adaptive response to low levels of alkylation damage *(148)*. Although both Tag and AlkA are effective in releasing 3MeA from DNA, their amino acid sequences are not similar *(88,147)*.

Both Tag and AlkA are able to repair several types of DNA damage in addition to 3MeA (*see* Table 1). However, the substrate range of AlkA is significantly more diverse than that of Tag. In fact, Tag has only been shown to repair 3-alkylpurines and is not able to repair O^2-alkylpyrimidines or major groove alkylated purines, which are substrates of AlkA *(8,73,90,141,175)*. In addition, Tag is substrate-inhibited by exposure to free 3MeA, whereas AlkA is not *(175)*. It is therefore likely that the enzyme mechanics of these two 3MeA DNA glycosylases are fundamentally different. In particular, it has been suggested that the substrate binding pocket for Tag is more restricted, thus narrowing the substrate range of this enzyme.

Early studies of AlkA revealed that this enzyme can repair both minor groove N3 and major groove N7 alkylated purines as well as O^2-alkylated pyrimidines *(119,175)*. It was originally thought that the primary function of AlkA is to protect against 3MeA lesions, since high levels of Tag can protect AlkA-deficient cells from killing by simple methylating agents *(88,147)*. This conclusion was also consistent with the observation that 7-methylguanine (Fig. 3), the major methylation product, appears to be a benign lesion *(80)*. However, subsequent research has revealed that the substrates of AlkA are quite diverse. Thus, it would appear that AlkA has many important in vivo substrates, since AlkA not only protects against killing by 3MeA, but also against the killing by more complex alkyl lesions and other modified bases not repaired by Tag.

Studies of the in vitro repair activity of AlkA have revealed the following types of substrates: a wide range of N3 and N7 alkylpurines, an intrastrand purine crosslink, cyclic purine lesions, deaminated adenine (hypoxanthine), and oxidized derivatives of thymine (*see* Table 1). It is difficult to imagine how one enzyme can repair such a diverse set of lesions, both in the major and minor grooves, and on purines and pyrimidines. In addition, not all of these substrates are the consequence of alkylation damage. Although many of these lesions are thought to potentially block DNA replication, some of the recently discovered substrates of AlkA, such as hypoxanthine, are known to promote mispairing during replication and are therefore mutagenic. It is noteworthy that *E. coli* lacking either AlkA and Tag have not been shown to have increased spontaneous mutation rates *(90)*. Some possible explanations for this observation are that:

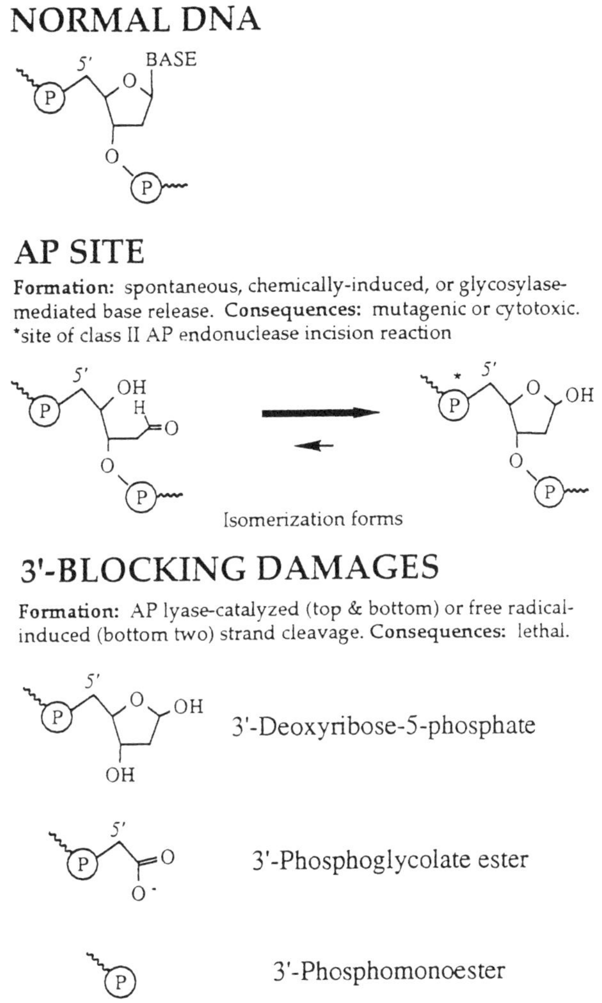

NORMAL DNA

5' BASE

AP SITE

Formation: spontaneous, chemically-induced, or glycosylase-mediated base release. **Consequences:** mutagenic or cytotoxic.
*site of class II AP endonuclease incision reaction

Isomerization forms

3'-BLOCKING DAMAGES

Formation: AP lyase-catalyzed (top & bottom) or free radical-induced (bottom two) strand cleavage. **Consequences:** lethal.

3'-Deoxyribose-5-phosphate

3'-Phosphoglycolate ester

3'-Phosphomonoester

Fig. 7. Structures of an AP site and several common 3'-blocking damages. The mechanisms of formation and the biological consequences of these damages are described here and in the text.

1. The repair of mutagenic lesions (e.g., hypoxanthine) by AlkA and Tag is so inefficient that the spontaneous mutation rate is not affected by the presence or absence of these enzymes;
2. The assay is not sensitive enough to detect subtle changes in spontaneous mutation rates; or
3. Repair of the mutagenic lesions is carried out by other repair pathways.

Thus, at present, it is thought that the major biological role of AlkA is to protect cells against killing. However, a role for AlkA in preventing mutation has not yet been definitively ruled out.

3. AP ENDONUCLEASES

Independent of DNA glycosylase-mediated base removal, AP sites (Fig. 7) arise in DNA via spontaneous hydrolysis of the *N*-glycosylic bond that links normal bases to DNA *(107)*. Moreover, DNA-damaging agents not only produce substrates for DNA

Table 2
***E. coli* AP Endonucleases**

	Exonuclease III	Endonuclease IV
Enzymatic activities	Class II AP endonuclease 3'-Diesterase 3'- to 5'-Exonuclease RNase H	Class II AP endonuclease 3'-Diesterase
General features	Stimulated by Mg^{2+} Requires double-stranded substrates	EDTA insensitive (in the presence of DNA) Contains essential metals
Null phenotype (sensitivities)	MMS, mitomysin C > *t*-$ButO_2H$ >> γ-rays, UV, bleomycin	*t*-$ButO_2H$, bleomycin > MMS, mitomycin C >> γ-rays, UV, H_2O_2

glycosylases, but they also generate certain base modifications that destabilize the *N*-glycosylic bond leading to a faster rate of hydrolytic base release. Studies have estimated that ~10,000 AP sites are generated per day per human cell under normal physiological conditions.

Abasic sites are noninstructional DNA lesions and are potentially mutagenic as well as cytotoxic *(111)*. In vitro and in vivo studies have shown that on encountering an AP site, both DNA and RNA polymerases of prokaryotes normally pause, but can bypass the lesion and insert preferentially an adenine opposite the AP site *(162,193)*. Repair of AP sites in DNA is initiated by class II AP endonucleases (Fig. 1). As noted before, these proteins cleave immediately 5' to an AP site in a hydrolytic reaction generating a normal 3'-OH group, an effective primer for DNA polymerase, and a 5'-abasic residue, which requires removal prior to DNA repair synthesis (Figs. 1 and 2). Moreover, class II AP endonucleases remove 3'-blocking damages from DNA (Fig. 1) to create normal 3'-OH ends. Nonconventional 3'-blocking termini (Fig. 7) are frequent products of attack of DNA by reactive oxygen species, and are refractory to DNA repair synthesis and, hence, potentially lethal. 3'-Blocking fragments are also formed by spontaneous or AP lyase-catalyzed β-elimination reactions at sites of base loss in DNA (Fig. 2). In *E. coli*, there exist two major AP endonucleases/3'-repair diesterases, exonuclease III and endonuclease IV (*see* Table 2), which represent the two evolutionarily conserved families of AP endonucleases discovered to date (reviewed in ref. *41*).

Exonuclease III was originally identified as a factor that removed 3'-phosphoryl groups from DNA, thus acting as a priming activity for DNA polymerase I *(142)*. This 3'-phosphatase activity copurified with a 3'- to 5'-exonuclease activity, and both activities were shown to reside in the same polypeptide *(143)*. Subsequently, exonuclease III, first termed endonuclease II *(56,57)*, was found to be the predominant AP endonuclease of *E. coli (183,190)*. Exonuclease III also demonstrates ribonuclease H activity that selectively degrades the RNA strand of an RNA–DNA duplex molecule *(146)*. These activities of exonuclease III require double-stranded substrates, are stimulated by the presence of Mg^{2+}, and are inhibited by the addition of ethylenediaminetetraacetic acid (EDTA). The 3'- to 5'-exonuclease and RNase H activities of exonuclease III account for >90% of these activities in bacteria, yet the biological roles of these functions are unknown. The biological contributions and in vivo importance of the 3'-phosphatase and AP endonuclease activities are discussed below.

Crystallographic studies have revealed that the 3-dimensional structure of exonuclease III, which consists of an αβ-sandwich motif, is strikingly similar to the overall structure of DNase I and RNase H, two other hydrolytic enzymes *(126)*. Structural analysis also located the active site and the divalent metal binding site in exonuclease III. Mol and colleagues *(126)* proposed that exonuclease III cleaves at abasic sites in DNA via a hydrolytic reaction that is facilitated by a single metal ion (presumably Mg^{2+}), which depolarizes the phosphate–oxygen bond 5' to the lesion permitting a nucleophilic attack by a hydroxide ion. This proposed mechanism of catalysis may apply to all the major activities (AP endonuclease, 3'-phosphatase, 3'- to 5'-exonuclease, and RNase H) of exonuclease III, consistent with there being a single active site for this 28-kDa protein, as first predicted by Weiss in 1976 *(183)*. Although site-directed mutagenesis has not been reported for exonuclease III, mutagenesis of the corresponding active site residues in its mammalian homolog, Hap1 or Ape *(43,145)*, indicate the importance of these amino acids in catalysis *(5)*.

The gene encoding exonuclease III, *xthA*, was identified by screening for *E. coli* mutants exhibited a defective nuclease activity *(123)*. Cells lacking exonuclease III display a modest hypersensitivity to the alkylating agent methylmethane sulfonate (MMS), a compound that indirectly generates a large number of abasic sites. Thus, exonuclease III functions in vivo in the repair of AP lesions, a finding consistent with the in vitro biochemical evidence. The role of exonuclease III in the initiation of AP site repair was further supported by experiments in which the dUTPase gene *(dut)* was deleted. The absence of dUTPase protein results in an increased intracellular concentration of dUTP and therefore a higher content of uracil in DNA through misincorporation during DNA replication. These uracil bases are removed by uracil DNA glycosylase producing a high number of abasic sites in DNA. Bacteria harboring both dUTPase and exonuclease III deficiencies are inviable, presumably a result of poor AP site repair *(169)*.

The role of exonuclease III in repairing oxidative DNA damage was highlighted by the demonstration that *xth* mutants are extremely sensitive to H_2O_2 *(40)*. Analysis of chromosomal DNA from H_2O_2-treated *xth* cells revealed an accumulation of 3'-fragments that are refractory to DNA polymerase activity. Presumably these 3'-blocking groups are 3'-phosphate and 3'-phosphoglycolate (3'-PG) residues (Fig. 7), which are products of free radical attack on DNA *(65,81)*. This conclusion is supported by the fact that exonuclease III comprises >99% of the cellular 3'-phosphatase activity *(123)*, and that exonuclease III demonstrates similar repair activities for 3'-phosphoglycol aldehyde residues (a synthetic analog to 3'-PG), 3'-phosphate groups, and AP sites in vitro *(44)*. Exonuclease III does display an endonucleolytic activity at urea residues in oxidized DNA *(98)*, but it would appear unlikely that this lesion contributes significantly to the hypersensitivity of *xth* mutants to H_2O_2 *(44)*. *xth* mutants demonstrate little, if any, increased sensitivity to UV light, ionizing radiation or bleomycin. Thus, in general, exonuclease III appears primarily to play a role in the repair of AP sites and 3'-oxidative damages in DNA.

Analysis of *xth* mutants revealed that these bacteria still retain significant levels (~10% of wild-type) of AP endonuclease activity. The residual endonuclease activity, later termed endonuclease IV, was shown to be heat-stable and EDTA-insensitive, clearly distinguishing it from exonuclease III *(109,110)*. The presence of a second AP endonuclease likely accounts for the fact that *xth* mutants are only mildly sensitive to MMS. Endonuclease IV also exhibits a 3'-repair diesterase activity capable of activating H_2O_2-damaged chromosomal DNA for DNA polymerase activity in vitro *(44)*. However, based on the extreme sensitivity of *xth* mutants to H_2O_2, it appears that the major enzyme responsible for the removal of 3'-phosphate damages produced by free radical attack of DNA in vivo is exonuclease III.

As noted, endonuclease IV was identified as an EDTA-resistant AP endonuclease activity. However, Levin et al. *(105)* discovered that, in the absence of substrate DNA, endonuclease IV protein can be inactivated by preincubation with chelating agents, such as EDTA or 1,10-phenanthroline, suggesting that the enzyme contains an essential metal component. Atomic absorption spectrometry *(106)* revealed that endonuclease IV contains multiple zinc atoms (2.4–3.1 atoms/endonuclease IV molecule) and substoichiometric amounts of manganese (0.6–0.8 atoms/protein molecule); the variation in the values probably stems from a mixed population of protein molecules. Enzyme that has been inactivated by treatment with a metal chelating agent can be reactivated to varying degrees by the reintroduction of different metals. The role of these metals in the mechanistic, structural, or regulatory functions of endonuclease IV awaits further study.

The gene encoding endonuclease IV, *nfo*, was isolated by identifying plasmids that contained *E. coli* genomic fragments expressing EDTA-resistant AP endonuclease activity *(37)*, and used to generate bacterial strains lacking endonuclease IV *(37)*. *nfo* mutants display an increased sensitivity to the alkylating agents MMS and mitomycin C, and to the oxidizing agents *tert*-butyl hydroperoxide (*t*-$ButO_2H$) and bleomycin. Deletion of *nfo* increased the killing of *xth* mutants by MMS, H_2O_2, *t*-$ButO_2H$, and γ-rays. Taken together, these findings indicate an involvement of endonuclease IV in the repair of AP sites and oxidative DNA damages, revealing that many of the repair activities of exonuclease III and endonuclease IV overlap. However, the differences in sensitivities of the *nfo* or *xth* single mutants to agents such as bleomycin or H_2O_2, indicate that these repair enzymes maintain distinct substrate specificities (Table 2). It was recently reported that bleomycin generates a specific DNA lesion in vivo that requires endonuclease IV for efficient repair *(104)*. Furthermore, comparison of the incision activity at synthetic AP site analogs clearly distinguishes these microbial enzymes *(168,188)*. Most notably, oxidized AP sites or complex damages (e.g., AP sites opposing strand breaks) that are generated by bleomycin or neocarzinostatin exposure, and the anoxic radiolysis product α-deoxyadenine, are clearly better substrates for endonuclease IV than exonuclease III *(77,84,136)*. Thus, although these two *E. coli* repair enzymes share many common substrates, they also possess distinct substrate preferences, likely explaining the different sensitivities of *nfo* and *xth* single mutants to various DNA-damaging agents. Exonuclease III and endonuclease IV homologs have been identified from several evolutionarily distinct organisms (reviewed in *42*).

Endonuclease IV is normally present at ~1/10th the amount of exonuclease III. However, Chan and Weiss *(27)* found that endonuclease IV is induced to levels comparable

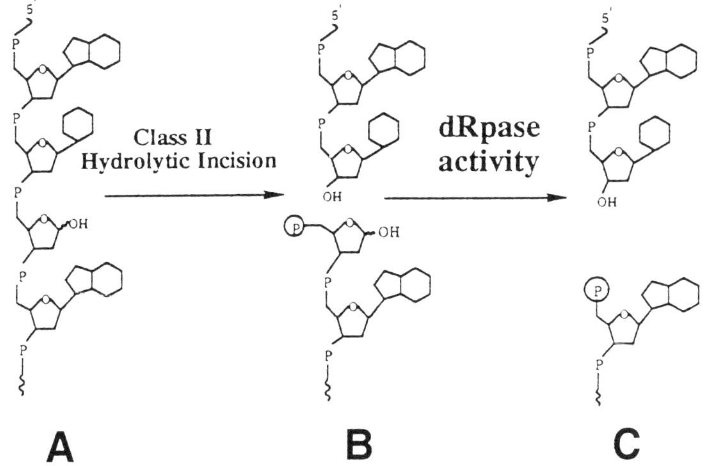

Fig. 8. Removal of dRp residues that arise from class II AP endonuclease catalyzed hydrolytic incision at AP sites. **(A)** The AP-containing strand of a duplex DNA molecule is depicted. **(B)** AP endonuclease cleavage generates a normal 3'-OH group and a 5'-abasic residue (dRp). **(C)** The 5'-dRp moeity can be removed by RecJ or Fpg/MutM to generate a single nucleotide gap. The complementary strand (not depicted) is used as the template strand for DNA repair synthesis, prior to ligation and the completion of BER.

to exonuclease III by superoxide-generating agents, which activate the *soxRS* response system (*see* Chapter 5). Similarly, nitric oxide can also stimulate the *soxRS* regulon and induce endonuclease IV levels *(130)*. Why endonuclease IV is the only DNA repair enzyme upregulated in such a response remains unknown, but it may reflect the types of damages generated under such conditions.

4. DEOXYRIBOPHOSPHODIESTERASE

Incision at an abasic site in DNA by a class II hydrolytic AP endonuclease, believed to be the major pathway for repair of AP sites in vivo *(38)*, generates a normal 3'-OH group and a 5'-terminal 2-deoxyribose-5-phosphate residue (dRp; Fig. 8). The primary candidate for the removal of these 5'-termini was initially the 5'- to 3'-exonuclease activity of *E. coli* DNA polymerase I (pol I). However, dRp residues were found to be poor substrates for this enzyme; indeed dRp residues hinder the progress of DNA pol I *(127)*. Thus, since dRp must be removed for efficient gap filling and ligation, a search for alternative bacterial proteins that would excise free dRp from DNA was launched.

Franklin and Lindahl *(55)* isolated an ~55-kDa *E. coli* protein that excises free dRp from DNA in a Mg^{2+}-dependent hydrolytic fashion. This activity, deoxyribophosphodiesterase (dRpase), was shown to release ~50 pmol of dRp min^{-1}/mg at 37°C under optimal reaction conditions and was later ascribed to the 5' to 3' single-strand-specific DNA exonuclease encoded by *recJ (48)*. RecJ exhibits dRpase activity in vitro on the "classical" 5'-abasic residues produced by endonuclease IV (class II) incision at AP sites in double-stranded DNA, but removes dRp residues 8–10 times faster from single-stranded DNA. Moreover, RecJ does not demonstrate exonuclease activity on duplex DNA, explaining the short patch repair observed at AP sites in DNA *(46,47)*. In vitro reconstitution experiments have indicated a role for RecJ in BER *(46)*.

The Fpg/MutM DNA glycosylase/AP lyase (*see* Section 2.1.) also promotes the release of dRp residues, but via a β-elimination reaction and at nearly two times the rate of RecJ *(67)*. In contrast to RecJ, the dRpase activity of Fpg/MutM is suppressed by Mg^{2+} *(48)*. Interestingly, *E. coli fpg/recJ* single or double mutants show nearly wild-type resistance to MMS, an agent that induces AP site formation *(48)*. Furthermore, these mutants are capable of maintaining and propagating λ phage DNA that contain numerous uracil bases that are readily removed by uracil DNA glycosylase. These findings suggest that *fpg/recJ* mutants are not defective in the repair of AP sites and that there probably exist alternative dRp-excision activities. Analysis of several other potential dRpase candidates, such as exonuclease III, V, and VII, and endonuclease I, III and IV, revealed that none of these enzymes remove free dRp from DNA *(48)*. Previous studies had reported that exonuclease I contains dRpase activity *(149)*, however, this finding was not reproducible and may have resulted from copurification of RecJ protein with exonuclease I *(48)*. Perhaps the nonenzymatic release of dRp residues by intracellular polyamines or basic proteins *(3)* in *fpg/recJ* double mutants can suffice for proficient BER. Alternatively, the ability of DNA pol I to release dRp in short oligonucleotides, although inefficient *(137)*, may compensate for the absence of normal dRp excision. Since dRpase is present at only 50 mol/cell, one might predict that there exist additional cellular components that remove 5'-dRp moieties.

5. DNA POLYMERASES

Incision by a class II AP endonuclease and removal of the 5'-abasic residue during BER leaves a one nucleotide gap that if left unrepaired can be lethal. *E. coli* possess three distinct DNA polymerases. DNA pol I is thought to be the major protein responsible for normal DNA repair synthesis, particularly at short gaps in DNA, but DNA polymerase II (pol II) and DNA polymerase III (pol III) contribute to specific and overlapping aspects of DNA repair (Table 3).

E. coli DNA pol I was so named because it was the first enzyme isolated with the ability to incorporate radiolabeled dNTPs into a template-primer substrate *(6,103)*. Pol I is a single polypeptide subunit with a molecular mass of 103 kDa that possesses three major activities: a 5'- to 3'-polymerase activity, a 3'- to 5'-exonuclease activity, and a 5'- to 3'-exonuclease activity *(96)*. Proteolytic digestion of pol I uncovered two functional units *(22)*: a carboxy-terminal fragment (the Klenow fragment) that contains the polymerase and 3'- to 5'-proofreading function, and a smaller amino-terminal fragment that possesses the 5'- to 3'-exonuclease activity involved in degrading the RNA portion of Okazaki fragments and displacing the 5'-strand during nick translation. Crystals of the Klenow fragment unveiled a small domain, which contains a tightly bound metal ion that interacts with both the protein and the 5'-phosphate of the deoxynucleoside monophosphate, and a large domain, which contains a deep crevice suitable for binding to double-stranded B-DNA *(134)*. Metal binding by both the enzyme and the DNA substrate is essential for the pol I catalytic reaction *(128)*. A flexible subdomain was also uncovered that appears to encircle the DNA substrate, enhancing processivity. Complementary site-directed mutagenesis studies have helped to identify several of the amino acids involved in the enzymatic reaction *(96)*.

Table 3
***E. coli* DNA Polymerases**

	Pol I[a]	Pol II[a]	Pol III, core[a]
General properties			
Structural gene	*polA*	*polB*	*polC* or *dnaE*
Size (kDa)	103	90	(130, 27.5, 10)[b]
Estimated mol/cell	400	40	10–20
Estimated turnover number (nts/min at 37°C)	700	30	>10,000
Enymatic activities			
5' to 3' polymerase	Yes	Yes	Yes
3' to 5' Exonuclease (proofreading)	Yes	Yes	Yes
5' to 3' Exonuclease	Yes	No	No
Utilization of Template-Primer Substrates[c]			
Single-stranded DNA with primer	Yes	No	No
Nicked duplex	Yes	No	No
Gapped duplex			
<100 nts	Yes	Yes	Yes
>100 nts	Yes	No	No
Primary in vivo role	Normal DNA gap repair	Adaptive mutagenesis	Replication and UV mutagenesis[a]

[a]"Yes" or "no" indicates the ability or inability to act in such capacity.
[b]Molecular mass of α, ε, and θ subunits of the core enzyme, respectively.
[c]None of the polymerases act on intact duplex DNA.

A unique feature of DNA pol I is its ability to promote replication at nicks independent of accessory proteins. Purified pol I exhibits an optimal turnover number of ~700 nucleotides polymerized/min at 37°C and displays a processivity of ~20 nucleotides *(95)*. The ability of pol I to synthesize DNA at nicks, and its low processivity, suggests a role for pol I in short patch DNA repair synthesis. Mutants lacking *polA*, the gene encoding pol I, are hypersensitive to several DNA-damaging agents, including UV light, X-rays and MMS *(34,68,118,177)*. In general, polA mutants grow normally, but are defective in a number of processes that require gap filling *(96)*. The relative abundance of pol I (200–400 mol/cell) as compared to pol II (40 mol/cell) and pol III (10–20 mol/cell) also suggests a role for pol I in DNA repair synthesis *(61)*. Nonetheless, all three DNA polymerases are able to function both in vivo and in vitro on gapped DNA substrates created by UvrABC endonuclease *(33,116)*. Thus, a preponderance of evidence implicates pol I as the major DNA repair polymerase, but the participation of the other enzymes cannot be ignored or excluded.

DNA pol II, the second polymerase identified and characterized *(39,186,187)*, is encoded by *polB (29)*, which was later shown to be allelic to *dinA*, a DNA-damage-inducible gene that is part of the error-prone SOS response in *E. coli (13,86)*. Indeed, pol II expression is elevated sevenfold on activation of the SOS regulon by UV irradiation *(14)*. Pol II is a single polypeptide with a molecular mass of 90 kDa and shows a high degree of homology to the replicative DNA polymerases of eukaryotes *(85)*. The enzyme prefers gapped DNA substrates and exhibits a poor processivity (~5 nucleotides/binding event) in the absence of accessory factors (which are protein subunits of the pol III holoenzyme) and single-stranded DNA binding protein (SSB). Addition of these factors with SSB can stimulate the processivity of pol II ~150–600-fold *(15,83)*. The in vivo function of pol II has remained elusive, since *polB* mutants are not defective in DNA replication or repair *(25,82)*. Moreover, it was surprising to find that despite being part of the SOS response, *polB* mutants display normal UV mutability as well as wild-type resistance to UV *(156)*. Thus, pol II does not appear to play a critical role in UV-induced SOS mutagenesis or in the process of nucleotide excision repair of DNA photoproducts in vivo. However, in the absence of pol I and pol III, pol II can participate in the repair of UV photodamages *(116,167)*. Studies using either a cell-free or a reconstituted in vitro system indicated that although pol III is the major polymerase involved in UV-induced mutagenesis, pol II can substitute, although poorly, for pol III in this capacity *(32,176)*. Perhaps pol II serves as a redundant polymerase molecule for such functions as error-prone repair synthesis, or alternatively, there may exist a yet unidentified specialized role for this protein. Recently, it was shown that *E. coli* harboring a proofreading-defective pol II, in place of the wild-type polymerase, exhibit an increase in adaptive mutations, suggesting a role in this phenomenon for pol II *(54)*.

DNA pol III is typically thought of as the replicative enzyme, but is capable of functioning in other areas of DNA metabolism *(96)*. The pol III holoenzyme is a dimer of two multiprotein complexes that each consist of 10 different subunits forming an assembly of two polymerase molecules with a molecular mass of 900 kDa; these various protein components are encoded by a series of *dna* genes (reviewed in *93*). The dimerization of two polymerase molecules is compatible with the model that two associated polymerases simultaneously conduct the coordinated synthesis of both strands of chromosomal DNA *(157)*. The pol III holoenzyme can be resolved into a smaller complex termed the core enzyme, which consists of three subunits (α, ε, θ), capable of performing DNA synthesis on short gapped substrates *(53)*; this smaller complex lacks the tremendous processivity associated with the pol III holoenzyme. On association of the other pol III accessory proteins, particularly the β-subunit, with the core enzyme, the polymerase exhibits processivity of up to 100,000 nucleotides/binding event at a rate of nearly 1000 nucleotides/s *(28,164)*. X-ray crystallographic studies have shown that the β-subunit dimerizes to form a ring-shaped structure that clamps around the DNA template allowing the holoenzyme to tether itself to DNA and to replicate in a "sliding clamp" mechanism *(164)*. Since pol III is present at only 10–20 mol/cell and since there exist multiple sites for DNA synthesis during replication, particularly along the lagging strand, pol III needs to be efficiently recycled. To accomplish this feat, pol III disengages from the β-clamp once it has completed synthesis of an Okazaki fragment and immediately reassociates with the preassembled sliding clamp of the upstream RNA primer *(165)*. Bacteria containing a temperature-sensitive mutation in the synthetic subunit of DNA Pol III (the α-subunit encoded by *dnaE*) are defective in normal DNA replication at nonpermissive temperatures (43°C) and are therefore less viable, clearly implicating pol III in the replicative process *(63)*.

The synthetic activity of DNA pol I can substitute for the α-subunit of DNA pol III when present with a *gyrB* mutation (*pcbA1*; *23*), a feature that has permitted further analysis into the physiological contribution(s) of DNA pol III. Using this and other approaches, *dnaE* mutants of pol III did not exhibit an increase in UV-induced SOS mutagenesis at restrictive temperatures, indicating that pol III is necessary for mutagenic bypass of UV photoproducts *(21,75,115,153)*. Reintroduction of a plasmid engineered to express wild-type Pol III restored normal UV-induced mutagenesis *(75)*. A smaller form of the β-subunit (β*) is synthesized in response to UV irradiation, and this factor may act as an alternative sliding clamp subunit for pol III during error-prone DNA synthesis *(135,158,159)*. One study has shown that the holoenzyme is required for the optimal repair of DNA damage generated by MMS and H_2O_2 *(74)*. Thus, although pol I is the major polymerase responsible for the repair of MMS- and H_2O_2-induced DNA damage *(74)*, there appears to be a role for Pol III in what is presumably BER. There may be a distinct group of lesions generated by MMS and H_2O_2 are that not corrected by BER and thus require pol III for effective repair, or for mutagenic bypass. It should be re-emphasized that, in the absence of one or two of the DNA polymerases, the remaining enzyme(s) can often acquire a surrogate replicative or repair function in vivo, and that several of the accessory components can functionally interchange between the different polymerases. Thus, clearly defined in vivo roles for the *E. coli* DNA polymerases remain unclear.

6. DNA LIGASE

Following DNA repair synthesis, a 5'-phosphoryl group and 3'-hydroxyl group remain juxtaposed in nicked duplex DNA. To seal this potentially harmful DNA strand break, *E. coli* DNA ligase catalyzes the synthesis of a phosphodiester linkage in a nicotinamide adenine dinucleotide (NAD)-dependent reaction (Fig. 9). This DNA end-joining activity of *E. coli* was originally identified as a function that rejoined DNA molecules during genetic recombination *(92,121,191)* or converted linear bacteriophage λ DNA to covalently closed circular duplexes *(9,191)*. In 1967, several laboratories simultaneously identified the single protein of *E. coli* that catalyzes the ligation of DNA ends *(35,62,64,133,184)*. It should be emphasized that while DNA ligase executes a vital step in the BER process, it also plays a critical role in many aspects of DNA metabolism, including the replication and recombination of DNA *(102)*.

E. coli DNA ligase is a single polypeptide chain of ~75 kDa and is present at 200–400 mol/cell *(102)*. Through a series of covalently linked intermediates, *E. coli* DNA ligase catalyzes the formation of a phosphodiester bond in nicked duplex DNA *(102)*. The first step of this reaction involves the nucleophilic attack of NAD by the ε-amino group of a single lysine residue in the protein, generating a ligase–adenylate complex and a nicotinamide mononucleotide product *(70)*. The adenyl group is then transferred from the enzyme to the 5'-phosphoryl group of nicked DNA, producing a DNA-adenylate intermediate that is very short-lived under steady-state conditions *(76,132)*. The final step entails the attack of the 5'-phosphate-adenylate by the opposing 3'-hydroxyl group to form the phosphodiester linkage. Steady-state kinetic analysis showed that *E. coli* DNA ligase has a K_m for single-strand breaks of 40–60 n*M* and a k_{cat} of 1.4 min^{-1} in the absence of NH_4^+ or 28 min^{-1} in its presence *(124)*. The manner in which monovalent cations, such as NH_4^+, Rb^+, or K^+, stimulate DNA ligase activity is unknown, but the activation would appear to occur prior to the formation of the ligase–adenylate intermediate. One theory is that monovalent cations induce a conformational change in the ligase protein producing a more catalytically efficient enzyme. However, initial

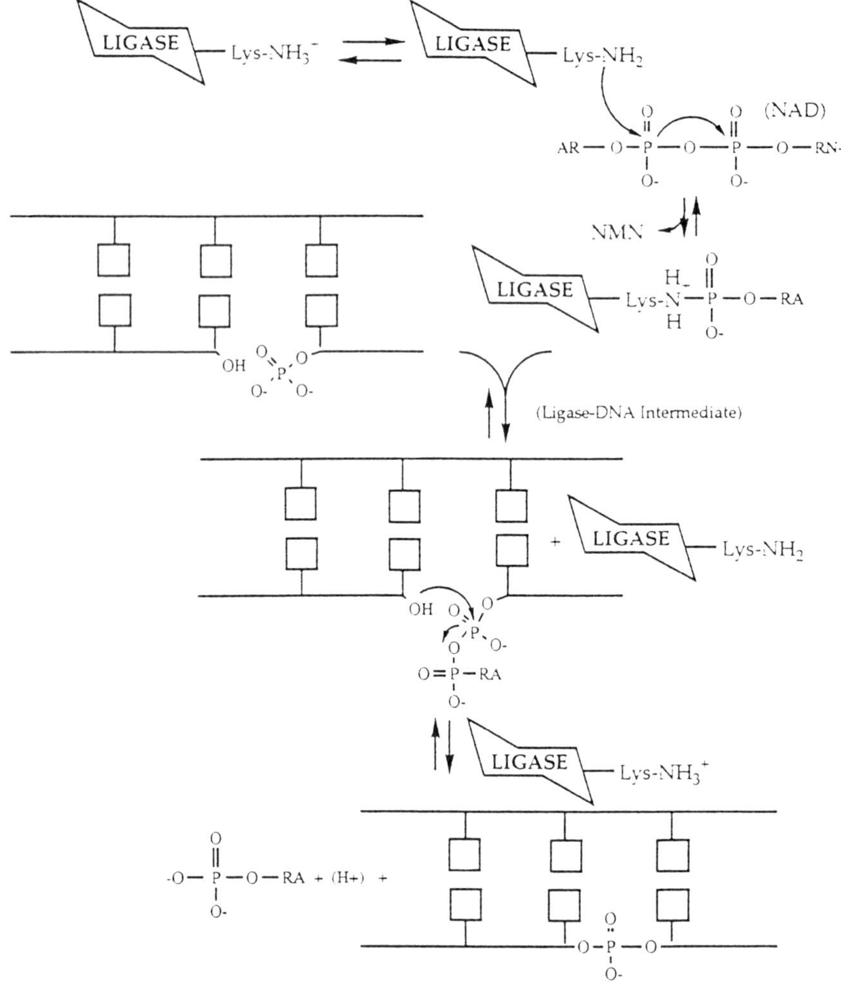

Fig. 9. Mechanism of *E. coli* DNA ligase. The various steps for ligation of a DNA nick with a 3'-OH and a 5'-phosphate; the steps are detailed in the text.

studies did not support this hypothesis *(124)*. Interestingly, the *E. coli* DNA ligase reaction is reversible *(125)*, a feature that helped to characterize the reaction mechanism *(102)*. The biological significance of this poor AMP-dependent endonuclease activity is unknown, and may simply reflect an in vitro phenomenon. Alternatively, it may serve a specialized role in vivo for relaxing superhelical structures in DNA.

Using a temperature-sensitive mutant in the *E. coli* DNA ligase gene (lig^{ts7}; *170*), which at permissive temperatures (25°C) exhibits 1–3% of the wild-type DNA ligase activity, it was found that cells grown at the nonpermissive temperature (42°C) lost viability much more rapidly than under permissive conditions *(66)*. Thus, DNA ligase I is clearly essential for viability. Treatment of lig^{ts7} mutants with the alkylating agent MMS or with UV at 25°C resulted in an increased cell killing, indicating the involvement of DNA ligase I in BER and nucleotide excision repair, respectively. Furthermore, these mutant cells were found to accumulate "Okazaki fragments" at both permissive and nonpermissive temperatures owing to an inability to ligate discontinuous replicated DNA strands (94). The ability of lig^{ts7} cells to grow normally at 25°C with ~2% of wild-type DNA ligase activity probably reflects the fact that wild-type cells possess an estimated >30-fold excess in DNA ligase activity *(102)*, accentuating the importance of this enzyme.

The crystal structure of the ATP-dependent DNA ligase of the bacteriophage T7 has recently been solved *(166)*. This structural information has provided insight into the mechanism employed by other nucleotidyltransferases. Furthermore, close examination of the T7 ligase structure revealed a domain with striking similarity to the DNA binding region of DNA methyltransferases. However, additional structural data, particularly of the protein–substrate complexes, is necessary to draw any further conclusions.

7. CONCLUSIONS

Each step of the multistep BER process has been presented here as if it is normally carried out independently of the other steps. Indeed, there is no doubt that in the test tube, each one of the BER enzymes is capable of functioning independently from the others. However, this does not rule out the possibility that in vivo a multienzyme complex is responsible and perhaps required for the efficient execution of BER. Currently, various groups are actively searching for interactions between the BER enzymes, using a combination of biochemical, immunological, molecular, and genetic approaches. It should soon be well established whether or not the BER enzymes (especially those downstream of the DNA glycosylases) must associate in vivo in order to carry out their important tasks efficiently.

In virtually all organisms, a vast array of improper DNA bases are constantly being eliminated and accurately being replaced via the BER pathway. Numerous DNA glycosylases feed into this pathway, and it is becoming increasingly clear that BER, as a whole, is a crucial player in maintaining genomic stability. However, unlike several other DNA repair pathways (e.g., nucleotide excision repair, DNA mismatch repair, and double-strand break repair), there are no documented cases of a human or rodent disease that find their origin in a BER defect. The conspicuous absence of a BER-deficient disease could be interpreted in a least three very different ways:

1. BER-deficient animals do exist, but we have not yet recognized their phenotype;
2. A BER defect has no obvious effect on the well being of an animal, perhaps because other DNA repair pathways can readily substitute for BER; or
3. BER is required for the viability of animals.

In support of the third interpretation, it was recently shown that β-polymerase homozygous null mice are embryonic lethal, and that cultured cells lacking β-polymerase are completely deficient in BER *(69,160)*. In further support, AP endonuclease homozygous null mice are also embryonic lethal *(189a)*. Thus, although BER is not essential for the survival of cells in culture, it appears that BER (at least in the context of the two enzymes discussed here) is essential for animal development. The embryonic lethality of the AP endonuclease- and β-polymerase-defective animals could be owing to the accumulation of large amounts of naturally occuring mutagenic or lethal lesions that would normally be removed from the genome by BER. Alternatively, the enzymes of BER may participate in other essential processes. It will be interesting to see, through the generation of tissue-specific or stage-specific knockouts, whether targeted deficiencies have any effect on maturation, aging, carcinogenesis, or other biological processes. Whether particular DNA glycosylases or the other BER enzymes are also required for development remains to be determined.

ACKNOWLEDGMENTS

D. M. W. III was supported by a National Research Service Award CA 62845 from the National Cancer Institute and, in part, from National Institutes of Health grant GM40000 to Bruce Demple, Harvard School of Public Health. B. P. E. and L. S. were supported by National Cancer Institute grant CA R01-55042 and National Institute of Environmental Health Sciences grant P01-ES03926. B. P. E. was the recipient of a Pharmaceutical Manufacturers Association Foundation Advanced Predoctoral Fellowship in Pharmacology/Toxicology and a Graduate Student Research Fellowship Award from the Society of Toxicology sponsored by Hoffmann-LaRoche, Inc. L. S. is a Burroughs Wellcome Toxicology Scholar. The authors would also like to thank Edy Y. Kim, a Harvard University undergraduate student, for his input.

REFERENCES

1. Asahara, H., P. M. Wistort, J. F. Bank, R. H. Bakerian, and R. P. Cunningham. 1988. Purification and characterization of *Escherichia coli* endonuclease III from the cloned *nth* gene. *Biochemistry* **28:** 4444–4449.

1a. Au, K. G., M. Cabrera, J. H. Miller, and P. Modich. 1988. *Escherichia coli mutY* gene product is required for specific A-G–C-G mismatch correction. *Proc. Natl. Acad. Sci. USA* **85:** 9163–9166.

2. Bailly, V. and W. G. Verly. 1987. *Escherichia coli* endonuclease III is not an endonuclease but a β-elimination catalyst. *Biochem. J.* **242:** 565–572.

3. Bailly, V. and W. G. Verly. 1988. Possible roles of β-elimination and δ-elimination reactions in the repair of DNA containing AP (apurinic/apyrimidinic) sites in mammalian cells. *Biochem. J.* **253:** 553–559.

4. Bailly, V., W. G. Verly, T. R. O'Connor, and J. Laval. 1989. Mechanism of DNA strand nicking at apurinic/apyrimidinic sites by *Escherichia coli* (formamidopyrimidine) DNA glycosylase. *Biochem. J.* **262:** 581–589.

5. Barzilay, G., C. D. Mol, C. N. Robson, L. J. Walker, R. P. Cunningham, J. A. Tainer, and I. D. Hickson. 1995. Identification of critical active-site residues in the multifunctional human DNA repair enzyme HAP1. *Nature Struct. Biol.* **2:** 561–568.

6. Bessman, M. J., I. R. Lehman, E. S. Simm, and A. Kornberg. 1958. Enzymatic synthesis of deoxyribonucleic acid. II. General properties of the reaction. *J. Biol. Chem.* **233:** 171–177.

7. Bjelland, S., N. K. Birkeland, T. Benneche, G. Volden, and E. Seeberg. 1994. DNA glycosylase activities for thymine residues oxidized in the methyl group are functions of the AlkA enzyme in *Escherichia coli*. *J. Biol. Chem.* **269:** 30,489–30,495.

8. Bjelland, S., M. Bjoras, and E. Seeberg. 1993. Excision of 3-methylguanine from alkylated DNA by 3-methyladenine DNA glycosylase I of *Escherichia coli*. *Nucleic Acids Res.* **21:** 2045–2049.

9. Bode, V. C. and A. D. Kaiser. 1965. Changes in the structure and activity of λ DNA in a superinfected immune bacterium. *J. Mol. Biol.* **14:** 399–417.

9a. Boiteux, S., M. Bichara, R. P. Fuchs, and J. Laval. 1989. Excision of the imidazole ring opened form of N-2-aminofluorene-C(8)-gaunine adduct in poly (dG-dc) by *Echerichia coli* formamidopyrimidine-DNA glycosylase. *Carcinogenesis* **10:** 1905–1909.

10. Boiteux, S. and O. Huisman. 1989. Isolation of a formamidopyrimidine-DNA glycosylase *(fpg)* mutant of *Escherichia coli* K12. *Mol. Gen. Genet.* **215:** 300–305.
11. Boiteux, S., O. Huisman, and J. Laval. 1984. 3-Methyladenine residues in DNA induce the SOS function *sfiA* in *Escherichia coli. EMBO J.* **3:** 2569–2573.
12. Boiteux, S. and J. Laval. 1983. Imidazole open ring 7-methylguanine: an inhibitor of DNA synthesis. *Biochemistry Biophys. Res. Commun.* **110:** 552–558.
13. Bonner, C. A., S. Hays, K. McEntee, and M. F. Goodman. 1990. DNA polymerase II is encoded by the DNA damage-inducible *dinA* gene of *Escherichia coli. Proc. Natl. Acad. Sci. USA* **87:** 7663–7667.
14. Bonner, C. A., S. K. Randall, C. Rayssiguier, M. Radman, R. Eritja, B. E. Kaplan, K. McEntee, and M. F. Goodman. 1988. Purification and characterization of an inducible *Escherichia coli* DNA polymerase capable of insertion and bypass at abasic lesions in DNA. *J. Biol. Chem.* **263:** 18,946–18,952.
15. Bonner, C. A., P. T. Stukenberg, M. Rajagopalan, R. Eritja, K. McEntee, H. Echols, and M. F. Goodman. 1992. Processive DNA synthesis by DNA polymerase II mediated by DNA polymerase III accessory proteins. *J. Biol. Chem.* **267:** 11,431–11,438.
16. Boorstein, R. J., T. P. Hilbert, J. Cadet, R. P. Cunningham, and G. W. Teebor. 1989. UV-induced pyrimidine hydrates in DNA are repaired by bacterial and mammalian DNA glycosylase activities. *Biochemistry* **28:** 6164–6170.
17. Breimer, L. H. 1984. Enzymatic excision from γ-irradiated polydeoxyribonucleotides of adenine residues whose imidazole rings have been ruptured. *Nucleic Acids Res.* **12:** 6359–6367.
18. Breimer, L. H. and T. Lindahl. 1984. DNA glycosylase activities for thymine residues damaged by ring saturation, fragmentation, or ring contraction are functions of endonuclease III in *Escherichia coli. J. Biol. Chem.* **259:** 5543–5548.
19. Breimer, L. H., and T. Lindahl. 1980. A DNA glycosylase from *Escherichia coli* that releases free urea from a polydeoxyribonucleotide containing fragments of base residues. *Nucleic Acids Res.* **8:** 6199–6211.
20. Breimer, L. H. and T. Lindahl. 1985. Thymine lesions produced by ionizing radiation in double-stranded DNA. *Biochemistry* **24:** 4018–4022.
21. Bridges, B. A. and H. Bates. 1990. Mutagenic DNA repair in *Escherichia coli.* XVIII. Involvement of DNA polymerase III α-subunit (DnaE protein) in mutagenesis after exposure to UV light. *Mutagenesis* **5:** 35–38.
22. Brutlag, D., M. R. Atkinson, P. Setlow, and A. Kornberg. 1969. An active fragment of DNA polymerase produced by proteolytic cleavage. *Biochem. Biophys. Res. Commun.* **37:** 982–989.
23. Bryan, S. K. and R. E. Moses. 1984. Map location of the *pcbA* mutation and physiology of the mutant. *J. Bacteriol.* **158:** 216–221.
24. Cabrera, M., Y. Nghiem, and J. H. Miller. 1988. *mutM*, a second mutator locus in *Escherichia coli* that generates GC—TA transversions. *J. Bacteriol.* **170:** 5405–5407.
25. Campbell, J. L., L. Soll, and C. C. Richardson. 1972. Isolation and partial characterization of a mutant of *Escherichia coli* deficient in DNA polymerase II. *Proc. Natl. Acad. Sci. USA* **69:** 2090–2094.
26. Carter, C. A., Y. Habraken, and D. B. Ludlum. 1988. Release of 7-alkylguanines from haloethylnitrosourea-treated DNA by *E. coli* 3-methyladenine-DNA glycosylase II. *Biochem. Biophys. Res. Commun.* **155:** 1261–1265.
27. Chan, E. and B. Weiss. 1987. Endonuclease IV of *Escherichia coli* is induced by paraquat. *Proc. Natl. Acad. Sci. USA* **84:** 3189–3193.
28. Chandler, M., R. E. Bird, and L. Caro. 1975. The replication time of the *Escherichia coli* K12 chromosome as a function of cell doubling time. *J. Mol. Biol.* **94:** 127–132.
29. Chen, H., S. K. Bryan, and R. E. Moses. 1989. Cloning the *polB* gene of *Escherichia coli* and identification of its product. *J. Biol. Chem.* **264:** 20,591–20,595.

29a. Chetsanga, C. F. and G. P. Frenette. 1983. Excision of aflatoxin B1-imidazole ring opened guanine adducts from DNA by formamidopyrimidine-DNA glycosylase. *Carcinogenesis* **4:** 997–1000.
30. Chetsanga, C. J. and T. Lindahl. 1979. Release of 7-methylguanine residues whose imidazole rings have been opened from damaged DNA by a DNA glycosylase from *Escherichia coli*. *Nucleic Acids Res.* **6:** 3673–3684.
31. Chung, M. H., H. Kasai, D. S. Jones, H. Inoue, H. Ishikawa, E. Ohtsuka, and S. Nishimura. 1991. An endonuclease activity of *Escherichia coli* that specifically removes 8-hydroxyguanine residues from DNA. *Mutat. Res.* **254:** 1–12.
32. Cohen-Fix, O. and Z. Livneh. 1994. *In vitro* UV mutagenesis associated with nucleotide excision-repair gaps in *Escherichia coli*. *J. Biol. Chem.* **269:** 4953–4958.
33. Cooper, P. K. 1982. Characterization of long patch excision repair of DNA in ultraviolet-irradiated *Escherichia coli*: an inducible function under *rec-lex* control. *Mol. Gen. Genet.* **185:** 189–197.
34. Cooper, P. K. and P. C. Hanawalt. 1972. Role of DNA polymerase I and the *rec* system in excision-repair in *Escherichia coli*. *Proc. Natl. Acad. Sci. USA* **69:** 1156–1160.
35. Cozzarelli, N. R. and N. E. Melechen, T. M. Jovin, A. Kornberg. 1967. Polynucleotide cellulose as a substrate for a polynucleotide ligase induced by phage T4. *Biochem. Biophys. Res. Commun.* **28:** 578–586.
36. Cunningham, R. P., H. Asahara, J. F. Bank, C. P. Scholes, J. C. Salerno, K. Surerus, E. Munck, J. McCracken, J. Peisach, M. H. and Emptage. 1989. Endonuclease III is an iron-sulfur protein. *Biochemistry* **28:** 4450–4455.
37. Cunningham, R. P., S. M. Saporito, S. G. Spitzer, and B. Weiss. 1986. Endonuclease IV *(nfo)* mutant of *Escherichia coli*. *J. Bacteriol.* **168:** 1120–1127.
38. Cunningham, R. P. and B. Weiss. 1985. Endonuclease III *(nth)* mutants of *Escherichia coli*. *Proc. Natl. Acad. Sci. USA* **82:** 474–478.
39. De Lucia, P. and J. Cairns. 1969. Isolation of an *Escherichia coli* strain with a mutation affecting DNA polymerase. *Nature* **224:** 1164–1166.
40. Demple, B., J. Halbrook, and S. Linn. 1983. *Escherichia coli xth* mutants are hypersensitive to hydrogen peroxide. *J. Bacteriol.* **153:** 1079–1082.
41. Demple, B. and L. Harrison. 1994. Repair of oxidative damage to DNA: enzymology and biology. *Ann. Rev. Biochem.* **63:** 915–948.
42. Demple, B., L. Harrison, D. M. Wilson, III, R. A. O. Bennett, T. Takagi, and A. G. Ascione. 1996. Regulation of eukaryotic abasic (AP) endonucleases and their role in genetic stability. *Environ. Health Perspect*, in press.
43. Demple, B., T. Herman, and D. S. Chen. 1991. Cloning and expression of APE, the cDNA encoding the major human apurinic endonuclease: definition of a family of DNA repair enzymes. *Proc. Natl. Acad. Sci. USA* **88:** 11,450–11,454.
44. Demple, B., A. Johnson, and D. Fung. 1986. Exonuclease III and endonuclease IV remove 3' blocks from DNA synthesis primers in H_2O_2-damaged *Escherichia coli*. *Proc. Natl. Acad. Sci. USA* **83:** 7731–7735.
45. Demple, B. and S. Linn. 1980. DNA N-glycosylase and UV repair. *Nature* **287:** 203–208.
46. Dianov, G. and T. Lindahl. 1994. Reconstitution of the DNA base excision-repair pathway. *Curr. Biol.* **4:** 1069–1076.
47. Dianov, G., A. Price, and T. Lindahl. 1992. Generation of single-nucleotide repair patches following excision of uracil residues from DNA. *Mol. Cell. Biol.* **12:** 1605–1612.
48. Dianov, G., B. Sedgwick, G. Daly, M. Olsson, S. Lovett, and T. Lindahl. 1994. Release of 5'-terminal deoxyribose-phosphate residues from incised abasic sites in DNA by the *Escherichia coli* RecJ protein. *Nucleic Acids Res.* **22:** 993–998.
48a. Dizdaroglu M., J. Laval, and S. Boiteux. 1993. Substrate specificity of the *Escherichia coli* endonuclease III: excision of thymine- and cytosine-derived lesions in DNA produced by radiation-generated free radicals. *Biochemistry* **32:** 12,105–12,111.

49. Dodson, M. L., M. L. Michaels, and R. S. Lloyd. 1994. Unified catalytic mechanism for DNA glycosylases. *J. Biol. Chem.* **269:** 32,709–32,712.
50. Doetsch, P. W. and R. P. Cunningham. 1990. The enzymology of apurinic/apyrimidinic endonucleases. *Mutat. Res.* **236:** 173–201.
51. Duncan, B. K. and B. Weiss. 1982. Specific mutator effects of *ung* (uracil-DNA glycosylase) mutations in *Escherichia coli. J. Bacteriol.* **151:** 750–755.
52. el-Hajj, H. H., L. Wang, and B. Weiss. 1992. Multiple mutant of *Escherichia coli* synthesizing virtually thymineless DNA during limited growth. *J. Bacteriol.* **174:** 4450–4456.
53. Fay, P. J., K. O. Johanson, C. S. McHenry, and R. A. Bambara. 1981. Size classes of products synthesized processively by DNA polymerase III and DNA polymerase III holocnzyme of *Escherichia coli. J. Biol. Chem.* **256:** 976–983.
54. Foster, P. L., G. Gudmundsson, J. M. Trimarchi, H. Cai, and M. F. Goodman. 1995. Proofreading-defective DNA polymerase II increases adaptive mutation in *Escherichia coli. Proc. Natl. Acad. Sci. USA* **92:** 7951–7955.
55. Franklin, W. A. and T. Lindahl. 1988. DNA deoxyribophosphodiesterase. *EMBO J.* **7:** 3617–3622.
56. Friedberg, E. C. and D. A. Goldthwait. 1969. Endonuclease II of *Escherichia coli.* I. Isolation and purification. *Proc. Natl. Acad. Sci. USA* **62:** 934–940.
57. Friedberg, E. C., S. M. Hadi, and D. A. Goldthwait. 1969. Endonuclease II of *Escherichia coli.* II. Enzyme properties and studies on the degradation of alkylated and native deoxyribonucleic acid. *J. Biol. Chem.* **244:** 5879–5889.
58. Friedberg, E. C. and J. J. King. 1969. Endonucleolytic cleavage of UV-irradiated DNA controlled by the *denV*$^+$ gene in phage T4. *Biochemistry Biophys. Res. Commun.* **37:** 649–651.
59. Friedberg, E. C., G. C. Walker, and W. Siede. 1995. *DNA Repair and Mutagenesis.* ASM Press, Washington, DC.
60. Gates, F. T. and S. Linn. 1977. Endonuclease from *Escherichia coli* that acts specifically upon duplex DNA damages by ultraviolet light, osmium tetroxide, acid or X rays. *J. Biol. Chem.* **252:** 2802–2807.
61. Gefter, M. L. 1975. DNA replication. *Ann. Rev. Biochem.* **44:** 45–78.
62. Gefter, M. L., A. Becker, and J. Hurwitz. 1967. The enzymatic repair of DNA. I. Formation of circular λ-DNA. *Proc. Natl. Acad. Sci. USA* **58:** 240–247.
63. Gefter, M. L., Y. Hirota, T. Kornberg, J. A. Wechsler, and C. Barnoux. 1971. Analysis of DNA polymerases II and III in mutants of *Escherichia coli* thermosensitive for DNA synthesis. *Proc. Natl. Acad. Sci. USA* **68:** 3150–3153.
64. Gellert, M. 1967. Formation of covalent circles of λ DNA by *Escherichia coli* extracts. *Proc. Natl. Acad. Sci. USA* **57:** 148–155.
65. Giloni, L., M. Takeshita, F. Johnson, C. Iden, and A. P. Grollman. 1981. Bleomycin-induced strand-scission of DNA. Mechanism of deoxyribose cleavage. *J. Biol. Chem.* **256:** 8608–8615.
66. Gottesman, M. M., M. L. Hicks, and M. Gellert. 1973. Genetics and function of DNA ligase in *Escherichia coli. J. Mol. Biol.* **77:** 531–547.
67. Graves, R. J., I. Felzenszwalb, J. Laval, and T. R. O'Connor. 1992. Excision of 5'-terminal deoxyribose phosphate from damaged DNA is catalyzed by the Fpg protein of *Escherichia coli. J. Biol. Chem.* **267:** 14,429–14,435.
68. Gross, J. and M. Gross. 1969. Genetic analysis of an *Escherichia coli* strain with a mutation affecting DNA polymerase. *Nature* **224:** 1166–1168.
69. Gu, H., J. D. Marth, P. C. Orban, H. Mossmann, and K. Rajewsky. 1994. Deletion of a DNA polymerase β gene segment in T cells using cell type-specific gene targeting. *Science* **265:** 103–106.
70. Gumport, R. I. and I. R. Lehman. 1971. Structure of the DNA ligase-adenylate intermediate: lysine (ε-amino)-linked adenosine monophosphoramidate. *Proc. Natl. Acad. Sci. USA* **68:** 2559–2563.

71. Habraken, Y., C. A. Carter, M. C. Kirk, and D. B. Ludlum. 1991. Release of 7-alkylguanines from *N*-(2-chloroethyl)-*N*'-cyclohexyl-*N*-nitrosourea-modified DNA by 3-methyladenine DNA glycosylase II. *Cancer Res.* **51:** 499–503.
72. Habraken, Y., C. A. Carter, M. Sekiguchi, and D. B. Ludlum. 1991. Release of *N*2,3-ethanoguanine from haloethylnitrosourea-treated DNA by *Escherichia coli* 3-methyladenine DNA glycosylase II. *Carcinogenesis* **12:** 1971–1973.
73. Habraken, Y. and D. B. Ludlum. 1989. Release of chloroethyl ethyl sulfide-modified DNA bases by bacterial 3-methyladenine-DNA glycosylases I and II. *Carcinogenesis* **10:** 489–492.
74. Hagensee, M. E., S. K. Bryan, and R. E. Moses. 1987. DNA polymerase III requirement for repair of DNA damage caused by methyl methanesulfonate and hydrogen peroxide. *J. Bacteriol.* **169:** 4608–4613.
75. Hagensee, M. E., T. L. Timme, S. K. Bryan, and R. E. Moses. 1987. DNA polymerase III of *Escherichia coli* is required for UV and ethyl methanesulfonate mutagenesis. *Proc. Natl. Acad. Sci. USA* **84:** 4195–4199.
76. Hall, Z. W. and I. R. Lehman. 1969. Enzymatic joining of polynucleotides. VI. Activity of a synthetic adenylylated polydeoxynucleotide in the reaction. *J. Biol. Chem.* **244:** 43–47.
77. Haring, M., H. Rudiger, B. Demple, S. Boiteux, and B. Epe. 1994. Recognition of oxidized abasic sites by repair endonucleases. *Nucleic Acids Res.* **22:** 2010–2015.
78. Haseltine, W. A., L. K. Gordon, C. P. Lindan, R. H. Grafstrom, N. L. Shaper, and L. Grossman. 1980. Cleavage of pyrimidine dimers in specific DNA sequences by a pyrimidine dimer DNA-glycosylase of *M. luteus*. *Nature* **285:** 634–641.
79. Hatahet, Z., Y. W. Kow, A. A. Purmal, R. P. Cunningham, and S. S. Wallace. 1994. New substrates for old enzymes. 5-Hydroxy-2'-deoxycytidine and 5-hydroxy-2'-deoxyuridine are substrates for *Escherichia coli* endonuclease III and formamidopyrimidine DNA *N*-glycosylase, while 5-hydroxy-2'-deoxyuridine is a substrate for uracil DNA *N*-glycosylase. *J. Biol. Chem.* **269:** 18,814–18,820.
80. Hendler, S., E. Furer, and P. R. Srinivasan. 1970. Synthesis and chemical properties of monomers and polymers containing 7-methylguanine and an investigation of the substrate or template properties for bacterial deoxyribonucleic acid or ribonucleic acid polymerases. *Biochemistry* **9:** 4141–4153.
81. Henner, W. D., S. M. Grunberg, and W. A. Haseltine. 1983. Enzyme action at 3' termini of ionizing radiation-induced DNA strand breaks. *J. Biol. Chem.* **258:** 15,198–15,205.
82. Hirota, Y., M. Gefter, and L. Mindich. 1972. A mutant of *Escherichia coli* defective in DNA polymerase II activity. *Proc. Natl. Acad. Sci. USA* **69:** 3238–3242.
83. Hughes, A. J., Jr., S. K. Bryan, H. Chen, R. E. Moses, and C. S. McHenry, C. S. 1991. *Escherichia coli* DNA polymerase II is stimulated by DNA polymerase III holoenzyme auxiliary subunits. *J. Biol. Chem.* **266:** 4568–4573.
84. Ide, H., K. Tedzuka, H. Shimzu, Y. Kimura, A. A. Purmal, S. S. Wallace, and Y. Kow. 1994. α-deoxyadenosine, a major anoxic radiolysis product of adenine in DNA, is a substrate for *Escherichia coli* endonuclease IV. *Biochemistry* **33:** 7842–7847.
85. Iwasaki, H., Y. Ishino, H. Toh, A. Nakata, and H. Shinagawa. 1991. *Escherichia coli* DNA polymerase II is homologous to a-like DNA polymerases. *Mol. Gen. Genet.* **226:** 24–33.
86. Iwasaki, H., A. Nakata, G. C. Walker, and H. Shinagawa. 1990. The *Escherichia coli polB* gene, which encodes DNA polymerase II, is regulated by the SOS system. *J. Bacteriol.* **172:** 6268–6273.
87. Jiang, D., Z. Hatahet, R. Melamede, A. Purmal, and S. S. Wallace. 1996. Cloning of the *Escherichia coli* gene for exonuclease VIII, *nei*, and further characterization of the protein, in preparation.
88. Kaasen, I., G. Evensen, and E. Seeberg. 1986. Amplified expression of the tag^+ and $alkA^+$ genes in *Escherichia coli*: identification of gene products and effects on alkylation resistance. *J. Bacteriol.* **168:** 642–647.

89. Kaplan, J. C., S. F. Kushner, and L. Grossman. 1969. Enzymatic repair of DNA. 1. Purification of two enzymes involved in the excision of thymine dimers from ultraviolet-irradiated DNA. *Proc. Natl. Acad. Sci. USA* **63:** 144–151.
90. Karran, P., T. Lindahl, I. Ofsteng, G. B. Evensen, and E. Seeberg. 1980. *Escherichia coli* mutants deficient in 3-methyladenine-DNA glycosylase. *J. Mol. Biol.* **140:** 101–127.
91. Katcher, H. L. and S. S. Wallace. 1983. Characterization of the *Escherichia coli* x-ray endonuclease, endonuclease III. *Biochemistry* **22:** 4071–4081.
92. Kellenberger, G., M. L. Zichichi, and J. J. Weigle. 1961. Exchange of DNA in the recombination of bacteriophage λ. *Proc. Natl. Acad. Sci. USA* **47:** 869–878.
93. Kelman, Z. and M. O'Donnell 1995. DNA polymerase III holoenzyme: structure and function of a chromosomal replicating machine. *Ann. Rev. Biochem.* **64:** 171–200.
94. Konrad, E. B., P. Modrich, and I. R. Lehman. 1973. Genetic and enzymatic characterization of a conditional lethal mutant of *Escherichia coli* K12 with a temperature-sensitive DNA ligase. *J. Mol. Biol.* **77:** 519–529.
95. Kornberg, A. 1969. Active center of DNA polymerase. *Science* **163:** 1410–1418.
96. Kornberg, A. and T. Baker. 1992. *DNA Replication*, 2nd ed. W. H. Freeman & Co., New York.
97. Kow, Y. W., H. Ide, R. J. Melamede, and S. S. Wallace. 1988. Comparative study of the mechanisms of action of the apurinic endonucleases of *Escherichia coli. J. Cell Biochem.* **12A(Suppl.):** 269.
98. Kow, Y. W. and S. S. Wallace. 1985. Exonuclease III recognizes urea residues in oxidized DNA. *Proc. Natl. Acad. Sci. USA* **82:** 8354–8358.
99. Kow, Y. W. and S. S. Wallace. 1987. Mechanism of action of *Escherichia coli* Endonuclease III. *Biochemistry* **26:** 8200–8206.
100. Kuo, C. F., D. E. McRee, C. L. Fisher, S. F. O'Handley, R. P. Cunningham, and J. A. Tainer. 1992. Atomic structure of the DNA repair [4Fe-4S] enzyme endonuclease III. *Science* **258:** 434–440.
101. Larson, K., J. Sahm, R. Shenkar, and B. Strauss. 1985. Methylation-induced blocks to *in vitro* DNA replication. *Mutat. Res.* **150:** 77–84.
102. Lehman, I. R. 1974. DNA ligase: structure, mechanism, and function. *Science* **186:** 790–797.
103. Lehman, I. R., M. J. Bessman, E. S. Simms, and A. Kornberg. 1958. Enzymatic synthesis of deoxyribonucleic acid. I. Preparation of substrates and partial purification of an enzyme from *E. coli. J. Biol. Chem.* **233:** 163–170.
104. Levin, J. D. and B. Demple. 1996. *In vitro* detection of endonuclease IV-specific DNA damage formed by bleomycin *in vivo. Nucleic Acids Res.* **24:** 885–889.
105. Levin, J. D., A. W. Johnson, and B. Demple. 1988. Homogeneous *Escherichia coli* endonuclease IV. Characterization of an enzyme that recognizes oxidative damage in DNA. *J. Biol. Chem.* **263:** 8066–8071.
106. Levin, J. D., R. Shapiro, and B. Demple. 1991. Metalloenzymes in DNA repair. *Escherichia coli* endonuclease IV and *Saccharomyces cerevisiae* Apn1. *J. Biol. Chem.* **266:** 22,893–22,898.
107. Lindahl, T. 1993. Instability and decay of the primary structure of DNA. *Nature* **362:** 709–715.
108. Lindahl, T. 1974. An *N*-glycosidase from *Escherichia coli* that releases free uracil from DNA containing deaminated cytosine residues. *Proc. Natl. Acad. Sci. USA* **71:** 3649–3653.
109. Ljungquist, S. 1977. A new endonuclease from *Escherichia coli* acting at apurinic sites in DNA. *J. Biol. Chem.* **252:** 2808–2814.
110. Ljungquist, S., T. Lindahl, and P. Howard-Flanders. 1976. Methyl methane sulfonate-sensitive mutant of *Escherichia coli* deficient in an endonuclease specific for apurinic sites in deoxyribonucleic acid. *J. Bacteriol.* **126:** 646–653.
111. Loeb, L. A. and B. D. Preston. 1986. Mutagenesis by apurinic/apyrimidinic sites. *Ann. Rev. Genet.* **20:** 201–230.

112. Lu, A. L., J. J. Tsai-Wu, and J. Cillo. 1995. DNA determinants and substrate specificities of *Escherichia coli* Mut Y. *J. Biol. Chem.* **270:** 23,582–23,588.
113. Luria, S. E. 1947. Reactivation of irradiated bacteriophage by transfer of self-reproducing units. *Proc. Natl. Acad. Sci. USA* **33:** 253–264.
114. Mackay, W. J., S. Han, and L. D. Samson. 1994. DNA alkylation repair limits spontaneous base substitution mutations in *Escherichia coli* [published erratum appears in *J. Bacteriol.* 1994. **176:** 5193]. *J. Bacteriol.* **176:** 3224–3230.
115. Maki, H., S. K. Bryan, T. Horiuchi, and R. E. Moses. 1989. Suppression of *dnaE* nonsense mutations by *pcbA1*. *J. Bacteriol.* **171:** 3139–3143.
116. Masker, W., P. Hanawalt, and H. Shizuya. 1973. Role of DNA polymerase II in repair replication in *Escherichia coli*. *Nature—New Biol.* **244:** 242,243.
117. Matijasevic, Z., M. Sekiguchi, and D. B. Ludlum. 1992. Release of N^2,3-ethenoguanine from chloroacetaldehyde-treated DNA by *Escherichia coli* 3-methyladenine DNA glycosylase II. *Proc. Natl. Acad. Sci. USA* **89:** 9331–9334.
118. Matson, S. W. and R. A. Bambara. 1981. Short deoxyribonucleic acid repair patch length in *Escherichia coli* is determined by the processive mechanism of deoxyribonucleic acid polymerase I. *J. Bacteriol.* **146:** 275–284.
119. McCarthy, T. V., P. Karran, and T. Lindahl. 1984. Inducible repair of O-alkylated DNA pyrimidines in *Escherichia coli*. *EMBO J.* **3:** 545–550.
120. Melamede, R. J., Z. Hatahet, Y. W. Kow, H. Ide, and S. S. Wallace. 1994. Isolation and characterization of endonuclease VIII from *Escherichia coli*. *Biochemistry* **33:** 1255–1264.
121. Meselson, M. and J. J. Weigle. 1961. Chromosome break accompanying genetic recombination in bacteriophage. *Proc. Natl. Acad. Sci. USA* **47:** 857–868.
122. Michaels, M. L., C. Cruz, A. P. Grollman, and J. H. Miller. 1992. Evidence that MutY and MutM combine to prevent mutations by an oxidatively damaged form of guanine in DNA. *Proc. Natl. Acad. Sci. USA* **89:** 7022–7025.
123. Milcarek, C. and B. Weiss. 1972. Mutants of *Escherichia coli* with altered deoxyribonucleases. I. Isolation and characterization of mutants for exonuclease III. *J. Mol. Biol.* **68:** 303–318.
124. Modrich, P. and I. R. Lehman. 1973. Deoxyribonucleic acid ligase: A steady state kinetic analysis of the reaction catalyzed by the enzyme from *E. coli*. *J. Biol. Chem.* **248:** 7502–7511.
125. Modrich, P., I. R. Lehman, and J. C. Wang. 1972. Enzymatic joining of polynucleotides. XI. Reversal of *Escherichia coli* deoxyribonucleic acid ligase reaction. *J. Biol. Chem.* **247:** 6370–6372.
126. Mol, C. D., C. F. Kuo, M. M. Thayer, R. P. Cunningham, and J. A. Tainer. 1995. Structure and function of the multifunctional DNA-repair enzyme exonuclease III. *Nature* **374:** 381–386.
127. Mosbaugh, D. W. and S. Linn. 1982. Characterization of the action of *Escherichia coli* DNA polymerase I at incisions produced by repair endodeoxyribonucleases. *J. Biol. Chem.* **257:** 575–583.
128. Mullen, G. P., E. H. Serpersu, L. J. Ferrin, L. A. Loeb, and A. S. Mildvan. 1990. Metal binding to DNA polymerase I, its large fragment, and two 3',5'-exonuclease mutants of the large fragment. *J. Biol. Chem.* **265:** 14,327–14,334.
129. Nghiem, Y., M. Cabrera, C. G. Cupples, and J. H. Miller. 1988. The *mutY* gene: a mutator locus in *Escherichia coli* that generates G:C—T:A transversions. *Proc. Natl. Acad. Sci. USA* **85:** 2709–2713.
130. Nunoshiba, T., T. deRojas-Walker, J. S. Wishnok, S. R. Tannenbaum, and B. Demple. 1993. Activation by nitric oxide of an oxidative-stress response that defends *Escherichia coli* against activated macrophages. *Proc. Natl. Acad. Sci. USA* **90:** 9993–9997.

131. O' Connor, T. and J. Laval. 1989. Physical association of the 2,6-diamino-4-hydroxy-5-(*N*-methyl) formamidopyrimidine-DNA glycosylase of *Escherichia coli* and an activity nicking DNA at apurinic/apyrimidinic sites. *Proc. Natl. Acad. Sci. USA* **86:** 5222–5226.
132. Olivera, B. M., Z. W. Hall, and I. R. Lehman. 1968. Enzymatic joining of polynucleotides, V. A DNA-adenylate intermediate in the polynucleotide-joining reaction. *Proc. Natl. Acad. Sci. USA* **61:** 237–244.
133. Olivera, B. M. and I. R. Lehman. 1967. Linkage of polynucleotides through phosphodiester bonds by an enzyme from *Escherichia coli. Proc. Natl. Acad. Sci. USA* **57:** 1426–1433.
134. Ollis, D. L., P. Brick, R. Hamlin, N. G. Xuong, and T. A. Steitz. 1985. Structure of large fragment of *Escherichia coli* DNA polymerase I complexed with dTMP. *Nature* **313:** 762–766.
135. Paz-Elizur, T., R. Skaliter, S. Blumenstein, and Z. Livneh. 1996. β*, a UV-inducible smaller form of the β subunit sliding clamp of DNA polymerase III of *Escherichia coli.* I. Gene expression and regulation. *J. Biol. Chem.* **271:** 2482–2490.
136. Povirk, L. F. and I. H. Goldberg. 1985. Endonuclease-resistant apyrimidinic sites formed by neocarzinostatin at cytosine residues in DNA: evidence for a possible role in mutagenesis. *Proc. Natl. Acad. Sci. USA* **82:** 3182–3186.
137. Price, A. 1992. Action of *Escherichia coli* and human 5'–3' exonuclease functions at incised apurinic/apyrimidinic sites in DNA. *FEBS Lett.* **300:** 101–104.
138. Radany, E. H. and E. C. Friedberg. 1980. A pyrimidine dimer-DNA glycosylase activity associated with the *denV* gene product of bacteriophage T4. *Nature* **286:** 182–185.
139. Radman, M. 1976. An endonuclease from *Escherichia coli* that introduces single polynucleotide chain scissions in ultraviolet-irradiated DNA. *J. Biol. Chem.* **251:** 1438–1445.
140. Rebeck, G. W. and L. Samson. 1991. Increased spontaneous mutation and alkylation sensitivity of *Escherichia coli* strains lacking the *Ogt* O6-methylguanine DNA repair methyltransferase. *J. Bacteriol.* **173:** 2068–2076.
141. Riazuddin, S. and T. Lindahl. 1978. Properties of 3-methyladenine-DNA glycosylase from *Escherichia coli. Biochemistry* **17:** 2110–2118.
142. Richardson, C. C. and A. Kornberg. 1964. A DNA phosphatase-exonuclease from *Escherichia coli.* I. Purification of the enzyme and characterization of the phosphatase activity. *J. Biol. Chem.* **239:** 242–250.
143. Richardson, C. C., I. R. Lehman, and A. Kornberg. 1964. A DNA phosphatase-exonuclease from *Escherichia coli.* II. Characterization of the exonuclease activity. *J. Biol. Chem.* **239:** 251–258.
144. Roberts, R. J. 1995. On base flipping. *Cell* **82:** 9–12.
145. Robson, C. N. and I. D. Hickson. 1991. Isolation of cDNA clones encoding a human apurinic/apyrimidinic endonuclease that corrects DNA repair and mutagenesis defects in *E. coli* xth (exonuclease III) mutants. *Nucleic Acids Res.* **19:** 5519–5523.
146. Rogers, S. G. and B. Weiss. 1980. Exonuclease III of *Escherichia coli* K-12, an AP endonuclease. *Methods Enzymol.* **65:** 201–211.
147. Sakumi, K., Y. Nakabeppu, Y. Yamamoto, S. Kawabata, S. Iwanaga, and M. Sekiguchi. 1986. Purification and structure of 3-methyladenine-DNA glycosylase I of *Escherichia coli. J. Biol. Chem.* **261:** 15,761–15,766.
148. Samson, L. and J. Cairns. 1977. A new pathway for DNA repair in *Escherichia coli. Nature* **267:** 281–283.
149. Sandigursky, M. and W. A. Franklin. 1992. DNA deoxyribophosphodiesterase of *Escherichia coli* is associated with exonuclease I. *Nucleic Acids Res.* **20:** 4699–4703.
150. Saparbaev, M., and J. Laval. 1994. Excision of hypoxanthine from DNA containing dIMP residues by the *Eschericia coli*, yeast, rat, and human alkylpurine DNA glycosylases. *Proc. Natl. Acad. Sci. USA* **91:** 5873–5877.

151. Schuster, H. 1960. The reaction of nitrous acid with dioxyribonucleic acid. *Biochemistry Biophys. Res. Commun.* **2:** 320–323.
152. Seiichi, Y., and M. Sekiguchi. 1970. T4 endonuclease involved in repair of DNA. *Proc. Natl. Acad. Sci. USA* **67:** 1839–1845.
153. Sharif, F., and B. A. Bridges. 1990. Mutagenic DNA repair in *Escherichia coli.* XVII. Effect of temperature-sensitive DnaE proteins on the induction of streptomycin-resistant mutations by UV light. *Mutagenesis* **5:** 31–34.
154. Shibutani, S., M. Takeshita, and A. P. Grollman. 1991. Insertion of specific bases during DNA synthesis past the oxidation-damaged base 8-oxodG. *Nature* **349:** 431–434.
155. Shimada, K., H. Nakayama, S. Okubo, M. Sekiguchi, and Y. Takagi. 1967. An endonucleolytic activity specific for ultraviolet-irradiated DNA in wild type and mutant strains of *Micrococcus lysodeikticus. Biochemistry Biophys. Res. Commun.* **27:** 539–545.
156. Shinagawa, H., H. Iwasaki, Y. Ishino, and A. Nakata. 1991. SOS-inducible DNA polymerase II of *E. coli* is homologous to replicative DNA polymerase of eukaryotes. *Biochimie* **73:** 433–435.
157. Sinha, N. K., C. F. Morris, and B. M. Alberts. 1980. Efficient in vitro replication of double-stranded DNA templates by a purified T4 bacteriophage replication system. *J. Biol. Chem.* **255:** 4290–4293.
158. Skaliter, R., M. Bergstein, and Z. Livneh. 1996. β*, a UV-inducible shorter form of the β subunit of DNA polymerase III of *Escherichia coli.* II. Overproduction, purification, and activity as a polymerase processivity clamp. *J. Biol. Chem.* **271:** 2491–2496.
159. Skaliter, R., T. Paz-Elizur, and Z. Livneh. 1996. A smaller form of the sliding clamp subunit of DNA polymerase III is induced by UV irradiation in *Escherichia coli. J. Biol. Chem.* **271:** 2478–2481.
160. Sobol, R. W., J. K. Horton, R. Kuhn, H. Gu, R. K. Singhal, R. Prasad, K. Rajewsky, and S. H. Wilson. 1996. Requirement of mammalian DNA polymerase-β in base-excision repair. *Nature* **379:** 183–186.
161. Strauss, B., T. Searashi, and M. Robbins 1966. Repair of DNA studied with a nuclease specific for UV-induced lesions. *Biochemistry* **56:** 932–939.
162. Strauss, B. S. 1991. The "A rule" of mutagen specificity: a consequence of DNA polymerase bypass of non-instructional lesions? *Bioessays* **13:** 79–84.
163. Strniste, G., and S. S. Wallace. 1975. Endonucleolytic incision of X-irradiated deoxyribonucleic acid by extracts of *Escherichia coli. Proc. Natl. Acad. Sci. USA* **72:** 1997–2001.
164. Stukenberg, P. T., P. S. Studwell-Vaughan, and M. O'Donnell. 1991. Mechanism of the sliding β-clamp of DNA polymerase III holoenzyme. *J. Biol. Chem.* **266:** 11,328–11,334.
165. Stukenberg, P. T., J. Turner, and M. O'Donnell. 1994. An explanation for lagging strand replication: polymerase hopping among DNA sliding clamps. *Cell* **78:** 877–887.
166. Subramanya, H. S., A. J. Doherty, S. R. Ashford, and D. B. Wigley. 1996. Crystal structure of an ATP-dependent DNA ligase from bacteriophage T7. *Cell* **85:** 607–615.
167. Tait, R. C., A. L. Harris, and D. W. Smith. 1974. DNA repair in *Escherichia coli* mutants deficient in DNA polymerases I, II and-or III. *Proc. Natl. Acad. Sci. USA* **71:** 675–679.
168. Takeuchi, M., R. Lillis, B. Demple, and M. Takeshita. 1994. Interactions of *Escherichia coli* endonuclease IV and exonuclease III with abasic sites in DNA. *J. Biol. Chem.* **269:** 21,907–21,914.
169. Taylor, A. F. and B. Weiss. 1982. Role of exonuclease III in the base-excision repair of uracil-containing DNA. *J. Bacteriol.* **151:** 351–357.
170. Taylor, A. L. and C. D. Trotter. 1967. Revised linkage map of *Escherichia coli. Bacteriol. Rev.* **31:** 332–353.
171. Tchou, J. and A. P. Grollman. 1995. The catalytic mechanism of Fpg protein. *J. Biol. Chem.* **270:** 11,671–11,677.

172. Tchou, J., H. Kasai, S. Shibutani, M. H. Chung, J. Laval, A. P. Grollman, and S. Nishimura. 1991. 8-oxoguanine (8-hydroxyguanine) DNA glycosylase and its substrate specificity. *Proc. Natl. Acad. Sci. USA* **88:** 4690–4694.
173. Tchou, J., M. L. Michaels, J. H. Miller, and A. P. Grollman. 1993. Function of the zinc finger in *Escherichia coli* Fpg protein. *J. Biol. Chem.* **268:** 26,738–26,744.
174. Teoule, R., C. Bert, and A. Bonicel. 1977. Thymine fragment damage retained in the DNA polynucleotide chain after gamma irradiation in aerated solutions. II. *Radiat. Res.* **72:** 190–200.

174a. Thayer, M. M., H. Ahern, D. Xing, R. P. Cunningham, and J. A. Tainer. 1995. Novel DNA binding motifs in the DNA repair enzyme endonuclease III crystal structure. *EMBO J.* **14:** 4108–4120.

175. Thomas, L., C. H. Yang, and D. A. Goldthwait. 1982. Two DNA glycosylases in *Escherichia coli* which release primarily 3-methyladenine. *Biochemistry* **21:** 1162–1169.
176. Tomer, G., O. Cohen-Fix, M. O'Donnell, M. Goodman, and Z. Livneh. 1996. Reconstitution of repair-gap UV mutagenesis with purified proteins from *E. coli*: A role for DNA polymerases III and II. *Proc. Natl. Acad. Sci. USA* **93:** 171–200.
177. Town, C. D., K. C. Smith, and H. S. Kaplan. 1971. DNA polymerase required for rapid repair of x-ray-induced DNA strand breaks *in vivo. Science* **172:** 851–854.
178. Tsai-Wu, J. J., H. F. Liu, and A. L. Lu. 1992. *Escherichia coli* MutY protein has both N-glycosylase and apurinic/apyrimidinic endonuclease activities on A-C and A-G mispairs. *Proc. Natl. Acad. Sci. USA* **89:** 8779–8783.

178a. Vassylyev, D. G., T. Kashiwagi, Y. Mikami, M. Ariyoshi, S. Iwai, E. Ohtsuka, and K. Morikawa. 1995. Atomic model of a pyrimidine dimer excision repair enzyme complexed with a DNA substrate: structural basis for damaged DNA recognition. *Cell* **83:** 773–782.

179. Verri, A., P. Mazzarello, S. Spadari, and F. Focher. 1992. Uracil-DNA glycosylases preferentially excise mispaired uracil. *Biochem. J.* **287:** 1007–1010.
180. Wallace, S. S. Oxidative damage to DNA and its repair, in *Oxidative Stress* (Scandalios, J., ed.), Cold Spring Harbor Press, Cold Spring Harbor, NY, in press.
181. Wallace, S. S. 1988. AP endonucleases and DNA glycosylases that recognize oxidative DNA damage. *Environ. Mol. Mutagen.* **12:** 431–477.
182. Warner, H. R., B. K. Duncan, C. Garrett, and J. Neuhard. 1981. Synthesis and metabolism or uracil-containing deoxyribonucleic acid in *Escherichia coli. J. Bacteriol.* **145:** 687–695.

182a. Warner, H. R., and P. A. Rochstroh. 1980. Incorporation and excision of 5-fluorouracil from deoxyribonucleic acid in *Escherichia coli. J. Bacteriol.* **141:** 680–686.

183. Weiss, B. 1976. Endonuclease II of *Escherichia coli* is exonuclease III. *J. Biol. Chem.* **251:** 1896–1901.
184. Weiss, B. and C. C. Richardson. 1967. Enzymatic breakage and joining of deoxyribonucleic acid, I. Repair of single-strand breaks in DNA by an enzyme system from *Escherichia coli* infected with T4 bacteriophage. *Proc. Natl. Acad. Sci. USA* **57:** 1021–1028.
185. Weiss, R. B., and N. J. Duker. 1986. Photoalkylated DNA and ultraviolet-irradiated DNA are incised at cytosines by endonuclease III. *Nucleic Acids Res.* **14:** 6621–6631.
186. Wickner, R. B., B. Ginsberg, I. Berkower, and J. Hurwitz. 1972. Deoxyribonucleic acid polymerase II of *Escherichia coli*. I. The purification and characterization of the enzyme. *J. Biol. Chem.* **247:** 489–497.
187. Wickner, R. B., B. Ginsberg, and J. Hurwitz. 1972. Deoxyribonucleic acid polymerase II of *Escherichia coli*. II. Studies of the requirements and the structure of the deoxyribonucleic acid product. *J. Biol. Chem.* **247:** 498–504.
188. Wilson, D. M., III, M. Takeshita, A. P. Grollman, and B. Demple. 1995. Incision activity of human apurinic endonuclease (Ape) at abasic site analogs in DNA. *J. Biol. Chem.* **270:** 16,002–16,007.

189. Wood, M. L., M. Dizdaroglu, E. Gajewski, and J. M. Essigmann. 1990. Mechanistic studies of ionizing radiation and oxidative mutagenesis: genetic effects of a single 8-hydroxy-guanine (7-hydro-8-oxoguanine) residue inserted at a unique site in a viral genome. *Biochemistry* **29:** 7024–7032.

189a. Xanthoudakis, S. R. J. Smeyne, J. D. Wallace, and T. Curran. 1996. The redox/DNA repair protein, Ref-1, is essential for early embryonic development in mice. *Proc. Natl. Acad. Sci. USA* **93:** 8919–8923.

190. Yajko, D. M. and B. Weiss. 1975. Mutations simultaneously affecting endonuclease II and exonuclease III in *Escherichia coli*. *Proc. Natl. Acad. Sci. USA* **72:** 688–692.

191. Young, E. T., II and R. L. Sinsheimer. 1964. Novel intracellular forms of λ DNA. *J. Mol. Biol.* **10:** 562–564.

192. Zastawny, T. H., P. W. Doetsch, and M. Dizdaroglu. 1995. A novel activity of *E. coli* uracil DNA *N*-glycosylase excision of isodialuric acid (5,6-dihydroxyuracil), a major product of oxidative DNA damage, from DNA. *FEBS Lett.* **364:** 255–258.

193. Zou, W. and P. W. Doetsch. 1993. Effects of abasic sites and DNA single-strand breaks on prokaryotic RNA polymerases. *Proc. Natl. Acad. Sci. USA* **90:** 6601–6605.

4

Oxidative DNA Damage and Mutagenesis

Terry G. Newcomb and Lawrence A. Loeb

1. INTRODUCTION

In 1952, Conger and Fairchild first demonstrated that hyperbaric oxygen produces chromosomal aberrations in pollen grains of *Tradescantia* flower buds *(24)*. Since that time, oxidative damage to DNA has been thought to be an important factor in mutagenesis in all aerobic organisms and has been postulated to contribute to a wide range of diseases. There is evidence that oxygen damage to DNA is a causative factor in human cancer induced by ionizing radiation *(19)*, transition metals *(38,39,64,65,99,100,111,112)*, and chronic inflammation *(43,125)*, and may play a role in hereditary syndromes with a proclivity to malignancy, such as ataxia telangiectasia, Bloom's syndrome, and Fanconi's anemia *(19)*. Oxidative damage may also contribute to the decline in function associated with normal aging *(2,52,53,92,81)* and the age-related increase in cancer risk *(1,3)*. Finally, oxidative damage has been considered to be a causative factor in a wide spectrum of other diseases, including atherosclerosis, strokes, and autoimmune syndromes *(59)*.

2. REACTIVE OXYGEN SPECIES (ROS)

2.1. *Magnitude of Oxygen Damage*

When assessing the deleterious effects of oxygen, it is instructive to consider quantitatively the number of oxygen species that are metabolized in an aerobic organism (Fig. 1). A typical human cell is approx 10 μm^3 in volume and weighs about 1 ng. This cell metabolizes roughly 10^{12} molecules of molecular oxygen/d and generates 3×10^9 molecules of hydrogen peroxide/h. The ROS produced have sufficient energy to damage any macromolecule in close proximity; Floyd has estimated that they damage 400 million protein mol and 14 million RNA mol/d *(42)*. Based on the excretion of oxygen-damaged DNA adducts, Ames and Shigenaga have estimated that approx 2×10^4 oxidative DNA lesions occur/human genome each day *(1)*. Assuming that resynthesis of each excised adduct involves a stretch of 10 nucleotides *(46)*, then oxygen-induced damage to DNA results in the replacement of 2×10^5 nucleotides/human cell/d.

We have focused on DNA as a critical target for oxygen damage based on the fact that essential genes are present in only one or two copies in both prokaryotes and eukaryotes, whereas there are multiple copies of RNA and protein species in each cell.

From: DNA Damage and Repair, Vol 1: DNA Repair in Prokaryotes and Lower Eukaryotes
Edited by: J. A. Nickoloff and M. F. Hoekstra © Humana Press Inc., Totowa, NJ

O_2 Metabolism	Steady State
Cell	Protein
$10\mu^3$	0.9% of O_2 consumed
1 ng	4 x 10^8 oxidized proteins / cell
Oxygen flux	RNA
~10^{12} O_2 / cell / day	0.03% of O_2 consumed
	1.4 x 10^7 lesions / total RNA / day
H_2O_2 production	DNA
3 x 10^9 H_2O_2 / cell / hour	0.008% of O_2 consumed
	2 x 10^4 lesions / genome / day
	2 x 10^5 nucleotides replaced / genome / day

Fig. 1. Oxygen metabolism in a typical human cell. The magnitude of oxygen damage is presented as estimated by Floyd *(42)*.

Damage to an essential gene, if not correctly repaired, alters the synthesis of all molecules coded by that gene throughout the life of the affected cell. Damage to DNA can be mutagenic and the resulting alteration in nucleotide sequence will be transferred to all progeny cells. The fact that DNA constitutes the genome of most organisms renders it by definition the critical target for mutagenesis and for the production of somatic mutations that accumulate during malignant progression *(74)*.

ROS cause a broad range of DNA damage. Historically, oxygen damage has often been described in terms of strand breaks, because the double-strand breaks formed can result in visible disruption of chromatin structure. It is now clear that, apart from the double-strand breaks, oxygen free radicals also induce single-strand breaks, base and sugar modifications, and DNA–protein crosslinks *(16)*. These lesions have the potential to cause base substitutions and, much less frequently, insertions and deletions. A major problem has been to identify lesions in DNA that are diagnostic of damage by oxygen free radicals. This has been an obstacle in establishing a causal relationship between oxygen damage to DNA and the different pathologies described in Section 1.

The oxidative adducts formed and the resulting mutations have been cataloged in both bacteria and mammalian cells. The mechanisms responsible for repairing oxidative damage have been characterized most extensively in *Escherichia coli*, facilitated by mutants in the oxidative repair pathways that exhibit enhanced sensitivity to oxidative insult. This chapter focuses on oxygen mutagenesis in prokaryotes with analogies drawn to mammalian systems whenever appropriate.

2.2. Fenton Reaction

There are several active or free radical oxygen moieties, collectively referred to as reactive oxygen species that directly or indirectly damage DNA. They are formed by both endogenous and exogenous processes in all aerobic organisms *(2)*. Superoxide anion (O_2^-), hydrogen peroxide (H_2O_2), and the hydroxyl radical (·OH) are generated in vivo by the stepwise reduction of dioxygen to water:

$$O_2 \rightarrow O_2^- \rightarrow H_2O_2 \rightarrow \cdot OH \rightarrow H_2O \quad (1)$$

Although cytochrome oxidase normally efficiently catalyzes this reaction series with no accumulation of the intermediates, some oxygen molecules are reduced monovalently to produce these active species *(43,49,66,83)*.

H_2O_2 and O_2^- are themselves poorly reactive, but can combine to form the ·OH radical in the Haber-Weiss reaction:

$$H_2O_2 + O_2^- \rightarrow \cdot OH + OH^- + O_2 \quad (2)$$

·OH formation occurs spontaneously, but is catalyzed by Fe^{2+} and other transition metals that are present in cells; this is referred to as the Fenton reaction:

$$Fe^{2+} + H_2O_2 \rightarrow \cdot OH + OH^- + Fe^{3+} \quad (3)$$

When Fe^{2+} is the transition metal catalyst, the production of ·OH can be enhanced by chelators, such as EDTA, that may function to keep the metalions in solution. In the case of Cu^{2+} ions, chelators decrease the rate of the metal-catalyzed reaction *(6)*. The Fenton reaction is biologically relevant because metal ions present in the chromatin can catalyze the production of ·OH in close proximity to the DNA *(4)*. It has been clearly demonstrated that the trace amounts of transition metals associated with DNA are required for H_2O_2-induced base damage of DNA *(11)*. The ·OH radical is capable of attacking the DNA directly and is generally assumed to be the critical reactive species that mediates damage via oxygen free radicals *(60,66,114)*.

2.3. Singlet Oxygen

The other radical that acts directly to modify DNA is singlet oxygen ($^1O_2$). This species is a strong electrophile with a relatively long lifetime compared to the ·OH radical *(84)*. Though it is less reactive than ·OH, it has a longer path of diffusion (1–2 μm) through the cell, so it is feasible that $^1O_2$ produced more distal to the chromatin might still reach the DNA and damage it (60). Dyes, such as methylene blue can be used to generate $^1O_2$ in vitro by excitation with visible light. This model system has been shown to be mutagenic in a variety of in vitro and in vivo assays (*see* Section 6.3.).

3. SOURCES OF ROS

3.1. Endogenous Sources

A number of cellular enzymes produce ROS. In particular, enzymes found in cells of the mammalian immune system generate ROS as part of an antimicrobial or antiviral response. This "respiratory burst" from activated immune cells has been demonstrated to cause mutations in DNA *(102)* or bacteria *(126)*. In neutrophils, myeloperoxidase converts H_2O_2 into hypochlorous acid (HOCl), a powerful oxidizing agent *(66)*. The NADPH oxidase associated with the plasma membrane of phagocytes and B-lymphocytes catalyzes the reduction of molecular oxygen to O_2^- at the expense of NADPH. Other ROS and $^1O_2$ are then formed secondarily *(21)*, although the mechanism involved is not clear.

Another enzyme that generates ROS for host defense is nitric oxide synthetase. It comprises part of the cytotoxic response in macrophages and functions to oxidize arginine to nitric oxide (NO·). The nitric oxide formed combines with the O_2^- to generate a peroxynitrile that spontaneously decomposes to form a strong radical that is similar to ·OH *(10)*. This enzyme is also present in vascular endothelial cells and has been postulated to be involved in oxidative injury during formation of atherosclerotic plaques.

Additional enzymatic reactions that form ROS as by-products include the well-studied xanthine oxidase and cytochrome P450. Xanthine oxidase produces O_2^- and H_2O_2 on converting xanthine or hypoxanthine to uric acid. This enzyme has been shown to modify DNA oxidatively *(5)* and produce mutagenic lesions *(73)* in the presence of iron. Cytochrome P450 generates singlet oxygen while metabolizing hydroperoxides and represents a major source of that species in mammals.

Gonzalez-Flecha and Demple *(47)* recently evaluated the sources of H_2O_2 and O_2^- in *E. coli* . The steady-state concentration of H_2O_2 is 0.15 μ*M*. Approximately 87% of the H_2O_2 produced comes from the respiratory chain. They demonstrated that both NADP hydrogenase and ubiquinone leak electrons that form O_2^-during aerobic growth. The O_2^- is converted to H_2O_2 by superoxide dismutase. The rate of production of H_2O_2 in exponentially growing cells in *E. coli* is 4 μ*M* s^{-1}, similar to that reported for mammalian cells. In eukaryotes, the mitochondrial respiratory chain has been demonstrated to be a significant source of O_2^- *(20)*.

The relationship between lipid peroxidation and DNA requires more thorough investigation. ·OH can abstract a hydrogen from a polyunsaturated fatty acid and produce a radical species that reacts with oxygen to produce a powerful peroxy radical. This species can remove a hydrogen atom from a nearby unsaturated fatty acid, thus initiating a radical chain reaction that could amplify an initial oxidative insult *(36,43)*. The lipid hydroperoxides formed are capable of producing double-strand breaks via a yet to be characterized mechanism that depends on the presence of transition metals *(88)*. They can also be degraded to aldehydes, such as malonaldehyde, that have been shown to be mutagenic in both *Salmonella typhimurium (80)* and in mammalian cells *(128)*. Lipid modifications of the bases in DNA could be a major contributor to the cytotoxicity and mutagenicity of ROS *(36)*.

3.2. *Exogenous Sources*

In addition to endogenous normal cellular process, ROS are also generated by exposure of cells to exogenous agents. The most well-studied example is ionizing radiation (*see* Chapter 5, vol. 2), which was the first demonstrated to be mutagenic by Muller in 1927 *(89)*. In fact, the concept that oxygen is mutagenic was fostered by the observation that chromosomal abnormalities induced by hyperbaric oxygen and those produced by ionizing radiation were indistinguishable *(24)*. Ionizing radiation generates the ·OH radical through water radiolysis and produces more than 100 distinct DNA adducts *(56)*. ·OH scavengers are capable of preventing most of the γ-irradiation-induced single-strand breaks in plasmid DNA *(104)* and also reduce the number of mutagenic events *(124)*.

Other exogenous sources of ROS include xenobiotic agents, such as bleomycin, neocarzinostatin, and mitomycin C. Each of these drugs damage DNA by an iron-dependent radical mechanism *(57)*. Both bleomycin and mitomycin C are used in the treatment of human cancer. Other chemical agents, including pesticides, also generate

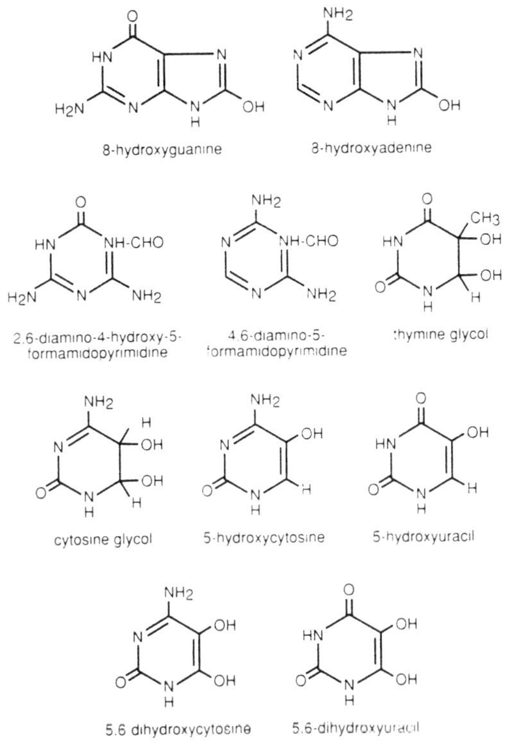

8-hydroxyguanine 8-hydroxyadenine

2,6-diamino-4-hydroxy-5-formamidopyrimidine 4,6-diamino-5-formamidopyrimidine thymine glycol

cytosine glycol 5-hydroxycytosine 5-hydroxyuracil

5,6 dihydroxycytosine 5,6-dihydroxyuracil

Fig. 2. Chemical structure of some of the major oxidized adducts resulting from ·OH radical attack on the base moieties of DNA.

ROS. For example, paraquat produces O_2^- anion by diverting electrons from NADPH *(62)*; it is highly cytotoxic and mutagenic in bacteria *(87)*.

4. DNA ADDUCTS FORMED BY ROS

4.1. *Types of DNA Damage*

ROS produce a multiplicity of alterations in DNA *(18,56)*. One approach to documenting these alterations has been to treat either cells or DNA with ROS generated by different protocols and then analyze the DNA adducts formed by gas chromatography-mass spectrometry or, in the case of adducts that have electrochemical activity, a combination of HPLC and electrochemical detection *(31,44,51)*. Oxidative attack on the deoxyribose moiety can result in the formation of sugar radicals and release of free bases from the DNA. ·OH can also attack the purine or pyrimidine bases directly at different sites (Fig. 2). For example, addition to positions C-4, C-5, or C-8 of guanine or adenine can result in ring opening, ring saturation, or hydroxylation. Attack on pyrimidines also yields a variety of alterations, including the production of thymine or cytosine glycols *(50,110,117)*. Oxidative damage to cytosine alone results in at least 40 modified species *(96)*. An alternative source of DNA adducts is the modification of nucleosides and nucleotides, and the subsequent incorporation of these modified substrates into DNA *(22,40)*. The major difficulty has been to establish which of the modifications in DNA is rate-limiting with respect to cytotoxicity or mutagenicity.

4.2. Base Damage

The most well-characterized oxygen free radical-induced alteration in DNA is 8-hydroxy-2'-deoxyguanosine (8-OH-dG). This adduct provides a biomarker of oxidative damage; it is usually quantitated by a combination of HPLC and electrochemical detection *(41,44,63,93)*. It is detected in DNA in increased amounts after exposure to a variety of agents that produce ROS, including ionizing radiation, the carcinogens 4-nitro-quinoline oxide and 2-nitro-propane *(32,59)* as well as $^1O_2$ *(12,115)* and Fe^{3+} or Cu^{2+} plus H_2O_2 *(6,7)*. In one of the initial studies, 8-OH-dG levels were shown to increase in the DNA of intact *S. typhimurium* treated with H_2O_2 *(63)*. The mutagenic potential of this adduct and its repair in *E. coli* have been well characterized *(13,22,48,69,114)*. 8-OH-dG is excised from DNA by the product of the *fpg* gene, formamidopyrimidine glycosylase (FAPY glycosylase). The *fpg* gene is identical to the *mutM* gene, and bacterial strains with mutations in *mutM* exhibit a mutator phenotype and an increase in G $\rightarrow$ T transversions. This is owing to 8-OH-dG mispairing with A during replication *(22,127)*. If the lesion remains unrepaired prior to replication, misinsertion of A opposite the adduct can occur. Cells have evolved an ancillary repair pathway to correct this mismatch by excising the deoxyadenosine *(85)*. The product of the *mutY* gene recognizes and specifically repairs the 8-OH-dG:dA mismatch.

8-OH-dG triphosphate can also form as a result of alteration of dGTP and the subsequent incorporation of 8-OH-dGTP into DNA. Thereafter, it frequently mispairs with A to yield A $\rightarrow$ C substitutions *(22,94)*. A third cellular defense mechanism against 8-OH-dG exists in *E. coli* to protect against this source of mutation. The product of the *mutT* gene catalyzes the conversion of the triphosphate adduct to the monophosphate, thus removing the premutagenic triphosphate from the precursor pool *(77)*. That such elaborate repair mechanisms are required to purge cells of 8-OH-dG reflects the highly mutagenic potential of this oxidative adduct.

The substrate specificity of FAPY glycosylase is not limited to 8-OH-dG; it also excises the oxidatively produced alteration, 2,6-diamino-4-hydroxy-5-formamidopyrimidine (FAPY Gua *[12]*). Using the combination of xanthine oxidase and the Fe–EDTA complex to generate ·OH, Aruoma et al. *(7)* showed that this imidazole ring-opened guanine was one of the major adducts formed in oxidatively damaged DNA. It is also one of the few adducts other than 8-OH-dG that have been demonstrated to result from damage by $^1O_2$ *(12)*. The analogous ring-opened adenine (4,6-diamino-5-formamidopyrimidine or FAPY Ade) is also formed by oxidant injury and repaired by FAPY glycosylase *(15)*. Both of these adducts are present at very low levels in undamaged DNA, but increase approx 10-fold in DNA exposed to Fe^{3+} or Cu^{2+} and H_2O_2 *(6)*. It remains to be determined whether FAPY glycosylase excises a wider spectrum of purine base alterations produced by ROS, i.e., whether FAPY glycosylase is a generic enzyme for the repair of purine adducts.

Altered pyrimidines are also produced by ROS. Ring-saturated products form as a result of ·OH radical attack on the C-5—C-6 double bonds of these residues. In the case of thymine, this leads to the formation of thymine glycol *(115)*, a major adduct formed by the combination of Fe^{3+}/H_2O_2 *(6)*. When present in a single-stranded DNA template, thymine glycol causes predominantly T $\rightarrow$ C mutations at a frequency of 0.3% *(9)*. When placed in double-stranded DNA, thymine glycol is not mutagenic; rather, it has a toxic effect because it is a strong block to replication *(16,37)*. The adduct is repaired by the product of the *nth* gene, endonuclease III *(29,34)*. Interpreting the in vivo significance of this was initially complicated by the fact that *nth* mutants do not show an increased sensitivity to oxidative damage *(25)*. However, another enzyme capable of excising thymine glycol has been described in *E. coli (82)*. This endonuclease, referred to as endonuclease VIII, displays functional redundancy with endonuclease III and may compensate for its loss in *nth* mutants.

A catalog of the mutations caused by oxygen free radicals indicates that cytosine residues are targets for the most frequent premutagenic lesions. Exposure of DNA to osmium tetroxide or ionizing radiation results in the formation of cytosine glycol and its dehydrated or deaminated derivatives, 5-hydroxycytosine (5-OH-dC) and 5-hydroxyuracil, respectively. 5-OH-dC is also formed, along with the deaminated derivative, 5,6-dihydroxyuracil *(31,33,116)*. Amounts of 5-OH-dC increase 50-fold in DNA exposed to Fe^{3+}/H_2O_2 and 150-fold in DNA treated with Cu^{2+}/H_2O_2 *(123)*. This altered base in DNA is a substrate for both endonuclease III and, unexpectedly, FAPY glycosylase *(54)*.

4.3. *Abasic Sites*

ROS can also cause the formation of apurinic/apyrimidinic (AP) sites in DNA via oxidation of the sugar moieties. For example, γ-irradiation can cause oxidation at C-1' or C-2' and concomitant release of the adjoining base *(97)*. Copying past abasic sites by DNA polymerases from both procaryotes *(106,108,113)* and eukaryotes *(71,106)* results predominantly in the insertion of dATP. However, incorporation of dATP is not the most frequent misinsertion in yeast lacking the major AP repair enzyme, Apn1 *(72)*. The specificity of incorporation at an abasic site has been reviewed *(76)*. These sites are repaired by several AP endonucleases. In *E. coli*, this includes the products of the *nth*, *xth*, and *nfo* genes *(35)*. One can argue that the findings of multiple repair pathways in bacteria for abasic sites in DNA indicate the potential deleterious effects of the unrepaired alterations.

4.4. *Strand Breaks*

Major structural alterations in the phosphodiester backbone of DNA caused by ROS are single- and double-strand breaks. These can result from the loss of a nucleoside moiety, leaving a phosphate group on both the free 5'- and 3'-termini, or from degradation of a sugar residue leaving the altered sugar molecule on the 3'-end of the break and a 5'-phosphate on the other. These breaks can be blocks to replication and are lethal if unrepaired. In *E. coli*, exonuclease III (the product of the *xth* gene) and endonuclease IV have been shown to remove the 3'-terminal phosphate and generate a 3'-hydroxyl terminus that can be extended by DNA polymerase I *(28,55)*. This explains the increased sensitivity of *xth* mutants to H_2O_2.

5. REVERSE CHEMICAL MUTAGENESIS

Although these and other premutagenic lesions in DNA have been extensively cataloged, the precise lesions that correlate directly with mutations in vivo have remained elusive. Once candidate adducts have been identified, they can be incorporated into a target sequence and assayed for mutagenesis *(109)*. The most informative methods involve site-specific mutagenesis. In this approach, an altered base is inserted into DNA either by the chemical synthesis of an oligonucleotide containing the altered base *(9)* or by incorporation of an altered deoxynucleoside triphosphate into DNA by a DNA polymerase *(22,129)*. Different methods are available to test the cytotoxicity and mutagenicity of these site-specific modifications in both bacteria or in mammalian cells in culture. The difficulty is that one needs first to identify the key lesion for investigation.

A novel approach to this problem, referred to as reverse chemical mutagenesis, has been described by Feig et al. *(40)*. ROS-mediated damage to DNA most frequently occurs at cytidine residues (*see* Section 6.2.). In the initial studies using reverse chemical mutagenesis, dCTP was damaged in vitro and fractionated by chromatography on HPLC columns. The individual fractions were incorporated into DNA in place of dCTP and then analyzed to determine which fractions were mutagenic. Incorporation into DNA was carried out with HIV reverse transcriptase using a gap in the *lacZ*α gene in circular M13mp2 DNA as a template and target site for mutagenesis. After transfecting the M13mp2 DNA into *E. coli* and selecting for *lacZ* mutants on X-gal (5-bromo-4-chloro-3-indolyl β-D-galactoside), the oxidized dCTP species in the most mutagenic fraction were identified by mass spectrometry.

The most mutagenic species identified was 5-hydroxy-2'-deoxycytidine. It causes C → T transitions in repair competent *E. coli* at a frequency of 2.5%, making it the most mutagenic oxidative adduct described to date. Wagner et al *(123)* have reported that 5-OH-dC levels increase in DNA exposed to ionizing radiation, or to H_2O_2 in the presence or absence of transition metals, and osmium tetroxide. It has also been reported that the Klenow fragment of *E. coli* DNA polymerase I will incorporate 5-hydroxy-deoxycytidine triphosphate with a greater efficiency than 8-OH-dGTP, leading to the proposal that a protein analogous to MutT (*see* Section 4.2.) may exist to remove 5-OH-dC triphosphate from the nucleotide precursor pool *(98)*. Taken together, these data argue strongly for a prominent role for 5-OH-dC in oxygen mutagenesis.

6. MUTATIONAL SPECTRA OF DNA EXPOSED TO ROS

6.1. Metal Ions

As is illustrated in the preceding discussion, ROS interact with DNA to produce an array of lesions that may affect the faithful transmission of genetic information. In aerobic organisms, mechanisms have evolved for the repair of many of these lesions (*see* refs. *17*, *27*, and *117* for comprehensive reviews). Those that escape repair, however, can lead to a variety of mutagenic changes. The spectrum of mutations generated at a particular target sequence of DNA depends in part on the agents used to generate ROS. Considerable data are available on mutations caused by exposure of DNA to ROS produced by metals in Fenton-type reactions.

The earliest demonstration of metal mutagenesis mediated by ROS was provided in studies with Fe^{2+} using ϕX174 am3 single-stranded DNA. The amber 3 mutation allows for scoring reversion mutations at nucleotide 587 (a T residue) in the amber codon, TAG. When ϕX174 am3 DNA was exposed to 5 μ*M* Fe^{2+} and transfected into spheroplasts from SOS-induced *E. coli*, 72% of the mutations were T → A, 22 % were T → G, and 6% were T → C. The overall reversion frequency was 28.3×10^{-6} (compared to a background reversion frequency of 2.4×10^{-6}). The addition of superoxide dismutase

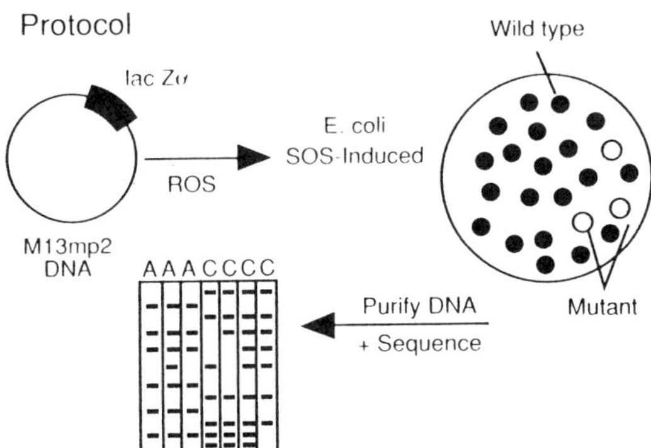

Fig. 3. Protocol for the M13mp2 forward mutation assay employed to determine the mutagenic spectrum of oxidative damage in SOS-induced *E. coli.*

had little effect on the frequency, indicating that the production of O_2^- was not a major factor, and the data are consistent with ·OH being the damaging species. Because the misincorporation of dATP is also characteristic for copying past an AP site, the Fe^{2+}-treated DNA was also exposed to a HeLa cell apurinic endonuclease prior to transfection. This abolished only 26% of the Fe^{2+}-induced mutations, suggesting that mutagenesis under these conditions is not primarily mediated by AP sites *(75)*.

Other transition metals have been tested in the ϕX174 reversion assay, as well as the combination of metal ions plus H_2O_2 *(120)*. Of nine metal ions assayed, $FeCl_2$ was the most mutagenic. Addition of 10 μ*M* H_2O_2 did not increase the reversion frequency over that produced by 10 μ*M* $FeCl_2$ alone. This contrasts with Cu^{2+}, which is by itself mutagenic, but Cu^{2+} mutagenicity is increased more than threefold by the addition of 10 μ*M* H_2O_2. This comparison is consistent with the data of Aruoma et al. *(6)*, who demonstrated that Cu^{2+}/H_2O_2 was more effective than Fe^{3+}/H_2O_2 at inducing different adducts measurable by gas chromatography-mass spectroscopy. Cu^{2+} is also a Fenton reactant *(107)*, but an oxygen species other than ·OH must be involved, since the addition of mannitol (which scavenges the ·OH radical) reduced mutagenesis by <50%. Furthermore, superoxide dismutase nearly abolished Cu^{2+} mutagenicity, indicating a role for O_2^-.

The mutagenic potential of these ROS-generating systems has been explored using a forward mutation assay that is capable of scoring for all possible single-base substitutions as well as deletions or insertions (Fig. 3). The M13mp2 assay described by Kunkel *(71)* utilizes single-stranded M13 DNA that contains a 474 nucleotide sequence coding for that portion of the *lacZ* gene necessary to produce a functional β-galactosidase when transfected into an *E. coli* strain carrying the remainder of the gene. The host *E. coli* is generally UV-irradiated at 50 J/m^2 prior to transfection. This induces the SOS response, which is believed to facilitate replicative bypass of lesions that are blocking in uninduced cells (reviewed in *90*; *see* Chapters 7 and 12). The complementation assay is scored by plating *E. coli* transfected with M13mp2 DNA onto the indicator substrate, X-Gal. Cells containing wild-type M13mp2 DNA produce dark blue colonies; cells containing mutations in the target sequence produce light blue or colorless colonies, depending on the extent to which the mutation renders the β-galactosidase inactive.

The background mutation frequency in this assay is 0.2×10^{-3}. However, the assay is sensitive, since mutations at a large number of sites are scorable. When M13mp2 DNA is treated with 10 μM Fe^{2+} and transfected into SOS-induced bacteria, the mutation frequency increased approx 75-fold. This increase can be inhibited by incubation of the treated DNA with catalase, mannitol, or superoxide dismutase. Single-base substitutions were overwhelmingly the predominant mutation observed (94% of those sequenced). The remaining 6% were deletions. The most common single base substitutions were G → C transversions, followed by C → T transitions, and G → T transversions *(78)*.

Cu^{2+} increased the mutation frequency of M13mp2 DNA in this assay about 66-fold *(120)*. Again, only a small percentage of the sequenced mutants were deletions (3.5%). One insertional event was detected. The remaining 95% of the mutations were single base substitutions. In this case, the predominant mutations were C → T transitions (59%), followed by G → T substitutions.

The mutational spectra of ROS generated by other protocols have also been assessed with the M13mp2 system. It has been observed that the tripeptide, glycine-glycine-histidine (GGH), enhances ROS production by Ni^{2+}/H_2O_2. Although the chemistry of the reaction is unclear, GGH has been proposed to behave as a chelator that allows Ni^{2+} to function as a Fenton catalyst *(122)*. In fact, the reaction may be more complicated, since in addition to $\cdot OH$, O_2^- and $^1O_2$ have also been demonstrated to form under these reaction conditions *(58)*. In vivo, histidine-containing proteins may serve this function and contribute to the carcinogenicity of Ni^{2+}. Ni^{2+}/H_2O_2 and GGH give rise to a wider spectrum of mutations than Fe^{2+} or Cu^{2+}. The most frequent types of substitutions were G → A and C → T transitions (38 and 32%, respectively). The fact that the mutation frequency of the Ni^{2+} system was reduced by the addition of scavengers of $\cdot OH$ and $^1O_2$ further indicates the involvement of ROS.

6.2. Ionizing Radiation

The mutagenic effect of ionizing radiation has been studied extensively in both bacteria and mammalian cells, although this discussion will focus on the former. Initial work on the mutagenic spectrum of ionizing radiation was done using ^{60}Co and M13 mp10 single-stranded phage DNA in an *E. coli lacZ* complementation assay similar to the one described (*see* Section 6.1. *[8]*). Of the mutants sequenced, 14 were single-base substitutions and 1 was an insertion mutation. The predominant mutations were C → T transitions, with 70% of all substitutions occurring at cytosine residues. Similar results were obtained when the λ phage cI repressor gene was used as a target for ^{60}Co mutagenesis *(118)*. Of 41 mutants sequenced, 85% of the mutations were base substitutions. Again, G:C → A:T transitions were predominant, followed by G:C → T:A. The predominance of these two mutations has also been documented in *S. typhimurium* exposed to ionizing radiation *(67)*.

One general conclusion that may be drawn from these studies is that C → T transitions are characteristic of damage by ROS induced by both metal ions and ionizing radiation. The most direct explanation for this is that ROS increase the rate of deamination of cytidine to uracil. However, when M13mp2 DNA was damaged with Cu^{2+} plus H_2O_2 and transfected into cells lacking a functional uracil glycosylase (ung^-), no increase in mutation frequency was seen over that observed when wild-type *E. coli* was used as the host. Uracil glycosylase removes uracil as a free base and leaves an abasic site that is recognized by cellular apurinic/apyrimidinic endonuclease. Therefore, if deamination of cytidine to uracil contributed significantly to oxygen mutagenesis, the mutation frequency should be increased. This was not the case *(101)*. The obverse was also observed; cells that lacked endonuclease III *(nth⁻)* did show an increased mutation frequency, indicating that the deamination of cytidine is not a major contributor to oxygen-induced mutagenesis.

6.3. *Methylene Blue Plus Light*

The mutational spectrum of singlet oxygen is quite different from that generated by Fenton reactants or ionizing radiation. When M13mp2 DNA was exposed to methylene blue plus light and transfected into uninduced *E. coli*, there was a >97% loss of viability *(79)*. This is consistent with the data of Piette et al. *(95)*, who demonstrated that proflavine plus light, which also produces $^1O_2$, causes the formation of blocking lesions at G-residues. Also in accordance with this, when the damaged DNA was transfected into SOS-induced cells, 88% of single base mutations occurred at template G-residues. The majority of these were G → C transversions. This suggests the formation of G:G mispairing, but the adduct responsible for this is unknown. The other mutation present in significant amounts was a G → T transversion. This mutation was predominant after $^1O_2$ treatment of double-stranded DNA in a forward mutation assay in *E. coli (26)* and in a single-stranded shuttle vector replicated in primate cells *(105)*. It likely results from the misinsertion of A opposite 8-OH-dG adducts, which were present in large amounts (600 8-OH-dG/10^5 dG residues) in this study and in other studies with $^1O_2$ generators *(30,45,79)*. The differences in these reports on mutational spectra could be owing to differences in the repair of the promutagenic lesion, 8-OH-dG, or in the unknown lesion responsible for the G → C transversion.

6.4. *Activated Leukemia Cells*

The M13mp2 forward mutation assay has also been used to catalog the mutations generated by a more complex biological source. When cells from the human leukemia line HL-60 are grown in culture and induced to differentiate along a granulocytic pathway, activation with phorbol esters stimulates them to release oxygen species. This mimics the respiratory burst of phagocytic cells in vivo (*see* Section 3.1.). The stimulated cells release both O_2^- and H_2O_2 *(102)*. When M13mp2 DNA was incubated with activated HL-60 cells, the mutation frequency increased sixfold. This increase was partially inhibited by catalase, superoxide dismutase, and mannitol, indicating the involvement of H_2O_2, O_2^-, and ·OH. One deletion and one insertional mutation were recovered in 72 mutant sequences. The remainder were single-base substitutions, predominately at G-residues. G → T substitutions were the most common, followed by G → C, T → C and C → T. The preponderance of G-substitutions may be owing to release of $^1O_2$ from these cells, as has been reported for eosinophils *(61)*, though release of this species was not measured directly in this study.

7. A SIGNATURE MUTATION FOR OXYGEN DAMAGE

In addition to the abundant single-base substitutions induced by oxygen damage, a rare double-tandem CC → TT mutation has also been reported and found to be characteristic of damage by ROS. In analyzing the sequence of M13mp2 mutations induced by metals or HL-60 cells, it became apparent that CC → TT mutations were a unique feature. The only other reported source of this mutation is UV irradiation *(14,86)*. The observation that oxidative damage to DNA causes CC → TT mutations was confirmed and extended using a reversion assay designed to recognize single- and double-tandem base substitutions at cytosine residues. A modification of the M13mp2 assay (referred to as M13G*1) was established. This reversion assay is based on a GCC → CCC alteration in the codon at positions 141–143 of the *lacZ*α gene that abolishes β-galactosidase activity. Any single mutation at positions 141 or 142 or a double-tandem mutation at 141–142 or 142–143 restores enzyme activity, and causes a reversion to the wild-type dark blue colonies on selective plates *(23)*. This assay has been used to measure the mutation frequency at cytosine residues as a result of damage with metal ions, metal ions plus H_2O_2, activated HL-60 cells, ionizing radiation, and UV light *(103,119,121)*.

The most common single-base substitutions produced by oxygen damage or UV irradiation are C → T substitutions. As noted above, this mutation has emerged as a general feature of oxygen damage. However, all other possible single-base substitutions were also recovered, with C → G being the least common. The overall single base mutation frequency within the CCC site was 1 in 2000 molecules. In comparison, the CC → TT mutations were observed at a frequency of about 1 in 50,000 when Cu^{2+} or Fe^{2+} plus H_2O_2 were used to damage DNA compared to about 1 in 20,000 for a 25 J/m^2 dose of UV *(103)*. The frequency of CC → TT mutations produced by Cu/H_2O_2 was 1000-fold higher than could be accounted for by two independent C → T mutations (Fig. 4). The addition of mannitol almost entirely abolished the Fe^{2+}-induced double tandem mutations, but had no effect on those induced by UV. These results could indicate that UV and metal ions do not generate a common oxygen radical that produces the CC → TT mutation. However, the same chemical alteration in DNA might be produced by the two agents. To date, CC → TT substitutions have been observed in DNA exposed to metal ions/H_2O_2, ionizing radiation, or activated HL-60 cells *(102,103,120,121)*. They have not been observed in DNA exposed to methylene blue plus light, which generates only singlet oxygen *(79)*, nor in DNA copied by a variety of DNA polymerases and reverse transcriptases.

The largest increase in CC → TT mutations was produced by the combination of Ni^{2+}/H_2O_2/GGH *(121)*. In this case, 20% of all mutants sequenced in the reversion assay had this double-tandem mutation. There was a definite bias towards mutations at the first two positions of the CCC target compared to the second and third positions. This may be because the Ni^{2+}/GGH complex binds preferentially to the first two cytosines. The affinity of Ni^{2+}/GGH for the site may also explain its efficiency at generating the double-tandem mutation.

The CC → TT mutation produced by oxygen damage has so far been described only in SOS-induced *E. coli*. It is relevant to question whether it occurs in mammalian cells, since they have not been clearly shown to exhibit an error-prone SOS pathway. Koch et al. *(67)* used a histidine reversion assay with a missense CCC codon in place of the wild-type CTC codon to score for CC → TT mutations in *S. typhimurium*. They demonstrated that UV-induced CC → TT mutations are also SOS-dependent in this system. However, CC → TT substitutions also occur in human *(14,91)* and in mouse *(68)* cells

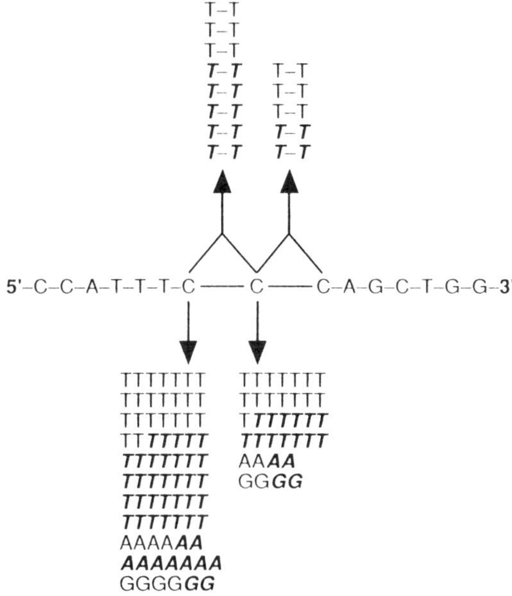

Fig. 4. Single- and double-tandem mutations induced by Fe^{2+}/H_2O_2 and Cu^{2+}/H_2O_2 in the M13G*1 reversion assay. The number and type of single-base substitutions at each of three cytosines residues at the target site are shown below the sequence; the double-tandem substitutions are shown above the sequence. Those mutations induced by Fe^{2+}/H_2O_2 are in italicized bold type; those induced by Cu^{2+}/H_2O_2 are in regular type *(103)*.

treated with UV radiation. Thus, SOS dependence in bacteria does not exclude the occurrence of a double-tandem mutation in mammalian cells.

CC → TT mutations are the result of both UV irradiation and ROS. They have been reported in the p53 gene in human skin cancer that presumably arose from exposure to UV irradiation *(14)*. The fact that these mutations in p53 are clonal indicates that the mutation imparted a proliferative advantage to the cancer cells. Since UV irradiation only penetrates the outermost cell layers, most internal tissues are not exposed to UV irradiation. The documentation of CC → TT mutations in internal cancers and in other pathologic conditions would implicate ROS in the pathogenesis of these conditions. In the case of nonskin cancers, there is no reason *a priori* to argue that CC → TT mutations in p53 would result in clonal proliferation. Thus, there is a need for assays to measure these signature mutations at low frequencies in cancers as well as in other diseases associated with oxygen free radicals.

8. CONCLUDING REMARKS

A number of ROS possessing the capacity to damage DNA are generated intracellularly. In response, cells have evolved multiple mechanisms to repair this oxidative damage. Although it now seems clear that a small fraction of adducts can escape repair and lead to mutagenesis, the precise chemical lesions that contribute most significantly to mutagenesis are only beginning to be identified. Additionally, the extent to which mutations attributable to oxidative damage accumulate over the lifetime of a cell or during the course of progression of disease is unknown. However, the observation that CC → TT mutations are diagnostic for oxygen damage (in the absence of UV irradiation) may provide a basis to quantitate the contribution of oxygen free radicals to mutagenesis, cancer, and other diseases, and perhaps normal aging.

ACKNOWLEDGMENTS

Support for our work comes from Grant 1-F32-CA67482-01 from the National Cancer Institute; Program Project Grant AG017151 from the National Aging Institute; and Training Grant 5-T32-AG00057 from the National Institutes of Health.

REFERENCES

1. Ames, B. N. and M. K. Shigenaga. 1992. Oxidants are a major contributor to aging. *Ann. NY Acad. Sci.* **663:** 85–96.
2. Ames, B. N. 1989. Mutagenesis and carcinogenesis: endogenous and exogenous factors. *Environ. Mol. Mutagen.* **16:** 66–77.
3. Ames, B. N. and L. S. Gold. 1991. Endogenous mutagens and the causes of aging and cancer. *Mutat. Res.* **250:** 3–16.
4. Andronikashvili, E. L., L. M. Mosulishvili, A. I. Belokobilski, N. E. Kharabadze, T. K. Tevzieva, and E. Y. Efremova. 1974. Content of some trace elements in sarcoma M-1 DNA in dynamics of malignant growth. *Cancer Res.* **34:** 271–274.
5. Aruoma, O. I., B. Halliwell, and M. Dizdaroglu. 1989. Iron ion-dependent modification of bases in DNA by the superoxide radical-generating system hypoxanthine/xanthine oxidase. *J. Biol. Chem.* **264:** 13024–13028.
6. Aruoma, O. I., B. Halliwell, E. Gajewski, and M. Dizdaroglu. 1991. Copper-ion-dependent damage to the bases in DNA in the presence of hydrogen peroxide. *Biochem. J.* **273:** 601–604.
7. Aruoma, O. I., B. Halliwell, E. Gajewski, and M. Dizdaroglu. 1989. Damage to the bases in DNA induced by hydrogen peroxide and ferric ion chelates. *J. Biol. Chem.* **264:** 20,509–20,512.
8. Ayaki, H., K. Higo, and O. Yamamoto. 1986. Specificity of ionizing radiation-induced mutagenesis in the *lac* region of single-stranded phage M13 mp10 DNA. *Nucleic Acids Res.* **14:** 5013–5018.
9. Basu, A. K., E. L. Loechler, S. A. Leadon, and J. M. Essigmann. 1989. Genetic effects of thymine glycol: site-specific mutagenesis and molecular modeling studies. *Proc. Natl. Acad. Sci. USA* **86:** 7677–7681.
10. Beckman, J. S., T. W. Beckman, J. Chen, P. A. Marshall, and B. A. Freeman. 1990. Apparent hydroxyl radical production by peroxynitrite: implications for endothelial injury from nitric oxide and superoxide. *Proc. Natl. Acad. Sci. USA* **87:** 1620–1624.
11. Blakely, W. F., A. F. Fuciarelli, B. J. Wegher, and M. Dizdaroglu. 1990. Hydrogen peroxide-induced base damage in deoxyribonucleic acid. *Radiat. Res.* **121:** 338–343.
12. Boiteux, S., E. Gajewski, J. Laval, and M. Dizdaroglu. 1992. Substrate specificity of the *Escherichia coli* Fpg protein (formamidopyrimidine-DNA glycosylase): excision of purine lesions in DNA produced by ionizing radiation or photosensitization. *Biochemistry* **31:** 106–110.
13. Boiteux, S., T. R. O' Conner, and J. Laval. 1987. Formamidopyrimidine-DNA glycosylase of *Escherichia coli*: cloning and sequencing of the *fpg* structural gene and overproduction of the protein. *EMBO J.* **6:** 3177–3183.
14. Brash, D. E., D. A. Rudolph, J. A. Simon, A. Lin, G. J. McKenna, H. P. Baden, A. J. Halperin, and J. Ponten. 1991. A role for sunlight in skin cancer: UV-induced p53 mutations in squamous cell carcinoma. *Proc. Natl. Acad. Sci. USA* **88:** 10,124–10,128.

15. Breimer, L. H. 1984. Enzymatic excision from gamma-irradiated polydeoxyribonucleotides of adenine residues whose imidazole rings have been ruptured. *Nucleic Acids Res.* **12:** 6359–6367.
16. Breimer, L. H. 1990. Molecular mechanisms of oxygen radical carcinogenesis and mutagenesis: the role of DNA base damage. *Mol. Carcinog.* **3:** 188–197.
17. Breimer, L. H. 1991. Repair of DNA damage induced by reactive oxygen species. *Free Radical Res. Commun.* **14:** 159–171.
18. Cadet, J. and M. Berger. 1985. Radiation-induced decomposition of the purine bases within DNA and related model compounds. *Int. J. Radiat. Biol. Relat. Stud. Phys. Chem. Med.* **47:** 127–143.
19. Cerutti, P. A. 1985. Prooxidant states and tumor promotion. *Science* **227:** 375–381.
20. Chance, B., H. Sies, and A. Boveris. 1979. Hydroperoxide metabolism in mammalian organs. *Physiol. Rev.* **59:** 527–605.
21. Chanock, S. J., J. el-Benna, R. M. Smith, and B. M. Babior. 1994. The respiratory burst oxidase. *J. Biol. Chem.* **269:** 24,519–24,522.
22. Cheng, K. C., D. S. Cahill, H. Kasai, S. Nishimura, and L. A. Loeb. 1992. 8-hydroxyguanine, an abundant form of oxidative DNA damage, causes G $\rightarrow$ T and A $\rightarrow$ C substitutions. *J. Biol. Chem.* **267:** 166–172.
23. Cheng, K. C., B. D. Preston, D. S. Cahill, M. K. Dosanjh, B. Singer, and L. A. Loeb. 1991. The vinyl chloride DNA derivative N2,3-ethenoguanine produces G–A transitions in *Escherichia coli*. *Proc. Natl. Acad. Sci. USA* **88:** 9974–9978.
24. Conger, A. D. and L. M. Fairchild. 1952. Breakage of chromosomes by oxygen. *Proc. Natl. Acad. Sci. USA* **38:** 289–299.
25. Cunningham, R. P. and B. Weiss. 1985. Endonuclease III *(nth)* mutants of *E. coli*. *Proc. Natl. Acad. Sci. USA* **82:** 474–478.
26. Decuyper-Debergh, D., J. Piette, and A. Van de Vorst. 1987. Singlet oxygen-induced mutations in M13 *lacZ* phage DNA. *EMBO J.* **6:** 3155–3161.
27. Demple, B. and L. Harrison. 1994. Repair of oxidative damage to DNA: enzymology and biology. *Annu. Rev. Biochem.* **63:** 915–948.
28. Demple, B., A. Johnson, and D. Fung. 1986. Exonuclease III and endonuclease IV remove 3' blocks from DNA synthesis primers in H_2O_2-damaged *Escherichia coli*. *Proc. Natl. Acad. Sci. USA* **83:** 7731–7735.
29. Demple, B. and S. Linn. 1980. DNA N-glycosylases and UV repair. *Nature* **287:** 203–208.
30. Devasagayam, T. P. A., S. Steenken, M. S. W. Obendorf, W. A. Schutz, and H. Sies. 1991. Formation of 8-hydroxy(deoxy)guanosine and generation of strand breaks at guanine residues in DNA by singlet oxygen. *Biochemistry* **30:** 6283–6289.
31. Dizdaroglu, M. 1985. Application of capillary gas chromatography-mass spectrometry to chemical characterization of radiation-induced base damage of DNA: implications for assessing DNA repair processes. *Anal. Biochem.* **144:** 593–603.
32. Dizdaroglu, M. 1985. Formation of an 8-hydroxyguanine moiety in deoxyribonucleic acid on gamma-irradiation in aqueous solution. *Biochemistry* **24:** 4476–4481.
33. Dizdaroglu, M., E. Holwitt, M. P. Hagan, and W. F. Blakely. 1986. Formation of cytosine glycol and 5,6-dihydroxycytosine in deoxyribonucleic acid on treatment with osmium tetroxide. *Biochem. J.* **235:** 531–536.
34. Dizdaroglu, M., J. Laval, and S. Boiteux. 1993. Substrate specificity of the *Escherichia coli* endonuclease III: excision of thymine- and cytosine-derived lesions in DNA produced by radiation-generated free radicals. *Biochemistry* **32:** 12,105–12,111.
35. Doetsch, P. W. and R. P. Cunningham. 1990. The enzymology of apurinic/apyrimidinic endonucleases. *Mutat. Res.* **236:** 173–201.
36. Esterbauer, H., P. Eckl, and A. Ortner. 1990. Possible mutagens derived from lipids and lipid precursors. *Mutat. Res.* **238:** 223–233.

37. Evans, J., M. Maccabee, Z. Hatahet, J. Courcelle, R. Bockrath, H. Ide, and S. Wallace. 1993. Thymine ring saturation and fragmentation products: lesion bypass, misinsertion and implications for mutagenesis. *Mutat. Res.* **299:** 147–156.
38. Feig, D. I. and L. A. Loeb. 1995. Endogenous mutagenesis, in *DNA Repair Mechanisms: Impact on Human Diseases and Cancer* (Vos, J.-M., ed.), R. G. Landes Company, New York, pp. 175–185.
39. Feig, D. I., T. M. Reid, and L. A. Loeb. 1994. Reactive oxygen species in tumorigenesis. *Cancer Res.* **54:** 1890s–1894s.
40. Feig, D. I., L. C. Sowers, and L. A. Loeb. 1994. Reverse chemical mutagenesis: Identification of the mutagenic lesions resulting from reactive oxygen species-mediated damage to DNA. *Proc. Natl. Acad. Sci. USA* **91:** 6609–6613.
41. Floyd, R. A. 1990. The development of a sensitive analysis for 8-hydroxy-2'-deoxyguanosine. *Free Radical Res. Commun.* **8:** 139–141.
42. Floyd, R. A. 1995. Measurement of oxidative stress in vivo, in *The Oxygen Paradox* (Davies, K. J. A. and Ursini, F., eds.), Cleup University Press, pp. 89–103.
43. Floyd, R. A. 1990. Role of oxygen free radicals in carcinogenesis and brain ischemia. *FASEB J.* **4:** 2587–2597.
44. Floyd, R. A., J. J. Watson, P. K. Wong, D. H. Altmiller, and R. C. Rickard. 1986. Hydroxyl free radical adduct of deoxyguanosine: sensitive detection and mechanisms of formation. *Free Radical Res. Commun.* **1:** 163–172.
45. Floyd, R. A., M. S. West, K. L. Eneff, and J. E. Schneider. 1989. Methylene blue plus light mediates 8-hydroxyguanine formation in DNA. *Arch. Biochem. Biophys.* **273:** 106–111.
46. Freidberg, E. C. 1985. Excision repair. I. DNA glycosylases and AP endonucleases, in *DNA Repair*. W. H. Freeman, New York.
47. Gonzalez-Flecha, B. and B. Demple. 1995. Metabolic sources of hydrogen peroxide in aerobically growing *Escherichia coli*. *J. Biol. Chem.* **270:** 13,681–13,687.
48. Grollman, A. P., F. Johnson, J. Tchou, and M. Eisenberg. 1994. Recognition and repair of 8-oxoguanine and formamidopyrimidine lesions in DNA. *Ann. NY Acad. Sci.* **726:** 208–213.
49. Gutteridge, J. M. 1994. Biological origin of free radicals, and mechanisms of antioxidant protection. *Chem. Biol. Interact.* **91:** 133–140.
50. Halliwell, B. and O. I. Aruoma. 1991. DNA damage by oxygen-derived species. Its mechanism and measurement in mammalian systems. *FEBS Lett.* **281:** 9–19.
51. Halliwell, B. and M. Dizdaroglu. 1992. The measurement of oxidative damage to DNA by HPLC and GC/MS techniques. *Free Radical Res. Commun.* **16:** 75–87.
52. Harman, D. 1956. Aging: a theory based on free radical and radiation chemistry. *J. Gerontol.* **11:** 298–300.
53. Harman, D. 1992. Role of free radicals in aging and disease. *Ann. NY Acad. Sci.* **673:** 126–141.
54. Hatahet, Z., Y. W. Kow, A. A. Purmal, R. P. Cunningham, and S. S. Wallace. 1994. New substrates for old enzymes: 5-hydroxy-2'-deoxycytidine and 5-hydroxy-2'-deoxyuridine are substrates for *Escherichia coli* endonuclease III and formamidopyrimidine DNA *N*-glycosylase, while 5-hydroxy-2'deoxyuridine is a substrate for uracil DNA *N*-glycosylase. *J. Biol. Chem.* **269:** 18,814–18,820.
55. Henner, W. D., S. M. Grunberg, and W. A. Haseltine. 1983. Enzyme action at 3'-termini of ionizing radiation-induced DNA strand breaks. *J. Biol. Chem.* **258:** 1683–1688.
56. Hutchinson, F. 1985. Chemical changes induced in DNA by ionizing radiation. *Prog. Nucleic Acid Res. Mol. Biol.* **32:** 115–154.
57. Imlay, J. A. and S. Linn. 1988. DNA damage and oxygen radical toxicity. *Science* **240:** 1302–1309.
58. Inoue, S. and S. Kawanishi. 1989. ESR evidence for superoxide, hydroxyl radicals and singlet oxygen produced from hydrogen peroxide and nickel (II) complex of glycylglycyl-L-histidine. *Biochem. Biophys. Res. Commun.* **159:** 445–451.

59. Janssen, Y. M. W., B. V. Houten, P. J. A. Borm, and B. T. Mossman. 1993. Cell and tissue response to oxidative damage. *Lab. Invest.* **261:** 261–274.
60. Joenje, H. 1989. Genetic toxicology of oxygen. *Mutat. Res.* **219:** 193–208.
61. Kanofsky, J. R., H. Hoogland, R. Wever, and S. J. Weiss. 1988. Singlet oxygen production by human eosinophils. *J. Biol. Chem.* **263:** 9692–9696.
62. Kappus, H. and H. Sies. 1981. Toxic drug effects associated with oxygen metabolism: redox cycling and lipid peroxidation. *Experientia* **37:** 1233–1241.
63. Kasai, H., P. F. Crain, Y. Kuchino, S. Nishimura, A. Ootsuyama, and H. Tanooka. 1986. Formation of 8-hydroxyguanine moiety in cellular DNA by agents producing oxygen radicals and evidence for its repair. *Carcinogenesis* **7:** 1849–1851.
64. Kasprak, K. 1991. The role of oxidative damage in metal carcinogenicity. *Chem. Res. Toxicol.* **4:** 604–615.
65. Kasprzak, K. S. 1995. Possible role of oxidative damage in metal-induced carcinogenesis. *Cancer Invest.* **13:** 411–430.
66. Klebanoff, S. J. 1988. Phagocytic cells: Products of oxygen metabolism, in *Inflammation: Basic Principles and Clinical Correlates* (Gallin, J. I., I. M. Goldstein, and R. Snyderman, eds.), Raven, New York, pp. 391–444.
67. Koch, W. H., E. N. Henrikson, E. Kupchella, and T. Cebula. 1994. Salmonella typhimurium strain TA100 differentiates several classes of carcinogens and mutagens by base substitution specificity. *Carcinogenesis* **15:** 79–88.
68. Kress, S., C. Sutter, P. T. Strickland, H. Mukhtar, J. Schweizer, and M. Schwarz. 1992. Carcinogen-specific mutational pattern in the p53 gene in ultraviolet B radiation-induced squamous cell carcinomas of mouse skin. *Cancer Res.* **52:** 6400–6403.
69. Kuchino, Y., F. Mori, H. Kasai, H. Inoue, S. Iwai, K. Miura, E. Ohtsuka, and S. Nishimura, S. 1987. Misreading of DNA templates containing 8-hydroxydeoxyguanosine at the modified base and at adjacent residues. *Nature* **327:** 77–79.
70. Kunkel, T. A., R. M. Schaaper, and L. A. Loeb. 1983. Depurination-induced infidelity of deoxyribonucleic acid synthesis with purified deoxyribonucleic acid replication proteins in vitro. *Biochemistry* **22:** 2378–2384.
71. Kunkel, T. A. 1984. Mutational specificity of depurination. *Proc. Natl. Acad. Sci.USA* **81:** 1494–1498.
72. Kunz, B. A., E. S. Henson, H. Roche, D. Ramotar, T. Nunoshiba, and B. Demple. 1994. Specificity of the mutator caused by deletion of the yeast structural gene *(APN1)* for the major apurinic endonucleases. *Proc. Natl. Acad. Sci. USA* **91:** 8165–8169.
73. Loeb, L. A. and S. J. Klebanoff. Unpublished results.
74. Loeb, L. A. 1991. Mutator phenotype may be required for multistage carcinogenesis. *Cancer Res.* **51:** 3075–3079.
75. Loeb, L. A., E. A. James, A. M. Waltersdorph, and S. J. Klebanoff. 1988. Mutagenesis by the autoxidation of iron with isolated DNA. *Proc. Natl. Acad. Sci. USA* **85:** 3918–3922.
76. Loeb, L. A. and B. D. Preston. 1986. Mutagenesis by apurinic/apyrimidinic sites. *Annu. Rev. Genet.* **20:** 201–230.
77. Maki, H. and M. Sekiguchi. 1992. MutT protein specifically hydrolyses a potent mutagenic substrate for DNA synthesis. *Nature* **355:** 273–275.
78. McBride, T. J., B. D. Preston, and L. A. Loeb. 1991. Mutagenic spectrum resulting from DNA damage by oxygen radicals. *Biochemistry* **30:** 207–213.
79. McBride, T. J., J. E. Schneider, R. A. Floyd, and L. A. Loeb. 1992. Mutations induced by methylene blue plus light in single-stranded M13mp2. *Proc. Natl. Acad. Sci. USA* **89:** 6866–6870.
80. Marnett, L. J., H. K. Hurd, M. C. Hollstein, D. E. Levin, H. Esterbauer, and B. H. Ames, B. H. 1985. Naturally occurring carbonyl compounds are mutagens in the Salmonella tester strain TA104. *Mutat. Res.* **148:** 25–34.

81. Martin, G.M. 1991. Genetic and environmental modulations of chromosomal stability: their roles in aging and oncogenesis. *Ann. NY Acad. Sci.* **621:** 401–417.
82. Melamede, R. J., Z. Hatahet, Y. Kow, H. Ide, and S. S. Wallace. 1994. Isolation and characterization of endonuclease VIII from Escherichia coli. *Biochemistry* **33:** 1255–1264.
83. Meneghini, R. 1988. Genotoxicity of active oxygen species in mammalian cells. *Mutat. Res.* **195:** 215–230.
84. Merkel, P. B., R. Nillson, and D. R. Kearns. 1972. Deuterium effects on singlet oxygen lifetimes in solutions. A new test of singlet oxygen reactions. *J. Am. Chem. Soc.* **94:** 1030–1031.
85. Michaels, M. L., J. Tchou, A. P. Grollman, and J. H. Miller. 1992. A repair system for 8-oxo-7,8-dihydrodeoxyguanine. *Biochemistry* **31:** 10,964–10,968.
86. Miller, J. H. 1985. Mutagenic specificity of ultraviolet light. *J. Mol. Biol.* **182:** 45–68.
87. Moody, C. S., and H. M. Hassan. 1982. Mutagenicity of oxygen free radicals. *Proc. Natl. Acad. Sci. USA* **79:** 2855–2859.
88. Morita, J. K., K. Ueda, K. Nakai, Y. Baba, and T. Komano. 1983. DNA strand breakage in vitro by autooxidized unsaturated fatty acids. *Agricultural Biol. Chem.* **47:** 2977–2979.
89. Muller, H. J. 1927. Artificial transmutation of the gene. *Science* **66:** 84–87.
90. Murli, S. and G. C. Walker. 1993. SOS mutagenesis. *Curr. Opinion Genet. Dev.* **3:** 719–725.
91. Nakazawa, H., D. English, P. L. Randell, K. Nakazawa, N. Martel, B. K. Armstrong, and H. Yamasaki. 1994. UV and skin cancer: Specific p53 gene mutation in normal skin as a biologically relevant exposure measurement. *Proc. Natl. Acad. Sci. USA* **91:** 360–364.
92. Pacifici, R. E. and K. J. A. Davies. 1991. Protein, lipid and DNA repair systems in oxidative stress: The free-radical theory of aging revisited. *Gerontology* **37:** 166–180.
93. Park, J. W., K. C. Cundy, and B. N. Ames. 1989. Detection of DNA adducts by high-performance liquid chromatography with electrochemical detection. *Carcinogenesis* **10:** 827–832.
94. Pavlov, Y. I., D. T. Minnick, S. Izuta, and T. A. Kunkel. 1994. DNA replication fidelity and 8-oxodeoxyguanosine triphosphate. *Biochemistry* **33:** 4685–4701.
95. Piette, J., M. P. Merville-Louis, and J. Decuyper. 1986. Damages induced in nucleic acids by photosensitization. *Photochem. Photobiol.* **44:** 793–802.
96. Polverelli, M. and R. Teoule. 1974. Gamma irradiation of cytosine in an aerated aqueous solution. I. Identification of radiolysis products of cytosine resulting from the deamination pathway. *Z. Naturforsch C.* **29:** 12–15.
97. Povirk, L. F. and R. J. Steighner. 1989. Oxidized apurinic/apyrimidinic sites formed in DNA by oxidative mutagens. *Mutat. Res.* **214:** 13–22.
98. Purmal, A. A., Y. W. Kow, and S. S. Wallace. 1994. Major oxidative products of cytosine, 5-hydroxycytosine and 5-hydroxyuracil, exhibit sequence context-dependent mispairing *in vitro*. *Nucleic Acids Res.* **22:** 72–78.
99. Reid, T. M., D. I. Feig, and L. A. Loeb. 1994. Mutagenesis by metal induced oxygen radicals. *Environ. Health Perspect* **102:** 57–61.
100. Reid, T. M., M. Fry, and L. A. Loeb. 1992. Endogenous mutations and cancer, in *Multistage Carcinogenesis*, vol. 22 (Harris, C. C., Hirohashi, S., Ito, N., Pitot, H. C., Sugimura, T., Terada, M., and Yokota, J., eds.), Jpn. Sci. Soc., Tokyo, Japan, pp. 221–229.
101. Reid, T. M. and L. A. Loeb. 1993. Effect of DNA-repair enzymes on mutagenesis by oxygen free radicals. *Mutat. Res.* **289:** 181–186.
102. Reid, T. M. and L. A. Loeb. 1992. Mutagenic specificity of oxygen radicals produced by human leukemia cells. *Cancer Res.* **52:** 1082–1086.
103. Reid, T. M. and L. A. Loeb. 1993. Tandem double CC → TT mutations are produced by reactive oxygen species. *Proc. Natl. Acad. Sci. USA* **90:** 3904–3947.

104. Repine, J. E., O. W. Pfenninger, D. W. Talmage, E. M. Berger, and D. E. Pettijohn. 1981. Dimethyl sulfoxide prevents DNA nicking mediated by ionizing radiation or iron/hydrogen peroxide-generated hydroxyl radical. *Proc. Natl. Acad. Sci. USA* **78:** 1001–1003.
105. Ribeiro, D. T., R. C. De-Oliveira, P. Di-Mascio, and C. F. Menck. 1994. Singlet oxygen induces predominantly G to T transversions on a single-stranded shuttle vector replicated in monkey cells. *Free Radical Res.* **21:** 75–83.
106. Sagher, D. and B. Strauss. 1983. Insertion of nucleotides opposite apurinic/apyrimidinic sites in deoxyribonucleic acid during in vitro synthesis: Uniqueness of adenine nucleotides. *Biochemistry* **22**:4518–4526.
107. Samuni, A., J. Aronovitch, D. Gadinger, M. Chevion, G. and Csapski. 1983. On the cytotoxicity of vitamin C and metal ions, a site-specific Fenton mechanism. *Eur. J. Biochem.* **137:** 119–129.
108. Schaaper, R. M., T. A. Kunkel, and L. A. Loeb. 1983. Infidelity of DNA synthesis associated with bypass of apurinic sites. *Proc. Natl. Acad. Sci. USA* **80**:487–491.
109. Singer, B. and J. M. Essigman. 1991. Site-specific mutagenesis: retrospective and prospective. *Carcinogenesis* **12:** 949–955.
110. Steenken, S. 1989. Structure, acid/base properties and transformation reactions of purine radicals. *Free Radical Res. Commun.* **6** 117–120.
111. Stohs, S. J. and D. Bagchi. 1995. Oxidative mechanisms in the toxicity of metal ions. *Free Radical Biol. Med.* **18:** 321–336.
112. Sunderman, F.W. 1984. Recent advances in metal carcinogenesis. *Ann. Clin. Lab. Sci.* **14:** 93–122.
113. Takeshita, M., C.-N. Chang, F. Johnson, S. Will, and A. Grollman. 1987. Oligonucleotides containing synthetic abasic sites. *J. Biol. Chem.* **262:** 10,171–10,179.
114. Tchou, J. and A. P. Grollman. 1993. Repair of DNA containing the oxidatively-damaged base, 8-oxoguanine. *Mutat. Res.* **299:** 277–287.
115. Teebor, G. W. 1995. Excision base repair, in *DNA Repair Mechanisms: Impact on Human Diseases and Cancer* (Vos, J.-M. H., ed.), R.G. Landes Company, NY, pp. 99–123.
116. Teebor, G.W., M. Goldstein, and K. Frenkel. 1984. Ionizing radiation and tritium transmutation both cause formation of 5-hydroxymethyl-2'-deoxyuridine in cellular DNA. *Proc. Natl. Acad. Sci. USA* **81**:318–321.
117. Teebor, G. W., R. J. Boorstein, and J. Cadet. 1988. The repairability of oxidative free radical mediated damage to DNA: a review. *Int. J. Radiat. Biol.* **54:** 131–150.
118. Tindall, K. R., J. Stein, and F. Hutchinson. 1988. Changes in DNA base sequence induced by gamma-ray mutagenesis of lambda phage and prophage. *Genetics* **118:** 551–560.
119. Tkeshelashvili, L. K., T. McBride, K. Spence, and L. A. Loeb. 1992. Mutation spectrum of copper-induced DNA damage. *J. Biol. Chem.* **267:** 13,778.
120. Tkeshelashvili, L. K., T. McBride, K. Spence, and L. A. Loeb. 1991. Mutation spectrum of copper-induced DNA damage [pub. erratum in *J. Biol. Chem.* 1992, **267:** 13,778]. *J. Biol. Chem.* **266:** 6401–6406.
121. Tkeshelashvili, L. K., T. M. Reid, T. J. McBride, and L. A. Loeb. 1993. Nickel induces a signature mutation for oxygen free radical damage. *Cancer Res.* **53:** 4172–4174.
122. Torreilles, J., M. C. Guerin, and A. Slaoui-Hasnaoui. 1990. Nickel (II) complexes of histidyl-peptides as Fenton-reaction catalysts. *Free Radical Res. Commun.* **11:** 159–166.
123. Wagner, J. R., C. C. Hu, and B. N. Ames. 1992. Endogenous oxidative damage of deoxycytidine in DNA. *Proc. Natl. Acad. Sci. USA* **89:** 3380–3384.
124. Ward, J. E. 1994. The complexity of DNA damage: relevance to biological consequences. *Int. J. Radiat. Biol.* **66:** 427–432.
125. Weitzman, S. A. and L. I. Gordon. 1990. Inflammation and cancer: role of phagocyte-generated oxidants in carcinogenesis. *Blood* **76:** 655–663.

126. Weitzman, S. A. and T. P. Stossel. 1982. Effects of oxygen radical scavengers and antioxidants on phagocyte-induced mutagenesis. *J. Immunol.* **128:** 2770–2772.
127. Wood, M. L., A. Esteve, M. L. Morningstar, G. M. Kuziemko, and J. M. Essigman. 1992. Genetic effects of oxidative DNA damage: comparative mutagenesis of 7,8,-dihydro-8-oxoguanine and 7,8-dihydro-8-oxoadenine in *Escherichia coli*. *Nucleic Acids Res.* **20:** 6023–6032.
128. Yau, T. M. 1979. Mutagenicity and cytotoxicity of malondialdehyde in mammalian cells. *Mech. Ageing Dev.* **11:** 137–144.
129. Zakour, R. A. and L. A. Loeb. 1982. Site-specific mutagenesis by error-directed DNA synthesis. *Nature* **295:** 708–710.

5

Regulation of Endonuclease IV as Part of an Oxidative Stress Response in *Escherichia coli*

Bernard Weiss

1. INTRODUCTION

In their natural environments, bacteria are continually subjected to both intracellular and extracellular oxidants *(13,42)*. During aerobic growth, H_2O_2 and superoxide anion radicals (O_2^-) may be generated when flavoproteins are auto-oxidized. The Fe^{2+} released by oxidatively damaged iron–sulfur proteins might then catalyze a Haber-Weiss reaction to produce the hydroxyl radical (·OH) *(13)*. The bacterial cell must also combat extracellular oxidants and free-radical sources that other organisms use as chemical weapons, some well-studied examples of which are nitric oxide, bleomycin, streptonigrin, plumbagin, juglone, and pyocyanine *(18)*. Because of its large size and small number of copies per cell, the DNA molecule is a prime target for lethal oxidative damage. However, among the known enzymes of DNA repair in *Escherichia coli*, only endonuclease IV was found to be induced specifically by oxidative stress.

It is not immediately obvious why endonuclease IV has been singled out in this fashion, why it is induced by some oxidants that are not considered to be primarily injurious to DNA, and what the intracellular effector is for this response. In an attempt to answer these questions, we must go far beyond discussing DNA repair and consider the induction of endonuclease IV within the context of the global response of which it is a part.

1.1. Properties of Endonuclease IV

What properties of endonuclease IV suit it for the repair of oxidative damage? Endonuclease IV is an AP endonuclease that cleaves 5' to an abasic site in double-stranded DNA, leaving 3'-hydroxyl and 5'-phosphoryl termini. It can remove from the 3'-termini of damaged DNA unusual end-groups that would otherwise block repair by most exonucleases and DNA polymerases *(26)*. These include 3'-phosphoryl groups, phosphoglycolates, and the unsaturated deoxyribose left by a β-elimination reaction. (For further discussion, *see* Chapter 3). All of these sites may be products of oxidative damage or intermediates in its repair.

Endonuclease IV is enzymatically similar to exonuclease III, the major AP endonuclease of *E. coli (26)*. It differs from the latter enzyme mainly in the lack of a strong associated exonuclease activity. Endonuclease IV is encoded by the *nfo* gene *(8,40)*.

From: DNA Damage and Repair, Vol 1: DNA Repair in Prokaryotes and Lower Eukaryotes
Edited by: J. A. Nickoloff and M. F. Hoekstra © *Humana Press Inc., Totowa, NJ*

An *nfo* mutation, either alone or in combination with an exonuclease III *(xth)* mutation, increases the cell's sensitivity to several DNA-damaging agents *(8)*. These agents include ionizing radiation, peroxides, and bleomycin, which mediate damage through direct oxidation or through the generation of free radicals. *nfo* and *xth* mutants display differences in sensitivity to different agents, and the purified enzymes differ in their abilities to attack certain types of oxidized abasic sites *(17,22)*. Nfo, for example, has a relatively strong preference for a type of oxidized abasic site produced by bleomycin, which might explain the sensitivity of *nfo* mutants to that agent. Therefore, although Nfo and Xth appear at first to be redundant, they each have unique specificities with respect to different lesions in oxidatively damaged DNA.

2. INDUCERS

When the *nfo* gene was inserted into a plasmid with a very high copy number, it was noted that the level of endonuclease IV activity in the cells was not commensurate with gene dose. Although this finding is still unexplained, it nevertheless led to the useful hypothesis that the gene is regulated and therefore inducible. Accordingly, many DNA-damaging agents were tested for their ability to induce endonuclease IV, including alkylating agents, oxidants, UV, and ionizing radiation. Only one set of agents were inducers *(5)*: paraquat (methyl viologen), plumbagin, menadione, and phenazine methosulfate, which are all superoxide generators. Induction could elevate *nfo* expression about 10-fold, bringing the level of the enzyme up to about that of exonuclease III.

The inducers were known to generate superoxide through univalent redox cycling in vivo *(18)*. They are reduced by NADPH either enzymatically (e.g., paraquat) or nonenzymatically (e.g., phenazine methosulfate) and are then auto-oxidized, yielding the superoxide anion radical as a byproduct. 4-Nitroquinoline-*N*-oxide is also an inducer, acting in an oxygen-dependent fashion, presumably through the production of O_2^- by redox cycling *(32)*. Endonuclease IV could be induced simply by growing a superoxide dismutase *(sodAB)* mutant in the presence of pure O_2 *(5)*, thereby confirming that O_2^- was indeed the inducer. It was later found that nitric oxide could induce endonuclease IV via the same genetic mechanism; however, it could do so anaerobically *(35)*. Therefore, the induction appeared to be mediated by one-electron redox reactions, and it was not specific for superoxide.

The trial of O_2^- generators in *nfo* induction experiments was prompted by several reports suggesting that O_2^- could damage DNA and induce a DNA repair system. Plumbagin was found to be mutagenic and to induce the host cell reactivation of λ bacteriophage that had been treated with riboflavin plus light (a source primarily of O_2^-), but not those treated with H_2O_2 *(10)*. *sodAB* mutants had a high rate of spontaneous mutations when grown aerobically *(9)*. In these studies, an *xth* mutation decreased (rather than increased) lethality and mutagenicity, suggesting that exonuclease III caused lethal breakage and stimulated mutagenic repair at oxidative lesions. However, a study suggesting that paraquat induced the SOS response *(2)* was not supported by related studies with plumbagin *(10)* or with *sod* mutants *(9)*, and lysogens of SOS-inducible λ bacteriophages were stable in both paraquat-treated *(43,46)* and plumbagin-treated *(10)* cells. These results suggested, therefore, that O_2^- causes DNA damage that is repaired by an SOS-independent pathway.

Table 1
Known Genes of the *soxRS* Regulon

Gene	Product	Function	Refs.
nfo	Endonuclease IV	Repair of oxidative damage to DNA	*15,43*
sodA	Mn^{2+}-superoxide dismutase	With catalase, renders O_2^- harmless by converting it to $O_2 + H_2O$	*15,43*
zwf	Glucose 6-phosphate dehydrogenase	Regenerates NADPH consumed during redox cycling of some inducers; might aid reduction of oxidized Fe–S proteins, including SoxR	
fumC	O_2^--resistant fumarase	Replaces two [4Fe–4S] isoenzymes	*28*
acnA	Aconitase	Replenishes damaged [4Fe–4S] enzyme	*16*
fpr	NADPH::ferredoxin (flavodoxin) oxidoreductase	May help to maintain Fe–S enzymes in a stable reduced form	
micF	Antisense mRNA inhibitor of *ompF* (porin gene)	Reduces permeability to xenobiotics	*6,15*
rimK	Modifies ribosomal protein S6	Resistance to xenobiotics	*15*

3. THE *sox*RS REGULON

3.1. Component Genes

The genes that regulate the induction of *nfo* were discovered in two ways: (1) by a search for mutants affecting the inducible expression of an *nfo-lacZ* fusion *(43)*, and (2) by a search for menadione-resistant mutants *(15)*. The mutations fell within two adjacent, divergently arranged genes, *soxR* and *soxS (1,46)*. Mutations in these genes affected the expression of at least 12 genes, including *nfo*, which were thus designated as the *soxRS* (or superoxide response) regulon. The term "regulon" refers to a set of dispersed bacterial genes that are coordinately controlled by the same regulatory gene or genes.

Table 1 lists those genes in the *soxRS* regulon for which the functions are known. Because the catalytic [4Fe–4S] clusters of dehydratases are readily inactivated through oxidation by O_2^- *(12,13)*, several products of the regulon (Zwf, Fpr, FumC, and AcnA) either compensate for or prevent this damage. Others protect directly against superoxide (SodA) or block nonspecifically the uptake of xenobiotics (*micF* RNA) or their effects on protein synthesis (RimK). The combined induction of *zwf* and of *fpr* has dual opposing effects. On the one hand, the gene products help to protect against O_2^-. They help:

1. To restore reducing equivalents (NADPH) that were lost by redox cycling;
2. To restore the balance between the oxidized and reduced forms of NAD(P), of ferredoxin, and of flavoproteins; and
3. To reduce unstable oxidized forms of [4Fe–4S] proteins.

On the other hand, they actually increase the production of O_2^- by paraquat: *zwf* generates NADPH, and Fpr has an NADPH:paraquat oxidoreductase activity *(29)*.

The regulon also contains some genes whose functions are unknown. One is downstream of *nfo* in a two-gene operon and encodes a protein resembling those of the PfkB family of carbohydrate kinases *(48)*. Other *soxRS*-regulated genes or gene products have been discovered by randomly screening random *lacZ* gene fusions *(24,25,31,39)* or by two-dimensional electrophoretic analysis of the total cellular proteins of *sox* mutants *(15)*. Some of these genes may turn out to be the same; if not, there may be as many as seven different genes yet to be characterized.

3.2. Relation to Other Regulons

The role of the *soxRS* regulon in overall oxidative stress responses has been recently reviewed *(21)*. The other well-studied oxidative stress regulon is the *oxyR* regulon of *E. coli* and *Salmonella typhimurium (21)*. The *oxyR* regulon is induced by peroxides, agents to which *nfo* mutants are unusually sensitive *(8)*; however, *nfo* is not a part of this regulon nor is it significantly induced by peroxides. Thus far, endonuclease IV is the only DNA repair enzyme known to be induced by oxidative stress.

Although many regulons overlap, *nfo* is not known to belong to any regulon other than *soxRS*. However, many other genes of the *soxRS* regulon do belong to other regulons as well. The *sodA* gene, for example, is controlled by five other global regulators *(7)*. As described elsewhere *(39)*, most of the genes of the *soxRS* regulon that are listed in Table 1 are also part of the *mar* (multiple antibiotic resistance) regulon, which can be induced by weak organic acids, antibiotics, and menadione (which is also an inducer of the *soxRS* regulon). In addition, SoxS has a great deal of overall homology with MarA, especially in the putative operator binding site (helix-turn-helix region). It is perhaps for this reason that overexpression of SoxS in strains containing a *soxR* regulon-constitutive mutation or a multicopy *soxS* plasmid caused an increase in *mar*-specific mRNA *(30)*. When families of regulatory proteins were first discovered, such crosstalk was thought to be a potential problem *(38)* in that a strong signal for one regulon might also induce unrelated regulons. However, perhaps in some cases, crosstalk exists by design in order to couple regulons having related functions, like those of *mar* and *soxRS*.

In addition to SoxS, AraC, and MarA, there are other AraC family members in *E. coli* that have an amino acid sequence similar to that of SoxS: (1) Rob binds to *oriC*, the origin of chromosomal replication *(41)*; and (2) TetD is product of the transposon Tn*10* *(14)*. It is not known what genes, if any, they regulate or what signals they recognize. It is possible, therefore, that they control some of the genes of the *soxRS* regulon.

3.3. Protective Role of Endonuclease IV

Although *soxRS* mutants are unusually sensitive to the lethal effects of superoxide generators *(43)*, *nfo* mutants are not very sensitive *(44)*. There is one oxidant, however, against which *nfo* has been clearly shown to have a protective function. Nitric oxide mediates the induction of *nfo* via the *soxRS* regulon *(35)*. *nfo* and Δ*soxRS* mutants had a similar decrease in survival in murine macrophages, with a kinetics of killing that paralleled intracellular NO· production *(35)*. The poor survival of the (Δ*soxRS* mutant in the macrophages could be reversed with an inhibitor of NO· synthesis *(34)*. It was proposed that *nfo* might not be recognizing damage to DNA by NO·, which mainly causes deaminations, but rather that caused by peroxynitrite (formed from NO· + O_2^-), which has a reactivity like that of the hydroxyl radical *(35)*. If so, peroxynitrite must still produce some unique lesions recognized by Nfo because an *nfo* mutant is not especially sensitive to γ-radiation, which forms hydroxyl radicals *(8)*.

The results with NO· provided the best evidence for the role of *soxRS*-dependent induction of *nfo* in protection against lethal DNA damage. However, these experiments,

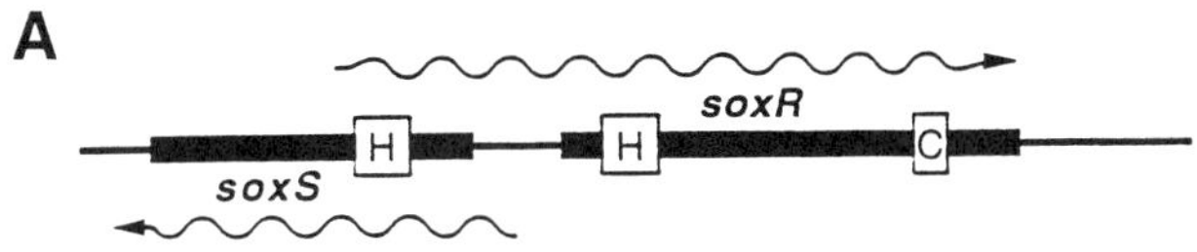

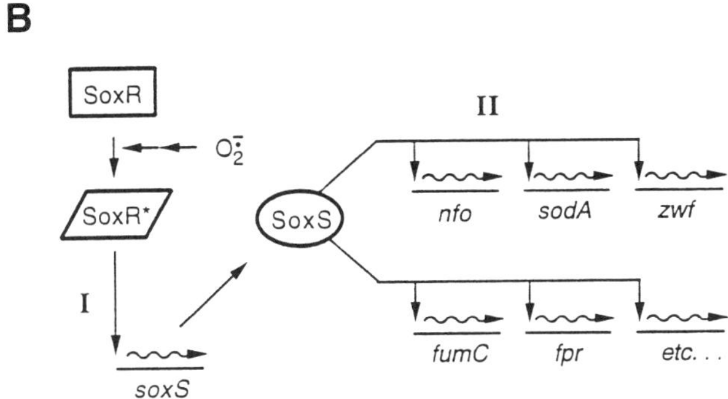

Fig. 1. Molecular organization of the *soxRS* region **(A)** and two-stage induction of the regulon **(B)**. A schema for the *soxRS* region (A) depicts the open reading frames (solid bars), overlapping transcripts (wavy lines), regions encoding the helix-turn-helix (HTH) motifs (H), and a region (C) encoding the four cysteines, which are probably coordinated to the iron-sulfur center. In B, the induction of the regulon is shown occurring in two steps. In step I, a univalent oxidant (in this case $O_2^{-}\cdot$) leads to the conversion of SoxR to SoxR*, which is a transcriptional activator of *soxS*. In step II, the resulting overproduced SoxS stimulates the transcription of *nfo* and at least 12 other genes of which seven produce known products (Table 1).

as well as some others *(8)*, must be interpreted with caution, because they employed an *nfo* insertion mutation. The insertion would also have affected the expression of the second gene of the *nfo* operon, and this gene of unknown function, although related by evolution to carbohydrate kinases *(48)*, might nevertheless be responsible for some of the properties attributed to *nfo* mutants.

3.4. The Regulatory Cascade

The *soxR* and *soxS* genes are divergently arranged (Fig. 1A). The *soxS* promoter is within the 85-nucleotide intergenic region, whereas that for *soxR* is within the *soxS* gene *(46)*. Therefore, their transcription is initially convergent and then divergent, and their mRNAs overlap. Despite this arrangement, the genes do not appear to be coordinately regulated. Induction of the regulon occurs in two stages (Fig. 1B). The SoxR protein, which contains iron–sulfur clusters, senses an unknown effector, perhaps through a redox reaction, and undergoes a conformational change. This activated SoxR then becomes a transcriptional activator for *soxS (36,47)*. SoxS may be induced up to 75-fold, judging from the expression of *soxS-lacZ* gene fusions. The induced native SoxS protein activates in turn the transcription of *nfo* and the other genes of the regulon. This outline was suggested by the initial observation that during induction, the level of *soxS* mRNA increased *(46)*. It was then proven by the following evidence based mainly on experiments performed in vivo with gene fusions *(36,47)*.

1. Mutations leading to the constitutive overexpression of the regulon were all in the *soxR* gene;
2. Induction of the regulon by agents like paraquat required both the *soxR* and *soxS* genes;
3. Induction was accompanied by *soxR*-dependent overexpression of SoxS;
4. Constitutive overexpression of SoxS activated the regulon even in the absence of an inducing agent and in the absence of a functional *soxR* gene; and
5. The spectrum of oxidants that induce *soxS* are similar to those that induce *nfo*, and the induction by superoxide generators is oxygen-dependent.

The known details of the system were gleaned largely from in vitro studies of DNA-protein binding (by DNA footprinting and by electrophoretic mobility shift assays) and of transcription. SoxR, even when purified from cells that have not been exposed to a regulon-inducing agent, was isolated in a transcriptionally active form in which its iron–sulfur clusters were in an oxidized state *(19,20,45)*. It binds specifically to the *soxS* promoter in the –10 to –35 region, which contains an 18-bp palindrome. It helps to recruit RNA polymerase to the operator, and it activates the initiation of transcription by promoting the formation of open complexes *(19,20)*. The SoxS protein binds to operator regions containing a consensus sequence ("soxbox") within a 30-nucleotide region just upstream or slightly overlapping the –35 hexamers. At promoters in which the soxbox overlaps the –35 hexamer (as with *nfo*), transcriptional activation does not require the C-terminal domain of the α-subunit of RNA polymerase; at other promoter sites, however, this domain is necessary *(23)*.

In vivo, nonactivated wild-type SoxR can inhibit the action of a mutated SoxR that is constitutively activated *(43,47)*. It is not known if the wild-type and mutant forms compete for operator sites or form inactive heterodimers. It is perhaps because of a similar competition between the activated and nonactivated forms of the wild-type protein that the cell produces a minimal amount of SoxR, thereby ensuring that all of it will be activated, even by a weak inducing signal. The production of SoxR *(43)* and of SoxR-LacZ fusion proteins *(47)* is therefore much lower than that of their SoxS counterparts in uninduced cells. The induction of *soxS* by SoxR results in signal amplification, permitting a limited amount of activated SoxR ultimately to affect transcription at many operator sites. In addition to increasing the sensitivity of a response, a regulatory cascade should also provide an opportunity for the temporal control of gene expression as is seen in many viral systems. Thus, genes activated directly by SoxR might be expressed earlier than SoxS-activated ones. To date, however, no genes (apart from *soxS*) have been found to be regulated by SoxR.

3.5. *Autoregulation*

The induction of the regulon may be autoregulated both at the level of SoxR activation and at the level of SoxS transcription. As discussed in Section 3.1., the products of the regulon help to restore the redox balance of the cell. If SoxR is activated by reversible oxidation, then the induction of the regulon might deactivate SoxR and turn off the regulon.

In addition, SoxS can bind weakly to its own promoter in vitro *(11,37)*, and it blocks its own induced transcription in vivo *(37)*. However, the presence of an intact *soxS*$^+$ gene did not affect the rate of decay in the expression of a truncated *soxS* gene (i.e., a *soxS'*::*lacZ* fusion) after the removal of paraquat. Therefore, it was proposed *(37)* that the effect of this autoregulation is not to turn off the regulon after induction, but rather to raise the threshold for induction. Thus, it might act like the spring in a trigger.

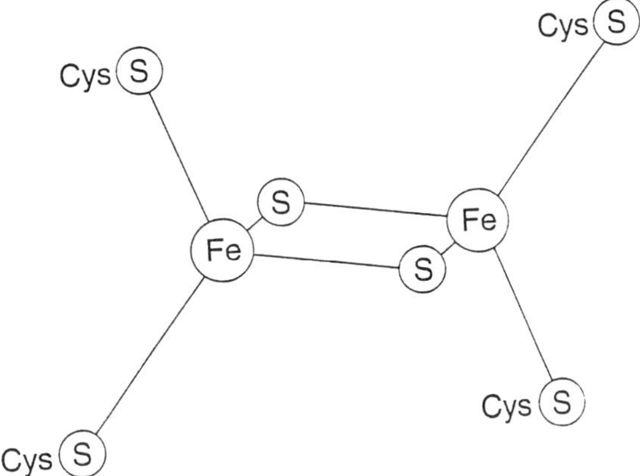

Fig. 2. Structure of a typical [2Fe–2S] center.

4. REGULATORY PROTEINS

4.1. SoxR

Because SoxR contains the sensor element for the regulon, its structure must contain important keys to understanding the mechanism of induction. The following information was obtained from an analysis of its predicted protein sequence *(1,46)* and from studies on protein purified from cells containing expression vectors *(19,20,45)*. SoxR protein is a globular homodimer of a 17-kDa polypeptide. The polypeptide possesses a helix-turn-helix (HTH) DNA-binding motif near its amino-terminus. Near its carboxyl end, it contains four cysteines that are probably involved in the formation of a [2Fe–2S] center, of which there are two per homodimer *(20,45)*. In most [2Fe–2S] proteins, the two iron atoms are coordinated to four cysteines (Fig. 2), and in all known cases, the coordinating amino acids are in the same polypeptide strand *(4)*. A [2Fe–2S] cluster is capable of undergoing univalent oxidation and reduction, and it is therefore probably a sensor element for the regulon. SoxR is thus related to the [2Fe–2S] ferredoxins, ubiquitous mediators of electron-transfer reactions. However, the spacing of its cysteines (CX_2CXCX_5C) is unique. SoxR is most closely related to the MerR protein of Tn*501* *(1)*. MerR binds Hg rather than Fe to its cysteines and activates the transcription of a Hg-responsive regulon. However, it contains only three cysteines per protomer, and they are spaced differently from those in SoxR; therefore, the proteins may be closer in evolution than in function.

In vitro, under aerobic conditions, the two Fe-S centers of the SoxR homodimer are in an oxidized ($[2Fe–2S]^{2+}$) state, and this oxidized SoxR is active in stimulating the transcription of *soxS* in vitro *(20,45)*. Although the Fe–S centers may be reduced anaerobically by dithionite, they are auto-oxidized in less than a minute after re-exposure to air. This extreme sensitivity to small amounts of oxygen has hindered efforts to test the reduced protein for transcriptional activity in vitro. The Fe–S centers are essential for activity; SoxR protein that has lost its Fe–S centers is still a dimer and is still capable of binding specifically to the *soxS* promoter region, but it no longer activates transcription *(20)*. In addition, SoxR can be inactivated by removal of its four cysteines by a deletion mutation *(47)*.

Some insights into mechanisms of activation have come from the DNA sequencing of SoxR-constitutive mutants, i.e., regulon-constitutive mutants in which SoxR is in its activated state even in the absence of inducer. There are 24 amino acids between the 4-cysteine cluster in SoxR and its carboxyl terminus. Constitutive mutations resulted from insertions or deletions that caused 7–19 terminal amino acids to be replaced with 8–148 new ones derived from adjacent vector or bacterial sequences *(33,44,46)*. A study of spontaneous constitutive mutants *(33)* revealed that two point mutations altered an amino acid 12 residues from the C-terminus. Two other constitutive mutants also lacked this region: a nonsense mutant and one with a 9-codon deletion. These results indicate that the dispensable C-terminal region must not play a positive role in activation. It appears instead to be needed for maintenance of SoxR in a nonactivated state. It might, for example, contain a recognition site for a SoxR reductase, or it might interfere with oxidation or with transcriptional activity by binding reversibly to another site in the protein. With regard to these hypotheses, it should be noted that two constitutive point mutations were in other regions: one was 24 amino acids proximal to the cysteine cluster and another within the helix-turn-helix region *(33)*. The full explanation, however, will probably be more complex because all the constitutive mutants *(33,46)* were still capable of further induction by paraquat.

4.2. SoxS

The SoxS protein is a member of the AraC family of transcriptional activators, with which it shares not only an overall homology, but also a characteristic sequence motif in its helix-turn-helix region *(1,14,46)* that is quite different from that of SoxR. Most of the AraC family members contain a sensor domain and an activator domain in different halves of the molecule. SoxS, however, is much smaller than its relatives, lacking a region corresponding to their sensor domains. Thus, SoxS is a simple transcriptional activator; the sensor function is supplied by SoxR. *soxS* mutants were found only among those that had a noninducible regulon *(46)*; to date, no regulon-constitutive *soxS* mutations have been isolated *(33,46)*.

In vitro studies of SoxS were difficult at first because the purified overexpressed protein was mostly insoluble, transcriptionally inactive, and probably incorrectly folded. Nevertheless, binding site studies could be performed *(27)*, and these were confirmed and extended after the isolation of a soluble MalE-SoxS fusion protein *(11)*. The fusion protein was active both in vivo and in vitro despite a size that was about four times that of native SoxS. The purified protein was used to identify a consensus operator site sequence ("soxbox") by means of DNA footprinting *(11)*, and it was found that the protein bound to the α-subunit of RNA polymerase at some promoters, but not that of *nfo* *(23)*. The SoxS protein lacks a region corresponding to the dimerization domain of its relative, AraC, and it therefore probably binds to DNA as a monomer *(3,14)*.

5. INDUCTION MECHANISM: THEORIES

Two experimental findings suggested that O_2^- is not the proximate effector for SoxR activation *(28)*. (1) The *soxRS*-dependent induction of *fumC* and *zwf* by paraquat was not affected by the prior overexpression of *sodA* from a synthetic *(tac)* promoter. (2) A *zwf* mutant, defective in the generation of NADPH needed for the production of O_2^- by redox cycling, was more sensitive rather than less sensitive to induction of *fumC* by paraquat. Thus, induction was not decreased under conditions that should have reduced the intracellular production of O_2^-. It was proposed, therefore, that the regulon responds primarily to a decrease in the NADPH/NADP$^+$ ratio brought about by the consumption of NADPH during redox cycling. However, O_2^- accumulation could induce the regulon in the absence of redox cycling compounds, as when a *sodAB* mutant was grown in the presence of O_2 *(43)*. This effect was explained by the ability of O_2^- to initiate free radical chain oxidations of NAD(P)H in the presence of Mn^{2+} *(13)*. Other evidence cited as consistent with this theory *(29)* included the anaerobic induction of *sodA* by diamide (a thiol oxidant) and by paraquat plus NO. However, this evidence is not definitive, because the induction was not shown to be *soxRS*-dependent, and *sodA* belongs to at least five regulons *(7)*. With the discoveries that the regulon governed a ferredoxin reductase (Fpr) and that SoxR was a ferredoxin-like protein, an additional link was forged in the chain of reasoning: the NADPH/NADP+ ratio affected in turn the ratio of reduced to oxidized flavodoxin and ferredoxin, and therefore perhaps the ratio of active to inactive forms of SoxR *(29)*.

In formulating a simple hypothesis for the activation of SoxR, we cannot ignore the observations that all of the known inducers are redox compounds that mediate one-electron transfers and that the SoxR homodimer contains as prosthetic groups two [2Fe–2S] centers that are each capable of univalent reduction and auto-oxidation. Peroxides, which primarily mediate two-electron transfers, are not effective inducers of the regulon. It is likely, therefore, that SoxR, like most ferredoxin-like compounds, is maintained in the cell in an inactive, reduced state by electrons derived from NAD(P)H. Aerobically, inducers mediate univalent redox cycling reactions that deplete the cell of reducing equivalents and permit the auto-oxidation of SoxR. Anaerobically, compounds like NO· might oxidize SoxR through an unknown electron chain, if not directly. This hypothesis led to two predictions. First, the oxidized, active form of SoxR should be reversibly inactivated by reduction; this was confirmed by experiments on the transcription of *soxS* in vitro *(14)*. Second, during induction of the regulon in vivo. The Fe–S centers of SoxR should shift from being predominantly reduced to being predominantly oxidized; this was recently confirmed by EPR *(14)*.

It has been also been proposed *(19)* that transitions of SoxR between nonactivated and activated states might be mediated through the reversible partial dissociation of iron–sulfur centers such as that which is commonly seen in [4Fe–4S] proteins *(12)*. This hypothesis is based on the observation of reversible inactivation of SoxR by very high concentrations of dithiothreitol, which is accompanied by subtle changes in the spectrum of the Fe-S centers; however, the levels of dithiothreitol needed for this in vitro effect exceeded those of intracellular thiol reductants.

6. CONCLUDING REMARKS

When we examine the components of the *soxRS* regulon, we are struck by its highly purposive nature. SodA, Fpr, AcnA, and FumC are specifically equipped to help the cell resist the effects of O_2^- and similar oxidants. Therefore, a reasonable guess is that the mechanism of *nfo* induction evolved to protect DNA from damage by univalent oxidants. However, to support this theory, we must find an inducer of the *soxRS* regulon to which *nfo* mutants are unusually sensitive, and it must produce DNA lesions that are specifically recognized by Nfo. This chain of evidence is currently incomplete with respect to any of the known inducers, and it is a worthy goal of future investigation.

In pursuing this goal, however, there may be several pitfalls. First, we may have not yet discovered an appropriate ubiquitous environmental oxidant (or stimulator of an endogenous oxidant) that will fulfill our criteria. Second, although such an agent might be harmful to DNA, it may kill cells primarily by damaging other cellular components; therefore, we might not observe an effect of *nfo* mutation on survival. This is probably what happened with the O_2^- generators that were used in previous studies; they may kill cells primarily by damaging [4Fe–4S] proteins. Third, there might be other repair enzymes, perhaps inducible, whose specificities overlap that of *nfo* such that the loss of *nfo* alone will not produce a decreased survival. This last problem, i.e., the redundancy and overlapping specificities of DNA repair pathways, resurfaces every time we try to use mutants to assess the role of an enzyme. In this vein, it should be noted that there are probably at least five other uncharacterized products of the regulon, some of which might be DNA repair enzymes.

Finally, before embarking on new adventures, it should be pointed out that as a result of the haste to explore these exciting control mechanisms, there has been no systematic study of the effect of *nfo* mutations on the lethality and mutagenicity of agents that induce the *soxRS* regulon. However, it was data on the DNA toxicity of these agents that led us to the discovery of *nfo* induction in the first place.

ACKNOWLEDGMENTS

I am especially grateful to the following colleagues who contributed to the timeliness of this chapter by sharing their findings in advance of publication: Bruce Demple, Judah L. Rosner, and Richard E. Wolf, Jr. Work in the author's laboratory was supported by grants from the National Science Foundation.

REFERENCES

1. Amábile-Cuevas, C. F. and B. Demple. 1991. Molecular characterization of the *soxRS* genes of *Escherichia coli*: two genes control a superoxide stress regulon. *Nucleic Acids Res.* **19:** 4479–4484.
2. Brawn, M. K. and I. Fridovich. 1985. Increased superoxide radical production evokes inducible DNA repair in *Escherichia coli. J. Biol. Chem.* **260:** 922–925.
3. Bustos, S. A. and R. F. Schleif. 1993. Functional domains of the AraC protein. *Proc. Natl. Acad. Sci. USA* **90:** 5638–5642.
4. Cammack, R. 1992. Iron-sulfur clusters in enzymes. *Adv. Inorg. Chem.* **38:** 281–322.
5. Chan, E. and B. Weiss. 1987. Endonuclease IV of *Escherichia coli* is induced by paraquat. *Proc. Natl. Acad. Sci. USA* **84:** 3189–3193.
6. Chou, J. H., J. T. Greenberg, and B. Demple. 1993. Posttranscriptional repression of *Escherichia coli* OmpF protein in reponse to redox stress: positive control of the *micF* antisense RNA by the *soxRS* locus. *J. Bacteriol.* **175:** 1026–1031.
7. Compan, I. and D. Touati. 1993. Interaction of six global transcription regulators in expression of manganese superoxide dismutase in *Escherichia coli* K-12. *J. Bacteriol.* **175:** 1687–1696.
8. Cunningham, R. P, S. M. Saporito, S. G. Spitzer, and B. Weiss. 1986. Endonuclease IV *(nfo)* mutant of *Escherichia coli. J. Bacteriol.* **168:** 1120–1127.
9. Farr, S. B., R. D'Ari, and D. Touati. 1986. Oxygen-dependent mutagenesis in *Escherichia coli* lacking superoxide dismutase. *Proc. Natl. Acad. Sci. USA* **83:** 8268–8272.
10. Farr, S. B., D. O. Natvig, and T. Kogoma. 1985. Toxicity and mutagenicity of plumbagin and the induction of a possible new DNA repair pathway in *Escherichia coli. J. Bacteriol.* **164:** 1309–1316.

11. Fawcett, W. P. and R. E. Wolf, Jr. 1994. Purification of a MalE-SoxS fusion protein and identification of the control sites of *Escherichia coli* superoxide-inducible genes. *Mol. Microbiol.* **14:** 669–679.
12. Flint, D. H., F. J. Tuminello, and M. H. Emptage. 1993. The inactivation of Fe-S cluster containing hydro-lyases by superoxide. *J. Biol. Chem.* **268:** 22,369–22,376.
13. Fridovich, I. 1994. A quarter century of superoxide dismutase: an exemplar of science triumphant, in *Frontiers of Reactive Oxygen Species in Biology and Medicine* (Asada, K. and T. Yoshikawa, eds.), Excerpta Medica, New York, pp. 1–9.
14. Gallegos, M.-T., C. Michán, and J. L. Ramos. 1993. The XylS/AraC family of regulators. *Nucleic Acids Res.* **21:** 807–810.
14a. Gaudu, P., N. Moon, W. R. Dunham, and B. Weiss. 1997. Regulation of the *soxRS* oxidative stress regulon: reversible oxidation of the F-S clusters of SoxR *in vivo*. *J. Biol. Chem.* **272:** in press.
14b. Gaudu, P. and B. Weiss. 1996. SoxR, a [2Fe-2S] transcription factor, is active only in its oxidized form. *Proc. Natl. Acad. Sci. USA* **93:** 10,094–10,098.
15. Greenberg, J. T., P. Monach, J. H. Chou, P. D. Josephy, and B. Demple. 1990. Positive control of a global antioxidant defense regulon activated by superoxide-generating agents in *Escherichia coli*. *Proc. Natl. Acad. Sci. USA* **87:** 6181–6185.
16. Gruer, M. J. and J. R. Guest. 1994. Two genetically-distinct and differentially-regulated aconitases (AcnA and AcnB) in *Escherichia coli*. *Microbiology* **140:** 2531–2541.
17. Häring, M., H. Rüdiger, B. Demple, S. Boiteux, and B. Epe. 1994. Recognition of oxidized abasic sites by repair endonucleases. *Nucleic Acids Res.* **22:** 2010–2015.
18. Hassan, H. M. and I. Fridovich. 1979. Intracellular production of superoxide radical and of hydrogen peroxide by redox active compounds. *Arch. Biochem. Biophys.* **196:** 385–395.
19. Hidalgo, E., J. M. Bollinger, Jr., T. M. Bradley, C. T. Walsh, and B. Demple. 1995. Binuclear [2Fe–2S] clusters in *Escherichia coli* SoxR protein and role of the metal centers in transcription. *J. Biol. Chem.* **270:** 20,908–20,914.
20. Hidalgo, E. and B. Demple. 1994. An iron-sulfur center essential for transcriptional activation by the redox-sensing SoxR protein. *EMBO J.* **13:** 138–146.
21. Hidalgo, E., and B. Demple. 1995. Adaptive responses to oxidative stress: the *soxRS* and *oxyR* regulons, in *Regulation of Gene Expression in* Escherichia coli (Lin, E. C. and A. S. Lynch, eds.), R. G. Landes, Austin, TX, pp. 433–450.
22. Ide, H., H. Shimizu, Y. Kimura, S. Sakamoto, K. Makino, M. Glackin, S. S. Wallace, H. Nakamuta, M. Sasaki, and N. Sugimoto. 1995. Influence of alpha-deoxyadenosine on the stability and structure of DNA. Thermodynamic and molecular mechanics studies. *Biochemistry* **34:** 6947–6955.
23. Jair, K.-W., P. W. Fawcett, N. Fujita, A. Ishihama, and R. E. Wolf, Jr. 1995. Ambidextrous transcriptional activation by SoxS: requirement for the C-terminal domain of the RNA polymerase alpha subunit in a subset of *Escherichia coli* superoxide-inducible genes. *Mol. Microbiol.* **19:** 307–317.
24. Kogoma, T., S. B. Farr, K. M. Joyce, and D. O. Natvig. 1988. Isolation of gene fusions *(soi::lacZ)* inducible by oxidative stress in *Escherichia coli*. *Proc. Natl. Acad. Sci. USA* **85:** 4799–4803.
25. Koh, Y.-S. and J.-H. Roe. 1995. Isolation of a novel paraquat-inducible *(pqi)* gene regulated by the *soxRS* locus in *Escherichia coli*. *J. Bacteriol.* **177:** 2673–2678.
26. Levin, J. D., A. W. Johnson, and B. Demple. 1988. Homogeneous *Escherichia coli* endonuclease IV. Characterization of an enzyme that recognizes oxidative damage in DNA. *J. Biol. Chem.* **263:** 8066–8071.
27. Li, Z. and B. Demple. 1994. SoxS, an activator of superoxide stress genes in *Escherichia coli*. *J. Biol. Chem.* **269:** 18,371–18,377.
28. Liochev, S. I. and I. Fridovich. 1992. Fumarase C, the stable fumarase of *Escherichia coli*, is controlled by the *soxRS* regulon. *Proc. Natl. Acad. Sci. USA* **89:** 5892–5896.

29. Liochev, S. I., A. Hausladen, W. F. Beyer, Jr., and I. Fridovich. 1994. NADPH:ferredoxin oxidoreductase acts as a paraquat diaphorase and is a member of the *soxRS* regulon. *Proc. Natl. Acad. Sci. USA* **91:** 1328–1331.
30. Miller, P. F., L. F. Gambino, M. C. Sulavik, and S. J. Gracheck. 1994. Genetic relationship between *soxRS* and *mar* loci in promoting multiple antibiotic resistance in *Escherichia coli*. *Antimicrob. Agents Chemother.* **38:** 1773–1779.
31. Mito, S., Q.-M. Zhang, and S. Yonei. 1993. Isolation and characterization of *Escherichia coli* strains containing new gene fusions *(soi::lacZ)* inducible by superoxide radicals. *J. Bacteriol.* **175:** 2645–2651.
32. Nunoshiba, T. and B. Demple. 1993. Potent intracellular oxidative stress exerted by the carcinogen 4-nitroquinoline-*N*-oxide. *Cancer Res.* **53:** 3250–3252.
33. Nunoshiba, T. and B. Demple. 1994. A cluster of constitutive mutations affecting the C-terminus of the redox-sensitive SoxR transcriptional activator. *Nucleic Acids Res.* **22:** 2958–2962.
34. Nunoshiba, T., T. deRojas-Walker, S. R. Tannenbaum, and B. Demple. 1995. Roles of nitric oxide in inducible resistance of *Escherichia coli* to activated murine macrophages. *Infect. Immun.* **63:** 794–798.
35. Nunoshiba, T., T. deRojas-Walker, J. S. Wishnok, S. R. Tannenbaum, and B. Demple. 1993. Activation by nitric oxide of an oxidative-stress response that defends *Escherichia coli* against activated macrophages. *Proc. Natl. Acad. Sci. USA* **90:** 9993–9997.
36. Nunoshiba, T., E. Hidalgo, C. F. Amábile-Cuevas, and B. Demple. 1992. Two-stage control of an oxidative stress regulon: the *Escherichia coli* SoxR protein triggers redox-inducible expression of the *soxS* regulatory gene. *J. Bacteriol.* **174:** 6054–6060.
37. Nunoshiba, T., E. Hidalgo, Z. Li, and B. Demple. 1993. Negative autoregulation by the *Escherichia coli* SoxS protein: a dampening mechanism for the *soxRS* redox stress response. *J. Bacteriol.* **175:** 7492–7494.
38. Ronson, C. W., B. T. Nixon, and F. M. Ausubel. 1987. Conserved domains in bacterial regulatory proteins that respond to environmental stimuli. *Cell* **49:** 579–581.
39. Rosner, J. L. and J. L. Slonczewski. 1994. Dual regulation of inaA by the multiple antibiotic resistance *(mar)* and superoxide *(soxRS)* stress response systems of *Escherichia coli*. *J. Bacteriol.* **176:** 6262–6269.
40. Saporito, S. M. and R. P. Cunningham. 1988. Nucleotide sequence of the *nfo* gene of *Escherichia coli* K-12. *J. Bacteriol.* **170:** 5141–5145.
41. Skarstad, K., B. Thöny, D. S. Hwang, and A. Kornberg. 1993. A novel binding protein of the origin of the *Escherichia coli* chromosome. *J. Biol. Chem.* **268:** 5365–5370.
42. Sies, H. 1991. *Oxidative Stress: Oxidants and Antioxidants.* Academic, London.
43. Tsaneva, I. R. and B. Weiss. 1990. *soxR*, a locus governing a superoxide reponse regulon in *Escherichia coli* K-12. *J. Bacteriol.* **172:** 4197–4205.
44. Weiss, B. Unpublished results.
45. Wu, J., W. R. Dunham, and B. Weiss. 1995. Overproduction and physical characterization of SoxR, a [2Fe-2S] protein that governs an oxidative response regulon in *Escherichia coli*. *J. Biol. Chem.* **270:** 10,323–10,327.
46. Wu, J., and B. Weiss. 1991. Two divergently transcribed genes, soxR and soxS, control a superoxide response regulon of *Escherichia coli*. *J. Bacteriol.* **173:** 2864–2871.
47. Wu, J. and B. Weiss. 1992. Two-stage induction of the *soxRS* (superoxide response) regulon of *Escherichia coli*. *J. Bacteriol.* **174:** 3915–3920.
48. Wu, L. F., A. Reizer, J. Reizer, B. Cai, J. M. Tomich, and M. H. Saier, Jr. 1991. Nucleotide sequence of the *Rhodobacter capsulatus fruK* gene, which encodes fructose-1-phosphate kinase: evidence for a kinase superfamily including both phosphofructokinases of *Escherichia coli*. *J. Bacteriol.* **173:** 3117–3127.

6
The "GO" System in *Escherichia coli*

Jeffrey H. Miller

1. INTRODUCTION—OXIDIZING AGENTS IN THE CELL

The repair of damage to DNA and DNA precursors is vital to living cells. One of the prime causes of spontaneous damage to cellular components, including DNA, is oxidation resulting from reactive oxygen species (Chapter 4). Failure to reverse the effects of oxidative damage leads to a higher rate of mutagenesis, carcinogenesis, aging, and a number of diseases in higher cells *(1,20,45)*. Reactive oxygen species are produced by normal cellular metabolism, as well as by exogenous sources, such as ionizing radiation. Reactive oxygen species that damage DNA include the OH^- radical *(20,53)* and singlet oxygen ($^1O_2$; *26*). Many different alterations in DNA are generated by oxidative damage (*21*; Chapter 4), including strand breaks, deamination, abasic sites, ring-opened purines, and deoxyribose modifications. Thymine and cytosine glycols are frequent pyrimidine oxidation products. At least two oxidative lesions have been implicated in causing mutation; 5-hydroxy-2'-deoxycytidine (5-OH-dC), which if not repaired can lead to C $\rightarrow$ T transitions (21), and 2'-deoxy-7,8-dihydro-8-oxoguanosine (8-oxoG or "GO"), which results in G $\rightarrow$ T transversions *(25,35)*.

The cell's first line of defense against oxidative damage is to neutralize reactive oxygen species before they damage DNA. For instance, *Escherichia coli* and *Salmonella typhimurium* contain at least three global regulons, the *oxyR*, *soxR*, and *katF* regulons, which control inducible enzymes involved in eliminating reactive oxygen species (*17*; *see* Chapter 5). The inducible enzymes include two superoxide dismutases that convert superoxides to hydrogen and two catalases that convert the hydrogen peroxide to water. However, when DNA is damaged by reactive oxygen species, repair enzymes act as the next line of defense to prevent mutation and its consequences. This chapter reviews the discovery of the "GO" system, which prevents mutation from one of the principal oxidative lesions in the DNA, 8-oxoG. This lesion is one of the most stable products of oxidative damage to DNA and is ubiquitous in nature *(16,20)*. Steady-state levels have been reported to be as high as 10^4/cell in humans *(23)*, although the validity of this measurement has been questioned *(31)*.

2. MUTATOR CELLS REVEAL THE GO SYSTEM

The analysis of different mutator strains together with a series of biochemical experiments has led to the elucidation of a three-component system geared toward

From: DNA Damage and Repair, Vol. 1: DNA Repair in Prokaryotes and Lower Eukaryotes
Edited by: J. A. Nickoloff and M. F. Hoekstra © *Humana Press Inc., Totowa, NJ*

preventing mutations resulting from oxidized guanine, specifically 8-oxodG, in the DNA *(25,35,36,39,51)* or in a DNA precursor *(33)*. Mutators are cells that have an increased rate of spontaneous mutation, and often result from a defect in a DNA repair enzyme *(11,40)*. The first mutator characterized in *E. coli*, *mutT (54)*, results in an increase in a specific transversion, the A:T → C:G change (*56)*, although it was not until recently that the biochemical basis of this specificity was understood (*33*; *see* Section 2.3.). Subsequently, mutators were found that helped to define the proofreading activity of DNA pol III, (*mutD*; *12,15,19*); the mismatch repair system (*mutU/uvrD, mutS, mutH, mutL*; *11,42,46*) and, among other processes, the in vivo repair of deamination by uracil DNA glycosylase (*ung*: *18*; *see* Chapter 3).

2.1. Detection of Mutators

The mismatch repair system corrects transition mispairs very efficiently in vivo, as revealed by the large increase in transition mutations in mismatch repair-deficient cells *(30,48)*. This raised the question of whether there were repair systems specific for transversions, and whether they could be detected by finding new mutators specific for transversions. Here we should consider how mutator cells are found. Since there is no selection for cells with higher spontaneous mutation rates, colonies need to be screened for mutator character. Initially, mutagenized cells were tested for increased antibiotic resistance or elevated reversion of auxotrophic markers *(15,53)*. Today, one widely used type of screen involves "papillation," the formation of microcolonies growing out of a larger colony *(34)*. In one version of this method, Lac^- cells are plated on medium with two carbon sources. The first carbon source allows growth of all the cells, which form colonies. After exhausting the first carbon source, the cells cannot grow on the second carbon source, lactose, since they are Lac^-. However, spontaneous revertants in the population of cells in each colony can grow, and they form papillae. The number of papillae is a rough measure of the reversion rate from Lac^- to Lac^+ for each colony *(27)*. A sensitive papillation test for Lac^+ microcolonies employs glucose as the first carbon source and phenyl-β,D-galactoside as the second *(43)*. In addition, the Lac^+ microcolonies are stained blue by 5-bromo-4-chloro-3-indolyl-β,D-galactoside. After mutagenizing a cell culture, the cells are plated on the specialized medium, and blue papillae are visualized after 3–4 d. Cells with significantly more blue papillae are candidate mutators.

A set of strains developed by Cupples and Miller *(14)* makes it possible to select for mutators that stimulate one of the six possible base substitutions, among other events, that might be elevated. The glutamic acid at position 461 in β-galactosidase is essential. Only glutamic acid at that point in the protein chain gives sufficient activity for the cells to grow on lactose (Lac^+). Therefore, strains with substitutions at the corresponding point in the gene need specific reversion events to restore the glutamic acid codon (GAG). Site-directed mutagenesis was used to construct six strains, each with a single base change at either the first or second position in the codon specifying Glu-461. Each of these Lac^- strains reverts back to Lac^+ by one of the six base substitutions, as shown in Table 1.

2.2. The mutY and mutM Genes

The abovementioned set of Lac^- strains and similar derivatives were used to detect mutators that could stimulate the G:C → T:A transversion, and this resulted in two new

Table 1
Altered Codon at Position 461 in *lacZ* and Changes That Restore the GAG Condon

Strain	Condon 461	Change
CC101	TAG	A:T → C:G
CC102	GGG	G:C → A:T
CC103	CAG	G:C → C:G
CC104	GCG	A:T → T:A
CC105	GTG	A:T → T:A
CC106	AAG	A:T → G:C

Table 2
Mutational Specificity of *mutM* and *mutY*[a]

Strain background	Base substitution	Control	*mutM*	*mutY*
CC101	A:T → C:G	0	0	0
CC102	G:C → A:T	6	7	4
CC103	G:C → C:G	0	0	0
CC104	G:C → T:A	4	55	155
CC105	A:T → T:A	0	0	0
CC106	A:T → G:C	0	0	0

[a]Data from refs. *8* and *43*.

mutators, *mutY* and *mutM (8,43)*. Table 2 shows the specificity of each of these mutators in reversion tests with the set of six strains. Both *mutY* and *mutM* stimulate only G:C → T:A transversions.

The cloning and sequencing of the *mutM* gene revealed that it encodes a previously described glycosylase *(6,37)*, the FAPY glycosylase, which was initially shown to remove ring-opened purines *(4,10)*. Because ring-opened purines block DNA synthesis *(5)* and do not lead to specific base substitutions, the FAPY lesion was not a good candidate for the mutagenic target of the MutM protein. However, 8-oxodG was a prime candidate, since the MutM protein removes this adduct from the DNA when paired with C *(52)*, and since this lesion results in frequent insertions of A opposite, in in vitro and in vivo systems *(49)*. Grollman and coworkers *(25,49)* have shown that polymerases involved in DNA replication, such as polymerases α and γ and polymerase III from *E. coli* incorporate A opposite 8-oxoG much more frequently than they incorporate C, with preferences ranging from 200:1 to 5:1 (Table 3). This eminates from the fact that guanine nucleotides substituted at the 8-position, can readily adopt the *syn* conformation *(13)* and pair with A *(28)*, but still can pair with C in the *anti* conformation *(44)*. The incorporation of A opposite 8-oxoG results in the G:C → T:A transversion *in vivo* at sites of these lesions in the DNA. This is shown most clearly by studies in which plasmid vectors containing a single 8-oxoG lesion are allowed to replicate in *E. coli (9,41,55)*.

The *mutY* gene was also cloned and sequenced, and shown to be partly homologous to endonuclease III, containing a series of evenly spaced cysteine residues that are

Table 3
Relative Incorporation of Nucleotides Opposite 8-oxoG by DNA Polymerases In Vitro[a]

DNA polymerase	Incorporation C:A
pol α	1 : 200
pol γ	1 : 5
pol III	1 : 8
pol β	4 : 1
pol I	7 : 1

[a]Data from refs. *24* and *47*.

found in iron–sulfur binding proteins, including endo III *(38)*. Initially, the MutY protein was characterized as a glycosylase that excised A residues from a G:A mispair *(2,3)*. However, it was also found that MutY can excise A residues from A:8-oxoG mispairs, leaving an AP site that is filled in by repair synthesis *(35)*. If repair synthesis had the same preference for inserting A across from 8-oxoG lesions, the strategy of MutY repair would be counterproductive. However, Grollman and coworkers (Table 3) have demonstrated that polymerases involved with repair, such as polymerase β and *E. coli* pol I, preferentially insert C across from 8-oxoG lesions, with respective biases of 4:1 and 7:1 *(25,49)*. These results lead to the sequence pictured in Fig. 1B for the in vivo concerted action of the MutM and MutY proteins in the repair of 8-oxoG lesions in the DNA *(36,39)*. As can be seen from this figure, 8-oxoG lesions are first excised by the MutM protein, with repair synthesis restoring the normal G:C base pair. The lesions that are missed result in insertion by the replicating polymerase of the incorrect A some of the time, but these As are excised by the MutY protein. Repair synthesis then inserts a C opposite the 8-oxoG lesion much of the time, returning it to the pool where it is a target once again for the MutM protein. This layered system of repair is very effective, since spontaneous G:C → T:A transversions are normally very rare, occurring on the order of 10^{-10}–10^{-9}/bp/cell/generation *(22,32)*.

It is likely that the principal function of the MutY protein in vivo is to repair 8-oxoG A mispairs, rather than G:A mispairs, since overproducing the MutM completely eliminates the mutator effect of a mutY strain *(35)*. If G:A mispairs that required the action of the MutY protein contributed to spontaneous mutagenesis, then these would not be affected by overproducing the MutM protein. Also, Grollman and coworkers (personal communication) have shown that purified MutY protein has a specificity constant (K_{cat}/K_m) for oligonucleotides containing A/8-oxoG that is two orders of magnitude higher than for oligos containing A/G.

The scheme in Fig. 1B is supported by genetic evidence. Mutants lacking either the MutM protein or the MutY protein, but retaining the other function have a significant, but moderate increase in mutations, all G:C → T:A transversions *(8,43)*, as seen in Table 4. However, the double mutant lacking both the MutM and MutY proteins has an enormous increase in mutations *(35)*, which is comparable to strains lacking both the proofreading and mismatch repair systems. Clearly, these two enzymes play a crucial role in avoiding high spontaneous mutation rates caused by the oxidative damage to DNA.

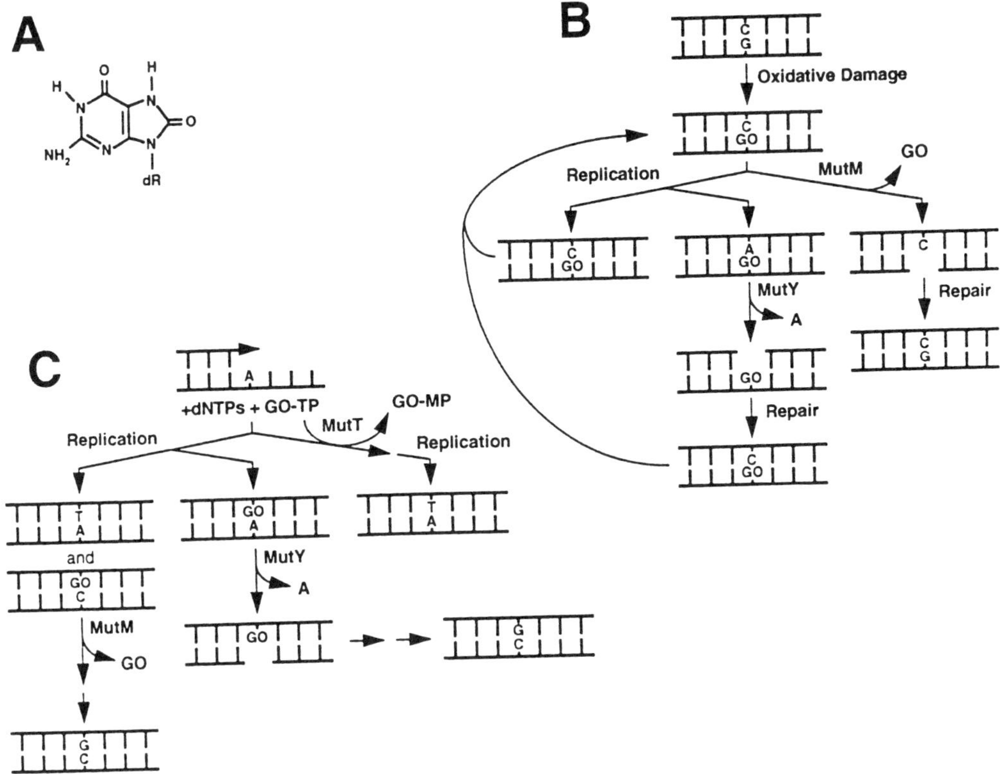

Fig. 1. The GO system *(34,37)*. **(A)** 2'-Deoxy-7,8dihydro-8-oxoguanosine (8-oxoG or "GO"). This is the structure of the predominant tautomeric form of the GO lesion. **(B)** Oxidative damage can lead to GO lesions in DNA that are repaired by the concerted action of the MutM and MutY proteins *K(see text)*. **(C)** Oxidative damage can also lead to 8-oxo-dGTP. Mut T is active on 8-oxo-dGTP and hydrolyzes it to 8-oxo-dGMP, effectively removing the triphosphate from the deoxynucleotide pool (*see text* for further details).

Table 4
Mutation Frequency of *mutM*, *mutY*, and *mutM mutY* Strains

Strain	No. of Lac$^+$	No. of Rif^r
CC104	3	5–10
CC104 *mutM*	25	151
CC104 *mutY*	62	290
CC104 *mutM mutY*	1900	8200
CSH115 *(mutS)*	ND	760
CSH116 *(mutD)*	ND	4900

2.3. *The* mutT *Gene*

The third protein involved in preventing 8-oxoG from causing mutations is the product of the *mutT* gene. This mutator causes A:T → C:G transversions *(56)*. The MutT protein is a hydrolase that converts 8-oxodGTP back to the monophosphate, 8-

oxodGMP, thus averting its incorporation into the DNA *(33)*. If 8-oxoG is incorporated into the DNA opposite an A, then it cannot be repaired and will result in an A:T → C:G transversion. In fact, the MutY protein will excise the A and lead to the insertion of a C opposite the resulting 8-oxoG lesion, leading to a more rapid fixation of the mutation. If the 8-oxoG lesion is incorporated opposite a C, then at the next round of replication, it could pair with A leading to a G:C → T:A transversion, although the MutM and MutY proteins prevent this from happening, thus producing the observed A:T → C:G specificity for *mutT* mutators.

3. FUTURE EXPERIMENTS

Significant insight into the detailed mechanism of action of the MutY, MutM, and MutT proteins will come from analysis of the crystal structures of these three enzymes. Such work is under way in several laboratories. Structure–function relationships for both MutM and MutY have been recently reviewed *(24)*. It is also of great interest to characterize the biological significance of the human counterpart of the GO system. The *mutT* and *mutY* genes have been cloned from human cells *(47,50)*, and the human MutT enzyme has been shown to complement the *E. coli mutT* gene *(47)*. Now that it has been shown that a deficiency in the human counterpart to the bacterial and yeast mismatch repair systems lead to colon cancer susceptibility *(7)*, it is of interest to determine whether other repair deficiencies leads to similar susceptibilities. Therefore, experiments with knockout mice lacking different combinations of the GO system genes will be of great interest. The availability of the human genes now makes these experiments possible.

ACKNOWLEDGMENTS

I would like to thank the members of my laboratory, and Dr. Arthur P. Grollman for helpful discussions during the writing of this manuscript. This work was supported by a grant from the NIH (GM32184).

REFERENCES

1. Ames, B. N., M. K. Shigenaga, and T. M. Hagen. 1993. Oxidants, antioxidants, and the degenerative diseases of aging. *Proc. Natl. Acad. Sci. USA* **90:** 7915–7922.
2. Au, K. G., M. Cabrera, J. H. Miller, and P. Modrich. 1988. The *Escherichia coli mutY* gene product is required for specific AG to CG mismatch correction. *Proc. Natl. Acad. Sci. USA* **85:** 9163–9166.
3. Au, K. G., L. Clark, J. H. Miller, and P. Modrich. 1989. *The Escherichia coli mutY* gene encodes an adenine glycosylase active on G-A mispairs. *Proc. Natl. Acad. Sci. USA* **86:** 5345–5349.
4. Boiteux, S., M. Bichara, R. P. Fuchs, and J. Laval. 1989. Excision of the imidazole ring-opened form of *N*-2–aminofluorene-C(8)-guanine adduct in poly (dG-dC) by *Escherichia coli* formamidopyrimidine-DNA glycosylase. *Carcinogenesis* **10:** 1905–1909.
5. Boiteux, S. and J. Laval. 1983. Imidazole ring-opened guanine: an inhibitor of DNA synthesis. *Biochem. Biophys. Res. Commun.* **110:** 625–631.
6. Boiteux, S., T. R. O'Conner, and J. Laval. 1987. Formamidopyrimidine-DNA glycosylase of *Escherichia coli*: cloning and sequencing of the fpg structural gene and overproduction of the protein. *EMBO J.* **6:** 3177–3183.

7. Bronner, C. E., S. M. Baker, P. T. Morrison, G. Warren, L. G. Smith, M. K. Lescoe, et al. 1994. Mutation in the DNA mismatch repair gene homologue hMLH1 is associated with hereditary nonpolyposis colon cancer. *Nature* **368:** 256–261.
8. Cabrera, M., Y. Nghiem, and J. H. Miller. 1988. *mutM*, a second mutator locus in *Escherichia coli* that generates GC → TA transversions. *J. Bacteriol.* **170:** 5405–5407.
9. Cheng, K. C., D. S. Cahill, H. Kasai, S. Nishimura, and L. A. Loeb. 1992. 8-hydroxyguanine, an abundant form of oxidative DNA damage, causes G → T and A → C substitutions. *J. Biol. Chem.* **267:** 166–172.
10. Chetsanga, C. J. and T. Lindahl. 1979. Release of 7-methylguanine residues whose imidazole rings have been opened from damaged DNA by a DNA glycosylase from *Escherichia coli. Nucleic Acids Res.* **6:** 3673–3683.
11. Cox, E. C. 1976. Bacterial mutator genes and the control of spontaneous mutation. *Ann. Rev. Genet.* **10:** 135–156.
12. Cox, E. C. and D. L. Horner. 1982. Dominant mutators in *Escherichia coli. Genetics* **100:** 7–18.
13. Culp, S. J., B. P. Cho, F. F. Kadulbar, and F. E. Evans. 1989. Structural and conformational analysis of 8-hydroxy-2'-deoxyguanosine. *Chem. Res. Toxicol.* **2:** 416–421.
14 Cupples, C. and J. H. Miller. 1989. A genetic system for the rapid determination of all six base substitutions in *Escherichia coli. Proc. Natl. Acad. Sci. USA* **86:** 8877–8881.
15. Degnen, G. E. and E. C. Cox. 1974. A conditional mutator gene in *Escherichia coli*: isolation, mapping and effector studies. *J. Bacteriol.* **117:** 477–487.
16. Dizdaroglu, M. 1985. Formation of an 8-hydroxyguanine moiety in deoxynucleic acid on gamma irradiation in aqueous solution. *Biochemistry* **24:** 4476–4481.
17. Demple, B. 1991. Regulation of bacterial oxidative stress genes. *Ann. Rev. Genet.* **25:** 315–337.
18. Duncan, B. and J. H. Miller. 1980. On the mutagenic determination of cytosine residues in DNA. *Nature* **287:** 560,561.
19. Echols, H., C. Lu, and P. M. J. Burgers. 1983. Mutator strains of *Escherichia coli, mutD5* and *dnaQ*, with defective exonucleolytic editing by DNA polymerase III holoenzyme. *Proc. Natl. Acad. Sci. USA* **80:** 2189–2192.
20. Farr, S. B. and T. Kogoma. 1991. Oxidative stress responses in *Escherichia coli* and *Salmonella typhimurium. Microbiol. Rev.* **55:** 561–585.
21. Feig, D. I., L. C. Sowers, and L. A. Loeb. 1994. Reverse chemical mutagenesis: identification of the mutagenic lesions resulting from reactive oxygen species-mediated damage to DNA. *Proc. Natl. Acad. Sci. USA* **91:** 6609–6613.
22. Fowler, R. G., G. E. Degnen, and E. C. Cox. 1974. Mutational specificity of a conditional *Escherichia coli* mutator, *mutD5. Mol. Gen. Genet.* **133:** 179–191.
23. Fraga, C. G., M. K. Shigenaga, J. W. Park, P. Degan, and B. N. Ames. 1990. Oxidative damage to DNA during aging: 8-hydroxy-2'-deoxyguanosine in rat organ DNA and urine. *Proc. Natl. Acad. Sci. USA* **87:** 4533–4537.
24. Grollman, A. P., F. Johnson, J. Tchou, and M. Eisenberg. 1994. Recognition and repair of 8-oxoguanine and formamidopyrimidine lesions in DNA. *Ann. NY Acad. Sci.* **726:** 208–214.
25. Grollman, A. P. and M. Moriya. 1993. Mutagenesis by 8-oxoguanine: an enemy within. *Trends Genet.* **9:** 246–249.
26. Joenje, H. 1989. Genetic toxicology of oxygen. *Mutat. Res.* **219:** 193–208.
27. Konrad, E. B. 1975. Isolation of an *Escherichia coli* K-12 *dnaE* mutation as a mutator. *J. Bacteriol.* **133:** 1197–1202.
28. Kouchakdjian, M., V. Bodepudi, S. Shibutant, M. Eisenber, F. Johnson, A. P. Grollman, and D. J. Patel. 1991. NMR structural studies of the ionizing radiation adduct 7-hydro-8-oxodeoxyguanosine (8-oxo-*7H*-dG) opposite deoxyadenosine in a DNA duplex. 8-Oxo-*7H*-dG(*syn*):dA(*anti*) alignments at lesion site. *Biochemistry* **30:** 1403–1412.
29. Liberfarb, R. M. and V. Bryson. 1970. Isolation, characterization and genetic analysis of mutator genes in *Escherichia coli* B and K-12. *J. Bacteriol.* **104:** 363–375.

30. Leong, P. M., H. C. Hsia, and J. H. Miller. 1986. Analysis of spontaneous base substitutions generated in mismatch repair-deficient strains of *Escherichia coli*. *J. Bacteriol.* **168:** 412–416.
31. Lindahl, T. 1993. Instability and decay of the primary structure of DNA. *Nature* **362:** 709–715.
32. Mackay, W. J., S. Han, and L. D. Samson. 1994. DNA alkylation repair limits spontaneous base substitution mutaitons in *Escherichia coli*. *J. Bacteriol.* **176:** 3224–3230.
33. Maki, H. and M. Sekiguchi. 1992. MutT protein specifically hydrolyzes a potent mutagenic substrate for DNA synthesis. *Nature (Lond.)* **355:** 273–275.
34. Massini, R. 1907. Über einen in biologischer Beziehun interessanten Kolistamm (Bacterium coli mutabile). Ein Beitrag sur Variationen bei Bakterien. *Archiv für Hygiene* **61:** 250–292.
35. Michaels, M. L., C. Cruz, A. P. Grollman, and J. H. Miller. 1992. Evidence that MutY and MutM combine to prevent mutations by an oxidatively damaged form of guanine in DNA. *Proc. Natl. Acad. Sci. USA* **89:** 7022–7025.
36. Michaels, M. L. and J. H. Miller. 1992. The GO system protects organisms from the mutagenic effect of the spontaneous lesion 8-hydroxyguanine (7,8-dihydro-8-oxoguanine). *J. Bacteriol.* **174:** 6321–6325.
37. Michaels, M. L., L. Pham, C. Cruz, and J. H. Miller. 1991. MutM, a protein that prevents G:C → T:A transversions, is formamidopyrimidine-DNA glycosylase. *Nucleic Acids Res.* **19:** 3629–3632.
38. Michaels, M. L., L. Pham, Y. Nghiem, C. Cruz, and J. H. Miller. 1990. MutY, an adenine glycosylase on G/A mispairs, has homology to endonuclease III. *Nucleic Acids Res.* **18:** 3841–3845.
39. Michaels, M. L., J. Tchou, A. P. Grollman, and J. H. Miller. 1992. A repair system for 8-oxo-7,8-dihydrodeoxyguanine. *Biochemistry* **31:** 10,964–10,968.
40. Miller, J. H. 1996. Spontaneous mutators in bacteria: insights into pathways of mutagenesis and repair. *Ann. Rev. Microbiol.* **50:** 625–643.
41. Moriya, M., C. Ou, V. Bodepudi, F. Johnson, M. Takeshita, and A. P. Grollman. 1991. Site specific mutagenesis using a gapped duplex vector: a study of translesion synthesis past 8–oxodeoxyguanosine in *E. coli*. *Mutat. Res.* **254:** 281–288.
42. Nevers, P. and H. Spatz. 1975. *Escherichia coli* mutants *uvrD uvrE* deficient in gene conversion of lambda heteroduplexes. *Mol. Gen. Genet.* **139:** 233–243.
43. Nghiem, Y., M. Cabrera, C. G. Cupples, and J. H. Miller. 1988. The *mutY* gene: a mutator locus in *Escherichia coli* that generates GC → TA transversions. *Proc. Natl. Acad. Sci. USA* **85:** 2709–2713.
44. Oda, Y., et al. 1991. NMR studies of a DNA containing 8-hydroxydeoxyguanosine. *Nucleic Acids Res.* **19:** 1407–14012.
45. Pacifici, R. E. and K. J. Davies. 1991. Protein, lipid and DNA repair systems in oxidative stress: the free radical theory of aging revisited. *Gerontology* **37:** 166–180.
46. Rydberg, B. 1978. Bromouracil mutagenesis and mismatch repair in mutator strains of *Escherichia coli*. *Mutat. Res.* **52:** 11–24.
47. Sakumi, K., M. Foruichi, T. Tsuzuki, S. Kawabata, H. Maki, M. Sekiguchi, et al. 1993. Cloning and expression of cDNA for a human enzyme that hydrolyzes 8–oxodGTP, a mutagenic substrate for DNA synthesis. *J. Biol. Chem.* **268:** 23,524–23,530.
48. Schaaper, R. M. and R. I. Dunn. 1987. Spectra of spontaneous mutations in *Escherichia coli* strains defective in mismatch correction: the nature of in vivo replication errors. *Proc. Natl. Acad. Sci. USA* **84:** 6220–6224.
49. Shibutani, S., M. Takeshita, A. P. Grollman. 1991. Insertion of specific bases during DNA synthesis past the oxidation-damaged base 8-oxodG. *Nature* **349:** 431–434.
50. Slupska, M. M., C. Baikalov, W. M. Luther, J. Chiang, Y. Wei, and J. H. Miller. 1996. Cloning and sequencing a human homolog (*hMYH*) of the *Escherichia coli mutY* gene

whose function is required for the repair of oxidative DNA damage. *J. Bacteriol.* **178:** 3885–3892.
51. Tchou, J. and A. P. Grollman. 1993. Repair of DNA containing the oxidatively-damaged base, 8–oxoguanine. *Mutat. Res.* **299:** 277–287.
52. Tchou, J., H. Kasai, S. Shibutani, M. H. Chung, J. Laval, A. P. Grollman, et al. 1991. 8-oxoguanine (8-hydroxyguanine) DNA glycosylase and its substrate specificity. *Proc. Natl. Acad. Sci. USA* **88:** 4690–4694.
53. Teebor, G. W., R. J. Boorstein, J. Cadet. 1988. The repairability of oxidative free radical mediated damage to DNA: a review. *Int. J. Radiat. Biol.* **54:** 131–150.
54. Treffers, H. P., V. Spinelli, and N. O. Belsser. 1954. A factor (or mutator gene) influencing mutation rates in *Escherichia coli. Proc. Natl. Acad. Sci. USA* **40:** 1064–1071.
55. Wood, M. L., M. Dizdaroglu, E. Gajewski, and J. M. Essigmann. 1990. Mechanistic studies of ionizing radiation and oxidative mutagenesis: genetic effects of a single 8-hydroxyguanine (7-hydro-8-oxoguanine) residue inserted at a unique site in a viral genome. *Biochemistry* 2**9:** 7024–7032.
56. Yanofsky, E. C., C. Cox, and V. Horn. 1966. The unusual mutagenic specificity of an *E. coli* mutator gene. *Proc. Natl. Acad. Sci. USA* **55:** 274–281.

7
The SOS Response

Walter H. Koch and Roger Woodgate

1. INTRODUCTION: THE *E. coli* PARADIGM

Escherichia coli, like many other living organisms, is often exposed to a variety of environmental agents, both natural and man-made, which perturb the integrity of DNA. These DNA adducts often block continued DNA replication and thereby pose immediate threats to continued cell survival. Perhaps as a consequence, *E. coli* (as well as many other prokaryotes) has evolved such that a number of unlinked genes involved in DNA repair, cell division, and damage tolerance are coordinately expressed after DNA damage. This global cellular response is often referred to as the "SOS response." Since its formal conception over 20 years ago *(132)*, our understanding of the SOS response has grown considerably and has been appropriately summarized in a number of excellent scientific reviews *(46,96,157)*. The goal of this chapter is to familiarize the reader with some recent advances in our understanding of the SOS response in *E. coli* and many other prokaryotic organisms.

2. REGULATION OF THE SOS RESPONSE

2.1. *Roles of RecA and LexA in Regulating the Response*

The complex SOS regulatory network is controlled by two proteins, RecA and the LexA transcriptional repressor, which are themselves SOS-inducible proteins. Accumulation of DNA damage or perturbation of DNA replication results in the activation of RecA coprotease activities (designated RecA*). In its activated state, RecA forms spectacular spiral nucleoprotein filaments on DNA *(56)*. Free LexA protein recognizes this structure and binds within the deep helical groove of the RecA nucleoprotein filament *(172)*. This interaction leads to the efficient autocatalytic cleavage of LexA at a scissile peptide bond located between residues Ala^{84} and Gly^{85} within a hinge region of LexA that connects its amino-terminal DNA binding and carboxyl-terminal dimerization domains (*see* Section 3.1., for a more detailed description of the mechanism of cleavage). As a consequence, the level of functionally active LexA protein able to bind specific sites found in the promoter/operator region of various LexA-regulated genes drops precipitously, thereby releasing the genes from negative transcriptional repression.

From: DNA Damage and Repair, Vol. 1: DNA Repair in Prokaryotes and Lower Eukaryotes
Edited by: J. A. Nickoloff and M. F. Hoekstra © Humana Press Inc., Totowa, NJ

2.2. *Differential Regulation of SOS Genes*

The exquisite molecular circuitry of the SOS network allows for a wide range of cellular responses in *E. coli*, based in part on the relative binding affinity of LexA for each individual binding site and the subsequent induction ratio of each SOS-encoded protein. More specifically, the graded response allows cells with minimal damage to induce error-free repair processes (such as nucleotide excision repair and recombinational repair) without inducing other repair pathways that are potentially error-prone *(46)*. Indeed, more extensive cellular DNA damage is necessary for the expression of the *umuDC* genes required for error-prone translesion synthesis of damaged DNA (*see* Section 5.3. and Chapter 12). Finally, the persistence of unrepaired DNA lesions leads to the induction of *sulA*-dependent filamentation, and allows prophage to escape a potentially inviable cell by entering into a lytic growth cycle *(46,157)*.

2.3. *Nature of the SOS-Inducing Signal*

The precise nature of the SOS-inducing signal in DNA-damaged cells has been the subject of some controversy. It is generally accepted that damaged DNA does not lead to RecA activation directly, but rather that single-stranded DNA (ssDNA), which accumulates when DNA replication is transiently inhibited by replication blocking lesions, provides the metabolic signal that activates the RecA coprotease *(142)*. Early evidence supporting this notion was provided by the demonstration that RecA can be activated in vitro in the presence of ssDNA and ATP *(27,28)*. More recent in vivo evidence supporting this hypothesis comes from studies using replication-defective episomal elements in *E. coli*. Mutations in the pR plasmid *bat* gene, which block production of ssDNA needed for pR plasmid replication by the rolling circle mechanism, also prevent the spontaneous induction of the *E. coli* SOS response by the pR plasmid *(49)*. Similarly, infection of *E. coli* with a mutant of the filamentous ssDNA phage f1 that is defective in initiation of minus-strand DNA synthesis induces the SOS response, whereas the wild-type phage does not *(58)*. Moreover, in the gram-positive organism *Bacillus subtilis* (which has an SOS response similar to that of *E. coli*) *(170)*, analysis of the SOS inducing signal has shown that it is DNA replication (and, by inference, the single-stranded gaps produced when a replication fork stops at DNA lesions) rather than DNA damage itself that serves as the inducing signal *(99)*.

2.4. *Genes Affecting RecA* Production and Induction of the SOS Response*

Genetic evidence indicates that optimal induction of the SOS response is dependent on *recF*, which encodes a protein involved in DNA recombination and repair *(26,142)*. More recently, it has been demonstrated that normal induction of the SOS response also requires *recO* and *recR*, which together with *recF* form the *recFOR* epistasis group and are required for the *recF* recombination pathway *(54,161)*; mutations in any one of these three genes delay the onset of SOS induction. It has been proposed that the RecFOR proteins facilitate SOS induction by improving the efficiency of RecA nucleation onto ssDNA, thereby intensifying the inducing signal for RecA activation *(102)*.

Induction of the SOS response by nalidixic acid, a gyrase inhibitor, requires the helicase activity of RecBCD, but not its nuclease activity *(23)*. This finding led to the suggestion that RecBCD unwinding activity in the presence of nalidixic acid produces the ssDNA required for RecA activation. The finding that the nonmutability by UV of *uvrD3 recB21* strains of *E. coli* was largely overcome by constitutive activation of RecA provided further evidence that helicase-dependent generation of ssDNA is required for RecA activation *(165)*. Similar processes appear to be involved in the RecBCD-dependent induction of the SOS response in F⁻ bacteria during interspecies mating *(106)*. Moreover, mutations in the UvrD helicase have been isolated that lead to expression of the SOS response in the absence of DNA damage, a process that may involve either excess unwinding of double-stranded DNA (dsDNA) or a blockade to unwinding during replication *(119)*.

It is likely that other proteins will also be identified that affect either the generation of the inducing signal or the ability of RecA to recognize it. Indeed, a recent study has identified a new mutation in *E. coli* called *isfA*, which inhibits several phenomena associated with the SOS response *(12)*. SOS mutagenesis, induced replisome reactivation (IRR, *see* Section 5.1.), prophage induction, and cell filamentation, phenomena that require RecA coprotease activation, are all inhibited by the *isfA* mutation, whereas conjugative recombination is unaffected. RecA-mediated proteolytic cleavage of UmuD is also apparently inhibited by the *isfA* mutation even in *E. coli recA730* and *recA730 lexA51*(Def) strains, which normally constitutively express coprotease activity towards UmuD *(13,166)*. The phenotype of the *isfA* mutation is somewhat similar to that exerted by the *psiB* gene, which inhibits SOS induction in recipient cells during conjugal transfer and is found on F factors and other conjugative plasmids *(5–7,36)*. Further studies into this phenomenon will likely determine whether the chromosomally encoded IsfA and plasmid encoded PsiB proteins are structurally as well as functionally related.

3. MECHANISM OF LexA, λ CI REPRESSOR, AND UmuD-LIKE CLEAVAGE REACTIONS

3.1. *Proteolytic Cleavage Site*

Earlier studies into the mechanism of LexA cleavage envisioned RecA as possessing proteolytic enzymatic activity towards LexA. However, a number of elegant studies from Little and colleagues revealed that LexA cleavage occurs via an intramolecular as well as intermolecular reaction in the absence of RecA protein *(71,93–95)*; RecA simply facilitates the efficiency of cleavage under physiological conditions. This autocatalytic cleavage reaction, which occurs efficiently in the absence of RecA at alkaline pH, has also been demonstrated in the functionally related λ/CI repressor protein *(50,146)* as well as in certain so-called mutagenesis proteins, like *E. coli* UmuD *(20)*, MucA *(53)*, and RumA($_{R391}$) *(86)*, that are structurally related. Key elements of the autocatalytic cleavage reaction include a conserved Ala(Cys)-Gly cleavage site and appropriately spaced Ser and Lys residues. Previous studies have suggested that the autocatalytic cleavage reaction is akin to that of serine proteases *(94)* or even signal peptidases *(101)*. More recently, however, it has been suggested that this reaction is more comparable to that of the β-lactamases *(46,94,123)*. Posttranslational processing of UmuD generates the shorter, but mutagenically active UmuD' protein, and recent structural analysis of UmuD' *(123)* suggests that the cleavage reaction is indeed similar to the hydrolysis of penicillin by the TEM1 β lactamase *(150)*. Although the cleavage site of UmuD is missing in the UmuD' molecule, UmuD' still retains the catalytically

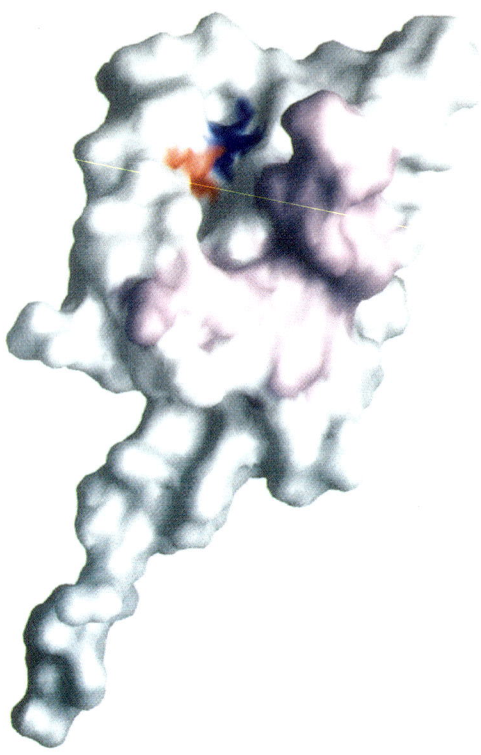

Fig. 1. A space-filling model of UmuD' monomer showing the cleft that leads to its catalytic active site: The surface contributed by the Ser^{60} residue is colored red, and the surface contributed by the Lys^{97} residue is colored blue. The residues that contribute to the molecular dimer interface of UmuD' have their surface contributions colored in violet. Reprinted with permission from *Nature (123)*.

active and conserved Ser and Lys residues required for the cleavage reaction. These residues are found at one end of a cleft in the UmuD' protein with the Ser^{60} residue residing only 2.8 Å away from the Lys^{97} residue (*123*; Fig. 1). Conceivably, the cleft in which Ser^{60} and Lys^{97} are found is the binding site for the amino-terminus of UmuD, thus ensuring the correct orientation of the cleavage site located between residues Cys^{24} and Gly^{25}. Support for the idea that this cleft may in fact form the catalytic active site is suggested by observations that certain *umuD* mutations, which detrimentally affect cleavage, are located along the sides of this cleft (*11,123*; Fig. 1). Furthermore, much of the cleft is occluded in the native UmuD' dimer (*123*; Fig. 1), perhaps explaining why cleavage of UmuD is relatively inefficient. In contrast, LexA (which autodigests much faster than UmuD) *(20)* is thought to exist primarily as a monomer in solution and only dimerizes on binding to its appropriate SOS-box in DNA *(70)*. Finally, certain mutations in the structurally related λ/CI repressor protein, which reduce dimerization, concomitantly led to an increase in RecA-mediated cleavage of λ/CI *(51)*. Taken together, these results strongly suggest that these structurally related proteins are in fact cleaved in a monomeric form at an active site that resembles that of the TEM1 β-lactamase *(123)*.

4. DAMAGE-INDUCIBLE PROTEIN EXPRESSION

4.1. Genes Identified as Part of the "Classic" SOS Regulon

Genes regulated in the classic sense of the SOS response are negatively regulated at the transcriptional level by LexA and become induced on damage-induced RecA-mediated autocatalytic cleavage of LexA. A hallmark of this type of gene is the appropriately positioned LexA binding site in the promoter/operator region of the inducible gene. Those SOS genes with a LexA binding site exhibiting the highest homology to the 20 bp palindromic consensus binding site (5'-TACTGTATATATATACAGTA-3') generally appear to be the most tightly regulated *(90)*. Variant nucleotides in the repeated $(TA)_5$ central region of the consensus sequence alter binding moderately, whereas those in the CTG and CAG triplet nucleotides flanking the central $(TA)_5$ region have a much more dramatic effect on the binding of LexA; mutations within this region often lead to an operator constitutive phenotype *(46)*. Indeed, recent computational analysis of the NMR-derived solution structure of LexA with DNA *(73)* has suggested that residues Asn^{41}, Glu^{44}, and Glu^{45} of the amino-terminal DNA binding domain of LexA actually form hydrogen bonds with the CTGT half-site sequence found in the consensus binding site, thus explaining the sequence specificity observed in LexA binding *(73)*.

Many of the earlier genes identified in the SOS regulon were determined experimentally through the increased damage-inducible expression of a β-galactosidase reporter gene fusion *(67)*. A number of these so-called *din* genes have now been identified, although their biochemical function often remains to be elucidated (Table 1). More recently, SOS-regulated genes have been identified through a computational search for putative LexA binding site sequences deposited in various genetic databases *(90)*. Using a mathematical approach, Lewis et al. *(90)* determined a heterology index for LexA binding sites. This index measures the degree by which any SOS box varies from the consensus. LexA boxes with a lower heterology index have a closer fit with the consensus sequence and are predicted to bind LexA more tightly that those with higher heterology indices. Interestingly, analysis revealed that those LexA binding sites that had been experimentally demonstrated to bind LexA protein physically had an index of 12.6 and lower, whereas those that did not bind LexA had an index of 15.0 or greater. Using such an approach, Lewis et al. *(90)* identified six new putative LexA-regulated genes called *sosA-F/dinJ-O* (Table 1). Although direct transcriptional induction of these genes has yet to be reported, it seems likely that they are indeed *bona fide* members of the SOS regulon.

Sequence analysis of the 5.5-min region of the *E. coli* chromosome has identified yet another *din* gene, designated *dinP (117)*. Somewhat surprisingly, DinP shares limited sequence homology to the *E. coli* UmuC, *Saccharomyces cerevisiae* Rev1p, and a *Caenorhabditis elegans* protein (YLW6-CAEEL), *(117)*. *dinP* is in fact allelic with the previously identified *dinB* gene (*67*; H. Ohmori, personal communication). Given the

Table 1
LexA-Regulated Damage Inducible Genes in *Escherichia coli*

Allele	Function/similarity/comments	Refs.
dinA/polB	DNA pol II	*14,63,67*
dinB/dinP	Identical to *dinP*; similarity to *E. coli* UmuC and *S. cerevisiae* REV1 proteins; role in untargeted λ mutagenesis	*18,67*; H. Ohmori, personal communication
dinD/pcsA	Cold-sensitive mutant *(pcsA68)*	*67,100,118*
dinE/uvrA	UvrA damage-recognition protein involved in nucleotide excision repair	*67,68*
dinF	Immediately downstream of the *lexA* gene	*55*
dinG	Noncannonical SOS box that binds LexA; putative helicase; similarity to *S. cerevisiae* Rad3 protein; multicopy suppressor of *pcsA68* cold-sensitive mutant	*82,91,92,171;* H. Ohmori, personal communication
dinH	Opertator binds LexA	*90,91*
dinI	Multicopy suppressor of *pcsA68* cold-sensitive mutant; similarity to plasmid TP110 ImpC protein; opertator binds LexA	*90,171;* H. Ohmori, personal communication
dinJ/sosA[a]	Predicted binding of LexA to operator based on similarity of SOS box to consensus sequence; map location 5.5 min	*90,117*
dinK/sosB[a]	Predicted binding of LexA to operator based on similarity of SOS box to consensus sequence; map location 86.6 min	*90*
dinL/sosC[a]	Operator binds LexA; map location 98.7	*90*
dinM/sosD[a]	Operator binds LexA; map location 38.0	*90*
dinN/sosE[a]	Predicted binding of LexA to operator based on similarity of SOS box to consensus sequence; map location 36.9 min	*90*
dinO/sosF[a]	Predicted binding of LexA to operator based on similarity of SOS box to consensus sequence; map location 47.2 min	*90*
dinP/dinB	Identical to *dinB*; similarity to *E. coli* UmuC and *S. cerevisiae* REV1 proteins; role in untargeted λ mutagenesis	*18,117*; H. Ohmori, personal communication
lexA	Transcriptional repressor	*46,156*
recA	Roles in recombination, regulation of the SOS response, and in SOS mutagenesis	*46,157*
uvrB	Protein involved in the incision step of nucleotide excision repair	*46,157*

uvrD	DNA helicase; involved in excision repair, mismatch repair, and the generation of an SOS inducing signal	*46,157*
sulA	Transient inhibition of cell division	*46,157*
umuDC	Recombinational repair	*46,157*
ruvAB	Recombinational repair	*46,156*
recN	Recombinational repair	*46,157*
sbmC	Resistance to Microcin B17	*10*
ssb	Single-stranded binding protein	*46,157*

[a]Although, technically, *sosA-F/dinJ-O* have not yet been shown to be damage-inducible, given the approriate positioning of the the LexA binding site in the operator/promoter regions of these genes, it seems highly likely that they are indeed *bona fide* members of the SOS regulon.

similarity to the UmuC protein, which is involved in SOS mutagenesis (*see* Section 6.2.), it is perhaps not too surprising that one of the few phenotypes of the DinB protein is a postulated role in untargeted phage λ mutagenesis *(18)*.

These *din/sos* genes, together with the independently characterized LexA-regulated genes *(46,157)*, now number at least 27 chromosomal *E. coli* genes (Table 1). This tally is likely to grow even larger once the nucleotide sequence of the entire *E. coli* genome has been determined. In addition to the 27 chromosomally encoded LexA-regulated genes described in Table 1, putative high-affinity LexA binding sites have been identified in the genomes of several *E. coli* bacteriophages, such as phage P1, T4, T5, and λ, as well as on several conjugative plasmids *(46,90)*. Thus, it would appear that regulation of gene expression via LexA-mediated transcriptional repression is quite widespread, at least in *E. coli*, its bacteriophages, and plasmids.

4.2. Damage-Inducible Protein Expression: Genes Not Directly Regulated by LexA

In addition to those genes that are negatively regulated by LexA in the classical sense of the SOS response, it is becoming increasingly clear that a number of proteins that do not possess even a putative LexA binding site within their operator/promoter regions are induced following cellular DNA damage. Indeed, using a two-dimensional gel electrophoresis assay, Lesca et al. *(88)*, observed the appearance of approx 22 new proteins in radiolabeled extracts obtained from UV-irradiated cells compared to their unirradiated controls. Further analysis has identified three of these proteins as the products of the *hga* gene *(22)* and the *dinY* locus *(126)*. Although the identity of the other proteins remains to be elucidated, it is likely that some of them will correspond to the products of the *dnaA*, *dnaB*, *dnaN*, *dnaQ*, *recQ*, and *phr* genes, all of which appear to be damage-inducible, yet are not directly regulated by LexA *(61,62,65,72,122,128–131)*.

Interestingly, many of these non-LexA-regulated proteins are involved in basic metabolic processes like the recovery of respiration following damage *(22)* and DNA replication. Indeed, although the level of many of the individual subunits that comprise DNA polymerase III (pol III) holoenzyme have yet to be experimentally established, it seems likely that all will exhibit a modest increase after DNA damage, since Bonner et al. *(15)* observed that the activity of DNA pol III was approx threefold higher in extracts from UV-irradiated cells than from nonirradiated cells. After UV irradiation, a variety of alternative replication pathways are induced (*see* Section 5.1.) and the increased activity of *E. coli* DNA polymerases after UV probably reflects their participation in these processes.

5. DNA REPLICATION AND THE SOS RESPONSE

5.1. Replication Restart After DNA Damage-Arrested DNA Synthesis

Induction of the SOS response by DNA damage is accompanied by a transient inhibition of DNA synthesis *(141)*. Approximately 30–45 min after UV radiation, DNA synthesis resumes in a discontinuous manner in which short nascent strands of DNA are produced, whose length corresponds to the average distance between dimers (reviewed in ref. *97*); this reinitiation of DNA synthesis has been termed replication restart or induced replisome reactivation (IRR). Replication restart requires RecA and at least one additional protein not expressed as part of the LexA-dependent SOS regulon *(69,164)*. The molecular basis for IRR is not well understood; however, existing evidence suggests that this recovery process is not caused by new DNA synthesis, but rather involves reinitiation of stalled replication complexes. Genetic analyses have shown that the process is independent of *recB*, *uvrA*, and *umuC* in *recA* wild-type cells; the latter finding suggests that IRR does not involve translesion synthesis *(69)*. A model explaining continuous DNA synthesis past DNA lesions in the absence of translesion synthesis postulates a RecA-mediated strand switch in which DNA pol III makes use of a newly synthesized daughter strand of similar polarity, and then switches back after passing the replication blocking lesion *(39,46)*.

5.2. Induced Stable DNA Replication

Chromosomal replication in *E. coli* normally requires transcription and translation and is initiated when the DnaA initiator protein binds to and opens duplex DNA at a unique origin of replication called *oriC (83)*. In cells induced for the SOS response by DNA damage or inhibition of DNA synthesis, a mode of chromosomal replication called inducible stable DNA replication (iSDR) occurs, which is independent of transcription, translation, DnaA, and *oriC* (reviewed in ref. *4*). iSDR makes use of at least three different replication origins (*oriM1A* and *oriM1B* within the minimal *oriC*, and *oriM2* within *terC*) and exhibits an absolute requirement for both the recombinase activity of RecA and the RecBCD helicase. The displacement loop (D-loop) model postulates that on induction of the SOS response, a double-strand break is introduced at or near the origin of iSDR. RecBCD helicase generates ssDNA that invades a homologous uncut duplex through the action of RecA, forming a D-loop and opening the duplex for initiation of replication. Support for this model is provided by the finding that when the RecBCD pathway is inactivated (i.e., in *recBC sbcC* or *recBC sbcBC* mutants), the *recE* or *recF* gene products, respectively, can mediate iSDR initiation *(2–4)*. iSDR differs further from normal DNA replication in that a separate primosome from the ABC priming system is employed; iSDR is abolished in strains carrying *priA* mutations, suggesting that the process relies on the PriA-type primosome assembly for initiation at oriM *(105)*. Interestingly, iSDR is less sensitive to UV radiation than normal DNA replication and may be error-prone. Thus, like *umuDC*-dependent translesion synthesis, iSDR may contribute to cell survival and increased genetic variability in bacterial populations under stress conditions.

5.3. Error-Prone Translesion Synthesis

Despite the fact that many *E. coli* SOS proteins are involved in error-free mechanisms of repair, it is clear that under certain conditions, perhaps when damage is more extensive, unrepaired lesions still remain in the genome. Both in vitro and in vivo studies have revealed that many of these lesions serve as blocks to continued DNA replication *(9,133)*. To avoid the immediate, and fatal consequences of this impasse, *E. coli* utilizes a number of damage-inducible proteins to facilitate a mode of replication that involves direct translesion DNA synthesis (*see* Chapter 12). Strictly speaking, this is not a repair pathway, since the offending lesion is not removed from the genome. It does, however, provide additional opportunities for the cell to repair the damage before the lesion is encountered in the next round of replication. Because the structure of the

DNA adduct frequently distorts the coding properties of the damaged nucleotide, this mode of replication is error-prone and is often referred to as "SOS mutagenesis."

6. SOS MUTAGENESIS

The field of SOS mutagenesis has also been the topic of a number of excellent reviews *(46,97,111,157,158,167,169)*.

6.1. Genetic Requirements For SOS Mutagenesis

Although at least 27 chromosomal genes are induced as part of the SOS response (Table 1), only the products of three LexA-regulated genes, namely RecA, UmuD, and UmuC are, in general, required for most chemical and UV-induced mutagenesis. Furthermore, basal levels of RecA appear to be adequate for SOS mutagenesis, and only UmuDC needs to become derepressed to achieve the cell's full mutagenic potential *(149)*. As discussed in Section 4.2., DNA pol III (which has also been implicated in the mutagenic process) is probably induced as a consequence of DNA damage, but is not directly regulated by LexA protein. As always, there are exceptions to these general requirements. For example, specific mutagenic pathways dealing with *N*-2-acetylaminofluorene (AAF) and ethenocytosine (εC) lesions do not always require the RecA and UmuDC proteins. In the case of AAF which can induce both –1 and –2 frameshift mutations, (depending on the sequence context in which the lesion occurs), the –1 frameshift pathway is both *recA*- and *umuDC*-dependent; a –2 frameshift pathway within alternating GC runs is, however, manifested even in the absence of *umuDC*. Instead, it requires another, as yet unidentified, LexA-regulated protein tentatively termed "Npf" (for *Nar*I processing factor; *64*). Likewise, mutagenesis of a single εC lesion located on the single-stranded phage, M13, increases after exposing the host cell to UV and alkylating agents *(120)*, yet does not require any LexA-regulated SOS protein *(120,121)*.

6.2. Possible Mechanisms of RecA-UmuDC-Dependent SOS Mutagenesis

Compared to many repair processes in *E. coli* that are well characterized, the biochemical mechanism of translesion DNA synthesis is poorly understood. Previous studies have shown that as part of the mutagenic process, the functionally inactive UmuD protein is processed to its mutagenically active form of UmuD' via a RecA-mediated reaction that is mechanistically similar to that of LexA cleavage (*see* Section 3.1.). Furthermore, by interacting with RecA, the acidic UmuD' protein appears to be specifically targeted to DNA *(8,45,151)*. UmuC, in contrast, is a basic protein that not only binds to a UmuD' dimer to form a UmuD'$_2$C complex, but also to regions of ssDNA *(19,127,168)*. The UmuD'$_2$C complex is believed to be mutagenically active since together with RecA and DNA pol III holoenzyme, they actually promote in vitro bypass of abasic lesions *(133)*. The mechanism by which this error-prone translesion DNA synthesis occurs still remains to be elucidated. The prevailing genetic model is based on a series of delayed photoreversal experiments reported by Bridges and Woodgate in the mid 1980s *(16,17)*. In this model, translesion synthesis was envisioned to occur in a two-step process. The first step, misincorporation of an incorrect nucleotide by DNA pol III opposite the DNA adduct, is mediated by the RecA protein (although its presence is not absolutely required for the misincorporation step to occur). The second step, bypass (or elongation from the misincorporated nucleotide), requires the Umu proteins. One of the major setbacks in testing this model has been the rather intractable nature of UmuC toward purification and characterization.

In an important step toward elucidating the mechanism of translesion DNA synthesis, Bruck et al. *(19)* have recently purified a soluble UmuD'$_2$C complex and have studied its ability to bind to DNA. Interestingly, the UmuD'$_2$C complex shows no affinity for dsDNA, but binds cooperatively to long stretches of ssDNA. Such binding might impede DNA replication and explain why overproduction of the Umu proteins leads to a cold-sensitive phenotype associated with the cessation of DNA replication *(103)*. Furthermore, DNA binding studies in the presence of RecA suggest that RecA is unable to outcompete UmuD'$_2$C on ssDNA once a UmuD'$_2$C–DNA complex has been formed. The latter observation may explain how the UmuD'C proteins inhibit RecA's recombinatorial activities when expressed at high cellular levels *(148)*; UmuD'$_2$C might inhibit formation of the extensive RecA nucleoprotein filament necessary for recombination by binding to the very end of the RecA filament *(148)*. Alternatively, UmuD'C may simply bind to the entire filament and occupy the same sites normally utilized by ssDNA *(136)*.

In contrast to UmuC, large quantities of UmuD' protein are readily purified. Crystals of UmuD' have recently been obtained *(125)* and its three-dimensional structure solved to a resolution of 2.5 Å by X-ray crystallography methods *(123)*. In addition to providing important information on the mechanism of autocatalytic cleavage, the structure provides some fascinating insights into the nature of the protein–protein interactions that are necessary for the mutagenic process. As well as forming molecular dimers, UmuD' also forms filament dimers within the crystal lattice. These filament dimers appear biologically important, since mutant UmuD' proteins unable to form these structures exhibit a greatly reduced affinity for a RecA-nucleoprotein filament and are completely defective in promoting mutagenesis in vivo *(124)*. Further analysis of UmuD' should identify which of its surfaces interacts with UmuC and/or subunits of DNA pol III holoenzyme.

6.3. *UmuDC Homologs in Organisms Other than* E. coli

Compared to *E. coli*, many bacteria are much less mutagenically responsive when exposed to DNA damage. This phenomenon led to the suggestion that the *umuDC* genes were perhaps limited to *E. coli* and very closely related organisms *(143)*. However, to date at least eight functional *umuDC* homologs have been isolated and sequenced *(46,59,81,86)*. Interestingly, these homologs reside either chromosomally or on certain conjugative R-plasmids. The reason for the dual location has yet to be identified, although their presence on transmissible plasmids means that organisms lacking a chromosomal copy of the *umuDC*-like genes (such as *Haemophilus influenzae [42]*) could acquire a copy in times of stress.

The observation that some bacteria carry *umuDC* homologs and yet are poorly mutable has led to the suggestion that these genes may be retained for participation in some other unknown processes *(144)*. *Salmonella typhimurium* species provide an interesting example, since they generally carry at least two different *umuDC*-like operons (chromosomally encoded *umuDC [147,155]* and virulence plasmid pSLT-encoded *samAB [113]*), and yet are generally much less responsive to SOS-dependent mutagens than *E. coli* K12 strains. Certain *S. typhimurium* strains have, however, also been found to carry the ColIa plasmid-encoded *impCAB* genes, which enhance UV mutagenesis in these strains *(74,78)*. Nevertheless, the limited mutability of most *S. typhimurium* can be ascribed to a defect in a small region of its UmuC protein *(80)* and the inactivity of pSLT-borne *samAB (76,114,116)*.

Bona fide umu-like genes are always found in a LexA-regulated operon with the *umuD*-like gene located immediately upstream of the *umuC*-like gene *(46)*. Surprisingly, putative *umu* homologs have been characterized that are apparently not located in an operon. These include the bacteriophage P1 *humD* gene, which encodes a LexA-regulated homolog of *umuD' (90)*, and the *E. coli dinB* gene, which appears to be a LexA-regulated homolog of *umuC (117)*. As one might expect, more distantly related homologs of UmuC/DinB have been identified in other organisms such as archaea *(85)* as well as in the yeast *S. cerevisiae* and the nematode *C. elegans (162)*. One of these homologs, the *S. cerevisiae* Rev1p protein, is required for damage-induced mutagenesis, suggesting that at least some of these homologs are functionally as well as structurally conserved *(46)*.

6.4. Mutational Specificity of SOS-Dependent Mutagenesis

As discussed in Section 6.2., mutagenesis arising from DNA lesions that are noninstructional (e.g., abasic sites) or misinstructional, is an active biochemical process, usually requiring the action of RecA* and UmuDC (or plasmid-encoded homologs) in conjunction with the major replicative DNA polymerase, pol III. In the almost 20 yr since the development of the elegant *lacI* system by J. Miller and coworkers, a large number of studies have examined the mutational specificity of UV and ionizing radiation, as well as of a wide variety of SOS-dependent chemical mutagens; many of these findings have recently been reviewed *(46,97)*. This discussion will be confined primarily to recent observations regarding SOS mutagenesis and the mutagenic specificity promoted by various Umu-like proteins.

Individual members of the Umu-like family differ significantly in their ability to promote SOS mutagenesis with a particular mutagen. For example, base substitution mutagenesis at several diverse loci in *E. coli* induced by the SOS-dependent chemical mutagen aflatoxin B1 (AFB1) is efficiently promoted by MucAB, but not the chromosomally encoded UmuDC *(44)*; similar observations have been made with AFB1 in *S. typhimurium* using the *hisG46* missense reversion assay *(156)*. Moreover, in *S. typhimurium* the plasmid-borne *samAB* genes are unable to promote UV mutagenesis unless expressed from a multicopy plasmid, and chromosomally encoded UmuDC is relatively inefficient at promoting UV mutagenesis *(40,76,79,116)*, whereas *S. typhimurium* strains carrying the pKM101-encoded *mucAB* genes are quite UV-mutable *(40,79)*. Indeed, incorporation of the R-factor pKM101 was in large part responsible for the success of the Ames *S. typhimurium* histidine reversion tester strains in detecting a wide variety of SOS-dependent mutagens and carcinogens *(1)*.

UmuDC-like homologs sometimes promote different kinds of mutations in response to a given mutagenic treatment. A set of *E. coli* strains carrying six mutant *lacZ* genes on F' plasmids, each of which are Lac^- unless a specific base substitution restores the wild-type Glu-461 residue, have been used to examine the kinds of base substitutions promoted by the *umuDC*, *mucAB*, *impCAB*, and *samAB* operons following exposure to UV radiation or chemicals *(32,159)*. With UV irradiation, *mucAB*, *impCAB*, and *umuDC* all preferentially promote reversion of the AT → GC transition indicator strain, and to a lesser extent, the GC → AT indicator *(32)*. In contrast, upon exposure to methyl methanesulfonate (MMS), transition reversions (AT → GC) are observed almost exclusively with chromosomally encoded *umuDC*, whereas multicopy *umuDC*, single or multicopy *impCAB*, and multicopy *mucAB* promote a dramatic increase in GC → TA and AT → TA indicator strain reversion. In a separate study using these *lacZ* alleles, introduction of multicopy plasmids encoding *umuDC* or homologous operons all preferentially enhanced GC → TA transversions induced by various mutagens with the order of potency being *mucAB* > *umuDC* ≥ *samAB*; *mucAB* also enhanced AT → CG and AT → TA transversion mutations *(159)*.

Similar results have been observed in *S. typhimurium* strains carrying the *hisG46* missense mutation and the *hisG428* ochre mutation. Here MMS-induced reversion of the *S. typhimurium hisG46* strains carrying or lacking functional chromosomal *umuDC* genes results in the exclusive recovery of GC → AT transitions, whereas strains carrying the plasmid pKM101-encoded *mucAB* operon revert primarily (about 75%) via GC → TA transversions *(79)*. Likewise, the presence of plasmid pKM101 results in a marked increase in spontaneous *hisG46* and *hisG428* reversion in an excision defective background (Δ*uvrB*), and dramatically increases the recovery of GC → TA transversion mutations relative to plasmid-deficient counterparts *(40,77,79)*. The recovery of these mutations from spontaneous populations suggests that cryptic lesions, perhaps apurinic sites *(109)*, are readily processed in the presence of MucAB after preferential insertion of adenosine opposite the DNA lesion (the "A rule") *(46)*.

The aforementioned *S. typhimurium* base substitution alleles and two frameshift reversion loci (*hisC3076*, a +1 frameshift and *hisD3052*, a –1 frameshift) have also revealed that plasmid pKM101, but not chromosomal *umuDC*, promotes multiple base substitutions, which are usually tandem, but sometimes within 1–10 bp, and complex frameshift/base pair substitution mutations *(75,77,79,89)*. The tandem and vicinal multiple base substitutions recovered in strains carrying the *hisG46* missense and *hisG428* ochre mutations have been proposed to arise as a targeted consequence of lesion-localized error-prone DNA polymerase activity, in which a misincorporation occurs not only opposite the damaged base, but also at an undamaged adjacent base *(79)*. In contrast, complex frameshift/base substitution mutations, which can account for as many as 40% of mutations induced by UV in *hisC3076* revertants carrying plasmid pKM101 *(75)*, may arise from a concerted primer-template slippage with concomitant misincorporation during lesion bypass synthesis.

The role of *umuDC*-homologs in frameshift mutational specificity has also been examined. Mutagenic treatment of a collection of *E. coli lacZ* frameshift mutants (+G, -G, +CG, +A, –A) carrying the *umuDC*, *mucAB*, or *samAB* operons revealed that *samAB* has no effect on frameshift mutagenesis induced by several classes of mutagens *(160)*, and that there are unique differences in the ability of the *umuDC* and *mucAB* genes to enhance particular frameshift mutations. In the alternating CG target sequences of the *S. typhimurium hisD3052* allele, *samAB* also fails to affect –2 frameshifts, whereas multicopy *umuDC* or *mucAB* genes promote -CG frameshifts with some mutagens, but not others *(115)*. Interestingly, deletion of the *S. typhimurium umuDC* genes eliminates CG deletions induced by 1-nitropyrene, but not *N*-hydroxyacetylamino-fluorene (N-OH-AAF), suggesting that here too, an "Npf" pathway may be operative *(64)*.

Single-stranded vectors containing uniquely placed single lesions have also been employed to examine the effect of UmuDC, MucAB, and RumAB proteins on translesion synthesis, and on the kinds of mutations recovered. T-T dimer bypass synthesis occurred with differing efficiencies in unirradiated cells expressing wild-type UmuDC, MucAB, and RumAB *(152)*. These differences were ascribed to the relative efficiencies with which the UmuD, MucA, and RumA proteins are poststranslationally activated, since recombinant constructs expressing the mutagenically active UmuD'C, MucA'B, and RumA'B proteins all promoted high-level bypass in UV-irradiated cells. In strains carrying *umuDC*, TA → AT transversions were preferentially induced over TA → CG transitions, whereas the opposite was true of strains carrying *mucAB* or its homolog *rumAB*. Similar results were obtained in studies using two specific abasic sites, in a single-stranded vector *(87)*. Here too, *mucAB* enhanced translesion synthesis relative to *umuDC* and also to *rumAB*. Although both *umuDC* and *rumAB* promoted bypass of the lesion with similar efficiencies, incorporation of dAMP opposite the abasic site was recovered somewhat more frequently with *rumAB*.

Mechanistically, the overall translesion synthesis efficiency (and mutagenesis) observed probably reflects the sum of various biochemical events that may include UmuD-like protein cleavage rate, Umu turnover rates, Umu influences on nucleotide insertion and/or proofreading by DNA polymerase, and the intrinsic ability of the different Umu proteins to promote elongation after nucleotide insertion opposite a DNA lesion. Taken together, the aforementioned studies have demonstrated a role for the susceptibility of UmuD-like proteins to undergo proteolytic cleavage and support the notion that translesion synthesis efficiency is also lesion-dependent. Dissection of the specific interactions and reactions involved in SOS mutagenesis awaits further biochemical analyses.

7. THE SOS RESPONSE IN OTHER PROKARYOTES

Although *E. coli* has served as the paradigm for studies into the SOS response, it is becoming increasingly clear that this type of damage-inducible global network is not limited to *E. coli*, but is widespread in both Gram-negative and Gram-positive eubacteria. Such a conclusion is based upon a variety of observations suggesting that the two key regulatory components of the SOS response have been conserved throughout evolution.

7.1. Conservation and Distribution of RecA-Like Proteins in Other Organisms

The RecA protein is a pivotal player not only in homologous recombination and as a regulatory factor in the induction of the SOS response, but also as a direct participant in several SOS-dependent DNA repair responses, including translesion bypass replication, and the repair of DNA double-strand breaks (DSBs) and of daughter strand gaps produced during replication of damaged DNA templates by homologous recombination processes (reviewed in ref. *46*). Indeed, the importance of RecA is evidenced by the more than 60 highly conserved homologs that have been characterized in a wide variety of bacteria representing Gram-negative, Gram-positive, and archaea *(66,110,140)*. Among bacterial RecA homologs, amino acid sequence conservation is generally quite high, yet the ability to promote particular RecA-dependent events such as prophage λ induction, LexA cleavage, or *umuDC*-dependent UV mutagenesis, is often variable (reviewed in ref. *110*).

Interestingly, several putative eukaryotic RecA homologs, such as the Rad51p, Rad55p, Rad57p, and Dmc1p proteins from *S. cerevisiae*, have been recently identified *(46)*. Rad51p, for example, is required for the repair of DSBs and for mitotic and meiotic recombination *(57)*, and not only shows that weak structural homology to the bacterial RecA proteins, but also that the gene is conserved from yeast to humans *(46,57,145)*. Although there are certain DNA repair and cell-cycle regulatory genes that are induced by DNA damage in eukaryotes, there is no evidence to date for a formal equivalent of the global negative transcriptional regulation characteristic of the SOS regulon in these organisms.

7.2. Identification of LexA-Like Proteins in Other Organisms

In addition to the wide spread conservation of the RecA protein, other components of the SOS response also appear to be widely conserved among bacteria. For example, evidence for *lexA*-like regulation in 30 species of bacteria, representing 20 Gram-negative genera, was revealed by the introduction of a plasmid-borne *recA-lacZ* fusion; most were able to induce or repress *recA* expression in the presence or absence of DNA damage, respectively *(41)*. Perhaps the most telling evidence that an SOS regulatory network similar to that found in *E. coli* exists in other organisms is the direct cloning of *lexA* genes from a number of different organisms, such as *S. typhimurium (47,112), Erwinia carotovora*, *Pseudomonas aeruginosa*, *Pseudomonas putida (47)*, *Providencia rettgeri (137)*, and *Aeromonas hydrophila (138)*. Many of these *lexA* genes have been cloned and characterized by Barbé and colleagues, who devised an elegant genetic assay that allowed for the direct selection and isolation of LexA-like encoding DNAs *(21)*. A putative *lexA* homolog called *dinR* has been described in *Bacillus subtilis (98,134, 135,163)* and more recently, genome sequencing projects have identified *lexA*-like genes in *H. influenzae (42)*, *Mycoplasma tuberculosis*, and *Mycoplasma leprae* (Genbank X91407, U00019). Searches of genetic data bases have also revealed the existence of putative -like LexA binding sites in the genomes of *Pseudomonas cepacia* (Genbank D90120), *Pseudomonas marginalis* (Genbank D32121), *Pseudomonas fluorescens* (Genbank M96558), *Proteus mirabilis* (Genbank X65079), *Bordella pertussis* (Genbank X53457), *Acidiphilium facilis* (Genbank D16538), *Yersina pestis* (Genbank Z54145), *Serratia marcescens* (Genbank X65080), and *Vibrio cholerae* (Genbank V10162), strongly suggesting that they too may regulate protein expression through a LexA-like repressor protein.

A great deal of homology exists between LexA repressors, reflecting conservation of the N-terminal domain involved in LexA DNA binding activity, the region flanking the Ala^{84}-Gly^{85} cleavage site, and C-terminal regions responsible for dimerization and catalysis of autodigestion (Fig. 2). As discussed in Section 3.1., the chemical mechanism whereby *E. coli* LexA is cleaved at the Ala^{84}-Gly^{85} bond requires appropriately spaced Ser and Lys residues; these cleavage and catalytic sites are conserved in both the Gram-negative and Gram-positive LexA homologs, suggesting that their mechanism of cleavage is similar (Fig. 2).

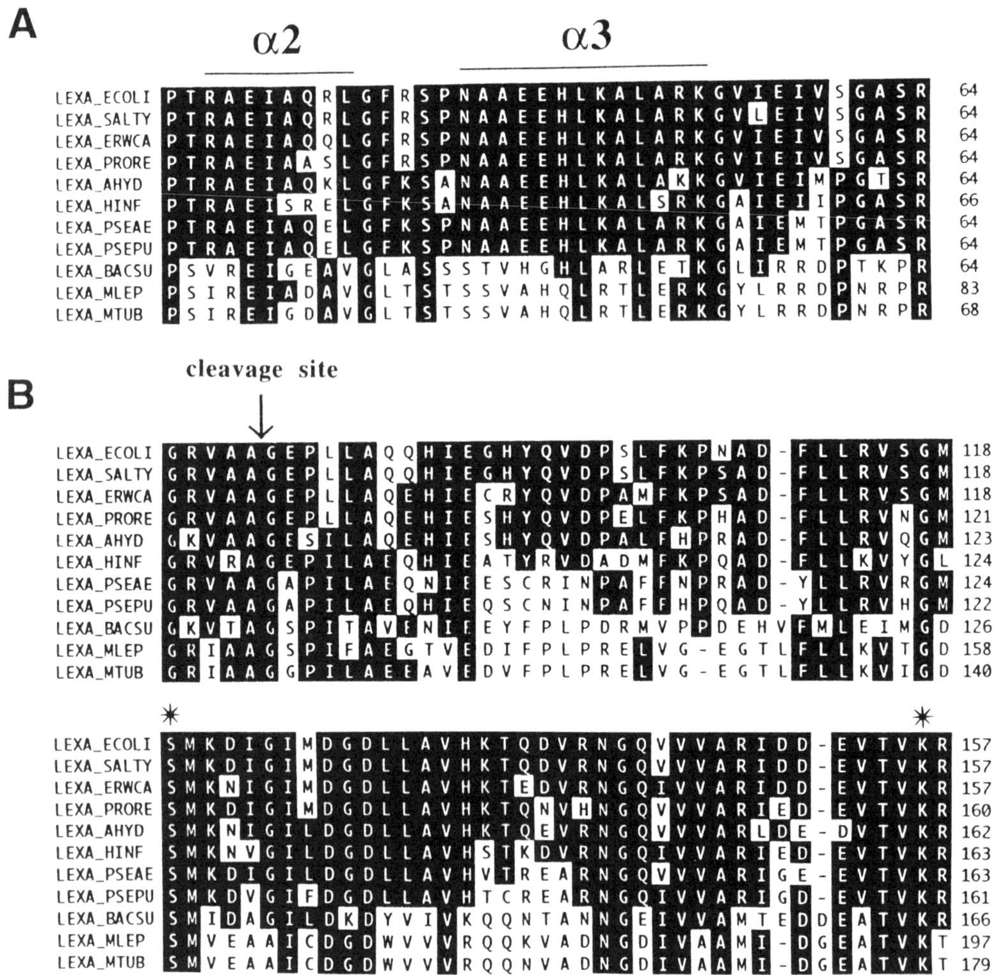

Fig. 2. Clustal multiple alignment of deduced amino acid sequence segments from the LexA family. Proteins are indicated by their Swiss-Prot database abbreviations except *A. hydrophila* (JC4042)-labeled LEXA_AHYD and *H. influenza* (HIN327529_1)-labeled LEXA_HINF. Amino acids that match a majority consensus are shaded in black. **(A)** N-terminal domain involved in DNA binding (*E. coli* residues 26–66). Regular α-helices that form a variant helix-turn-helix DNA binding motif in *E. coli* LexA are indicated as per Fogh et al. *(43)*. **(B)** Homology of LexA cleavage site and C-terminal residues that form the active site for intramolecular cleavage (*E. coli* residues 80–159 are shown). Ala-Gly cleavage site indicated by arrow. *Conserved Ser and Lys residues required for cleavage. Reproduced with permission from *J. Bacteriol.*

The *E. coli* LexA binding site is a 20-bp palindrome with the consensus sequence of (5'-TACTG-N_{10}-CAGTA-3') *(90)*. This differs from the consensus LexA binding site found in *B. subtilis*, which has been identified as 5'GAAC-N_4-GTTC3' *(24,25)*. In addition, studies on the damage inducibility of the *recA* genes from the α-group proteobacteria, *Agrobacterium tumefaciens*, *Rhizobium meliloti*, *Rhizobium phaseoli*, and *Rhodobacter sphaeroides* strongly suggest that their LexA binding site also differs dramatically from that found in *E. coli (139)*. At the present time, it is not known whether the LexA-binding site in these bacteria resembles that of *B. subtilis* or will perhaps be representative of yet another consensus sequence.

Various structural analyses have also suggested that the LexA repressor is a member of the catabolite gene activator protein (CAP)-like DNA binding domain superfamily *(43,60,73)*. Moreover, NMR data have revealed clear evidence of a helix-turn-helix (HTH) domain at *E. coli* LexA residues 28–48, a region highly conserved among all LexA homologs except those of *B. subtilis* and the two *Mycoplasma* species (Fig. 2) *(43)*. This difference probably reflects the binding site recognition divergence among the *E. coli* and *B. subtilis* LexA homologs described above.

7.3. Interaction Between the SOS Response and Other Global Stress Responses

There is accumulating evidence of interactions between the SOS response and other stress response pathways, which vary with the organism and stress to which it is exposed. In *E. coli*, phosphate starvation, as well as low temperature can transcriptionally induce certain LexA-controlled genes (e.g., *sulA*) *(33)*. In fact, a partial induction of some SOS genes may even occur as a function of growth phase on complex media *(33)*. More recently it was demonstrated that changes of intracellular pH can reversibly affect both cellular LexA levels and the rate of transcription of the *sulA* gene; acidic conditions reduce *sulA* expression, whereas alkaline conditions enhance expression *(34)*. These findings imply some interaction between the SOS response and cellular metabolism. Further evidence to support this notion involves the return to steady state after the SOS response, which requires induction of 2-keto-4-hydroxyglutarate aldolase, an enzyme involved in the resumption of respiratory metabolism *(22)*. This process requires RecA* and at least one unidentified LexA-regulated protein, but not LexA itself (Table 2).

Further examples of interactions between the regulation of cellular metabolism and the SOS response have recently emerged. In *E. coli* resting in controlled structured environments (e.g., on agar plates), the SOS response is induced in a cAMP-dependent manner *(153)*. Another example of a stationary phase-induced SOS gene in *E. coli* is the *sbmC* gene, which encodes a product that protects cells from the DNA replication inhibitor microcin B17 *(10)*. The *sbmC* SOS box is identical to that of *umuDC*, and the gene is strongly derepressed in the presence of mitomycin C and at the onset of stationary phase. These findings suggest that the SOS response is induced not only in response to DNA damage and arrest of DNA replication, but also in response to environmental growth conditions. Interestingly, the induction of the SOS response in resting bacterial populations is accompanied by an increase in mutation *(153)*, perhaps providing an adaptive advantage (increased fitness) to the bacteria during periods of starvation as first proposed by Echols *(38)*.

There are also interactions between elements of the SOS and heat-shock responses. UV induces expression of the *groEL* and *dnaK* genes, members of the cellular heat-shock regulon, in a LexA-independent manner *(84)*. In particular, the GroEL and GroES heat-shock proteins (HSPs) are required for *umuDC*-dependent SOS mutagenesis, where they appear to help fold UmuC *(30,31)*. Indeed, the Hsp60 (GroEL and GroES) and Hsp70 (DnaK, DnaJ, and GrpE) chaperone complexes have recently been shown to help fold UmuC in vitro in such a way as to promote translesion DNA synthesis *(126)*.

Table 2
LexA-Independent Damage Inducible Genes in *Escherichia coli*

Allele	Function/similarity	Refs.
dnaA	Chromosomal initiator protein	*128,129*
dnaB	DNA helicase; *lacZ* transcriptional fusion induced after DNA damage; inducible in a *recA*13 mutant	72
dnaN	β-subunit of DNA pol III; regulation at the posttranslational level; not induced in *recA*13 or lexA3 mutants	*130,154*
dnaQ	ε-subunit of DNA polymerase III	*130,131*
phr	DNA photolyase; operator possesses two SOS boxes that do not appear to bind LexA	*61,122*
hga	2-Keto-4-hydroxyglutarate aldolase; required for the recovery of respiration following DNA damage; induction requires activated RecA and at least one other LexA-regulated protein.	*22*
dinY	*lacZ* transcriptional fusion induced after DNA damage; mutant defective in Weigle reactivation of UV-irradiated bacteriophage λ; activated RecA required for induction	*126*
nrdAB	Ribonucleotide reductase	*48*
recQ	Recombinational repair	*62*
sfiC	Cell division	*29*
himA	Site-specific recombination	*107,108,157*

In *Lactococcus lactis*, the *recA* gene is involved in the regulation of cellular responses to heat and oxidative stresses *(37)*. *L. lactis recA* mutants grow poorly at elevated temperatures and are deficient in the DnaK, GroEL, and GrpE HSPs. This deficiency may be caused by elevated levels of HflB, an HSP that in *E. coli* downregulates the heat-shock response by promoting degradation of the heat shock transcription factor σ^{32}. *L. lactis recA* mutants also grow poorly in aerated cultures and exhibit significant DNA degradation. This effect is alleviated by removal of Fenton reaction substrates (i.e., H_2O_2, and Fe[II]), indicative of a hypersensitity to hydroxyl radical-induced damage in the absence of RecA.

In some Gram-positive species, a natural transformation process occurs as a function of growth phase. The development of competence, defined as the ability to take up exogenous DNA, is controlled as part of a global regulon involving the expression of over a dozen proteins (for review, *see* refs. *35* and *52*). Interestingly, in *B. subtilis*, expression of *recA* can be induced both as part of competence-specific gene activation and as a member of the SOS regulon (reviewed in ref. *170*). Similarly, induction of the SOS response and the *recA* gene during the development of competence has also been observed for *Streptococcus pneumoniae (104)*. In view of the high degree of conservation of RecA among families of bacteria, this type of dual regulation may be a universal phenomenon among many naturally competent bacteria.

8. SPECULATIONS FOR FUTURE WORK

Although the basic molecular circuitry of the SOS response was determined in *E. coli* approx 15 yr ago, much still remains to be learned. One area that will almost certainly be active will be the identification of damage-inducible genes/proteins that are not transcriptionally regulated by LexA. Analysis has already revealed that these damage-inducible proteins can be regulated at the transcriptional as well as the post-translational levels. As our understanding of SOS gene expression increases, the mechanism of exactly how these proteins are induced will presumably be elucidated. With the completion of the *E. coli* genome sequencing project, it is also likely that additional members of the LexA-regulated SOS regulon will be identified.

The biochemical activities of several damage-inducible proteins also remain uncharacterized. Given that most of the well-characterized SOS proteins are involved in a variety of DNA repair pathways, molecular and genetic analyses aimed at identifying the function of *dinB*, *dinF*, and *dinH-P* will most likely provide further insights into established, and perhaps novel, repair pathways.

LexA plays a central role in regulating the SOS response, and much has been learned recently from NMR studies about its ability to interact with its consensus binding site. Crystal structure analysis of the UmuD' protein has also provided insights into the catalytic cleavage site of these structurally related proteins. Certainly determination of the LexA crystal structure will provide a more complete picture of both RecA-regulated autodigestion catalysis and the unique HTH DNA binding domain.

It is clear that an *E. coli*-like SOS response has been conserved in many prokaryotes, although LexA recognition sites vary. Although it does not appear that a comparable global negative transcriptional control circuit exists in eukaryotes, it would be extremely interesting to determine if SOS regulation extends to archaea, many of which have evolved to grow under constant environmental stress.

The molecular mechanism of SOS mutagenesis also is still poorly understood. Although it is generally accepted that the Umu-like proteins are required for translesion DNA synthesis, how they promote bypass remains to be resolved. Studies that shed light on the interactions between the Umu-like proteins and subunits of DNA pol III holoenzyme will be a significant step towards achieving this goal.

Finally, the recent discovery that some SOS genes are induced in cells entering stationary phase, with concomitant increases in mutagenesis, again raises the possibility that the SOS response serves prokaryotic cells not only in DNA damage avoidance and tolerance, but also by playing a role in the survival, adaptation, and evolution of bacteria during the periods of extreme starvation they so frequently must endure.

ACKNOWLEDGMENTS

We would like to thank Tom Peat and Wayne Hendrickson for stimulating discussions about the structure of UmuD' and for kindly supplying the picture used in Fig. 1; Arthur S. Levine, Eugene LeClerc, and Tom Cebula for their comments on the manuscript; and Haruo Ohmori for communicating data prior to publication.

REFERENCES

1. Ames, B. N., J. McCann, and E. Yamasaki. 1975. Methods for detecting carcinogens and mutagens with the Salmonella/mammalian-microsome mutagencity test. *Mutat. Res.* **31:** 347–364.
2. Asai, T., M. Imai, and T. Kogoma. 1994. DNA damage-inducible replication of the *Escherichia coli* chromosome is initiated at separable sites within the minimal *oriC*. *J. Mol. Biol.* **235:** 1459–1469.

3. Asai, T. and T. Kogoma. 1994. The RecF pathway of homologous recombination can mediate the initiation of DNA damage-inducible replication of the *Escherichia coli* chromosome. *J. Bacteriol.* **176:** 7113,7114.
4. Asai, T. and T. Kogoma. 1994. D-loops and R-loops: alternative mechanisms for the initiation of chromosome replication in *Escherichia coli. J. Bacteriol.* **176:** 1807–1812.
5. Bagdasarian, M., A. Bailone, J. F. Angulo, P. Scholz, M. Bagdasarian, and R. Devoret. 1992. PsiB, an anti-SOS protein, is transiently expressed by the F sex factor during its transmission to an *Escherichia coli* K-12 recipient. *Mol. Microbiol.* **6:** 885–893.
6. Bagdasarian, M., A. Bailone, M. M. Bagdasarian, P. A. Manning, R. Lurz, K. N. Timmis, and R. Devoret. 1986. An inhibitor of SOS induction, specified by a plasmid locus in *Escherichia coli. Proc. Natl. Acad. Sci. USA* **83:** 5723–5726.
7. Bailone, A., A. Backman, S. Sommer, J. Celerier, M. M. Bagdasarian, M. Bagdasarian, and R. Devoret. 1988. PsiB polypeptide prevents activation of RecA protein in *Escherichia coli. Mol. Gen. Genet.* **214:** 389–395.
8. Bailone, A., S. Sommer, J. Knezevic, M. Dutreix, and R. Devoret. 1991. A RecA protein mutant deficient in its interaction with the UmuDC complex. *Biochimie* **73:** 479–484.
9. Banerjee, S. K., R. B. Christensen, C. W. Lawrence, and J. E. LeClerc. 1988. Frequency and spectrum of mutations produced by a single *cis-syn* thymine-thymine dimer in a single-stranded vector. *Proc. Natl. Acad. Sci. USA* **85:** 8141–8145.
10. Baquero, M. R., M. Bouzon, J. Varea, and F. Moreno. 1995. *sbmC*, a stationary-phase induced SOS *Escherichia coli* gene, whose product protects cells from the DNA replication inhibitor microcin B17. *Mol. Microbiol.* **18:** 301–311.
11. Battista, J. R., T. Ohta, T. Nohmi, W. Sun, and G. C. Walker. 1990. Dominant negative *umuD* mutations decreasing RecA-mediated cleavage suggest roles for intact UmuD in modulation of SOS mutagenesis. *Proc. Natl. Acad. Sci. USA* **87:** 7190–7194.
12. Bebenek, A. and I. Pietrzykowska. 1995. A new mutation in *Escherichia coli* K12, *isfA*, which is responsible for inhibition of SOS functions. *Mol. Gen. Genet.* **248:** 103–113.
13. Bebenek, A. and I. Pietrzykowska. 1996. The *isfA* mutation inibits mutator activity and processing of UmuD protein in *Escherichia coli recA*730 strains. *Mol. Gen. Genet.* **250:** 674–680.
14. Bonner, C. A., S. Hays, K. McEntee, and M. F. Goodman. 1990. DNA polymerase II is encoded by the DNA damage-inducible *dinA* gene of *Escherichia coli. Proc. Natl. Acad. Sci. USA* **87:** 7663–7667.
15. Bonner, C. A., S. K. Randall, C. Rayssiguier, M. Radman, R. Erijta, B. E. Kaplan, K. McEntee, and M. F. Goodman, 1988. Purification and characterization of an inducible *Escherichia coli* DNA polymerase capable of insertion and bypass at abasic lesions in DNA. *J. Biol. Chem.* **263:** 18,946–18,952.
16. Bridges, B. A. and R. Woodgate. 1984. Mutagenic repair in *Escherichia coli*, X. The *umuC* gene product may be required for replication past pyrimidine dimers but not for the coding error in UV mutagenesis. *Mol. Gen. Genet.* **196:** 364–366.
17. Bridges, B. A. and R. Woodgate. 1985. Mutagenic repair in *Escherichia coli*: products of the *recA* gene and of the *umuD* and *umuC* genes act at different steps in UV-induced mutagenesis. *Proc. Natl. Acad. Sci. USA* **82:** 4193–4197.
18. Brotcorne-Lannoye, A. and G. Maenhaut-Michel. 1986. Role of RecA protein in untargeted UV mutagenesis of bacteriophage λ: evidence for the requirement for the dinB gene. *Proc. Natl. Acad. Sci. USA* **83:** 3904–3908.
19. Bruck, I., R., Woodgate, K. McEntee, and M. F. Goodman. 1996. Purification of a soluble UmuD'C complex from *Escherichia coli*: cooperative binding of UmuD'C to single-stranded DNA. *J. Biol. Chem.* **271:** 10,767–10,774.
20. Burckhardt, S. E., R. Woodgate, R. H. Scheuermann, and H. Echols. 1988. UmuD mutagenesis protein of *Escherichia coli*: overproduction, purification and cleavage by RecA. *Proc. Natl. Acad. Sci. USA* **85:** 1811–1815.

21. Calero, S., X. Garriga, and J. Barbé. 1991. One-step cloning system for isolation of bacterial *lexA*-like genes. *J. Bacteriol.* **173:** 7345–7350.
22. Cayrol, C., C. Petit, B. Raynaud, J. Capdevielle, J. C. Guillemot, and M. Defais. 1995. Recovery of respiration following the SOS response of *Escherichia coli* requires RecA-mediated induction of 2-keto-4-hydroxyglutarate aldolase. *Proc. Natl. Acad. Sci. USA* **92:** 11,806–11,809.
23. Chaudhury, A. M. and G. R. Smith. 1985. Role of *Escherichia coli* RecBC enzyme in SOS induction. *Mol. Gen. Genet.* **201:** 525–528.
24. Cheo, D. L., K. W. Bayles, and R. E. Yasbin. 1991. Cloning and characterization of DNA damage-inducible promoter regions from *Bacillus subtilis*. *J. Bacteriol.* **173:** 1696–1703.
25. Cheo, D. L., K. W. Bayles, and R. E. Yasbin. 1993. Elucidation of regulatory elements that control damage induction and competence induction of the *Bacillus subtilis* SOS system. *J. Bacteriol.* **175:** 5907–5915.
26. Clark, A. J. and S. J. Sandler. 1994. Homologous genetic recombination: the pieces begin to fall into place. *Crit. Rev. Microbiol.* **20:** 125–142.
27. Craig, N. L. and J. W. Roberts. 1980. *E. coli recA* protein-directed cleavage of phage lambda repressor requires polynucleotide. *Nature* **283:** 26–30.
28. Craig, N. L. and J. W. Roberts. 1981. Function of nucleoside triphosphate and polynucleotide in *Escherichia coli recA* protein-directed cleavage of phage lambda repressor. *J. Biol. Chem.* **256:** 8039–8044.
29. D'Ari, R. and O. Huisman. 1983. Novel mechanism of cell division inhibition associated with the SOS response in *Escherichia coli*. *J. Bacteriol.* **156:** 243–250.
30. Donnelly, C. E. and G. C. Walker. 1989. *groE* mutants of *Escherichia coli* are defective in *umuDC*-dependent UV mutagenesis. *J. Bacteriol.* **171:** 6117–6125.
31. Donnelly, C. E. and G. C. Walker. 1992. Coexpression of UmuD' with UmuC suppresses the UV mutagenesis deficiency of *groE* mutants. *J. Bacteriol.* **174:** 3133–3139.
32. Doyle, N. and P. Strike. 1995. The spectra of base substitutions induced by the *impCAB*, *mucAB* and *umuDC* error-prone DNA repair operons differ following exposure to methyl methansulfonate. *Mol. Gen. Genet.* **247:** 735–741.
33. Dri, A.-M. and P. L. Moreau. 1993. Phosphate starvation and low temperature as well as ultraviolet irradiation transcriptionally induce the *Escherichia coli* LexA-controlled gene *sfiA*. *Mol. Microbiol.* **8:** 697–706.
34. Dri, A.-M. and P. L. Moreau. 1994. Control of the LexA regulon by pH: evidence for a reversible inactivation of the LexA repressor during the growth cycle of *Escherichia coli*. *Mol. Microbiol.* **12:** 621–629.
35. Dubnau, D. 1991. Genetic competence in *Bacillus subtilis*. *Microbiol. Rev.* **55:** 395–424.
36. Dutreix, M., A. Backman, J. Celerier, M. M. Bagdasarian, S. Sommer, A. Bailone, R. Devoret, and M. Bagdasarian. 1988. Identification of *psiB* genes of plasmids F and R6-5. Molecular basis for *psiB* enhanced expression in plasmid R6-5. *Nucleic Acids Res.* **16:** 10,669–10,679.
37. Duwat, P., S. D. Ehrlich, and A. Gruss. 1995. The *recA* gene of *Lactococcus lactis*: characterization and involvement in oxidative and thermal stress. *Mol. Microbiol.* **17:** 1121–1131.
39. Echols, H. and M. F. Goodman. 1990. Mutation induced by DNA damage: a many protein affair. *Mutat. Res.* **236:** 301–311.
40. Eisenstadt, E., J. K. Miller, L.-S. Kahng, and W. M. Barnes. 1989. Influence of *uvrB* and pKM101 on the spectrum of spontaneous, UV- and γ-ray-induced base substitutions that revert hisG46 in Salmonella typhimurium. *Mutat. Res.* **210:** 113–125.
41. Fernéndez de Henestrosa, A. R., S. Calero, and J. Barbé. 1991. Expression of the *recA* gene of *Escherichia coli* in several species of gram-negative bacteria. *Mol. Gen. Genet.* **226:** 503–506.
42. Fleischmann, R. D., M. D. Adams, O. White, R. A. Clayton, E. F. Kirkness, A. R. Kerlavage, C. J. Bult, J. F. Tomb, B. A. Dougherty, J. M. Merrick, K. McKenney, G.

Sutton, W. FitzHugh, C. A. Fields, J. D. Gocayne, J. D. Scott, R. Shirley, L.-I. Liu, A. Glodek, J. M. Kelley, J. F. Weidman, C. A. Phillips, T. Spriggs, E. Hedblom, M. D. Cotton, T. R. Utterback, M. C. Hanna, D. T. Nguyen, D. M. Saudek, R. C. Brandon, L. D. Fine, J. L. Fritchman, J. L. Fuhrmann, N. S. M. Geoghagen, C. L. Gnehm, L. A. McDonald, K. V. Small, C. M. Fraser, H. O. Smith, and J. C. Venter. 1995. Whole-genome random sequencing and assembly of *Haemophilus influenzae* Rd. *Science* **269:** 496–512.

43. Fogh, R. H., G. Ottleben, H. Rüterjans, M. Schnarr, R. Boelens, and R. Kaptein. 1994. Solution structure of the LexA repressor DNA binding domain determined by ^{1}H NMR spectroscopy. *EMBO J.* **13:** 3936–3944.
44. Foster, P. L., J. D. Groopman, and E. Eisenstadt. 1988. Induction of base substitution mutations by aflatoxin B1 is *mucAB* dependent in *Escherichia coli*. *J. Bacteriol.* **170:** 3415–3420.
45. Frank, E. G., J. Hauser, A. S. Levine, and R. Woodgate. 1993. Targeting of the UmuD, UmuD' and MucA' mutagenesis proteins to DNA by RecA protein. *Proc. Natl. Acad. Sci. USA* **90:** 8169–8173.
46. Friedberg, E. C., G. C. Walker, and W. Siede. 1995. *DNA Repair and Mutagenesis*. American Society of Microbiology, Washington, DC.
47. Garriga, X., S. Calero, J. and Barbé. 1992. Nucleotide sequence analysis and comparison of the *lexA* genes from *Salmonella typhimurium*, *Erwinia carotovora*, *Pseudomonas aeruginosa* and *Pseudomonas putida*. *Mol. Gen. Genet.* **236:** 125–134.
48. Gibert, I., S. Calero, and J. Barbé. 1990. Measurement of in vivo expression of *nrdA* and *nrdB* genes of *Escherichia coli* by using *lacZ* gene fusions. *Mol. Gen. Genet.* **220:** 400–408.
49. Gigliani, F., C. Ciotta, M. F. Del Grosso, and P. A. Battaglia. 1993. pR plasmid replication provides evidence that single-stranded DNA induces the SOS system in vivo. *Mol. Gen. Genet.* **238:** 333–338.
50. Gimble, F. S. and R. T. Sauer. 1986. λ repressor inactivation: properties of purified *ind-* proteins in the autodigestion and RecA-mediated cleavage reactions. *J. Mol. Biol.* **192:** 39–47.
51. Gimble, F. S. and R. T. Sauer. 1989. λ repressor mutants that are better substrates for RecA-mediated cleavage. *J. Mol. Biol.* **206:** 29–39.
52. Grossman, A. D. 1995. Genetic networks controlling the initiation of sporulation and the development of genetic competence in *Bacillus subtilis*. *Ann. Rev. Genet.* **29:** 477–508.
53. Hauser, J., A. S. Levine, D. G. Ennis, K. M. Chumakov, and R. Woodgate. 1992. The enhanced mutagenic potential of the MucAB proteins correlates with the highly efficient processing of the MucA protein. *J. Bacteriol.* **174:** 6844–6851.
54. Hegde, S., S. J. Sandler, A. J. Clark, and M. V. V. S. Madiraju. 1995. *recO* and *recR* mutations delay induction of the SOS response in *Escherichia coli*. *Mol. Gen. Genet.* **246:** 254–258.
55. Heide, L., M. Melzer, M. Siebert, A. Bechthold, J. Schroder, and K. Severin. 1993. Clarification of the *Escherichia coli* genetic map in the 92-minute region containing the *ubiCA* operon and the *plsB*, *dgk*, *lexA*, and *dinF* genes. *J. Bacteriol.* **175:** 5728,5729.
56. Heuser, J. and J. Griffith. 1989. Visualization of RecA protein and its complexes with DNA by quick-freeze/deep-etch electron microscopy. *J. Mol. Biol.* **210:** 473–484.
57. Heyer, W. D. 1994. The search for the right partner: homologous pairing and DNA strand exchange proteins in eukaryotes. *Experientia* **50:** 223–233.
58. Higashitani, N., A. Higashitani, and K. Horiuchi. 1995. SOS induction in *Escherichia coli* by single-stranded DNA of mutant filamentous phage: monitoring by cleavage of LexA repressor. *J. Bacteriol.* **177:** 3610–3612.
59. Ho, C., O. I. Kulaeva, A. S. Levine, and R. Woodgate. 1993. A rapid method for cloning mutagenic DNA repair genes: isolation of *umu*-complementing genes from multidrug resistance plasmids R391, R446b, and R471a. *J. Bacteriol.* **175:** 5411–5419.
60. Holm, L., C. Sander, H. Rüterjans, M. Schnarr, R. Fogh, R. Boelens, and R. Kaptein. 1994. LexA repressor and iron uptake regulator from *Escherichia coli*: new members of the CAP-like DNA binding domain superfamily. *Protein Eng.* **7:** 1449–1453.

61. Ihara, M., K. Yamamoto, and T. Ohnishi. 1987. Induction of phr gene expression by irradiation of ultraviolet light in *Escherichia coli*. *Mol. Gen. Genet.* **209:** 200–202.
62. Irino, N., K. Nakayama, and H. Nakayama. 1986. The *recQ* gene of *Escherichia coli* K12: primary structure and evidence for SOS regulation. *Mol. Gen. Genet.* **205:** 298–304.
63. Iwasaki, H., Y. Ishino, H. Toh, A. Nakata, and H. Shinagawa. 1991. *Escherichia coli* DNA polymerase II is homologous to alpha-like DNA polymerases. *Mol. Gen. Genet.* **226:** 24–33.
64. Janel-Bintz, R., G. Maenhaut-Michel, and R. P. P. Fuchs. 1994. MucAB but not UmuDC proteins enhance-2 frameshift mutagenesis induced by *N*-2-acetylaminofluorene at alternating GC sequences. *Mol. Gen. Genet.* **245:** 279–285.
65. Kaasch, M., J. Kaasch, and A. Quiñones. 1989. Expression of the *dnaN* and *dnaQ* genes of *Escherichia coli* is inducible by mitomycin C. *Mol. Gen. Genet.* **219:** 187–192.
66. Karlin, S., G. M. Weinstock, and V. Brendel. 1995. Bacterial classifications derived from *recA* protein sequence comparisons. *J. Bacteriol.* **177:** 6881–6893.
67. Kenyon, C. J. and G. C. Walker. 1980. DNA-damaging agents stimulate gene expression at specific loci in *Escherichia coli*. *Proc. Natl. Acad. Sci. USA* **77:** 2819–2823.
68. Kenyon, C. J. and G. C. Walker. 1981. Expression of the *E. coli uvrA* gene is inducible. *Nature* **289:** 808–810.
69. Khidhir, M. A., S. Casaregola, and I. B. Holland. 1985. Mechanism of transient inhibition of DNA synthesis in ultraviolet-irradiated: inhibition is independent of RecA whilst recovery requires RecA protein itself and an additional, inducible SOS function. *Mol. Gen. Genet.* **199:** 133–140.
70. Kim, B. and J. W. Little. 1992. Dimerization of a specific DNA-binding protein on the DNA. *Science* **255:** 203–206.
71. Kim, B. and J. W. Little. 1993. LexA and l CI repressors as enzymes: specific cleavage in an intermolecular reaction. *Cell* **73:** 1165–1173.
72. Kleinsteuber, S. and A. Quiñones. 1995. Expression of the *dnaB* gene of *Escherichia coli* is inducible by replication-blocking DNA damage in a *recA*-independent manner. *Mol. Gen. Genet.* **248:** 695–702.
73. Knegtel, R. M. A., R. H. Fogh, G. Ottleben, H. Rüterjans, P. Dumoulin, M. Schnarr, R. Boelens, and R. Kaptein. 1995. A model for the LexA repressor DNA complex. *Proteins* **21:** 226–236.
74. Koch, W. H., T. A. Cebula, and E. Eisenstadt., unpublished results.
75. Koch, W.H. and T. A. Cebula, unpublished results.
76. Koch, W. H., T. A. Cebula, P. L. Foster, and E. Eisenstadt. 1992. UV mutagenesis in Salmonella typhimurium is *umuDC* dependent despite the presence of *samAB*. *J. Bacteriol.* **174:** 2809–2815.
77. Koch, W. H., E. Henrikson, and T. A. Cebula. 1996. Molecular analyses of *Salmonella hisG428* ochre revertants for rapid characterization of mutational specificity. *Mutagenesis* **11:** 341–348.
78. Koch, W. H., E. Henrikson, E. Eisenstadt, and T. A. Cebula. 1995. *Salmonella typhimurium* LT7 and LT2 strains carrying the *imp* operon on ColIa. *J. Bacteriol.* **177:** 1903–1905.
79. Koch, W. H., E. Henrikson, E. Kupchella, and T. A. Cebula. 1994. *Salmonella typhimurium strain* TA100 differentiates several classes of carcinogens and mutagens by base substitution specificity. *Carcinogenesis* **15:** 79–88.
80. Koch, W. H., G. Kopsidas, B. Meffle, A. S. Levine, and R. Woodgate. 1996. Analysis of chimeric UmuC proteins: identification of regions in *Salmonella typhimurium* UmuC important for mutagenic activity. *Mol. Gen. Genet.* **251:** 121–129.
81. Koch, W. H. and R. Woodgate. Identification of new *umuC* homologs by degenerate primer PCR amplification, unpublished results.
82. Koonin, E. V. 1993. *Escherichia coli dinG* gene encodes a putative DNA helicase related to a group of eukaryotic helicases including Rad3 protein. *Nucleic Acids Res.* **21:** 1497.

83. Kornberg, A. and T. A. Baker. 1992. *DNA replication.* W. H. Freeman & Co, New York.
84. Krueger, J. H., S. J. Elledge, and G. C. Walker. 1983. Isolation and characterization of Tn*5* insertion mutations in the *lexA* gene of *Escherichia coli. J. Bacteriol.* **153:** 1368–1379.
85. Kulaeva, O. I., E. V. Koonin, J. P. McDonald, S. K. Randall, N. Rabinovich, J. F. Connaughton, A. S. Levine, and R. Woodgate. Identification of a DinB/UmuC homolog in the archeon *Sulfolobus solfataricus. Mutat. Res.* **357:** 245–253.
86. Kulaeva, O. I., J. C. Wootton, A. S. Levine, and R. Woodgate. 1995. Characterization of the *umu*-complementing operon from R391. *J. Bacteriol.* **177:** 2737–2743.
87. Lawrence, C. W., A. Borden, and R. Woodgate. Analysis of the mutagenic properties of the UmuDC, MucAB and RumAB proteins, using a site specific abasic lesion. *Mol. Gen. Genet.***251:** 493–498.
88. Lesca, C., C. Petit, and M. Defais. 1991. UV induction of LexA independent proteins which could be involved in SOS repair. *Biochimie* **73:** 407–409.
89. Levine, J. G., R. M. Schaaper, and D. M. DeMarini. 1994. Complex frameshift mutations mediated by plasmid pKM101: mutational mechanisms deduced from 4-aminobiphenyl-induced mutation spectra in Salmonella. *Genetics* **136:** 731–746.
90. Lewis, L. K., G. R. Harlow, L. A. Gregg-Jolly, and D. W. Mount. 1994. Identification of high affinity binding sites for LexA which define new DNA damage-inducible genes in *Escherichia coli. J. Mol. Biol.* **241:** 507–523.
91. Lewis, L. K., M. E. Jenkins, and D. W. Mount. 1992. Isolation of DNA damage-inducible promoters in *Escherichia coli*: regulation of *polB (dinA)*, *dinG*, and *dinH* by LexA repressor. *J. Bacteriol.* **174:** 3377–3385.
92. Lewis, L. K. and D. W. Mount. 1992. Interaction of LexA repressor with the asymmetric *dinG* operator and complete nucleotide sequence of the gene. *J. Bacteriol.* **174:** 5110–5116.
93. Little, J. W. 1984. Autodigestion of LexA and phage repressors. *Proc. Natl. Acad. Sci. USA* **81:** 1375–1379.
94. Little, J. W. 1993. LexA cleavage and other self-processing reactions. *J. Bacteriol.* **175:** 4943–4950.
95. Little, J. W., B. Kim, K. L. Roland, M. H. Smith, L.-L. Lin, and S. N. Slilaty. 1994. Cleavage of LexA repressor. *Methods Enzymol.* **244:** 266–284.
96. Little, J. W. and D. W. Mount. 1982. The SOS regulatory system of *Escherichia coli. Cell* **29:** 11–22.
97. Livneh, Z., O. Cohen-Fix, R. Skaliter, and T. Elizur. 1993. Replication of damaged DNA and the molecular mechanism of ultraviolet light mutagenesis. *Crit. Rev. Biochem. Mol. Biol.* **28:** 465–513.
98. Lovett, C. M. Jr., K. C. Cho, and T. M. O'Gara. 1993. Purification of an SOS repressor from *Bacillus subtilis. J. Bacteriol.* **175:** 6842–6849.
99. Lovett, C. M. Jr., T. M. O'Gara, and J. N. Woodruff. 1994. Analysis of the SOS inducing signal in *Bacillus subtilis* using *Escherichia coli* LexA as a probe. *J. Bacteriol.* **176:** 4914–4923.
100. Lundegaard, C. and K. F. Jensen. 1994. The DNA damage-inducible *dinD* gene of *Escherichia coli* is equivalent to *orfY* upstream of *pyrE. J. Bacteriol.* **176:** 3383–3385.
101. Maarten van Dijl, J., A. de Jong, G. Venema, and S. Bron. 1995. Identification of the potential active site of the signal peptidase SipS of *Bacillus subtilis*. Structural and functional similarities with LexA-like proteases. *J. Biol. Chem.* **270:** 3611–3618.
102. Madiraju, M. V. V. S., A. Templin, and A. J. Clark. 1988. Properties of a *recA*-encoded protein reveal a possible role for *Escherichia coli recF*-encoded protein in genetic recombination. *Proc. Natl. Acad. Sci. USA* **85:** 6592–6596.
103. Marsh, L. and G. C. Walker. 1985. Cold sensitivity induced by overproduction of *umuDC* in *Escherichia coli. J. Bacteriol.* **162:** 155–161.

104. Martin, B., P. Garcia, M. P. Castanie, and J. P. Claverys. 1995. The *recA* gene of *Streptococcus pneumoniae* is part of a competence-induced operon and controls lysogenic induction. *Mol. Microbiol.* **15:** 367–379.
105. Masai, H., T. Asai, Y. Kubota, K. Arai, and T. Kogoma. 1994. *Escherichia coli* PriA protein is essential for inducible and constitutive stable DNA replication. *EMBO J.* **13:** 5338–5345.
106. Matic, I., C. Rayssiguier, and M. Radman. 1995. Interspecies gene exchange in bacteria: the role of SOS and mismatch repair systems in evolution of species. *Cell* **80:** 507–515.
107. Mechulam, Y., G. Fayat, and S. Blanquet. 1985. Sequence of the *Escherichia coli pheST* operon and identification of the *himA* gene. *J. Bacteriol.* **163:** 787–791.
108. Miller, H. I., M. Kirk, and H. Echols. 1981. SOS induction and autoregulation of the himA gene for site-specific recombination in *Escherichia coli. Proc. Natl. Acad. Sci. USA* **78:** 6754–6758.
109. Miller, J. H. and K. B. Low. 1984. Specificity of mutagenesis resulting from the induction of SOS system in the absence of mutagenic treatment. *Cell* **37:** 675–682.
110. Miller, R. V. and T. A. Kokjohn. 1990. General microbiology of *recA*: environmental and evolutionary significance. *Ann. Rev. Microbiol.* **44:** 365–394.
111. Murli, S. and G. C. Walker. 1993. SOS mutagenesis. *Curr. Opinion Genet. Dev.* **3:** 719–725.
112. Mustard, J. A., A. T. Thliveris, and D. W. Mount. 1992. Sequence of the *Salmonella typhimurium* LT2 *lexA* gene and its regulatory region. *Nucleic Acids Res.* **20:** 1813.
113. Nohmi, T., A. Hakura, Y. Nakai, M. Watanabe, S. Y. Murayama, and T. Sofuni. 1991. *Salmonella typhimurium* has two homologous but different *umuDC* operons: cloning of a new *umuDC*-like operon *(samAB)* present in a 60-megadalton cryptic plasmid of *S. typhimurium*. *J. Bacteriol.* **173:** 1051–1063.
114. Nohmi, T., M. Yamada, K. Matsui, M. Watanabe, and T. Sofuni. 1995. Specific disruption of *samAB* genes in a 60-megadalton cryptic plasmid of *Salmonella typhimurium*. *Mutat. Res.* **329:** 1–9.
115. Nohmi, T., M. Yamada, M. Matsui, M. Watanabe, and T. Sofuni. 1995. Involvement of *umuDCST* genes in nitropyrene-induced-CG frameshift mutagenesis at repetitive CG sequence in the *hisD3052* allele of *Salmonella typhimurium*. *Mol. Gen. Genet.* **247:** 7–16.
116. Nohmi, T., M. Yamada, M. Watanabe, S. Y. Murayama, and T. Sofuni. 1992. Roles of *Salmonella typhimurium umuDC* and *samAB* in UV mutagenesis and UV sensitivity. *J. Bacteriol.* **174:** 6948–6955.
117. Ohmori, H., E. Hatada, Y. Qiao, M. Tsuji, and R. Fukuda. 1995. *dinP*, a new gene in *Escherichia coli*, whose product shows similarities to UmuC and its homologues. *Mutat. Res.* **347:** 1–7.
118. Ohmori, H., M. Saito, T. Yasuda, T. Nagata, T. Fujii, M. Wachi, and K. Nagai. 1995. The *pcsA* gene is identical to *dinD* in *Escherichia coli. J. Bacteriol.* **177:** 156–165.
119. Ossanna, N. and D. W. Mount. 1989. Mutations in *uvrD* induce the SOS response in *Escherichia coli*. *J. Bacteriol.* **171:** 303–307.
120. Palejwala, V. A., G. A. Pandya, O. S. Bhanot, J. J. Solomon, H. S. Murphy, P. M. Dunman, and M. Z. Humayun. 1994. UVM, an ultraviolet-inducible RecA-independent mutagenic phenomenon in *Escherichia coli*. *J. Biol. Chem.* **269:** 27,433–27,440.
121. Palejwala, V. A., G. E. Wang, H. S. Murphy, and M. Z. Humayun. 1995. Functional *recA*, *lexA*, *umuD*, *umuC*, *polA*, and *polB* genes are not required for the *Escherichia coli* UVM response. *J. Bacteriol.* **177:** 6041–6048.
122. Payne, N. S. and A. Sancar. 1989. The LexA protein does not bind specifically to the two SOS box-like sequences immediately 5' to the *phr* gene. *Mutat. Res.* **218:** 207–210.
123. Peat, T. S., E. G. Frank, J. P. McDonald, A. S. Levine, R. Woodgate, and W. A. Hendrickson. 1996. Structure of the UmuD' protein and its regulation in response to DNA damage. *Nature* **380:** 727–730.

124. Peat, T. S., E. G. Frank, J. P. McDonald, A. S. Levine, R. Woodgate, and W. A. Hendrickson. The UmuD' protein filament and its role in damage induced mutagenesis, unpublished results.
125. Peat, T. S., E. G. Frank, R. Woodgate, and W. A. Hendrickson. 1996. Production and crystallization of a selenomethionyl variant of UmuD', an *Escherichia coli* SOS response protein. *Proteins* **25:** 506–509.
126. Petit, C., C. Cayrol, C. Lesca, P. Kaiser, C. Thompson, and M. Defais. 1993. Characterization of *dinY*, a new *Escherichia coli* DNA repair gene whose products are damage inducible even in a *lexA*(Def) background. *J. Bacteriol.* **175:** 642–646.
127. Petit, M. A., W. Bedale, J. Osipiuk, C. Lu, M. Rajagopalan, P. McInerney, M. F. Goodman, and H. Echols. 1994. Sequential folding of UmuC by the Hsp70 and Hsp60 chaperone complexes of *Escherichia coli*. *J. Biol. Chem.* **269:** 23,824–23,849.
128. Quiñones, A., W. R. Jueterbock, and W. Messer. 1991. DNA lesions that block DNA replication are responsible for the *dnaA* induction caused by DNA damage. *Mol. Gen. Genet.* **231:** 81–87.
129. Quiñnones, A., W. R. Juterbock, and W. Messer. 1991. Expression of the *dnaA* gene of *Escherichia coli* is inducible by DNA damage. *Mol. Gen. Genet.* **227:** 9–16.
130. Quiñones, A., J. Kaasch, M. Kaasch, and W. Messer. 1989. Induction of *dnaN* and *dnaQ* gene expression in *Escherichia coli* by alkylation damage to DNA. *EMBO J.* **8:** 587–593.
131. Quiñones, A., R. Piechocki, and W. Messer. 1988. Expression of the *Escherichia coli dnaQ (mutD)* gene is inducible. *Mol. Gen. Genet.* **211:** 106–112.
132. Radman, M. 1974. Phenomenology of an inducible mutagenic DNA repair pathway in *Escherichia coli*: SOS repair hypothesis, in *Molecular and Environmental Aspects of Mutagensis* (L. Prakash, F. Sherman, M. Miller, C. W. Lawrence, and H. W. Tabor, eds.), Charles C. Thomas, Springfield, Ill, pp. 128–142.
133. Rajagopalan, M., C. Lu, R. Woodgate, M. O'Donnell, M. F. Goodman, and H. Echols. 1992. Activity of the purified mutagenesis proteins UmuC, UmuD' and RecA in replicative bypass of an abasic DNA lesion by DNA polymerase III. *Proc. Natl. Acad. Sci. USA* **89:** 10,777–10,781.
134. Raymond-Denise, A. and N. Guillen. 1991. Identification of *dinR*, a DNA damage-inducible regulator gene of *Bacillus subtilis*. *J. Bacteriol.* **173:** 7084–7091.
135. Raymond-Denise, A. and N. Guillen. 1992. Expression of the *Bacillus subtilis dinR* and *recA* genes after DNA damage and during competence. *J. Bacteriol.* **174:** 3171–3176.
136. Rehrauer, W. M. and S. C. Kowalczykowski. 1996. The DNA binding site(s) of the *Escherichia coli* RecA protein. *J. Biol. Chem.* **271:** 11,996–12,002.
137. Riera, J. and J. Barbé. 1993. Sequence of the *Providencia rettgeri lexA* gene and its control region. *Nucleic Acids Res.* **21:** 2256.
138. Riera, J. and J. Barbé. 1995. Cloning, sequence and regulation of expression of the *lexA* gene of *Aeromonas hydrophila*. *Gene* **154:** 71–75.
139. Riera, J., A. R. Fernéndez de Henestrosa, X. Garriga, A. Tapias, and J. Barbé. 1994. Interspecies regulation of the *recA* gene of gram-negative bacteria lacking an-like SOS operator. *Mol. Gen. Genet.* **245:** 523–527.
140. Roca, A. I. and M. M. Cox. 1990. The RecA protein: structure and function. *Crit. Rev. Biochem. Mol. Biol.* **25:** 415–456.
141. Rupp, W. D. and P. Howard-Flanders. 1968. Discontinuities in the DNA synthesized in an excision-defective strain of *Escherichia coli* following ultraviolet radiation. *J. Mol. Biol.* **31:** 291–304.
142. Sassanfar, M. and J. W. Roberts. 1990. *Nature* of the SOS-Inducing signal in *Escherichia coli*: The involvement of DNA replication. *J. Mol. Biol.* **212:** 79–96.
143. Sedgwick, S. G. and P. Goodwin. 1985. Differences in mutagenic and recombinational repair in enterobacteria. *Proc. Natl. Acad. Sci. USA* **82:** 4172–4176.

144. Sedgwick, S. G., C. Ho, and R. Woodgate. 1991. Mutagenic DNA repair in Enterobacteria. *J. Bacteriol.* **173:** 5604–5611.
145. Shinohara, A., H. Ogawa, and T. Ogawa. 1992. Rad51 protein involved in repair and recombination in S. cerevisiae is a RecA-like protein. *Cell* **69:** 457–470.
146. Slilaty, S. N., J. A. Rupley, and J. W. Little. 1986. Intramolecular cleavage of LexA and phage λ repressors. Dependence of kinetics on repressor concentration, pH, temperature and solvent. *Biochemistry* **25:** 6866–6875.
147. Smith, C. M., W. H. Koch, S. B. Franklin, P. L. Foster, T. A. Cebula, and E. Eisenstadt. 1990. Sequence analysis of the *Salmonella typhimurium* LT2 *umuDC* operon. *J. Bacteriol.* **172:** 4964–4978.
148. Sommer, S., A. Bailone, and R. Devoret. 1993. The appearance of the UmuD'C protein complex in *Escherichia coli* switches repair from homologous recombination to SOS mutagenesis. *Mol. Microbiol.* **10:** 963–971.
149. Sommer, S., J. Knezevic, A. Bailone, and R. Devoret. 1993. Induction of only one SOS operon, *umuDC*, is required for SOS mutagenesis in *Escherichia coli. Mol. Gen. Genet.* **239:** 137–144.
150. Strynadka, N. C., S. E. Jensen, K. Johns, H. Blanchard, M. Page, A. Matagne, J. M. Frere, and M. N. James. 1994. Structural and kinetic characterization of a beta-lactamase-inhibitor protein. *Nature* **368:** 657–660.
151. Sweasy, J. B., E. M. Witkin, N. Sinha, and V. Roegner-Maniscalco. 1990. RecA protein of *Escherichia coli* has a third essential role in SOS mutator activity. *J. Bacteriol.* **172:** 3030–3036.
152. Szekeres, E. S. Jr., R. Woodgate, and C. W. Lawrence. 1996. Substitution of *mucAB* or *rumAB* for *umuDC* alters the relative frequencies of the two classes of mutations induced by a site-specific T-T cyclobutane dimer and the efficiency of translesion DNA synthesis. *J. Bacteriol.* **178:** 2559–2563.
153. Taddei, F., I. Matic, and M. Radman. 1995. cAMP-dependent SOS induction and mutagenesis in resting bacterial populations. *Proc. Natl. Acad. Sci. USA* **92:** 11,736–11,740.
154. Tadmor, Y., M. Bergstein, R. Skaliter, H. Shwartz, and Z. Livneh. 1994. β subunit of DNA polymerase III holoenzyme is induced upon ultraviolet irradiation or nalidixic acid treatment of *Escherichia coli. Mutat. Res.* **308:** 53–64.
155. Thomas, S. M., H. M. Crowne, S. C. Pidsley, and S. G. Sedgwick. 1990. Structural characterization of the *Salmonella typhimurium* LT2 *umu* operon. *J. Bacteriol.* **172:** 4979–4987.
156. Urios, A., G. Herrera, V. Aleixandre, S. Sommer, and M. Blanco. 1994. Mutability of *Salmonella* tester strains TA1538 *(his D3052)* and TA1535 *(hisG46)* containing the UmuD' and UmuC prokins of *Escherichia coli. Environ. Mol. Mutagen.* **23:** 281–285.
157. Walker, G. C. 1984. Mutagenesis and inducible responses to deoxyribonucleic acid damage in *Escherichia coli. Microbiol. Rev.* **48:** 60–93.
158. Walker, G. C. 1995. SOS-regulated proteins in translesion DNA synthesis and mutagenesis. *Trends Biochem. Sci.* **20:** 416–420.
159. Watanabe, M., T. Nohmi, and T. Ohta. 1994. Effects of the *umuDC*, *mucAB*, and *samAB* operons on the mutational specificity of chemical mutagenesis in *Escherichia coli*: II. Base substitution mutagenesis. *Mutat. Res.* **314:** 39–49.
160. Watanabe, M., T. Nohmi, and T. Ohta. 1994. Effects of the *umuDC*, *mucAB*, and *samAB* operons on the mutational specificity of chemical mutagenesis in *Escherichia coli*: I. Frameshift mutagenesis. *Mutat. Res.* **314:** 27–37.
161. Whitby, M. C. and R. G. Lloyd. 1995. Altered SOS induction associated with mutations in *recF*, *recO* and *recR. Mol. Gen. Genet.* **246:** 174–179.
162. Wilson, R., R. Ainscough, K. Anderson, C. Baynes, M. Berks, J. Bonfield, J. Burton, M. Connell, T. Copsey, J. Cooper, A. Coulson, M. Craxton, S. Dear, Z. Du, R. Durbin, A.

Favello, A. Fraser, L. Fulton, A. Gardner, P. Green, T. Hawkins, L. Hillier, M. Jier, L. Johnston, M. Jones, J. Kershaw, J. Kirsten, N. Laisster, P. Latreille, J. Lightning, C. Lloyd, B. Mortimore, M. O'Callaghan, J. Parsons, C. Percy, L. Rifken, A. Roopra, D. Saunders, R. Shownkeen, M. Sims, N. Smaldon, A. Smith, M. Smith, E. Sonnhammer, R. Staden, J. Sulston, J. Thierry-Mieg, K. Thomas, M. Vaudin, K. Vaughan, R. Waterston, A. Watson, L. Weinstock, J. Wilkinson-Sproat, and P. Wohldman. 1994. 2.2 Mb of contiguous nucleotide sequence from chromosome III of *C. elegans*. *Nature* **368:** 32–38.

163. Winterling, K. W., A. S. Levine, R. E. Yasbin, and R. Woodgate. 1997. Characterization of DinR, the *Bacillus subtilis* SOS Repressor. *J. Bacteriol.*, in press.
164. Witkin, E. M., V. Roegner-Maniscalco, J. B. Sweasy, and J. O. McCall. 1987. Recovery from ultraviolet light-inhibition of DNA synthesis requires *umuDC* gene products in *recA718* mutant strains but not in $recA^+$ strains of *Escherichia coli*. *Proc. Natl. Acad. Sci. USA* **84:** 6804–6809.
165. Woodgate, R., B. A. Bridges, and C. Kelly. 1989. Non-mutability by ultraviolet light in *uvrD recB* derivatives of *Escherichia coli* WP2 *uvrA* is due to inhibition of RecA protein activation. *Mutat. Res.* 22**6:** 141–144.
166. Woodgate, R. and D. G. Ennis. 1991. Levels of chromosomally encoded Umu proteins and requirements for in vivo UmuD cleavage. *Mol. Gen. Genet.* **229:** 10–16.
167. Woodgate, R. and A. S. Levine. 1996. Damage inducible mutagenesis: recent insights into the activities of the Umu family of mutagenesis proteins, in *Cancer Surveys: Genetic Instability in Cancer*, vol 28. (T. Lindahl, ed.), Cold Spring Harbor Laboratory, Cold Spring Harbor, NY, pp. 117–140.
168. Woodgate, R., M. Rajagopalan, C. Lu, and H. Echols. 1989. UmuC mutagenesis protein of *Escherichia coli*: purification and interaction with UmuD and UmuD'. *Proc. Natl. Acad. Sci. USA* **86:** 7301–7305.
169. Woodgate, R. and S. G. Sedgwick. 1992. Mutagenesis induced by bacterial UmuDC proteins and their plasmid homologues. *Mol. Microbiol.* **6:** 2213–2218.
170. Yasbin, R. E., D. L. Cheo, and K. W. Bayles. 1992. Inducible DNA repair and differentiation in *Bacillus subtilis*: interactions between global regulons. *Mol. Microbiol.* **6:** 1263–1270.
171. Yasuda, T., T. Nagata, and H. Ohmori. 1996. Multicopy suppressors of the cold-sensitive phenotype of the *pcsA68 (dinD68)* mutation in *Escherichia coli*. *J. Bacteriol.* **178:** 3854–3859.
172. Yu, X. and E. H. Egelman. 1993. The LexA repressor binds within the deep helical groove of the activated RecA filament. *J. Mol. Biol.* **231:** 29–40.

8

DNA Double-Strand Break Repair and Recombination in *Escherichia coli*

Gerald R. Smith

1. INTRODUCTION

If left unrepaired, double-strand breaks (DSBs) in DNA can be lethal, since they result in faulty segregation of chromosomes and the likely loss of essential genes. Faithful repair of such breaks cannot occur by simple excision and resynthesis, as can happen for single-stranded (ss) lesions in DNA, because at a DSB, no intact template is present. Instead, the template for repair of a DSB appears to be an intact homologous DNA molecule, either a sister chromosome or a homolog. Interaction of broken and intact molecules can result in exchange of genetic information, i.e., homologous recombination, between them. As expected from this view, there is a close connection between DSB repair and recombination. In fact, part of the evidence that DSB repair frequently occurs by interaction with homologous DNA is the coordinate loss of DSB repair and homologous recombination in mutant organisms. For this reason, DSB repair is often called recombinational repair*.

Many genes and enzymes required for DSB repair and recombination have been well characterized in numerous organisms, but no cell-free system catalyzing these complete processes has, to my knowledge, been reproducibly reported for any organism. Molecular mechanisms are therefore based on inferences from the activities of individual components, many of which are well characterized in *Escherichia coli*.

At least part of the time the chromosome of *E. coli* is circular and therefore lacks double-strand (ds) ends *(2)*. DNA with ds ends can arise in *E. coli* after enzymatic, chemical, or physical damage to the DNA, as a consequence of attempted repair or replication of ss lesions, or after the introduction of linear DNA during transformation, conjugation, or phage infection. Repair or recombination of these DSBs requires numerous gene products. This chapter reviews the sources of DSBs in *E. coli*, general models for their repair, and the *E. coli* genes and enzymes necessary for their repair. It

*DNA metabolism at certain ss lesions, such as daughter strand gaps remaining after attempted replication of DNA containing a UV-induced lesion, is also referred to as recombinational repair, but the mechanisms envisaged do not remove the lesion, i.e., repair the DNA. Instead, they offer a way of tolerating the lesion by allowing synthesis of one lesion-free DNA molecule and, thereby, dilution of the lesion in the replicating DNA population.

From: DNA Damage and Repair, Vol. 1: DNA Repair in Prokaryotes and Lower Eukaryotes
Edited by: J. A. Nickoloff and M. F. Hoekstra © Humana Press Inc., Totowa, NJ

then discusses experimental means for measuring DSB repair. Although some of these experiments were designed to study recombination, a common mechanism appears to unite recombination with DSB repair.

2. SOURCES OF DS BREAKS IN *E. COLI*

DSBs can be introduced into the circular *E. coli* chromosome in numerous ways. Ionizing radiation, such as X-rays or γ-rays, creates DSBs (and other damages) at essentially random positions in DNA (*see* Section 6.1.). Attempted repair of randomly located UV light-induced lesions, such as cyclobutane dimers, is thought to leave ss gaps; enzymatic cutting of the intact strand before repair is completed could produce a DSB. Alternatively, replication of the gapped DNA (or of any nicked DNA) could produce a DSB. It has been hypothesized that stalled replication forks, for example, those at the *ter* sites near the terminus of replication, may break, converting the transient ssDNA in the fork into a DSB (*see* Section 7.). DSBs at more precisely defined locations on the chromosome have been produced by expression of site-specific endonucleases (for example, *Eco*RI and λ terminase; *see* Section 6.2.). Linear DNA, with its ds ends, can be introduced into *E. coli* by conjugation or phage infection (*see* Section 7.), or by transformation with naked DNA after electroporation or Ca^{2+} treatment of the cells (*see* Sections 6.4., 6.5., and 8.). Transformation with linear DNA allows manipulation of the DNA in ways not otherwise attainable to study the substrate requirements for DSB (or gap) repair and recombination, and to elucidate the types of products. Randomly located DSBs, from ionizing and UV radiation, conjugation, and transduction, have been useful in identifying the genes and enzymes required for DSB repair and recombination. The more precisely defined DSBs have been especially fruitful in elucidating molecular mechanisms.

3. GENERAL MODELS FOR DSB REPAIR AND RECOMBINATION

As noted in Section 1., restoration of the original nucleotide sequence at a DSB appears to require interaction with an intact homologous DNA molecule. Such an interaction is facilitated by the multiple chromosomes generally present in growing *E. coli* cells. Homology can also be provided by repeated sequences within a DNA molecule. Although some of the models discussed here were based on observations in other organisms, the features of these models are readily extended to *E. coli*.

For the repair of a DSB in DNA without repeats, Resnick *(68)* proposed a model of limited ss degradation of the ds ends, formation of hybrid DNA between the resultant 3' ss tails and the intact homolog, and DNA synthesis primed by the 3' ends and templated by the intact homolog. In alternative schemes (Fig. 1A and A') the tails anneal sequentially with or without the formation of a Holliday junction and with or without crossover products (those containing DNA from opposite parents flanking the repaired site). Resnick's model proposed precise homologous interaction by the formation of hybrid DNA, postulated by Holliday *(25)* to be an intermediate in recombination. Resnick noted that alternative resolutions of this intermediate, coupled with base-mismatch repair, would produce either reciprocal or nonreciprocal recombinants (crossing over and gene conversion, respectively). He supported his model by noting that X-rays induce recombination in *Saccharomyces cerevisiae*, and that *rad51* and *rad52* mutants are highly

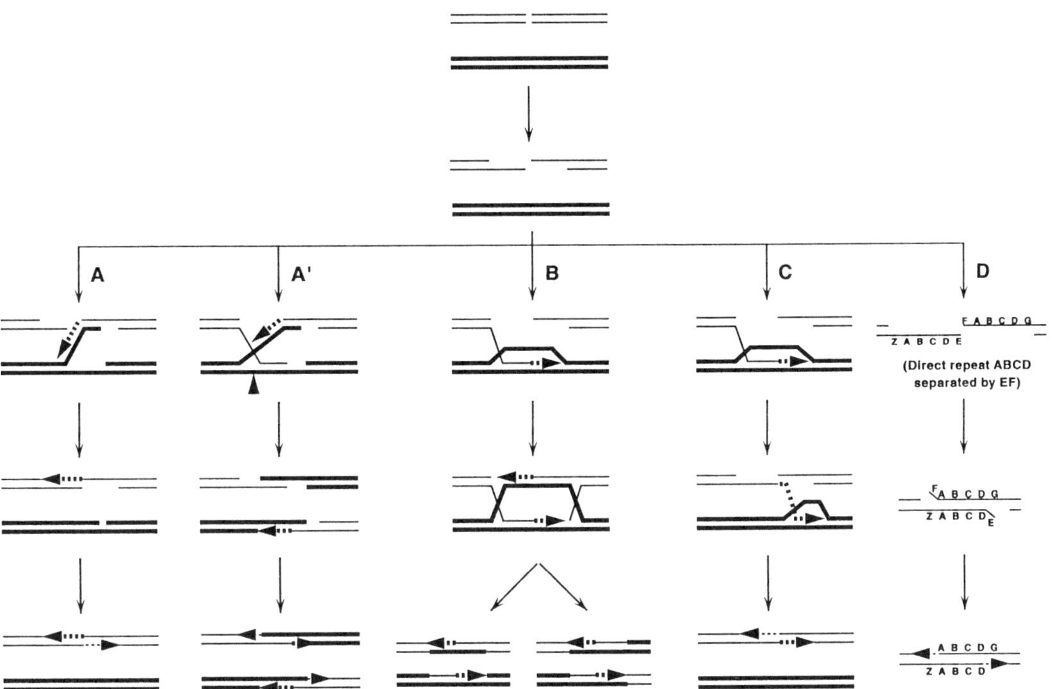

Fig. 1. General models for DSB repair and recombination. Each line represents a single strand of DNA, thick and thin lines being from homologs. Dotted lines represent newly synthesized DNA, thick or thin according to their templates. At each dsDNA end, a nuclease digests one strand to generate 3' ssDNA tails, which invade an intact homologous duplex **(A–C)** or anneal with each other **(D)**. (A) After Resnick *(68)*. The homologous (intact) duplex is nicked and partially unwound; the unwound strand anneals with one tail whose 3'-end primes DNA synthesis using the unwound strand as template. This product is in turn unwound, and the newly synthesized strand anneals with the other tail, which primes synthesis. Ligation restores intact duplexes. In an alternative scheme (A'), allowing reciprocal exchange (crossover), the second tail anneals with the ss gap in the initially intact duplex before the product is unwound. This creates a nicked Holliday junction, not present in scheme (A). Nicking of the Holliday junction (▲) resolves it into crossover duplexes, which are made intact by DNA synthesis and ligation. (B) After Szostak et al. *(88)*. One tail invades the intact homolog and primes DNA synthesis. When the D-loop is large enough, it anneals with the second tail, which also primes DNA synthesis. Joining of the newly synthesized 3'-ends to the 5'-ends on the broken DNA creates two Holliday junctions. Resolution of these junctions in the same or opposite planes creates noncrossover or crossover products, respectively. (C) After Formosa and Alberts *(18)*. One tail invades the intact homolog and primes DNA synthesis. The D-loop, instead of enlarging as in (B), migrates along the DNA as the newly synthesized DNA unwinds from its template. When the newly synthesized DNA is long enough, it dissociates from the homolog and anneals with the second tail, which also primes DNA synthesis. Note that no Holliday junction or crossover product is formed, and that no coordination of the two tails' invasion is required. (D) After Lin, et al. *(45)*. If the DSB occurs within or between direct repeats (ABCD), the tails can anneal without involving a homolog or strand invasion. Nonhomologous ends are removed by nucleases, and gaps are filled by DNA polymerase.

sensitive to X-rays and are recombination-deficient. He also calculated that one X-ray-induced DSB results in one gene conversion event. An attractive feature of his model was its reliance on known or easily visualized enzymatic processes and its avoidance of complex structures, such as histones and membranes, previously postulated to be involved in DSB repair. He speculated that his model could account for meiotic recombination if DSBs occur in meiosis.

Szostak et al. *(88)* amplified a similar scheme to account for meiotic recombination in fungi. Their model was inspired, in part, by their observations on transformation of mitotically dividing *S. cerevisiae* with linear dsDNA lacking a segment of an *S. cerevisiae* gene carried by the plasmid *(64)*. The ds gap in these plasmids could be as large as 1.2 kb and was apparently restored by interaction with the homologous chromosomal gene. They supposed that the ds ends at DSBs are typically degraded to yield ds gaps, which are subsequently repaired as shown in Fig. 1B. In the model of Szostak et al. *(88)*, the ds ends bracketing the ds gap separately invade the intact homolog to yield two separate regions of heteroduplex DNA and two Holliday junctions. Alternative resolutions of the two Holliday junctions yield crossover or noncrossover products. A novel feature of their model was the formation of gene convertants by ds gap formation and resynthesis of both strands using the intact homolog as template; in other words, no heteroduplex DNA is formed in the interval containing the gap, and no mismatch repair need be involved in gene conversion. Subsequent work has shown that during *S. cerevisiae* meiotic recombination DSBs are formed and degraded to generate 3' ss tails that form heteroduplex DNA and double Holliday junctions, but ds gaps may be typically no more than a couple of base pairs long, and gene conversion apparently occurs primarily by mismatch repair (reviewed in *44*; *see* Chapter 16).

Based on their studies of phage T4 replication, Formosa and Alberts *(18)* proposed a model (Fig. 1C) in which one 3'-tail invades an intact homolog and primes DNA synthesis templated by the homolog, as in the model of Szostak et al. *(88)*. However, here the D-loop migrates with the point of synthesis, and the newly synthesized DNA unwinds from the homolog and eventually anneals with the second tail. No Holliday junction or crossover product is formed.

An intramolecular repair scheme is possible when the DNA contains repeats bracketing the DSB *(45)*. As in the preceding models, there is limited ss digestion of the ds ends, but instead of invading an intact homologous duplex, the exposed 3'-tails anneal in the region of the repeat (Fig. 1D). Trimming of ss tails, filling of ss gaps, and ligation restore an intact duplex, but this duplex has a deletion of one repeat and any DNA between the initial repeats. This mechanism is called ss annealing (SSA).

These models have prompted many experimental tests of the role of DSBs in the initiation of recombination (*see* Sections 6. and 7.). In many cases, only recombinant formation, rather than DSB repair, was examined. One can imagine that a recombinant is formed by action at just one end of the DSB (Fig. 1C, top panels) and that a combined action with the other end would repair the DSB. The combined action might require coordination of the two events; the relative frequency of one-ended vs two-ended events remains undetermined (*see* Section 9.). The models in Fig. 1 provide a useful framework for describing enzymes required for DSB repair and for thinking about how they mediate repair.

Table 1
***E. coli* Gene Products Involved in DSB Repair and Recombination**

Gene(s)	Biochemical activity of gene product(s)
Presynapsis[a]	
recBCD	ATP-dependent ds and ssDNA exonuclease and DNA unwinding; interaction with Chi sites
recE	dsDNA exonuclease digesting 5' → 3'
recJ	ssDNA exonuclease digesting 5' → 3'
recQ	DNA helicase
Synapsis[a]	
recA	Homologous DNA strand exchange
recF	Binds ss and dsDNA and ATP
recO	Stimulates RecA under suboptimal conditions
recR	Stimulates RecA under suboptimal conditions
recT	Homologous ssDNA annealing
ssb	ssDNA binding
Postsynapsis[a]	
recG	Dissociates Holliday junctions
ruvA	Binds Holliday junctions and RuvB
ruvB	Branch migrates Holliday junctions
ruvC	Cleaves Holliday junctions
priA	Primosome assembly factor
Other	
recN	Unknown
sbcA[b]	Activates *recE* and *recT* genes of rac prophage
sbcB[c]	ssDNA exonuclease digesting 3' → 5'
subCD	ATP-dependent dsDNA exonuclease
lexA	Repressor of numerous genes in DNA repair

[a]These stages of action are not clearly established, especially for *recF*, *recO*, *recR*, and *recN*. Some activities may act in more than one stage. *See* Section 4.
[b]*sbcA* are regulatory mutations that do not define a gene.
[c]Also designated *xonA*.

4. GENES AND ENZYMES PROMOTING DSB REPAIR

Enzymes required for DSB repair and recombination can be divided into three sets, depending on their inferred stage of action. The first (presynaptic) set converts dsDNA ends into proper substrates for the second (synaptic) set, which pairs homologous molecules. The third (postsynaptic) set converts the paired molecules into repaired or recombined products by a combination of branch migration, DNA polymerization and ligation, and cleavage into separate DNA molecules. These three sets are listed in Table 1 and described more fully below. (For more extensive reviews, *see 37,80,104*).

4.1. RecBCD Enzyme

RecBCD enzyme, also called exonuclease V, is composed of three polypeptides, encoded by the *recB*, *recC*, and *recD* genes *(92)*. RecBCD enzyme is active as a monomer (B_1 C_1 D_1) with an M_r of 330 kDa *(97)* and is one of a few *E. coli* enzymes with a high affinity for ds ends. The dissociation constant K_d is about $10^{-10}M$; the enzyme is, therefore, expected to act at its maximal rate with just one ds end per *E. coli* cell (about $10^{-9}M$) *(97)*. In the presence of ATP and Mg^{2+} ions, the enzyme unwinds DNA from a ds end, with the production of one ss loop and two ss tails. The enzyme unwinds DNA at about 350 bp/s, and the loop grows and moves along the DNA *(69,94)*. This mechanism may help assure that the unwound DNA does not immediately rewind behind the enzyme, but instead remains as a potent ss substrate for RecA protein (*see* Section 4.5.).

During its unwinding of DNA, RecBCD enzyme interacts with Chi (5' GCTGGTGG 3' or its complement or the duplex), initially identified as a hot spot of recombination *(62,83)*, to produce an ssDNA tail with Chi near its 3'-end. This "3' Chi tail" is a potent substrate for homologous pairing by RecA protein *(13,101)*. Depending on the reaction conditions (primarily the ratio of the ATP and Mg^{2+} ion concentrations), there are two modes of interaction of RecBCD enzyme with Chi (Fig. 2). With [ATP] > [Mg^{2+}], the enzyme unwinds DNA and nicks the strand containing 5' GCTGGTGG 3' a few nucleotides to the 3'-side of Chi *(93)*. With [Mg^{2+}] > [ATP], the enzyme's ds exonuclease activity degrades the strand that has a 3'-end at which the enzyme initiated attack; degradation is attenuated at or a few nucleotides to the 3'-side of Chi *(14,15,98)*. This action, plus continued unwinding, produces a "3' Chi tail" essentially identical to that produced with [ATP] > [Mg^{2+}], but in addition, with [Mg^{2+}] > [ATP], RecBCD enzyme nicks the complementary DNA strand within or near Chi *(98)*. Whether RecBCD enzyme creates DNA ends on both strands of the same DNA molecule (to translate the ds end from its initial position to Chi) or only on separate DNA molecules (to form ds molecules with 3' or 5'-tails) is unclear. The consequences of these two modes of Chi–RecBCD enzyme interaction are discussed elsewhere *(98)*. Molecules with both 5' and 3' ss tails have also been proposed as recombination intermediates *(67,70)*.

Under both reaction conditions and in *E. coli* cells, RecBCD enzyme interacts with Chi only if it approaches Chi from the right (as written above) *(31,93)*. With continued unwinding, RecBCD enzyme produces ssDNA to the left of Chi, and in *E. coli* cells recombination is stimulated to the left of Chi *(6,85)*. Since Chi occurs, on the average, about once per 5 kb or about 1000 times on the *E. coli* chromosome *(17)*, it is likely to be encountered in its active orientation by RecBCD enzyme within about 10 kb from any ds end made on the chromosome. This outcome assures that repair will be effected within a short distance (≈10 kb or ≈0.2% of the chromosome length) from the DSB.

After interacting with Chi, RecBCD enzyme loses its ability to interact with a subsequently encountered Chi, on either the same or a separate DNA molecule *(96)*. This feature predicts single recombinational exchanges or simple repair events from each ds end–RecBCD enzyme–Chi set, and avoids more complex DNA–DNA interactions. The physical change of RecBCD enzyme at Chi is unclear, but has been hypothesized to be the loss or inactivation of the RecD subunit *(12,32,61)*.

RecBCD enzyme has strong exonuclease activities on either dsDNA, as noted above, or ssDNA. The latter activity could digest long ss tails on otherwise dsDNA molecules and render their ends flush or nearly flush, as required for the unwinding activity of RecBCD enzyme *(95)*. The strong ds exonuclease activity, perhaps in conjunction with the unwinding and ss exonuclease activities, degrades DNA to oligonucleotides *(92)*. This action destroys foreign DNA and DNA that cannot recombine owing, for example,

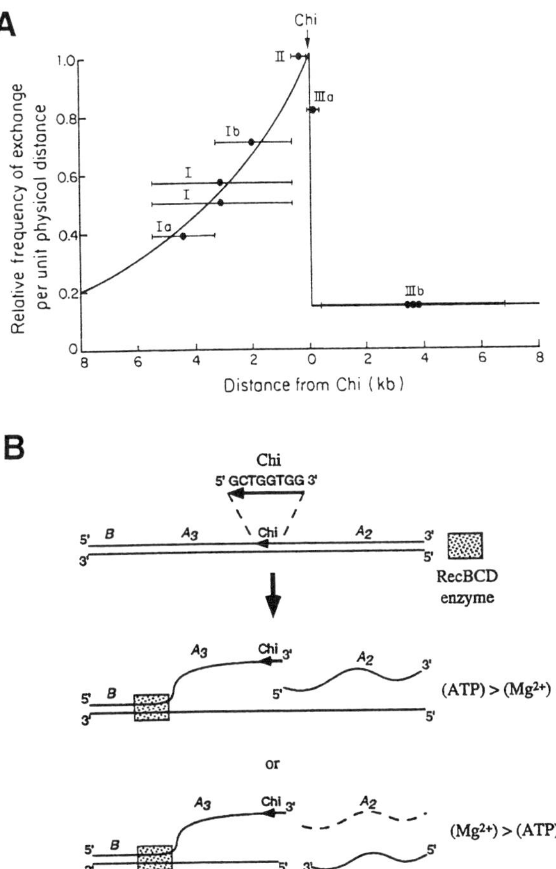

Fig. 2. Action of Chi sites and RecBCD enzyme. **(A)** Stimulation of recombination by the RecBCD pathway in phage λ vegetative crosses. The frequency of recombination per unit physical distance in genetic intervals I, Ia, and so forth (normalized to interval II = 1) is plotted as a function of the distance of the midpoint of the interval (●) from a Chi site oriented to be active with RecBCD entering from the right. From ref. *(6)*. **(B)** Reactions of purified RecBCD enzyme on DNA with a Chi site. RecBCD enzyme (stippled box) binds to a dsDNA end. With [ATP] > [Mg^{2+}], the enzyme unwinds the DNA with the production of ssDNA loops (not shown), nicks the "upper" strand a few nt to the right of Chi, and continues unwinding *(93)*. With [Mg^{2+}] > [ATP], the enzyme degrades the "upper" strand up to a few nt from Chi and continues unwinding with reduced degradation *(14)*; the "bottom" strand is also nicked near Chi, although not necessarily on the same DNA molecule on which the "upper" strand was partially degraded *(98)*. Note that under both conditions, RecBCD enzyme produces an ssDNA fragment with Chi near its 3'-end, a substrate for homologous pairing by RecA protein *(13)*. From ref. *(84)*.

to lack of a homolog or RecA protein *(43)*. The nuclease activity is reduced by Chi, as noted above, and by ssDNA binding protein (SSB; *see* Section 4.6.).

Two additional activities of RecBCD enzyme are a strong ATPase, which is dependent on ds or ssDNA and is, in turn, required for the enzyme's unwinding and exonuclease activities, and a weak ssDNA endonuclease activity, which may be a manifestation of the enzyme's Chi-nicking activity noted above. RecBCD enzyme has no detectable activity on circular dsDNA, whether supercoiled, relaxed, or nicked *(95)*. Therefore, in the absence of DNA damage, or exogenous linear DNA, intracellular RecBCD enzyme is presumably inactive.

4.2. RecE and λ Exonucleases

RecE exonuclease, also called exonuclease VIII, is encoded by the *recE* gene, a part of the cryptic λ-like prophage designated rac, present in wild-type *E. coli* strain K12 but not in some of its derivatives. The *recE* gene and the closely associated *recT* gene (*see* Section 4.7.) are activated by *sbcA* mutations, which suppress the DNA damage sensitivity and recombination deficiency of *recBC* mutations (*21* and references therein). The *recE* gene encodes a 96-kDa polypeptide, which as a trimer or tetramer has strong ATP-independent exonuclease activity on ds, but not ssDNA *(29)*. From a given ds end, the enzyme digests one strand, from its 5'-end, in a highly processive manner and produces long 3' ss tails. These 3'-tails, like those produced by RecBCD enzyme (*see* Section 4.1.), are likely substrates for homologous annealing or synapsis (*see* Section 4.7.).

The analogous enzyme encoded by the *exo* (or *redA* or *redα*) gene of λ has a similar enzymatic activity, but the λ exonuclease is considerably smaller (a dimer of 24-kDa polypeptides) *(46)*. The role of the extra mass of RecE exonuclease is unclear.

4.3. RecJ Exonuclease

RecJ protein (63 kDa) is an ATP-independent nuclease highly active on ss, but not ds, linear DNA and digests from the 5'-end *(52)*. In conjunction with the RecQ helicase (*see* Section 4.4.), RecJ exonuclease could produce dsDNA molecules with long 3'-tails, presumably appropriate substrates for synapsis by RecA protein. Alternatively, RecJ exonuclease could remove ss tails after synapsis.

4.4. RecQ DNA Helicase

RecQ protein, in the presence of ssDNA binding protein (SSB) (*see* Section 4.6.), unwinds dsDNA that has a 3' ss tail, with concomitant ATP hydrolysis *(102)*. SSB is presumably required to prevent immediate rewinding of the DNA. Coupling of RecQ's helicase activity with RecJ's exonuclease activity could produce 3' ss tails as substrates for RecA protein.

4.5. RecA Homologous Pairing and Strand-Exchange Protein

Although it is a relatively small protein (38 kDa), RecA protein has numerous activities. One set of activities pairs homologous DNA molecules and exchanges strands between them, whereas another set stimulates the cleavage of LexA repressor to induce expression of genes whose products repair DNA (*see* Section 4.13.), and yet another set appears to be involved in mutagenic replication past DNA damages. The presence of multiple activities in one protein underscores the central role of RecA protein in sensing and repairing DNA damage. For reviews and further references, *see* refs. *(3)* and *(37)*.

To promote interaction between homologous DNA molecules, RecA protein first forms an extensive filament on an ss portion of one of the substrate DNAs. This obligate ssDNA is formed adjacent to a DSB by the enzymes described in Sections 4.1.–4.4. Filament formation requires either ATP or an analog, such as the nonhydrolyzable ATP-γ-S. The ssDNA in the filament is stretched nearly to its limit—5Å/nucleotide (nt)—and is fully coated with RecA protein—1 RecA monomer/3 nt. This filament finds a homologous dsDNA, perhaps by diffusion through the solvent or along the dsDNA, and forms a structure in which the two DNA molecules intertwine within the filament. If the length of homology is sufficient, the ssDNA in the filament exchanges place with its equal in the dsDNA to form new Watson-Crick base pairs. This strand exchange can continue throughout the length of homology for many kilobases. The newly formed ssDNA, derived from the original dsDNA, may be sequestered by being bound with SSB (*see* Section 4.6.).

RecA's reaction to this point is isoenergetic—1 base pair is formed for each base pair dissociated—and does not require ATP hydrolysis. The role of its DNA-dependent ATPase has been puzzling. One role is to enable strand exchange to pass through regions of nonhomology, an energy-requiring process *(71)*. Another role may be to propel strand exchange in one direction (forward). One would expect the reaction in the absence of proteins to be bidirectional, but the asymmetry of the reaction—RecA-coated ssDNA in and SSB-coated ssDNA out—may provide directionality even without ATP hydrolysis.

If the RecA filament was formed on ssDNA attached to dsDNA, a four-strand reaction occurs beyond the ss–ds junction. The symmetric transfer of strands, also isoenergetic, forms a Holliday junction (Fig. 1A',B). Continued action of RecA or of RuvAB protein (*see* Section 4.8.) can move this junction along the DNA, a process called branch migration, until the junction is resolved (*see* Section 4.9.).

The three-dimensional structure of an inactive form of RecA filaments in the absence of DNA at atomic resolution, coupled with mutant analyses, has revealed amino acid domains responsible for binding DNA, binding ATP, forming RecA–RecA contacts in filaments, and forming RecA–LexA contacts *(86)*.

4.6. ssDNA Binding Protein

The SSB of *E. coli* is a tetramer of 19-kDa polypeptides with a higher affinity for ssDNA than for dsDNA (reviewed in *57*). In limited amounts, it stimulates RecA's strand-exchange reaction by effectively removing the displaced ssDNA and by removing intrastrand pairings in the substrate ssDNA, thereby facilitating RecA filament formation. In excess amounts, however, SSB inhibits RecA's strand-exchange reaction, presumably by so thoroughly coating the substrate ssDNA that RecA filaments cannot form.

SSB is also important for DNA replication; conditional lethal mutations show that SSB is essential for cell viability. Certain *ssb* mutants are sensitive to ionizing and UV radiation, and are recombination-deficient, showing that SSB is involved in recombination and repair as well.

4.7. RecT and λ β Strand Annealing Proteins

The rac prophage and λ phage encode the RecT and λ β proteins, which promote annealing of homologous, complementary ssDNA (reviewed in *35*). The genes encoding these proteins are adjacent to the *recE* and λ *exo* genes encoding exonucleases (*see* Section 4.2.). The purified RecT and λ β proteins can anneal two complementary ssDNA molecules, but can exchange ssDNA and dsDNA (the three-strand reaction promoted by RecA protein) only if the dsDNA has an ss tail to allow an initial two-strand reaction. The RecT and λ β proteins may, therefore, anneal ssDNA with 3'-tails produced by the phage-encoded exonucleases. This reaction would require a ds end in both homologous DNAs (Fig. 1D). Alternatively, or in addition, the RecE-T or λ exo–β complexes may promote a three-strand reaction, which would require a ds end in just one parental DNA. Studying the action of these complexes is complicated by their nuclease activities, which can destroy the DNA.

4.8. RuvA-RuvB and RecG Branch Migration Proteins

RuvA and RuvB appear to act together to promote migration of Holliday junctions (reviewed in *105* and *107*). *ruvA* encodes a 22-kDa protein that as a tetramer preferentially binds four-way dsDNA junctions. *ruvB* encodes a 37-kDa protein that forms hexameric rings on DNA and has weak DNA-dependent ATPase activity. Together, RuvA and RuvB promote rapid branch migration of Holliday junctions, for example, those formed by RecA protein. Migration can occur in either direction, forward to products (with more extensive hybrid DNA) or reverse to the initial substrates. Presumably, a RuvA tetramer targets to the junction a RuvB hexamer, which forms a ring around the DNA, moves along the DNA while hydrolyzing ATP, and forces the branch to migrate.

RecG protein is a 76-kDa protein that, like RuvAB, binds to synthetic Holliday junctions and dissociates them in an ATP-dependent manner *(50)*. RecG also promotes branch migration of RecA-formed Holliday junctions, but the reaction is weak: more than 1 mol of RecG is required to dissociate or branch migrate 1 mol of DNA in 15 min *(106)*.

4.9. RuvC Holliday Junction Resolving Protein

RuvC is a dimer of 19-kDa polypeptides that preferentially binds Holliday junctions and, in the presence of Mg^{2+} ions, makes two symmetrically placed nicks on strands of like polarity (reviewed in ref. *105*). This is the "biologically sensible" cleavage that produces recombinant molecules, either "patch" or "splice," depending on the pair of strands cleaved (Fig. 1B, left and right, respectively). Cleavage produces simple nicks, with 5'-phosphate (5'-PO_4^-) and 3'-hydroxyl (3'-OH) groups that are substrates for DNA ligase (*see* Section 4.11.). Cleavage is preferentially at the arrow in the sequence 5' (A/T) T T ↓(G/C) 3'. This result predicts hot spots of resolution, but because the preferred sequence occurs so frequently in *E. coli* DNA, such hot spots may not be detected genetically. Although RuvC cleaves Holliday junctions with high specificity, the activity is weak: under currently optimal conditions, more than 1 mol of RuvC is required to cleave 1 mol of junction DNA in a 30-min reaction. The clustering of the *ruvA, B,* and *C* genes and the activities of their products suggest that, as a complex in *E. coli*, they may be more active than the separate proteins.

4.10. DNA Polymerases

If nucleotides are lost during DSB repair, fully homologous recombination requires that these nucleotides be restored during the repair event. Nucleotides might be lost at the initial stages by nucleases acting on the DSB (*see* Sections 4.1.–4.3.) or at the resolving stages by aberrant RuvC cleavage (*see* Section 4.9.), or by other nucleases acting on the RuvC-produced nicks. Lost nucleotides may be replaced by DNA polymerase I or II. Alternatively, recombination intermediates may be converted into full-fledged replication forks by the action of DNA polymerase III and its associated proteins (*see* Section 7.).

4.11. PriA Protein

PriA protein, also called n', helps assemble the primosome for DNA polymerase III holoenzyme (reviewed in *36*). The primosome contains DnaG primase and DnaB helicase; other assembly factors include PriB, PriC, DnaC, and DnaT. PriA protein has DNA-dependent ATPase and ATP-dependent helicase and translocase activities. It is essential for replication of phage ϕX174 DNA with purified components, but not for *E. coli oriC*-dependent replication. However, *priA* mutants grow poorly and are radiation-sensitive and recombination-deficient *(34,74)*. This last property strongly implicates DNA replication in recombination and DSB repair (reviewed in *33*).

4.12. DNA Ligase

DNA ligase, a monomeric 75-kDa protein, seals nicks with 5'-PO_4^- and 3'-OH groups, for example, those left by RuvC cleavage or by DNA polymerase filling of ss gaps (*see* Sections 4.9. and 4.10.).

4.13. Rec F, O, R, and N Proteins

The *recF*, *O*, and *R* genes were identified by mutations that inactivate the RecF pathway of recombination (*see* Section 5.4.). Cloning and sequencing of the genes revealed some features of the encoded proteins, but their activities are still unclear (*see 49* for a review). The phenotypes of *recF*, *O*, and *R* mutants, singly or in double or triple combinations, are similar; these observations, plus the ability of the phage λ *ninR* gene *orf* to suppress *recF*, *O*, and *R* mutations in λ vegetative crosses *(76)*, are consistent with the RecF, O, and R proteins acting as a complex.

The available evidence suggests that the RecF, O, and R proteins aid RecA protein's synapsis or strand-exchange activity (*see 103* and references therein). Certain dominant *recA* mutations suppress *recF*, *O*, and *R* mutations, and these mutant RecA proteins have greater activity than wild-type RecA protein under certain suboptimal conditions, e.g., with excess SSB protein. Furthermore, under such conditions, RecO + RecR proteins stimulate the activity of wild-type RecA protein (i.e., they overcome the inhibition by excess SSB). RecF protein does not, under the conditions tested, stimulate RecA, but it does bind DNA. In the presence of ATP-γS, binding of RecF is tighter to circular duplex DNA containing an ss gap than to fully ds or ssDNA *(22)*. This result suggests that RecF has a preference for the junction between ds and ssDNA, perhaps to direct RecA protein to the junction.

The *recN* gene was identified by its induced expression after DNA damage; *recN* mutations render the RecF pathway inactive. The nucleotide sequence of the gene suggests that RecN protein (64 kDa) binds ATP, but no biochemical activity for RecN has been reported (*72* and references therein).

4.14. LexA Protein

The LexA protein is a repressor of a set of genes, the LexA regulon, whose expression is important for recovery from DNA damage (for a review and references, *see 19*). Among these genes are *recA*, *recN*, *ruvA*, *ruvB* (but not *ruvC*), and *lexA* itself. DNA damage leads to a signal, probably ssDNA, that activates RecA protein to bind to LexA protein and to stimulate LexA autocleavage, thereby inactivating it as a repressor. The LexA regulon is induced, and its products effect DNA repair, after which newly synthesized LexA returns the regulon to its basal level of expression. Note that RecA stimulates its own synthesis after DNA damage and reduces it by repairing the DNA.

Certain *lexA* and *recA* mutations have opposing effects on expression of the LexA regulon. *lexA* null mutations, called *lexA* (Def), have constitutively high levels of expression; these are partially mimicked by special *recA* mutations, formerly called *tif*, that behave as though they activated RecA protein in the absence of DNA damage. Special *lexA* mutations, called *lexA* (Ind$^-$), render LexA protein noncleavable; in these mutants, the LexA regulon is not inducible. These mutations are mimicked by special *recA* mutations, formerly called *lexB*, that are inactive for LexA cleavage, but are recombination-proficient. *recA* null mutations are inactive for both LexA cleavage and recombination proficiency.

Other *rec* genes are also involved in regulating the LexA regulon, via generation of the RecA-activating signal. Following nalidixic acid treatment, which inhibits DNA gyrase and may lead to DSBs, RecBCD enzyme is required for induction of the LexA regulon. Presumably, RecBCD enzyme attacks the dsDNA end, and its unwinding activity produces ssDNA, which activates RecA. Following UV irradiation, RecF function is required for induction, but RecF's mode of action is unclear. Because these and perhaps other Rec proteins affect expression of other genes, a *rec* mutation may abolish repair and recombination either directly by inactivating the encoded enzyme, or indirectly by blocking induction of another required gene. This complexity is evident for RecA and may also apply to other Rec proteins, such as RecBCD and RecF.

Induction of the LexA regulon is important for DNA repair because some of its gene products are required at high level after extensive DNA damage. Most notable is RecA protein, which is required in stoichiometric amounts for effective presynaptic filament formation on ssDNA and strand exchange (*see* Section 4.5.). Less clear is why the apparently enzymatic RuvAB proteins are induced. There may be additional, unidentified genes in the LexA regulon that are important for DNA repair. For example, there are suggestions that DNA damage induces an inhibitor of RecBCD enzyme, perhaps analogous to those encoded by phages (e.g., λ Gam protein).

5. PATHWAYS OF DSB REPAIR

Clark *(7)* introduced the concept of pathways of genetic recombination based on studies of suppressors of *recB* and *recC* mutations. By his definition, based on recombination proficiency in Hfr conjugational crosses, the RecBC pathway is active in wild-type *E. coli*; this was renamed the RecBCD pathway on the discovery of a third subunit of RecBCD enzyme *(1)*. *recB* and *recC* mutations inactivate the RecBCD enzyme and pathway. Suppressors of *recBC* mutations activate either the RecE pathway (*sbcA*, suppressor of *recB recC* mutations) or the RecF pathway (*sbcB* suppressor mutations); for high-level suppression, the *sbcB* mutations require the subsequently identified *sbcC* or *sbcD* mutations. Support for the idea of separate pathways came from the phenotypes of *recE, F, J, N, O, R* and other mutations that inactivate the RecE or RecF pathway (or both), but not the RecBCD pathway. Complexities arose when recombination other than conjugational recombination was studied (for example, plasmid recombination) *(81)*. Nevertheless, the pathways described below are useful for organizing the roles of gene functions in recombination and repair.

5.1. RecBCD Pathway

The RecBCD pathway, active in wild-type *E. coli*, appears to act only when dsDNA ends are available, for example, after X-irradiation or conjugation; it appears to play only a minor role in most plasmid recombination. This pathway requires the RecBCD enzyme, which acts with high efficiency at dsDNA ends to produce ssDNA for RecA protein. ssDNA with 3'-ends are produced at high frequency at Chi sites (Section 4.1. and Fig. 2) and at lower frequency at Chi-like sites differing from Chi at one or more base pairs; Chi-like sites may account for recombination by the RecBCD pathway of DNA lacking Chi *(5)*. Strand exchange with an intact homologous duplex, promoted by RecA and SSB proteins, produces joint molecules. The enzymes resolving these joint molecules are unclear, but may be either RuvC or an unknown enzyme acting with RecG: *ruvC* and *recG* single mutations reduce recombination by a factor of 3–4, but *ruvC* and *recG* double mutations reduce recombination by a factor of about 1000 *(47)*.

5.2. ‡ (RecD⁻) Pathway

Null mutations in *recD* and special mutations in *recC* abolish RecBCD enzyme's nuclease activity, including Chi cutting, but leave the cells recombination-proficient and resistant to DNA-damaging agents *(1,4)*. This phenotype is designated ‡ ("double dagger") *(4)* or RecBC(D⁻) *(100)*. The recombination-promoting activity retained by the ‡ mutants was hypothesized to be DNA unwinding *(4)*, but reconstituted RecBC enzyme (i.e., lacking RecD) has only about 1% as much unwinding activity as holoenzyme (*12;* A. Taylor, personal communication). A high level of linearized Chi-containing DNA in *E. coli* cells or treatment with bleomycin, which presumably linearizes the Chi-containing chromosome, converts these cells into partial ‡ phenocopies *(32,61)*. This observation supports an earlier hypothesis that at Chi RecBCD enzyme loses the RecD subunit *(99)*. In this view the ‡ enzyme is equivalent to native enzyme after its interaction with Chi, and the ‡ pathway is equivalent to the RecBCD pathway on Chi-containing DNA. However, conjugational recombination by the ‡ pathway requires RecJ and, to a lesser extent, RecN and is reduced by a *lexA* (Ind⁻) mutation; *recJ, recN,* and *lexA* mutations have less effect on the RecBCD pathway *(53)*. Further knowledge of the activities of the ‡ enzyme may clarify the relation of the two pathways.

5.3. RecE and λ Red Pathways

The *sbcA* mutations activate the *recE* and *recT* genes of the rac prophage. Acting at ds ends, RecE exonuclease makes ssDNA substrates for the RecT ss annealing protein or RecA strand-exchange protein. The analogous λ exonuclease and β protein act similarly in the λ Red pathway. Resolution of the joint molecules, at least by the RecE pathway, apparently occurs by RuvC protein, since *recBC sbcA ruvC* mutants are Rec⁻ *(48)*. The RecE and Red pathways require RecA protein in some cases (e.g., conjugational crosses and λ crosses in which replication is blocked) but not in others (e.g., in plasmid recombination and in freely replicating λ crosses). The basis of this differential requirement is unknown, but may reflect formation of recombinants by the SSA pathway (Fig. 1D) in some situations, but not others. The SSA pathway may require strand annealing by RecT or β, but not strand exchange by RecA.

5.4. RecF Pathway

The *sbcB* mutations alter exonuclease I (ExoI), a 3' → 5' ssDNAse, and were thought to allow survival of recombination intermediates with 3' ss ends, such as substrates for RecA protein *(80)*. At best, this view is incomplete *(37)*. *xonA* mutations that delete the gene for ExoI and adjacent genes do not suppress the Rec⁻ phenotype of *recB recC* mutations. Full suppression requires not only an *sbcB* mutation but also a mutation in *sbcC* or *D*, which inactivates the SbcCD ATP-dependent nuclease *(9)*. This nuclease may destroy substrates or intermediates of recombination in the RecF pathway but evidently not in the other pathways discussed above.

Resolution of intermediates in the RecF pathway also appears to differ from that in the RecBCD pathway. Individual mutations in *ruvA, B, C*, or *recG* reduce recombination by the RecF pathway *(48)*, but only a combination of a *ruv* mutation and a *recG* mutation reduces recombination by the RecBCD pathway *(47)*. These results suggest that both Ruv and RecG functions are required for the RecF pathway and that either is sufficient for the RecBCD pathway. The DNA intermediates in the two pathways presumably differ in an as-yet-unknown manner.

6. EXPERIMENTAL MEASURES OF DSB REPAIR

6.1. Ionizing Radiation

Ionizing radiation, such as X-rays and γ-rays, produces DSBs (and other damages) in DNA. The breakage and subsequent repair can be followed by extracting the DNA and estimating its size by the rate of sedimentation during centrifugation. Rapidly growing *E. coli* cells, with multiple genomes per cell, repair the breaks efficiently, but slowly growing cells, with closer to 1 genome/cell, do not *(38)*. This result indicates that repair requires an intact homolog. Repair is defective in *recA*, *recB*, *recC*, *recN*, and *lexA* (Ind⁻) mutants, and these mutants as well as *priA* mutants are hypersensitive to ionizing radiation *(34,38,66,75)*. Induction of the LexA regulon is important, to give high levels of RecA and RecN (and perhaps other) proteins *(39)*. Repair can occur by the RecBCD, ‡, or RecF pathway; others have not been tested.

6.2. Cutting of the E. coli Chromosome by Endogenous Nucleases

DSBs can be made at defined positions on the chromosome by expressing site-specific nucleases in *E. coli*. The λ terminase (Ter) protein cuts the λ *cos* site to generate linear DNA in preparation for packaging the DNA into phage particles. The *E. coli* strain K-12 chromosome contains *cos*-like sites, presumably in cryptic prophages, and is sensitive to Ter: expression of Ter kills *recA* or *recB* mutants, but not rec⁺ cells or *recD* mutants *(60)*. *E. coli* strain C apparently contains no *cos*-like site and is resistant to Ter expression even when made *recA* mutant, but these mutants become sensitive when a λ *cos* site is placed in the chromosome. These results provide one of the simplest demonstrations of DSB repair by the RecBCD and ‡ pathways.

From the preceding observations, one would expect cutting of the *E. coli* chromosome by *Eco*RI restriction enzyme to be lethal in Rec⁻ cells, but this is not the case. At semipermissive temperatures, *E. coli* cells harboring a mutant thermosensitive *Eco*RI enzyme are not further sensitized by *recA*, *recB*, *recN*, or *lexA* (Ind⁻) mutations, but they are further sensitized by *lig* mutations *(23)*. The investigators suggested that cleaved *Eco*RI sites are readily sealed by DNA ligase, but ionizing radiation leaves nonligatable DNA termini, and λ Ter remains bound to one of the cut ends and presumably blocks ligation. Presumably as a consequence, repair of the latter damages, but not that from *Eco*RI cleavage, requires recombination. It is unclear, however, how the correct *Eco*RI ends are rejoined.

6.3. Cutting of Phage Chromosomes or Plasmids by Homing Intron Endonucleases

Group I introns encode site-specific endonucleases that cleave intronless chromosomes at the allelic site. Repair of the DSB by recombination with the intron-containing chromosome allows the intron to move to the initially intronless chromosome, a process called homing (Fig. 3A). Belfort and colleagues have studied the requirements for homing of the intron in the thymidylate synthetase *(td)* gene of phage T4. Homing requires the intron's endonuclease I-*Tev*I, its recognition site on the recipient (intronless) DNA flanked by homology to the DNA flanking the intron on the donor DNA, and many of T4's recombination and replication functions *(8,59)*. These functions include UvsX and Y (a strand-exchange protein complex similar to RecA protein), gene 46 and gene 47 proteins (a putative exonuclease), gene 32 protein (similar to SSB), gene 43 protein (DNA polymerase), and replication accessory proteins. These requirements indicate a coupling of recombination and replication, as deduced earlier for T4 recombination *(58)* and are consistent with the models in Fig. 1A–C. Since gene 49 protein (similar to RuvC) is not absolutely required, Mueller et al. *(59)* suggested that the I-*Tev*I-induced DSB may be repaired without the formation of Holliday junctions by a mechanism like that in Fig. 1C. This mechanism, unlike that in Fig. 1A' or B, does not predict "crossover resolution," which would integrate the entire donor DNA into the recipient DNA. Indeed, in only about 20% of the homing events does the entire donor integrate (Fig. 3A), perhaps by mechanisms similar to those in Fig. 1A' or B.

To study homing in other contexts, the *td* gene with and without the cutting site has been transferred to phage λ and to a plasmid *(65)*. When the recipient λ phage contains the cutting site and I-*Tev*I is induced from a second plasmid, in a single infection cycle, about half of the λ progeny phage incorporate the surrogate intron *(kan)* initially on the donor plasmid; again, only about 20% of these events form cointegrates, suggesting primarily "noncrossover resolution." This "homing" requires RecA protein, λ exonuclease, λ β, and either RecBCD enzyme or exonuclease III (a dsDNA 3' → 5' exonuclease). These requirements suggest a complex exonucleolytic digestion from the dsDNA end followed by strand annealing and strand exchange.

In a modification of these experiments, Eddy and Gold *(16)* showed that *Eco*RI restriction enzyme could replace I-*Tev*I. They constructed a plasmid with the *Eco*RI restriction and modification genes flanked by parts of *lacZ*. During infection with λgt11, bearing an *Eco*RI cleavage site in *lacZ*, 3–10% of the phage incorporated part or all of the plasmid. Incorporation required active *Eco*RI from the plasmid and an *Eco*RI recognition site on the phage, and could proceed by the RecE or λ Red pathway, but not

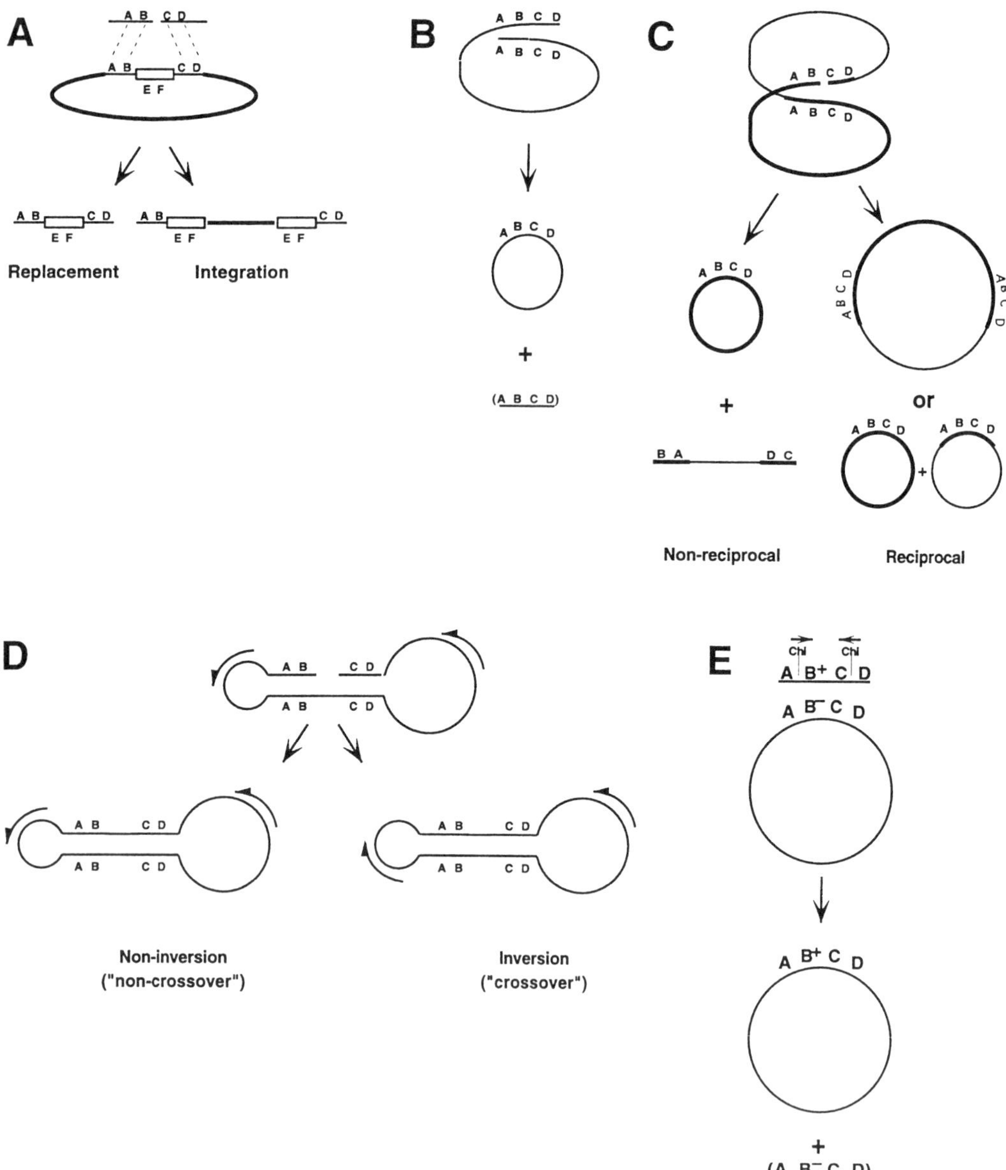

Fig. 3. Substrates for measuring DSB repair and recombination with plasmid-sized (≈10 kbp) molecules. Each line represents dsDNA. **(A)** Recipient (phage) DNA cut intracellularly by I-*Tev*I homing endonuclease or *Eco*RI restriction endonuclease recombines with a resident donor (plasmid) sharing homology with the recipient DNA flanking the cleavage site. The recipient can incorporate part of the donor (replacement) or all of it with duplication of the DNA between the flanking homology (integration). **(B)** Linear DNA containing terminal direct repeats is introduced into Ca^{2+}-treated cells or generated intracellularly by *Eco*RI restriction enzyme after infection with λ phage, the DNA of which contains the substrate between *Eco*RI sites. Recombination produces a heritable circular plasmid and, if the reaction is reciprocal, a nonheritable linear fragment. **(C)** Linear DNA containing two direct repeats in three segments is introduced or generated in *E. coli* as in (B). Here, reciprocal recombination produces one dimeric plasmid or two monomeric plasmids, and nonreciprocal recombination produces a

the RecBCD pathway. Failure of the RecBCD pathway may have been owing to the lack of appropriate Chi sites in the DNA substrates (*see* Section 4.1.).

6.4. Transformation with Linear DNA Containing Homologous Repeats

DNA containing direct or inverted repeats can undergo intramolecular recombination (Fig. 3B–D). Circular plasmids containing repeats do recombine in *E. coli*, but the frequency is elevated by a DSB. With inverted repeats, the products contain the repeats of the parent, i.e., the break is repaired without loss of a repeat (Fig. 3D). With direct repeats, the products usually contain only one repeat; i.e., the break is typically repaired with the loss of one repeat (Fig. 1D, 3B,C). In both cases, allelic exchange (recombination) between the repeats often accompanies the repair. Analysis of the circular products and the gene functions required for their formation has given insight into the mechanisms of DSB repair.

Wild-type *E. coli* is transformed with linear, direct repeat DNA (Fig. 3B) at much lower frequency (transformants/μg of DNA) than with the corresponding circular dimer. The ratio of these frequencies (L/C) is taken as a measure of repair potential, which varies with the recombination pathway. The ratio is low (~4%) by the RecBCD pathway, but high (40–140%) by the ‡, RecE, and RecF pathways *(53–55,87)*. The low ratio by the RecBCD pathway may be owing to the absence of appropriately placed Chi sites in the DNA (*see* Section 4.1.). The high ratio by other pathways indicates, on the assumption that linear and circular DNA enter *E. coli* equally frequently, that nearly every linear DNA is repaired. Repair by the ‡ pathway is reduced by mutations in *recA*, *recC*, *recJ*, *recN*, *recO*, and *lexA* (Ind$^-$). Repair by the RecE pathway appears to require only RecE or RecT, or both, whereas that by the RecF pathway requires RecA, F, J, O, and Q and LexA, but not RecN or RuvB. These requirements differ from those for other types of repair and recombination by these pathways (*see* Sections 6.1., 6.3., and 7.) and indicate that the functional requirements for a pathway depend on the substrates.

A simple model for the repair of linear, direct repeat DNA by the RecE pathway employs RecE exonuclease generating 3'-ended ssDNA at one or both ends of the substrate. These ends then invade homologous dsDNA (Fig. 1A, B, or C) or anneal with homologous ssDNA (Fig. 1D). However, several kb of nonhomologous DNA on one or both ends do not significantly reduce the L/C ratio, arguing against all of these models *(55)*. The authors prefer a model in which a protein enters the dsDNA end and promotes a four-strand reaction in the interior of the homologous repeats.

In related studies, linear DNA with direct repeats was generated in *E. coli* expressing the RecE pathway and *Eco*RI; cells were infected with phage λ bearing the DNA with repeats bracketed by *Eco*RI sites *(79)*. Recombination between the repeats is

---

heritable circular plasmid and a nonheritable linear fragment. **(D)** Linear DNA containing two inverted repeats in three segments is introduced into *E. coli*. An intramolecular reciprocal ("conservative") reaction repairs the gap (≈300 bp) without or with inversion ("crossover") of the DNA flanking the repeats. **(E)** Linear DNA with homology to the chromosome is introduced into *E. coli*. Two exchange reactions replace the chromosomal allele *(B$^-$)* with that from the introduced DNA *(B$^+$)* ("gene targeting"). A nonheritable linear fragment is produced if the exchange reactions are reciprocal. Two Chi sites, oriented to act in the interval between them, stimulate the reaction by the RecBCD pathway.

strongly stimulated by *Eco*RI cutting; at low multiplicity of infection, 10% of the infected cells yield circular plasmids with one repeat. Physical and genetic analyses showed the formation of heteroduplex DNA in the repeated DNA with the DNA strand polarity expected for 5' → 3'-digestion by RecE and strand annealing as in Fig. 1D. The paucity of one recombinant type expected by the model in Fig. 1B from a related substrate with direct repeats (Fig. 3C) led Silberstein et al. *(79)* to argue against two-end invasion models, such as that in Fig. 1B, and to favor alternatives, such as those in Fig. 1A, C, and D.

Recombination of direct repeats in a more complex substrate (Fig. 3C) by the RecF pathway is also stimulated 10- to 100-fold by a DSB or an ~300-bp gap within one repeat *(91)*. Repair is efficient (30–100% measured as L/C as above) and requires RecA, F, J, and N. When the DNA contains two origins of replication, allowing the recovery of the two reciprocal recombinant plasmids, only one monomeric plasmid is obtained at high frequency. This result indicates that the overall reaction is nonreciprocal, or in the authors' nomenclature "nonconservative," meaning that two parental DNA duplexes produce only one heritable product DNA duplex (Fig. 3C, left). Alternatively, the reaction at each DNA end may be reciprocal, but if the reactions at the two ends occur independently (i.e., without coordination), the reciprocal products might only rarely occur in the same cell. Successive (presumably uncoordinated) rounds of recombination of uncut established plasmids with inverted repeats (Fig. 3D; *see below*) have been invoked to explain the types of plasmids generated by the RecF pathway *(109)*.

Repair of an ~300 bp dsDNA gap in one of two inverted repeats on a plasmid (Fig. 3D) has been extensively studied (reviewed in *30*). Recombination between the repeats generates an intact neo^+ gene, which confers kanamycin resistance (Kan^R). In one interval between the repeats is an intact *amp* gene, conferring ampicillin resistance (Amp^R). Amp^R transformants of *E. coli* are produced with or without recombination, provided that the gap is repaired to produce circular DNA. The gap stimulates recombination (measured as Kan^R transformants/μg of DNA) 10- to 100-fold, indicating the recombinogenicity of dsDNA ends. Repair efficiency, measured as the ratio of the Amp^R transformant frequency with linear DNA to that with circular DNA (L/C as above), is about 5% by the RecE and Red pathways, but 50–100% in cells overexpressing RecE and T *(41)*. This result suggests that these enzymes are limiting for repair of DSBs in *recBC sbcA* mutants.

Repair of the gap in this plasmid (Fig. 3D) by the RecE and Red pathways appears to occur by the mechanism in Fig. 1A or B. The reaction can be intramolecular (monomolecular) *(110)* and requires RecE and T, or Redα and β, for the two pathways, respectively, but does not require RecA, F, G, J, N, O, R, or Q, RuvC, or an inducible LexA regulon *(40,41,89,90)*. These requirements, as far as tested, are the same as those for recombination of linear DNA with direct repeats (Fig. 3B) by the RecE pathway (*see above* in this Section). However, terminal heterologous DNA blocks recombination (repair) of the inverted repeat DNA (cited in *41*), but not that of direct repeat DNA *(55)*. These results suggest that homologous DNA ends are required for the invasions shown in Fig. 1B for repair of the gap in the inverted repeat, but that another reaction, such as SSA (Fig. 1D), may allow direct repeat recombination. As expected for an SSA reaction (*see* Sections 3., 4.7., and 5.3.), direct repeat recombination does not require RecA protein, but unexpectedly neither does inverted repeat recombination.

Inversion of the DNA between the repeats (Fig. 3D), called "crossing-over," occurs in about 70% of the *neo*$^+$ recombinants by the RecE pathway. Although *recQ* and *recJ* mutations do not greatly alter the repair efficiency, they do reduce the frequency of inversion among the *neo*$^+$ recombinants to 40 and 10%, respectively, when present singly or to 40% when present together *(40)*. These mutations have parallel effects on the sensitivity of the cells to γ-rays, suggesting that crossing over or a related reaction is important for repair of DSBs in the chromosome.

Repair of a gap (Fig. 3D) occurs at low or undetectable frequency by the RecBCD, ‡, and RecF pathways *(30)*. Lack of properly placed Chi sites in the DNA may account for lack of repair by the RecBCD pathway. It is unclear why repair does not occur by the ‡ pathway; Kobayashi *(30)* has suggested that the RecBC complex binds to a 3'-end and blocks its priming of the DNA synthesis necessary for gap repair. The RecF pathway appears unable to effect gap repair for a different reason. Established (uncut) inverted repeat plasmids (Fig. 3D) recombine to *neo*$^+$ by the RecF pathway, requiring RecA, F, J, and N and RuvC, but do so without inversion of the flanking DNA ("noncrossover") *(109)*. The authors invoke the "nonconservative" mechanism discussed above for direct repeat recombination by the RecF pathway: at least two rounds of single nonreciprocal exchanges, necessarily intermolecular, would be required to generate a circular *neo*$^+$ derivative of the inverted repeat plasmid (Fig. 3D), and on transformation only one linear DNA may typically enter a cell. In other words, the RecF pathway cannot repair the gap in a monomolecular reaction because of its apparent nonreciprocal ("nonconservative") nature.

6.5. *Transformation with Linear DNA Homologous to the* E. coli *Chromosome ("Gene Targeting")*

Bacterial transformation was first demonstrated with chromosomal DNA extracted from *Streptococcus pneumoniae* and added to specially grown ("competent") recipient cells *(20)*. With Ca^{2+}-treated *E. coli*, transformation with chromosomal DNA occurs, but at low frequency: about 10^{-8}–10^{-6} of the recipient cells are transformed. Transformation can occur by the RecBCD pathway or, at about 10-fold higher frequency, by the RecE or RecF pathway *(10,24)*. The DNA used in these experiments was about 100 kbp long and therefore likely contained multiple Chi sites enabling the RecBCD pathway.

Shorter linear DNA (≤10 kbp) prepared from plasmids has been used to introduce into the chromosome mutations generated on the plasmid ("gene targeting"; Fig. 3E). Transformation with replacement of the chromosomal allele occurs at a low frequency (1–30 recombinants/μg of DNA) by the RecF pathway *(28,108)*, and at a similar frequency (1/μg; ref. *78*) or much higher frequency (100–3000/μg; ref. *73*) by the ‡ pathway. The efficiency of transformation (L/C as in Section 6.4.) is 10^{-4} by the ‡ pathway *(73)*. Terminal heterology on the linear DNA does not seem to alter the frequency of transformation, but reducing the length of homology from 4.3 kb to 1.9 or 0.3 kb reduces the frequency by a factor of 5 or >50, respectively *(73)*.

The RecBCD pathway can also effect gene replacement provided the linear DNA has a Chi site near each end properly oriented to promote recombination in the homologous region between the Chi sites (Fig. 3E) *(11)*. The frequency of transformation is increased from 1.4/μg of linear DNA to 64/μg by the presence of both Chi sites; single Chi sites give intermediate frequencies. The efficiency (L/C) with the double Chi substrate is about 10^{-4}, the same as for the ‡ pathway. Gene replacement by the RecBCD pathway requires RecA protein and RecBCD enzyme. These results support the model for the RecBCD pathway discussed in the next section.

7. EXTENSION TO HOMOLOGOUS RECOMBINATION

Homologous recombination appears to occur most frequently in *E. coli* when there is linear DNA in the cell (for examples, *see* Sections 6.3.–6.5.). This outcome supports the view, amplified below, that recombination is a reflection of DSB repair. Stahl and his colleagues have extensively studied the effect of DSBs on phage λ recombination by each of the pathways described in Section 5. (For reviews *see 62* and *100*). They studied breaks at defined sites induced by *Eco*RI restriction enzyme or by λ Ter protein. In each case, the breaks increased the recombinant frequency among the surviving molecules. Except for the RecBCD pathway, stimulation of recombination is near the break (≤5 kb away), as expected by the invasion of an intact homolog by a broken end. For the RecBCD pathway, stimulation is at and "downstream" of a Chi site encountered by RecBCD enzyme entering the dsDNA end (*see* Section 4.1.). The stimulation of recombination by DSBs is consistent with events diagramed in Fig. 1 occurring after formation of the break. The multiplicity of λ chromosomes in the cell and the possibility of multiple events, however, make it difficult to show that the DSB is repaired (i.e., that the two broken fragments of a single substrate are rejoined). Rejoining may require a coordinated, and hence rare, action of the two ends with an intact homolog.

Recombination of *E. coli* chromosomal fragments injected into a cell during phage P1-mediated transduction appears to occur primarily near the ends of the injected DNA; conjugational DNA fragments, after their conversion to dsDNA by lagging strand synthesis, also appear to recombine primarily near their ends (*see 82* for a review). By definition, conjugational recombination can occur by each pathway described in Section 5, as can transduction. Processing of these ends, by RecBCD or other enzymes (Sections 4.1.–4.4.), is postulated to generate 3' ss tails that invade an intact homolog (Sections 4.5.–4.7.) (Fig. 4A). These 3'-invading ends are further postulated to prime DNA synthesis that leads to two replication forks proceeding in opposite directions around the circular chromosome (Fig. 4B, top). Completion of this bidirectional replication yields two intact chromosomes joined by two Holliday junctions, which are resolved (Sections 4.8. and 4.9.) into one dimeric or two monomeric chromosomes. The dimer can be converted to monomers by the XerCD site-specific recombination at *dif* sites by the XerCD enzyme (reviewed in *77*).

Repair of a DSB is topologically identical to conjugational or transductional recombination by this mechanism (Fig. 4B, bottom). A broken chromosome is equivalent to a fragment nearly 100% of the length of a chromosome. These views are consistent with the similar gene functions required for repair of DNA broken by X-rays (Section 6.1.) and for conjugational recombination (Section 5.). The finding that PriA is required for each process, at least by the RecBCD pathway, supports the hypothesis that both involve DNA replication *(34)*.

The following observations have suggested that blockage of the *E. coli* replication fork near the terminus of replication leads to a DSB at which RecBCD enzyme enters and promotes recombination. In the absence of RNase H, *E. coli* chromosomal replication can apparently be initiated by transcripts in R-loops at many places on the chromo-

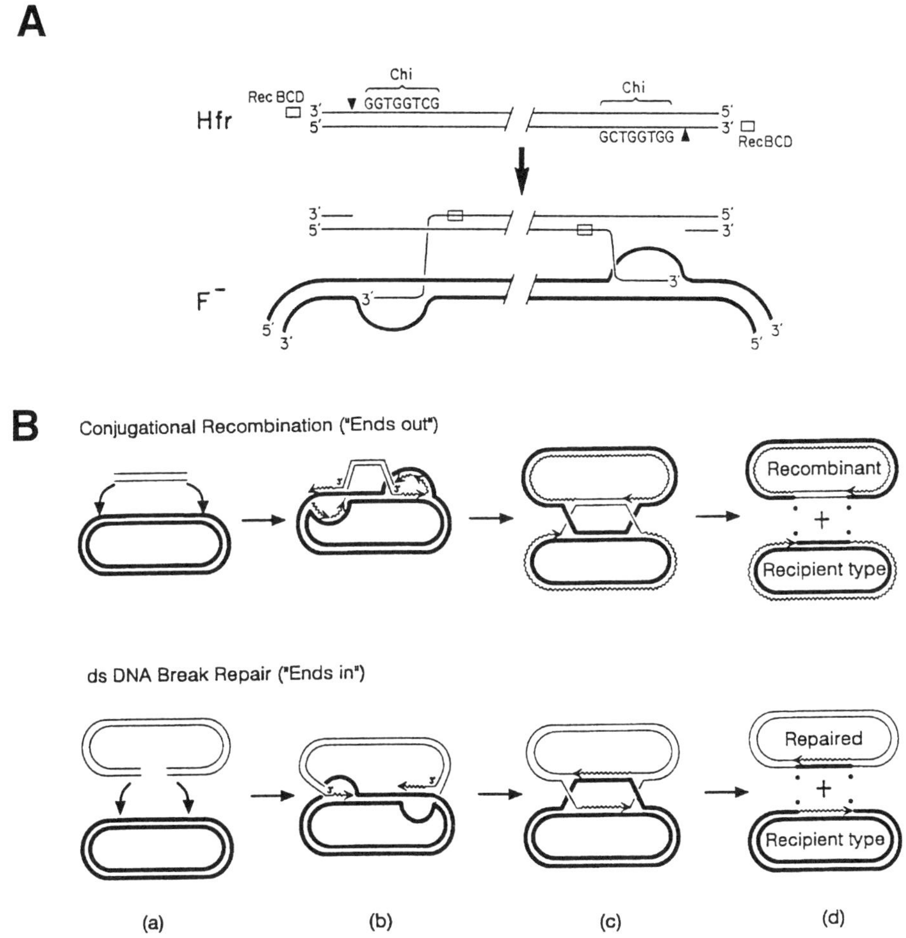

Fig. 4. Model for RecBCD enzyme-dependent recombination and DSB repair in *E. coli*. **(A)** Two RecBCD enzymes (open boxes) attack the ends of a linear DNA molecule (thin lines). Encounter with a properly oriented Chi sequence and continued unwinding generate at each end a 3' ss tail (Fig. 2B) that, aided by RecA and SSB proteins, invades an intact homologous chromosome (thick lines). These ends prime DNA replication as shown in **(B)**. (B) The reactions in (A) lead to two replication forks advancing in opposite directions around the chromosome. Completion of replication and joining of lagging strand 3'-ends to the initially linear DNA generate two complete chromosomes joined by two Holliday junctions. Resolution of these junctions in the horizontal plane produces the recombinant (or repaired) and recipient-type chromosomes diagramed. Alternative resolutions produce dimeric chromosomes or reciprocals of the types diagrammed. Bullets (●) indicate potential points of hybrid DNA. The bottom scheme is essentially identical to that in Fig. 1A,B *(68,88)*. From ref. *(82)* with permission.

some. Nishitani et al. *(63)* sought these origins on *E. coli* DNA fragments ligated to an origin-less *kan* fragment. The fragments found, as those giving high-frequency Kan^R transformants containing circular *kan* DNA ("Hot" assay), were mostly from the terminus of replication and contained Chi. These fragments were not "Hot" in a *recA* mutant or in cells with a chromosomal deletion of the fragment. From these observations, the authors deduced that the initial circular DNA integrates into the chromosome to form a direct repeat and is excised at high frequency, in a Chi-dependent way, to form the observed circles. "Hot" activity depends on Chi (in at least the one case tested by site-directed mutagenesis) and on Tau, a protein that halts replication complexes at sites *(Ter)* flanking the terminus of replication *(27)*. A *Ter* site-Hot Chi^+ fragment pair behaves similarly when placed at *lac*, far from the normal replication terminus *(26)*. Growth of such a strain depends on RecA and RecBCD enzyme. The authors propose that a replication fork blocked at *Ter* frequently leads to a DSB, for example, by scission of the ss gap at the fork, and that this DSB must be repaired by RecA-RecBCD-Chi-dependent reactions, perhaps similar to those in Fig. 4. Louarn et al. *(51)* also observed high-frequency excision between direct repeats near the terminus of replication and postulated a role for homologous recombination in the final stages of chromosome replication. For further discussion of the role of recombination in the repair of broken replication forks, *see* Kuzminov *(42)*.

8. DS BREAK REPAIR IN THE ABSENCE OF HOMOLOGY

Linear DNA lacking extensive homology within the molecule or with the chromosome circularizes at low frequency in *E. coli* (*56* and references therein). The frequency of transformation with linear DNA compared to that with circular DNA (L/C ratio or efficiency, as discussed in Sections 6.4. and 6.5.) is about 10^{-3}. In nearly all cases, DNA from one or both sides of the DSB is deleted in the recovered plasmids. The deletions generally occur between direct repeats of 2–11 bp, which may be from 4 to > 1500 bp apart. Mutations inactivating exonuclease III, RecF, RecR (but not RecO), RuvA, RuvB (but not RuvC), and RecG reduce the L/C ratio by more than a factor of 100 relative to that in wild-type cells. These results suggest degradation of the dsDNA end by exonuclease III and annealing of the resultant ssDNA at points of short homology, perhaps followed by trimming and ligation to produce plasmids with deletions. This postulated mechanism is similar to that in Fig. 1D.

9. CONCLUSION

There are multiple pathways by which DSBs can be repaired in *E. coli*. A major mechanism appears to be that by which homologous recombinants are produced, but even here, there are multiple potential pathways depending on the genotype of the *E. coli* cells and the structure of the substrates. A common feature of the recombinational mechanisms appears to be the generation, at the DSB, of 3'-ended ssDNA that invades an intact homolog and primes DNA synthesis (Fig. 1). The reassociation of the two initial ends would require that both ends invade the same homolog. Since in many of the cases discussed here there are multiple homologs (e.g., multiple chromosomes or plasmid copies), the two ends may invade different homologs. Such independent invasions would, by further processing, remove the ends from the cell, but not restore the initial chromosome. Invasion of one end but not the other could also generate a recombinant, but not restore the initial chromosome. The relative frequencies of these three events (two ends invading one homolog, two ends invading two homologs, and one end invading) remain undetermined. If the first event were to occur at a high frequency, it might reflect a higher-order structure that coordinates DSB repair to ensure its high fidelity.

ACKNOWLEDGMENTS

For fruitful discussions and helpful comments on the manuscript, I am grateful to Sue Amundsen, Jerry Bedoyan, Marlene Belfort, Amikam Cohen, Joe Farah, Ichizo Kobayashi, Richard Kolodner, Andrew Taylor, Steve West, and a reviewer. I thank Karen Brighton for skillfully and patiently preparing the manuscript, and Jerry Bedoyan for Figs. 1 and 3. Research in my laboratory is supported by research grants GM31693 and GM32194 from the National Institutes of Health.

REFERENCES

1. Amundsen, S. K., A. F. Taylor, A. M. Chaudhury, and G. R. Smith. 1986. *recD*: The gene for an essential third subunit of exonuclease V. *Proc. Natl. Acad. Sci. USA* **83:** 5558–5562.
2. Cairns, J. 1963. Bacterial chromosome and its manner of replication as seen by autoradiography. *J. Mol. Biol.* **6:** 208–213.
3. Camerini-Otero, R. D. and P. Hsieh. 1993. Parallel DNA triplexes, homologous recombination, and other homology-dependent DNA interactions. *Cell* **73:** 217–223.
4. Chaudhury, A. M. and G. R. Smith. 1984. A new class of *Escherichia coli recBC* mutants: Implications for the role of RecBC enzyme in homologous recombination. *Proc. Natl. Acad. Sci. USA* **81:** 7850–7854.
5. Cheng, K. C. and G. R. Smith. 1987. Cutting of Chi-like sequences by the RecBCD enzyme of *Escherichia coli*. *J. Mol. Biol.* **194:** 747–750.
6. Cheng, K. C. and G. R. Smith. 1989. Distribution of Chi-stimulated recombinational exchanges and heteroduplex endpoints in phage lambda. *Genetics* **123:** 5–17.
7. Clark, A. J. 1971. Toward a metabolic interpretation of genetic recombination of *E. coli* and its phages. *Annu. Rev. Microbiol.* **25:** 437–464.
8. Clyman, J. and M. Belfort. 1992. *Trans* and *cis* requirements for intron mobility in a prokaryotic system. *Genes Dev.* **6:** 1269–1279.
9. Connelly, J. C. and D. R. F. Leach. 1996. The *sbcC* and *sbcD* genes of *Escherichia coli* encode a nuclease involved in palindrome inviability and genetic recombination. *Genes to Cells* **1:** 285–291.
10. Cosloy, S. D. and M. Oishi. 1973. Genetic transformation in *Escherichia coli* K12. *Proc. Natl. Acad. Sci. USA* **70:** 84–87.
11. Dabert, P. and G. R. Smith. 1997. Gene replacement in wild-type *Escherichia coli*: enhancement by Chi sites. *Genetics* **145:** 877–889.
12. Dixon, D. A., J. J. Churchill, and S. C. Kowalczykowski. 1994. Reversible inactivation of the *Escherichia coli* RecBCD enzyme by the recombination hotspot chi *in vitro*: Evidence for functional inactivation or loss of the RecD subunit. *Proc. Natl. Acad. Sci. USA* **91:** 2980–2984.
13. Dixon, D. A. and S. C. Kowalczykowski. 1991. Homologous pairing in vitro stimulated by the recombination hotspot, Chi. *Cell* **66:** 361–371.
14. Dixon, D. A. and S. C. Kowalczykowski. 1993. The recombination hotspot chi is a regulatory sequence that acts by attenuating the nuclease activity of the E. coli RecBCD enzyme. *Cell* **73:** 87–96.
15. Dixon, D. A. and S. C. Kowalczykowski. 1995. Role of the *Escherichia coli* recombination hotspot, χ, in RecABCD-dependent homologous pairing. *J. Biol. Chem.* **270:** 16,360–16,370.
16. Eddy, S. R. and L. Gold. 1992. Artificial mobile DNA element constructed from the *Eco*RI endonuclease gene. *Proc. Natl. Acad. Sci. USA* **89:** 1544–1547.
17. Faulds, D., N. Dower, M. M. Stahl, and F. W. Stahl. 1979. Orientation-dependent recombination hotspot activity in bacteriophage lambda. *J. Mol. Biol.* **131:** 681–695.

18. Formosa, T. and B. M. Alberts. 1986. DNA synthesis dependent on genetic recombination: characterization of a reaction catalyzed by purified bacteriophage T4 proteins. *Cell* **47:** 793–806.
19. Friedberg, E. C., G. C. Walker, and W. Siede 1995. *DNA Repair and Mutagenesis.* ASM Press, Washington, DC.
20. Griffith, F. 1928. The significance of pneumococcal types. *J. Hyg.* **27:** 113–159.
21. Hall, S. D. and R. D. Kolodner. 1994. Homologous pairing and strand exchange promoted by the *Escherichia coli* RecT protein. *Proc. Natl. Acad. Sci. USA* **91:** 3205–3209.
22. Hegde, S. P., M. Rajagopalan, and M. V. V. S. Madiraju. 1996. Preferential binding of *Escherichia coli* RecF protein to gapped DNA in the presence of adenosine (γ-thio) triphosphate. *J. Bacteriol.* **178:** 184–190.
23. Heitman, J., N. D. Zinder, and P. Model. 1989. Repair of the *Escherichia coli* chromosome after *in vivo* scission by the *Eco*RI endonuclease. *Proc. Natl. Acad. Sci. USA* **86:** 2281–2285.
24. Hoekstra, W. P. M., J. E. N. Bergmans, and E. M. Zuidweg. 1980. Role of *recBC* nuclease in *Escherichia coli* transformation. *J. Bacteriol.* **143:** 1031,1032.
25. Holliday, R. 1964. A mechanism for gene conversion in fungi. *Genet. Res.* **5:** 282–304.
26. Horiuchi, T. and Y. Fujimura. 1995. Recombinational rescue of the stalled DNA replication fork: a model based on analysis of an *Escherichia coli* strain with a chromosome region difficult to replicate. *J. Bacteriol.* **177:** 783–791.
27. Horiuchi, T., Y. Fujimura, H. Nishitani, T. Kobayashi, and M. Hidaka. 1994. The DNA replication fork blocked at the *Ter* site may be an entrance for the RecBCD enzyme into duplex DNA. *J. Bacteriol.* **176:** 4656–4663.
28. Jasin, M. and P. Schimmel. 1984. Deletion of an essential gene in *Escherichia coli* by site-specific recombination with linear DNA fragments. *J. Bacteriol.* **159:** 783–786.
29. Joseph, J. W. and R. Kolodner. 1983. Exonuclease VIII of *Escherichia coli.* I. Purification and physical properties. *J. Biol. Chem.* **258:** 10,411–10,417.
30. Kobayashi, I. 1992. Mechanisms for gene conversion and homologous recombination: The double-strand break repair model and the successive half crossing-over model. *Adv. Biophys.* **28:** 81–133.
31. Kobayashi, I., H. Murialdo, J. M. Crasemann, M. M. Stahl, and F. W. Stahl. 1982. Orientation of cohesive end site *cos* determines the active orientation of chi sequence in stimulating recA-recBC mediated recombination in phage lambda lytic infections. *Proc. Natl. Acad. Sci. USA* **79:** 5981–5985.
32. Köppen, A., S. Krobitsch, B. Thoms, and W. Wackernagel. 1995. Interaction with the recombination hot spot χ *in vivo* converts the RecBCD enzyme of *Escherichia coli* into a χ-independent recombinase by inactivation of the RecD subunit. *Proc. Natl. Acad. Sci. USA* **92:** 6249–6253.
33. Kogoma, T. 1996. Recombination by replication. *Cell* **85:** 625–627.
34. Kogoma, T., G. W. Cadwell, K. G. Barnard, and T. Asai. 1996. The DNA replication priming protein, PriA, is required for homologous recombination and double-strand break repair. *J. Bacteriol.* **178:** 1258–1264.
35. Kolodner, R., S. D. Hall, and C. Luisi-DeLuca. 1994. Homologous pairing proteins encoded by the *Escherichia coli recE* and *recT* genes. *Mol. Microb.* **11:** 23–30.
36. Kornberg, A. and T. A. Baker. 1992. *DNA Replication.* W. H. Freeman and Company, New York.
37. Kowalczykowski, S. C., D. A. Dixon, A. K. Eggleston, S. D. Lauder, and W. M. Rehrauer. 1994. Biochemistry of homologous recombination in *Escherichia coli. Microbiol. Rev.* **58:** 401–465.
38. Krasin, F. and F. Hutchinson. 1977. Repair of DNA double-strand breaks in *Escherichia coli*, which requires *recA* function and the presence of a duplicate genome. *J. Mol. Biol.* **116:** 81–98.

39. Krasin, F. and F. Hutchinson. 1981. Repair of DNA double-strand breaks in *Escherichia coli* cells requires synthesis of proteins that can be induced by UV light. *Proc. Natl. Acad. Sci. USA* **78:** 3450–3453.
40. Kusano, K., Y. Sunohara, N. Takahashi, H. Yoshikura, and I. Kobayashi. 1994. DNA double-strand break repair: Genetic determinants of flanking crossing-over. *Proc. Natl. Acad. Sci. USA* **91:** 1173–1177.
41. Kusano, K., N. K. Takahashi, H. Yoshikura, and I. Kobayashi. 1994. Involvement of RecE exonuclease and RecT annealing protein in DNA double-strand break repair by homologous recombination. *Gene* **138:** 17–25.
42. Kuzminov, A. 1995. Collapse and repair of replication forks in *Escherichia coli. Mol. Microb.* **16:** 373–384.
43. Kuzminov, A., E. Schabtach, and F. W. Stahl. 1994. χ sites in combination with RecA protein increase the survival of linear DNA in *Escherichia coli* by inactivating exoV activity of RecBCD nuclease. *EMBO J.* **13:** 2764–2776.
44. Lichten, M. and A. S. H. Goldman. 1995. Meiotic recombination hotspots. *Annu. Rev. Genet.* **29:** 423–444.
45. Lin, F.-L., K. Sperle, and N. Sternberg. 1984. Model for homologous recombination during transfer of DNA into mouse L cells: Role for DNA ends in the recombination process. *Mol. Cell. Biol.* **4:** 1020–1034.
46. Little, J. W. 1967. An exonuclease induced by bacteriophage lambda. II. Nature of the enzymic reaction. *J. Biol. Chem.* **242:** 679–686.
47. Lloyd, R. G. 1991. Conjugational recombination in resolvase-deficient *ruvC* mutants of *Escherichia coli* K-12 depends on *recG. J. Bacteriol.* **173:** 5414–5418.
48. Lloyd, R. G., C. Buckman, and F. E. Benson. 1987. Genetic analysis of conjugational recombination in *Escherichia coli* K12 strains deficient in RecBCD enzyme. *J. Gen. Microbiol.* **133:** 2531–2538.
49. Lloyd, R. G. and G. J. Sharples. 1992. Genetic analysis of recombination in prokaryotes. *Curr. Opinions Genet. Dev.* **2:** 683–690.
50. Lloyd, R. G. and G. J. Sharples. 1993. Dissociation of synthetic Holliday junctions by *E. coli* RecG protein. *EMBO J.* **12:** 17–22.
51. Louarn, J.-M., J. Louarn, V. François, and J. Patte. 1991. Analysis and possible role of hyperrecombination in the termination region of the *Escherichia coli* chromosome. *J. Bacteriol.* **173:** 5097–5104.
52. Lovett, S. T. and R. D. Kolodner. 1989. Identification and purification of a single-stranded-DNA-specific exonuclease encoded by the *recJ* gene of *Escherichia coli. Proc. Natl. Acad. Sci. USA* **86:** 2627–2631.
53. Lovett, S. T., C. Luisi-DeLuca, and R. D. Kolodner. 1988. The genetic dependence of recombination in *recD* mutants of *Escherichia coli. Genetics* **120:** 37–45.
54. Luisa-DeLuca, C., S. T. Lovett, and R. D. Kolodner. 1989. Genetic and physical analysis of plasmid recombination in *recB recC sbcB* and *recB recC sbcA Escherichia coli* K-12 mutants. *Genetics* **122:** 269–278.
55. Luisi-DeLuca, C. and R. D. Kolodner. 1992. Effect of terminal non-homology on intramolecular recombination of linear plasmid substrates in *Escherichia coli. J. Mol. Biol.* **227:** 72–80.
56. McFarlane, R. J. and J. R. Saunders. 1996. Molecular mechanisms of intramolecular recombination-dependent recircularization of linearized plasmid DNA in *Escherichia coli*: requirements for the *ruvA, ruvB, recG, recF* and *recR* gene products. *Gene* **177:** 209–216.
57. Meyer, R. R. and P. S. Laine. 1990. The single-stranded DNA-binding protein of *Escherichia coli. Microbiol. Rev.* **54:** 342–380.
58. Mosig, G. 1994. Homologous recombination, in *Molecular Biology of Bacteriophage T4* (Karam, J. D., ed.), American Society for Microbiology, Washington, DC, pp. 54–82.

59. Mueller, J. E., J. Clyman, Y.-J. Huang, M. M. Parker, and M. Belfort. 1996. Intron mobility in phage T4 occurs in the context of recombination-dependent DNA replication by way of multiple pathways. *Genes Dev.* **10:** 351–364.
60. Murialdo, H. 1988. Lethal effect of lambda DNA terminase in recombination-deficient *Escherichia coli. Mol. Gen. Genet.* **213:** 42–49.
61. Myers, R. S., A. Kuzminov, and F. W. Stahl. 1995. The recombination hot spot χ activates RecBCD recombination by converting *Escherichia coli* to a *recD* mutant phenocopy. *Proc. Natl. Acad. Sci. USA* **92:** 6244–6248.
62. Myers, R. S. and F. W. Stahl. 1994. χ and the RecBCD enzyme of *Escherichia coli. Annu. Rev. Genet.* **28:** 49–70.
63. Nishitani, H., M. Hidaka, and T. Horiuchi. 1993. Specific chromosomal sites enhancing homologous recombination in *Escherichia coli* mutants defective in RNase H. *Mol. Gen. Genet.* **240:** 307–314.
64. Orr-Weaver, T. L., J. W. Szostak, and R. J. Rothstein. 1981. Yeast transformation: A model system for the study of recombination. *Proc. Natl. Acad. Sci. USA* **78:** 6354–6358.
65. Parker, M. M., D. A. Court, K. Preiter, and M. Belfort. 1996. Homology requirements for double-strand break-mediated recombination in a phage λ-*td* intron model system. *Genetics* **143:** 1057–1068.
66. Picksley, S. M., P. Attfield V, and R. G. Lloyd. 1984. Repair of DNA double-strand breaks in *Escherichia coli* K12 requires a functional *recN* product. *Mol. Gen. Genet.* **195:** 267–274.
67. Razavy, G., S. K. Szigety, and S. M. Rosenberg. 1996. Evidence for both 3' and 5' single-strand DNA ends in intermediates in Chi-stimulated recombination *in vivo*. *Genetics* **142:** 333–339.
68. Resnick, M. A. 1976. The repair of double-strand breaks in DNA: a model involving recombination. *J. Theor. Biol.* **59:** 97–106.
69. Roman, L. J. and S. C. Kowalczykowski. 1989. Characterization of the helicase activity of *Escherichia coli* RecBCD enzyme using a novel helicase assay. *Biochemistry* **28:** 2863–2873.
70. Rosenberg, S. M. and P. J. Hastings. 1991. The split-end model for homologous recombination at double-strand breaks and at Chi. *Biochimie* **73:** 385–397.
71. Rosselli, W. and A. Stasiak. 1991. The ATPase activity of RecA is needed to push the DNA strand exchange through heterologous regions. *EMBO J.* **10:** 4391–4396.
72. Rostas, K., S. J. Morton, S. M. Picksley, and R. G. Lloyd. 1987. Nucleotide sequence and LexA regulation of the *Escherichia coli recN* gene. *Nucleic Acids Res.* **15:** 5041–5049.
73. Russell, C. B., D. S. Thaler, and F. W. Dahlquist. 1989. Chromosomal transformation of *Escherichia coli recD* strains with linearized plasmids. *J. Bacteriol.* **171:** 2609–2613.
74. Sandler, S. J., H. S. Samra, and A. J. Clark. 1996. Differential suppression of *priA2::kan* phenotypes in *Escherichia coli* K-12 by mutations in *priA*, *lexA*, and *dnaC*. *Genetics* **143:** 5–13.
75. Sargentini, N. J. and K. C. Smith. 1986. Quantitation of the involvement of the *recA*, *recB*, *recC*, *recJ*, *recN*, *lexA*, *radA*, *radB*, *uvrD*, and *umuC* genes in the repair of DNA double-strand breaks in *Escherichia coli*. *Radiat. Res.* **107:** 58–72.
76. Sawitzke, J. A. and F. W. Stahl. 1994. The phage λ *orf* gene encodes a *trans*-acting factor that suppresses *Escherichia coli recO*, *recR*, and *recF* mutations for recombination of λ but not of *E. coli*. *J. Bacteriol.* **176:** 6730–6737.
77. Sherratt, D. J., L. K. Arciszewska, G. Blakely, S. Colloms, K. Grant, N. Leslie, and R. McCulloch. 1995. Site-specific recombination and circular chromosome segregation. *Phil. Trans. R. Soc. (London)* **347:** 37–42.
78. Shevell, D. E., A. M. Abou-Zamzam, B. Demple, and G. C. Walker. 1988. Construction of an *Escherichia coli* K-12 *ada* deletion by gene replacement in a *recD* strain reveals a second methyltransferase that repairs alkylated DNA. *J. Bacteriol.* **170:** 3294–3296.

79. Silberstein, Z., M. Shalit, and A. Cohen. 1993. Heteroduplex strand-specificity in restriction-stimulated recombination by the RecE pathway of *Escherichia coli*. *Genetics* **133:** 439–448.
80. Smith, G. R. 1988. Homologous recombination in procaryotes. *Microbiol. Rev.* **52:** 1–28.
81. Smith, G. R. 1989. Homologous recombination in E. coli: multiple pathways for multiple reasons. *Cell* **58:** 807–809.
82. Smith, G. R. 1991. Conjugational recombination in E. coli: myths and mechanisms. *Cell* **64:** 19–27.
83. Smith, G. R. 1994. Hotspots of homologous recombination. *Experientia* **50:** 234–241.
84. Smith, G. R., S. K. Amundsen, P. Dabert, and A. F. Taylor. 1995. The initiation and control of homologous recombination in *Escherichia coli*. *Phil. Trans. R. Soc. (London)* **347:** 13–20.
85. Stahl, F. W., M. M. Stahl, R. E. Malone, and J. M. Crasemann. 1980. Directionality and nonreciprocality of Chi-stimulated recombination in phage lambda. *Genetics* **94:** 235–248.
86. Story, R. M., I. T. Weber, and T. A. Steitz. 1992. The structure of the *E. coli* recA protein monomer and polymer. *Nature* **355:** 318–325.
87. Symington, L. S., P. Morrison, and R. Kolodner. 1985. Intramolecular recombination of linear DNA catalyzed by the *Escherichia coli RecE* recombination system. *J. Mol. Biol.* **186:** 515–525.
88. Szostak, J. W., T. L. Orr-Weaver, R. J. Rothstein, and F. W. Stahl. 1983. The double-strand-break repair model for recombination. *Cell* **33:** 25–35.
89. Takahashi, N. and I. Kobayashi. 1990. Evidence for the double-strand break repair model of bacteriophage λ recombination. *Proc. Natl. Acad. Sci. USA* **87:** 2790–2794.
90. Takahashi, N. K., K. Kusano, T. Yokochi, Y. Kitamura, H. Yoshikura, and I. Kobayashi. 1993. Genetic analysis of double-strand break repair in *Escherichia coli*. *J. Bacteriol.* **175:** 5176–5185.
91. Takahashi, N. K., K. Yamamoto, Y. Kitamura, S.-Q. Luo, H. Yoshikura, and I. Kobayashi. 1992. Nonconservative recombination in *Escherichia coli*. *Proc. Natl. Acad. Sci. USA* **89:** 5912–5916.
92. Taylor, A. F. 1988. RecBCD enzyme of *Escherichia coli*, in *Genetic Recombination* (Kucherlapati, R. and G. R. Smith, eds.), American Society for Microbiology, Washington, DC, pp. 231–263.
93. Taylor, A. F., D. W. Schultz, A. S. Ponticelli, and G. R. Smith. 1985. RecBC enzyme nicking at Chi sites during DNA unwinding: Location and orientation dependence of the cutting. *Cell* **41:** 153–163.
94. Taylor, A. F. and G. R. Smith. 1980. Unwinding and rewinding of DNA by the RecBC enzyme. *Cell* **22:** 447–457.
95. Taylor, A. F. and G. R. Smith. 1985. Substrate specificity of the DNA unwinding activity of the RecBC enzyme of *Escherichia coli*. *J. Mol. Biol.* **185:** 431–443.
96. Taylor, A. F. and G. R. Smith. 1992. RecBCD enzyme is altered upon cutting DNA at a Chi recombination hotspot. *Proc. Natl. Acad. Sci. USA* **89:** 5226–5230.
97. Taylor, A. F. and G. R. Smith. 1995. Monomeric RecBCD enzyme binds and unwinds DNA. *J. Biol. Chem.* **270:** 24,451–24,458.
98. Taylor, A. F. and G. R. Smith. 1995. Strand specificity of nicking of DNA at Chi sites by RecBCD enzyme: modulation by ATP and magnesium levels. *J. Biol. Chem.* **270:** 24,459–24,467.
99. Thaler, D. S., E. Sampson, I. Siddiqi, S. M. Rosenberg, F. W. Stahl, and M. Stahl. 1988. A hypothesis: Chi-activation of RecBCD enzyme involves removal of the RecD subunit, in *Mechanisms and Consequences of DNA Damage Processing* (Friedberg, E. and P. Hanawalt, eds.), Alan R. Liss, New York, pp. 413–422.
100. Thaler, D. S. and F. W. Stahl. 1988. DNA double-chain breaks in recombination of phage lambda and of yeast. *Annu. Rev. Genet.* **22:** 169–197.

101. Tracy, R. B. and S. C. Kowalczykowski. 1996. In vitro selection of preferred DNA pairing sequences by the *Escherichia coli* RecA protein. *Genes Dev.* **10:** 1890–1903.
102. Umezu, K. and H. Nakayama. 1993. RecQ DNA helicase of *Escherichia coli*: characterization of the helix-unwinding activity with emphasis on the effect of single-stranded DNA binding protein. *J. Mol. Biol.* **230:** 1145–1150.
103. Webb, B. L., M. M. Cox, and R. B. Inman. 1995. An interaction between the *Escherichia coli* RecF and RecR proteins dependent on ATP and double-stranded DNA. *J. Biol. Chem.* **270:** 31,397–31,404.
104. West, S. C. 1992. Enzymes and molecular mechanisms of genetic recombination. *Annu. Rev. Biochem.* **61:** 603–640.
105. West, S. 1996. The RuvABC proteins and Holliday junction processing in *Escherichia coli*. *J. Bacteriol.* **178:** 1237–1241.
106. Whitby, M. C., L. Ryder, and R. G. Lloyd. 1993. Reverse branch migration of Holliday junctions by RecG protein: a new mechanism for resolution of intermediates in recombination and DNA repair. *Cell* **75:** 341–350.
107. Whitby, M. C., G. J. Sharples, and R. G. Lloyd 1995. The RuvAB and RecG proteins of *Escherichia coli*, in *Nucleic Acids and Molecular Biology*, vol. 9 (Eckstein, F. and D. M. J. Lilley, eds.), Springer-Verlag, Berlin Heidelberg, pp. 66–83.
108. Winans, S. C., S. J. Elledge, J. H. Krueger, and G. C. Walker. 1985. Site-directed insertion and deletion mutagenesis with cloned fragments in *Escherichia coli*. *J. Bacteriol.* **161:** 1219–1221.
109. Yamamoto, K., K. Kusano, N. K. Takahashi, H. Yoshikura, and I. Kobayashi. 1992. Gene conversion in the *Escherichia coli* RecF pathway: a successive half crossing-over model. *Mol. Gen. Genet.* **234:** 1–13.
110. Yokochi, T., K. Kusano, and I. Kobayashi. 1995. Evidence for conservative (two-progeny) DNA double-strand break repair. *Genetics* **139:** 5–17.

9

Transcription Repair Coupling in *Escherichia coli*

Richard Bockrath

1. INTRODUCTION

The broad specificity of nucleotide excision repair for a variety of lesions in DNA is commonly attributed to recognition of distortion in the double helix. The repair process removes from one strand a fragment of DNA containing damage and uses the other strand as template for repair synthesis to restore integrity of the duplex DNA (*18,39*; Chapter 2).

The founding ideas about this widespread repair process appreciated the redundancy of information in duplex DNA, which was essential for repair of damage in one strand, without regard for which strand contained the damage. Since the two sugar-phosphate repeating polymers of DNA run antiparallel and therefore the basic nature of the duplex is twofold symmetrical, the possibility of strand-specific repair received little attention.

In the past few years, however, these ideas have been unsettled by realization that lesions in one strand of DNA can be preferentially repaired. The symmetry taken for granted for so long has been broken, and gene expression using one strand of DNA as template for RNA synthesis during transcription is apparently the defining feature. There is, in fact, a fundamental aspect of nucleotide excision repair that couples it with transcription, very likely in all cells, to account for selective repair in the transcribed strand.

Transcription-coupled excision repair was discovered in eukaryotic cells. It was then demonstrated in *Escherichia coli*, where it was linked to a particular gene by a combination of contemporary biochemical studies and a lineage of UV mutagenesis reports going back to the 1950s.

2. DISCOVERING THE TRANSCRIPTION REPAIR COUPLING FACTOR

2.1. *Eukaryotic Studies and Initial In Vitro Studies with* E. coli *Components*

Hanawalt *(16)* has recently reviewed studies of excision repair at select genetic regions in eukaryotic cells. Variations in damage and repair throughout the eukaryotic genome attracted interest for many years, but emerged with experimental definition only in the past 10–15 yr.

Zolan et al. *(57)* first reported differential repair activity, finding repair of furocoumarin adducts in highly repetitive α-DNA only 30% as efficient as in bulk DNA. Bohr et al. *(6)* then described much more efficient repair of UV-induced

From: DNA Damage and Repair, Vol. 1: DNA Repair in Prokaryotes and Lower Eukaryotes
Edited by: J. A. Nickoloff and M. F. Hoekstra © Humana Press Inc., Totowa, NJ

cyclobutane pyrimidine dimers (CPDs) in the expressed dihydrofolate reductase gene *(DHFR)* than in nonexpressed DNA just outside the gene or in bulk DNA.

Repair after UV irradiation was measured at specific genetic regions by a Southern transfer and quantitative hybridization with ^{32}P-labeled nick-translated probes *(6,16)*. Mellon et al. *(27)*, modifying this protocol to use RNA probes specific for one or the other strand, demonstrated a clear bias for excision of CPDs from the transcribed strand of the *DHFR* gene in rodent and human cells.

Their kinetic data showed similar, efficient rates of repair in the transcribed strand for either rodent or human cells. However, repair in the nontranscribed strand was essentially zero in rodent cells but moderate in human cells. Since the sensitivity to UV inactivation of the two cell types is similar even though the bulk rate of repair in rodent cells is about half that in human cells, these results underscored the possible significance to overall survival of strand-specific repair in the transcribed strands of expressed genes. In addition, this work considered repair in a region adjacent to *DHFR* that was transcribed in the divergent direction. Preferential repair was again found in the transcribed strand arguing against the possibility that the asymmetry arose from DNA replication.

Mayne and Lehmann *(23)* had previously noticed unusually slow recovery of RNA synthesis after UV irradiation of fibroblasts from a patient with Cockayne's syndrome. Normally, RNA synthesis is an early recovery response. They suggested that these cells might be defective in repair at active genes. Conversely, cells from patients with xeroderma pigmentosum (complementation group C) seem to be generally defective for excision repair, except for expressed genes of the nuclear matrix *(28)*. Transcription repair coupling in eukaryotic cells has now emerged as a sophisticated process, and genetic analysis has identified individual proteins serving both transcription and nucleotide excision repair (*15,16*; Chapters 10 and 18, vol. 2).

Soon after the first demonstration in eukaryotic cells, Mellon and Hanawalt *(25)* found preferential excision in the transcribed strand of the lactose operon in *E. coli*. This underscored the fundamental nature of this type of mechanism and allowed a test for the role of gene activity. The lactose operon can be repressed or derepressed. When the cells were grown with IPTG to induce expression, repair kinetics for CPDs in the nontranscribed strand were unaltered, but repair in the transcribed DNA strand was strikingly more rapid. The authors proposed a model in which the stalled RNA polymerase at a lesion served to recruit repair proteins to the damaged site.

Selby and Sancar *(41,42)* then set up an in vitro system of *E. coli* components to examine selective excision repair in transcribed strands. A region of DNA in a plasmid between asymmetrical restriction endonuclease sites could be monitored by electrophoresis on a sequencing gel to quantify the amount of repair synthesis specifically in the transcribed or the nontranscribed strand. The system included UvrA, UvrB, UvrC, UvrD, DNA polymerase I, and ligase, in addition to RNA polymerase for transcription.

Initially, only damage in the transcribed strand was found to stall transcription, and this actually resulted in less repair synthesis than did damage in the nontranscribed strand *(41)*. Presumably stalled RNA polymerase encumbered access of repair proteins. However, when a complex extract from *E. coli* was added, this not only relieved the inhibition, but resulted in preferential repair of damage in the transcribed strand *(42)*. This indicated that an additional activity from *E. coli* was required to reconstruct the sort of strand bias observed in vivo.

2.2. *UV Mutagenesis and Mutation Frequency Decline (MFD)*

Early UV mutagenesis experiments with *E. coli* carried some mystery because the quantitative results varied depending on certain aspects of cell preparation or the assay medium. Thus, in the 1950s, the precision expected of discrete genetic transitions was not easily separated from the adaptive nature of bacteria. Two types of experiment helped greatly to make sense of this ambiguity.

Typically, auxotrophic cells requiring a particular amino acid were assayed for mutations on defined medium lacking the required amino acid. Reversion mutations would produce colonies. However, small supplements of nutrient broth in the assay medium, which allowed limited growth of the parental cells, could produce greater numbers of revertant colonies from cells exposed to the same fluence of UV light. Initially, such semienriched medium was thought to enhance the mutation frequency because mutagenesis was associated with residual cell divisions on the assay plate. UV damage was thought to alter the spontaneous mutation process producing a greater probability of mutation per cell division. Hence, more cell divisions on the assay plate would produce more mutations.

In a series of incisive experiments, Witkin *(50)* found instead that the semienriched medium altered some mutation process occurring during an initial "sensitive period" of about 1 h. Over a limited range, increasing amounts of nutrient supplement increased the resulting mutation frequency, but this did not result from increased numbers of cell divisions on the assay plate. Cells were shifted from one medium surface to another during early incubation to show that the amount of supplement present during the sensitive period determined the outcome, not the number of divisions possible after that period. These experiments were done before excision repair was discovered, but they clearly showed that metabolic activity in the cells sustaining UV damage influenced the consequence of that damage.

Doudney and Haas *(12,13)* tested numerous types of physiological variation to cells during incubation immediately after UV irradiation. The cells were incubated in liquid medium so that samples could easily be removed and assayed. Thus, cells could be challenged after UV by incubation in different biochemical environments to see what variations in mutation frequency might result during the first 60–90 min after irradiation. They introduced the phrase "mutation frequency decline" to describe the rapid decrease observed in the measured mutation frequency (viable revertants per total viable titer) during certain post-UV incubations. These incubations had essentially no effect on the viable titer but could decrease the mutation frequency 10-fold in about 10 min.

Once excision repair was discovered in *E. coli*, and a mutant strain defective in excision repair was available *(7,17,37,49)*, it was shown that MFD was absent in cells incapable of excision repair *(29,52)* and also that a strain defective in MFD, with the *mfd-1* allele, had a "markedly lower" rate of excision repair *(52)*. Although several genetic defects greatly affected excision repair and had a large effect on the ability of cells to survive UV inactivation (e.g., *uvrA* or *uvrB*), the *mfd-1* defect had no effect on UV inactivation of viability (Section 3.1.). It did, however, entirely block MFD.

In other early studies, Osborn et al. *(33)* described how *E. coli* reversion mutations could arise by nonsense-suppressing tRNA gene mutations. This supported an earlier report implicating a suppressor mutation *(51)*. Auxotrophs, defective because of amber or ochre nonsense codons in particular biosynthetic genes, could revert either by mutations that replaced the nonsense with sense or acceptable missense (backmutations) or by mutations that changed the anticodon of a particular tRNA to the inverse complement of nonsense (tRNA suppressor mutations). This led directly to the realization that glutamine tRNA suppressor mutations were the majority of mutations studied in many UV mutagenesis reversion assays *(10,32,33)*. Moreover, Bridges et al. *(11)* found that MFD only affected glutamine tRNA suppressor mutations. The rapid, extensive decrease in mutation frequency during certain types of post-UV incubation was specific for a particular mutation event that happened to be the most common event among a variety that accounted for reversion mutation.

2.3. *Targeting Photoproducts in Mutagenesis and MFD*

Before DNA sequencing was possible, several tRNAs were sequenced and the corresponding gene sequences could be inferred. This helped Osborn et al. *(33)* explain the nonsense-suppressing tRNA mutations and allowed Person et al. *(35)* to consider the likely base pair substitutions produced by UV that were responsible for glutamine tRNA suppressor mutations. The DNA sequences in the transcribed strand for the anticodon loops of the two species of glutamine-charging tRNA that recognize the glutamine codons CAA or CAG would be: 5'-A-T-C-A-A-A-A- and 5'-A-T-C-A-G-A-A- at *glnU* or *glnV*, respectively (anticodons are underlined). Therefore, C-to-T transitions could produce tRNAs altered at the 3'-end of anticodons that recognize ochre or amber codons (UAA or UAG, respectively); and T<>C CPDs could form where these transitions occur *(14)*.

A working model for UV mutagenesis required induction of a generalized SOS response by damage throughout the genome and photoproducts at the sites of DNA alterations to "target" the mutations *(53)*. Thus, it was possible that cytosine-containing CPDs targeted glutamine tRNA suppressor mutations and accounted for the efficient production of these particular mutations *(3)*.

Person and Osborn *(36)* noticed that an amber glutamine tRNA suppressor mutation could be converted to an ochre glutamine tRNA suppressor mutation by a transition at the 5'-end of the anticodon (a G-to-A transition in the sequence for the transcribed strand, above). *De novo* production of the amber suppressor mutation could involve a T<>C CPD in the transcribed DNA strand, and the converted suppressor mutation could be targeted by a T<>C CPD in the nontranscribed strand. Bockrath and Palmer *(5)* found that both of these suppressor mutations were efficiently produced by UV mutagenesis, and, most important to this discussion, that only the *de novo* suppressor mutation was affected by MFD. The mutation frequency for converted glutamine tRNA suppressor mutations, possibly targeted by T<>C dimers in the nontranscribed strand of the same gene, was relatively unaffected.

MFD then could be explained with a fairly simple model. Glutamine tRNA suppressor mutations targeted by UV photoproducts in the transcribed DNA strand were susceptible to an efficient form of transcribed strand-specific excision repair. This model was supported in 1987 *(2)*. When these suppressor mutations were targeted by O^6-ethylguanine after alkylation mutagenesis, which could form at the G-residue opposite the C-residue of the T-C sequence, the converted suppressor mutation was sensitive to MFD, whereas the *de novo* suppressor mutation was not. Strand-specific excision repair could remove targeting lesions from the transcribed strand at particular sites in glutamine tRNA genes to account for MFD and explain much of the variability in ultimate mutation frequency characteristic of UV mutagenesis studies with *E. coli*.

2.4. *Mfd Protein is Transcription-Repair Coupling Factor (TRCF)*

In the in vitro studies of Selby and Sancar (Section 2.1.), the addition of cell extract to their defined system produced the sort of strand-biased repair seen in whole cells. To identify the responsible agent(s), they took advantage of the slowly developing MFD story and found that an extract from *mfd-1*-defective cells did not contain the necessary factor(s) to restore transcription-repair coupling *(47)*. They then isolated the wild-type *mfd* allele and purified the *mfd* gene product, which they called TRCF *(43)*. Purified TRCF was necessary and sufficient to produce transcription-repair coupling in their defined in vitro system.

Selby and Sancar *(44)* have noted the significant contribution TRCF makes to understanding MFD. The idea of targeting lesions in the transcribed strand and their repair by a special form of excision repair interlocks quite nicely with the clear demonstration of transcription repair coupling in *E. coli* and discovery of TRCF. A recent paper supports this integration *(4)*, and another comments on MFD historically *(54)*.

3. STRUCTURE AND FUNCTION OF THE TRCF

3.1. *The* mfd *Gene and Product in* E. coli

The *mfd-1* allele, first isolated in *E. coli* B/r *(52)*, was transduced into *E. coli* K-12 and mapped at 25.3 min on the genome *(43)*. The wild-type allele was cloned from a Kohara library. Although the *mfd-1* allele in B/r had a negligible effect on cell sensitivity to UV inactivation, in K-12 or in K-12 with an *recA*-defect, *mfd-1* conferred moderate UV sensitivity *(30,43)*. Therefore, UV sensitivity could serve as a phenotype, and sensitive cells with the defective genomic allele were complemented to wild-type resistance when transformed with a plasmid expressing the *mfd*$^+$ allele *(43)*.

The DNA sequence of the *mfd* gene encodes a 130-kDa protein of 1148 amino acids *(43)*. Initial consideration suggested three regions of particular interest. Toward the amino-terminal end, residues 82–219 have 22–25% identity with UvrB protein of *E. coli*, *Micrococcus luteus* and *Streptococcus pneumoniae*, the three prokaryotes from which a UvrB protein has been characterized. A central region of 387 amino acids contains seven helicase motifs with 38% identity to a region of *E. coli* RecG protein that also contains helicase motifs. Finally, there is an indication of a leucine zipper motif near the carboxy-terminus (*see* Fig. 1 of ref. *45*).

The *mfd* gene and its product (~500 mol/cell, more than any of the Uvr proteins) are not essential for viability *(45)*. Cells containing a null allele constructed by gene replacement (inserting a diminished *mfd::kan* allele, strain UNCNOMFD) had normal growth characteristics.

3.2. *Activity of TRCF (Mfd Protein)*

Purified TRCF was examined with an in vitro assay employing damaged DNA transcribed either by *E. coli* or T7 RNA polymerase *(43)*. The test lesion, a psoralen-thymine monoadduct, stalled transcription by either polymerase only when in the transcribed strand. As visualized with footprint analysis, TRCF had no effect on transcription initiation, but released RNA polymerase stalled during elongation by a process requiring ATP hydrolysis. Nascent RNA was released at the same time and was not altered in length from that formed during stalled transcription. However, the release was specific for *E. coli* RNA polymerase. TRCF did not release stalled T7 polymerase.

When UvrABC components were added to the system, TRCF did not affect repair if there was no transcription. An affinity column with TRCF attached to the matrix bound UvrA and, if presented with $UvrA_2B_1$, still bound UvrA but not UvrB. It did not bind *E. coli* RNA polymerase. Thus, since $UvrA_2B_1$ is thought to deliver UvrB to the damage site, TRCF may facilitate such delivery. Recall that TRCF has a region of similarity with UvrB (Section 3.1.), which might mediate an association with UvrA in displacing UvrB.

When repair of the damaged DNA was measured in this system, TRCF was required. However, although it restored repair when transcription was "on" (which otherwise was inhibited by stalled RNA polymerase), it did not produce a rate of repair greater than that when transcription was "off." That is, this in vitro system demonstrated the benefit of released polymerase, but did not show the enhanced rate of repair characteristic of in vivo transcription-repair coupling.

A larger DNA substrate was then tried perhaps better to detect both release of inhibition and stimulation of repair by TRCF. Also, UV damage was used rather than a particular psoralen-adduct. This gave results with assays for nicking or repair synthesis that demonstrated RNA polymerase release and stimulation of repair by TRCF. A completely defined system of *E. coli* components produced the hallmark characteristics of transcription repair coupling *(4,30,46)*.

3.2.1. TRCF Domains for Interaction with UvrABC Repair and Stalled RNA Polymerase

Brenner et al. *(8)* have outlined the logical necessities for protein complexes attending gene activities. There must be a DNA binding region that recognizes an address on DNA, and patches for interprotein recognition. Selby and Sancar *(45,46)* considered this sort of organization to locate the domains in the primary sequence of TRCF responsible for particular interactions by using altered TRCFs, including one truncated form lacking the carboxy-terminus (composed of the amino acids 1–938) and five fusion proteins between maltose binding protein (MBP) and different regions of the TRCF: N-V, V-C, N-X, X-C, and P-C (Fig. 1).

UvrA protein binds the amino-terminal N-X domain, but no other. Since the region of similarity with UvrB is entirely within N-X, this supports the idea that the interaction between TRCF and UvrA is similar to that between UvrB and UvrA. UvrB has a second binding site for UvrA near the carboxy-terminus. The first, near the amino-terminus, presumably binds to the same region of UvrA as TRCF, and one can imagine a staged exchange as TRCF displaces UvrB from $UvrA_2B_1$.

The binding of TRCF to DNA seems to require concurrent binding of ATP, and release occurs on ATP hydrolysis. For example, a stable complex formed in the presence of ATPγS. Various forms of DNA and RNA were tested, including constructions containing a small bubble or an RNA:DNA hybrid, and simple double-stranded DNA was the best substrate. Footprint analysis revealed "alternating protected and hyper-

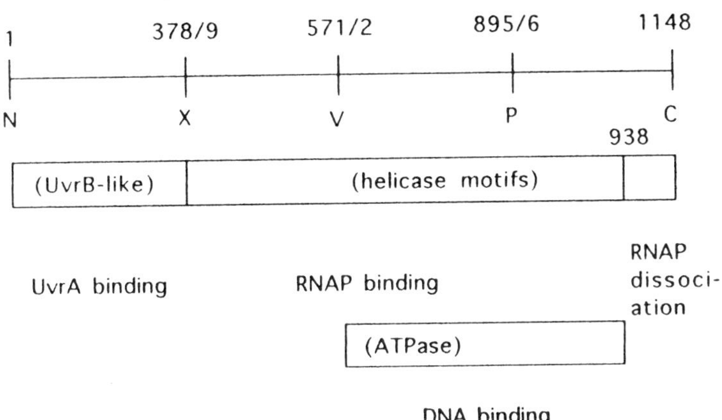

Fig. 1. Summary sketch for interactive regions of TRCF. Boxes represent regions of TRCF, between the amino- (1, N) and carboxyl- (1148, C) terminal ends, indicated by tests with fragment constructions for the activities given in the labels (not drawn to scale). Adapted from Fig. 1 of ref. *45*.

sensitive regions," suggesting that double-stranded DNA wrapped around TRCF with repeats phased from a particular sequence or structure. Tests with the various TRCF fragments located the DNA binding domain to the 571–938 region (Fig. 1). This contains the helicase motifs, which, if ineffective in producing actual helicase activity *(43,46)*, may nevertheless account for the association of TRCF with DNA. This region also contains the site for ATPase activity. Presumably, bound ATP activates the helicase region of TRCF for binding to DNA.

Although binding of TRCF to RNA polymerase could not be demonstrated with an affinity column, some binding was demonstrated in incubations of TRCF-MBP linked to amylose resin and separated by centrifugation (but RNA polymerase bound less well than UvrA). Tests with the fusions of TRCF fragments located RNA polymerase binding to amino acids 379–571. In addition, RNA polymerase bound to DNA (prepared as a stalled elongation complex by nucleotide starvation) was analyzed by gel shift with the TRCF fusion fragments and the truncated form. The N-V fragment did not bind to this complex perhaps because the electrophoresis assay conditions were too severe. The X-C fragment caused dissociation of stalled polymerase just as did whole TRCF, indicating that amino acids 1–378, which contain the binding site for UvrA, were not essential for release of stalled polymerase. Only the truncated form of TRCF bound to the stalled elongation complex, suggesting that the terminal residues 896–1148 play a role in the ultimate dissociation of stalled RNA polymerase, nascent RNA, DNA, and TRCF.

3.2.2. Dissociation of a Robust Transcription Complex

Recently the strength of transcription by *E. coli* RNA polymerase was measured *(55)*. A single active complex, incubated in vitro with saturating concentrations of nucleotide triphosphates, required a force of 14 pN to stop it, indicating an elongation force greater than that of the cytoskeletal motors kinesin and myosin. For this reason, the verb "stall" has been used here, rather than "block," to connote a subtle disruption of a rather imposing process. Hence, a stalled transcription complex is not to be taken lightly *(38)*. It should probably be regarded as a substantial molecular association of many close interactions jammed by a critical misalignment where the template normally receives the next monomeric unit.

Since release and stimulation of repair by TRCF at UV damage in linear double-stranded DNA require an interval of about 90 nucleotides on the template strand downstream of the stalling lesion, TRCF may bind DNA just beyond the site of stalled transcription *(46)*. From that position TRCF could effect a conformational change in the template strand where it enters the transcription complex and essentially derail the apparatus. The elongation forces would then be redirected to dissociation.

Selby and Sancar *(46)* commented, "TRCF is the only known protein in *E. coli* other than ρ that is able to dissociate a ternary RNA Pol complex," and compared the activities of TRCF and ρ factor. Unlike ρ, TRCF did not release complexes at ρ-dependent terminators, did not bind RNA and had no helicase activity on RNA:DNA hybrids. However, in one respect, TRCF and ρ were similar. They both released the transcription complex stalled at a protein "roadblock," which in vivo might be a bound repressor or certain types of DNA binding protein *(48)*. A defective form of *Eco*RI bound at an *Eco*RI site was used, which previously had been shown to arrest transcription. ρ was known to release the transcription components from this blockage *(34)*, and TRCF effected a similar release.

Zhou et al. *(56)* have noted an interesting fact about transcription and damaged template. Nicks and small gaps in the transcribed strand are accommodated, at least by T7 RNA polymerase. The nascent RNA is continuous as though the break in the transcribed strand did not exist. This underscores a likely role for the nontranscribed strand as a close and sustaining component of the transcription complex. There may be a direct association of RNA polymerase with the sugar–phosphate backbone of the nontranscribed strand that is insensitive to chemical alterations of the bases. In any case, the transcription complex is robust, the process initiated by certain template damage causing elongation to stall must be clever, and the molecular mechanism by which TRCF disassociates the stalled complex must be equally clever.

3.3. Bacillus subtilis *Mfd Protein*

An open reading frame for a protein of 1177 amino acids in *B. subtilis* has been termed an *mfd* gene because of similarities of the predicted and actual protein product with the TRCF of *E. coli (1)*. The amino acid sequence has 37% identity with that of TRCF. The protein is monomeric in solution like TRCF, having a size of about 135 kDa. Although it lacks the leucine zipper motif of TRCF in the carboxyl-terminus, it has regions of similarity with UvrB and RecG, and contains helicase motifs as does TRCF.

B. subtilis carrying a null allele of the gene are viable, sensitive to inactivation by MMS or 4NQO, and apparently defective in aspects of homologous recombination. In an analysis of forward mutations induced by UV that inactivated the *phl* gene (confers resistance to phleomycin), cells with the null allele acquired a relatively greater number of mutations that could be attributed to photoproducts at adjacent pyrimidines in the transcribed strand of the *phl* gene. This was similar to results by Oller et al. *(30)* for UV-induced mutations in the *lacI* gene of mfd^+ and mfd^- *E. coli*. The shift in mutation spectra supports the idea that functional Mfd protein allows enhanced repair of photoproducts in the transcribed strand, which reduces their contribution to the resulting mutation spectrum.

Additional in vitro studies of the *B. subtilis* Mfd protein suggested a modest ATPase activity unaffected by addition of single- or double-stranded DNA, an inability to form a stable complex with various DNA substrates (even in the presence of ATPγS), and no helicase activity. *B. subtilis* Mfd released stalled *B. subtilis* RNA polymerase and less efficiently, *E. coli* RNA polymerase. The nascent RNA was released as well.

Thus, the two Mfd proteins are similar in several respects pertaining to the role for clearing stalled transcription complexes. The absence of a leucine zipper motif in the *B. subtilis* variety and the ability of the truncated *E. coli* variety to form a stable complex with the stalled transcription apparatus suggest that this motif is not important to the association with RNA polymerase (*see* ref. *45*). The region for association with RNA polymerase in *E. coli* Mfd protein has a corresponding region conserved in the *B. subtilis* protein *(1)*. The two proteins differ in their ability to bind DNA, and the *B. subtilis* Mfd protein appears to play a significant role in recombination, which involves more overlapping systems in *B. subtilis* than in *E. coli*. It was suggested that this stems from the similarity with RecG protein and/or an ability possessed by the *B. subtilis* protein to facilitate introduction of certain recombination proteins at recombinogenic abnormalities in DNA *(1)*.

E. coli and *B. subtilis* are not closely related phylogenetically *(31)*. Although additional comparative studies will be revealing, the work by Ayora et al. *(1)* establishes the likely significance of transcription repair coupling across a range of bacteria by a rather large monomeric protein that can associate with RNA polymerase but is not an integral and required component of the elongation complex.

4. CELLULAR COMPLEXITIES OF TRANSCRIPTION REPAIR COUPLING

4.1. Transcription Repair Coupling and Mismatch Repair

Mellon and Champe *(24)* recently made a remarkable observation using the assay for removal of CPDs from specific strands of the lactose operon after UV irradiation (Section 2.1.). They found that *E. coli* carrying certain defective alleles for mismatch repair no longer performed preferential repair at the transcribed strand of the derepressed operon. Until this time, nucleotide excision repair and mismatch repair, in dealing respectively with DNA damage and editing of incorrect bases inserted during semiconservative DNA replication (or mismatches formed during homologous recombination), were considered quite distinct repair processes with different purposes. However, the dramatic repair rate at the transcribed strand during active expression *(25)* was absent in cells with an *mutS* or *mutL* defect.

This result was attributed to a possible interplay between the mechanism mediated by TRCF and functions of MutS and MutL that normally recognize a mismatched base pair and communicate this to a site of endonuclease activation (*see* Chapter 11). While a role for mismatch repair proteins could not be substantiated in vitro *(46)*, the loosely condensed state of DNA associated with histone-like proteins in the whole cell may require activities of mismatch repair to achieve the sort of biased repair indicated in dilute in vitro assays (*see* Section 5.). An interplay of mismatch repair and transcription-repair coupling in vivo was also supported by slower kinetics for MFD with cells having mismatch repair defects *(21)*.

Mellon et al. *(26)* then confirmed the fundamental nature of an interaction between transcription coupled excision repair and mismatch repair by extending the observation to human cells. This is particularly noteworthy because defective genes for human mismatch repair have been linked to hereditary nonpolyposis colorectal cancer (*see* Chapter 20, vol. 2). A fundamental interdependence between transcription repair coupling and mismatch repair could mean that individuals with this hereditary disposition would be sensitive to genotoxic environmental agents as well as to errors in DNA replication.

4.2. Other In Vivo Characteristics of Transcription Repair Coupling

A recent study of MFD noted a detail of that phenomenon that had been misunderstood for several years *(4)*. As previously mentioned, certain post-UV incubation environments produce a rapid decline in the ultimate mutation frequency without affecting overall viability (Sections 2.2. and 2.3.). Commonly, auxotrophic cells were incubated with glucose, but without the amino acids required for growth. This unbalanced metabolic condition might then be associated with transcription repair coupling for lesions that targeted certain suppressor mutations.

However, it was shown that MFD required only a change in temperature (also noted in 1963, but neglected thereafter) *(4)*. Cells in buffer at room temperature were stable but underwent MFD if simply shifted to 37°C without the addition of glucose. Thus, the UvrABC excision mediated by TRCF affecting a certain set of photoproducts may well have a critical temperature dependence.

In addition, during an MFD experiment, plating for mutation frequency on semienriched agar medium (containing a small supplement of mixed amino acids) must immediately arrest the repair process *(4)*. As a possible explanation, Selby and Sancar *(44)* have suggested that rapid transcription at the tRNA operon stimulated by this supplement could actually interfere with transcription repair coupling. What effect a second elongation complex may have on an initial stalled complex already in some stage of the release process is an interesting question.

The actual benefit of transcription repair coupling to successful expression of a UV-damaged gene in *E. coli* has been examined *(22)*. After incremental exposures to UV, cells were incubated with IPTG to induce the lactose operon and assayed to monitor the rate of enzyme synthesis during the ensuing 40 min. The rates of synthesis decreased exponentially with UV exposure, and the exponential coefficients could be used to compare the effectiveness of UV damage in strains fully proficient in repair, defective in excision repair generally (*uvrA* or *uvrC*), or defective in transcription coupled excision repair (*mfd::kan*). For expression of β-galactosidase, UV inactivation was two to three times more effective when either general excision repair or transcription-coupled repair was defective compared to wild-type cells indicating that general excision repair made no contribution to expression recovery apart from that made by transcription-coupled repair.

Kunala and Brash *(19,20)* measured in vivo repair of CPDs at individual bases of the *lacI* and *lacZ* genes in *E. coli*. DNA extracted from UV-irradiated cells was digested with a restriction endonuclease, and a tail of radiolabeled nucleotides was added to a specific fragment. Subsequent digestion with T4 endonuclease V then indicated the positions of CPDs on a sequencing gel. The results for the *lacI* gene corroborated the inference drawn from mutation spectra by Oller et al. *(30)* about preferential excision at the transcribed strand. In *mfd*$^+$ cells, repair was more rapid at CPDs in the transcribed strand than in the nontranscribed strand, and in *mfd*-defective cells there was little or no repair in the transcribed strand *(19)*. The results for the *lacZ* gene, considering the first ~150 bp just past the transcription start site, showed a transition in the repair kinetics about 32 nucleotides into the transcribed strand where transcription would shift from an initiation to an elongation complex *(20)*. This is consistent with in vitro data showing that TRCF facilitates repair at a stalled transcription elongation complex but not at the initiation complex *(43,46)*. The high-resolution study also demonstrated preferential repair in the transcribed strand of the extreme proximal region of *lacZ* during active expression in *mfd*-defective cells. This contrasts with results for repair throughout the *lacZ* gene showing similar repair in the two strands in *mfd*-defective cells *(24)* and with results for the *lacI* gene indicating substantially less repair in the transcribed strand than in the nontranscribed strand in *mfd*-defective cells *(19,30)*. Moreover, MFD is not observed in *mfd*-defective cells (presumably with no repair of targeting lesions in the transcribed strand of the relevant tRNA gene) as long as transcription is not inhibited *(21)*.

Bridges *(9)* noted an unusual phenotype of *mfd*-defective cells in a study of starvation-associated mutation in *E. coli*. Contrasting results were observed between strain WU3610-45 *(mfd-1)*, the original strain in which the defective allele for TRCF was isolated, and strains CM1279 and CM1281. The latter two strains were recently constructed by transducing the defective alleles *mfd::kan* or *mfd-1*, respectively, into WU3610, the immediate precursor used to isolate WU3610-45 *(52)*. There appears to be an additional defect in WU3610-45 that interferes with starvation-associated mutation (unless the cells are also *uvrA*-defective) and causes the strain to die on starvation. This latter observation is included here as a practical note (warning) for those attempting to maintain strain WU3610-45 on agar medium. Furthermore, it is not clear whether the appearance of the additional defect is wholly independent of *mfd-1*.

5. CONCLUSIONS

Two conclusions may be drawn about transcription repair coupling with some confidence. First, DNA damage interfering with the ongoing precision of RNA synthesis can be a fundamental cellular problem. For many years, DNA damage that stopped DNA replication was of major concern to investigators. Recent discoveries now bring attention to damaged templates that stop RNA synthesis and to repair of this damage. Second, transcription repair coupling in bacteria seems to involve a particular protein essential and sufficient for release of the transcription apparatus stalled during elongation. The protein is not a required part of the normally functioning transcription apparatus.

Less clear is the sequence of events and the time-course of individual steps in the process. One outline for this was proposed by Selby and Sancar (Fig. 7 of ref. *43*) based on results with their in vitro defined systems:

1. A lesion in the transcribed strand stalls elongation of RNA polymerase.
2. The stalled complex interacts with TRCF and ATP leading to release of RNA polymerase, nascent RNA, ADP, and P_i.

3. TRCF remains attached to the damaged DNA and mediates an interchange with $UvrA_2B_1$ (and ATP) that leaves UvrB bound with the damaged DNA and releases UvrA, TRCF, ADP, and P_i.
4. Nucleotide excision repair is effected with the arrival of UvrC and subsequent proteins.

In vitro evidence suggests that repair stimulation arises from delivery of UvrB aided by TRCF, although the mechanism for this is not understood *(46)*. Since the transcription elongation complex is thought to bend DNA in its association with DNA *(38)* and damaged DNA bound to UvrB also is thought to be bent *(40)*, it is possible that rapid kinetics result from release of the stalled transcription complex by TRCF that preserves the bent conformation of DNA. This could reduce the time required for successful delivery of UvrB. Moreover, DNA conformation has been implicated in experiments showing an effect by mismatch repair defects on transcription repair coupling (Section 4.1.).

Mellon and Champe *(24)* suggest that mismatch repair proteins MutS and MutL, essential for transcription-repair coupling in vivo, may introduce a DNA loop where damage has stalled transcription. The loop, in turn, could stabilize DNA topology so that, after release of RNA polymerase, $UvrA_2B_1$ might more readily deliver UvrB. However, in the minimal in vitro system, cell free extracts from *mutS* or *mutL* cells with or without added MutS produced no change in the level of transcription repair coupling, and topology was not an obvious factor in the rates of reaction *(46)*. Nevertheless, the demonstration of an analogous influence by mismatch repair defects in human cells *(26)* raises questions about the contrasting in vivo and in vitro results with *E. coli*. A better understanding of how MutS and MutL, or the human analogs, interact with DNA will clarify these issues.

Finally, in reviewing the work on transcription repair coupling in *E. coli*, one notes a lopsided contribution. Essentially all of the work has been done by groups interested in DNA repair. It seems that understanding of transcription repair coupling could benefit from considered insights by those studying transcription.

ACKNOWLEDGMENTS

The author is grateful for helpful comments by A. Ganesan, P. C. Hanawalt, C. Selby, and M. Zolan during the course of this writing. Also, fresh news of ongoing developments in this area from I. Mellon during her research is greatly appreciated. Funding by N. I. H. grant GM 21788.

NOTE ADDED IN PROOF

A paper by George and Witkin (*Mutat. Res.* [1975] **28:** 347–354), which quantified excision kinetics in an *mfd* mutant, was overlooked in Section 2.2.

REFERENCES

1. Ayora, S., F. Rojo, N. Ogasawara, S. Nakai, and J. C. Alonso. 1996. The Mfd protein of *Bacillus subtilis* 168 is involved in both transcription-coupled DNA repair and DNA recombination. *J. Mol. Biol.* **256:** 301–318.
2. Bockrath, R., A. Barlow, and J. Engstrom. 1987. Mutation frequency decline in *Escherichia coli* B/r after mutagenesis with ethyl methanesulfonate. *Mutat. Res.* **183:** 241–247.

3. Bockrath, R. and M. Chung. 1973. The role of nutrient broth supplementation in UV mutagenesis of *E. coli. Mutat. Res.* **19:** 23–32.
4. Bockrath, R. and B.-H. Li. 1995. Mutation frequency decline in *Escherichia coli*. II. Kinetics support the involvement of transcription-coupled excision repair. *Mol. Gen. Genet.* **249:** 591–599.
5. Bockrath, R., and J. Palmer. 1977. Differential repair of premutational UV-lesions at tRNA genes in *E. coli. Mol. Gen. Genet.* **156:** 133–140.
6. Bohr, V. A., C. A. Smith, D. S. Okumoto, and P. C. Hanawalt. 1985. DNA repair in an active gene: removal of pyrimidine dimers from the DHFR gene of CHO cells is much more efficient than in the genome overall. *Cell* **40:** 359–369.
7. Boyce, R. P. and P. Howard-Flanders. 1964. Release of ultraviolet light-induced thymine dimers from DNA in *E. coli* K-12. *Proc. Natl. Acad. Sci. USA* **51:** 293–300.
8. Brenner, S., W. Dove, I. Herskowitz, and R. Thomas. 1990. Genes and development: molecular and logical themes. *Genetics* **126:** 479–486.
9. Bridges, B. A. 1995. Starvation-associated mutation in *Escherichia coli* strains defective in transcription repair coupling factor. *Mutat. Res.* **329:** 49–56.
10. Bridges, B. A., R. E. Dennis, and R. J. Munson. 1967. Mutation in *Escherichia coli* B/r WP2 *tyr*⁻ by reversion or suppression of a chain-terminating codon. *Mutat. Res.* **4:** 502–504.
11. Bridges, B. A., R. E. Dennis, and R. J. Munson. 1967. Differential induction and repair of ultraviolet damage leading to true reversion and external suppressor mutations of an ochre codon in *Escherichia coli* B/r WP2. *Genetics* **57:** 897–908.
12. Doudney, C. O. and F. L. Haas. 1958. Modification of ultraviolet-induced mutation frequency and survival in bacteria by post-irradiation treatment. *Proc. Natl. Acad. Sci. USA* **44:** 390–401.
13. Doudney, C. O. and F. L. Haas. 1959. Mutation induction and macromolecular synthesis in bacteria. *Proc. Natl. Acad. Sci. USA* **45:** 709–721.
14. Engstrom, J. E., S. Larsen, S. Rodgers, and R. Bockrath. 1984. UV-mutagenesis at a cloned target sequence: Converted suppressor mutation is insensitive to mutation frequency decline regardless of the gene orientation. *Mutat. Res.* **132:** 143–152.
15. Friedberg, E. C., G. C. Walker, and W. Siede. 1995. *DNA Repair and Mutagenesis.* ASM, Washington, DC.
16. Hanawalt, P. C. 1995. Intragenomic DNA repair heterogenety, in *DNA Repair Mechanisms: Impact on Human Diseases and Cancer* (Vos, J.-M. H., ed.), R. G. Landes Co., Austin, TX, pp. 161–174.
17. Hill, R. 1958. A radiation-sensitive mutant of *Escherichia coli. Biochem. Biophys. Acta* **30:** 636,637.
18. Huang, J.-C., D. L. Svoboda, J. T. Reardon, and A. Sancar. 1992. Human nucleotide excision nuclease removes thymine dimers from DNA by incising the 22nd phosphodiester bond 5' and the 6th phophodiester bond 3' to the photodimer. *Proc. Natl. Acad. Sci. USA* **89:** 3664–3668.
19. Kunala, S. and D. E. Brash. 1992. Excision repair at individual bases of the *Escherichia coli lacI* gene: relation to mutation hot spots and transcription coupling activity. *Proc. Natl. Acad. Sci. USA* **89:** 11,031–11,035.
20. Kunala, S., and D. E. Brash. 1995. Intragenic domains of strand-specific repair in *Escherichia coli. J. Mol. Biol.* **246:** 264–272.
21. Li, B.-H. and R. Bockrath. 1995. Mutation frequency decline in *Escherichia coli*. I. Effects by mismatch repair defects. *Mol. Gen. Genet.* **249:** 585–590.
22. Li, B.-H. and R. Bockrath. 1995. Benefit of transcription-coupled nucleotide excision repair for gene expression in u. v.-damaged *Escherichia coli. Mol. Micro.* **18:** 615–622.
23. Mayne, L. V. and A. R. Lehmann. 1982. Failure of RNA synthesis to recover after UV irradiation: an early defect in cells from individuals with Cockayne's syndrome and xeroderma pigmentosum. *Cancer Res.* **42:** 1473–1478.

24. Mellon, I. and G. N. Champe. 1996. Products of DNA mismatch repair genes *mutS* and *mutL* are required for transcription-coupled nucleotide excision repair of the lactose operon in *Escherichia coli*. *Proc. Natl. Acad. Sci. USA* **93:** 1292–1297.
25. Mellon, I. and P. C. Hanawalt. 1989. Induction of the *Escherichia coli* lactose operon selectively increases repair of its transcribed DNA strand. *Nature* **342:** 95–98.
26. Mellon, I., D. K. Rajpal, M. Koi, R. Boland and G. N. Champe. 1996. Transcription-coupled repair deficiency and mutations in human mismatch repair genes. *Science* **272:** 557–560.
27. Mellon, I., G. Spivak, and P. C. Hanawalt. 1987. Selective removal of transcription-blocking DNA damage from the transcribed strand of the mammalian DHFR gene. *Cell* **51:** 241–249.
28. Mullenders, L. H. F., A. C. van Kesteren van Leeuwen, A. A. van Zeeland and A. T. Natarajan. 1988. Nuclear matrix associated DNA is preferentially repaired in normal human fibroblasts, exposed to a low dose of ultraviolet light but not in Cockayne's syndrome fibroblasts. *Nucleic Acids Res.* **16:** 10,607–10,622.
29. Munson, R. J. and B. A. Bridges. 1966. Non-photoreactivating repair of mutational lesions inducd by ultraviolet and ionizing radiation in *Escherichia coli*. *Mutat. Res.* **3:** 461–469.
30. Oller, A. R., I. J. Fijalkowska, R. L. Dunn, and R. M. Schaaper. 1992. Transcription-repair coupling determines the strandedness of ultraviolet mutagenesis in *Escherichia coli*. *Proc. Natl. Acad. Sci. USA* **89:** 11,036–11,040.
31. Olsen, G. J. and C. R. Woese. 1993. Ribosomal RNA: a key to phylogeny. *FASEB J.* **7:** 113–123.
32. Osborn, M. and S. Person. 1967. Characterization of revertants of *E. coli* WU36-10 and WP2 using amber mutants and an ochre mutant of bacteriophage T4. *Mutat. Res.* **4:** 504–507.
33. Osborn, M., S. Person, S. Phillips, and F. Funk. 1967. A determination of mutagen specificity in bacteria using nonsense mutants of bacteriophage T4. *J. Mol. Biol.* **26:** 437–447.
34. Pavco, P. A. and D. A. Steege. 1990. Elongation by *Escherichia coli* RNA polymerase is blocked in vitro by a site-specific DNA binding protein. *J. Biol. Chem.* **265:** 9960–9969.
35. Person, S., J. A. McCloskey, W. Snipes, and R. C. Bockrath. 1974. Ultraviolet mutagenesis and its repair in an *Escherichia coli* strain containing a nonsense codon. *Genetics* **78:** 1035–1049.
36. Person, S. and M. Osborn. 1968. The conversion of amber suppressors to ochre suppressors. *Proc. Natl. Acad. Sci. USA* **60:** 1030–1037.
37. Pettijohn, D. and P. Hanawalt. 1964. Evidence for repair-replication of ultraviolet damaged DNA in bacteria. *J. Mol. Biol.* **9:** 395–410.
38. Richardson, J. P. and J. Greenblatt. 1996. Control of RNA chain elongation and termination, in *Escherichia coli* and *Salmonella typhimurium. Cellular and Molecular Biology*, vol. 1 (Neidhardt, F. C., ed.), ASM, Washington, DC, pp. 822–848.
39. Sancar, A. and W. D. Rupp. 1983. A novel repair enzyme: UvrABC excision nuclease of *Escherichia coli* cuts a DNA strand on both sides of the damaged region. *Cell* **33:** 249–260.
40. Sancar, A. and M. S. Tang. 1993. Nucleotide excision repair. *Photochem. Photobiol.* **57:** 905–921.
41. Selby, C. P. and A. Sancar. 1990. Transcription preferentially inhibits nucleotide excision repair of the template DNA strand in vitro. *J. Biol. Chem.* **265:** 21,330–21,336.
42. Selby, C. P. and A. Sancar. 1991. Gene- and strand-specific repair in vivo: partial purification of a transcription-repair coupling factor. *Proc. Natl. Acad. Sci. USA* **88:** 8232–8236.
43. Selby, C. P. and A. Sancar. 1993. Molecular mechanism of transcription-repair coupling. *Science* **260:** 53–58.
44. Selby, C. P. and A. Sancar. 1993. Transcription-repair coupling and mutation frequency decline. *J. Bacteriol.* **175:** 7509–7514.
45. Selby, C. P. and A. Sancar. 1995. Structure and function of transcription-repair coupling factor. I. Structural domains and binding properties. *J. Biol. Chem.* **270:** 4882–4889.

46. Selby, C. P. and A. Sancar. 1995. Structure and function of transcription-repair coupling factor. II. Catalytic properties. *J. Biol. Chem.* **270:** 4890–4895.
47. Selby, C. P., E. M. Witkin, and A. Sancar. 1991. *Escherichia coli mfd* mutant deficient in "Mutation Frequency Decline" lacks strand-specific repair: in vitro complementation with purified coupling factor. *Proc. Natl. Acad. Sci. USA* **88:** 11,574–11,578.
48. Sellitti, M. A., P. A. Pavco, and D. A. Steege. 1987. *lac* repressor blocks in vivo transcription of *lac* control region DNA. *Proc. Natl. Acad. Sci. USA* **84:** 3199–3203.
49. Setlow, R. B., and W. L. Carrier. 1964. The disappearance of thymine dimers from DNA: an error-correcting mechanism. *Proc. Natl. Acad. Sci. USA* **51:** 226–231.
50. Witkin, E. M. 1956. Time, temperature and protein synthesis: a study of ultraviolet-induced mutation in bacteria. *Cold Spring Harbor Symp. Quant. Biol.* **21:** 123–240.
51. Witkin, E. M. 1963. One step reversion to prototrophy in a selected group of multi-auxotrophic substrains of *Escherichia coli*. *Genetics* **48:** 916.
52. Witkin, E. M. 1966. Radiation-induced mutations and their repair. *Science* **152:** 1345–1353.
53. Witkin, E. M. 1976. Ultraviolet mutagenesis and inducible DNA repair in *Escherichia coli*. *Bact. Rev.* **40:** 869–907.
54. Witkin, E. M. 1994. Mutation frequency decline revisited. *BioEssays* **16:** 437–444.
55. Yin, H., M. D. Wang, K. Svoboda, R. Landick, S. M. Block, and J. Gelles. 1995. Transcription against an applied force. *Science* **270:** 1653–1657.
56. Zhou, W., D. Reines, and P. W. Doetsch. 1995. T7 RNA polymerase bypass of large gaps on the template reveals a critical role of the nontemplate strand in elongation. *Cell* **82:** 577–585.
57. Zolan, M. E., G. A. Cortopassi, C. A. Smith, and P. C. Hanawalt. 1982. Deficient repair of chemical adducts in alpha DNA of monkey cells. *Cell* **28:** 613–619.

10

Branched DNA Resolving Enzymes (X-Solvases)

Börries Kemper

1. INTRODUCTION

Double-stranded DNA is a nonbranched linear or circular macromolecule. However, DNA can become transiently branched when, for example, replication forks are initiated and three-arm Y-junctions move through the DNA or when genetic recombination joins DNA molecules creating Y-junctions or four-arm cross strand structures (X-junctions). Branches in DNA are potentially lethal, and may interfere with the distribution of chromosomes to progeny cells or viruses. DNA debranching systems are therefore expected to be present, and it is not surprising that debranching systems have been detected in numerous organisms.

Several endonucleases with specificity for branched DNAs have been described. Involvement in genetic recombination is implied by terms like Holliday structure resolving enzyme or cruciform resolving (or cutting) enzyme. The less biased term "X-solvase" (cross-solvase) is used throughout this chapter. X-solvases share their ability to recognize four-arm junctions with cruciform binding proteins (CBPs), which do not execute DNA cleavage. A brief survey of these proteins is therefore included.

2. BRANCHED DNAS

"True" Y- and X-branches arise when one DNA duplex replicates to form two molecules or when genetic recombination joins two molecules into one—in either case, at some point, two separate DNA duplexes are involved. Replication forks are usually removed from the DNA by the replication machinery itself. The Y-junctions are either transferred to a free end where they resolve spontaneously or they are resolved by proteins with topoisomerase activity after encountering a termination sequence. Recombination branches are removed from the DNA by members of the recombination machinery. The X-junctions are either actively moved by branch migration to a free end or are resolved by special endonucleases while the junction is still intact *(52)*.

"Pseudobranching" occurs occasionally at inverted repeats (IR) when palindromic sequences form cruciform structures. These branches do not involve two distinct DNA molecules and apparently require no resolution. These pseudobranches may instead require protection from attack by debranching enzymes like X-solvases. CBPs are good candidates for this function.

From: DNA Damage and Repair, Vol. 1: DNA Repair in Prokaryotes and Lower Eukaryotes
Edited by: J. A. Nickoloff and M. F. Hoekstra © Humana Press Inc., Totowa, NJ

2.1. Four-Way X-Structures

2.1.1. Holliday Structures

The Holliday structure is the best known X-structure in DNA. It is formed as a key intermediate in genetic recombination through a series of distinct steps. These steps include homologous pairing of chromosomes, mutual exchange of strands, branch migration, and isomerization of crossovers. Resolution of Holliday structures and repair of remaining breaks complete genetic recombination (*52*; Fig. 1).

Holliday structures were isolated from in vivo reactions and visualized by electron microscopy (*see 106* and references therein). Theoretical considerations and model building predict a high degree of structural flexibility, a requisite for strand movement during isomerization (*see 127* and references therein). According to the original model, such movement occurs with arms of the Holliday structure aligned in parallel. However, recent experiments with synthetic Holliday structure analogs (synthetic cruciforms) suggest that the arms of a four-way junction are aligned in an antiparallel orientation *(10,128)*.

2.1.2. Cruciforms

The cruciform structure (cruciform) by definition is a four-armed structure. It occurs naturally at palindromes in supercoiled DNA (supercoil cruciforms or SC cruciforms) or it is generated artificially by annealing oligonucleotides in vitro (synthetic cruciforms). Different types of cruciforms are shown in Fig. 2.

2.1.2.1. SC Cruciforms

Shortly after the structure of B-DNA was elucidated, SC cruciforms were predicted to exist at sites with inverted repeated sequences. The possibility that SC cruciforms control gene expression by structural means was discussed at that time. Convincing evidence for the existence of SC cruciforms was obtained only much later (reviewed in ref. *69*).

Pioneering work from several laboratories revealed how energy stored in supercoiled DNA drives cruciform formation from palindromes (reviewed in ref. *41*). Release of supercoils results in energy release and loss of the cruciform structure. Appearance and disappearance of these structures were monitored indirectly by physical, enzymatic, and chemical methods *(1,24,98)*. Sudden changes in superhelical density and transient appearance of single-stranded regions in the center of palindromes are measurable events related to the formation and reversion of SC cruciforms, respectively *(23,41)*.

Direct measurements of SC cruciforms became available when T4 endonuclease VII (Endo VII) was isolated and shown to cleave across the junction at the base of the cruciform. The enzyme does not attack relaxed palindromes. Through experiments with Endo VII, SC cruciforms have been easily mapped in many DNAs in vitro.

The question whether SC cruciforms also exist in vivo is closely related to the question whether supercoils exist in vivo. Several positive indications for supercoils in vivo were obtained experimentally (reviewed in *25*). For example, transient induction of the X-solvase T7 endonuclease I (Endo I) in *Escherichia coli* causes rapid linearization of colE1 plasmid DNA. The sites of linearization were mapped to the same locations in supercoiled colE1 plasmid DNA after treatment with T7 Endo I in vitro *(90)*.

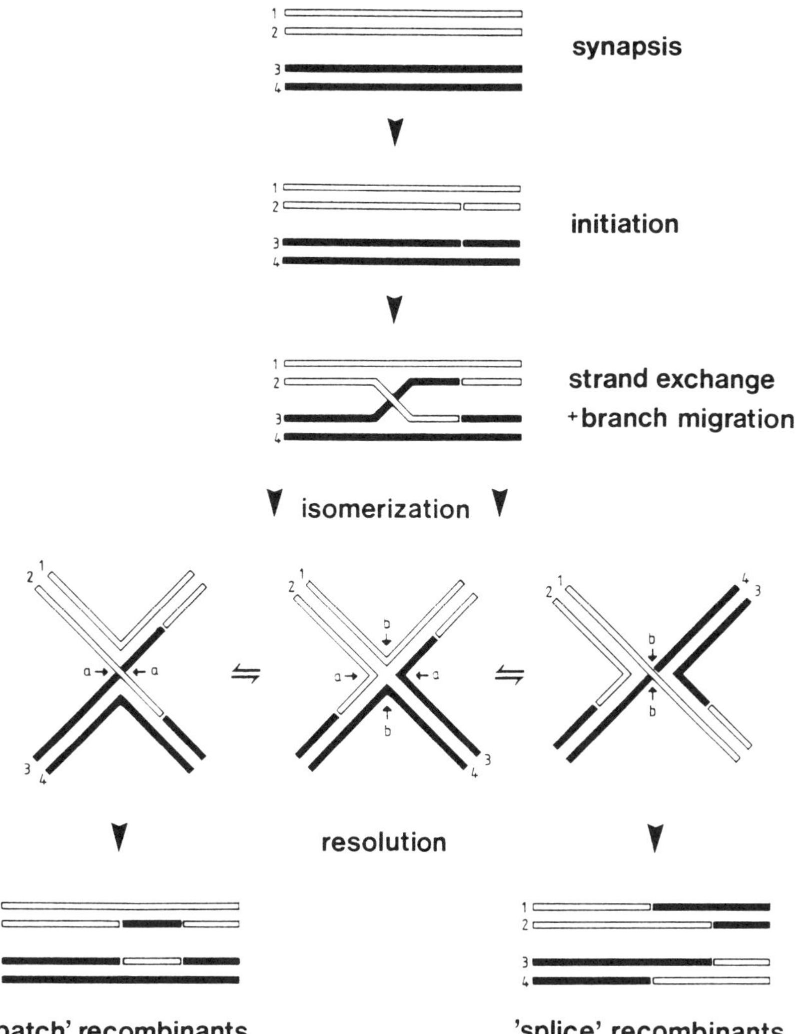

Fig. 1. Model of genetic recombination. The model as originally proposed by Holliday *(52)* requires exchanges between strands of like polarity after synapsis of two double-stranded DNA molecules is established. The resulting crossover leads to the Holliday structure as the central intermediate. This structure has the ability to "branch migrate" into either direction, away from the original site of exchange, thereby extending the region of hDNA. During this time, isomerization can occur, which allows those strands that were not originally broken and exchanged to cross over. It was assumed that resolution of a Holliday structure occurs at crossing strands, affecting the two pairs of homologs with an equal probability. Successful resolution would then lead to genetically distinguishable "patch" and "splice" recombinants. The resolution of Holliday structures is, therefore, a key event during genetic recombination and was assumed to be enzymatically driven.

2.1.2.2. Synthetic Cruciforms

Synthetic cruciforms are made in the test tube by simply annealing oligonucleotides *(22,60)*. These cruciforms can be made in larger amounts with variable nucleotide sequences and in different isomeric conformations. Synthetic cruciforms were exten-

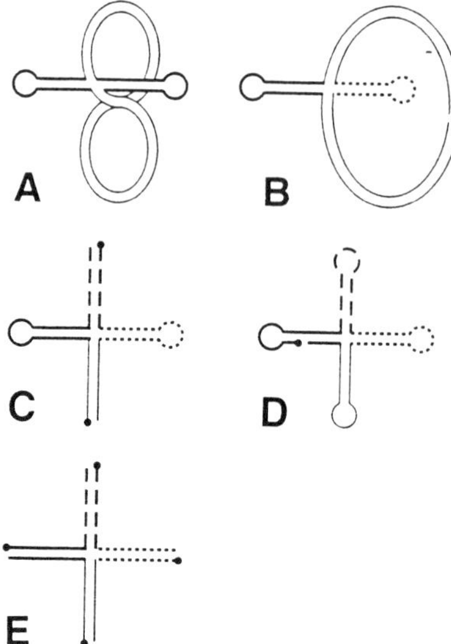

Fig. 2. Cruciform structures. **(A)** Natural cruciform DNA in supercoiled plasmid DNA. **(B)** Synthetic hybrid cruciform DNA made by annealing purified plus and minus strands of two phage M13 strains, each with a different palindrome in the same location. **(C)** Synthetic cruciform made from four oligonucleotides. **(D)** Synthetic cruciform made from two, partially self-complementary oligonucleotides. **(E)** Synthetic cloverleaf cruciform made from a single snap-back oligonucleotide. Nonhomologous regions are shown by different line types. Free 5'-termini are marked by dots.

sively used in analyses of X-solvases (*17,29*; Fig. 3), and as analogs of Holliday structure isomers in studies of structural parameters *(10,31,107)*.

Synthetic cruciforms adopt a defined conformation in solution. Two monoclonal antibodies (MAb) to cruciform DNA bind to the genuine structures used as antigens in vitro. The antibodies bind to structural determinants at or near the base of synthetic cruciforms, and protect two of the arms near the junction from cleavage by X-solvases. They do not protect the termini of the arms from cleavage by single-strand specific mung bean endonuclease *(42,43)*. One antibody binds also to a Y-junction and to a synthetic cruciform with a sequence differing from the original construct. Neither of the antibodies binds to linear double-stranded or single-stranded DNA, or to single-stranded DNA with terminal loops.

Structural parameters of synthetic cruciforms were further investigated in a series of physical and chemical experiments. The results support a model in which in the presence of Mg^{2+} ions, a synthetic cruciform adopts an X-shaped scissor-like conformation with antiparallel arms opened in a 60 and a 120° angle, respectively (*10,22,29,87*; Fig. 4). This model is supported by the observation that T4 Endo VII preferentially cleaves swapping strands in antiparallel constructs of synthetic cruciforms *(44,48,75)*.

Antiparallel arms in synthetic cruciforms contrast with parallel arms postulated for true Holliday structures *(52)*. This conflict was recently resolved when a Holliday structure analog was found to adopt a square-shaped, open conformation (central isomer in Fig. 1) after the *E. coli* branch migration proteins RuvA and RuvB were loaded onto

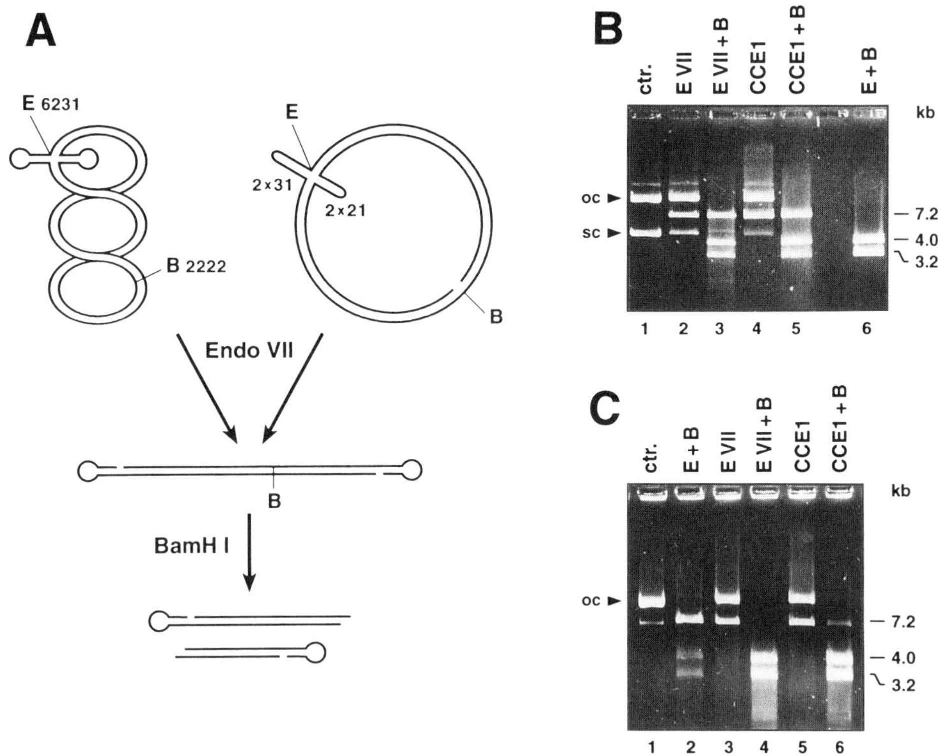

Fig. 3. Assay for X-solvase activity. **(A)** Supercoil cruciform DNA (left) is from phage M13mp2IR62E^{cc} *(51)*. The phage were obtained after inserting a 62-bp perfect palindrome into the unique *Eco*RI restriction site (E) of M13mp2. The relaxed hybrid cruciform DNA M13mp2IR62E/42E^{oc} (right) is made by annealing circular plus strands from M13mp2IR62E to linearized minus strands from M13mp2IR42E, which carries a 42-bp palindrome in the same *Eco*RI restriction site. Cleavages across the junction of the cruciform structure by an X-solvase (e.g., Endo VII) converts the circular molecules to their linear form with fold-back hairpins at both ends, including the IR. Further digestion with a unique restriction enzyme (e.g., *Bam*HI [B]) delivers two diagnostic fragments of predicted sizes, in this example, 4 and 3.2 kbp, respectively. **(B)** Analysis of the reaction products obtained from M13mp2IR62E^{cc} with Endo VII (EVII) or Cce1 as analyzed on 1% agarose gels. One hundred nanograms of supercoiled M13mp2IR62E^{cc}-DNA were treated with 10 U of Endo VII (lanes 2 and 3) or 2 U of Cce1 (lanes 4 and 5). Controls include one mock treated sample (lane 1) and one sample treated with *Eco*RI (E) (lane 6). Additional treatment with restriction enzyme *Bam*HI (B) gave the predicted fragments of 4 and 3.2 kbp (lanes 3, 5, and 6). **(C)** Analysis of the reaction products obtained from relaxed hybrid M13mp2IR62E/42E^{oc}-DNA treated in parallel and analyzed as for (B).

the junction. These proteins promote branch migration and transfer the DNA across the junction from one arm to the next *(50,93)*. Parallel and antiparallel conformations in cruciforms may therefore be relevant only for protein-free structures.

3. X-SOLVASES

X-solvases have the ability to recognize and resolve branches in DNA by introducing staggered nicks on either side of the junction. X-solvases apparently are ubiquitous

Table 1
X-Solvases and Cruciform Binding Proteins

Enzyme	Source	Gene	Refs.
Prokaryotic X-solvases			
Endo VII	Phage T4	49	*81*
Endo I	Phage T7	3	*26*
Int	Phage λ	*int*	*53*
Cre	Phage P1	*cre*	*51*
RuvC	*E. coli*	*ruv*C	*19*
RusA	*E. coli*/prophage DLP12	rus	*77*
XerC/XerD	*E. coli*/plasmid	*xer*C, *xer*D	*4*
RecBCD	*E. coli*	*recB, recC, recD*	*122*
Eukaryotic X-solvases			
Endo X1	*S. cerevisiae*	Unknown	*130*
Endo X2	*S. cerevisiae*	*CCE1*	*119*
Endo X3	*S. cerevisiae*	Unknown	*57*
Endo X4	*S. cerevisiae*	*CCE1*	*64*
Flp protein	*S. cerevisiae*	*flp*	*79*
No name	Human placenta	Unknown	*59*
No name	Calf thymus	Unknown	*35*
RuvC analog	Calf thymus	Unknown	*54*
RC-1	Calf thymus	Unknown	*58*
EMX1	Mouse B-cells	Unknown	*116*
No name	Poxvirus	Unknown	*118*
Topoisomerase I	Poxvirus	*Topo*	*28*
CBPs			
HMG1	Rat liver	*Hmg1*	*11*
H1	Chicken erythocytes	h1	*124*
HU	*E. coli*	*hupA hupB*	*14*
No name	Human lymphoblasts	Unknown	*34*
No name	*Ustilago maydis*	Unknown	*66*
RecG	*E. coli*	*recG*	*73*
RuvA	*E. coli*	*ruvA*	*91*

enzymes, having been isolated from many prokaryotic and eukaryotic systems (Table 1). Some X-solvases have been purified to homogeneity, and their activities studied extensively in vitro. Others are still in early stages of analysis. Most X-solvases were detected in assays using SC-cruciform substrates (Fig. 3). The major cleavage sites introduced by a number of the X-solvases at the junction of a synthetic cruciform are shown in Fig. 4.

The biological functions of some X-solvases are quite well understood (e.g., Endo VII from phage T4 and Endo I from phage T7). However, the functions of most X-solvases are not clear. Many of the known X-solvases react with the same type of SC cruciform, suggesting that they share a functional relationship. The proteins, however, differ considerably in their physical properties and in their amino acid sequences. T4

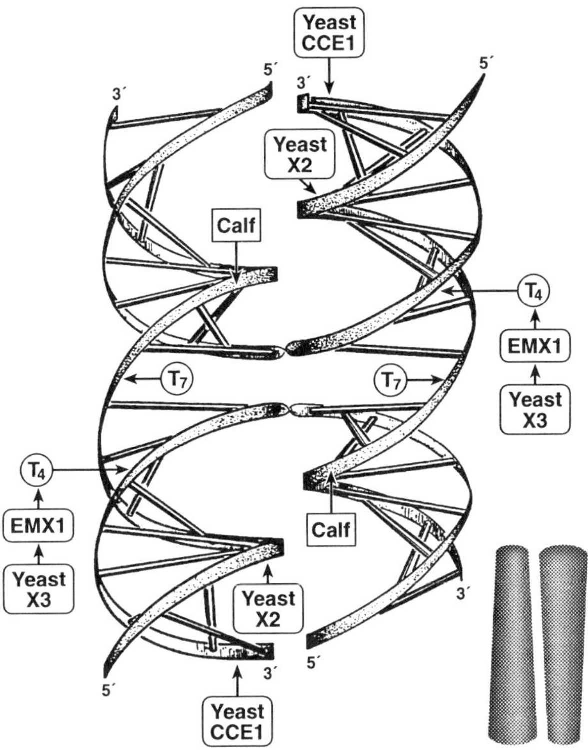

Fig. 4. Location of X-solvase cleavage sites for five enzymes in cruciform DNA. A view from the minor groove side of an antiparallel cruciform structure (Fig. 1C in ref. *10*). The arrows point to the major cleavage sites of six different X-solvases. The enzymes are T4 Endo VII (T4-EVII), T7 EndoI (T7), yeast Endo X2 (Yeast X2), yeast Endo X3 (Yeast EX3), yeast Endo X4 and Cce1 (Yeast *CCE1*), an X-solvase from calf thymus (calf), and X-solvase from mouse B-cells (EMX1). Note that only the X-solvases T4 Endo VII, yeast Endo X1, and the enzyme from calf thymus were tested on the same cruciform structure *(7)*. Predicted cleavage positions of the other enzymes are based on results with different cruciforms and T4 Endo VII as a reference enzyme *(24,27,52,67)*. The drawing in the lower right corner shows a view of the two DNA molecules of a Holliday structure to clarify how the strands cross in space forming a small 60° angle.

Endo VII is the prototype X-solvase, because it was the first enzyme shown to resolve Holliday structures in vitro.

Other enzymes with similar activities include Endo I of phage T7 *(26)*, endonucleases 1, 2, 3, and 4 from yeast *Saccharomyces cerevisiae (57,64,119,129)*, and endonucleases from *E. coli (3,19,55,77)*, human placenta *(59)*, HeLa cells *(126)*, calf thymus *(35)*, and nuclear extracts of mouse B-cells *(116)*. Some of the well-known site-specific recombination proteins also resolve Holliday structures, provided that these include authentic signal sequences. Examples of these proteins are the Flp protein of the 2-μm plasmid from *S. cerevisiae (2,68,103)*, the Int protein of bacteriophage λ *(53)*, the Cre protein of bacteriophage P1 *(51)*, and the XerD/XerC proteins of *E. coli* (*4*; Section 3.2.3.).

3.1. BACTERIOPHAGE X-SOLVASES

3.1.1. Endo VII from Phage T4

3.1.1.1. Physical and Enzymatic Properties

Endo VII is the product of gene 49 (gp49) of bacteriophage T4. The enzyme was initially detected in and purified from phage-infected *E. coli*. After gene 49 was isolated *(6,123)*, the protein was purified to homogeneity from overexpressing bacteria *(46,65)*. The 18-kDa protein contains 157 amino acids with the potential for several alternative Zn-finger motifs. The protein exhibits 48% sequence similarity between amino acids 111 and 138 to T4 endonuclease V (Endo V), a T4-encoded DNA repair enzyme with high specificity for thymine dimers. This particular region can be replaced in T4 Endo VII by the corresponding T4 Endo V region without affecting the Endo VII cruciform cleavage activity *(45)*.

Endo VII executes its activity as a homodimer *(65)*. The enzyme degrades very fast sedimenting phage DNA (VFS-DNA) in vitro by introducing a limited number of double-strand breaks. VFS-DNA is the natural substrate for T4 Endo VII and can be isolated from gene-49$^-$-defective phage-infected cells. Molecules of approximately one-third the size of unit-length T4 DNA were found in a limit digest, indicating a limited number of cleavage sites per phage genome.

T4 Endo VII also shows activity with a large number of additional DNA substrates, including true Holliday structures from in vivo recombined figure-eight plasmid molecules, in vitro generated recA-mediated Holliday structures *(85)*, supercoil cruciforms, hybrid cruciforms (*see also* Fig. 3), synthetic cruciforms *(57,61,87,88,92,99,108)*, Y-structures *(56,61,101)*, heteroduplex loops *(12,63,117)*, five-arm and six-arm branched DNAs *(44)*, mismatches in double-stranded DNA *(12,61,117)*, single-strand overhangs *(61,101)*, single-stranded loops longer than 5 bases at the ends of hairpins *(61)*, nicks and gaps in double-stranded DNA *(61,101)*, curved DNA *(10)*, bulky adducts in double-stranded DNAs *(9,89)*, and apurinic sites in double-stranded DNA *(48)*.

In all cases tested to date, the enzyme introduces asynchronous nicks in both strands 3' to and 2–6 bases from the structural perturbation of the double-stranded target DNA *(61,101)*. The cleavage pattern varies depending on the local sequence. Therefore, homologous sequences exhibit highly symmetrical cleavage patterns. The time delay between nick and counternick is such that DNA polymerase and DNA ligase may repair mismatched nucleotides. Repair is effected by the 3' → 5' exonuclease activity of DNA polymerase entering at the nick and excising the nonmatching nucleotides from one DNA strand, resynthesis of the resulting gap, and rejoining by DNA ligase *(117)*. The repair tracts at a C-C-mismatch were very short, only three to four bases in length. DNA loops of up to twenty bases in length were completely removed from the DNA and replaced by short repair tracts of similar length in vitro *(12)*. The complete repair reactions are surprisingly efficient and suggest that T4 Endo VII may function in vivo primarily as a versatile repair protein, which flags primary defects by introducing a single-strand break, thus preparing the defect for further processing, and preventing double-strand breakage *(12,117)*. Repair was found in either strand of the constructs depending on where the first nick was introduced. A strand bias was found to depend on the individual sequences flanking the mismatching nucleotides *(63,101)*.

Gel retardation and footprint analyses were used to demonstrate the selective binding of T4 Endo VII to a synthetic cruciform. The hydroxyl radical cleavage pattern indicates protection of approx 5 bases in two strands that are diametrically opposed across the junction point *(92)*.

When synthetic cruciforms are made in parallel and antiparallel conformation by tethering appropriate arms, T4 Endo VII shows marked preference for crossing strands in antiparallel constructs *(44,87,88)*. These observations and conformational studies of free synthetic cruciforms suggest that T4 Endo VII cleaves at phosphodiester bonds that are presented on one side of the cruciform molecule (*10*; Fig. 4).

3.1.1.2. Biological Function of T_4 Endo VII

T4 Endo VII removes branches from DNA prior to packaging; consequently, gene 49^- phage are deficient in packaging their newly synthesized DNA into phage heads *(62)*. This result of the mutation led to the gene originally being classified as a morphopoietic gene. The DNA in gene 49^- phage infections accumulates in highly branched, VFS complexes. Many sites of this DNA still support simultaneous initiation of packaging, but completion of the process is defective. The defect is not abortive, however, and the resolution of VFS-DNA continues if, for example in gene 49ts infections permissive conditions are re-established. In the electron microscope, VFS-DNA appears as compact and complex structures made of tightly packed DNA *(39)*. VFS-DNA, like all phage DNA, contains hydroxymethyl cytosine (HMC), which renders the DNA resistent to cleavage by many restriction enzymes including *Cla*I. Replacement of HMC by unmodified cytosine allows digestion with *Cla*I. Digestion fragments are X- and Y-branched structures when viewed in the electron microscope *(39)*. All branches were sensitive to digestion with purified T4 Endo VII in vitro (*39*; Fig. 5).

Surprisingly, gene 49^- mutants show little reduction of recombination frequencies measured among progeny that arise despite defective packaging *(80,84,114)*. In fact, the formation of heteroduplex DNA (hDNA) is remarkably increased *(84)*, suggesting that genetic recombination involves the repair functions of T4 Endo VII rather than the Holliday structure-resolving function. The crucial role of debranching functions of T4 Endo VII was demonstrated in situations when T4 DNA synthesis starts from secondary replication start sites. T4 creates these secondary start sites frequently during middle- and late-stage of infection. Recombination between homologous sequences yields single-stranded 3'-termini capable of priming DNA synthesis that replicates the displaced strands *(40)*. Functional gp49 is required to resolve the resulting D-loops and Y-structures prior to further processing and packaging of the newly replicated DNA *(82,83)*.

T4 Endo VII is a highly active enzyme that completely and rapidly destroys all bacterial host DNA *(65)*. The gene is transcribed during early and late phases of infection from different promoters. In the early transcript of gene 49, the Shine-Dalgarno sequence of the first ribosome binding site is sequestered in a hairpin structure, thereby preventing efficient translation *(6)*. Late transcripts are initiated from a late promoter immediately upstream of the first ribosome binding site containing a sequence that cannot form a hairpin. This arrangement results in a more efficient translation of the 18-kDa protein during packaging at times when there is an increased demand for Endo VII activity *(61)*. Initiation from a proximal AUG and an internal GUG allows two

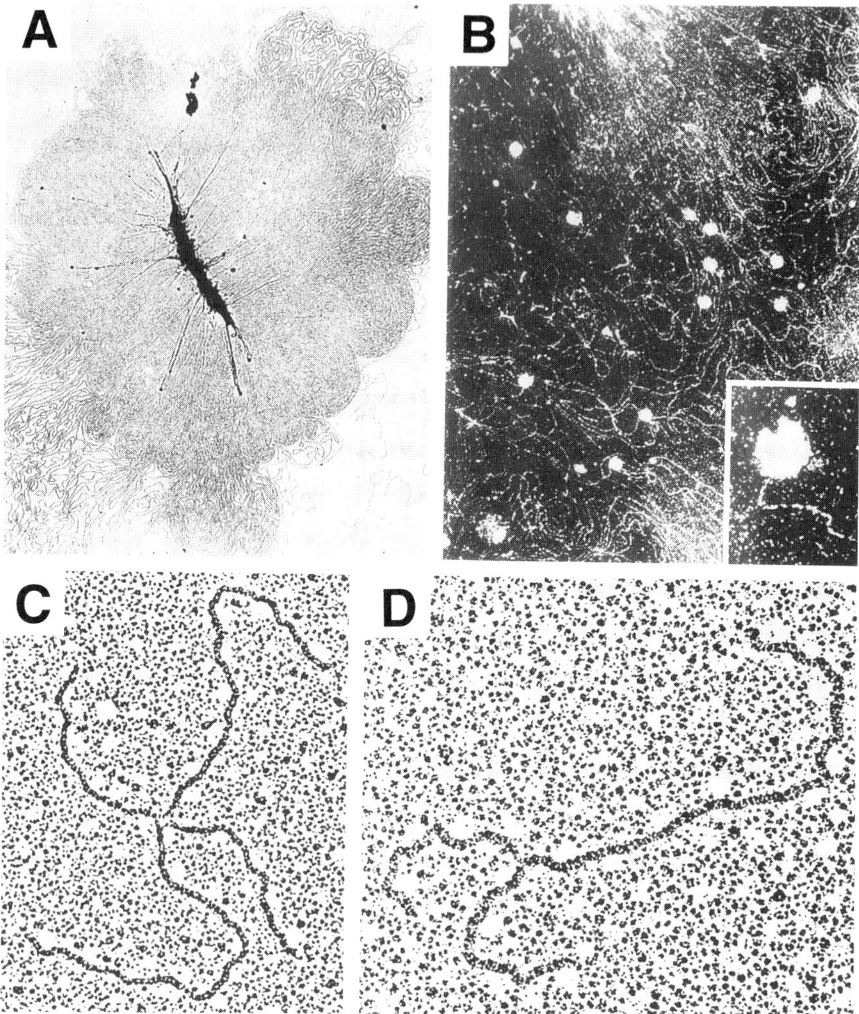

Fig. 5. Analyses of VFS-DNA from T4 gene 49⁻-infected cells. Electron micrographs show different stages of the analysis of VFS-DNA. **(A)** VFS-DNA purified from gene 49⁻ infections. The complex contains on average 300 chromosomes of phage T4. **(B)** VFS-DNA after crosslinking with formaldehyde. Partially filled heads are seen attached to free ends. The inset shows a magnification of a partially filled head structure. **(C,D)** Two *Cla*I fragments isolated from cytosine containing VFS-DNA showing typical X- and Y-structures.

in-frame overlapping peptides of 18 and 12 kDa to be made. The 18-kDa protein is T4 Endo VII. The function of the 12-kDa protein is still unknown *(6)*. Purified 12 kDa protein has no detectable endonucleolytic activity *(49)*.

3.1.2. Endo I from Phage T7

3.1.2.1. Physical and Enzymatic Properties

Endo I is the product of gene 3 (gp3) of bacteriophage T7 *(16)*. The protein has 143 amino acids and a mol wt of 18 kDa. Endo I is maximally expressed at 7.5 min after infection at 37°C. The X-solvase was originally partially purified from extracts of T7$^+$ phage-infected *E. coli* B *(15,104)*. T7 Endo I cleaves both double- and single-stranded DNA and is at least 100 times more active on single-stranded DNA than duplex DNA. Both the single- and double-strand cleavage activities are associated with a single protein *(15)*. The hydrolysis of single-stranded DNA results in the formation of acid-soluble 5'-phosphate and 3'-hydroxyl-terminated oligonucleotides with a chain length of about 10 bases. The enzyme introduces both single-strand and double-strand breaks having 5'-phosphate and 3'-hydroxyl-termini in duplex DNA, but does not produce acid-soluble material. The limit product obtained from digestion of double-stranded DNA consists of fragments of duplex DNA approx 125 bp long. The enzyme shows some preference for pyrimidines, generating fragments with predominantly pyrimidine-containing nucleotides at the 5'-termini. The enzyme acts equally well on DNAs from *E. coli*, T4, T7, and phage λ *(104)*.

The reported preference of T7 Endo I for single-stranded DNA suggested that it is a single-strand specific endonuclease. When T7 Endo I was used to map SC cruciforms in supercoiled plasmid DNA, cleavage sites were therefore misidentified in the single-stranded loops of cruciforms. Fine mapping revealed later that T7 Endo I cleaves at the base of cruciforms similarly to T4 Endo VII. The enzyme cuts two opposing strands at or near the branch point, resolving these substrates into linear molecules. However, although T4 Endo VII cleaves 3' to cruciform junctions, T7 Endo I cleaves 5' to these junctions (Fig. 4). T7 Endo I cuts at branched X- and Y-structures in relaxed as well as supercoiled DNAs *(26,27)*. Even the purest preparation of T7 Endo I is still more active on single-stranded DNA than on double-stranded DNA. This is presumably owing to the high content of secondary structures in single-stranded DNA; these are also substrates of T4 Endo VII *(102)*.

In the absence of divalent cations, T7 Endo I forms a distinct protein–DNA complex with a synthetic cruciform structure. Probing this complex with hydroxyl radicals reveals that the nuclease binds all four strands at the junction point *(31,94,99)*.

Mutants of T7 Endo I have been characterized that retain full structural selectivity of binding to four-way junctions, but that are completely inactive as nucleases. These mutations are clustered in the second quarter of the primary sequence, which displays some sequence similarity with T4 Endo VII *(31)*. The addition of T7 Endo I to strand exchange reactions catalyzed by RecA protein of *E. coli* in vitro leads to the formation of duplex products that correspond to "patch" and "splice" type recombinants (*86*; Fig. 1).

3.1.2.2. Biological Function of T7 Endo I

Amber mutants of phage T7 defective in gene 3 synthesize only a limited amount of DNA, they are defective in carrying out efficient degradation of host DNA *(16)*, and they exhibit low recombination frequencies. Gene 3 product together with the products of gene 4 (DNA replication protein), gene 5 (T7 DNA polymerase), and gene 6 (T7 exonuclease) are all involved in phage T7 genetic recombination. The effect of mutations in any of these genes on recombination frequencies in two-factor crosses, however, was comparably moderate; reductions of between 5- (gene 3) and 10-fold (gene 4) were observed, suggesting the existence of backup recombination mechanisms. For T7, it appears that recA, recB, or recC is not involved in such a backup system, since mutations in any of these genes do not reduce recombination in phage two-factor crosses.

As seen for phage T4 infections, newly synthesized T7 DNA accumulates in complexes containing up to several hundred phage equivalents of DNA per infected cell. The complexes are normally converted to unit-size DNA during packaging. They are stable to treatment with ionic detergents, protease, and phenol, and are resistant to elevated temperatures. These complexes consist of looping DNA molecules anchored in a dense central core. The DNA is rich in single-stranded regions. During infection by gene 3^- phage, complexes form normally, but maturation of the complex to unit-size DNA is blocked. Physical studies of the gene 3^-–DNA complexes uncovered numerous X- and Y-branched structures containing two parental DNA molecules that resulted from incomplete recombination events. Such structures were not found in gene 3^+ infections, indicating that Endo I is directly involved in processing these structures in vivo. This phenotype is very similar to the phenotype of T4 gene 49^- mutations described in Section 3.1.1.2.

3.1.3. Other Bacteriophage X-Solvases

To date, no systematic search for X-solvases from other phage has been done. However, two examples of well-known proteins with highly specialized functions in site-specific recombination were tested for their ability to resolve Holliday structures in vitro. These are the Int protein (product of phage λ gene *int*) and the Cre protein (product of phage P1 gene *cre*). These proteins catalyze recombination reactions between either the *att* sites for phage λ or the *lox* sites for phage P1. The breakage and rejoining of partner molecules during this recombination process are highly concerted. Integration and excision reactions are initiated by single-strand breaks followed by reciprocal strand exchange and the formation of Holliday structures. The integrity of the recombined molecules is fully restored by resolving the branches and relegating the nicks. Synthetic Holliday structure analogs made with *att* sites for phage λ and *lox* sites for phage P1 were clearly resolved in vitro by Int and Cre, respectively *(51,53)*. Another protein, RusA, was recently identified as a cruciform-resolving enzyme in *E. coli*. It is produced by a defective lambdoid prophage, DLP12, which is probably derived from a phage related to Φ82 and PA-2 (*77*; Section 3.2.2.).

3.2. Bacterial X-Solvases

3.2.1. RuvC from E. coli

3.2.1.1. Physical and Enzymatic Properties

When recombinant figure-eight structures made by RecA protein in vitro were transformed into *recA*⁻ cells, they were resolved into patch and splice recombinant products with nearly an equal probability. The transformed structures were treated in *E. coli* as true recombinant intermediates without need for generation in vivo or functional RecA protein.

The same synthetic Holliday junctions were also resolved in vitro by fractionated extracts from wild-type *E. coli*, but not by *ruvC* mutant extracts *(20,21)*. The *ruvC* deficiency may be restored by a multicopy plasmid carrying a $ruvC^+$ gene that overexpresses the junction-resolving activity *(20,21)*. The purified RuvC protein is an 18-kDa protein. It is active as a dimer and resolves both recombination intermediates, when made from fully homologous DNA molecules, and synthetic cruciforms containing a stretch of at least 12 homologous base pairs. SC cruciforms also serve as substrates. In contrast to earlier observations *(32)*, recent experiments revealed that RuvC can cleave three-way junctions with high efficiency if they contain a core of homology *(120)*. RuvC initially binds nonproductively to four-way junctions, which does not require divalent metal ions, and subsequently cleaves the junctions by a mechanism dependent on a divalent cation and a particular topological conformer that is induced by the sequences at both the mobile X- and Y-junctions *(120)*. Branches are resolved by the introduction of litagable nicks into strands of same polarity *(7,33,55,120)*.

RuvC cleaves the consensus sequence 5'- (A)/TTT ↓ (G)/(C)-3' in duplex DNA in one strand, suggesting that Holliday junctions only need to be translocated to specific cleavage sites. Mutation of the consensus site in synthetic Holliday junctions abolishes or significantly reduces the efficiency of cleavage. The strand bias of this sequence can also affect the outcome of recombinational crosses in vivo by directing resolution to either "patch" or "splice" recombinant products *(110)*. When RuvC is bound to a synthetic cruciform, two of the four strands at the junction become hypersensitive to hydroxyl radical attack *(7)*. Three point mutations were reported in *ruvC* that affect either the nuclease activity of RuvC or both the nuclease activity and the cruciform binding activity of the protein *(113)*, suggesting these functions reside in distinct domains.

The activity of RuvC protein should be considered in conjunction with the activities of RuvA and RuvB proteins. These proteins are important in the formation of hDNA during genetic recombination and recombinational repair processes. RuvA binds to Holliday junctions in vitro and directs RuvB to the junction. RuvA is then sandwiched between two ring-shaped hexameric molecules of RuvB. Together, RuvA and RuvB promote branch migration in an ATP-dependent reaction that increases the length of the hDNA *(1,93)*. The interactions between RuvC and Holliday junctions in complex with these proteins has yet to be elucidated.

3.2.1.2. Biological Role of RuvC

The *E. coli ruvC* gene is arranged together with four open reading frames in the order orf17-orf26-*ruvC*-orf23. The open reading frames are located immediately upstream of the *ruvAB* operon. Only *ruvC* is involved in the repair of UV-damaged DNA, and in contrast to the *ruvAB* operon, *ruvC* is not regulated by the SOS system *(121)*. Mutations in any of the *ruv* genes confer increased sensitivity to DNA-damaging agents, such as UV- and γ-irradiation, and certain mutagens *(111)*. *ruv* single mutants have little effect on conjugal recombination, but confer recombination deficiency and extreme sensitivity to ionizing radiation in *recBC sbcBC* strains *(71)*, *recBC sbcA* strains *(72,76)*, and *recG* strains *(70)*. These results indicate that *ruvC* is multifunctional, participating in several DNA protection pathways supported by genes *recBC*, *sbcABC* and *recG*. Whether it functions as a Holliday structure resolving activity in vivo remains uncertain, since *ruvC* mutants show nearly normal recombination frequencies in genetic crosses. Recent findings indicate that *recG* overlaps functionally with *ruvC*. The 76-kDa RecG protein binds to synthetic cruciform DNA as well Y-junctions. It has helicase activity that dissociates recombination intermediates into mature products by catalyzing branch migration of Holliday structures in vitro *(73,131)*. RuvC and RecG belong to a network of multilayered DNA rescue mechanisms, which complement each other.

3.2.2. E. coli *RusA*

Suppressor mutants of the UV-sensitive phenotype of *ruvC* mutants were isolated in *E. coli* and named *rus-1 (78)*. Recent studies have revealed that the *rus* gene is part of the defective prophage related to phage 82, and an identical protein is encoded by phage 82 itself *(77)*. Whether *rus-1* is expressed under normal cell growth is not yet clear. A 14-kDa protein, named RusA, was purified from overexpressing clones. RusA cleaves synthetic cruciforms as well as RecA-induced Holliday junctions, introducing symmetrical cuts in two strands to give nicked duplex products. The cleavage activity on cruciform DNA is remarkably similar to that of RuvC; both proteins preferentially cut the same two strands at the same location in a specific cruciform *(112)*.

3.2.3. E. coli *XerC and XerD*

The *E. coli* Xer site-specific recombination system functions in the normal segregation of multicopy plasmids and the bacterial chromosome by converting circular multimers to monomers. Both proteins are closely related to the λ integrase family and use specific sites on the DNA, for example *cer* in plasmid ColE1 and *dif* in the terminus of *E. coli (13)*. This site-specific recombination system is unique, since it involves two recombination proteins instead of one as with phage λ Int/*attP*/*attB*, the phage P1 Cre/*lox*, and the yeast 2-μm plasmid Flp/frt systems. A synthetic Holliday junction-containing molecule that was free to branch migrate through the 28-bp recombination core region of the *cer6* recombination site is a substrate for Xer recombination in vitro. Both XerC and XerD proteins were required for the resolution of the preformed recombination intermediate and termination of the recombination event *(3)*. No evidence has been reported to date that Xer participates in recombination reactions other than sequence-specific ones.

3.2.4. E. coli *recBCD*

The reactivity of purified RecBCD enzyme to resolve pre-existing Holliday structure analogs was examined using an open-ended DNA cruciform with limited ability to branch migrate *(122)*. The enzyme cleaved two strands of the cruciform near its base with two nicks introduced separately. Cruciforms whose four arms were blocked by fold-back loop structures (Fig. 2D) were not detectably nicked by the enzyme. Use of varying combinations of open and loop-protected arms in the same construct led to the conclusion that RecBCD enzyme molecules must enter the termini of duplex DNA and approach the cruciform from more than one direction in order to cleave it into recombinant products. It was therefore inferred that intracellular RecBCD enzyme cannot cleave pre-existing Holliday junctions into recombinants, but may cleave Holliday junctions in whose formation it participates. This is apparently a special case restricted to the *recBCD*-dependent conjugal recombination pathway in *E. coli*.

3.3. Yeast X-Solvases

Four X-solvases have been detected in cell extracts from yeast *S. cerevisae (57,64,119,129)*. The activities were arbitrarily named yeast endonucleases X1 (Endo X1), X2 (Endo X2), X3 (Endo X3) and X4 (Endo X4). Endo X4 was synonymous with cruciform cutting endonuclease 1 (*CCE1*). Since these activities were detected using basically the same assay procedure (Fig. 3), isolates were expected to overlap in activities. Evidence (cited in *64*) suggests that Endo X2 and Endo X4 are indeed the same protein.

3.3.1. Yeast Endo X1

Activity of yeast Endo X1 increases in cells after treatment with mechlorethamine, a DNA-damaging nitrogen mustard. The activity was purified through several chromatographic steps. The purest fraction has a protein with an apparent mol wt of about 200 kDa, as judged from gel-filtration analysis. The enzyme requires Mg^{2+} for optimal activity. Neither Ca^{2+}, Co^{2+}, nor Mn^{2+} substitutes for Mg^{2+}. The activity linearizes supercoiled cruciforms by cleaving diagonally across the junction. The enzyme shows no activity with single-stranded DNA, nicked circular, or linear duplex DNA *(130)*. The high molecular weight of the protein in the active fractions is more consistent with an aggregation of proteins rather than a single polypeptide. All more extensively studied X-solvases exhibit small mol wts of 18–40 kDa. Further characterization of the enzymatic activities and studies of the role of Endo X1 in DNA recombination and repair requires further purification of the protein(s).

3.3.2. Yeast Endo X2

Yeast Endo X2 was purified 500- to 1000-fold from yeast cell extracts by a series of chromatographic steps. The activity linearizes supercoiled cruciforms by cleaving across the junction of the structure. It also resolves true Holliday structures in recombinant figure-eight DNA from phage G4 into linear products. Endo X2 is not active on single-stranded, linear, or circular duplex DNA *(119)*. A fully synthetic cruciform structure is not cleaved, but it is bound by the enzyme, as shown by gel-mobility shift assays *(36)*. Three-armed Y-structures are also not substrates for Endo X2. However, synthetic cruciforms made from wild-type λ attachment sites are resolved by pairs of single-strand nicks in opposite strands across the junction. Single nucleotide changes in the sequence located immediately adjacent to the junction, but still within the 15-bp common core of the λ *att*-site, had dramatic effects on both the directionality and rate of resolution *(37)*. Since this enzyme is the same as yeast Endo X4 or CCE1, *see* Section 3.3.4. for further discussion of its function.

3.3.3. Yeast Endo X3

Yeast Endo X3 was detected in crude extracts from mitotically growing cells of yeast *S. cerevisiae*. The activity was purified more than 1000-fold. On denaturing SDS PAGE, the protein has an apparent mol wt of 18 kDa. Gel filtration under native conditions revealed that the enzymatic activity eluted together with a protein of a mol wt of 43 kDa, suggesting that the active enzyme may exist as a dimer. The enzyme introduces staggered nicks flanking the junction in branched DNAs of different SC cruciforms, hybrid cruciforms, synthetic cruciforms with arms of 9 bp and longer, and synthetic Y-junctions. It also cleaves at DNA loops and mismatches in synthetic hDNAs. Cleavage patterns were indistinguishable from those seen with T4 Endo VII in the same substrates (Section 3.1.1.). Anti-Endo VII antibodies inhibit yeast Endo X3 activity with the same efficiency as they inhibit the phage T4 enzyme *(57)*. This feature is shared with mouse enzyme EMX1 (Section 3.4.). Further evaluation of these data require further tests with more purified enzymes. Their existence, however, raises the question of the evolution of X-solvases in different organisms, and one expects the proteins to share common structure/function motifs.

3.3.4. Yeast Endo X4 (Cce1p)

Yeast Endo X4 was detected in crude extracts from yeast *S. cerevisiae*. Although missing in a temperature-sensitive mutant *(64)*, the ts phenotype was separable from the X-solvase mutation. The CCE1 mutation was mapped to the left arm of chromosome XI within 3 c*M* of the centromere. *CCE1* has an open reading frame for a 41-kDa protein. The protein was overexpressed in *E. coli* as native *(64)* and fusion proteins *(132)*. After purification to near homogeneity, Cce1p has an apparent mol wt of 38 kDa *(67)* or 40 kDa *(132)*. The enzymatic activity of Cce1p was analyzed with SC cruciforms and a series of sequence-related synthetic cruciforms. Cce1p linearizes SC cruciforms by cutting at the base of the cruciform structure. It does not require a supercoiled substrate, since a relaxed hybrid cruciform was cleaved with the same efficiency. Cce1p is active with fully synthetic cruciforms, provided an *Eco*RI restriction sequence was maintained at the junction. A fully randomized sequence at the junction of the same construct prevented cleavage by Cce1p. A synthetic three-armed Y-junction was also cleaved by Cce1p, but with a lower efficiency than the related four-armed construct. The wild-type protein is a monomer in solution as determined by gel filtration *(67)*. In contrast, the fusion protein Cce1p and the fusion-released Cce1p protein form dimers with each other and with themselves when run in native polyacrylamide gels *(132)*. Also a pronounced sequence selectivity for cleavage at 5'-CT was reported for fusion-released Cce1p *(132)*.

The amino acid sequence of Cce1p shows no homology to prokaryotic resolvases or to any known protein. A *cce1* null mutation has no obvious growth defect, and shows no defect in meiotic or mitotic recombination. The only phenotype observed for *cce1* mutants is a higher than normal frequency of petite cells, suggesting that Cce1p is important for the maintenance of mitochondrial DNA *(64)*. This was supported by the finding that Cce1p activity is associated with the mitochondrial inner membrane *(38)*.

CCE1 was found to be allelic with *MGT1*, a gene required for the highly biased transmission of ρ^- mitochondrial DNA in matings between ρ^- and ρ^+ cells. This segregation advantage is abolished by mutations in *MGT1/CCE1 (133)*. In accordance with its function as a Holliday structure resolving enzyme, these mutants accumulate mitochondrial DNA in aggregation complexes with molecules linked together by recombination junctions *(74)*. This phenotype is reminiscent of the phenotypes of mutant genes 49 and 3 of phages T4 and T7, respectively (Sections 3.1.1. and and 3.1.2.).

3.4. X-Solvases from Higher Eukaryotes

Using a SC cruciform assay, X-solvase activities have been detected in cellular and nuclear extracts from several types of eukaryotic cells, including human placenta *(59)*, HeLa cells *(126)*, calf thymus *(35)*, and mouse B-cells *(116)*.

The X-solvase activity from human placenta was partially purified and shown to cleave SC cruciforms as well as synthetic cruciforms *(59)*. The X-solvase activity from calf thymus was also partially purified, exhibiting a mol wt of approx 75 kDa as determined by gel filtration *(35)*. An X-solvase activity purified from cytoplasmic extracts of HeLa cells cleaves diagonally across the central junction of a synthetic cruciform *(126)*.

The X-solvase activity EMX1 was detected in nuclear extracts from mouse B-cells using a covalently closed synthetic cruciform structure as substrate. Loops of five T-residues attached to the ends of each arm of the cruciform structure provided resistance to exonucleases in extracts (Fig. 2D). EMX1 shows surprising similarities to T4 Endo VII: the enzyme cleaves at 3'-positions across the junction, and recognizes loops and mismatches in synthetic heteroduplexes *(116)*. EMX1 is strongly inhibited by anti-Endo VII antibodies, a feature shared with Endo X3 from yeast *S. cerevisae (57)*.

A Holliday structure resolving activity was identified in crude protein extracts from vaccinia virus-infected cells at late times during the replication cycle *(118)*. The activity accurately resolved poxviral inverted repeat replicative intermediates. The substrate used for investigating the activity was a cloned version of the Shope fibroma virus replicated telomere in an inverted repeat configuration, which presumably adopts the secondary structure of an SC-cruciform similar to that shown in Fig. 3. The resolved linear products are identical to resolution products in vivo. They possess symmetrical nicks, which mapped at the borders of the inverted repeats. The cleavage of poxviral telomeric substrates resembles the resolution of true Holliday junctions, and it was shown that nonviral Holliday junction analogs were also cleaved by the extract. These results suggest that the resolution activity may also play a role in general viral recombination during late stages of infection.

Recently, it was shown that highly purified vaccinia virus topoisomerase I is capable of precisely resolving synthetic cruciform constructs if they contain the topoisomerase I recognition sequence 5'-YCCTT (Y = C or T) at their junctions. The reaction is very efficient and proceeds by transesterification of two recognition sites in opposing strands generating two linear duplexes *(109)*. Although nothing is known to date about viral or host proteins involved in vaccinia virus recombination, topoisomerase I is a good candidate for a Holliday structure resolving enzyme. More than 1000 topoisomerase cleavage sites are within the 192-kbp vaccinia genome, providing ample locations for the resolution reactions. In addition to RuvC from *E. coli*, vaccinia topoisomerase I is an example of a sequence-dependent Holliday structure resolving enzyme. Whether it is related to the telomere resolving activity described in the preceding section *(118)* remains to be seen.

3.5. Recognition of Holliday Structures by X-Solvases

To understand the molecular recognition of branched DNAs by X-solvases, many experiments were performed with synthetic cruciform constructs in solution. Electrophoretic and fluorescence resonance energy transfer analyses of pure DNA in solution indicate that in the absence of Mg^{2+} ions, synthetic four-arm junctions adopt a square planar conformation, while at a sufficient Mg^{2+} concentration, the junction folds into an antiparallel-stacked X-structure with angles of about 60 and 120°, respectively, between the arms *(18,30)*.

The antiparallel conformation of a cruciform structure is sterically very attractive for enzymatic resolution, since dimeric X-solvase proteins can selectively bind to one face of the junction and simultaneously contact phosphodiester bonds in the swapping strands located 3' and 5' immediately adjacent to the branch point (Fig. 4). This model was further supported by observations that T4 Endo VII resolves cruciform structures that were forced into an antiparallel conformation by appropriate tethering, with high relative efficiencies and high precision. Structures having identical sequences, but tethered into a parallel conformation were not resolved at all, nor were they cleaved with high relative efficiencies *(10,44,48)*. Tethered constructs have limited freedom to switch between isomeric conformations. These results therefore suggest that Endo VII recognizes antiparallel molecules without an ability to change their global structure. Recent evidence obtained with RuvC, however, suggests that after protein binding, the cruciform is converted to an open planar configuration *(7)*, which enables it to cleave the continuous, rather than the crossover, pair of strands *(8)*.

RuvC is the first Holliday structure resolving protein for which the crystal 3D-structure has been determined at 2.5 Å resolution *(5)*. The active enzyme forms a dimer with 19-kDa subunits related by a dyad axis. The catalytic center, comprising four acidic residues, lies at the bottom of a cleft into which duplex DNA fits. The structural features of the dimer, with a 3-nm space between the two catalytic centers, provide a substantially defined image of the Holliday junction architecture. A model of a RuvC protein dimer binding to a cruciform structure was obtained by computer fitting two DNA molecules into the clefts of each protein subunit. The DNA molecules were then covalently linked by connecting nucleotides and energetically optimized. The best-fit was obtained with a stacked antiparallel X-structure, which is in principle similar to the structure of a synthetic cruciform in solution described before *(29,125)*.

Holliday structures are from a geneticist's point of view best described as parallel aligned homologous chromosomes connected by mutually exchanging strands *(52)*. The conformation of a true Holliday structure in solution has to date not been determined. Synthetic cruciforms are generally designed free from extensive homologies for stability reasons, and they must therefore be considered Holliday structure analogs. It is important to note that the results obtained with synthetic cruciforms can not be applied to natural Holliday junctions without further experimental evidence.

4. CRUCIFORM BINDING PROTEINS

CBPs belong to a heterogeneous class of proteins whose members are operationally defined by their ability to bind to cruciform DNA as determined by either electromobility gel-shift assays or affinity chromatography Table 1. CBPs do not hydrolyze DNA, which clearly distinguishes them from X-solvases. CBPs have been isolated from rat liver *(11)*, human lymphoblasts *(34)*, HeLa cells *(95)*, and *Ustilago maydis (66)*. A few proteins well known for their cellular functions other than recombination were also found to bind to cruciforms. Among these were HU protein from *E. coli (14)*, histone H1 from chicken erythrocytes *(124)*, and adenosine diphosphoribosyl transferase (ADPRT) from calf thymus *(105)*.

CBP from rat liver was the first protein shown to recognize cruciform DNA selectively. It was identified as the nonhistone high mobility group protein 1 (HMG1), which is an evolutionarily conserved, essential, and abundant nuclear component *(11)*. Histone H1, another abundant nuclear protein that is also a major constituent of chromatin and helps to organize nucleosomes into higher-order structures, was also found to bind to cruciform DNA. The binding may reflect its natural binding to the incoming and outgoing DNA duplex crossing over itself on the surface of the nucleosome.

Hydroxyl radical footprinting experiments with the *E. coli* CBP revealed that the bases protected by HU are localized at the junction point. These bases are primarily located in two of the four oligonucleotides constituting the cruciform DNA. These two oligonucleotides are unpaired and opposite each other. These results support a model where two HU protein dimers bind to two sides of the cruciform junction with almost no dimer–dimer interactions *(14)*. HU protein appears to inhibit cruciform extrusion from supercoiled inverted repeat DNA, either by constraining supercoiling or by trapping a metastable interconversion intermediate. This behavior of HU is reminiscent of the active protection that cruciforms may need from cleavage by X-solvases in vivo suggested in Section 2. All these properties are analogous to the properties of the mammalian chromatin protein HMG1 and suggest that HU is a prokaryotic HMG1-like protein *(97,100)*.

CBP from HeLa cells has an apparent mol wt of 66 kDa *(95)*. Hydroxyl radical footprinting analyses revealed that the protein binds to the four-way junction of the base of the cruciform. On CBP binding, structural distortions were observed in the cruciform stems and in a DNA region adjacent to the junction *(96)*. CBP from human lymphoblasts forms complexes with synthetic X- and Y-junctions. The binding is specific for DNA structure rather than sequence, since cruciforms with different sequences bind equally well. Linear molecules containing the same sequences are not bound, and one X-junction successfully competes with another for binding whereas linear duplexes do not compete *(34)*.

The protein from *U. maydis* has a mol wt of 11 kDa *(66)*. The protein recognizes synthetic as well as supercoiled cruciforms. Its physical and biochemical properties in conjunction with its amino acid composition are reminiscent of properties observed for the HMG proteins *(66)*.

Adenosine diphosphoribosyl transferase, another eukaryotic nuclear protein, was found to bind specifically to SC cruciforms in plasmid DNA in vitro *(105)*. The enzyme catalyzes the transfer of ADP-ribose moiety of NAD^+ to itself and to other cellular proteins. The protein binds to one side of the base of the cruciform, leaving the other side and the tip of the structure open to nuclease attack *(105)*.

5. CONCLUDING REMARKS

X-solvases are widespread in prokaryotes and eukaryotes, demonstrating general importance. RuvC from *E. coli* and Endo VII from phage T4 are the best-studied examples so far, and their functions in general genetic recombination and repair of DNA have been demonstrated. Future studies should reveal how RuvC and Endo VII interact with Holliday structures in the presence of other system-related recombination proteins, such as strand transfer protein RecA, and branch migration proteins RuvA and RuvB from *E. coli* or the strand transfer protein UvsX and the helper protein UvsY from phage T4.

Resolving Holliday junctions, however, may be only one aspect of X-solvase activity. For example, RuvC is involved in several recombinational DNA repair pathways as is T4 Endo VII, which was shown to function in mismatch repair *(47)*. Since this enzyme detects many different lesions in DNA, Endo VII may be a representative from an ancient class of repair enzymes with broad substrate specificities. This possibility requires more detailed investigation. The small effect that gene 49^- mutants have on genetic recombination frequencies in two-factor crosses suggests that phage T4 may employ another T4 Holliday structure resolving enzyme or a host enzyme. With the wide variety of suitable substrates and mutant phage and host strains available, this question is amenable to experiment.

Many X-solvases are small proteins with a mol wt of around 18 kDa (e.g., T4 Endo VII, T7 Endo I, *E. coli* RuvC, *E. coli* RusA, yeast Endo X3, mouse EMX1) that may facilitate crystallization and structure analyses. Some of these proteins are already available in pure form and their crystallization may be a question of time. It would be very interesting to see whether the X-solvase proteins have a (new) DNA binding domain in common, allowing specific recognition of their branched targets, probably via kinks or sharp bends in DNA.

CBPs like HMG1, histone, H1, or HU may be involved in the intracellular organization of DNA. Binding to cruciforms mimics their binding to arms of kinked DNA present at turns of looping molecules tightly packed in chromosomal structures. RuvA may be an exceptional CBP. It binds as a tetramer to cruciforms and moves (with the help of two hexameric rings of RuvB) the junction of a Holliday structure by translocating the DNA duplexes of pairing molecules (reviewed in ref. *115*). RuvA/B and RecG are Holliday junction specific helicases.

It remains to be seen whether CBPs are also involved in stabilizing SC cruciforms and protecting them, for example, from attack by X-solvases in vivo. CBPs may prove useful for studying the recognition of branched DNAs by proteins.

ACKNOWLEDGMENTS

The studies on enzymes Endo VII of phage T4, Endo X3, and *CCE1* of yeast in this laboratory are supported by grants from the Deutsche Forschungsgeinschaft (Ke 188/8-1, Ke 188/8-2), through SFB 274 (TP A2), and by Avitech Corporation, Malvern PA. The author is grateful for the helpful comments provided by A. Huff and the editors.

REFERENCES

1. Adams, D. E. and S. C. West. 1995. Unwinding of closed circular DNA by the *Escherichia coli* RuvA and RuvB recombination/repair proteins. *J. Mol. Biol.* **247:** 404–417.
2. Amin, A., H. Roca, K. Luetke, and P. D. Sadowski. 1991. Synapsis, strand scission, and strand exchange induced by the FLP recombinase: analysis with half-FRT sites. *Mol. Cell. Biol.* **11:** 4497–4508.
3. Arciszewska, L., I. Grainge, and D. Sherratt. 1995. Effects of Holliday junction position on Xer-mediated recombination in vitro. *EMBO J.* **14:** 2651–2660.
4. Arciszewska, L. K. and D. J. Sherratt. 1995. Xer site-specific recombination in vitro. *EMBO J.* **14:** 2112–2120.
5. Ariyoshi, M., D. G. Vassylyev, H. Iwasaki, H. Nakamura, H. Shinagawa, and K. Morikawa. 1994. Atomic structure of the RuvC resolvase: A Holliday junction-specific endonuclease from *E. coli*. *Cell* **78:** 1063–1072.
6. Barth, K. A., D. Powell, M. Trupin, and G. Mosig. 1988. Regulation of two nested proteins from gene 49 (recombination endonuclease VII) and of a lambda RexA-like protein of bacteriophage T4. *Genetics* **120:** 329–343.
7. Bennett, R. J., H. J. Dunderdale, and S. C. West. 1993. Resolution of Holliday junctions by RuvC resolvase: cleavage specificity and DNA distortion. *Cell* **74:** 1021–1031.
8. Bennett, R. J. and S. C. West. 1995. RuvC protein resolves Holliday junctions via cleavage of the continuous (noncrossover) strands. *Proc. Natl. Acad. Sci. USA* **92:** 5635–5639.
9. Bertrand-Burggraf, E., B. Kemper, and R. P. P. Fuchs. 1994. Endonuclease VII of phage T4 nicks N-2 acetylaminofluorene induced DNA structures in vitro. *Mutat. Res.* **314:** 287–295.

10. Bhattacharyya, A., A. I. H. Murchie, E. V. Kitzing, S. Diekmann, B. Kemper, and D. M. J. Lilley. 1991. Model for the interaction of DNA junctions and resolving enzymes. *J. Mol. Biol.* **221:** 1191–1207.
11. Bianchi, M. E., M. Beltrame, and G. Paonessa. 1989. Specific recognition of cruciform DNA by nuclear protein HMG1. *Science* **243:** 1056–1059.
12. Birkenkamp, K. and B. Kemper. 1995. In vitro processing of heteroduplex loops and mismatches by endonuclease VII. *DNA Res.* **2:** 9–14.
13. Blakely, G., G. May, R. McCulloch, L. K. Arciszewska, M. Burke, S. T. Lovett, and D. J. Sherratt. 1993. Two related recombinases are required for site-specific recombination at dif and cer in *E. coli* K12. *Cell* **75:** 351–361.
14. Bonnefoy, E., M. Takahashi, and J. R. Yaniv, Jr. 1994. DNA-binding parameters of the HU protein of *Escherichia coli* to cruciform DNA. *J. Mol. Biol.* **242:** 116–129.
15. Center, M. S. and C. C. Richardson. 1970. An endonuclease induced after infection of *Escherichia coli* with bacteriophage T7. II. Specificity of the enzyme toward single- and double-stranded deoxyribonucleic acid. *J. Biol. Chem.* **245:** 6292–6299.
16. Center, M. S., F. W. Studier, and C. C. Richardson. 1970. The structural gene for a T7 endonuclease essential for phage DNA synthesis. *Proc. Natl. Acad. Sci. USA* **65:** 242–248.
17. Chen, J. H., M. E. Churchill, T. D. Tullius, N. R. Kallenbach, and N. C. Seeman. 1988. Construction and analysis of monomobile DNA junctions. *Biochemistry* **27:** 6032–6038.
18. Clegg, R. M., A. I. Murchie, and D. M. Lilley. 1994. The solution structure of the four-way DNA junction at low-salt conditions: a fluorescence resonance energy transfer analysis. *Biophys. J.* **66:** 99–109.
19. Connolly, B., C. A. Parsons, F. E. Benson, H. J. Dunderdale, G. J. Sharples, R. G. Lloyd, and S. C. West. 1991. Resolution of Holliday junctions in vitro requires the *Escherichia coli* ruvC gene product. *Proc. Natl. Acad. Sci. USA* **88:** 6063–6067.
20. Connolly, B., C. A. Parsons, F. E. Benson, H. J. Dunderdale, G. J. Sharples, R. G. Lloyd, and S. C. West. 1991. Resolution of Holliday junctions in vitro requires the *Escherichia coli* ruvC gene product. *Proc. Natl. Acad. Sci. USA* **88:** 6063–6067.
21. Connolly, B. and S. C. West. 1990. Genetic recombination in *Escherichia coli*: Holliday junctions made by RecA protein are resolved by fractionated cell-free extracts. *Proc. Natl. Acad. Sci. USA* **87:** 8476–8480.
22. Cooper, J. P. and P. J. Hagerman. 1987. Gel electrophoretic analysis of the geometry of a DNA four-way junction. *J. Mol. Biol.* **198:** 711–719.
23. Courey, A. J. and J. C. Wang. 1983. Cruciform formation in a negatively supercoiled DNA may be kinetically forbidden under physiological conditions. *Cell* **33:** 817–829.
24. Cozzarelli, N. R., T. C. Boles, and J. H. White. 1990. Primer on the topology and geometry of DNA supercoiling, in *DNA Topology and Its Biological Effects* (Cozzarelli, N. R. and J. D. Wang, eds.), Cold Spring Harbor Laboratory, Cold Spring Harbor, NY, pp. 139–184.
25. Cozzarelli, N. R. and J. C. Wang. 1990. *DNA Topology and Its Biological Effects.* Cold Spring Harbor Laboratory, Cold Spring Harbor, NY.
26. de Massy, B., R. A. Weisberg, and F. W. Studier. 1987. Gene 3 endonuclease of bacteriophage T7 resolves conformationally branched structures in double-stranded DNA. *J. Mol. Biol.* **193:** 359–376.
27. Dickie, P., A. R. Morgan, and G. McFadden. 1988. Conformational isomerization of the Holliday junction associated with a cruciform during branch migration in supercoiled plasmid DNA. *J. Mol. Biol.* **201:** 19–30.
28. Dickie, P., A. R. Morgan, and G. McFadden. 1987. Cruciform extrusion in plasmids bearing the replicative intermediate configuration of a poxvirus telomere. *J. Mol. Biol.* **196:** 541–558.

29. Duckett, D., A. I. H. Murchie, S. Diekmann, E. v.Kitzing, B. Kemper, and D. M. J. Lilley. 1988. The structure of the Holliday junction and its resolution. *Cell* **55:** 79–89.
30. Duckett, D. R., A. I. Murchie, M. J. Giraud Panis, J. R. Pohler, and D. M. Lilley. 1995. Structure of the four-way DNA junction and its interaction with proteins. *Philos. Trans. R. Soc. Lond. B. Biol. Sci.* **347:** 27–36.
31. Duckett, D. R., I. H. Murchie, A. Bhattacharyya, R. M. Clegg, S. Diekmann, E. V. Kitzing, and D. M. J. Lilley. 1992. The structure of DNA junctions and their interaction with enzymes. *Eur. J. Biochem.* **207:** 285–295.
32. Duckett, D. R., M. J. E. G. Panis, and D. M. J. Lilley. 1995. Binding of the junction-resolving enzyme bacteriophage T7 endonuclease I to DNA: Separation of binding and catalysis by mutation. *J. Mol. Biol.* **246:** 95–107.
33. Dunderdale, H. J., F. E. Benson, C. A. Parsons, G. J. Sharples, R. G. Lloyd, and S. C. West. 1991. Formation and resolution of recombination intermediates by *E. coli* RecA and RuvC proteins. *Nature* **354:** 506–510.
34. Elborough, K. M. and S. C. West. 1988. Specific binding of cruciform DNA structures by a protein from human extracts. *Nucleic Acids Res.* **16:** 3603–3616.
35 Elborough, K. M. and S. C. West. 1990. Resolution of synthetic Holliday junctions in DNA by an endonuclease activity from calf thymus. *EMBO J.* **9:** 2931–2936.
36. Evans, D. H. and R. Kolodner. 1987. Construction of a synthetic Holliday junction analog and characterization of its interaction with a *Saccharomyces cerevisiae* endonuclease that cleaves Holliday junctions. *J. Biol. Chem.* **262:** 9160–9165.
37. Evans, D. H. and R. Kolodner. 1988. Effect of DNA structure and nucleotide sequence on Holliday junction resolution by a *Saccharomyces cerevisiae* endonuclease. *J. Mol. Biol.* **201:** 69–80.
38. Ezekiel, U. R. and H. P. Zassenhaus. 1993. Localization of a cruciform cutting endonuclease to yeast mitochondria. *Mol. Gen. Genet.* **240:** 414–418.
39. Flemming, M., B. Deumling, and B. Kemper. 1993. Function of gene 49 of bacteriophage T4 III. Isolation of Holliday-structures from very fast-sedimenting DNA. *Virology* **196:** 910–913.
40. Formosa, T. and B. M. Alberts. 1986. DNA synthesis dependent on genetic recombination: characterization of a reaction catalyzed by purified bacteriophage T4 proteins. *Cell* **47:** 793–806.
41. Frank-Kamenetskii, M. D. 1990. DNA supercoiling and unusual structures, in *DNA Topology and Its Biological Effects* (Cozzarelli, N. R. and J. C. Wang, eds.), Cold Spring Harbor Laboratory, Cold Spring Harbor, NY, pp. 185 –215.
42. Frappier, L., G. B. Price, R. G. Martin, and M. Zannis-Hadjopoulos. 1987. Monoclonal antibodies to cruciform DNA structures. *J. Mol. Biol.* **193:** 751–758.
43. Frappier, L., G. B. Price, R. G. Martin, and M. Zannis-Hadjopoulos. 1989. Characterization of the binding specificity of two anticruciform DNA monoclonal antibodies. *J. Biol. Chem.* **264:** 334–341.
44. Fu, T. J., B. Kemper, and N. C. Seeman. 1994. Cleavage of double-crossover molecules by T4 endonuclease VII. *Biochemistry* **33:** 3896–3905.
45. Giraud-Panis, M. J. E., D. R. Duckett, and D. M. J. Lilley. 1995. The modular character of a DNA junction-resolving enzyme: A zinc-binding motif in bacteriophage T4 endonuclease VII. *J. Mol. Biol.* **252:** 596–610.
46. Golz, S., R. Birkenbihl, and B. Kemper. 1996. Improved large scale preparation of phage T4 endonuclease VII overexpressed in *E. coli*. *DNA Res.* **2:** 277–284.
47. Grebenshchikova, S. M., L. A. Plugina, and V. P. Shcherbakov. 1994. The role of T4-bacteriophage endonuclease-VII in correction of mismatched regions. *Genetika* **30:** 622–626.
48. Greger, B. and B. Kemper. 1995. Unpublished results.
49. Hindermann, S. and B. Kemper. 1996. Unpublished results.

50. Hiom, K. and S. C. West. 1995. Branch migration during homologous recombination: Assembly of a RuvAB-Holliday junction complex in vitro. *Cell* **80:** 787–793.
51. Hoess, R., A. Wierzbicki, and K. Abremski. 1987. Isolation and characterization of intermediates in site-specific recombination. *Proc. Natl. Acad. Sci. USA* **84:** 6840–6844.
52. Holliday, R. 1964. A mechanism for gene conversion in fungi. *Genet. Res.* **5:** 282–304.
53. Hsu, P. L. and A. Landy. 1984. Resolution of synthetic att-site Holliday structures by the integrase protein of bacteriophage lambda. *Nature* **311:** 721–726.
54. Hyde, H., A. A. Davies, F. E. Benson, and S. C. West. 1994. Resolution of recombination intermediates by a mammalian activity functionally analogous to *Escherichia coli* RuvC resolvase. *J. Biol. Chem.* **269:** 5202–5209.
55. Iwasaki, H., M. Takahagi, T. Shiba, A. Nakata, and H. Shinagawa. 1991. *Escherichia coli* RuvC protein is an endonuclease that resolves the Holliday structure. *EMBO J.* **10:** 4381–4389.
56. Jensch, F. and B. Kemper. 1986. Endonuclease VII resolves Y-junctions in branched DNA in vitro. *EMBO J.* **5:** 181–189.
57. Jensch, F., H. Kosak, N. C. Seeman, and B. Kemper. 1989. Cruciform cutting endonucleases from Saccharomyces cerevisiae and phage T4 show conserved reactions with branched DNAs. *EMBO J.* **8:** 4325–4334.
58. Jessberger, R., G. Chui, S. Linn, and B. Kemper. 1996. Analysis of the mammalian recombination protein complex RC-1. *Mutat. Res. Fundam. Mol. Mech. Mutagen.* **350:** 217–227.
59. Jeyaseelan, R. and G. Shanmugam. 1988. Human placental endonuclease cleaves Holliday junctions. *Biochem. Biophys. Res. Commun.* **156:** 1054–1060.
60. Kallenbach, N. R., R.-I. Ma, and C. Seeman. 1983. An immobile nucleic-acid junction constructed from oligonucleotides. *Nature* **305:** 829–831.
61. Kemper, B., S. Pottmeyer, P. Solaro, and H. Kosak. 1990. Resolution of DNA-secondary structures by endonuclease VII (Endo VII) from phage T4, in *Structure and Methods*, vol. 1. *Human Genome Initiative and DNA Recombination* (Sarma, R. H. and M. H. Sarma, eds.), Adenine, Schenectady, NY, pp. 215–229.
62. Kemper, B. and D. T. Brown. 1976. Function of gene 49 of bacteriophage T4. II. Analysis of intracellular development and the structure of very fast-sedimenting DNA. *J. Virol.* **18:** 1000–1015.
63. Kleff, S. and B. Kemper. 1988. Initiation of heteroduplex-loop repair by T4-encoded endonuclease VII in vitro. *EMBO J.* **7:** 1527–1535.
64. Kleff, S., B. Kemper, and R. Sternglanz. 1992. Identification and characterization of yeast mutants and the gene for a cruciform cutting endonuclease. *EMBO J.* **11:** 699–704.
65. Kosak, H. G. and B. Kemper. 1990. Large-scale preparation of T4 endonuclease VII from over-expressing bacteria. *Eur. J. Biochem.* **194:** 779–784.
66. Kotani, H., E. B. Kmiec, and W. K. Holloman. 1993. Purification and properties of a cruciform DNA binding protein from Ustilago maydis. *Chromosoma* **102:** 348–354.
67. Kupfer, C. and B. Kemper. 1996. Reactions of mitochondrial cruciform cutting endonuclease 1 *(CCE1)* of yeast *S. cerevisiae* with branched DNAs in vitro. *Eur. J. Biochem.* **238:** 77–87.
68. Lee, J. and M. Jayaram. 1995. Role of partner homology in DNA recombination—Complementary base pairing orients the 5'-hydroxyl for strand joining during Flp site-specific recombination. *J. Biol. Chem.* **270:** 4042–4052.
69. Lilley, D. M. J. 1984. DNA: sequence, structure and supercoiling. *Biochem. Soc. Transact.* **12:** 127–140.
70. Lloyd, R. G. 1991. Conjugational recombination in resolvase-deficient *ruvC* mutants of *Escherichia coli* K-12 depends on recG. *J. Bacteriol.* **173:** 5414–5418.
71. Lloyd, R. G., F. E. Benson, and C. E. Shurvinton. 1984. Effect of ruv mutations on recombination and DNA repair in *Escherichia coli* K12. *Mol. Gen. Genet.* **194:** 303–309.

72. Lloyd, R. G., C. Buckman, and F. E. Benson. 1987. Genetic analysis of conjugational recombination in *Escherichia coli* K12 strains deficient in RecBCD enzyme. *J. Gen. Microbiol.* **133:** 2531–2538.
73. Lloyd, R. G. and G. J. Sharples. 1993. Dissociation of synthetic Holliday junctions by *E. coli* RecG protein. *EMBO J.* **12:** 17–22.
74. Lockshon, D., S. G. Zweifel, L. L. Freeman-Cook, H. E. Lorimer, B. J. Brewer, and W. L. Fangman. 1995. A role for recombination junctions in the segregation of mitochondrial DNA in yeast. *Cell* **81:** 947–955.
75. Lu, M., Q. Guo, N. C. Seeman, and N. R. Kallenbach. 1991. Parallel and antiparallel Holliday junctions differ in structure and stability. *J. Mol. Biol.* **221:** 1419–1432.
76. Luisi-DeLuca, C., S. T. Lovett, and R. D. Kolodner. 1989. Genetic and physical analysis of plasmid recombination in recB recC sbcB and recB recC sbcA *Escherichia coli* K-12 mutants. *Genetics* **122:** 269–278.
77. Mahdi, T. N., G. J. Sharples, T. N. Mandal, and R. G. Lloyd. 1996. Holliday junction resolvases encoded by homologous *rusA* genes in *Escherichia coli* K-12 and phage 82. *J. Mol. Biol.* **258:** 561–573.
78. Mandal, T. N., A. A. Mahdi, G. J. Sharples, and R. G. Lloyd. 1993. Resolution of Holliday intermediates in recombination and DNA repair: indirect suppression of *ruvA*, *ruvB*, and *ruvC* mutations. *J. Bacteriol.* **175:** 4325–4334.
79. Meyer-Leon, L., L. C. Huang, S. W. Umlauf, M. M. Cox, and R. B. Inman. 1988. Holliday intermediates and reaction by-products in FLP protein-promoted site-specific recombination. *Mol. Cell. Biol.* **8:** 3784–3796.
80. Miyazaki, J., Y. Ryo, and T. Minagawa. 1983. Involvement of gene 49 in recombination of bacteriophage T4. *Genetics* **104:** 1–9.
81. Mizuuchi, K., B. Kemper, J. Hays, and R. A. Weisberg. 1982. T4 endonuclease VII cleaves Holliday structures. *Cell* **29:** 357–365.
82. Mosig, G. 1994. Homologous recombination, in *Molecular Biology of Bacteriophage T4* (Karam, J. D. ed). ASM American Society for Microbiology, Washington, DC, pp. 54–82.
83. Mosig, G., A. Luder, A. Ernst, and N. Canan. 1991. Bypass of a primase requirement for bacteriophage T4 DNA replication in vivo by a recombination enzyme, endonuclease VII. *The New Biologist* **3:** 1195–1205.
84. Mosig, G., M. Shaw, and G. M. Garcia. 1984. On the role of DNA replication, endonuclease VII, and rII proteins in processing of recombinational intermediates in phage T4. *Cold Spring Habor Symp. Quant. Biol.* **49:** 371–382.
85. Mueller, B., C. Jones, B. Kemper, and S. C. West. 1990. Enzymatic formation and resolution of Holliday junctions in vitro. *Cell* **60:** 329–336.
86. Mueller, B., C. Jones, and S. C. West. 1990. T7 endonuclease I resolves Holliday junctions formed in vitro by RecA protein. *Nucleic Acids Res.* **18:** 5633–5636.
87. Mueller, J. E., B. Kemper, R. P. Cunningham, N. R. Kallenbach, and N. C. Seeman. 1988. T4 endonuclease VII cleaves the crossover strands of Holliday junction analogs. *Proc. Natl. Acad. Sci. USA* **85:** 9441–9445.
88. Mueller, J. E., C. J. Newton, F. Jensch, B. Kemper, R. Cunningham, N. R. Kallenbach, and N. C. Seeman. 1990. Resolution of Holliday junction analogs by T4 Endonuclease VII can be directed by substrate structure. *J. Biol. Chem.* **265:** 13,918–13,924.
89. Murchie, A. I. H. and D. M. J. Lilley. 1993. T4 endonuclease VII cleaves DNA containing a cisplatin adduct. *J. Mol. Biol.* **233:** 77–85.
90. Panayotatos, N. and A. Fontaine. 1987. A native cruciform DNA structure probed in bacteria by recombinant T7 endonuclease. *J. Biol. Chem.* **262:** 11,364–11,368.
91. Parson, C. A., I. Tsaneva, R. G. Lloyd, and S. C. West. 1992. Interaction of *Escherichia coli* RuvA and RuvB proteins with synthetic Holliday junctions. *Proc. Natl. Acad. Sci. USA* **89:** 5452–5456.

92. Parsons, C. A., B. Kemper, and S. C. West. 1990. Interaction of a four-way junction in DNA with T4 endonuclease VII. *J. Biol. Chem.* **265:** 9285–9289.
93. Parsons, C. A., A. Stasiak, R. J. Bennett, and S. C. West. 1995. Structure of a multisubunit complex that promotes DNA branch migration. *Nature* **374:** 375–378.
94. Parsons, C. A. and S. C. West. 1990. Specificity of binding to four-way junctions in DNA by bacteriophage T7 endonuclease I. *Nucleic Acids Res.* **18:** 4377–4384.
95. Pearson, C. E., M. T. Ruiz, G. B. Price, and M. Zannis Hadjopoulos. 1994. Cruciform DNA binding protein in HeLa cell extracts. *Biochemistry* **33:** 14,185–14,196.
96. Pearson, C. E., M. Zannis Hadjopoulos, G. B. Price, and H. Zorbas. 1995. A novel type of interaction between cruciform DNA and a cruciform binding protein from HeLa cells. *EMBO J.* **14:** 1571–1580.
97. Pettijohn, D. E. 1988. Histone-like proteins and bacterial chromosome structure. *J. Biol. Chem.* **263:** 12,793–12,796.
98. Pettijohn, D. E., R. R. Sinden, and S. S. Broyles. 1988. Cruciform transition assayed using psoralen crosslinking method: applications to measurements of DNA torsional tension, in *Unusual DNA Structures* (Wells, R. D. and S. C. Harvey, eds.), Springer Verlag, Heidelberg, pp. 103–113.
99. Picksley, S. M., C. A. Parsons, B. Kemper, and S. C. West. 1990. Cleavage specificity of bacteriophage T4 endonuclease VII and bacteriophage T7 endonuclease I on synthetic branch migratable Holliday junctions. *J. Mol. Biol.* **212:** 723–735.
100. Pontiggia, A., A. Negri, M. Beltrame, and M. E. Bianchi. 1993. Protein HU binds specifically to kinked DNA. *Mol. Microbiol.* **7:** 343–350.
101. Pottmeyer, S. and B. Kemper. 1992. T4 endonuclease VII resolves cruciform DNA with nick and counter-nick and its activity is directed by local nucleotide sequence. *J. Mol. Biol.* **223:** 607–615.
102. Pottmeyer, S. and B. Kemper. 1994. Unpublished results.
103. Qian, X. H., R. B. Inman, and M. M. Cox. 1992. Reactions between half- and full-FLP recombination target sites. A model system for analyzing early steps in FLP protein-mediated site-specific recombination. *J. Biol. Chem.* **267:** 7794–7805.
104. Sadowski, P. 1992. Bacteriophage T7 Endonuclease I. Properties of the enzyme from T7 phage-infected *Escherichia coli*. *J. Biol. Chem.* **246:** 200–216.
105. Sastry, S. S. and E. Kun. 1990. The interaction of adenosine diphosphoribosyl transferase (ADPRT) with a cruciform DNA. *Biochem. Biophys. Res. Commun.* **167:** 842–847.
106. Schwacha, A. and N. Kleckner. 1994. Identification of joint molecules that form frequently between homologs but rarely between sister chromatids during yeast meiosis. *Cell* **76:** 51–63.
107. Seeman, N. C. and N. R. Kallenbach. 1994. DNA branched junctions. *Annu. Rev. Biophys. Biomol. Struct.* **23:** 53–86.
108. Seeman, N. C., J. E. Mueller, J. Chen, M. E. A. Churchill, A. Kimball, T. D. Tullius, B. Kemper, R. P. Cunningham, and N. R. Kallenbach. 1990. Immobile junctions suggest new features of the structural chemistry of recombination, in *Human Genome Initiative and DNA Recombination* (Sarma, R. H. and M. H. Sarma, eds.), Adenine, Schenectady, NY, pp. 137–156.
109. Sekiguchi, J., N. C. Seeman, and S. Shuman. 1996. Resolution of Holliday junctions by eukaryotic DNA topoisomerase. *Proc. Natl. Acad. Sci. USA* **93:** 785–789.
110. Shah, R., R. J. Bennett, and S. C. West. 1994. Genetic recombination in E-coli: RuvC protein cleaves Holliday junctions at resolution hotspots in vitro. *Cell* **79:** 853–864.
111. Sharples, G. J., F. E. Benson, G. T. Illing, and R. G. Lloyd. 1990. Molecular and functional analysis of the *ruv* region of *Escherichia coli* K-12 reveals three genes involved in DNA repair and recombination. *Mol. Gen. Genet.* **221:** 219–226.
112. Sharples, G. J., S. N. Chan, A. A. Mahdi, M. C. Whitby, and R. G. Lloyd. 1994. Processing of intermediates in recombination and DNA repair: indentification of a new endonuclease that specifically cleaves Holliday junctions. *EMBO J.* **13:** 6133–6142.

113. Sharples, G. J. and R. G. Lloyd. 1993. An *E. coli ruvC* mutant defective in cleavage of synthetic Holliday junctions. *Nucleic Acids Res.* **21:** 3359–3364.
114. Shcherbakov, V. P. and L. A. Plugina. 1991. Marker-dependent recombination in T4 bacteriophage. III. Structural prerequisites for marker discrimination. *Genetics* **128:** 673–685.
115. Shinagawa, H. and H. Iwasaki. 1996. Processing the Holliday junction in homologous recombination. *TIBS* **21:** 107–111.
116. Solaro, P., B. Greger, and B. Kemper. 1995. Detection and partial purification of a cruciform-resolving activity (X-solvase) from nuclear extracts of mouse B-cells. *Eur. J. Biochem.* **230:** 926–933.
117. Solaro, P. C., K. Birkenkamp, P. Pfeiffer, and B. Kemper. 1993. Endonuclease VII of phage T4 triggers mismatch correction in vitro. *J. Mol. Biol.* **230:** 868–877.
118. Stuart, D., K. Ellison, K. Graham, and G. McFadden. 1992. In vitro resolution of poxvirus replicative intermediates into linear minichromosomes with hairpin termini by a virally induced Holliday junction endonuclease. *J. Virol.* **66:** 1551–1563.
119. Symington, L. S. and R. Kolodner. 1985. Partial purification of an enzyme from Saccharomyces cerevisiae that cleaves Holliday junctions. *Proc. Natl. Acad. Sci. USA* **82:** 7247–7251.
120. Takahagi, M., H. Iwasaki, and H. Shinagawa. 1994. Structural requirements of substrate DNA for binding to and cleavage by RuvC, a Holliday junction resolvase. *J. Biol. Chem.* **269:** 15,132–15,139.
121. Takahaki, M., H. Iwasaki, A. Nakata, and H. Shinagawa. 1991. Molecular analysis of the *Escherichia coli ruvC* gene which encodes a Holliday junction-specific endonuclease. *J. Bacteriol.* **173:** 5747–5753.
122. Taylor, A. F. and G. R. Smith. 1990. Action of RecBCD enzyme on cruciform DNA. *J. Mol. Biol.* **211:** 117–134.
123. Tomaschewski, J. and W. Rueger. 1987. Nucleotide sequence and primary structures of gene products coded for by T4 genome between map positions 48.266 Kb and 39.166 Kb. *Nucleic Acids Res.* **15:** 3632,3633.
124. Varga-Weisz, P., K. Van Holde, and J. Zlatanova. 1993. Preferential binding of histone H1 to four-way helical junction DNA. *J. Biol. Chem.* 2**68:** 20,699–20,700.
125. von Kitzing, E., D. M. Lilley, and S. Diekmann. 1990. The stereochemistry of a four-way DNA junction: a theoretical study. *Nucleic Acids Res.* **18:** 2671–2683.
126. Waldman, A. S. and R. M. Liskay. 1988. Resolution of synthetic Holliday structures by an extract of human cells. *Nucleic Acids Res.* **16:** 10,249–10,266.
127. West, S. C. 1992. Enzymes and molecular mechanisms of genetic recombination. *Ann. Rev. Biochem.* **61:** 603–640.
128. West, S. C. 1994. The processing of recombination intermediates: mechanistic insights from studies of bacterial proteins. *Cell* **76:** 9–15.
129. West, S. C. and A. Koerner. 1985. Cleavage of cruciform DNA structures by an activity from Saccharomyces cerevisiae. *Proc. Natl. Acad. Sci. USA* **82:** 6445–6449.
130. West, S. C., C. A. Parsons, and S. M. Picksley. 1987. Purification and properties of a nuclease from Saccharomyces cerevisiae that cleaves DNA at cruciform junctions. *J. Biol. Chem.* **262:** 12,752–12,758.
131. Whitby, M. C., S. D. Vincent, and R. G. Lloyd. 1994. Branch migration of Holliday junctions: Identification of RecG protein as a junction specific: DNA helicase. *EMBO J.* **13:** 5220–5228.
132. White, M. F. and D. M. J. Lilley. 1996. The structure-selectivity and sequence-preference of the junction-resolving enzyme CCE1 of *Saccharomyces cerevisiae*. *J. Mol. Biol.* **257:** 330–341.
133. Zweifel, S. G. and W. L. Fangman. 1991. A nuclear mutation reversing a biased transmission of yeast mitochondrial DNA. *Genetics* **128:** 241–249.

11

Dam-Directed DNA Mismatch Repair

Lene Juel Rasmussen, Leona Samson, and M. G. Marinus

1. INTRODUCTION

Base mismatches in duplex DNA are repaired by a variety of systems in *Escherichia coli*, including:

1. Very short patch (VSP) repair;
2. MutY-dependent repair; and
3. DNA adenine methyltransferase-directed DNA mismatch repair (Dam-directed DNA mismatch repair [DDMR]).

The first two systems have a restricted substrate specificity (T·G and A·G mismatches, respectively), whereas the third system can repair 11 out of the 12 possible base mispairs with only the C·C mismatch being refractory. There is no known repair pathway for C·C mismatches in duplex DNA. The DDMR pathway also acts on insertions/deletions ("loops") of up to 4 bases.

In this chapter are reviewed the VSP repair and DDMR systems with emphasis on the latter. The MutY-pathway is discussed in Chapter 6.

2. MECHANISM OF DAM-DIRECTED DNA MISMATCH REPAIR IN *E. COLI*

Studies of DDMR in cell-free extracts of *E. coli* pioneered by Modrich and coworkers have led to a detailed understanding of the mechanism underlying the repair process *(50,60)* as diagrammed in Fig. 1 *(76)*. The DNA in enterobacteria is normally fully methylated at the N^6-position of adenine in the sequence d(GATC) by the Dam methyltransferase, and this methylation pattern directs the mismatch repair proteins to correct errors in the newly synthesized strand *(92)*. Immediately after a new round of replication, the DNA is hemimethylated at d(GATC) sites where MutH endonuclease binds *(1,16,29,62,66,109)*. The MutS protein recognizes and binds to sequences containing errors introduced into the DNA by the replisome *(83,103,104)*. The exact function of the MutL protein is unknown, but is believed to create a contact between MutH and MutS to facilitate bending of the DNA into an α-shaped looped structure *(34)*. The MutH endonuclease activity is hereby activated, and a single-stranded nick is introduced into the newly synthesized strand at the GATC site closest to the mismatch. The DNA is unwound by the UvrD (also called UvrE, MutU, or RecL) helicase, and

From: DNA Damage and Repair, Vol. 1: DNA Repair in Prokaryotes and Lower Eukaryotes
Edited by: J. A. Nickoloff and M. F. Hoekstra © *Humana Press Inc., Totowa, NJ*

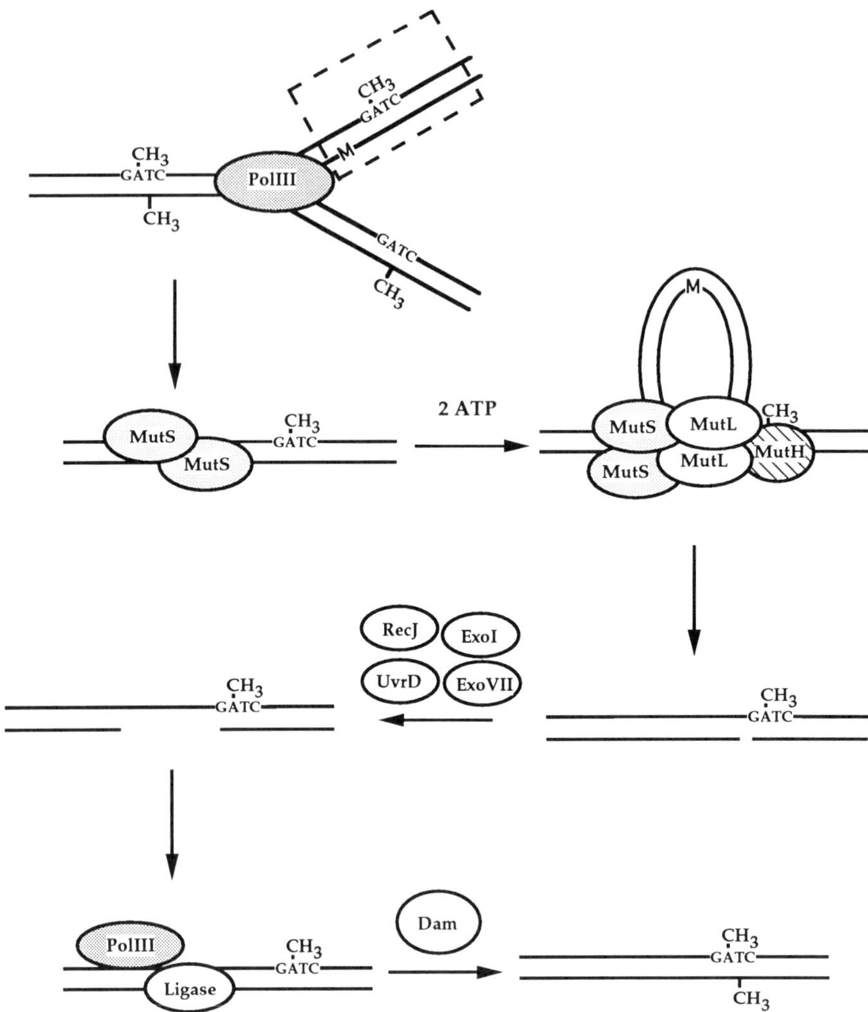

Fig. 1. Model for initiation of DDMR. This model is based on in vitro properties of the MutS, MutL, and MutH proteins. The stoichiometry and the nature of protein–protein interactions are not understood in detail and are shown here for illustrative purposes only. **M:** base–base mismatch.

depending on whether the nick is located 3' or 5' to the mismatch, exonucleases ExoI, ExoVII, or RecJ removes the bases on the newly synthesized strand *(19,33,49,59)*. In *E. coli*, the repair tracts in mismatched DNA heteroduplexes can extend over a distance of 1–3 kbp *(11,108)*. DNA polymerase III fills in the excision tract and DNA ligase closes the nick. The final step in the repair process is methylation of the GATC sites on the newly synthesized strand by Dam methyltransferase. When the DNA template is fully methylated, the repair of the newly synthesized strand is inhibited, since MutH cannot bind *(23)*.

The substrate specificity of the DDMR system has been studied extensively, and has been shown to correlate with the binding affinity of the MutS protein to various mismatches and insertions/deletions both in vivo and in vitro. Genetic and biochemical experiments indicate that the G·T mismatch is repaired most efficiently and C·C not at all, whereas the efficiency of correction of the other mismatches lies between these extremes *(15,47,76)*. Loops of 1, 2 or 3 bases are repaired at high efficiency both in vitro *(51)* and in vivo *(83)*, and MutS has a higher affinity for these than the G·T mismatch *(83)*. The fidelity of repair not only depends on the type of mispaired bases but is also influenced by the base composition of the flanking nucleotide sequence *(43,103)*.

The types of mutations that are generated in *dam*, *mutH*, *mutL*, and *mutS* strains have also been studied extensively. These indicate that AT to GC and GC to AT transition mutations are most common. Frameshift mutations resulting from single base insertions/deletions, however, show the highest recovery when calculated on the basis of available mutable sites. These results are consistent with the known specificity of MutS recognition. In contrast, *mutD (dnaQ)* mutator strains with decreased proofreading activity of polymerase III holoenzyme show an increase in mutations resulting from transversion mutations (T·T, A·A, G·G, and C·C mispairs). Under certain growth conditions, the *mutD* mutant strain shows a pattern of mutagenesis like the *mutH*, *mutL*, and *mutS* mutant strains due to saturation of the mismatch repair system *(30)*.

3. GENES INVOLVED IN DDMR

Marinus and Morris *(69)* isolated mutants of the Dam methyltransferase, which was later shown to methylate the adenine residue in the sequence -GATC- in double-stranded DNA *(38,68,69)*. The Dam methyltransferase is involved in the control of several cellular events, such as:

1. Initiation of chromosome replication;
2. DDMR;
3. Transposition; and
4. Gene regulation.

The *E. coli* origin of replication *(oriC)* remains hemimethylated for about 30% of the cell cycle *(13,79)*, thus preventing premature initiation. Dam methylation also ensures that the initiation of all replication origins in the cell occurs simultaneously *(9)*. DNA methylation at the replication fork lags behind synthesis, such that DNA is in a hemimethylated configuration for a short period *(13)*. During this interval, the hemimethylated DNA undergoes mismatch repair for the removal of replication errors *(76*; Fig. 1). Genetic evidence has indicated a role for the Dam methyltransferase in the correction of mismatches. Mutants deficient in Dam methyltransferase activity are hypermutable, as are strains that overproduce the methylase *(40,68,70)*. The state of DNA methylation also provides an elegant means of cell-cycle-specific gene expression, because some promoters such as that regulating the transposase gene of Tn*10*, are more active in the hemimethylated configuration *(96)*.

In contrast to our knowledge about the biochemical functions of proteins involved in DDMR, little is known about regulation. The best-characterized are the *dam* and *mutL* genes. The *dam* gene is part of a transcription unit containing at least seven genes *(57,61)*. The *dam* gene is subject to growth rate control, thus linking its expression to the rate at which the chromosome is replicated *(93,94)*. It was suggested that the level of Dam methyltransferase must be tightly controlled because too much or too little methylase leads to increased mutagenesis, asynchronous initiation of chromosome replication, and altered frequency of transposition *(9,68,70,96)*. The *mutL* gene is also part of a superoperon consisting of at least five genes. The genes in this operon seem to be regulated by multiple mechanisms (also operative in the regulation of the *dam* operon), including growth rate control, degree of cotranscription from upstream genes, modulation of internal promoter strength, and mRNA stability mediated by RNaseE activity *(17,18,105–107)*. Both the *mutS* and *mutH* genes have been cloned and sequenced, but little is known about their regulation *(32,36,64,81,100)*. It has been shown that during exponential growth in rich medium, the cell contains between 100 and 200 molecules of MutH, MutL, MutS, and Dam *(10,27)*.

The *hex* genes of *Streptococcus pneumoniae* show a high degree of homology to their microbial (Fig. 2A,B) as well as to their mammalian (Chapter 20, vol. 2) counterparts *(3,72,87,88,90)*. Surprisingly, the *hexA* gene fails to complement a *mutS* mutation of *E. coli*, and both the HexA protein and the human MutS homolog, *hMSH2*, induce a mutator phenotype when expressed in *E. coli*, indicating that these proteins interfere with the *E. coli* DDMR pathway, perhaps by forming nonfunctional repair complexes *(28,91)*. DDMR homologs from several microbial organisms have been identified (Fig. 2A,B), and many others are likely to be found during genome sequencing projects.

Clearly, a great deal remains unknown about the regulation of DDMR genes in the bacterial cell cycle. Such knowledge may prove helpful in understanding the mechanism underlying tissue specific expression of DNA mismatch repair genes in mammalian cells *(2,110)*.

4. DAM-DIRECTED DNA MISMATCH REPAIR PROTEINS

4.1. MutS DNA Mismatch Recognition Protein

The MutS protein with a monomeric mol wt of 97 kDa binds as a dimer to heteroduplexes containing DNA mismatches and small loops of 1–4 unpaired bases *(83,104)*. The binding of MutS to a mismatch protects about 22 bp of DNA from digestion with DNase I *(104)*.

The MutS protein contains a Walker A nucleotide binding motif (P-loop motif). Mutational analysis has shown that this sequence is essential for the binding and hydrolysis of ATP, and genetic alteration of this sequence results in a mutator phenotype. However, MutS protein mutated in the P-loop is able to bind heteroduplexes specifically, but does so with a reduced affinity *(37)*. In the presence of ATP, the MutS protein forms α-shaped looped structures which are stabilized at the junctions by the protein (reviewed in refs. *76* and *78)*. The formation of loops has been suggested to facilitate the interaction among the MutS, MutH, and MutL proteins during the initiation of the repair process.

In order to identify the functional domains of MutS, Wu and Marinus *(114)* used a genetic screen to isolate dominant negative mutator mutations. They found that one class of mutants mapped to the P-loop motif and mapped to conserved MutS sequences of unknown function. The dominant mutations in the P-loop motif caused a reduced repair of heteroduplex DNA. Interestingly, several *mutS* mutant isolates showed phenotypic reversal on increasing the gene dosage of *mutL* or *mutH*. These findings indicate an in vivo interaction between MutS, MutL, and MutH.

4.2. MutL Protein

The *E. coli* MutL protein (monomeric mol wt of 70 kDa) exists as a dimer in solution *(34)*. It has been shown that the heteroduplex region protected by MutS from DNAse I digestion increases from 20 to >100 bp on addition of MutL and ATP, suggesting an interaction between these proteins *(34)*. MutL and ATP are also required for activating the latent endonucleolytic activity of MutH *(1)*, suggesting that MutL also interacts with MutH. No enzymatic activity has been found associated with MutL in the initiation phase of repair. However, this does not preclude an active catalytic role in the excision or resynthesis phases. Purified MutL has been reported to bind both single- and double-stranded DNA *(4)*, though the physiological significance of such binding remains unclear.

4.3. MutH Endonuclease

The 25-kDa MutH protein possesses a weak Mg^{2+}-dependent endonuclease activity that cleaves 5' to the G of both un- and hemimethylated GATC sites *(109)*. The specific activity of MutH endonuclease increases 20 to 70-fold in the presence of ATP, MutS, and MutL, indicating that the latter two proteins are required for full activity of MutH. Fully methylated GATC sites are resistant to cleavage by the MutH endonuclease, hemimethylated sites are cleaved only on the unmethylated strand, and unmethylated GATC sites are cleaved randomly on either strand. MutH endonuclease is not present in most bacterial species (e.g., *S. pneumoniae*), but is present in *Haemophilus influenza* and *E. coli* and its relatives. The mechanism for strand discrimination in bacteria that lack MutH is unknown to date. Candidates for strand discrimination in organisms lacking MutH, such as methylation of DNA, single-stranded breaks, and strand discontinuities in newly replicated DNA, have been proposed and are described in Section 4.4.

4.4. Dam Methyltransferase

Methyltransferases are generally associated with restriction-modification systems in bacteria. Methylation can take place at the C^5- or N^4-position of cytosine, or the N^6-position of adenine (reviewed in ref. *67)*. The Dam methyltransferase catalyzes the transfer of a single methyl group from *S*-adenosylmethionine to the N^6 position of adenine in the sequence GATC, and the rate of methylation is markedly influenced by the flanking sequences *(6,40)*. The Dam methyltransferase has a molecular mass of 32 kDa and acts as a monomer. It has been shown that *S*-adenosylmethionine not only serves as a substrate for the methylation reaction, but also serves as an allosteric activator to stimulate Dam binding to its target sequence GATC *(5)*.

The amino acid sequence of Dam methyltransferase shows considerable similarity with other GATC recognizing adenine methyltransferases, such as bacteriophage T4 Dam methyltransferase, *Dpn*II methyltransferase, as well as with the GATATC methylating enzyme *Eco*RV methyltransferase. *Dpn*II DNA methylase of *S. pneumoniae* is

A

MutS Homologs

```
(Sp) hexA   IQLVTGPNMSGKSTYMRQLAMTAVMAQ.LGSYVPAESAHLPIFDAIFTRIGAADDLVSGQSTFMVEMMEANNAISHATKNSLILFDELGRGTATYDGMALAQSIIEYIHEHIGA.KTLFATHYH
(Hi) mutS   LLVITGPNMGGKSTYMRQTALITLLAYYIGSFVPADSARIGPIDRIFTRIGASDDLASGRSTFMVEM......LHQATAQSLVLIDEIGRGTSTYDGLSLAWACAEWLSKKIRS.LTLFATHYF
(Avi) mutS  MLIITGPNMGGKSTYMRQTALIVLLAH.IGSFVPAQSCELSLVDRIFTRIGSSDDLAGGRSTFMVEMSETANILHNASERSLVLMDEVGRGTSTFDGLSLAWAAAEHLAGLAAW..TLFATHYF
(Ec) mutS   MLIITGPNMGGKSTYMRQTALIALMAY.IGSYVPAQKVEIGPIDRIFTRVGAADDLASGRSTFMVEMTETANILHNATEYSLVLMDEIGRGTSTYDGLSLAWACAENLANKIKA.LTLFATHYF
(St) mutS   MLIITGPNMGGKSTYMRQTALIALLAY.IGSYVPAQNVEIGPIDRIFTRVGAADDLASGRSTFMVEMTETANILHNATENSLVLMDEIGRGTSTYDGLSLAWACAENLANKIKA.LTLFATHYF
(Bs) mutS   MLLITGPNMSGKSTYMRQIALISIMAQ.IGCFVPAKKAVLPIFDQIFTRIGAADDLISGQSTFMVEMLEAKNAIVNATKNSLILFDEIGRGTSTYDGMALAQAIIEYVHDHIGA.KTLFSTHYH
(Ta) mutS   LVLITGPNMAGKSTFLRQTALIALLAQ.VGSFVPAEEAHLPLFDGIYTRIGASDDLAGGKSTFMVEMEEVALILKEATENSLVLLDEVGRGTSSLDGVAIATAVAEALHERRA..YTLFATHYF
(Tt) mutS   LVLVTGPNMAGKSTFLRQTALIALLAQ.IGSFVPAEEAELPLFDGIYTRIGASDDLAGGKSTFMVEMEEVALVLKEATERSLVLLDEVGRGTSSLDGVAIATALAEALHERRC..YTLFATHYF

yMSH1       LWVITGPNMGGKSTFLRQNAIIVILAQ.IGCFVPCSKARVGIVDKLFSRVGSADDLYNEMSTFMVEMIETSFILQGATERSLAILDEIGRGTSGKEGISLAYATLKYLLENNQC.RTLFATHFG
yMSH2       FLIITGPNMGGKSTYIRQVGVISLMAQ.IGCFVPCEEAEIAIVDAILCRVGAGDSQLKGVSTFMVEILETASILKNASKNSLIIVDELGRGTSTYDGFGLAWAIAEHIASKIGC.FALFATHFH
yMSH3       INIITGPNMGGKSSYIRQVALLTIMAQ.IGSFVPAEEIRLSIFENVLTRIGAHDDIINGDSTFKVEMLDILHILKNCNKRSLLLLDEVGRGTGTHDGIAISYALIKYFSELSDCPLILFTTHFP
yMSH4       LQIITGCNMSGKSVYLKQVALICIMAQ.MGSGIPALYGSFPVFKRLHARV.CNDSMELTSSNFGFEMKEMAYFLDDINTETLLILDELGRGSSIADGFCVSLAVTEHLLRTEATVF.L.STHF
yMSH5       IIVVTGANASGKSVYLTQNGLIVYLAQ.IGSFVPAERARIGIADKILTRIRTQETVYKTQSSFLLDSQQMAKSLSLATEKSLILIDEYGKGTDILDGPSLFGSIMLNMSKSEKCPRIIACTHFH
yMSH6       LGLLTGANAAGKSTILRMACIAVIMAQ.MGCYVPCESAVLTPIDRIMTRLGANDNIMQGKSTFFVELAETKKILDMATNRSLLVVDELGRGGSSSDGFAIAESVLHHVATHIQS.LGFFATHYG

swi4        CLLITGPNMGGKSSFVKQLALSAIMAQ.SGCFVPAKSALLPIFDSILIRMGSSDNLSVNMSTFMVEMLETKEVLSKATEKSMVIIDELGRGTSTIDGEAISYAVLHYLNQYIKS.YLLFVTHFP
```

B

MutL Homologs

```
(St) mutL  ..MPIQVLP.PQLANQIAAGEVVERPASVVKELVENSLDAGATRVDIDIERGGAKLIRIRDNGCGIKKEELALALARHATSKIASLDDLEAIISLGFRGEALASISSVSRLTLTSR
(Ec) mutL  ..MPIQVLP.PQLANQIAAGEVVERPASVVKELVENSLDAGATRIDIDIERGGAKLIRIRDNGCGIKKDELALALARHATSKIASLDDLEAIISLGFRGEALASISSVSRLTLTSR
(Sp) hexB  ..MSHIIELPEMLANQIAAGEVIERPASVCKELVENAIDAGSSQIIIEIEEAGLKKVQITDNGHGIAHDEVELALRRHATSKIKNQADLFRIRTLGFRGEALPSIASVSVLTLLTA
(Hi) mutL  ..MPIKILS.PQLANQIAAGEVVERPASVVKELVENSLDAG......DIENGGANLIRIRDNGCGIPKEELSLALARHATSKIADLDDLEAILSLGFRG......SSVSRLTLTSR
(Bs) mutL  ..MAKVIQLSDELSNKIAAGEVVERPASVVKELVENAIDADSTVIEIDIEEAGLASIRVLDNGEGMENEDCKRAFRRHATSKIKDENDLFRVRTLGFRGEALPSIASVSHLEITTS

yMLH1      .MSLRIKALDASVVNKIAAGEIIISPVNALKEMMENSIDANATMIDILVKEGGIKVLQITDNSGGINKADLPILCERFTTSKLQKFEDLSQIQTYGFRGEALASISHVARVTVTTK
yPMS1      .SMTQIHQINDIDVHRITSGQVITDLTTAVKELVDNSIDANANQIEIIFKDYGLESIECSDNGDGIDPSNYEFLALKHYTSKIAKFQDVAKVQTLGFRGEALSSLCGIAKLSVITT

(St) mutL  TAEQAEAWQAYAE.GRDMDVTVKPAAHPVGTTLEVLDLFYNTPARRKFMRTEKTEFNHIDEIIRRIALARFDVTLNLSHNGKLVRQYRAVAKDGQKERRLGAICGTPFLEQALAIE
(Ec) mutL  TAEQQEAWQAYAE.GRDMNVTVKPAAHPVGTTLEVLDLFYNTPARRKFLRTEKTEFNHIDEIIRRIALARFDVTINLSHNGKIVRQYRAVPEGGQKERRLGAICGTAFLEQALAIE
(Sp) hexB  VDGASHGTKLVAR.GGEVE.EVIPATSPVGTKVCVEDLFFNTPARLKYMKSQQAELSHIIDIVNRLGLAHPEISFSLISDGKEMTRTAGTGQLRQAIAGIYGLVSAKKMIEIENSD
(Hi) mutL  TEEQTEAWQVYAQ.GRDMETTIKPASHPVGTTVEVANLFFNTLRTDKTEFSHIDEVIRRIALTKFNTAFTLTHNGKIIRQYRPAEAINQQLKRVDFVKNALRIEWKHDDLHLSGWV
(Bs) mutL  TGEGAGTKLVL.QGGNIISESRSS..SRKGTEIVVSNLFFNTPARLKYMKTVHTELGNITDVVNRIALAHPEVSIRLRHHGKNLLQTNGNGDVRHVLAAIYGTAVAKKMLPLHVSS

yMLH1      VKEDRCAWRVSYAEGKMLESPKPVAGKD.GTTILVEDLFFNIPSRLRALRSHNDEYSKILDVVGRYAIHSKDIGFSCKKFGDSNYSLSVKPSYTVQDRIRTVFNKSVASNLITFHI
yPMS1      TSPPKADKLEYDMVGHITSKTTTSRNK..GTTVLVSQLFHNLPVRQKEFSKTFKRQFTKCLTVIQGYAIINAAIKFSVWNITPKGKKNLILSTMRNSSMRKNISSVFGAGGMRGLE

(St) mutL  WQHGDLTLRGWVADPNHTTTALTEIQYCYVNGRMMRDRLINHAIRQACEDKLGADQQPAFVL.........................YLEIDPHQV DVNVHPAKHE
(Ec) mutL  WQHGDLTLRGWVADPNHTTPALAEIQYCYVNGRMMRDRLINHAIRQACEDKLGADQQPAFVL.........................YLEIDPHQV DVNVHPAKHE
(Sp) hexB  LDFEISGFVSLPELTTRANRNYISLFINGRYIKNFLLNRAILDGFGSKLMVGRFPLAVIHIH...........................IDPYLA DVNVHPTKQE
(Hi) mutL  ATPNFSRTQNDLS............YCYINGRMVRDKVISHAQYLPTDAY......PAFVL.........................FIDLNPHDV DVNVHPTKHE
(Bs) mutL  LDFEVKGYIALPEITRASRNYMSSVVNGRYIKNFPLVKAVHEGYHTLLPIGRHPITFIEITM............................DPILV DVNVHPSKLE

yMLH1      SKVEDLNLESVDGKVCNLNFISKKSISLIFFINNRLVTCDLLRRALNSVYSNYLPKGFRPFIYLGIV.......................IDPAAV DVNVHPTKRE
yPMS1      EVDLVLDLNPFKNRMLGKYTDDPDFLDLDYKIRVKGYISQNSFGCGRNSKDRQFIYVNKRPVEYSTLLKCCNEVYKTFNNVQFPAVFLNLELPMSLI DVNVTPDKRV
```

Fig. 2. Alignment of microbial DDMR genes. All sequences shown in this figure were obtained from GENBANK. **(A)** Alignment of the C-terminal regions of the proteins. **(Sp) hexA:** *S. pneumoniae* MutS, **(Hi) mutS:** *H. influenzae* MutS, **(Avi) mutS**: *Azotobacter vinelandii* MutS, **(Ec) mutS:** *E. coli* MutS, **(St) mutS**: *S. typhimurium* MutS, **(Bs) mutS**: *Bacillus subtilis* MutS, **(Ta) mutS**: *Thermus aquaticus* MutS, **(Tt) mutS**: *Termus thermophilus* MutS, **yMSH1-6**: *Saccharomyces cerevisiae* MutS homologs, **swi4**: *Schizosaccharomyces pombe* MutS homolog. **(B)** Alignment of the N-terminal regions of the proteins. **(St) mutL**: *S. typhimurium* MutL, **(Ec) mutL**: *E. coli* MutL, **(Sp) hexB**: *S. pneumoniae* MutL, **(Hi) mutL**: *H. influenzae* MutL, **(Bs) mutL**: *B. subtilis* MutL, **yMLH1** and **yPMS1**: *S. cerevisiae* MutL homologs, **ScpPMS1**: *S. pombe* MutL homolog.

homologous to the *E. coli* Dam methylase, but there is no evidence that this restriction methyltransferase plays any role in DNA mismatch repair *(65)*.

An alternative mechanism for strand discrimination has been proposed based on the presence of free ends on the donor DNA within heteroduplex recombination intermediates in *S. pneumoniae (35)*. The hypothesis that strand discrimination during DNA replication involves the recognition of strand breaks is based on the fact that the lagging strand of DNA is synthesized discontinuously. The ends of the Okazaki fragments could serve to target the *S. pneumoniae* Hex DNA mismatch repair system to the daughter strand, and such a strand-discriminating system would also be a good candidate for how eukaryotic cells discriminate parental from newly synthesized strands. However, this mechanism does not account for repair in leading strands. It has been speculated that breaks in newly synthesized strands could arise from incorporation of uracil followed by removal by uracil-DNA glycosylase, and cleavage at the resulting abasic sites. However, this possibility has been ruled out, since an *S. pneumoniae ung* mutant strain, deficient in uracil-DNA glycosylase activity, shows normal mutation avoidance *(14)*.

5. VSP REPAIR SYSTEM

VSP repair was discovered through the unusual distribution of recombinants ("high negative interference") in genetic crosses using specific amber mutations in the *cI* gene of bacteriophage λ *(55)*. Since genetic markers less than five nucleotides away from the amber mutations were not affected, the repair tracts were very short (hence the designation VSP). The amber codons involved occurred in the sequence -CTAGG- (or -NTAGN-), which is derived from the wild-type sequence -CCAGG-. Thus, high negative interference in genetic crosses was explained by the unidirectional bias in the repair effecting TA to CG transition mutations in heteroduplex molecules.

The wild-type -CCAGG- sequence is naturally methylated to form 5-methylcytosine at the second cytosine; on spontaneous deamination, this 5-methylcytosine yields thymine. These T·G mismatches in double-stranded DNA need to be repaired to prevent the occurrence of CG to TA mutations. It had previously been shown *(20)* that such mutations indeed occur at amber sites. It was appealing therefore to model VSP repair as counteracting the potential mutagenic effect of spontaneous deamination of 5-methylcytosine *(56)*. Indeed, bacteria that lacked this methylated base had vastly reduced spontaneous amber mutations at -CCAGG- sites *(20)*.

Optimal VSP repair requires the action of the *vsr*, *mutS*, *mutL*, and *polA* gene products *(7,25,54)*. The *vsr* gene encodes an 18-kDa Mg^{2+}-dependent endonuclease, which cleaves 5' to the T·G mismatch *(39)*. Thus, a simple model for repair is that after deamination of 5-methylcytosine to form the T·G mismatch, Vsr binds to and cleaves at the mismatch followed by nick repair by DNA polymerase I (the *polA* product) and DNA ligase. Although the action of a helicase might be expected, the *uvrD* gene product is not required *(54)*.

The simple model above does not take into account the role(s) of the *mutS* and *mutL* gene products. In their absence, VSP activity is significantly reduced, but not eliminated *(54)*. An alternative model can be proposed that takes into account the action of MutS, MutL, and ATP on mismatched DNA *(78)*. After binding to the mismatch, MutS

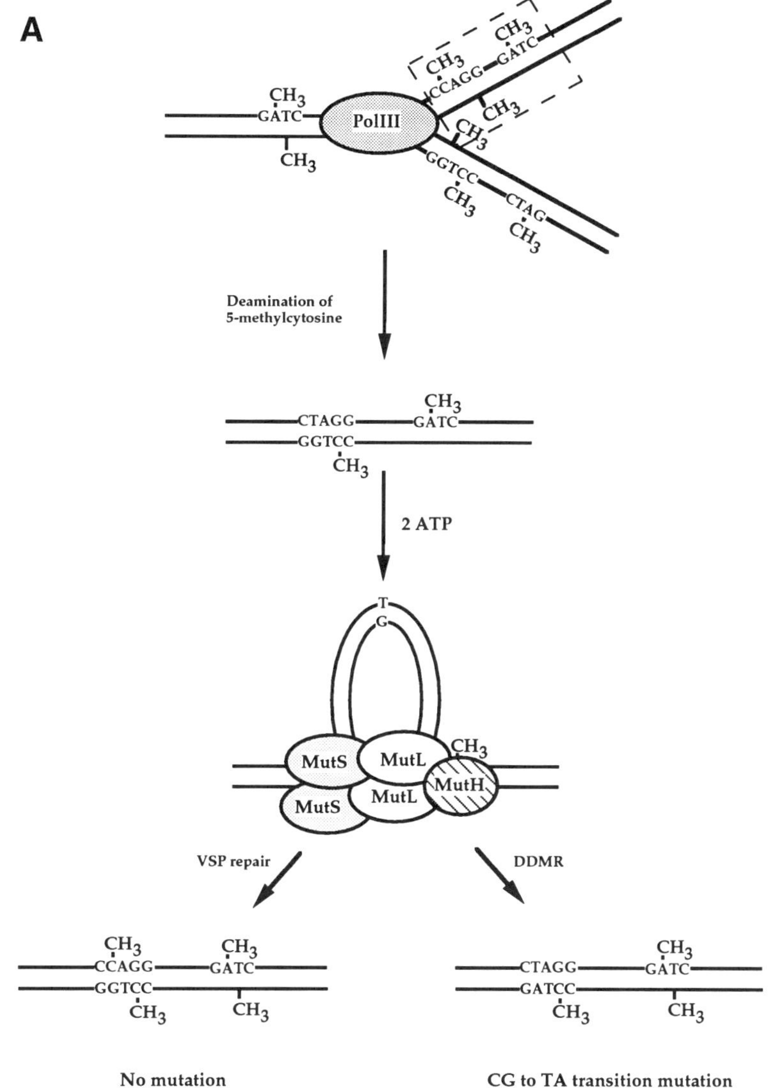

Fig. 3. Model for how the VSP repair and DDMR systems interact with G·T mismatches resulting from deamination of 5-methylcytosine in DNA. **(A)** Deamination of 5-methylcytosine in the parental strand when the adjacent GATC site is hemimethylated.

in the presence of ATP is able to form a loop in which MutS and MutL remain bound at the base of the loop (Fig. 3). This leaves the mismatch exposed in the loop and available for Vsr cleavage. Figure 3 illustrates how the DDMR system can affect the processing of G·T mismatches arising from deamination of 5-methylcytosine in DNA. Figure 3A and B shows two different situations where the deamination of 5-methylcytosine has occurred right after replication and when the adjacent GATC site is still hemimethylated. In this situation, the resulting G·T mismatch is subject to repair by the DDMR system. Depending on whether deamination of 5-methylcytosine occurs on the

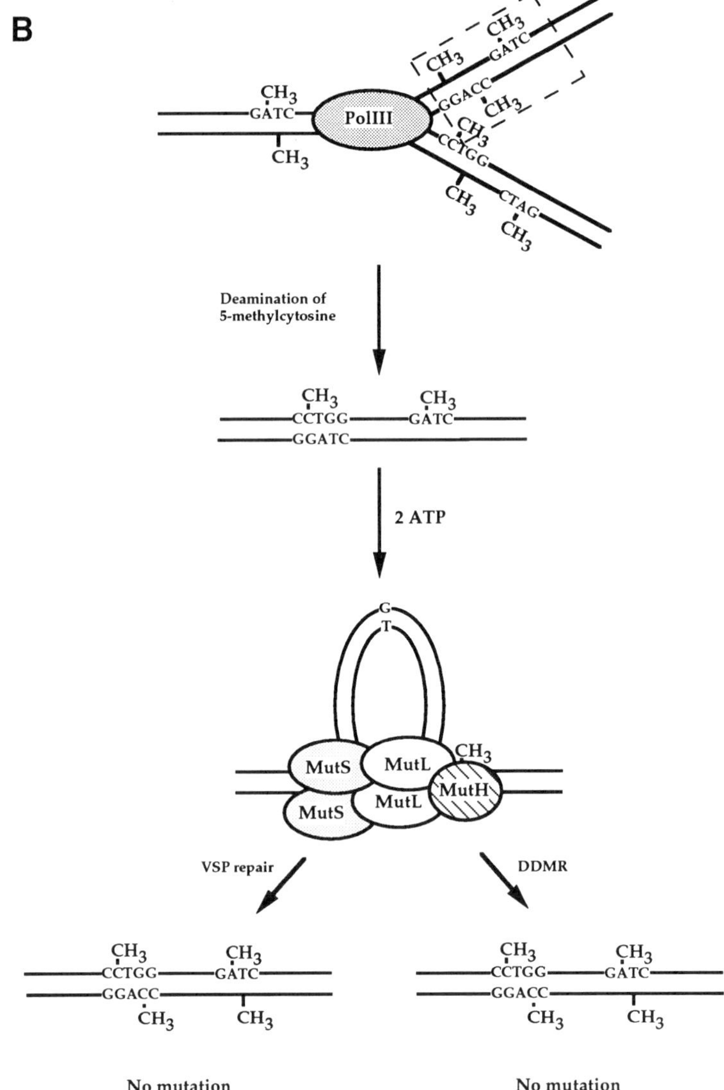

Fig. 3. (B) Deamination of 5-methylcytosine in the newly synthesized strand when the adjacent GATC site is hemimethylated.

parental or the newly synthesized strand, DDMR will result in either a CG to TA transition mutation (Fig. 3A) or no mutation (Fig. 3B). A third possibility is shown in Fig. 3C, where deamination of 5-methylcytosine occurs when the adjacent GATC site is methylated on both strands. This may occur in either slow-growing or resting cells. When the GATC site is fully methylated, the G·T mismatch is refractory to repair by the DDMR system, although the MutS-MutL-mediated α-shaped looped structure can still be formed. Therefore, the G·T mismatch is only processed by the VSP repair system, preventing mutations from being introduced into the genome. This model, if correct, suggests either that MutS affinity for G·T mismatches is higher than that of Vsr

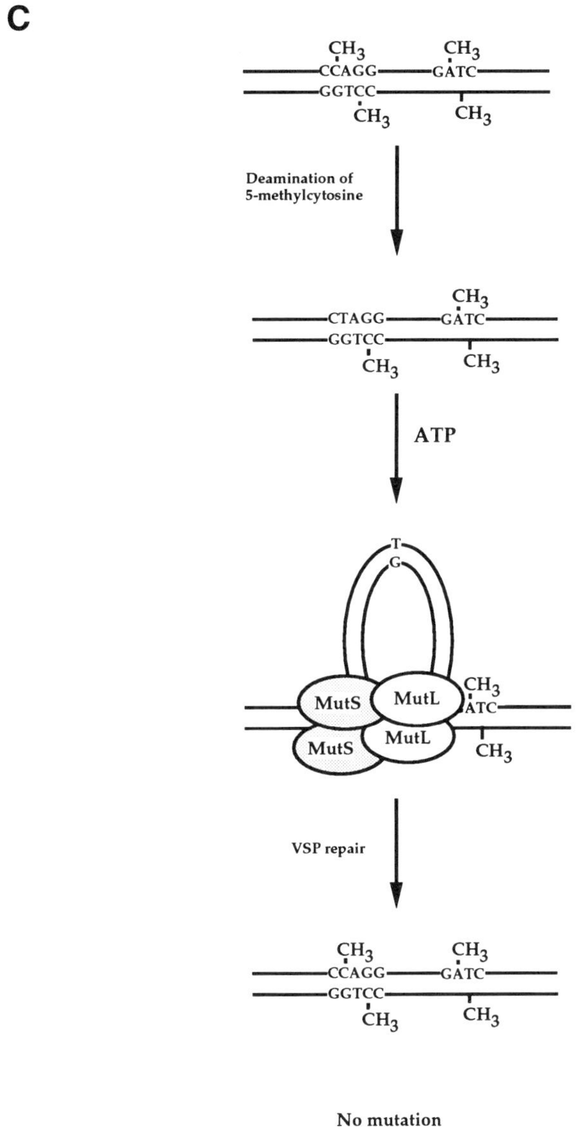

C

CH_3 CH_3
CCAGG GATC
GGTCC
CH_3 CH_3

Deamination of 5-methylcytosine

CH_3
CTAGG GATC
GGTCC
CH_3 CH_3

ATP

T
G
MutS MutL
MutS MutL
CH_3
ATC
CH_3

VSP repair

CH_3 CH_3
CCAGG GATC
GGTCC
CH_3 CH_3

No mutation

Fig. 3. (C) Deamination of 5-methylcytosine when the adjacent GATC site is methylated on both strands.

and/or that an α-shaped looped structure containing mismatched DNA is a better substrate for cleavage. These predictions can be verified experimentally. The model in Fig. 3 also accommodates the competition between the VSP and DDMR systems for the T·G mismatch, and it raises the possibility that processing exclusively by the DDMR system is responsible, at least in part, for the mutagenic event by removal of the G resulting in a CG to TA base pair change.

The *vsr* gene is located immediately downstream of the *dcm* gene and is part of the same transcriptional unit *(102)*. The *dcm* gene encodes the DNA cytosine methyltransferase responsible for the formation of 5-methylcytosine at certain sequences. This gene arrangement is present in several bacterial species in addition to *E. coli* (M. Lieb, personal communication). A possible explanation is that *vsr*-dependent mutagenesis has a selective value in evolution to generate diversity *(31)*. Indeed, there is a good correlation between the efficiency of Vsr cleavage and the under-representation of -NTAGN- sequences in different strains of *E. coli*. This could come about by the conversion of replication-generated T·G mismatches by VSP repair at -NTAGN- sequences to -NCAGN- *(31)*.

A widely used mutation, *dcm-6*, inactivates both the *dcm* and *vsr* genes *(22)* and explains why a requirement of the DNA methyltransferase for VSP repair was originally inferred *(54)*. It was later shown, as expected, that the Dcm protein is not required for repair *(102)*.

6. DDMR AND TRANSCRIPTION-COUPLED NUCLEOTIDE-EXCISION REPAIR (TCR)

In *E. coli* and other organisms, there appear to be two kinds of nucleotide excision repair *(30)* that operate on transcriptionally active and inactive DNA. The former has been termed transcription coupled repair (TCR). The two forms of repair are mechanistically similar, but TCR requires additional proteins to target excision repair to the transcribed strand. Defects in DDMR reduce the efficiency of TCR. An example of TCR in *E. coli* is the reduction in UV-induced mutation frequency at nonsense codons on inhibition of protein synthesis and was termed "mutation frequency decline" *(111)*. The basis of this phenomenon appears to be the rapid strand-specific repair of the photoproduct in the genes coding for glutamine tRNA generating nonsense suppressor tRNAs *(8)*. Note that tRNA genes are constantly transcribed. *E. coli mfd* mutants have lost strand discrimination owing to lack of the Mfd protein to recruit nucleotide-excision repair enzymes at sites of DNA damage where the RNA polymerase has stalled and, therefore, show no UV-induced mutation frequency decline. *E. coli mutS* and *mutL* mutant strains show mutation frequency decline intermediate between wild type and *mfd* mutant strains *(53)*. The authors suggested that the concentration of nucleotide-excision repair enzyme complexes ($[UvrA]_2UvrB$; Chapter 2) may be limiting at stalled transcription complexes in DDMR-deficient cells owing to their binding at DNA mismatches.

TCR has also been demonstrated in the *E. coli lacZ* gene; genomic DNA isolated at various times after UV irradiation was probed on Southern blots for the presence of damage in the transcribed vs the nontranscribed strands. Induction of transcription dramatically increases the rate of thymine-dimer excision from the transcribed strand *(75)*. This increase is abolished in *mutL*, *mutS*, and *mfd* mutant strains, resulting in an equal rate of dimer removal from each strand *(74)*. The authors prefer a model in which MutL and MutS are bound directly to some feature of the stalled transcription complex to allow formation of an α-shaped looped structure. This altered topology then allows for enhanced binding of the nucleotide-excision enzyme complex. Whatever the correct explanation, these results show a connection between TCR and DDMR. This was not predicted based on the known substrate specificities and the proteins involved in the two processes.

7. STRAND-SLIPPAGE MUTAGENESIS

Duplicated regions that have identical or similar sequences in chromosomes can present a problem, since they may recombine, causing either a deletion of the intervening material or a duplication. Recent studies indicate that recombination-independent events at the replication fork may also lead to such chromosome alterations, and these changes are either enhanced or reduced in the absence of DDMR.

Deletions between two 101-bp tandem repeats with 4% sequence divergence occurs at high frequency in *E. coli (58)*. Inactivation of homologous recombination by mutation in *recA* does not alter this frequency. In *mutH*, *mutL*, *mutS*, *uvrD*, or *dam* mutants, however, the frequency increases almost 100-fold. Deletions between perfectly homologous repeats are not affected by inactivation of DDMR. These results indicate that a heteroduplex intermediate between the two repeats is formed at or near the replication fork, which is susceptible to DDMR in a *dam*-dependent manner. Such an intermediate could be formed by strand slippage where pairing occurs out of register either in the newly synthesized DNA strand or the template strand. Although Lovett and Feschenko *(58)* measured only deletions between repeats, the similarity in mechanism would predict that duplications are also influenced by DDMR.

Dinucleotide and trinucleotide repeats occur in eukaryotic DNA, and are expected to present a potential problem during replication owing to strand slippage leading to repeat deletion or expansion. The frequency of deletions and expansions in an $(AC)_{20}$ dinucleotide repeat is increased at least 13-fold in *E. coli mutL* and *mutS* strains but not at all in a *recA* background *(52)*. This result correlates with the subsequent finding of dinucleotide repeat instability in mismatch repair-deficient human cell lines (reviewed in ref. *78)*.

In contrast, $(CTG)_{180}$ trinucleotide repeats behave quite differently; these repeats are actually more stable in *mutH*, *mutL*, or *mutS* mutants *(42)*. Inactivation of homologous recombination by mutation in the *recA* gene had no effect. These results suggest that DDMR promotes deletion (and probably expansion) of trinucleotide repeats at the replication fork owing to strand slippage. It is unclear how many other triplet repeats may also be affected, since DDMR had no influence on the stability of $(CGG)_{80}$ *(42)*. Trinucleotide instability in human mismatch repair-deficient cells has not been reported. Differences in secondary structure and stability of the individual repeats may account for the variation in DDMR-dependent processing of di-/tri-nucleotide repeats.

8. HOMEOLOGOUS RECOMBINATION

Recombination between related, interspersed, and diverged DNA sequences is a potential source of genomic instability, as first proposed by Karimova et al. in 1985 *(44)*. This phenomenon has been studied extensively since then by interspecies matings between related, but diverged organisms like *E. coli* and *Salmonella typhimurium*, and by following the ability of bacteriophages to inject DNA into their hosts *(30)*. These studies indicate that the DDMR proteins can act to prevent interchromosomal exchanges between homologous DNAs and that the recombinational events are RecA-dependent. The exact mechanism underlying homeologous recombination is not fully understood, but it has been shown that the MutS protein inhibits RecA-mediated strand exchange (*113*; Fig. 4).

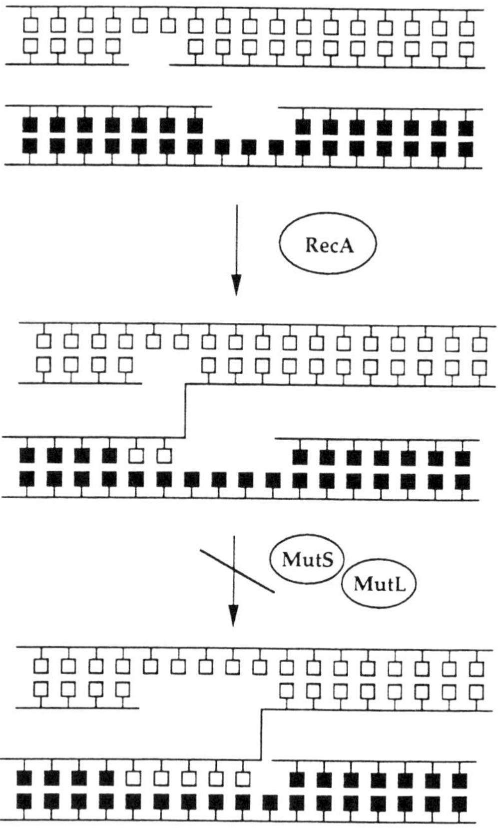

Fig. 4. Mechanism of homeologous recombination. The Mut proteins block the branch migration step of the reaction, probably in response to the occurrence of mismatches within the recombination heteroduplex. Adapted from ref. *77* with permission.

The DDMR enzymes protect the cell against mutation as well as from interspecies recombination, whereas the SOS response (Chapter 7) increases both *(12,73,99)*. That a connection between these two phenomenona exists was shown by Matic et al. *(73)*, who suggested that homeologous recombination may play an important role in the evolution of species. They hypothesized that the opposing activities of DDMR and SOS systems could determine the rate at which sequences diverge, which is a key component in the evolution of species. Homeologous recombination has also been shown to occur in yeast *(101)*. From an evolutionary perspective, it would be interesting to see whether an SOS response exists in mammalian cells and, if so, whether there would be any similarity to the bacterial SOS system. Possible candidates for proteins involved in an SOS-like response in eukaryotic cells are the checkpoint proteins. In contrast to the results with *E. coli*, Humbert et al. *(41)* showed that the Hex system is not a barrier to interspecies recombination in *S. pneumoniae*. They found that the Hex DNA mismatch repair system was completely saturated during heterospecific transformation of chromosomal DNA from the two related streptococcal species, *S. oralis* and *S. mitis*. Only by overexpressing the HexA and HexB proteins could a decrease in recombination between the two genomes be obtained. The DDMR pathway of *E. coli* and *S. typhimurium* appears different from the Hex mismatch repair system of *S. pneumoniae* in that the Gram-negative repair pathways do not saturate easily. Saturation of the Gram-negative DDMR pathway is observed in strains carrying a mutation in the *mutD* gene, which renders the cells deficient in DNA polymerase III proofreading activity, and in the presence of multicopy single-stranded DNA that adopts structures containing mismatches *(21,63,98)*. These results may indicate that the DDMR pathway is not a general barrier against recombination between related, but diverged DNA. Alternatively, the Hex system could represent a special case among DDMR systems, because *S. pneumoniae* is a naturally transformable species.

A saturation of DNA mismatch repair systems may explain why some individuals heterozygous for the human *mutS* homolog, *hMSH2*, are predisposed to hereditary nonpolyposis colorectal carcinoma (HNPCC) *(84)*. Alternatively, these individuals may carry a dominant negative mutation in one allele of *hMSH2*. It remains to be seen whether the presence of only a single intact copy of *hMSH2* in the germline of these individuals results in a lower concentration of the corresponding protein and whether this affects repair capacity.

Homologous recombination is also increased by about 2- to 20-fold in the absence of DDMR in *E. coli (26,114)*. The increase is highly marker-dependent, which is expected since only those susceptible to DDMR will be affected. The molecular basis for the increase in homologous recombination is unknown, but may be related to that for homeologous recombination.

9. RECOGNITION OF O^6-METHYLGUANINE (O^6MEG) DNA LESIONS BY THE MutS PROTEIN

The finding that DDMR proteins recognize mispaired bases has raised the question whether these proteins also recognize modified bases in DNA. Wood *et al. (112)* showed that the *E. coli* MutS protein does bind certain guanine analogs but failed to bind certain adenine analogs. The presence of O^6-methylguanine (O^6MeG) lesions in double-stranded DNA have been associated with mutation and lethality (Chapter 6, vol. 2). These lesions can, in principle, be produced by at least three different mechanisms: direct alkylation of G·C base pairs in double-stranded DNA; alkylation of guanine residues in single-stranded regions of DNA associated with replication forks; and alkylation of the DNA precursor pool followed by incorporation of O^6-methyl deoxyguanosine triphosphate during DNA replication. The modified base O^6MeG is one of the lesions introduced into DNA after treatment with methylating agents, such as *N*-methyl-*N'*-nitro-*N*-nitrosoguanidine (MNNG). In *E. coli*, this lesion is repaired by the DNA repair methyltransferases (MTases), *Ada* and *Ogt*, which remove the methyl group from the O^6 position of guanine (reviewed in ref. *97)*. The O^6MeG DNA lesion is mutagenic because it can base pair with thymine, causing GC to AT transition mutations. However, O^6MeG is also capable of pairing with cytosine, in which case no mutation results. In addition to being mutagenic, this lesion is cytotoxic in both prokaryotic and eukaryotic cells, but the mechanism by which O^6MeG causes cell death is not fully understood. Two lines of evidence suggest that processing of O^6MeG by the DDMR system may cause cell death. Karran and Marinus *(46)* reported that *E. coli dam*⁻ strains, which are proficient in DDMR but deficient in strand discrimination, showed a marked sensitivity to killing by MNNG, but not by dimethyl sulfate. The major difference between these alkylating agents is that MNNG induces significantly higher levels of O^6MeG in DNA than does dimethyl sulfate. The hypersensitivity of *dam*⁻ cells to MNNG was suppressed by mutations in either *mutS* or *mutL*, which inactivate the DDMR pathway. These observations led Karran and Marinus *(46)* to suggest that the cytotoxicity may be mediated by the interaction of the DDMR system with O^6MeG sites on DNA. Such an interaction of DDMR with alkylation damage was proposed to occur on replication of DNA containing O^6MeG; the insertion of either C or T opposite O^6MeG (the only two bases that efficiently pair with O^6MeG) would result in repeated futile rounds of DDMR (Fig. 5). This would ultimately cause cell death either owing to persistent DNA breaks or because DNA polymerase III would be sequestered at the repair site and, therefore, not be available for DNA replication.

This phenomenon is not unique for *E. coli*, but has also been discovered in other bacteria and in mammalian cells (reviewed in refs. *45* and *48)*. In contrast, it was recently shown that DNA mismatch repair-deficient mutants do not rescue MNNG-induced genotoxicity in DNA repair MTase-deficient yeast cells *(115)*. The proposal that a cellular mechanism exists whereby unrepaired O^6MeG DNA lesions may trigger cell death rather than persisting as potentially mutagenic lesions in the genome has been confirmed by many groups. Some studies showed that the presence of O^6MeG lesions in DNA is more mutagenic in an *E. coli mutS* mutant strain, indicating that the DDMR pathway processes such lesions *(85,86)*. Further evidence has come from biochemical studies that show that MutS has a specific affinity for O^6MeG in DNA oligonucleotides *(95)*. It remains to be shown whether the binding of O^6MeG·C DNA lesions by MutS or any of the mammalian MutS homologs is capable of initiating mismatch repair, leading to an accumulation of DNA strand breaks. Such evidence would validate the proposal that in some instances, the DNA mismatch repair system may convert a potentially mutagenic lesion into a cytotoxic lesion in both prokaryotic and eukaryotic cells.

10. CONCLUDING REMARKS

The recent findings that several human genetic diseases, such as HNPCC, are characterized by a defect in the DNA mismatch repair pathway (Chapter 20, vol. 2) have generated widespread interest in this research area. The genes involved in DNA mismatch repair in mammalian cells were identified primarily on the basis of high homology of the predicted gene products to their yeast and bacterial counterparts. In contrast to bacteria, eukaryotic cells contain several homologs of both the MutL and MutS proteins, which have been shown to act together in a complex in the DNA repair process (Fig. 6) *(24,80,82,89)*. Thus far, six MutS homologs (Msh1–6) and two MutL homologs (Mlh1 and Pms1) have been identified in *Saccharomyces cerevisiae* (Fig. 2A,B). In humans, three MutS homologs (hMSH2, hMSH3, and p160/GTBP) and three MutL homologs (hMLH1, PMS1, and PMS2) have been identified. The finding that eukaryotic cells have several MutL and MutS homologs raises some very interesting questions. For instance, how does a single protein like MutS in prokaryotic cells perform

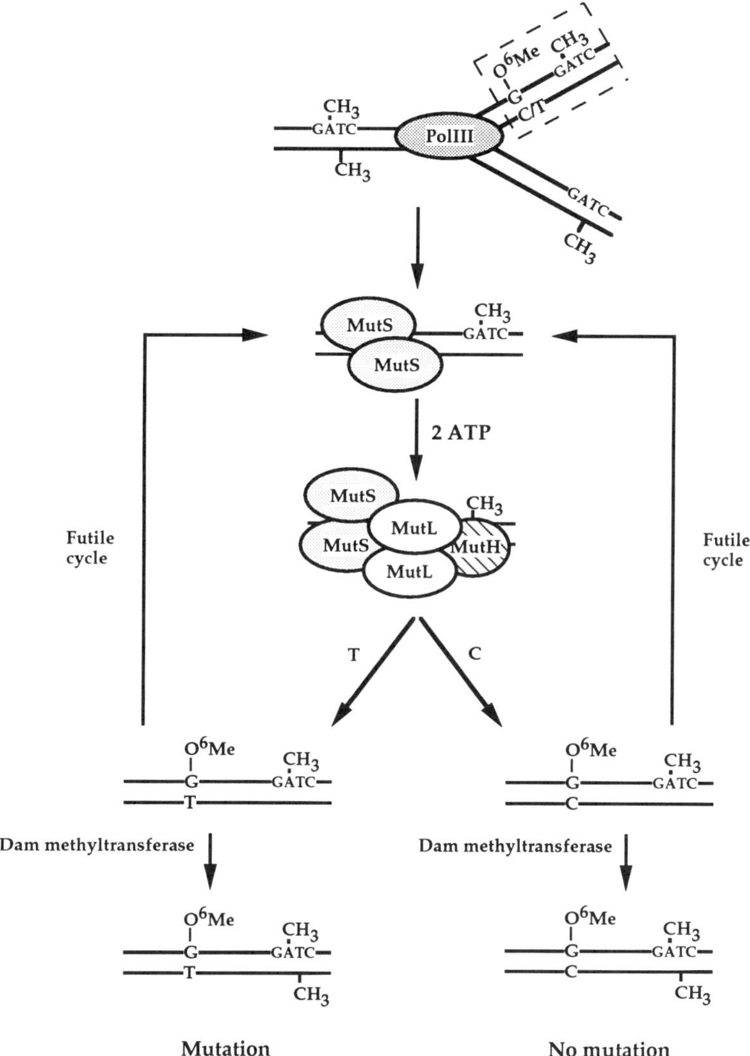

Fig. 5. Model for how the cytotoxicity of O^6MeG DNA lesions may be mediated by the interaction of the DDMR system with O^6MeG sites on DNA. The model illustrates how the insertion of either C or T opposite O^6MeG could result in repeated futile rounds of DNA mismatch repair.

functions similar to those carried out by a multiprotein complex like $_{\text{Msh}}2\text{–}3/_{\text{Msh}}6$ in yeast *(71)* or hMSH2-GTBP in human cells?

The similarity in prokaryotes and eukaryotes of the biochemical steps carried out by MutS and MutL proteins and their homologs argues for continued studies of this pathway in both systems. A detailed characterization of the structural domains of the prokaryotic Mut proteins, and characterization of their functions in VSP and TCR should shed light on their eukaryotic counterparts.

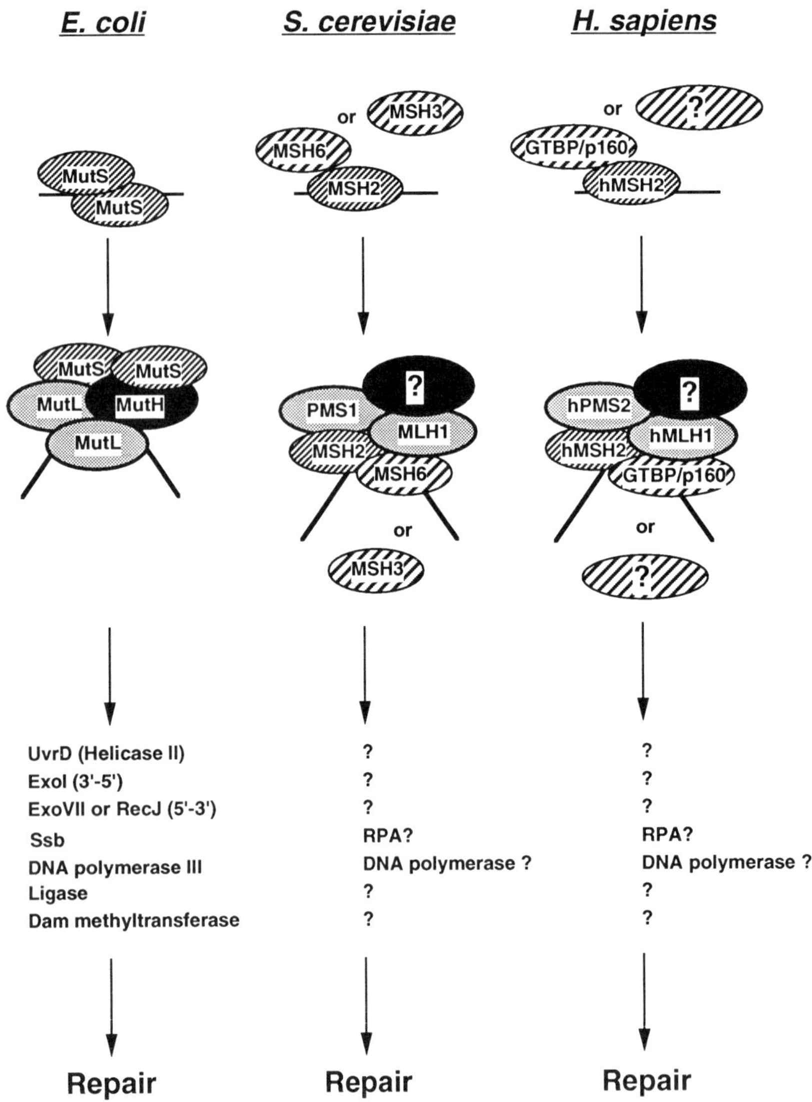

Fig. 6. Comparison among DNA mismatch repair pathways in *E. coli*, *S. cerevisiae*, and *Homo sapiens*. The stoichiometry and the nature of protein–protein interactions are not understood in detail, and are shown here for illustration only.

ACKNOWLEDGMENT

We would like to thank D. Raychaudhuri for critical reading of the manuscript.

REFERENCES

1. Au, K. G., K. Welsh, and P. Modrich. 1992. Initiation of methyl-directed mismatch repair. *J. Biol. Chem.* **267:** 12,142–12,148.
2. Baker, S. M., C. E. Bronner, L. Zhang, A. W. Plug, M. Robatzek, G. Warren, E. A. Elliott, J. A. Yu, T. Ashley, N. Arnheim, R. A. Flavall, and R. M. Liskay. 1995. Male mice defec-

tive in the DNA mismatch repair gene *PMS2* exhibit abnormal chromosome synapsis in meiosis. *Cell* **82:** 309–319.

3. Balganesh, T. S. and S. A. Lacks. 1985. Heteroduplex DNA mismatch repair system of *Streptococcus pneumoniae*: cloning and expression of the *hexA* gene. *J. Bacteriol.* **162:** 979–984.
4. Bende, S. M. and R. H. Grafstrom. 1991. The DNA binding properties of the MutL protein isolated from *Escherichia coli. Nucleic Acids Res.* **19:** 1549–1555.
5. Bergerat, A. and W. Guschlbauer. 1990. The double role of methyl donor and allosteric effector of *S*-adenosyl-methionine for Dam methylase of *E. coli. Nucleic Acids Res.* **18:** 4369–4375.
6. Bergerat, A., A. Kriebardis, and W. Guschlbauer. 1989. Preferential site-specific hemimethylation of GATC sites in pBR322 DNA by Dam methyltransferase from *Escherichia coli. J. Biol. Chem.* **264:** 4064–4070.
7. Bhagwat, A. S., A. Sohail, and M. Lieb. 1988. A new gene involved in mismatch correction in *Escherichia coli. Gene* **74:** 155,156.
8. Bockrath, R. and J. Palmer. 1977. Differential repair of pre-mutational UV-lesions at tRNA genes in *E. coli. Mol. Gen. Genet.* **156:** 133–140.
9. Boye, E. and A. Løbner-Olesen. 1990. The role of *dam* methyltransferase in the control of DNA replication in *E. coli. Cell* **62:** 981–989.
10. Boye, E., M. G. Marinus, and A. Løbner-Olesen. 1992. Quantitation of Dam methyltransferase in *Escherichia coli. J. Bacteriol.* **174:** 1682–1685.
11. Bruni, R., D. Martin, and J. Jiricny. 1988. d(GATC) sequences influence *Escherichia coli* mismatch repair in a distance-dependent manner from positions both upstream and downstream of the mismatch. *Nucleic Acids Res.* **16:** 4875–4890.
12. Caillet-Fauquet, P., G. Maenhaut-Michel, and M. Radman. 1984. SOS mutator effect in *E. coli* mutants deficient in mismatch repair correction. **3:** 707–712.
13. Campbell, J. L. and N. Kleckner. 1990. *E. coli oriC* and the *dnaA* gene promoter are sequestered from *dam* methyltransferase following the passage of the chromosomal replication fork. *Cell* **62:** 967–979.
14. Chen, J. D. and S. A. Lacks. 1991. Role of uracil-DNA glycosylase in mutation avoidance by *Streptococcus pneumoniae. J. Bacteriol.* **173:** 283–290.
15. Claverys, J. P. and S. A. Lacks. 1986. Heteroduplex deoxyribonucleic acid base mismatch repair in bacteria. *Microbiol. Rev.* **50:** 133–165.
16. Claverys, J. P. and V. Mejean. 1988. Strand targeting signal(s) for in vivo mutation avoidance by post-replication mismatch repair in *Escherichia coli. Mol. Gen. Genet.* **214:** 574–578.
17. Connolly, D. M., and M. E. Winkler. 1989. Genetic and physiological relationships among the *miaA* gene, 2-methylthio-N^6-(delta 2-isopentenyl)-adenosine tRNA modification, and spontaneous mutagenesis in *Escherichia coli* K-12. *J. Bacteriol.* **171:** 3233–3246.
18. Connolly, D. M. and M. E. Winkler. 1991. Structure of *Escherichia coli* K-12 *miaA* and characterization of the mutator phenotype caused by *miaA* insertion mutations. *J. Bacteriol.* **173:** 1711–1721.
19. Cooper, D. L., R. S. Lahue, and P. Modrich. 1993. Methyl-directed mismatch repair is bidirectional. *J. Biol. Chem.* **268:** 11,823–11,829.
20. Coulondre, C., J. H. Miller, P. J. Farabaugh, and W. Gilbert. 1978. Molecular basis of substitution hotspots in *Escherichia coli. Nature* **274:** 775–780.
21. Damagnez, V., M. P. Doutriaux, and M. Radman. 1989. Saturation of mismatch repair in the *mutD5* mutator strain of *Escherichia coli. J. Bacteriol.* **171:** 4494–4497.
22. Dar, M. E. and A. S. Bhagwat. 1993. Mechanism of expression of DNA repair gene vsr, an *Escherichia coli* gene that overlaps the DNA cytosine methylase gene, dcm. *Mol. Microbiol.* **9:** 823–833.
23. Dohet, C., R. Wagner, and M. Radman. 1986. Methyl-directed repair of frameshift mutations in heteroduplex DNA. *Proc. Natl. Acad. Sci. USA* **83:** 3395–3397.

24. Drummond, J. T., G.-M. Li, M. J. Longley, and P. Modrich. 1995. Isolation of an h*MSH2*-p160 heterodimer that restores DNA mismatch repair to tumor cells. *Science* **268:** 1909–1912.
25. Dzidic, S. and M. Radman. 1989. Genetic requirements for hyper- recombination by very short patch mismatch repair: involvement of *Escherichia coli* DNA polymerase I. *Mol. Gen. Genet.* **217:** 254–256.
26. Feinstein, S. I. and K. B. Low. 1986. Hyper-recombining recipient strains in bacterial conjugation. *Genetics* **113:** 13–33.
27. Feng, G., H.-C. T. Tsui, and M. E. Winkler. 1996. Depletion of the cellular amounts of the MutS and MutH methyl-directed mismatch repair proteins in stationary-phase *Escherichia coli* K-12 cells. *J. Bacteriol.* **178:** 2388–2396.
28. Fishel, R., M. K. Lescoe, M. R. Rao, N. G. Copeland, N. A. Jenkins, J. Garber, M. Kane, and R. Kolodner. 1993. The human mutator gene homolog *MSH2* and its association with hereditary nonpolyposis colon cancer. *Cell* **75:** 1027–1038.
29. Fishel, R. A., E. C. Siegel, and R. Kolodner. 1986. Gene conversion in *Escherichia coli*. Resolution of heteroallelic mismatched nucleotides by co-repair. *J. Mol. Biol.* **188:** 147–157.
30. Friedberg, E. C., G. C. Walker, and W. Siede. 1995. *DNA Repair and Mutagenesis.* ASM, Washington, D. C.
31. Glasner, W., R. Merkl, V. Schellenberger, and H.-J. Fritz. 1995. Substrate preferences of Vsr DNA mismatch endonuclease and their consequenses for the evolution of the *Escherichia coli* K-12 genome. *J. Mol. Biol.* **245:** 1–7.
32. Grafstrom, R. H.,and R. H. Hoess. 1987. Nucleotide sequence of the *Escherichia coli mutH* gene. *Nucleic Acids Res.* **15:** 3073–3084.
33. Grilley, M., J. Griffith, and P. Modrich. 1993. Bidirectional excision in methyl-directed mismatch repair. *J. Biol. Chem.* **268:** 11,830–11,837.
34. Grilley, M., K. M. Welsh, S. S. Su, and P. Modrich. 1989. Isolation and characterization of the *Escherichia coli mutL* gene product. *J. Biol. Chem.* **264:** 1000–1004.
35. Guild, W. R., and N. B. Shoemaker. 1976. Mismatch correction in pneumococcal transformation: donor length and *hex*-dependent marker efficiency. *J. Bacteriol.* **125:** 125–135.
36. Haber, L. T., P. P. Pang, D. I. Sobell, J. A. Mankovich, and G. C. Walker. 1988. Nucleotide sequence of the *Salmonella typhimurium mutS* gene required for mismatch repair: homology of MutS and HexA of *Streptococcus pneumoniae*. *J. Bacteriol.* **170:** 197–202.
37. Haber, L. T. and G. C. Walker. 1991. Altering the conserved nucleotide binding motif in the *Salmonella typhimurium* MutS mismatch repair protein affects both its ATPase and mismatch binding activities. *EMBO J.* **10:** 2707–2715.
38. Hattman, S., J. E. Brooks, and M. Masurekar. 1978. Sequence specificity of the P1 modification methylase (M. Eco P1) and the DNA methylase (M. Eco dam) controlled by the *Escherichia coli dam* gene. *J. Mol. Biol.* **126:** 367–380.
39. Hennecke, F., H. Kolmar, K. Brundl, and H.-J. Fritz. 1991. The *vsr* gene product of *E. coli* K-12 is a strand- and sequence-specific DNA mismatch endonuclease. *Nature* **353:** 776–778.
40. Herman, G. E., and P. Modrich. 1982. *Escherichia coli dam* methylase. Physical and catalytical properties of the homogeneous enzyme. *J. Biol. Chem.* **257:** 2605–2612.
41. Humbert, O., M. Prudhomme, R. Hakenbeck, C. G. Dowson, and J. P. Claverys. 1995. Homeologous recombination and mismatch repair during transformation in *Streptococcus pneumoniae*: saturation of the Hex mismatch repair system. *Proc. Natl. Acad. Sci. USA* **92:** 9052–9056.
42. Jarworski, A., W. A. Rosche, R. Gellibolian, S. Kang, M. Shimizu, R. P. Bowater, R. R. Sinden, and R. D. Wells. 1995. Mismatch repair in *Escherichia coli* enhances instability of $(CTG)_n$ triplet repeats from human hereditary diseases. *Proc. Natl. Acad. Sci. USA* **92:** 11,019–11,023.
43. Jones, M., R. Wagner, and M. Radman. 1987. Repair of a mismatch is influenced by the base composition of the surrounding nucleotide sequence. *Genetics* **115:** 605–610.

44. Karimova, G. A., P. S. Grigor'ev, and V. N. Rybchin. 1985. The role of genes of the system of mismatched base correction in genetic recombination of *Escherichia coli. Mol. Genetika, Microbiol. Virus* **10:** 29–34.
45. Karran, P. and M. Bignami. 1994. DNA damage tolerance, mismatch repair and genome instability. *Bioessays* **16:** 833-839.
46. Karran, P. and M. G. Marinus. 1982. Mismatch correction at O^6-methylguanine residues in *E. coli* DNA. *Nature* **296:** 868,869.
47. Kramer, B., W. Kramer, and H. J. Fritz. 1984. Different base/base mismatches are corrected with different efficiencies by the methyl-directed DNA mismatch-repair system of *E. coli. Cell* **38:** 879–887.
48. Kumaresan, K. R., S. S. Springhorn, and S. A. Lacks. 1995. Lethal and mutagenic actions of *N*-methyl-*N*-nitro-*N*-nitrosoguanidine potentiated by oxidized glutathione, a seemingly harmless substance in the cellular environment. *J. Bacteriol.* **177:** 3641–3646.
49. Laengle-Rouault, F., G. Maenhaut-Michel, and M. Radman. 1987. GATC sequences, DNA nicks and the MutH function in *Escherichia coli* mismatch repair. *EMBO J.* **6:** 1121–1127.
50. Lahue, R. S., K. G. Au, and P. Modrich. 1989. DNA mismatch correction in a defined system. *Science* **245:** 160–164.
51. Learn, B. A. and R. H. Grafstrom. 1989. Methyl-directed repair of frameshift heteroduplexes in cell extracts from *Escherichia coli. J. Bacteriol.* **171:** 6473–6481.
52. Levinson, G. and G. A. Gutman. 1987. High frequencies of short frameshifts in poly-CA/TG tandem repeats borne by bacteriophage M13 in *Escherichia coli. Nucleic Acids Res.* **15:** 5323–5338.
53. Li, B.-H. and R. Bockrath. 1995. Mutation frequency decline in *Escherichia coli.* I. Effects of defects in mismatch repair. *Mol. Gen. Genet.* **249:** 585–590.
54. Lieb, M. 1987. Bacterial genes *mutL*, *mutS* and *dcm* participates in repair of mismatches at 5-methylcytosine sites. *J. Bacteriol.* **169:** 5241–5246.
55. Lieb, M. 1983. Specific mismatch correction in bacteriophage lambda crosses by very short patch repair. *Mol. Gen. Genet.* **191:** 118–125.
56. Lieb, M. 1991. Spontaneous mutation at a 5-methylcytosine hotspot is prevented by very short patch (VSP) mismatch repair. *Genetics* **128:** 23–27.
57. Løbner-Olesen, A., E. Boye, and M. G. Marinus. 1992. Expression of the *Escherichia coli dam* gene. *Mol. Microbiol.* **6:** 1841–1851.
58. Lovett, S. T. and V. V. Feschenko. 1996. Stabilization of diverged tandem repeats by mismatch repair: evidence for deletion formation via a misaligned replication intermediate. *Proc. Natl. Acad. Sci. USA* **93:** 7120–7124.
59. Lu, A. L. 1987. Influence of GATC sequences on *Escherichia coli* DNA mismatch repair in vitro. *J. Bacteriol.* **169:** 1254–1259.
60. Lu, A. L., S. Clark, and P. Modrich. 1983. Methyl-directed repair of DNA base-pair mismatches in vitro. *Proc. Natl. Acad. Sci. USA* **80:** 4639–4643.
61. Lyngstadaas, A., A. Løbner-Olesen, and E. Boye. 1995. Characterization of three genes in the *dam*-containing operon of *Escherichia coli. Mol. Gen. Genet.* **247:** 546–554.
62. Lyons, S. M. and P. F. Schendel. 1984. Kinetics of methylation in *Escherichia coli* K-12. *J. Bacteriol.* **159:** 421–423.
63. Maas, W. K., C. Wang, T. Lima, G. Zubay, and D. Lim. 1994. Multicopy single-stranded DNAs with mismatched base pairs are mutagenic in *Escherichia coli. Mol. Microbiol.* **14:** 437–441.
64. Mankovich, J. A., C. A. McIntyre, and G. C. Walker. 1989. Nucleotide sequence of the *Salmonella typhimurium mutL* gene required for mismatch repair: homology of MutL to HexB of *Streptococcus pneumoniae* and to PMS1 of the yeast *Saccharomyces cerevisiae. J. Bacteriol.* **171:** 5325–5331.
65. Mannarelli, B. M., T. S. Balganesh, B. Greenberg, S. S. Springhorn, and S. A. Lacks. 1985. Nucleotide sequence of the *DpnII* DNA methylase gene of *Streptococcus*

pneumoniae and its relationship to the *dam* gene of *Escherichia coli*. *Proc. Natl. Acad. Sci. USA* **82:** 4468–4472.
66. Marinus, M. G. 1976. Adenine methylation of Okazaki fragments in *Escherichia coli*. *J. Bacteriol.* **128:** 853,854.
67. Marinus, M. G. 1996. Methylation of DNA, in *Escherichia coli and* Salmonella: *Cellular and Molecular Biology*, 2nd ed. (Neidhardt, F. C. et al., eds.), ASM, Washington, DC.
68. Marinus, M. G. and N. R. Morris. 1974. Biological function for 6- methyladenine residues in the DNA of *Escherichia coli* K12. *J. Mol. Biol.* **85:** 309–322.
69. Marinus, M. G., and N. R. Morris. 1973. Isolation of deoxyribonucleic acid methylase mutants of *Escherichia coli* K-12. *J. Bacteriol.* **114:** 1143–1150.
70. Marinus, M. G., A. Poteete, and J. A. Arraj. 1984. Correlation of DNA adenine methylase activity with spontaneous mutability in *Escherichia coli* K-12. *Gene* **28:** 123–125.
71. Marsischky, G. T., N. Filosi, M. F. Kane, and R. Kolodner. 1996. Redundancy of *Saccharomyces cerevisiae MSH3* and *MSH6* in *MSH2*- dependent mismatch repair. *Genes Dev.* **10:** 407–420.
72. Martin, B., H. Prats, and J. P. Claverys. 1985. Cloning of the *hexA* mismatch-repair gene of *Streptococcus pneumoniae* and identification of the product. *Gene* **34:** 293–303.
73. Matic, I., C. Rayssiguier, and M. Radman. 1995. Interspecies gene exchange in bacteria: the role of SOS and mismatch repair systems in evolution of species. *Cell* **80:** 507–515.
74. Mellon, I. and G. N. Champe. 1995. Products of DNA mismatch repair genes *mutS* and *mutL* are required for transcription-coupled nucleotide-excision repair of the lactose operon in *Escherichia coli*. *Proc. Natl. Acad. Sci. USA* **93:** 1292–1297.
75. Mellon, I. and P. C. Hanawalt. 1989. Induction of the *Escherichia coli* lactose operon selectively increases repair of its transcribed DNA strand. *Nature* **342:** 95–98.
76. Modrich, P. 1991. Mechanisms and biological effects of mismatch repair. *Ann. Rev. Genet.* **25:** 229–253.
77. Modrich, P. 1995. Mismatch repair, genetic stability and tumour avoidance. *Phil. Trans. R. Soc. Lond. B.* **347:** 89–95.
78. Modrich, P. and R. S. Lahue. 1996. Mismatch repair in replication fidelity, genetic recombination, and cancer biology. *Ann. Rev. Biochem.* **65:** 101–133.
79. Ogden, G. B., M. J. Pratt, and M. Schaechter. 1988. The replicative origin of the *E. coli* chromosome binds to cell membranes only when hemimethylated. *Cell* **54:** 127–135.
80. Palombo, F., P. Gallinari, I. Iaccarino, T. Lettieri, M. Hughes, A. D'Arrigo, O. Truong, J. J. Hsuan, and J. Jiricny. 1995. GTBP, a 160-kilodalton protein essential for mismatch-binding activity in human cells. *Science* **268:** 1912–1914.
81. Pang, P. P., A. S. Lundberg, and G. C. Walker. 1985. Identification and characterization of the *mutL* and *mutS* gene products of *Salmonella typhimurium* LT2. *J. Bacteriol.* **163:** 1007–1015.
82. Papadopoulos, N., N. C. Nicolaides, B. Liu, R. Parsons, C. Lengauer, F. Palombo, A. D'Arrigo, S. Markowitz, J. K. V. Willson, K. W. Kinzler, J. Jiricny, and B. Vogelstein. 1995. Mutations of GTBP in genetically unstable cells. *Science* **268:** 1915–1917.
83. Parker, B. O. and M. G. Marinus. 1992. Repair of DNA heteroduplexes containing small heterologous sequences in *Escherichia coli*. *Proc. Natl. Acad. Sci. USA* **89:** 1730–1734.
84. Parsons, R., G. M. Li, M. J. Longley, W. H. Fang, N. Papadopoulos, J. Jen, A. de la Chapelle, K. W. Kinzler, B. Vogelstein, and P. Modrich. 1993. Hypermutability and mismatch repair in RER+ tumor cells. *Cell* **75:** 1227–1236.
85. Pauly, G. T., S. H. Hughes, and R. C. Moschel. 1995. Mutagenesis in *Escherichia coli* by three O^6-substituted guanines in double-stranded or gapped plasmids. *Biochemistry* **34:** 8924–8930.
86. Pauly, G. T., S. H. Hughes, and R. C. Moschel. 1994. Response of repair- competent and repair-deficient *Escherichia coli* to three O^6-substituted guanines and involvement of

methyl-directed mismatch repair in the processing of O^6-methylguanine residues. *Biochemistry* **33:** 9169–9177.
87. Prats, H., B. Martin, and J. P. Claverys. 1985. The *hexB* mismatch repair gene of *Streptococcus pneumoniae*: characterization, cloning and identification of the product. *Mol. Gen. Genet.* **200:** 482–489.
88. Priebe, S. D., S. M. Hadi, B. Greenberg, and S. A. Lacks. 1988. Nucleotide sequence of the *hexA* gene for DNA mismatch repair in *Streptococcus pneumoniae* and homology of *hexA* to *mutS* of *Escherichia coli* and *Salmonella typhimurium*. *J. Bacteriol.* **170:** 190–196.
89. Prolla, T. A., Q. Pang, E. Alani, R. D. Kolodner, and R. M. Liskay. 1994. MLH1, PMS1, and *MSH2* interactions during the initiation of DNA mismatch repair in yeast. *Science* **265:** 1091–1093.
90. Prudhomme, M., B. Martin, V. Mejean, and J. P. Claverys. 1989. Nucleotide sequence of the *Streptococcus pneumoniae hexB* mismatch repair gene: homology to *mutL* of *Salmonella typhimurium* and to *PMS1* of *Saccharomyces cerevisiae*. *J. Bacteriol.* **171:** 5332–5338.
91. Prudhomme, M., V. Mejean, B. Martin, and J. P. Claverys. 1991. Mismatch repair genes of *Streptococcus pneumoniae*: HexA confers a mutator phenotype in *Escherichia coli* by negative complementation. *J. Bacteriol.* **173:** 7196–7203.
92. Pukkila, P. J., J. Peterson, G. Herman, P. Modrich, and M. Meselson. 1983. Effects of high levels of DNA adenine methylation on methyl-directed mismatch repair in *Escherichia coli*. *Genetics* **104:** 571–582.
93. Rasmussen, L. J., A. Løbner-Olesen, and M. G. Marinus. 1995. Growth-rate-dependent transcription initiation from the *dam* P2 promoter. *Gene* **157:** 213–215.
94. Rasmussen, L. J., M. G. Marinus, and A. Løbner-Olesen. 1994. Novel growth rate control of *dam* gene expression in *Escherichia coli*. *Mol. Microbiol.* **12:** 631–638.
95. Rasmussen, L. J. and L. Samson. 1996. The *Escherichia coli* MutS DNA mismatch binding protein specifically binds O^6-methylguanine DNA lesions. *Carcinogenesis* **17:** 2085–2088.
96. Roberts, D., B. C. Hoopes, W. R. McClure, and N. Kleckner. 1985. IS10 transposition is regulated by DNA adenine methylation. *Cell* **43:** 117–130.
97. Samson, L. D. 1992. The repair of DNA alkylation damage by methyltransferases and glycosylases. *Essays Biochem.* **27:** 69–78.
98. Schaaper, R. M. and M. Radman. 1989. The extreme mutator effect of *Escherichia coli mutD5* results from saturation of mismatch repair by excessive DNA replication errors. *EMBO J.* **8:** 3511–3516.
99. Schellhorn, H. E. and K. B. Low. 1991. Indirect stimulation of recombination in *Escherichia coli* K-12: dependence on *recJ*, *uvrA*, and *uvrD*. *J. Bacteriol.* **173:** 6192–6198.
100. Schlensog, V. and A. Bock. 1991. The *Escherichia coli* fdv gene probably encodes *mutS* and is located at minute 58.8 adjacent to the *hyc-hyp* gene cluster. *J. Bacteriol.* **173:** 7414,7415.
101. Selva, E. M., L. New, G. F. Crouse, and R. S. Lahue. 1995. Mismatch correction acts as a barrier to homeologous recombination in *Saccharomyces cerevisiae*. *Genetics* **139:** 1175–1188.
102. Sohail, A., M. Lieb, M. Dar, and A. Bhagwat. 1990. A gene required for very short patch repair in *Escherichia coli* is adjacent to the DNA cytosine methylase gene. *J. Bacteriol.* **172:** 4214–4221.
103. Su, S. S., R. S. Lahue, K. G. Au, and P. Modrich. 1988. Mispair specificity of methyl-directed DNA mismatch correction *in vitro*. *J. Biol. Chem.* **263:** 6829–6835.
104. Su, S. S. and P. Modrich. 1986. *Escherichia coli mutS*-encoded protein binds to mismatched DNA base pairs. *Proc. Natl. Acad. Sci. USA* **83:** 5057–5061.
105. Tsui, H. C., H. C. Leung, and M. E. Winkler. 1994. Characterization of broadly pleiotropic phenotypes caused by an *hfq* insertion mutation in *Escherichia coli* K-12. *Mol. Microbiol.* **13:** 35–49.

106. Tsui, H. C., G. Zhao, G. Feng, H. C. Leung, and M. E. Winkler. 1994. The *mutL* repair gene of *Escherichia coli* K-12 forms a superoperon with a gene encoding a new cell-wall amidase. *Mol. Microbiol.* **11:** 189–202.
107. Tsui, H. C. T. and M. E. Winkler. 1994. Transcriptional patterns of the MutL-MiaA superoperon of *Escherichia coli* K-12 suggest a model for posttranscriptional regulation. *Biochimie* **76:** 1168–1177.
108. Wagner, R. J. and M. Meselson. 1976. Repair tracts in mismatched DNA heteroduplexes. *Proc. Natl. Acad. Sci. USA* **73:** 4135–4139.
109. Welsh, K. M., A. L. Lu, S. Clark, and P. Modrich. 1987. Isolation and characterization of the *Escherichia coli mutH* gene product. *J. Biol. Chem.* **262:** 15,624–15,629.
110. Wilson, T. M., R. Ewel, J. R. Duguid, J. N. Eble, M. K. Lescoe, R. Fishel, and M. R. Kelley. 1995. Differential cellular expression of the human *MSH2* repair enzyme in small and large intestine. *Cancer Res.* **55:** 5146–5150.
111. Witkin, E. M. 1956. Time, temperature and protein synthesis: a study of ultra-violet mutation in bacteria. *Cold Spring Harbor Symp. Quant. Biol.* **21:** 123–140.
112. Wood, S. G., A. Ubasawa, D. Martin, and J. Jiricny. 1986. Guanine and adenine analogues as tools in the investigation of the mechanisms of mismatch repair in *E. coli*. *Nucleic Acids Res.* **14:** 6591–6602.
113. Worth, L. J., S. Clark, M. Radman, and P. Modrich. 1994. Mismatch repair proteins MutS and MutL inhibit RecA-catalyzed strand transfer between diverged DNAs. *Proc. Natl. Acad. Sci. USA* **91:** 3238–3241.
114. Wu, T. and M. G. Marinus. 1994. Dominant negative mutator mutations in the *mutS* gene of *Escherichia coli*. *J. Bacteriol.* **176:** 5393–5400.
115. Xiao, W., L. Rathgeber, T. Fontainie, and S. Bawa. 1995. DNA mismatch repair mutants do not increase N-methyl-N'-nitro-N-nitrosoguanidine tolerance in O^6-methylguanine DNA methyltransferase-deficient yeast cells. *Carcinogenesis* **16:** 1933–1939.

12

Translesion DNA Synthesis

Zafer Hatahet and Susan S. Wallace

1. INTRODUCTION

At the DNA level, the biological consequences of individual DNA lesions are the product of the efficiency of repair of the lesion and the interaction of any unrepaired damages with the DNA replication apparatus. Whether or not a DNA polymerase can bypass a particular cellular DNA damage (translesion synthesis) and which base it inserts if it does are the ultimate determining factors governing the potential mutagenicity or carcinogenicity of that lesion. Clearly, the most readily bypassed lesions that have the greatest propensity to miscode are the most important lesions from a biological point of view. The classic example is uracil derived from deamination of cytosine, which is bypassed with 100% efficiency and miscodes (as T not C) with 100% efficiency; fortunately, it is efficiently repaired (reviewed in ref. *218*).

DNA synthesis past a wide variety of DNA damages has been studied both in vitro and in vivo (for reviews *see* refs. *87*, *153*, and *258*) and has led to the conclusion that the ability of a DNA polymerase to transverse a lesion is complex and depends on the structure of the particular lesion, the DNA polymerase used, and the sequence context within which the lesion is presented. These three aspects of translesion synthesis will be discussed in the three sections of this chapter. Although all three parameters influence whether or not a particular lesion will be bypassed and which base is inserted, the major determining factors for bypass efficiency are the presentation of the lesion and the polymerase used. In general lesions that have a poor ability to form base-pairing interactions with an incoming nucleotide, or lesions that cause significant structural distortion or alter the base-stacking properties of the DNA molecule can be strong blocks to DNA polymerases, thus reducing translesion synthesis. However, exceptions to this generality are widespread and polymerase-dependent. Lesion structure is also a primary determinant governing the coding properties of the lesion, which again is modulated by the polymerase and the surrounding sequence. The relationships between lesion structure and translesion synthesis are discussed in Section 2. Because the ternary complexes formed among the lesion-containing template/primer, the incoming dNTP, and the DNA polymerase active site vary from polymerase to polymerase, it is easy to envisage that the efficiency of translesion synthesis would depend on the particular polymerase used. The various parameters governing DNA polymerase interactions with normal template/primers, mismatched base pairs, and lesions are important

From: DNA Damage and Repair, Vol. 1: DNA Repair in Prokaryotes and Lower Eukaryotes
Edited by: J. A. Nickoloff and M. F. Hoekstra © Humana Press Inc., Totowa, NJ

for model building and are discussed in Section 3. Finally, the microstructure of the surrounding DNA sequence is influenced by the structure of the lesion, which in turn influences the structure of the DNA polymerase-primer/template–dNTP ternary complex. Accordingly, surrounding DNA sequences affect the efficiency of translesion synthesis past the lesion as well as the coding characteristics of the lesion, and these context effects are discussed in Section 4.

2. LESION STRUCTURE PERSPECTIVE

2.1. Background

With the advent of sophisticated nuclear magnetic resonance (NMR) methods that are able to resolve the structures of oligonucleotides containing specific modified bases, this past decade has supplied a wealth of information on the solution structures of a variety of important mutagenic and carcinogenic DNA lesions. These data provide a structural framework for the biochemical and biological observations made over the past two decades with damage-containing DNA. Although it is not the purview of this chapter to document all of the structural data that have accumulated, or to survey all the lesions and their biological consequences, pertinent information relevant to several well-studied lesions, for which structural information is available and which have been examined with respect to translesion synthesis, will be discussed. Both NMR and X-ray crystallographic determination of lesion structures in DNA are extremely valuable as a point of departure for understanding interactions between damage-containing DNA molecules and DNA polymerases. NMR studies are particularly useful because they provide solution structures of the molecules that may approximate the lesion-containing DNA in the cell. One important caveat with respect to the structures solved thus far is that, with one exception, the lesions have been placed in the center of small 9–12 bp duplex DNA molecules, which is not the structure that the DNA polymerase encounters when it reaches a lesion during replication. In contrast, in the biochemical studies of translesion synthesis, an oligonucleotide or DNA molecule is used in single-stranded form, with the polymerase bound to the newly synthesized duplex primer terminus region and the single-stranded DNA template downstream. Although in theory, the latter structure more closely approximates the expected structure at the cellular replication fork, comparisons to structural studies are compromised. Also, because of technical limitations, the NMR structures are determined in the absence of the DNA polymerase molecule. However, and very importantly, structural analysis of lesion-containing duplex molecules provides insight into the miscoding properties of the lesion. Despite the shortcomings of the structural approaches, both NMR and X-ray crystallography are extremely important tools to guide the thinking of biochemists and biologists who are interested in the consequences of particular DNA lesions.

2.2. Free Radical-Induced DNA Base Damages and Sites of Base Loss

The solution structures, and in some cases the crystal structures, of several well-studied oxidative base lesions in DNA have been determined. The thymine products investigated include the ring-saturation product, thymine glycol (Tg), the ring fragmentation product, urea, and 5-hydroxymethyl-uracil (5-HMU). The structures of the major purine oxidation products, 7,8-dihydro-8-oxoguanine (8-oxoG) and 7,8-dihydro-

8-oxoadenine (8-oxoA), have also been determined. These represent only a small fraction of the 100+ different oxidation products, yet in all cases, the structures determined in DNA are consistent with the interpretations of both the biochemical and biological experimental data where translesion synthesis was measured.

2.2.1. *Thymine Glycol*

Tg is produced by ionizing radiation (*19,91,268*; *see* Chapter 5, vol. 2) and a variety of chemical oxidants, such as potassium permanganate and osmium tetroxide. For the latter two chemicals, the *cis*-stereoisomer (5R, 6S) of Tg is the predominant product of DNA oxidation *(72,73)*. Tgs are also a frequently produced biological lesion; Ames and coworkers have estimated that during normal oxidative metabolism, about 320 Tgs are formed/human genome/day *(26)*. Of all the oxidatively modified DNA bases investigated to date, Tg produces the most distortion to the DNA molecule, although it pairs correctly with A. Proton NMR spectroscopy of a duplex oligonucleotide containing Tg shows that the base portion of Tg and the adenine opposite appear to be extrahelical, indicating that the Tg moiety exerts a major structural perturbation to the DNA molecule *(121)*. Molecular modeling has also suggested that Tg exerts a significant effect on DNA structure. Early studies showed that Tg destabilized the base pair on the 5'-side of the lesion but the Tg-A pair was energetically stable *(36)*. More recent molecular dynamics simulations bear out this original observation and clarify it further. Simulations of 5-hydroxy-5,6-dihydrothymine, 6-hydroxy-5,6-dihydrothymine, 5,6-dihydrothymine, and Tg lend support to the idea that the hydroxy group at the 5 position causes a pseudo-axial-oriented methyl group that results in a distortion of the base pair 5' to both Tg and 5-hydroxy, 5,6-dihydrothymine *(169,172)*. In fact, termination of primer extension occurs at the lesion site with A inserted opposite Tg *(35,93,104,221)*. In the presence of proofreading, extensions can proceed no further, supporting the structural studies showing the major distortions to occur at the base pair 5' to Tg. Structural distortion is substantially less for 6-hydroxy-5,6-dihydrothymine and 5,6-dihydrothymine *(169,172)*, which do not contain a C^5 hydroxyl substituent, suggesting that the substituent at the 5 position plays a major role in perturbing DNA rather than the half-chair conformation assumed by all pyrimidine ring-saturation products. This is in keeping with the experimental data showing that, unlike Tg, which strongly blocks translesion synthesis, the other ring-saturation products examined, dihydrothymine, dihydrouracil, and uracil glycol, are readily bypassed *(107,206)*. There is, however, suggestive evidence that 6-hydroxy-5,6-dihydrothymine inhibits *Escherichia coli* DNA polymerase I (pol I) *(77)*. It should be noted that, in contrast to the NMR studies *(121)*, the molecular mechanics and dynamics simulations show Tg to remain intrahelical pairing with A *(169,172)*. Although the biochemical and biological data are consistent with Tg being a structurally distorting lesion, they are not consistent with it being in an extrahelical form, as determined by its solution structure (at least during replication). To clarify, no (or very few) single base deletions are observed during in vitro and in vivo translesion synthesis past Tg, which would be expected if the extrahelical configuration of the Tg-A pair predominated during synthesis. In contrast, acetylaminofluorine DNA adducts stabilize an extrahelical bulge, and one and two base deletions predominate (*see* Section 2.3.3.). The apparently contradictory observations between the structural and functional observations with Tg exemplify the issue pointed out in Section 1., that is, the solution structure was determined with Tg in a duplex structure, which is not the structure encountered during DNA synthesis.

Because Tg is a strong block to DNA polymerases, it is likely to be a potentially lethal lesion. This has been verified by measuring the inactivation efficiency of Tg in (unrepairable) single-stranded DNA *(1,94)*; Tg is also lethal in duplex DNA *(141,182)*. Because Tg pairs with A, it is a poor premutagenic lesion. When the block to replication is lifted, as in SOS-induced *E. coli* (*see* Chapter 7), no mutations are observed at Tg sites *(94)*. However, when Tg was inserted into a bypass sequence (*see* Section 4.), a low level of misincorporation of G opposite Tg occurred giving rise to mutations *(5)*.

2.2.2. *Urea*

The solution structure of urea, a fragmentation product of oxidized thymine, has also been determined *(82)*. Although urea and abasic (AP) sites (Section 2.2.6.) share many similar features, including lack of stacking, urea, by virtue of its size, is more capable of forming hydrogen bonds than an AP site. NMR spectroscopic analysis of a duplex dodecamer containing urea opposite T shows that both urea and its partner T are stacked into the helix *(82)*. These authors *(82)* predict that the *trans*-isomer of urea would pair with T with two hydrogen bonds, whereas the *cis*-isomer is highly unfavorable for the formation of hydrogen bonds. The urea-T structure also exhibits greater thermal stability than a similar structure containing an AP site instead of urea, lending further support to this suggestion. Guanine can potentially form the same pattern of hydrogen bonding with urea as thymine; however, the NMR spectrum of this combination could not be resolved. Biochemical and biological studies have shown that urea, a poorly coding and stacking lesion, is a strong block to translesion synthesis *(104)* and is lethal in vivo *(141)*. In vitro studies demonstrate that when urea is bypassed, purines are the preferred inserted bases *(107)*. Mutational studies of urea derived from thymine *(158)* show that guanine is the major residue inserted opposite urea during translesion synthesis, although some thymine pairs were observed. (Adenine insertions are not mutagenic and therefore were not detected in this assay.)

2.2.3. *5-Hydroxymethyluridine*

Another common oxidative product of thymine, 5-HMU *(74,91,268)*, has been analyzed by NMR *(165)*. When paired with A, 5-HMU assumes normal Watson-Crick geometry, and the global conformation of the duplex is normal B form DNA with no distortion observed. This is expected, since 5-HMU completely substitutes for T in phage SPO-1 *(256)*. When 5-HMU is paired with G at acidic and neutral pH, the pair is in a wobble configuration; at basic pH, 5-HMU ionizes and the pair assumes a Watson-Crick configuration. The expected infrequent ionization of the 5-HMU-G pair at neutral pH is in keeping with the observation that 5-HMU pairs correctly in vitro *(97)* and is a poor premutagenic lesion *(147)*.

2.2.4. *Other Pyrimidine Oxidation Products*

Translesion synthesis past other pyrimidine oxidation products has been examined biochemically, but no structures have been resolved to date. Dihydrothymine (DHT) *(107)*, uracil glycol (Ug) *(206)*, dihydrouracil (DHU) *(206)*, 5-hydroxycytosine (5-OHC) *(207)*, and 5-hydroxyuracil (5-OHU) *(207)* are readily bypassed by DNA polymerases. All pair primarily with their cognate base, although mispairing of 5-OHC and 5-OHU occurs with A and C, and C, respectively. 5-OHC has been shown to be mutagenic *(66)*, whereas DHT is not *(158)*. Cytosine products oxidatively deaminated to uracil products, such as DHU, Ug, and 5-OHU, should be potent premutagenic lesions, since they all pair with A. The nucleoside triphosphate forms of DHT *(105,106)*, DHU *(206)*, 5-OHC *(207,208)*, and 5-OHU *(207,208)* are all readily incorporated into DNA with pairing properties similar to those determined when they are present in template DNA. In contrast, the ring-open product of DHT, β-ureidoisobutyric acid (UBA), is a strong block to DNA polymerases in vitro *(107,163)*. When bypassed in vitro, purines and T are inserted *(107)*; in vivo, when mutations are measured, T is also inserted opposite with T → A transversions predominating *(158)*. This is in contrast to urea (*see* Section 2.2.2.), where purines are inserted both in vitro *(107)* and in vivo *(158)*, and is in keeping with structural differences between these two lesions, that is, UBA is larger and contains more hydrogen bonding potential than urea.

2.2.5. 8-Oxopurines

8-OxoG and 8-oxoA are the major stable purine products produced by oxidation and radiolysis (*91,122,268; see* Chapter 6, this vol., and Chapter 5, vol. 2). Metabolic oxygen has been estimated to be responsible for the formation of 168 8-oxoG lesions/human genome/d *(243)*. The structural characteristics of both 8-oxoG and 8-oxoA have been examined by NMR spectroscopy and X-ray crystallography. As determined by NMR, 8-oxoG opposite C in a self-complementary dodecanucleotide *(191)* exhibits an overall structure that is very similar to the unmodified duplex. The base exists in a 6,8-diketo tautomeric form, assuming an *anti* conformation and is appropriately base paired to cytosine. NMR analysis of 8-oxoG opposite adenine *(133)* shows 8-oxoG to assume a *syn* conformation forming a Hoogsteen base pair with adenine in the *anti* alignment. In the case of both C and A opposite 8-oxoG, the stable base pairs formed are intrahelical. Essentially identical conclusions were reached after X-ray crystallographic analysis of similar constructs containing 8-oxoG paired with either A or C *(152,164)*, and molecular mechanics calculations *(203)* are also in basic agreement with the NMR and crystallographic results. The structural data showing that stable interactions are possible between 8-oxoG and both C and A are in keeping with biochemical data showing that 8-oxoG is readily bypassed during DNA synthesis, and that C or A is inserted opposite with insertion and extension rates dependent on the polymerase used *(236)*. Incorporation of 8-oxodGTP into DNA has also been observed and the modified nucleotide, again, pairs with A or C *(32,162,208)*. As these data predict, 8-oxoG is a potent premutagenic lesion giving rise to the G → T transversions from template 8-oxoG found in both bacterial *(32,184,280)* and mammalian *(127,185)* cells, as well as A → C transversions *(32,198)* from misincorporation of 8-oxodGTP opposite A.

NMR spectra of 8-oxoA in a duplex nonamer revealed that the modified base exists predominately in the keto form and is in the *anti* position when paired with T *(90)*. T pairings with 8-oxoA are the predominant pairing events found when translesion synthesis is measured *(90)*. The crystal structure of a duplex dodecomer with 8-oxoA opposite G has also been determined *(146)*. Here, with G *anti* and 8-oxoA *syn,* a wobble configuration is adopted, and it was suggested that the G-8-oxoA base pair is held together by four hydrogen bonds *(146)*. 8-OxoA-G pairs are formed at a very low frequency during in vitro DNA synthesis *(90,238)* and in one sequence context in *E. coli*

(281). However, in NIH3T3 cells, 8-oxoA was as mutagenic as 8-oxoG *(120)* at least in a single-sequence context.

2.2.6. Sites of Base Loss

The AP site, the most common DNA lesion produced by depurination at the level of 10,000/human genome/d *(148,149)*, has been investigated by proton and phosphorous NMR. The earliest three-dimensional structure determined by NMR used a nonamer duplex containing a stable AP site analog, tetrahydrofuran (furan) *(45)*. These data showed that the furan opposite unpaired adenine assumed all of the aspects of classical B-DNA with both the furan and the unpaired adenine lying inside the helix. Furthermore, little distortion and no change in the melting temperature of the furan-A nonamer compared to the nonamer containing the T-A pair were observed. When the structure of duplex oligomers containing furan opposite G, T, and C were examined *(46)*, furan-G was also intrahelical; furan-T was present in two species, intrahelical and extrahelical; whereas furan-C was extrahelical. An independent NMR analysis of furan opposite A *(116)* also showed the duplex nonamer to assume a normal B conformation with both the deoxyribose and the A opposite being inserted into the helix; however, significant distortion was observed. A more recent NMR study investigated the conformational and dynamic properties of a thermally stable 11-bp duplex containing a naturally occurring aldehydic AP site opposite A or G *(86,277)*. Again, the opposed bases were stacked into the helix. In contrast to one of the studies with furan *(116)*, there was little distortion observed with the natural AP site, with disruption extending only to the adjacent base pairs. The observation that aldehydic AP sites appear to be less distorting than furan may explain observed differences found during translesion synthesis past these lesions (*see* Section 4.).

AP site interactions with DNA polymerases were among the earliest studied *(235)*. In general, AP sites are strong blocks to in vitro *(98,136,224,225)* and in vivo *(229)* DNA synthesis (reviewed in ref. *87* and *154*). In vitro, in the presence of proofreading, termination sites are observed one base prior to the lesion *(224,225)*; this steady-state observation presumably results from the exonucleolytic removal of the unpaired nucleotide opposite the AP site, since in the absence of proofreading, termination sites occur opposite the lesion. Similar observations have been made with furan *(261)*. The extent of bypass of AP sites derived from purines has been correlated to mutation frequency at that site *(136,137)*. AP sites are potentially lethal *(63,141,228,229)*; however, when AP sites are bypassed in vitro and in vivo, A is preferentially inserted followed by G or T, depending on the sequence context and polymerase used, thus resulting in mutations *(16,23,24,136,143,187,211,224,225,228,229)*. The relative preference of these insertions is in agreement with the structural data where purines opposite an AP site give rise to stable intrahelical structures, with T opposite producing a less stable structure; and C opposite, an unstable structure, although in the yeast, *Saccharomyces cerevisiae*, C is the preferred insertion *(85)*. Thus the original "A rule" hypothesis, which suggested that polymerases prefer to insert A opposite "noninstructional" lesions (*257*; reviewed in ref. *260*), appears to be superceded by a modified "structural hypothesis" suggesting that "noninstructional" lesions are indeed instructional with the inserted base being governed by the stability of the "lesion pair" in a particular surrounding sequence in the polymerase-primer/template-dNTP ternary complex.

2.3. Alkylation Products, Bipyrimidine Dimers, and Bulky Chemical Adducts

There is a wealth of information on the structures of the plethora of so-called bulky mutagenic and carcinogenic lesions, as well as information on their interactions with DNA polymerases. For this chapter, the discussion will be limited to the less bulky alkylation products, O^6-methylguanine (O^6-meG) and O^4-methylthymine (O^4-meT); the principal UV-induced photoproduct, the cyclobutane pyrimidine dimer (CPD); and several products of guanine-reactive bulky carcinogens, benzo(*a*)pyrene diol epoxide (BPDE) and the arylamines, 2-aminofluorene (AF) and 2-acetyl-2-aminofluorene (AAF). A more complete discussion of carcinogen adducts and their interaction with DNA can be found in Chapter 4, vol. 2.

2.3.1. O^6-*meG and* O^4-*meT*

The structural characteristics of the important mutagenic and carcinogenic products formed by alkylation of guanine O^6 and thymine O^4 have been elucidated. NMR analysis of the O^6-meG-C pair in a self-complementary dodecanucleotide shows slight distortions in the vicinity of the lesion with both bases in the *anti* conformation *(194)*. The data suggest that the O^6-meG forms a wobble pair with C, with the O^6-meG sliding toward the major groove and the C toward the minor groove. Wobble pairing between O^6-ethylG and C was found by X-ray crystallographic analysis of a similar dodecamer *(252)*. Although the O^6-meG -T mispair appears to be somewhat less thermodynamically favorable than the O^6-meG-C pair *(76)*, the X-ray crystallographic results show that O^6-meG forms a normal Watson-Crick pair with thymine *(145)*. In the minor groove, the O^6-meG -T pair is indistinguishable from a normal G-C pair, whereas the O^6-methyl group protrudes somewhat into the major groove. Both NMR *(195)* and X-ray crystallography *(145)* show that both bases are in the *anti* configuration, and only minor perturbations are observed at the damage site. Similarly, when the O^4-meT-A correct pair was analyzed by NMR *(117)*, both bases adopted the normal *anti* configuration, but formed a wobble base pair. In contrast, for the O^4-meT-G mispair *(118)*, the data suggest that a Watson-Crick pair is formed with one short hydrogen bond between the N^2 of G and the 2-carbonyl of O^4-meT.

The *O*-alkyl pyrimidines are major players in mutagenesis and carcinogenesis (for reviews, *see* refs. *4, 6,* and *245*). The structural data compiled from studies with O^6-meG-C or -T pairs and O^4-meT-A or -G pairs demonstrate that little distortion is associated with the presence of the methyl group. However, the mispairs O^6-meG-T and O^4-meT-G form Watson-Crick hydrogen bonds in the correct B-DNA geometry, whereas the "correct" pairs form wobble base pairs, suggesting a propensity for O^6-meG and O^4-meT to mispair. Neither O^6-meG nor O^4-meT are strong blocking lesions to most DNA polymerases *(56,247)*, although pause sites are observed *(167,276)* and DNA polymerase β (pol β) is completely blocked by O^6-meG in a single sequence context *(276)*. Interestingly, in some sequence contexts (*see* Section 4.), the above "mispairs" are favored 10-fold over the correct pairs using the Klenow fragment of *E. coli* DNA polymerase I (K_f), whereas in other contexts, they are not, with misinsertions favored in some cases and extensions of mispairs in others *(53,54,246,247)*. Bypass of O^6-meG in vivo also leads to a significant number of T insertions and thus mutations *(13,28,61,62,155)*, although replication blockage leading to strand bias has been observed *(27,197)*. A similar story can be told for O^4-meT, only here the mutagenic G is inserted opposite giving rise to T → C transitions *(55,204,205)*. As has been demonstrated for other nonblocking lesions, the nucleotide triphosphate forms of O^6-meG and O^4-meT are readily incorporated into DNA by DNA polymerases pairing with the predicted correct and incorrect bases *(244,249,272)*.

2.3.2. Cyclobutane Pyrimidine Dimers

The CPD is the prototypic DNA lesion that has been studied for many years using both biochemical and cellular approaches, and was long considered to perturb significantly DNA structure and miscode. However, the NMR solution structures of a duplex octamer containing a thymine dimer produced by UV irradiation *(124,125)* and a duplex decamer containing a synthesized *cis-syn* thymine dimer *(265)* show that the naturally occurring *cis-syn* thymine dimer produces rather small distortions in the B-DNA structure with the main conformational change occurring at the dimer site. Furthermore, both thymines in the dimer remain hydrogen bonded to their complimentary As, although these base pairs are substantially weakened. When chemically synthesized *cis-syn* and *trans-syn* dimers were compared using NMR *(265)*, the small perturbations produced by the *cis-syn* product are more to the 3'-side of the dimer, whereas the *trans-syn* causes a much greater perturbation, possibly in the form of a kink, on the 5' side of the dimer. Molecular mechanics analysis of a complementary dodecamer also showed no gross distortion in the double helix produced by the *cis-syn* thymine dimer *(212)*, although a second molecular mechanics analysis of a *cis-syn* thymine dimer, using constraints from the X-ray crystallographic analysis of the dinucleotide, gave a significantly different result, that is, a substantial kink was observed at the site of the lesion *(199)*.

CPDs are effective blocks to DNA synthesis catalyzed by a variety of DNA polymerases with termination sites occurring primarily 1 base prior to the lesion *(29,178,180,275)*. Although the structural studies using duplex oligomers suggest that correct pairing occurs with CPDs opposite As, some local distortion occurs, perhaps explaining its ability to block DNA polymerases. It certainly appears that the encountered CPD on single-stranded DNA must present a significant obstacle to polymerization. In keeping with the structural observations that CPDs pair correctly, when bypass is forced in vitro, 95% of the time A is incorporated opposite *(209, 266)*. T-T CPDs are also poor premutagenic lesions. In *E. coli*, a *cis-syn* T-T dimer, site-specifically introduced, coded accurately for A 93% of the time *(2)*. Results with the *trans-syn* T-T dimer showed that translesion synthesis was more efficient than with the *cis-syn* dimer, with replication being more accurate in the absence of SOS induction and less accurate in its presence *(3)*. The fact that the *cis-syn* dimer blocks DNA synthesis more efficiently than the *trans-syn* dimer suggests that perturbation 3' to the lesion prevents the formation of the appropriate ternary complex more effectively than distortion 5' to the lesion. Both the extent and accuracy of T-T dimer bypass vary between yeast and *E. coli (83)*, again emphasizing the role of the polymerase (*see* Section 3.).

Most UV-induced mutations occur at C-containing bipyrimidine sites, leading to C → T transitions *(69,151,264,282)*; however, no C-containing bipyrimidine structures have been solved yet. The T-C CPD has been postulated to give rise to mutation during translesion synthesis by two potential mechanisms, the imino tautomer of C pairing with A or the deamination of C to U, which then pairs with A *(112,200,266,269)*. Another UV-induced bipyrimidine product, T-C pyrimidine (6-4) pyrimidone [(6-4)PD], is a potent premutagenic lesion that blocks translesion synthesis in vitro *(29)* but when bypassed, miscodes leading to C → T transitions *(101)*. Interestingly, the T-T (6-4)PD product is highly mutagenic in *E. coli (144)* and mammalian cells *(81)*, but much less so in yeast *(84)*.

2.3.3. Bulky Chemical Adducts

The bulky chemical adducts, BPDE, AF, and AAF exhibit a preference for binding to G. The solution structures of four stereoisomers of *anti* benzo(*a*)pyrene-G adducts in duplex oligomers *(39,40,49,70)* and in oligomers where the adduct was positioned opposite a deletion site *(41,42)* have been determined by NMR. The carbon-10 (C^{10}) of BPDE, the activated form of the carcinogen, binds to the N^2 extracyclic amino group of guanine either *cis* or *trans* relative to the OH at C^9. The particular adduct formed depends on the reaction conditions, and there are also specific sequence context effects. For example, a G within a run of Gs is severalfold more reactive toward anti-BPDE than a G with non-G neighbors *(111)*. The benzopyrene moiety of the *trans* adduct, the common adduct, is located in a widened minor groove of the DNA duplex with a pyrene ring oriented towards the 5' (the strong carcinogen, [+]-*trans-anti*, C^{10} S) end of the modified strand or the 3' (the weak carcinogen, [–]-*trans-anti*, C^{10} R) end of the modified strand depending on which enantiomer is examined. A minimally perturbed B-DNA helix and correct Watson-Crick pairing for BPDE-G-C and surrounding pairs were observed for both *trans-anti* adducts *(39,49,70)*. In contrast, when positioned opposite a deletion site, the (+)-*trans-anti*-BPDE-G adduct was intercalated into the B-DNA duplex with the G displaced into the major groove *(41)*. In the *cis* adducts, the pyrenyl group is quasi-intercalated with the guanine ring of the adduct displaced into the minor groove *(40)*. A similar structure was observed for (+)-*cis-anti*-BPDE opposite a deletion site *(42)*. Recently, the solution structure of the (+)- and (–)-*trans-anti*-BPDE-G adducts was determined in a DNA-template-primer junction (13/9 nucleotides [nt]) *(43)*. The primer (9 nt) did not include the base opposite the adduct. In the 9-bp duplex, all the base pairs were in Watson-Crick hydrogen-bonded alignments. Conformational heterogeneity was observed 5' to the adduct with the adducted guanine assuming a *syn* conformation and displaced into the major groove, no longer stacking over the adjacent base pair, which now stacks with one face of the pyrenyl ring. This study, although an important undertaking, points out the major stumbling block to solving structure–function relationships: the lack of structural information on lesion-containing DNA in the presence of the polymerase molecule.

BPDE also adducts to the N^2 of adenine, but much less frequently than to guanine. One of the stereoisomers of a BPDE-A adduct was positioned opposite G in a nonanucleotide duplex and the structure solved by NMR *(234)*. The carcinogen intercalated into the major groove oriented toward the 5'-side of the adducted strand. Both the adducted A and the mispaired G are in the *anti* form.

Because BPDE forms bulky DNA adducts that protrude into the minor groove of the DNA molecule, they might be expected to block DNA synthesis. This was shown in early studies with randomly introduced BPDE *(178)*. Major termination sites occurred 1 base 3' to the lesion. More recent studies, using a single BPDE-adducted G ([+]-*cis-anti* and [+]-*trans-anti*) showed that the *cis* BPDE-G adduct was a more effective block to Sequenase and human pol α than the *trans* adduct *(102)*. When primed up to the block, C was preferentially added opposite both adducted Gs, although traces of A were incorporated opposite the *trans* adduct *(102)*. The *trans* adduct is more effectively bypassed than the *cis* adduct, which is in keeping with the structural data showing that the base pairing face of the *trans* adduct is intact. When the four stereoisomers of *anti*-BPDE-G adducts were examined with *E. coli* pol I, all inhibited the polymerase, but when bypassed, the extent of deletions and substitutions opposite the lesion depended on the isomer *(241)*. The insertion of A opposite and the frequency of extension from the (–)-*trans-anti*-BPDE-A pair were much greater than from the other adducts. When four *trans* adducts of A were examined using primer extension *(34)*, all were effective blocks to DNA synthesis, again, with an efficiency dependent on the polymerase. In the cases where insertion occurred, T or A was incorporated depending on the polymerase. Another primer extension study with six BPDE stereoisomers of A adducts *(31)* showed that the extent of blockage as well as the inserted base was again polymerase-dependent with T4 pol most strongly inhibited and 3' → 5' exonuclease-free Klenow fragment (K_f exo$^-$) the least inhibited. Other than the DNA polymerases used, the R or S configuration of C^{10} of the pyrenyl ring was the principal determinant of the outcome. However, these two studies were not in agreement on either the extent of blockage or the base incorporated opposite the adduct.

In keeping with both their structural characteristics and their ability to block translesion synthesis in vitro, BPDE adducts are lethal *(103)*. In vivo mutational spectra with BPDE adducts have generally employed racemic mixtures of both (+) and (–)-*anti*-BPDE. The most prevalent mutations observed were G → T and A → T transversions *(12,20,60)* in agreement with the observed misinsertions in vitro. Interestingly, when optically pure enantiomers were used, the extent of mutagenicity of the enantiomer differed between bacterial and mammalian cells *(255,279)*. With the site-specifically introduced (+)-*trans-anti*-BPDE-G *(159)*, targeted G → T transversions predominated and were the only mutations produced in the 5'-TG-3' sequence *(159)*. Six stereoisomers of BPDE-A adducts introduced into codon 61 of N-*ras* generated exclusively A → G transitions with all isomers being mutagenic *(30)*. With both site-specifically introduced BPDE-G and BPDE-A adducts, >97% of the time that the adduct is bypasssed in vivo the correct base is inserted, suggesting flexibility of the polymerase molecule *(159)* when the hydrogen-bonding face of the adducted base is not disrupted.

The solution structures of the carcinogenic AF *(33,58,188)*, and AAF *(171,190)* (for a review, *see* ref. *95*) have been solved by NMR. Both AF and AAF adduct to the C^8 of guanine. NMR analysis of duplexes containing AF-G opposite C shows that in the major conformer, the AF moiety is positioned in the major groove of a slightly disturbed B-DNA duplex leaving the AF-C pair intact *(33,58)*. In a second conformer *(58)*, the AF is stacked within the helix displacing the modified G and disrupting the Watson-Crick pairs. NMR analysis of the AF-G-A mispair shows the AF-G to assume a *syn* conformation opposite A *anti,* with the aminofluorene positioned in the minor groove and Watson-Crick pairs on either side of the lesion *(188)*. Both duplex and heteroduplex structures of AAF-bound adducts have been determined by NMR. For the duplex structure, there was a predominant conformer (70%), which showed stacking of the fluorene moiety on the adjacent base pair with the distal ring of the fluorene moiety protruding into the minor groove with no evidence of base pairing between the AAF-G and C *(190)*. In the heteroduplex containing a bulged guanine, the AAF was external to the helix *(171)*. However, the AAF-modified heteroduplex appeared to be much more stable than the unmodified bulged duplex in agreement with its stability measured by melting temperatures *(78)*.

Termination of primer extension on AAF-modified DNA occurs one base prior to the adducted site, and strong termination bands are observed *(179,180,259)* even with pol III holoenzyme *(10)*. When bypass is forced, the specificity of incorporation depends on the polymerase and sequence *(210)*. In contrast, AF is not as strong a block to in vitro synthesis *(51,170,181,259)*. With T7 pol, the AF adduct is a block *(170)*; with T4 pol, a base is inserted opposite *(181)*; with K_f, AF is completely bypassed *(51,170)*; and with pol III holoenzyme, AF is partially bypassed *(51)*. C is incorporated opposite the AF-G adduct *(170)*.

In vivo, AAF adducts are lethal, whereas AF adducts are less so *(156,263)*. In fact, in duplex DNA transfected into nucleotide excision repair-deficient cells, over 80% of the AF adducts did not inactivate the DNA *(157,220)*. More than 90% of randomly introduced AAF mutations are frameshift mutations, whereas AF adducts, the more effective premutagenic lesion *(96)*, produce primarily (85%) base substitutions, mostly G $\rightarrow$ T transversions *(14,160,226)*. With site-specific adducts of AAF, both deletions and base substitutions are observed *(183)*. In the *Nar*I hotspot (GGCGCC), AAF adducts in the third G produce –2 base frameshifts *(21,129,267)*, whereas AF induces base substitutions *(267)*. In other site-specific studies with AF adducts, deletions were observed *(173)*. These data are consistent with the structural results that show more distortion of the DNA duplex by AAF than AF. Also, a bulge containing G-AAF is more stable than a bulged G *(171)*, suggesting that an inserted nucleotide opposite AAF is not elongated, but that the adduct promotes slippage and deletion *via* a stabilized intermediate *(78)*.

3. DNA POLYMERASE PERSPECTIVE

3.1. *Background*

Detailed understanding of translesion bypass would not be possible without accurate structural and mechanistic models of DNA synthesis in general. Fortunately, resolution of the crystal structures of several DNA polymerases, including K_f *(8)*, human immunodeficiency virus reverse transcriptase (HIVRT) *(132)*, rat DNA pol β *(48,227)*, and *Thermus aquaticus (Taq)* DNA polymerase *(126)*, has greatly contributed to a unified model for all members of the polymerase superfamily, including RNA polymerases. These proteins share a tertiary structure that is analogous to an half-open right hand with three distinct subdomains resembling the palm, thumb, and fingers *(114,115,254)*. In conjunction with protein mutagenesis studies, the palm subdomain is shown to contain the polymerase catalytic center, the thumb is responsible for binding double-stranded DNA upstream of the primer terminus, whereas the fingers bind the single-stranded template downstream of the primer terminus. Other functions, such as dNTP binding, involve amino acids from both the palm and finger subdomains. Although polymerases share functional analogy in their general organization, large structural variations exist between the subdomains of the crystallized polymerases, which may explain differences in their properties, such as template preference and level of fidelity.

Similar advances have been made in developing a mechanistic model of DNA synthesis largely owing to transient-state kinetic analysis of several DNA polymerases, including K_f *(47,59,135,174–176)*, T7 DNA polymerase (T7 pol) *(52,196,278)*, T4 DNA polymerase (T4 pol) *(25)*, and HIVRT *(123)*. Briefly, the fidelity of DNA synthesis has been shown to be the combined product of preferential incorporation of a correctly pairing nucleotide, preferential extension of a properly paired primer terminus, and when present, exonucleolytic proofreading of a mispaired primer terminus (for reviews, *see* refs. *57* and *113*). Misincorporation can result from failure in one or more of these steps. It has also been shown that, for several polymerases, different amino acids are involved in each of these steps, and in some cases, a single step may be modulated differently by several amino acids in the polymerase molecule. For instance, amino acid changes in either the polymerase or proofreading 3' → 5' exonuclease domains of T4 pol result in mutator or antimutator phenotypes *(75,215,251)*. Similarly, mutator and antimutator strains of *E. coli* can be produced by amino acid changes in *dnaE (67,68,161,177,192,233)* and *dnaQ (134,248)*, the genes coding for the polymerase (α) and proofreading exonuclease (ε) subunits of DNA pol III, respectively. More interestingly, two *dnaE* alleles, with different single amino acid substitutions, result in an antimutator phenotype by increasing the fidelity of DNA synthesis. One substitution reduces A → G transitions, whereas the other reduces G → A transitions *(233)*.

Although much work has focused on translesion DNA synthesis during the past two decades, detailed mechanistic studies are still lacking for most lesions and most DNA polymerases. Two exceptions are AP sites and the arylamines, AF and AAF. These have been studied both in vivo *(23,79,80,85,128,139,142,143,183,187,222,232,262,267)* and in vitro, and with both prokaryotic *(10,11,17,108,109,225,239,261,274)* and eukaryotic DNA polymerases *(202,271)*. Steady-state *(24,211,239,240)* and transient *(150)* kinetic analyses are available as well. Therefore, using these lesions as models, in conjunction with existing mechanisms of nucleotide misincorporation on natural DNA templates, it is possible to reconcile the large body of translesion synthesis literature and derive several basic rules that govern DNA polymerase interactions with template lesions.

3.2. The Ability to Bypass a Particular Lesion Is an Intrinsic Property of a DNA Polymerase

The function of a DNA polymerase is to arrange the various substrate molecules, namely the 3'-hydroxyl of a primer terminus, the α-phosphate of the incoming dNTP, and a pair of divalent metal ions, in the appropriate geometry that renders a phosphoryl transfer reaction energetically favorable *(253)*. The geometry was initially proposed to be that of B-DNA *(253)*. However, recent crystallographic data favor a conformation more similar to A-DNA *(7)*. This arrangement, however, can be provided by several protein structures that involve, to varying degrees, different amino acids in the palm, thumb, and fingers subdomains. Indeed, DNA polymerases have been grouped, based on primary sequence alignments, into three major families, A, B, and C, named for their homology to, respectively, DNA pol I and II, and the α-subunit of DNA pol III of *E. coli (18,110)*. When lesions are involved, on the template strand or as dNTPs, it is presumed that the efficiency of translesion synthesis greatly depends on how well the polymerase–primer/template–incoming nucleotide ternary complex approximates that in the absence of a lesion. Distortion in DNA structure introduced by a lesion may be offset by flexibility in the catalytic center of the polymerase. Since polymerases differ in the geometry of their active centers, it stands to reason that those with structurally homologous active centers may interact with lesions in a similar manner. The small number of available crystal structures, however, precludes verification of this hypothesis. On the other hand, there is strong experimental evidence to support the idea that the ability to bypass a particular lesion is strongly influenced by the intrinsic properties of the polymerase. Hoffmann et al. *(99)* have shown that a single d(GpG)-cisplatin adduct placed on codon 13 of the human *H-ras* gene strongly blocked in vitro DNA synthesis by calf thymus DNA pol α, δ, and ε. Pol β, on the other hand, was able both to bypass the lesion and rescue the termination products generated by the other polymerases. In another study, pol β was also shown to bypass several different platinum (II) monoadducts, which blocked synthesis by *E. coli* pol I, *Taq* polymerase, and calf thymus pol α *(100)*. In contrast, a single O^6-meG residue completely blocked in vitro synthesis by human pol β, but only partially blocked synthesis by human DNA pol α *(276)*. Templates containing one of several intrastrand crosslinked adducts induced by *cis*-diamminedichloroplatinum (II) were also shown to block in vitro synthesis by T4 pol or pol III holoenzyme, but only partially block synthesis by K_f or Sequenase *(38)*. Similarly, aminofluorine-modified guanine blocked in vitro synthesis by T7 pol, but not by pol I *(170,259)*. Finally, using gapped duplex shuttle vectors containing unique lesions in a single-stranded region, Lawrence and colleagues have recently shown that bypass of both thymine dimers and AP sites in vivo is much more efficient in *S. cerevisiae* than in *E. coli (83,85)*. These data are consistent with the proposal that structurally and functionally related polymerases interact with lesions in a similar manner. For example, the marked difference in the outcome of translesion synthesis between several polymerases and pol β is consistent with the latter polymerase's distinct structure *(115)* and primary sequence *(110)*.

3.3. The Mutagenic Potential of a Lesion Is Partially Determined by DNA Polymerase

A strong correlation between the coding properties of a lesion and the nature of the polymerase used in translesion DNA synthesis is both predicted by and consistent with the physical and mechanistic models that were briefly described in Section 3.1. If the success of translesion synthesis depends on how well the polymerase catalytic center offsets distortions (deviations from the catalytically favored DNA conformation) caused by a lesion in the polymerase–primer/template–dNTP ternary complex, then it would stand to reason that the dNTP in the complex also plays an important role. Interestingly, although the replicative T7 pol and T4 pol DNA polymerases discriminate against the binding of an incorrect dNTP during synthesis on natural templates by a factor of ~10^3 *(25,278)*, correct and incorrect dNTPs bind equally well in the ternary complex formed by the K_f *(59)*, a repair polymerase. Although largely dependent on chemical structure, the coding properties, and therefore mutagenic potential of several lesions have been shown to vary with different polymerases, both in vivo and in vitro. When Feig and Loeb *(65)* monitored the accuracy of in vitro DNA synthesis on oxygen radical-damaged phage M13 using a *lacZ* forward mutation assay, striking differences in the type, frequency, and sites of mutation were observed when either pol α or pol β was used. Also, the nucleotide incorporated opposite 8-oxoG or 8-oxoA during in vitro synthesis varied depending on the polymerase used. K_f favors incorporation of C opposite 8-oxoG, Pol δ and pol α favor the incorporation of A, whereas Sequenase, pol β and pol γ incorporate both A and C to varying degrees *(202,236,237,242)*. Similarly, although 8-oxoA codes exclusively for T when most polymerases are used, pol β also incorporates G, albeit to a much lesser extent *(237,238)*. Recent studies have shown that both T-T CPDs and T-T (6-4)PDs have different coding properties in *E. coli, S. cerevisiae,* and mammalian cells *(81,83,84)*, when copied in vivo from gapped duplex vectors. The *cis-syn* dimers and (6-4)PDs were less mutagenic in yeast than in *E. coli* or mammalian cells, whereas the reverse was true for the *trans-syn* dimers. More interestingly, T-T (6-4) PD replicated in mammalian cells induced semitargeted mutations at the base before the 5' T of the photoproduct *(81)*. When a similar system was used to address the coding properties of AP sites, striking differences were observed as well *(85,143)*. While A is preferentially incorporated opposite AP sites in *E. coli*, with some variability depending on sequence context (*see* Section 4.), C is highly favored in yeast independent of sequence context. Although these pairing properties of AP sites could be unique to the polymerase responsible for gap filling in yeast, they still demonstrate the ability of the polymerase to modulate the mutagenic potential of a lesion. AP sites introduced site specifically or randomly into both single-stranded and double-stranded shuttle vectors propagated in mammalian cells show a preference for the nucleotide inserted opposite the lesion, which is different from either yeast or *E. coli*. When the AP site was introduced site-specifically, all four nucleotides were inserted opposite it with equal frequency *(23,80)*, whereas G was most frequently inserted opposite the lesions which were randomly introduced *(187)*. Here again, in spite of the observed difference in coding properties between the site-specific and randomly introduced AP sites, which are likely owing to sequence context effects, it is clear that the polymerase plays a major role in determining the mutagenic potential of the lesion.

Finally, kinetic studies have shown that the role of a polymerase in modulating the mutagenic potential of a lesion is played at one or both of two steps in the DNA synthesis pathway. The first step involves the insertion of a mutagenic nucleotide opposite the lesion, and the second involves the extension of that pair. Although K_f incorporates A less efficiently than C opposite 8-oxoG, it extends the mutagenic pair more efficiently than the nonmutagenic one *(236)*. Pol α, on the other hand, favors both the formation and extension of 8-oxoG-A pairs over 8-oxoG-C pairs.

3.4. The Efficiency of Translesion Synthesis Is Influenced by Polymerase Processivity and Proofreading Exonuclease Activity

In the presence of certain lesions, a polymerase–template–dNTP complex that approximates that of B-DNA may not be possible, and such lesions would represent absolute blocks to most DNA polymerases. This could be owing in part to the absence of coding information, as in the case in AP sites where the polymerase fails to form a stable pair with the lesion *(24,109,211)*, or to structural distortions, which deem the nucleotide incorporated opposite the lesion unextendable, as in the case of Tg *(104,107,168,169)*. Kinetically, this is expressed as a reduction in the rate of phosphodiester bond formation, which represents forward movement relative to the rate of 3' → 5' exonuclease activity (reflecting the strength of proofreading) and the rate of dissociation from the template (reflecting processivity). Although different DNA polymerases exhibit variable levels of proofreading and processivity on natural templates, productive DNA synthesis requires that the rate of phosphodiester bond formation be higher than that of proofreading or dissociation. It is presumed, therefore, that any factor that maintains a favorable rate of phosphodiester bond formation over either proofreading or dissociation would effectively enhance the chance of lesion bypass. This prediction is largely borne out by experimental data. In vitro bypass of *cis-syn* and *trans-syn* T-T CPDs by pol δ is higher in the presence of PCNA, the accessory protein that confers higher processivity on the enzyme *(189)*. Similarly, the *E. coli* β,γ, single-strand binding protein complex, that confers processivity on pol III, enhances the ability of pol II to bypass AP sites in vitro *(17)*.

The negative effect of the proofreading 3' → 5' exonuclease activity on the efficiency of translesion synthesis has been demonstrated by comparing in vitro DNA synthesis past template AF adducts by T7 pol DNA polymerase deficient (Sequenase) or proficient in proofreading *(259)*. Although the latter enzyme was completely blocked by the lesion, the former readily bypassed it. Recently, an in vitro reconstituted T4 pol holoenzyme system has been used to address the influence of both proofreading, using polymerase subunits proficient or deficient in the 3' → 5' exonuclease activity, and processivity, by comparing holoenzyme with the core polymerase, on the efficiency of bypass of several oxidative lesions, including Tg, urea, and AP sites. The data showed that, as predicted, higher processivity and lack of proofreading enhanced the efficiency of lesion bypass with the latter having a slightly greater effect than the former *(92)*.

4. SEQUENCE CONTEXT PERSPECTIVE

4.1. Background

It has been argued, so far, that successful DNA synthesis, in the absence or presence of a lesion depends on the ability of the polymerase to anchor a primer/template-incoming dNTP in a catalytically favorable geometry. DNA conformation, however, is greatly influenced by nucleotide sequence *(50,217)*, and as a result, the degree of polymerase fidelity varies with sequence context. Biologically, this variation in spontaneous mutation can be observed as hot- and coldspots *(71,140,230,231)* and mechanistically as alterations in either or both of the kinetic parameters, K_m and V_{max}, of forming and extending a proper base pair vs a mispair *(166,201,216)*. Consequently, the extent to which a lesion affects local DNA structure is not expected to be independent of sequence context, which might further exaggerate or offset any deviation from a B-DNA conformation caused by the lesion. Sequence context effects on lesion presentation need to be addressed both when the DNA is in solution and when it is in the confines of the polymerase active site. The importance of the latter effect is obvious, yet the former effect is also important as it relates to polymerase binding during replication restarts, i.e., reinitiation of synthesis at a primer terminus blocked by a template lesion.

Sequence context effects on both the efficiency and mutagenic outcome of lesion bypass have been reported for all classes of DNA damage. Several studies have identified sequence contexts for successful bypass of several base oxidation products, which generally block DNA synthesis, including AP sites, urea, β-ureidoisobutyric acid and Tg *(37,63,92,93,104,107)*. With Tg and K_f, the effect is dramatic. In all but one sequence context, 5' C(Tg)Pu3', Tg constitutes an absolute block to DNA synthesis. In the bypassable context, bypass is 100% *(63,93)*. In addition, single-stranded phage containing randomly distributed oxidative lesions showed a nonrandom distribution of mutations at mutable sites when replicated either in vivo *(63,66,158)* or in vitro *(64)*. Similar results were seen with other types of DNA lesions, including AAF adducts *(222,232)*, polycyclic hydrocarbon adducts *(15,34,219)*, alkylation damage *(22,89,131,250)*, and UV damage *(193,223)*. Analysis of sequence context effects on lesion bypass have generally taken two approaches: studying the influence of the lesion's nearest neighbors, and searching for consensus sequences common to hot- or coldspots within large targets containing randomly distributed lesions.

4.2. Nearest-Neighbor Effects on Translesion Synthesis

The extent to which a lesion distorts local DNA structure depends on its ability to form hydrogen bonding interactions with a base on the complementary strand and, perhaps more importantly, to form proper stacking interactions with adjacent bases on the same strand. The latter consideration has prompted several studies that monitored translesion synthesis as a function of 3' or 5' nearest neighbors. Goodman and colleagues determined the kinetic parameters of synthesis past tetrahydrofuran, an AP site analog, as influenced by the 3' *(211)* or 5' neighbor *(24)*. Although the two studies used different DNA polymerases, Pol α in the former and HIVRT in the latter, it was clearly demonstrated that nearest neighbors influenced both the efficiency of the bypass and the choice of nucleotide incorporated opposite the lesion. This influence becomes even greater when lesion bypass follows a misalignment mechanism *(138)*. Similar effects were seen with AF- and AAF-modified guanine. These lesions have been shown to cause frameshift mutations through a misalignment mechanism in the hotspot sequence 5'GGCGCC and in runs of Gs. Interestingly, lesions at the third G in the hotspot sequence *(11)* or at the 3'-terminal position in a run of Gs *(186)* were much more mutagenic than in neighboring Gs. In addition, only the 3' neighbor modulated the mutagenic outcome when bases on both sides of the lesion were systematically altered *(130,186)*, consistent with the fact that the 3' neighboring pair contributes to base stacking interactions in the ternary complex. Direct correlations were elegantly demonstrated between the ability of AF and AAF to induce mutations in these hotspots and lesion effects on DNA conformation as judged by melting temperatures *(78)*, reactivity to modifying chemical agents *(9,11)*, and circular dichroism spectroscopy *(129)*. 5' Nearest-neighbor effects on the ability of several BPDE-modified G isomers to induce frameshift -1 deletions through a misalignment mechanism were also reported *(241)*. This lesion mispairs with A and, not surprisingly, induces –1 deletions when the 5' nearest neighbor is T. In vitro bypass of BPDE modified A by K_f, Sequenase, HIVRT, and pol α has also been shown to vary within two positions in SupF *(34)*. 8-OxoG introduced into two positions of *H-ras* codon 12 and transfected into NIH3T3 cells coded exclusively as T when at position 1 and both as G and as T when at position 2 *(120)*. Finally, the 3' neighbor of O^6-meG strongly modulated the K_m and V_{max} for both misincorporation of T opposite the lesion and extension of the resulting mispair *(246)*.

4.3. Larger Sequence Context Effects on Translesion Synthesis

In spite of the valuable information provided by studies of nearest-neighbor effects on lesion bypass, especially when kinetic data are generated, such studies fall short of defining sequence contexts that influence the mutagenic potential of a lesion. The range at which a lesion has an effect on local DNA structure and, conversely, the range at which neighboring nucleotides have an effect on lesion presentation undoubtedly varies depending on the chemical structure of the lesion. However, it is reasonable to presume that, at least in the template–polymerase–dNTP ternary complex, nucleotides that are in contact with the polymerase play a role in modulating bypass outcome. Physical and biochemical studies have indicated that single polypeptide polymerases, such as pol I and T4 pol, cover 6–8 nucleotides when the primer terminus is in the polymerase active site *(44,88,114)*. Larger polymerase complexes such as *E. coli* pol III holoenzyme, footprint three helical turns of the DNA *(213)* with the α-subunit covering 13 nucleotides upstream of the 3' primer terminus *(214)*. Analysis of sequences surrounding mutagenic hot- and coldspots obtained from targets with randomly introduced lesions confirmed that sequence context effects extend beyond nearest neighbors. For instance, a study employing in vivo-replicated single-stranded phage DNA containing randomly distributed β-ureidoisobutyric acid revealed two hotspots within the sequence GTG where 25 independent T → A transversions were scored *(158)*. However, only one mutation was scored at two other mutable T sites that were also flanked by Gs on both sides. Similar conclusions can be drawn from studies using single-stranded phage containing multiple 5-OHC residues replicated in *E. coli (66)*, double-stranded phagemid containing multiple O^6-meG *(273)*, or UV-irradiated phage DNA replicated in human cell extracts *(270)*. Interestingly, in spite of the comparable overall frequency of UV-induced C → T transitions when the damaged template was copied as the leading or lagging strand in the latter study, certain sites showed a higher substitution frequency when copied as the leading strand, whereas others showed the opposite effect.

In an attempt to address large sequence context effects on translesion synthesis in a systematic way, oligonucleotides that contain unique oxidative lesions surrounded by four randomized nucleotides on both the 3' and 5' sides were used as templates for in vitro DNA synthesis. The aim of these studies was to define sequence contexts that render blocking lesions bypassable as well as address sequence context effects on the mispairing properties of oxidative lesions. The initial study used templates containing either AP sites or furan and versions of T4 pol holoenzyme proficient or deficient in proofreading *(92)*. Under conditions where polymerase rebinding was not permitted, successful bypass products displayed clear biases for and against certain nucleotides at several of the eight randomized positions. Analysis of approx 100 independent bypass products for each lesion/polymerase combination revealed strong consensus sequences which enhanced or impeded lesion bypass, especially in the case of furan. The consensus covered on average 3–4 nt and was more frequently found 3' of the lesion, consistent with the polymerase subunit footprint *(88)*. Also, sequences on the 3'-end of the lesion are in a double-stranded conformation in the polymerase–template–dNTP ternary complex and therefore might contribute more to stacking interactions than sequences on the 5' side. More interestingly, the furan consensus sequence, which promoted bypass, could be translated over several positions to the 3'-side of the lesion and in the opposite orientation on the 5' side of the lesion. In addition, contexts that promoted the incorporation of G, rather than A, opposite AP sites were defined. In a parallel study, contexts that promoted the misincorporation of A opposite 8-oxoG were defined *(92)*. Although both studies resulted in consensus sequences that did not cover the entire randomized region of the template, it may be necessary to expand the randomized region in order to define more accurately the distance at which neighboring nucleotides have an effect on the lesion.

5. SUMMARY

Although translesion synthesis is clearly a complex process, several statements can be made:

1. Lesion structure plays the primary role in predicting bypass, that is, nonblocking lesions form hydrogen bonds with an incoming nucleotide and produce little distortion to the DNA molecule. 8-oxoA, 8-oxoG, O^6-meG, and O^4-meT readily pair and mispair, show little distortion to the DNA molecule, and are readily bypassed with pausing dependent on the polymerase and the sequence context. Interestingly, the T-T CPD, the *trans* adducts of *anti*-BPDE-G, and one conformer of the AF-G adduct all have intact hydrogen bonding faces as determined by NMR and can be bypassed, although inefficiently, by DNA polymerases.
2. Lesion structure is also the primary determinant for which particular base is inserted opposite the damage during translesion synthesis. This is particularly exemplified by the solution structures of O^6-meG, O^4-meT, and 8-oxoG, where the mispairs form highly favorable structures and for abasic sites that form stable structures with purines inserted opposite. Again, however, the particular polymerase and surrounding sequence context affect the outcome. For example, although not favored structurally, C is the favored insertion opposite abasic sites in yeast, and all four bases are inserted at equal frequency in a mammalian cell system.
3. Whether or not a DNA-distorting or poorly pairing lesion is a strong block to DNA synthesis is highly dependent on the particular polymerase used and its surrounding DNA sequence. Strongly blocking lesions, such as Tg, BPDE, and AAF, can all be bypassed by particular polymerases and in particular sequence contexts.

In order to develop a comprehensive model for translesion synthesis, data obtained through divergent biophysical, biochemical, and biological approaches need to be integrated and used to guide new studies. For instance, interactions between amino acid residues known to contribute to DNA polymerase fidelity and DNA lesions need to be addressed. Such studies should be guided by the structural data available for polymerases and lesion-containing DNAs, as well as biological data on polymerase mutants with altered fidelity. The biochemical mechanism of translesion DNA synthesis needs to be delineated, and should be guided by the transient-state kinetic studies, which have been used successfully to define the mechanism of DNA synthesis on natural templates. Once structural and mechanistic models are developed and verified for at least one member of each family of lesions and polymerases, predictions could be made and tested much more readily based on structural and functional homology between molecules. Ultimately, it is hoped that, with the help of comprehensive models, the exponentially growing power of computers could be utilized to make predictions about the outcome of interactions between base damages and DNA polymerases.

ACKNOWLEDGMENTS

The work reported from this laboratory was supported by NIH grants CA 52040 and CA 33657 awarded by the National Cancer Institute and a grant from the U. S. Department of Energy. The authors are grateful to Dr. Suse Broyde for critically reading the manuscript.

REFERENCES

1. Achey, P. M. and C. F. Wright. 1983. Inducible repair of thymine ring saturation damage in phi X174 DNA. *Radiat. Res.* **93:** 609–612.
2. Banerjee, S. K., R. B. Christensen, C. W. Lawrence, and J. E. LeClerc. 1988. Frequency and spe32ctrum of mutations produced by a single cis-syn thymine-thymine cyclobutane dimer in a single-stranded vector. *Proc. Natl. Acad. Sci. USA* **85:** 8141–8145.
3. Banerjee, S. K., A. Borden, R. B. Christensen, J. E. LeClerc, and C. W. Lawrence. 1990. SOS-dependent replication past a single trans-syn T-T cyclobutane dimer gives a different mutation spectrum and increased error rate compared with replication past this lesion in uninduced cells. *J. Bacteriol.* **172:** 2105–2112.
4. Basu, A. K. and J. M. Essigmann. 1988. Site-specifically modified oligodeoxynucleotides as probes for the structural and biological effects of DNA-damaging agents. *Chem. Res. Toxicol.* **1:** 1–18.
5. Basu, A. K., E. L. Loechler, S. A. Leadon, and J. M. Essigmann. 1989. Genetic effects of thymine glycol: site-specific mutagenesis and molecular modeling studies. *Proc. Natl. Acad. Sci. USA* **86:** 7677–7681.
6. Basu, A. K. and J. M. Essigmann. 1990. Site-specifically alkylated oligodeoxynucleotides: probes for mutagenesis, DNA repair and the structural effects of DNA damage. *Mutat. Res.* **233:** 189–201.
7. Beese, L. S. Personal communication.
8. Beese, L. S., J. M. Friedman, and T. A. Steitz. 1993. Crystal structures of the Klenow fragment of DNA polymerase I complexed with deoxynucleoside triphosphate and pyrophosphate. *Biochemistry* **32:** 14,095–14,101.
9. Belguise-Valladier, P. and R. P. Fuchs. 1991. Strong sequence-dependent polymorphism in adduct-induced DNA structure: analysis of single N-2-acetylaminofluorene residues bound within the *NarI* mutation hot spot. *Biochemistry* **30:** 10,091–10,100.
10. Belguise-Valladier, P., H. Maki, M. Sekiguchi, and R. P. Fuchs. 1994. Effect of single DNA lesions on *in vitro* replication with DNA polymerase III holoenzyme. Comparison with other polymerases. *J. Mol. Biol.* **236:** 151–164.
11. Belguise-Valladier, P. and R. P. Fuchs. 1995. *N*-2-aminofluorene and *N*-2 acetylaminofluorene adducts: the local sequence context of an adduct and its chemical structure determine its replication properties. *J. Mol. Biol.* **249:** 903–913.
12. Bernelot-Moens, C., B. W. Glickman, and A. J. Gordon. 1990. Induction of specific frameshift and base substitution events by benzo[*a*]pyrene diol epoxide in excision-repair-deficient *Escherichia coli. Carcinogenesis* **11:** 781–785.
13. Bhanot, O. S. and A. Ray. 1986. The *in vivo* mutagenic frequency and specificity of O^6-methylguanine in ϕX174 replicative form DNA. *Proc. Natl. Acad. Sci. USA* **83:** 7348–7352.
14. Bichara, M. and R. P. Fuchs. 1985. DNA binding and mutation spectra of the carcinogen N-2-aminofluorene in *Escherichia coli.* A correlation between the conformation of the premutagenic lesion and the mutation specificity. *J. Mol. Biol.* **183:** 341–351.
15. Bigger, C. A., J. St. John, H. Yagi, D. M. Jerina, and A. Dipple. 1992. Mutagenic specificities of four stereoisomeric benzo[c]phenanthrene dihydrodiol epoxides. *Proc. Natl. Acad. Sci. USA* **89:** 368–372.

16. Boiteux, S. and J. Laval. 1982. Coding properties of poly(deoxycytidylic acid) templates containing uracil or apyrimidinic sites: *in vitro* modulation of mutagenesis by deoxyribonucleic acid repair enzymes. *Biochemistry* **21:** 6746–6751.
17. Bonner, C. A., P. T. Stukenberg, M. Rajagopalan, R. Eritja, M. O'Donnell, K. McEntee, H. Echols, and M. F. Goodman. 1992. Processive DNA synthesis by DNA polymerase II mediated by DNA polymerase III accessory proteins. *J. Biol. Chem.* **267:** 11,431–11,438.
18. Braithwaite, D. K. and J. Ito. 1993. Compilation, alignment, and phylogenetic relationships of DNA polymerases. *Nucleic Acids Res.* **21:** 787–802.
19. Breimer, L. H. and T. Lindahl. 1985. Thymine lesions produced by ionizing radiation in double-stranded DNA. *Biochemistry* **24:** 4018–4022.
20. Brookes, P. and M. R. Osborne. 1982. Mutation in mammalian cells by stereoisomers of anti-benzo[a] pyrene-diolepoxide in relation to the extent and nature of the DNA reaction products. *Carcinogenesis* **3:** 1223–1226.
21. Burnouf, D., P. Koehl, and R. P. Fuchs. 1989. Single adduct mutagenesis: strong effect of the position of a single acetylaminofluorene adduct within a mutation hot spot (published erratum appears in *Proc. Natl. Acad. Sci. USA*, 1989) *Proc. Natl. Acad. Sci. USA* **86:** 4147–4151.
22. Burns, P. A., A. J. Gordon, and B. W. Glickman. 1987. Influence of neighbouring base sequence on *N*-methyl-*N*'-nitro-*N*-nitrosoguanidine mutagenesis in the lacI gene of *Escherichia coli*. *J. Mol. Biol.* **194:** 385–390.
23. Cabral Neto, J. B., R. E. Cabral, A. Margot, F. Le Page, A. Sarasin, and A. Gentil. 1994. Coding properties of a unique apurinic/apyrimidinic site replicated in mammalian cells. *J. Mol. Biol.* **240:** 416–420.
24. Cai, H., L. B. Bloom, R. Eritja, and M. F. Goodman. 1993. Kinetics of deoxyribonucleotide insertion and extension at abasic template lesions in different sequence contexts using HIV-1 reverse transcriptase. *J. Biol. Chem.* **268:** 23,567–23,572.
25. Capson, T. L., J. A. Peliska, B. F. Kaboord, M. W. Frey, C. Lively, M. Dahlberg, and S. J. Benkovic. 1992. Kinetic characterization of the polymerase and exonuclease activities of the gene 43 protein of bacteriophage T4. *Biochemistry* **31:** 10,984–10,994.
26. Cathcart, R., E. Schwiers, R. L. Saul, and B. N. Ames. 1984. Thymine glycol and thymidine glycol in human and rat urine: a possible assay for oxidative DNA damage. *Proc. Natl. Acad. Sci. USA* **81:** 5633–5637.
27. Ceccotti, S., E. Dogliotti, J. Gannon, P. Karran, and M. Bignami. 1993. O^6-methylguanine in DNA inhibits replication *in vitro* by human cell extracts. *Biochemistry* **32:** 13,664–13,672.
28. Chambers, R. W. 1991. Site-specific mutagenesis in cells with normal DNA repair systems: transitions produced from DNA carrying a single O^6-alkylguanine. *Nucleic Acids Res.* **19:** 2485–2488.
29. Chan, G. L., P. W. Doetsch, and W. A. Haseltine. 1985. Cyclobutane pyrimidine dimers and (6-4) photoproducts block polymerization by DNA polymerase I. *Biochemistry* **24:** 5723–5728.
30. Chary, P., G. J. Latham, D. L. Robberson, S. J. Kim, S. Han, C. M. Harris, T. M. Harris, and R. S. Lloyd. 1995. *In vivo* and *in vitro* replication consequences of stereoisomeric benzo[*a*]pyrene-7,8-dihydrodiol 9,10-epoxide adducts on adenine N6 at the second position of N-ras codon 61. *J. Biol. Chem.* **270:** 4990–5000.
31. Chary, P. and R. S. Lloyd. 1995. *In vitro* replication by prokaryotic and eukaryotic polymerases on DNA templates containing site-specific and stereospecific benzo[*a*] pyrene-7,8-dihydrodiol-9,10-epoxide adducts. *Nucleic Acids Res.* **23:** 1398–1405.
32. Cheng, K. C., D. S. Cahill, H. Kasai, S. Nishimura, and L. A. Loeb. 1992. 8-Hydroxyguanine, an abundant form of oxidative DNA damage, causes G $\rightarrow$ T and A $\rightarrow$ C substitutions. *J. Biol. Chem.* **267:** 166–172.

33. Cho, B. P., F. A. Beland, and M. M. Marques. 1994. NMR structural studies of a 15-mer DNA duplex from a ras protooncogene modified with the carcinogen 2-aminofluorene: conformational heterogeneity. *Biochemistry* **33:** 1373–1384.
34. Christner, D. F., M. K. Lakshman, J. M. Sayer, D. M. Jerina, and A. Dipple. 1994. Primer extension by various polymerases using oligonucleotide templates containing stereoisomeric benzo[*a*]pyrene-deoxyadenosine adducts. *Biochemistry* **33:** 14,297–14,305.
35. Clark, J. M. and G. P. Beardsley. 1986. Thymine glycol lesions terminate chain elongation by DNA polymerase I *in vitro. Nucleic Acids Res.* **14:** 737–749.
36. Clark, J. M., N. Pattabiraman, W. Jarvis, and G. P. Beardsley. 1987. Modeling and molecular mechanical studies of the cis-thymine glycol radiation damage lesion in DNA. *Biochemistry* **26:** 5404–5409.
37. Clark, J. M. and G. P. Beardsley. 1989. Template length, sequence context, and 3'-5' exonuclease activity modulate replicative bypass of thymine glycol lesions *in vitro. Biochemistry* **28:** 775–779.
38. Comess, K. M., J. N. Burstyn, J. M. Essigmann, and S. J. Lippard. 1992. Replication inhibition and translesion synthesis on templates containing site-specifically placed cis-diamminedichloroplatinum(II) DNA adducts. *Biochemistry* **31:** 3975–3990.
39. Cosman, M., C. de los Santos, R. Fiala, B. E. Hingerty, S. B. Singh, V. Ibanez, L. A. Margulis, D. Live, N. E. Geacintov, S. Broyde, and et ale. 1992. Solution conformation of the major adduct between the carcinogen (+)-anti-benzo[*a*]pyrene diol epoxide and DNA. *Proc. Natl. Acad. Sci. USA* **89:** 1914–1918.
40. Cosman, M., C. de los Santos, R. Fiala, B. E. Hingerty, V. Ibanez, E. Luna, R. Harvey, N. E. Geacintov, S. Broyde, and D. J. Patel. 1993. Solution conformation of the (+)-*cis-anti*-[BP]dG adduct in a DNA duplex: intercalation of the covalently attached benzo[*a*]pyrenyl ring into the helix and displacement of the modified deoxyguanosine. *Biochemistry* **32:** 4145–4155.
41. Cosman, M., R. Fiala, B. E. Hingerty, S. Amin, N. E. Geacintov, S. Broyde, and D. J. Patel. 1994. Solution conformation of the (+)-*trans-anti*-[BP]dG adduct opposite a deletion site in a DNA duplex: intercalation of the covalently attached benzo[*a*]pyrene into the helix with base displacement of the modified deoxyguanosine into the major groove. *Biochemistry* **33:** 11,507–11,517.
42. Cosman, M., R. Fiala, B. E. Hingerty, S. Amin, N. E. Geacintov, S. Broyde, and D. J. Patel. 1994. Solution conformation of the (+)-*cis-anti*-[BP]dG adduct opposite a deletion site in a DNA duplex: intercalation of the covalently attached benzo[*a*]pyrene into the helix with base displacement of the modified deoxyguanosine into the minor groove. *Biochemistry* **33:** 11,518–11,527.
43. Cosman, M., B. E. Hingerty, N. E. Geacintov, S. Broyde, and D. J. Patel. 1995. Structural alignment of (+)- and (–)-trans-anti-benzo[a]pyrene-dG adducts positioned at a DNA template-primer junction. *Biochemistry* **34:** 15,334–15,350.
44. Cowart, M., K. J. Gibson, D. J. Allen, and S. J. Benkovic. 1989. DNA substrate structural requirements for the exonuclease and polymerase activities of procaryotic and phage DNA polymerases. *Biochemistry* **28:** 1975–1983.
45. Cuniasse, P., L. C. Sowers, R. Eritja, B. Kaplan, M. F. Goodman, J. A. Cognet, M. Le Bret, W. Guschlbauer, and G. V. Fazakerley. 1989. Abasic frameshift in DNA. Solution conformation determined by proton NMR and molecular mechanics calculations. *Biochemistry* **28:** 2018–2026.
46. Cuniasse, P., G. V. Fazakerley, W. Guschlbauer, B. E. Kaplan, and L. C. Sowers. 1990. The abasic site as a challenge to DNA polymerase. A nuclear magnetic resonance study of G, C and T opposite a model abasic site. *J. Mol. Biol.* **213:** 303–314.
47. Dahlberg, M. E. and S. J. Benkovic. 1991. Kinetic mechanism of DNA polymerase I (Klenow fragment): identification of a second conformational change and evaluation of the internal equilibrium constant. *Biochemistry* **30:** 4835–4843.

48. Davies, J. F. N., R. J. Almassy, Z. Hostomska, R. A. Ferre, and Z. Hostomsky. 1994. 2.3 A crystal structure of the catalytic domain of DNA polymerase beta. *Cell* **76:** 1123–1133.
49. de los Santos, C., M. Cosman, B. E. Hingerty, V. Ibanez, L. A. Margulis, N. E. Geacintov, S. Broyde, and D. J. Patel. 1992. Influence of benzo[*a*]pyrene diol epoxide chirality on solution conformations of DNA covalent adducts: the (–)-trans-anti-[BP]G.C adduct structure and comparison with the (+)-trans-anti-[BP]G.C enantiomer. *Biochemistry* **31:** 5245–5252.
50. Dickerson, R. E. 1992. DNA structure from A to Z. *Methods Enzymol.* **211:** 67–111.
51. Doisy, R. and M. S. Tang. 1995. Effect of aminofluorene and (acetylamino)fluorene adducts on the DNA replication mediated by *Escherichia coli* polymerases I (Klenow fragment) and III. *Biochemistry* **34:** 4358–4368.
52. Donlin, M. J., S. S. Patel, and K. A. Johnson. 1991. Kinetic partitioning between the exonuclease and polymerase sites in DNA error correction. *Biochemistry* **30:** 538–546.
53. Dosanjh, M. K., J. M. Essigmann, M. F. Goodman, and B. Singer. 1990. Comparative efficiency of forming m4T.G versus m4T.A base pairs at a unique site by use of *Escherichia coli* DNA polymerase I (Klenow fragment) and Drosophila melanogaster polymerase alpha-primase complex. *Biochemistry* **29:** 4698–4703.
54. Dosanjh, M. K., G. Galeros, M. F. Goodman, and B. Singer. 1991. Kinetics of extension of O^6-methylguanine paired with cytosine or thymine in defined oligonucleotide sequences. *Biochemistry* **30:** 11,595–11,599.
55. Dosanjh, M. K., B. Singer, and J. M. Essigmann. 1991. Comparative mutagenesis of O^6-methylguanine and O^4-methylthymine in *Escherichia coli. Biochemistry* **30:** 7027–7033.
56. Dosanjh, M. K., P. Menichini, R. Eritja, and B. Singer. 1993. Both O^4-methylthymine and O^4-ethylthymine preferentially form alkyl T.G pairs that do not block *in vitro* replication in a defined sequence. *Carcinogenesis* **14:** 1915–1919.
57. Echols, H. and M. F. Goodman. 1991. Fidelity mechanisms in DNA replication. *Ann. Rev. Biochem.* **60:** 477–511.
58. Eckel, L. M. and T. R. Krugh. 1994. 2-Aminofluorene modified DNA duplex exists in two interchangeable conformations. *Nat. Struct. Biol.* **1:** 89–94.
59. Eger, B. T. and S. J. Benkovic. 1992. Minimal kinetic mechanism for misincorporation by DNA polymerase I (Klenow fragment). *Biochemistry* **31:** 9227–9236.
60. Eisenstadt, E., A. J. Warren, J. Porter, D. Atkins, and J. H. Miller. 1982. Carcinogenic epoxides of benzo[*a*]pyrene and cyclopenta[cd]pyrene induce base substitutions via specific transversions. *Proc. Natl. Acad. Sci. USA* **79:** 1945–1949.
61. Ellison, K. S., E. Dogliotti, T. D. Connors, A. K. Basu, and J. M. Essigmann. 1989. Site-specific mutagenesis by O^6-alkylguanines located in the chromosomes of mammalian cells: influence of the mammalian O^6-alkylguanine-DNA alkyltransferase. *Proc. Natl. Acad. Sci. USA* **86:** 8620–8624.
62. Essigmann, J. M., K. W. Fowler, C. L. Green, and E. L. Loechler. 1985. Extrachromosomal probes for mutagenesis by carcinogens: studies on the mutagenic activity of O^6-methylguanine built into a unique site in a viral genome. *Environ. Health Perspect.* **62:** 171–176.
63. Evans, J., M. Maccabee, Z. Hatahet, J. Courcelle, R. Bockrath, H. Ide, and S. Wallace. 1993. Thymine ring saturation and fragmentation products: lesion bypass, misinsertion and implications for mutagenesis. *Mutat. Res.* **299:** 147–156.
64. Feig, D. I. and L. A. Loeb. 1993. Mechanisms of mutation by oxidative DNA damage: reduced fidelity of mammalian DNA polymerase beta. *Biochemistry* **32:** 4466–4473.
65. Feig, D. I. and L. A. Loeb. 1994. Oxygen radical induced mutagenesis is DNA polymerase specific. *J. Mol. Biol.* **235:** 33–41.
66. Feig, D. I., L. C. Sowers, and L. A. Loeb. 1994. Reverse chemical mutagenesis: identification of the mutagenic lesions resulting from reactive oxygen species-mediated damage to DNA. *Proc. Natl. Acad. Sci. USA* **91:** 6609–6613.

67. Fijalkowska, I. J., R. L. Dunn, and R. M. Schaaper. 1993. Mutants of *Escherichia coli* with increased fidelity of DNA replication. *Genetics* **134:** 1023–1030.
68. Fijalkowska, I. J. and R. M. Schaaper. 1993. Antimutator mutations in the alpha subunit of *Escherichia coli* DNA polymerase III: identification of the responsible mutations and alignment with other DNA polymerases. *Genetics* **134:** 1039–1044.
69. Fix, D. and R. Bockrath. 1983. Targeted mutation at cytosine-containing pyrimidine dimers: studies of *Escherichia coli* B/r with acetophenone and 313-nm light. *Proc. Natl. Acad. Sci. USA* **80:** 4446–4449.
70. Fountain, M. A. and T. R. Krugh. 1995. Structural characterization of a (+)-trans-anti-benzo[*a*]pyrene-DNA adduct using NMR, restrained energy minimization, and molecular dynamics. *Biochemistry* **34:** 3152–3161.
71. Fowler, R. G., R. M. Schaaper, and B. W. Glickman. 1986. Characterization of mutational specificity within the *lacI* gene for a *mutD5* mutator strain of *Escherichia coli* defective in 3' → 5' exonuclease (proofreading) activity. *J. Bacteriol.* **167:** 130–137.
72. Frenkel, K., M. S. Goldstein, N. J. Duker, and G. W. Teebor. 1981. Identification of the cis-thymine glycol moiety in oxidized deoxyribonucleic acid. *Biochemistry* **20:** 750–754.
73. Frenkel, K., M. S. Goldstein, and G. W. Teebor. 1981. Identification of the cis-thymine glycol moiety in chemically oxidized and gamma-irradiated deoxyribonucleic acid by high-pressure liquid chromatography analysis. *Biochemistry* **20:** 7566–7571.
74. Frenkel, K., A. Cummings, J. Solomon, J. Cadet, J. J. Steinberg, and G. W. Teebor. 1985. Quantitative determination of the 5-(hydroxymethyl)uracil moiety in the DNA of gamma-irradiated cells. *Biochemistry* **24:** 4527–4533.
75. Frey, M. W., N. G. Nossal, T. L. Capson, and S. J. Benkovic. 1993. Construction and characterization of a bacteriophage T4 DNA polymerase deficient in 3' → 5' exonuclease activity. *Proc. Natl. Acad. Sci. USA* **90:** 2579–2583.
76. Gaffney, B. L. and R. A. Jones. 1989. Thermodynamic comparison of the base pairs formed by the carcinogenic lesion O^6-methylguanine with reference both to Watson-Crick pairs and to mismatched pairs. *Biochemistry* **28:** 5881–5889.
77. Ganguly, T. and N. J. Duker. 1992. Inhibition of DNA polymerase activity by thymine hydrates. *Mutat. Res.* **293:** 71–77.
78. Garcia, A., I. B. Lambert, and R. P. Fuchs. 1993. DNA adduct-induced stabilization of slipped frameshift intermediates within repetitive sequences: implications for mutagenesis. *Proc. Natl. Acad. Sci. USA* **90:** 5989–5993.
79. Gentil, A., A. Margot, and A. Sarasin. 1984. Apurinic sites cause mutations in simian virus 40. *Mutat. Res.* **129:** 141–147.
80. Gentil, A., G. Renault, C. Madzak, A. Margot, J. B. Cabral-Neto, J. J. Vasseur, B. Rayner, J. L. Imbach, and A. Sarasin. 1990. Mutagenic properties of a unique abasic site in mammalian cells. *Biochem. Biophys. Res. Commun.* **173:** 704–710.
81. Gentil, A., F. Le Page, A. Margot, C. W. Lawrence, A. Borden, and A. Sarasin. 1996. Mutagenicity of a unique thymine-thymine dimer or thymine-thymine pyrimidine pyrimidone (6-4) photoproduct in mammalian cells. *Nucleic Acids Res.* **24:** 1837–1840.
82. Gervais, V., A. Guy, R. Teoule, and G. V. Fazakerley. 1992. Solution conformation of an oligonucleotide containing a urea deoxyribose residue in front of a thymine. *Nucleic Acids Res.* **20:** 6455–6460.
83. Gibbs, P. E., B. J. Kilbey, S. K. Banerjee, and C. W. Lawrence. 1993. The frequency and accuracy of replication past a thymine-thymine cyclobutane dimer are very different in *Saccharomyces cerevisiae* and *Escherichia coli*. *J. Bacteriol.* **175:** 2607–2612.
84. Gibbs, P. E., A. Borden, and C. W. Lawrence. 1995. The T-T pyrimidine (6-4) pyrimidinone UV photoproduct is much less mutagenic in yeast than in *Escherichia coli*. *Nucleic Acids Res.* **23:** 1919–1922.
85. Gibbs, P. E. and C. W. Lawrence. 1995. Novel mutagenic properties of abasic sites in *Saccharomyces cerevisiae*. *J. Mol. Biol.* **251:** 229–236.

86. Goljer, I., S. Kumar, and P. H. Bolton. 1995. Refined solution structure of a DNA heteroduplex containing an aldehydic abasic site. *J. Biol. Chem.* **270**: 22,980–22,987.
87. Goodman, M. F., S. Creighton, L. B. Bloom, and J. Petruska. 1993. Biochemical basis of DNA replication fidelity. *Crit. Rev. Biochem. Mol. Biol.* **28:** 83–126.
88. Gopalakrishnan, V. and S. J. Benkovic. 1994. Spatial relationship between polymerase and exonuclease active sites of phage T4 DNA polymerase enzyme. *J. Biol. Chem.* **269:** 21,123–21,126.
89. Gordon, A. J., P. A. Burns, and B. W. Glickman. 1990. *N*-methyl-*N*'-nitro-*N*-nitrosoguanidine induced DNA sequence alteration; non-random components in alkylation mutagenesis. *Mutat. Res.* **233:** 95–103.
90. Guschlbauer, W., A. M. Duplaa, A. Guy, R. Teoule, and G. V. Fazakerley. 1991. Structure and *in vitro* replication of DNA templates containing 7,8-dihydro-8-oxoadenine. *Nucleic Acids Res.* **19:** 1753–1758.
91. Hagen, U. 1986. Current aspects on the radiation induced base damage in DNA. *Radiat Environ. Biophys.* **25:** 261–271.
92. Hatahet, Z. and S. S. Wallace. Unpublished results.
93. Hayes, R. C. and J. E. LeClerc. 1986. Sequence dependence for bypass of thymine glycols in DNA by DNA polymerase I. *Nucleic Acids Res.* **14:** 1045–1061.
94. Hayes, R. C., L. A. Petrullo, H. M. Huang, S. S. Wallace, and J. E. LeClerc. 1988. Oxidative damage in DNA. Lack of mutagenicity by thymine glycol lesions. *J. Mol. Biol.* **201:** 239–246.
95. Heflich, R. H. and R. E. Neft. 1994. Genetic toxicity of 2-acetylaminofluorene, 2-aminofluorene and some of their metabolites and model metabolites. *Mutat. Res.* **318:** 73–114.
96. Heller, E. P., E. J. Rosenkranz, E. C. McCoy, and H. S. Rosenkranz. 1984. Comparative mutagenesis by aminofluorene derivatives. A possible effect of DNA configuration. *Mutat. Res.* **131:** 89–95.
97. Herrala, A. M. and J. A. Vilpo. 1989. Template-primer activity of 5-(hydroxymethyl) uracil-containing DNA for prokaryotic and eukaryotic DNA and RNA polymerases. *Biochemistry* **28:** 8274–8277.
98. Hevroni, D. and Z. Livneh. 1988. Bypass and termination at apurinic sites during replication of single-stranded DNA *in vitro*: a model for apurinic site mutagenesis. *Proc. Natl. Acad. Sci. USA* **85:** 5046–5050.
99. Hoffmann, J. S., M. J. Pillaire, G. Maga, V. Podust, U. Hubscher, and G. Villani. 1995. DNA polymerase beta bypasses *in vitro* a single d(GpG)-cisplatin adduct placed on codon 13 of the *H-ras* gene. *Proc. Natl. Acad. Sci. USA* **92:** 5356–5360.
100. Holler, E., R. Bauer, and F. Bernges. 1992. Monofunctional DNA-platinum(II) adducts block frequently DNA polymerases. *Nucleic Acids Res.* **20:** 2307–2312.
101. Horsfall, M. J. and C. W. Lawrence. 1994. Accuracy of replication past the T-C (6-4) adduct. *J. Mol. Biol.* **235:** 465–471.
102. Hruszkewycz, A. M., K. A. Canella, K. Peltonen, L. Kotrappa, and A. Dipple. 1992. DNA polymerase action on benzo[*a*]pyrene-DNA adducts. *Carcinogenesis* **13:** 2347–2352.
103. Hsu, W. T., E. J. Lin, R. G. Harvey, and S. B. Weiss. 1977. Mechanism of phage phiX174 DNA inactivation by benzo(a)pyrene-7,8-dihydrodiol-9,10-epoxide. *Proc. Natl. Acad. Sci. USA* **74:** 3335–3339.
104. Ide, H., Y. W. Kow, and S. S. Wallace. 1985. Thymine glycols and urea residues in M13 DNA constitute replicative blocks *in vitro*. *Nucleic Acids Res.* **13:** 8035–8052.
105. Ide, H., R. J. Melamede, and S. S. Wallace. 1987. Synthesis of dihydrothymidine and thymidine glycol 5'-triphosphates and their ability to serve as substrates for *Escherichia coli* DNA polymerase I. *Biochemistry* **26:** 964–969.
106. Ide, H. and S. S. Wallace. 1988. Dihydrothymidine and thymidine glycol triphosphates as substrates for DNA polymerases: differential recognition of thymine C5-C6 bond saturation and sequence specificity of incorporation. *Nucleic Acids Res.* **16:** 11,339–11,354.

107. Ide, H., L. A. Petrullo, Z. Hatahet, and S. S. Wallace. 1991. Processing of DNA base damage by DNA polymerases. Dihydrothymine and beta-ureidoisobutyric acid as models for instructive and noninstructive lesions. *J. Biol. Chem.* **266:** 1469–1477.
108. Ide, H., H. Murayama, A. Murakami, T. Morii, and K. Makino. 1992. Effects of base damages on DNA replication—mechanism of preferential purine nucleotide insertion opposite abasic site in template DNA. *Nucleic Acids Symp. Ser.* **27:** 167,168.
109. Ide, H., H. Murayama, S. Sakamoto, K. Makino, K. Honda, H. Nakamuta, M. Sasaki, and N. Sugimoto. 1995. On the mechanism of preferential incorporation of dAMP at abasic sites in translesional DNA synthesis. Role of proofreading activity of DNA polymerase and thermodynamic characterization of model template-primers containing an abasic site. *Nucleic Acids Res.* **23:** 123–129.
110. Ito, J., and D. K. Braithwaite. 1991. Compilation and alignment of DNA polymerase sequences. *Nucleic Acids Res.* **19:** 4045–4057.
111. Jernstrom, B. and A. Graslund. 1994. Covalent binding of benzo[*a*]pyrene 7,8-dihydrodiol 9,10-epoxides to DNA: molecular structures, induced mutations and biological consequences. *Biophys. Chem.* **49:** 185–199.
112. Jiang, N. and J. S. Taylor. 1993. *In vivo* evidence that UV-induced C → T mutations at dipyrimidine sites could result from the replicative bypass of cis-syn cyclobutane dimers or their deamination products. *Biochemistry* **32:** 472–481.
113. Johnson, K. A. 1993. Conformational coupling in DNA polymerase fidelity. *Ann. Rev. Biochem.* **62:** 685–713.
114. Joyce, C. M. and T. A. Steitz. 1994. Function and structure relationships in DNA polymerases. *Ann. Rev. Biochem.* **63:** 777–822.
115. Joyce, C. M. and T. A. Steitz. 1995. Polymerase structure and function: variations on a theme? *J. Bacteriol.* **177:** 6321–6329.
116. Kalnik, M. W., C. N. Chang, A. P. Grollman, and D. J. Patel. 1988. NMR studies of abasic sites in DNA duplexes: deoxyadenosine stacks into the helix opposite the cyclic analogue of 2-deoxyribose. *Biochemistry* **27:** 924–931.
117. Kalnik, M. W., M. Kouchakdjian, B. F. Li, P. F. Swann, and D. J. Patel. 1988. Base pair mismatches and carcinogen-modified bases in DNA: an NMR study of A.C and A.O^4meT pairing in dodecanucleotide duplexes. *Biochemistry* **27:** 100–108.
118. Kalnik, M. W., M. Kouchakdjian, B. F. Li, P. F. Swann, and D. J. Patel. 1988. Base pair mismatches and carcinogen-modified bases in DNA: an NMR study of G.T and G.O^4meT pairing in dodecanucleotide duplexes. *Biochemistry* **27:** 108–115.
119. Kamiya, H. and H. Kasai. 1995. Formation of 2-hydroxydeoxyadenosine triphosphate, an oxidatively damaged nucleotide, and its incorporation by DNA polymerases. Steady-state kinetics of the incorporation. *J. Biol. Chem.* **270:** 19,446–19,450.
120. Kamiya, H., N. Murata-Kamiya, S. Koizume, H. Inoue, S. Nishimura, and E. Ohtsuka. 1995. 8-Hydroxyguanine (7,8-dihydro-8-oxoguanine) in hot spots of the c-Ha-ras gene: effects of sequence contexts on mutation spectra. *Carcinogenesis* **16:** 883–889.
121. Kao, J. Y., I. Goljer, T. A. Phan, and P. H. Bolton. 1993. Characterization of the effects of a thymine glycol residue on the structure, dynamics, and stability of duplex DNA by NMR. *J. Biol. Chem.* **268:** 17,787–17,793.
122. Kasai, H., P. F. Crain, Y. Kuchino, S. Nishimura, A. Ootsuyama, and H. Tanooka. 1986. Formation of 8-hydroxyguanine moiety in cellular DNA by agents producing oxygen radicals and evidence for its repair. *Carcinogenesis* **7:** 1849–1851.
123. Kati, W. M., K. A. Johnson, L. F. Jerva, and K. S. Anderson. 1992. Mechanism and fidelity of HIV reverse transcriptase. *J. Biol. Chem.* **267:** 25,988–25,997.
124. Kemmink, J., R. Boelens, T. Koning, G. A. van der Marel, J. H. van Boom, and R. Kaptein. 1987. 1H NMR study of the exchangeable protons of the duplex d(GCGTTGCG). d(CGCAACGC) containing a thymine photodimer. *Nucleic Acids Res.* **15:** 4645–4653.

125. Kemmink, J., R. Boelens, T. M. Koning, R. Kaptein, G. A. van der Marel, and J. H. van Boom. 1987. Conformational changes in the oligonucleotide duplex d(GCGTTGCG) × d(CGCAACGC) induced by formation of a *cis-syn* thymine dimer. A two-dimensional NMR study. *Eur. J. Biochem.* **162:** 37–43.
126. Kim, Y., S. H. Eom, J. Wang, D. S. Lee, S. W. Suh, and T. A. Steitz. 1995. Crystal structure of *Thermus aquaticus* DNA polymerase. *Nature* **376:** 612–616.
127. Klein, J. C., M. J. Bleeker, C. P. Saris, H. C. Roelen, H. F. Brugghe, H. van den Elst, G. A. van der Marel, J. H. van Boom, J. G. Westra, E. Kriek, and A. J. M. Berns. 1992. Repair and replication of plasmids with site-specific 8-oxodG and 8-AAFdG residues in normal and repair-deficient human cells. *Nucleic Acids Res.* **20:** 4437–4443.
128. Klinedinst, D. K. and N. R. Drinkwater. 1992. Mutagenesis by apurinic sites in normal and ataxia telangiectasia human lymphoblastoid cells. *Mol. Carcinog.* **6:** 32–42.
129. Koehl, P., P. Valladier, J. F. Lefevre, and R. P. Fuchs. 1989. Strong structural effect of the position of a single acetylaminofluorene adduct within a mutation hot spot. *Nucleic Acids Res.* **17:** 9531–9541.
130. Koffel-Schwartz, N. and R. P. Fuchs. 1995. Sequence determinants for -2 frameshift mutagenesis at *Nar*I-derived hot spots. *J. Mol. Biol.* **252:** 507–513.
131. Kohalmi, S. E. and B. A. Kunz. 1988. Role of neighbouring bases and assessment of strand specificity in ethylmethanesulphonate and *N*-methyl-*N'*-nitro-*N*-nitrosoguanidine mutagenesis in the *SUP4*-o gene of *Saccharomyces cerevisiae*. *J. Mol. Biol.* **204:** 561–568.
132. Kohlstaedt, L. A., J. Wang, J. M. Friedman, P. A. Rice, and T. A. Steitz. 1992. Crystal structure at 3.5 A resolution of HIV-1 reverse transcriptase complexed with an inhibitor. *Science* **256:** 1783–1790.
133. Kouchakdjian, M., V. Bodepudi, S. Shibutani, M. Eisenberg, F. Johnson, A. P. Grollman, and D. J. Patel. 1991. NMR structural studies of the ionizing radiation adduct 7-hydro-8-oxodeoxyguanosine (8-oxo-7H-dG) opposite deoxyadenosine in a DNA duplex. 8-Oxo-7H-dG*(syn)* · dA*(anti)* alignment at lesion site. *Biochemistry* **30:** 1403–1412.
134. Krishnaswamy, S., J. A. Rogers, R. J. Isbell, and R. G. Fowler. 1993. The high mutator activity of the dnaQ49 allele of *Escherichia coli* is medium-dependent and results from both defective 3' → 5' proofreading and methyl-directed mismatch repair. *Mutat. Res.* **288:** 311–319.
135. Kuchta, R. D., V. Mizrahi, P. A. Benkovic, K. A. Johnson, and S. J. Benkovic. 1987. Kinetic mechanism of DNA polymerase I (Klenow). *Biochemistry* **26:** 8410–8417.
136. Kunkel, T. A., R. M. Schaaper, and L. A. Loeb. 1983. Depurination-induced infidelity of deoxyribonucleic acid synthesis with purified deoxyribonucleic acid replication proteins *in vitro*. *Biochemistry* **22:** 2378–2384.
137. Kunkel, T. A. 1984. Mutational specificity of depurination. *Proc. Natl. Acad. Sci. USA* **81:** 1494–1498.
138. Kunkel, T. A. 1990. Misalignment-mediated DNA synthesis errors. *Biochemistry* **29:** 8003–8011.
139. Kunz, B. A., E. S. Henson, H. Roche, D. Ramotar, T. Nunoshiba, and B. Demple. 1994. Specificity of the mutator caused by deletion of the yeast structural gene *(APN1)* for the major apurinic endonuclease. *Proc. Natl. Acad. Sci. USA* **91:** 8165–8169.
140. Lai, M. D. and K. L. Beattie. 1988. Influence of DNA sequence on the nature of mispairing during DNA synthesis. *Biochemistry* **27:** 1722–1728.
141. Laspia, M. F. and S. S. Wallace. 1988. Excision repair of thymine glycols, urea residues, and apurinic sites in *Escherichia coli*. *J. Bacteriol.* **170:** 3359–3366.
142. Laspia, M. F. and S. S. Wallace. 1989. SOS processing of unique oxidative DNA damages in *Escherichia coli*. *J. Mol. Biol.* **207:** 53–60.
143. Lawrence, C. W., A. Borden, S. K. Banerjee, and J. E. LeClerc. 1990. Mutation frequency and spectrum resulting from a single abasic site in a single-stranded vector. *Nucleic Acids Res.* **18:** 2153–2157.

144. LeClerc, J. E., A. Borden, and C. W. Lawrence. 1991. The thymine-thymine pyrimidine-pyrimidone(6-4) ultraviolet light photoproduct is highly mutagenic and specifically induces 3' thymine-to-cytosine transitions in *Escherichia coli*. *Proc. Natl. Acad. Sci. USA* **88:** 9685–9689.

145. Leonard, G. A., J. Thomson, W. P. Watson, and T. Brown. 1990. High-resolution structure of a mutagenic lesion in DNA. *Proc. Natl. Acad. Sci. USA* **87:** 9573–9576.

146. Leonard, G. A., A. Guy, T. Brown, R. Teoule, and W. N. Hunter. 1992. Conformation of guanine-8-oxoadenine base pairs in the crystal structure of d(CGCGAATT(O8A)GCG). *Biochemistry* **31:** 8415–8420.

147. Levy, D. D. and G. W. Teebor. 1991. Site directed substitution of 5-hydroxymethyluracil for thymine in replicating ϕX174am3 DNA via synthesis of 5-hydroxymethyl-2'-deoxyuridine-5'-triphosphate. *Nucleic Acids Res.* **19:** 3337–3343.

148. Lindahl, T. and B. Nyberg. 1972. Rate of depurination of native deoxyribonucleic acid. *Biochemistry* **11:** 3610–3618.

149. Lindahl, T. 1993. Instability and decay of the primary structure of DNA. *Nature* **362:** 709–715.

150. Lindsley, J. E. and R. P. Fuchs. 1994. Use of single-turnover kinetics to study bulky adduct bypass by T7 DNA polymerase. *Biochemistry* **33:** 764–772.

151. Lippke, J. A., L. K. Gordon, D. E. Brash, and W. A. Haseltine. 1981. Distribution of UV light-induced damage in a defined sequence of human DNA: detection of alkaline-sensitive lesions at pyrimidine nucleoside-cytidine sequences. *Proc. Natl. Acad. Sci. USA* **78:** 3388–3392.

152. Lipscomb, L. A., M. E. Peek, M. L. Morningstar, S. M. Verghis, E. M. Miller, A. Rich, J. M. Essigmann, and L. D. Williams. 1995. X-ray structure of a DNA decamer containing 7,8-dihydro-8-oxoguanine. *Proc. Natl. Acad. Sci. USA* **92:** 719–723.

153. Livneh, Z., O. Cohen-Fix, R. Skaliter, and T. Elizur. 1993. Replication of damaged DNA and the molecular mechanism of ultraviolet light mutagenesis. *Crit. Rev. Biochem. Mol. Biol.* **28:** 465–513.

154. Loeb, L. A. and B. D. Preston. 1986. Mutagenesis by apurinic/apyrimidinic sites. *Ann. Rev. Genet.* **20:** 201–230.

155. Loechler, E. L., C. L. Green, and J. M. Essigmann. 1984. *In vivo* mutagenesis by O^6-methylguanine built into a unique site in a viral genome. *Proc. Natl. Acad. Sci. USA* **81:** 6271–6275.

156. Lutgerink, J. T., J. Retel, and H. Loman. 1984. Effects of adduct formation on the biological activity of single- and double-stranded ϕX174 DNA, modified by N-acetoxy-N-acetyl-2-aminofluorene. *Biochim. Biophys. Acta* **781:** 81–91.

157. Lutgerink, J. T., J. Retel, J. G. Westra, M. C. Welling, H. Loman, and E. Kriek. 1985. Bypass of the major aminofluorene-DNA adduct during *in vivo* replication of single- and double-stranded ϕX174 DNA treated with *N*-hydroxy-2-aminofluorene. *Carcinogenesis* **6:** 1501–1506.

158. Maccabee, M., J. S. Evans, M. P. Glackin, Z. Hatahet, and S. S. Wallace. 1994. Pyrimidine ring fragmentation products. Effects of lesion structure and sequence context on mutagenesis. *J. Mol. Biol.* **236:** 514–530.

159. Mackay, W., M. Benasutti, E. Drouin, and E. L. Loechler. 1992. Mutagenesis by (+)-anti-B[a]P-N2-Gua, the major adduct of activated benzo[*a*]pyrene, when studied in an *Escherichia coli* plasmid using site-directed methods. *Carcinogenesis* **13:** 1415–1425.

160. Mah, M. C., J. Boldt, S. J. Culp, V. M. Maher, and J. J. McCormick. 1991. Replication of acetylaminofluorene-adducted plasmids in human cells: spectrum of base substitutions and evidence of excision repair. *Proc. Natl. Acad. Sci. USA* **88:** 10,193–10,197.

161. Maki, H., J. Y. Mo, and M. Sekiguchi. 1991. A strong mutator effect caused by an amino acid change in the alpha subunit of DNA polymerase III of *Escherichia coli*. *J. Biol. Chem.* **266:** 5055–5061.

162. Maki, H. and M. Sekiguchi. 1992. MutT protein specifically hydrolyses a potent mutagenic substrate for DNA synthesis. *Nature* **355:** 273–275.
163. Matray, T. J., K. J. Haxton, and M. M. Greenberg. 1995. The effects of the ring fragmentation product of thymidine C5-hydrate on phosphodiesterases and klenow (exo-) fragment. *Nucleic Acids Res.* **23:** 4642–4648.
164. McAuley-Hecht, K. E., G. A. Leonard, N. J. Gibson, J. B. Thomson, W. P. Watson, W. N. Hunter, and T. Brown. 1994. Crystal structure of a DNA duplex containing 8-hydroxydeoxyguanine-adenine base pairs. *Biochemistry* **33:** 10,266–10,270.
165. Mellac, S., G. V. Fazakerley, and L. C. Sowers. 1993. Structures of base pairs with 5-(hydroxymethyl)-2'-deoxyuridine in DNA determined by NMR spectroscopy. *Biochemistry* **32:** 7779–7786.
166. Mendelman, L. V., M. S. Boosalis, J. Petruska, and M. F. Goodman. 1989. Nearest neighbor influences on DNA polymerase insertion fidelity. *J. Biol. Chem.* **264:** 14,415–14,423.
167. Menichini, P., M. M. Mroczkowska, and B. Singer. 1994. Enzyme-dependent pausing during *in vitro* replication of O^4-methylthymine in a defined oligonucleotide sequence. *Mutat. Res.* **307:** 53–59.
168. Miaskiewicz, K., J. Miller, and R. Osman. 1993. *Ab initio* theoretical study of the structures of thymine glycol and dihydrothymine. *Int. J. Radiat. Biol.* **63:** 677–686.
169. Miaskiewicz, K., J. Miller, R. Ornstein, and R. Osman. 1995. Molecular dynamics simulations of the effects of ring-saturated thymine lesions on DNA structure. *Biopolymers* **35:** 113–124.
170. Michaels, M. L., T. M. Reid, C. M. King, and L. J. Romano. 1991. Accurate *in vitro* translesion synthesis by *Escherichia coli* DNA polymerase I (large fragment) on a site-specific, aminofluorene-modified oligonucleotide. *Carcinogenesis* **12:** 1641–1646.
171. Milhe, C., C. Dhalluin, R. P. Fuchs, and J. F. Lefevre. 1994. NMR evidence of the stabilisation by the carcinogen *N*-2-acetylaminofluorene of a frameshift mutagenesis intermediate. *Nucleic Acids Res.* **22:** 4646–4652.
172. Miller, J., K. Miaskiewicz, and R. Osman. 1994. Structure–function studies of DNA damage using ab initio quantum mechanics and molecular dynamics simulation. *Ann. NY Acad. Sci.* **726:** 71–91.
173. Mitchell, N. and G. Stohrer. 1986. Mutagenesis originating in site-specific DNA damage. *J. Mol. Biol.* **191:** 177–180.
174. Mizrahi, V., R. N. Henrie, J. F. Marlier, K. A. Johnson, and S. J. Benkovic. 1985. Rate-limiting steps in the DNA polymerase I reaction pathway. *Biochemistry* **24:** 4010–4018.
175. Mizrahi, V., P. Benkovic, and S. J. Benkovic. 1986. Mechanism of DNA polymerase I: exonuclease/polymerase activity switch and DNA sequence dependence of pyrophosphorolysis and misincorporation reactions. *Proc. Natl. Acad. Sci. USA* **83:** 5769–5773.
176. Mizrahi, V., P. A. Benkovic, and S. J. Benkovic. 1986. Mechanism of the idling-turnover reaction of the large (Klenow) fragment of *Escherichia coli* DNA polymerase I. *Proc. Natl. Acad. Sci. USA* **83:** 231–235.
177. Mo, J. Y., H. Maki, and M. Sekiguchi. 1991. Mutational specificity of the dnaE173 mutator associated with a defect in the catalytic subunit of DNA polymerase III of *Escherichia coli*. *J. Mol. Biol.* **222:** 925–936.
178. Moore, P. and B. S. Strauss. 1979. Sites of inhibition of *in vitro* DNA synthesis in carcinogen- and UV-treated ϕX174 DNA. *Nature* **278:** 664–666.
179. Moore, P. D., S. D. Rabkin, and B. S. Strauss. 1980. Termination of vitro DNA synthesis at AAF adducts in the DNA. *Nucleic Acids Res.* **8:** 4473–4484.
180. Moore, P. D., K. K. Bose, S. D. Rabkin, and B. S. Strauss. 1981. Sites of termination of *in vitro* DNA synthesis on ultraviolet- and N-acetylaminofluorene-treated ϕX174 templates by prokaryotic and eukaryotic DNA polymerases. *Proc. Natl. Acad. Sci. USA* **78:** 110–114.
181. Moore, P. D., S. D. Rabkin, A. L. Osborn, C. M. King, and B. S. Strauss. 1982. Effect of acetylated and deacetylated 2-aminofluorene adducts on *in vitro* DNA synthesis. *Proc. Natl. Acad. Sci. USA* **79:** 7166–7170.

182. Moran, E. and S. S. Wallace. 1985. The role of specific DNA base damages in the X-ray-induced inactivation of bacteriophage PM2. *Mutat. Res.* **146:** 229–241.
183. Moriya, M., M. Takeshita, F. Johnson, K. Peden, S. Will, and A. P. Grollman. 1988. Targeted mutations induced by a single acetylaminofluorene DNA adduct in mammalian cells and bacteria. *Proc. Natl. Acad. Sci. USA* **85:** 1586–1589.
184. Moriya, M., C. Ou, V. Bodepudi, F. Johnson, M. Takeshita, and A. P. Grollman. 1991. Site-specific mutagenesis using a gapped duplex vector: a study of translesion synthesis past 8-oxodeoxyguanosine in *E. coli. Mutat. Res.* **254:** 281–288.
185. Moriya, M. 1993. Single-stranded shuttle phagemid for mutagenesis studies in mammalian cells: 8-oxoguanine in DNA induces targeted G.CÆT.A transversions in simian kidney cells. *Proc. Natl. Acad. Sci. USA* **90:** 1122–1126.
186. Napolitano, R. L., I. B. Lambert, and R. P. Fuchs. 1994. DNA sequence determinants of carcinogen-induced frameshift mutagenesis. *Biochemistry* **33:** 1311–1315.
187. Neto, J. B., A. Gentil, R. E. Cabral, and A. Sarasin. 1992. Mutation spectrum of heat-induced abasic sites on a single-stranded shuttle vector replicated in mammalian cells. *J. Biol. Chem.* **267:** 19,718–19,723.
188. Norman, D., P. Abuaf, B. E. Hingerty, D. Live, D. Grunberger, S. Broyde, and D. J. Patel. 1989. NMR and computational characterization of the N-(deoxyguanosin-8-yl)aminofluorene adduct [(AF)G] opposite adenosine in DNA: (AF)G*[syn]* · A*[anti]* pair formation and its pH dependence. *Biochemistry* **28:** 7462–7476.
189. O'Day, C. L., P. M. Burgers, and J. S. Taylor. 1992. PCNA-induced DNA synthesis past cis-syn and trans-syn-I thymine dimers by calf thymus DNA polymerase delta *in vitro*. *Nucleic Acids Res.* **20:** 5403–5406.
190. O'Handley, S. F., D. G. Sanford, R. Xu, C. C. Lester, B. E. Hingerty, S. Broyde, and T. R. Krugh. 1993. Structural characterization of an *N*-acetyl-2-aminofluorene (AAF) modified DNA oligomer by NMR, energy minimization, and molecular dynamics. *Biochemistry* **32:** 2481–2497.
191. Oda, Y., S. Uesugi, M. Ikehara, S. Nishimura, Y. Kawase, H. Ishikawa, H. Inoue, and E. Ohtsuka. 1991. NMR studies of a DNA containing 8-hydroxydeoxyguanosine. *Nucleic Acids Res.* **19:** 1407–1412.
192. Oller, A. R. and R. M. Schaaper. 1994. Spontaneous mutation in *Escherichia coli* containing the dnaE911 DNA polymerase antimutator allele. *Genetics* **138:** 263–270.
193. Parris, C. N., D. D. Levy, J. Jessee, and M. M. Seidman. 1994. Proximal and distal effects of sequence context on ultraviolet mutational hotspots in a shuttle vector replicated in xeroderma cells. *J. Mol. Biol.* **236:** 491–502.
194. Patel, D. J., L. Shapiro, S. A. Kozlowski, B. L. Gaffney, and R. A. Jones. 1986. Structural studies of the O6meG.C interaction in the d(C-G-C-G-A-A-T-T-C-O6meG-C-G) duplex. *Biochemistry* **25:** 1027–1036.
195. Patel, D. J., L. Shapiro, S. A. Kozlowski, B. L. Gaffney, and R. A. Jones. 1986. Structural studies of the O6meG.T interaction in the d(C-G-T-G-A-A-T-T-C-O6meG-C-G) duplex. *Biochemistry* **25:** 1036–1042.
196. Patel, S. S., I. Wong, and K. A. Johnson. 1991. Pre-steady-state kinetic analysis of processive DNA replication including complete characterization of an exonuclease-deficient mutant. *Biochemistry* **30:** 511–525.
197. Pauly, G. T., S. H. Hughes, and R. C. Moschel. 1995. Mutagenesis in *Escherichia coli* by three O^6-substituted guanines in double-stranded or gapped plasmids. *Biochemistry* **34:** 8924–8930.
198. Pavlov, Y. I., D. T. Minnick, S. Izuta, and T. A. Kunkel. 1994. DNA replication fidelity with 8-oxodeoxyguanosine triphosphate. *Biochemistry* **33:** 4695–4701.
199. Pearlman, D. A., S. R. Holbrook, D. H. Pirkle, and S. H. Kim. 1985. Molecular models for DNA damaged by photoreaction. *Science* **227:** 1304–1308.
200. Person, S., J. A. McCloskey, W. Snipes, and R. C. Bockrath. 1974. Ultraviolet mutagenesis and its repair in an *Escherichia coli* strain containing a nonsense codon. *Genetics* **78:** 1035–1049.

201. Petruska, J. and M. F. Goodman. 1985. Influence of neighboring bases on DNA polymerase insertion and proofreading fidelity. *J. Biol. Chem.* **260:** 7533–7539.
202. Pinz, K. G., S. Shibutani, and D. F. Bogenhagen. 1995. Action of mitochondrial DNA polymerase gamma at sites of base loss or oxidative damage. *J. Biol. Chem.* **270:** 9202–9206.
203. Poltev, V. I., S. L. Smirnov, O. V. Issarafutdinova, and R. Lavery. 1993. Conformations of DNA duplexes containing 8-oxoguanine. *J. Biomol. Struct. Dyn.* **11:** 293–301.
204. Preston, B. D., B. Singer, and L. A. Loeb. 1986. Mutagenic potential of O4-methylthymine *in vivo* determined by an enzymatic approach to site-specific mutagenesis. *Proc. Natl. Acad. Sci. USA* **83:** 8501–8505.
205. Preston, B. D., B. Singer, and L. A. Loeb. 1987. Comparison of the relative mutagenicities of O-alkylthymines site-specifically incorporated into ϕX174. *J. Biol. Chem.* **262:** 13,821–13,827.
206. Purmal, A. A. and S. S. Wallace. Unpublished observations.
207. Purmal, A. A., Y. W. Kow, and S. S. Wallace. 1994. Major oxidative products of cytosine, 5-hydroxycytosine and 5-hydroxyuracil, exhibit sequence context-dependent mispairing *in vitro*. *Nucleic Acids Res.* **22:** 72–78.
208. Purmal, A. A., Y. W. Kow, and S. S. Wallace. 1994. 5-Hydroxypyrimidine deoxynucleoside triphosphates are more efficiently incorporated into DNA by exonuclease-free Klenow fragment than 8-oxopurine deoxynucleoside triphosphates. *Nucleic Acids Res.* **22:** 3930–3935.
209. Rabkin, S. D., P. D. Moore, and B. S. Strauss. 1983. *In vitro* bypass of UV-induced lesions by *Escherichia coli* DNA polymerase I: specificity of nucleotide incorporation. *Proc. Natl. Acad. Sci. USA* **80:** 1541–1545.
210. Rabkin, S. D. and B. S. Strauss. 1984. A role for DNA polymerase in the specificity of nucleotide incorporation opposite N-acetyl-2-aminofluorene adducts. *J. Mol. Biol.* **178:** 569–594.
211. Randall, S. K., R. Eritja, B. E. Kaplan, J. Petruska, and M. F. Goodman. 1987. Nucleotide insertion kinetics opposite abasic lesions in DNA. *J. Biol. Chem.* **262:** 6864–6870.
212. Rao, S. N., J. W. Keepers, and P. Kollman. 1984. The structure of d(CGCGAAT^TCGCG) · d(CGCGAATTCGCG); the incorporation of a thymine photodimer into a B-DNA helix. *Nucleic Acids Res.* **12:** 4789–4807.
213. Reems, J. A. and C. S. McHenry. 1994. *Escherichia coli* DNA polymerase III holoenzyme footprints three helical turns of its primer. *J. Biol. Chem.* **269:** 33,091–33,096.
214. Reems, J. A., S. Wood, and C. S. McHenry. 1995. *Escherichia coli* DNA polymerase III holoenzyme subunits alpha, beta, and gamma directly contact the primer-template. *J. Biol. Chem.* **270:** 5606–5613.
215. Reha-Krantz, L. J. and R. L. Nonay. 1994. Motif A of bacteriophage T4 DNA polymerase: role in primer extension and DNA replication fidelity. Isolation of new antimutator and mutator DNA polymerases. *J. Biol. Chem.* **269:** 5635–5643.
216. Ricchetti, M. and H. Buc. 1990. Reverse transcriptases and genomic variability: the accuracy of DNA replication is enzyme specific and sequence dependent. *EMBO J.* **9:** 1583–1593.
217. Rich, A. 1993. DNA comes in many forms. *Gene* **135:** 99–109.
218. Richards, R. G., L. C. Sowers, J. Laszlo, and W. D. Sedwick. 1984. The occurrence and consequences of deoxyuridine in DNA. *Adv. Enzyme Regul.* **22:** 157–185.
219. Ross, H., C. A. Bigger, H. Yagi, D. M. Jerina, and A. Dipple. 1993. Sequence specificity in the interaction of the four stereoisomeric benzo[c]phenanthrene dihydrodiol epoxides with the supF gene. *Cancer Res.* **53:** 1273–1277.
220. Ross, J., R. Doisy, and M. S. Tang. 1988. Mutational spectrum and recombinogenic effects induced by aminofluorene adducts in bacteriophage M13. *Mutat. Res.* **201:** 203–212.

221. Rouet, P. and J. M. Essigmann. 1985. Possible role for thymine glycol in the selective inhibition of DNA synthesis on oxidized DNA templates. *Cancer Res.* **45:** 6113–6118.
222. Roy, A. and R. P. Fuchs. 1994. Mutational spectrum induced in Saccharomyces cerevisiae by the carcinogen *N*-2-acetylaminofluorene. *Mol. Gen. Genet.* **245:** 69–77.
223. Sage, E., E. Cramb, and B. W. Glickman. 1992. The distribution of UV damage in the lacI gene of *Escherichia coli*: correlation with mutation spectrum. *Mutat. Res.* **269:** 285–299.
224. Sagher, D. and B. Strauss. 1983. Insertion of nucleotides opposite apurinic/apyrimidinic sites in deoxyribonucleic acid during *in vitro* synthesis: uniqueness of adenine nucleotides. *Biochemistry* **22:** 4518–4526.
225. Sagher, D. and B. Strauss. 1985. Abasic sites from cytosine as termination signals for DNA synthesis. *Nucleic Acids Res.* **13:** 4285–4298.
226. Sahm, J., E. Turkington, D. LaPointe, and B. Strauss. 1989. Mutation induced *in vitro* on a C-8 guanine aminofluorene containing template by a modified T7 DNA polymerase. *Biochemistry* **28:** 2836–2843.
227. Sawaya, M. R., H. Pelletier, A. Kumar, S. H. Wilson, and J. Kraut. 1994. Crystal structure of rat DNA polymerase beta: evidence for a common polymerase mechanism. *Science* **264:** 1930–1935.
228. Schaaper, R. M. and L. A. Loeb. 1981. Depurination causes mutations in SOS-induced cells. *Proc. Natl. Acad. Sci. USA* **78:** 1773–1777.
229. Schaaper, R. M., T. A. Kunkel, and L. A. Loeb. 1983. Infidelity of DNA synthesis associated with bypass of apurinic sites. *Proc. Natl. Acad. Sci. USA* **80:** 487–491.
230. Schaaper, R. M., B. N. Danforth, and B. W. Glickman. 1986. Mechanisms of spontaneous mutagenesis: an analysis of the spectrum of spontaneous mutation in the *Escherichia coli lacI* gene. *J. Mol. Biol.* **189:** 273–284.
231. Schaaper, R. M. and R. L. Dunn. 1987. Spectra of spontaneous mutations in *Escherichia coli* strains defective in mismatch correction: the nature of *in vivo* DNA replication errors. *Proc. Natl. Acad. Sci. USA* **84:** 6220–6224.
232. Schaaper, R. M., N. Koffel-Schwartz, and R. P. Fuchs. 1990. *N*-acetoxy-*N*-acetyl-2-aminofluorene-induced mutagenesis in the lacI gene of *Escherichia coli*. *Carcinogenesis* **11:** 1087–1095.
233. Schaaper, R. M. 1993. The mutational specificity of two *Escherichia coli* dnaE antimutator alleles as determined from lacI mutation spectra. *Genetics* **134:** 1031–1038.
234. Schurter, E. J., H. J. Yeh, J. M. Sayer, M. K. Lakshman, H. Yagi, D. M. Jerina, and D. G. Gorenstein. 1995. NMR solution structure of a nonanucleotide duplex with a dG mismatch opposite a 10R adduct derived from trans addition of a deoxyadenosine N6-amino group to (–)-(7S,8R,9R,10S)-7,8-dihydroxy-9,10-epoxy-7,8,9,10- tetrahydrobenzo[*a*] pyrene. *Biochemistry* **34:** 1364–1375.
235. Shearman, C. W. and L. A. Loeb. 1979. Effects of depurination on the fidelity of DNA synthesis. *J. Mol. Biol.* **128:** 197–218.
236. Shibutani, S., M. Takeshita, and A. P. Grollman. 1991. Insertion of specific bases during DNA synthesis past the oxidation-damaged base 8-oxodG. *Nature* **349:** 431–434.
237. Shibutani, S. 1993. Quantitation of base substitutions and deletions induced by chemical mutagens during DNA synthesis *in vitro*. *Chem. Res. Toxicol.* **6:** 625–629.
238. Shibutani, S., V. Bodepudi, F. Johnson, and A. P. Grollman. 1993. Translesional synthesis on DNA templates containing 8-oxo-7,8-dihydrodeoxyadenosine. *Biochemistry* **32:** 4615–4621.
239. Shibutani, S. and A. P. Grollman. 1993. Nucleotide misincorporation on DNA templates containing *N*-(deoxyguanosin-*N*2-yl)-2-(acetylamino)fluorene. *Chem. Res. Toxicol.* **6:** 819–824.
240. Shibutani, S. and A. P. Grollman. 1993. On the mechanism of frameshift (deletion) mutagenesis *in vitro*. *J. Biol. Chem.* **268:** 11,703–11,710.

241. Shibutani, S., L. A. Margulis, N. E. Geacintov, and A. P. Grollman. 1993. Translesional synthesis on a DNA template containing a single stereoisomer of dG-(+)- or dG-(–)-anti-BPDE (7,8-dihydroxy-anti-9,10-epoxy-7,8,9,10-tetrahydrobenzo[*a*]pyrene). *Biochemistry* **32:** 7531–7541.
242. Shibutani, S. and A. P. Grollman. 1994. Miscoding during DNA synthesis on damaged DNA templates catalysed by mammalian cell extracts. *Cancer Lett.* **83:** 315–322.
243. Shigenaga, M. K., C. J. Gimeno, and B. N. Ames. 1989. Urinary 8-hydroxy-2'-deoxyguanosine as a biological marker of *in vivo* oxidative DNA damage. *Proc. Natl. Acad. Sci. USA* **86:** 9697–9701.
244. Singer, B., J. Sagi, and J. T. Kusmierek. 1983. *Escherichia coli* polymerase I can use O2-methyldeoxythymidine or O4-methyldeoxythymidine in place of deoxythymidine in primed poly(dA-dT).poly(dA-dT) synthesis. *Proc. Natl. Acad. Sci. USA* **80:** 4884–4888.
245. Singer, B. 1986. O-alkyl pyrimidines in mutagenesis and carcinogenesis: occurrence and significance. *Cancer Res.* **46:** 4879–4885.
246. Singer, B., F. Chavez, M. F. Goodman, J. M. Essigmann, and M. K. Dosanjh. 1989. Effect of 3' flanking neighbors on kinetics of pairing of dCTP or dTTP opposite O6-methylguanine in a defined primed oligonucleotide when *Escherichia coli* DNA polymerase I is used. *Proc. Natl. Acad. Sci. USA* **86:** 8271–8274.
247. Singer, B. and M. K. Dosanjh. 1990. Site-directed mutagenesis for quantitation of base-base interactions at defined sites. *Mutat. Res.* **233:** 45–51.
248. Slater, S. C., M. R. Lifsics, M. O'Donnell, and R. Maurer. 1994. *holE*, the gene coding for the theta subunit of DNA polymerase III of *Escherichia coli*: characterization of a holE mutant and comparison with a *dnaQ* (epsilon-subunit) mutant. *J. Bacteriol.* **176:** 815–821.
249. Snow, E. T., R. S. Foote, and S. Mitra. 1984. Kinetics of incorporation of O^6-methyl-deoxyguanosine monophosphate during *in vitro* DNA synthesis. *Biochemistry* **23:** 4289–4294.
250. Sockett, H., S. Romac, and F. Hutchinson. 1991. Mutagenic specificity of N-methyl-N'-nitro-N-nitrosoguanidine in the *gpt* gene on a chromosome of Chinese hamster ovary cells and of *Escherichia coli* cells. *Mol. Gen. Genet.* **227:** 252–259.
251. Spacciapoli, P. and N. G. Nossal. 1994. A single mutation in bacteriophage T4 DNA polymerase (A737V, tsL141) decreases its processivity as a polymerase and increases its processivity as a 3' → 5' exonuclease. *J. Biol. Chem.* **269:** 438–446.
252. Sriram, M., G. A. van der Marel, H. L. Roelen, J. H. van Boom, and A. H. Wang. 1992. Conformation of B-DNA containing O6-ethyl-G-C base pairs stabilized by minor groove binding drugs: molecular structure of d(CGC[e6G]AATTCGCG complexed with Hoechst 33258 or Hoechst 33342. *EMBO J.* **11:** 225–232.
253. Steitz, T. A. and J. A. Steitz. 1993. A general two-metal-ion mechanism for catalytic RNA. *Proc. Natl. Acad. Sci. USA* **90:** 6498–6502.
254. Steitz, T. A., S. J. Smerdon, J. Jager, and C. M. Joyce. 1994. A unified polymerase mechanism for nonhomologous DNA and RNA polymerases. *Science* **266:** 2022–2025.
255. Stevens, C. W., N. Bouck, J. A. Burgess, and W. E. Fahl. 1985. Benzo[a]pyrene diol-epoxides: different mutagenic efficiency in human and bacterial cells. *Mutat. Res.* **152:** 5–14.
256. Stewart, C. 1988. Phage SPO-1, in *The Bacteriophages*, vol. 1 (Calendar, R., ed.), Plenum, New York, p. 477.
257. Strauss, B., S. Rabkin, D. Sagher, and P. Moore. 1982. The role of DNA polymerase in base substitution mutagenesis on non-instructional templates. *Biochimie* **64:** 829–838.
258. Strauss, B. S. 1985. Translesion DNA synthesis: polymerase response to altered nucleotides. *Cancer Surveys* **4:** 439–516.
259. Strauss, B. S. and J. Wang. 1990. Role of DNA polymerase 3 '× 5' exonuclease activity in the bypass of aminofluorene lesions in DNA. *Carcinogenesis* **11:** 2103–2109.

260. Strauss, B. S. 1991. The "A rule" of mutagen specificity: a consequence of DNA polymerase bypass of non-instructional lesions? *Bioessays* **13:** 79–84.
261. Takeshita, M., C. N. Chang, F. Johnson, S. Will, and A. P. Grollman. 1987. Oligodeoxynucleotides containing synthetic abasic sites. Model substrates for DNA polymerases and apurinic/apyrimidinic endonucleases. *J. Biol. Chem.* **262:** 10,171–10,179.
262. Takeshita, M. and W. Eisenberg. 1994. Mechanism of mutation on DNA templates containing synthetic abasic sites: study with a double strand vector. *Nucleic Acids Res.* **22:** 1897–1902.
263. Tang, M., M. W. Lieberman, and C. M. King. 1982. *uvr* Genes function differently in repair of acetylaminofluorene and aminofluorene DNA adducts. *Nature* **299:** 646–648.
264. Tang, M. S., J. Hrncir, D. Mitchell, J. Ross, and J. Clarkson. 1986. The relative cytotoxicity and mutagenicity of cyclobutane pyrimidine dimers and (6-4) photoproducts in *Escherichia coli* cells. *Mutat. Res.* **161:** 9–17.
265. Taylor, J. S., D. S. Garrett, I. R. Brockie, D. L. Svoboda, and J. Telser. 1990. 1H NMR assignment and melting temperature study of *cis-syn* and *trans-syn* thymine dimer containing duplexes of d(CGTATTATGC).d(GCATAATACG). *Biochemistry* **29:** 8858–8866.
266. Taylor, J. S. and C. L. O'Day. 1990. *cis-syn* thymine dimers are not absolute blocks to replication by DNA polymerase I of *Escherichia coli in vitro*. *Biochemistry* **29:** 1624–1632.
267. Tebbs, R. S. and L. J. Romano. 1994. Mutagenesis at a site-specifically modified *Nar*I sequence by acetylated and deacetylated aminofluorene adducts. *Biochemistry* **33:** 8998–9006.
268. Teoule, R. 1987. Radiation-induced DNA damage and its repair. *Int. J. Radiat. Biol. Relat. Stud. Phys. Chem. Med.* **51:** 573–589.
269. Tessman, I., S. K. Liu, and M. A. Kennedy. 1992. Mechanism of SOS mutagenesis of UV-irradiated DNA: mostly error-free processing of deaminated cytosine. *Proc. Natl. Acad. Sci. USA* **89:** 1159–1163.
270. Thomas, D. C., D. C. Nguyen, W. W. Piegorsch, and T. A. Kunkel. 1993. Relative probability of mutagenic translesion synthesis on the leading and lagging strands during replication of UV-irradiated DNA in a human cell extract. *Biochemistry* **32:** 11,476–11,482.
271. Thomas, D. C., X. Veaute, T. A. Kunkel, and R. P. Fuchs. 1994. Mutagenic replication in human cell extracts of DNA containing site-specific N-2-acetylaminofluorene adducts. *Proc. Natl. Acad. Sci. USA* **91:** 7752–7756.
272. Toorchen, D. and M. D. Topal. 1983. Mechanisms of chemical mutagenesis and carcinogenesis: effects on DNA replication of methylation at the O^6-guanine position of dGTP. *Carcinogenesis* **4:** 1591–1597.
273. Topal, M. D., J. S. Eadie, and M. Conrad. 1986. O^6-methylguanine mutation and repair is nonuniform. Selection for DNA most interactive with O^6-methylguanine. *J. Biol. Chem.* **261:** 9879–9885.
274. Veaute, X. and R. P. Fuchs. 1993. Greater susceptibility to mutations in lagging strand of DNA replication in *Escherichia coli* than in leading strand. *Science* **261:** 598–600.
275. Villani, G., S. Boiteux, and M. Radman. 1978. Mechanism of ultraviolet-induced mutagenesis: extent and fidelity of *in vitro* DNA synthesis on irradiated templates. *Proc. Natl. Acad. Sci. USA* **75:** 3037–3041.
276. Voigt, J. M. and M. D. Topal. 1995. O^6-methylguanine-induced replication blocks. *Carcinogenesis* **16:** 1775–1782.
277. Withka, J. M., J. A. Wilde, P. H. Bolton, A. Mazumder, and J. A. Gerlt. 1991. Characterization of conformational features of DNA heteroduplexes containing aldehydic abasic sites. *Biochemistry* **30:** 9931–9940.
278. Wong, I., S. S. Patel, and K. A. Johnson. 1991. An induced-fit kinetic mechanism for DNA replication fidelity: direct measurement by single-turnover kinetics. *Biochemistry* **30:** 526–537.

279. Wood, A. W., R. L. Chang, W. Levin, H. Yagi, D. R. Thakker, D. M. Jerina, and A. H. Conney. 1977. Differences in mutagenicity of the optical enantiomers of the diastereomeric benzo[*a*]pyrene 7,8-diol-9,10-epoxides. *Biochem. Biophys. Res. Commun.* **77:** 1389–1396.
280. Wood, M. L., M. Dizdaroglu, E. Gajewski, and J. M. Essigmann. 1990. Mechanistic studies of ionizing radiation and oxidative mutagenesis: genetic effects of a single 8-hydroxyguanine (7-hydro-8-oxoguanine) residue inserted at a unique site in a viral genome. *Biochemistry* **29:** 7024–7032.
281. Wood, M. L., A. Esteve, M. L. Morningstar, G. M. Kuziemko, and J. M. Essigmann. 1992. Genetic effects of oxidative DNA damage: comparative mutagenesis of 7,8-dihydro-8-oxoguanine and 7,8-dihydro-8-oxoadenine in *Escherichia coli*. *Nucleic Acids Res.* **20:** 6023–6032.
282. Wood, R. D. 1985. Pyrimidine dimers are not the principal pre-mutagenic lesions induced in lambda phage DNA by ultraviolet light. *J. Mol. Biol.* **184:** 577–585.

13

DNA Repair and Mutagenesis in *Streptococcus pneumoniae*

Sanford A. Lacks

1. INTRODUCTION

There are several reasons why *Streptococcus pneumoniae* (also known as pneumococcus and previously classified as *Diplococcus pneumoniae*) has been a particularly interesting organism for the study of DNA damage and repair. Until the advent of antibiotics in the 1940s, this pathogenic bacterium, which is responsible for human lobar pneumonia, was a leading cause of death throughout the world. As an outcome of the intense scrutiny of this pathogen, genetic transformation was discovered in this species in 1928 *(38)* and shown to result from the transfer of DNA in 1944 *(3)*. Because the transformation phenomenon allowed controlled treatment of DNA outside the cell, the system was useful for assessing the effect of different types of DNA damage in vitro on biological activity in vivo. Both inactivating and mutagenic effects could be examined.

Although the investigation of DNA damage and repair in *S. pneumoniae* contributed to an understanding of the mechanism of transformation, more often, the reverse was true. A preliminary discussion of the molecular mechanisms of transformation in *S. pneumoniae,* therefore, will be presented. Studies of pneumococcal transformation in the 1960s led to the discovery of the Hex DNA mismatch repair system of *S. pneumoniae (21,22,56,57,66)*. In the 1970s, the Mut DNA mismatch repair system was elucidated in *Escherichia coli (38,150,155)*, and in the 1980s, these systems were shown to be closely related *(39,61,107,109)*, and homologous systems of mismatch repair were found in eukaryotes, as well *(51,156)*. In the 1990s, genetic defects in the human homolog of these mismatch repair systems were shown to contribute to the causation of human colon cancer *(25,74)*. Discovery of these human genes depended on the DNA sequence information from the bacterial systems, a ramification that would have been difficult to foresee at the time of the original investigations of mismatch repair in transformation.

In addition to mismatched DNA, the effects on transformation of various other kinds of alterations in DNA, particularly those that can be characterized in vitro, are discussed. The known cellular mechanisms of DNA repair in *S. pneumoniae* will be examined. Several unusual mechanisms of mutagenesis found in this species will be

From: DNA Damage and Repair, Vol. 1: DNA Repair in prokaryotes and Lower Eukaryotes
Edited by: J. A. Nickoloff and M. F. Hoekstra © Humana Press Inc., Totowa, NJ

discussed. Finally, two cases demonstrating the interaction of two or more repair systems in *S. pneumoniae* will be presented.

2. MOLECULAR MECHANISM OF GENETIC TRANSFORMATION

2.1. *Competence for DNA Uptake*

Cultures of *S. pneumoniae* are most competent in the late logarithmic phase of growth, when a 17-aa polypeptide *(42)* secreted into the growth medium *(8,47)* reaches a sufficient concentration *(144)* and elicits expression of a number of cellular proteins *(97)*, one of which binds single-stranded DNA *(96)*. There is no evidence for direct involvement of the competence polypeptide in DNA uptake; rather, it appears to stimulate expression of proteins needed for DNA binding.

Inasmuch as an analysis of DNA damage or genetic recombination often requires comparisons of different cultures, variations in culture competence could perturb such analyses. This problem was solved by using a single reference marker, usually for a drug resistance, in the DNA to be analyzed for other markers, to measure the competence in a particular culture *(66)*.

2.2. *Processing of DNA on Binding and Entry into the Recipient Cell*

The mechanism of DNA uptake in pneumococcal transformation consists of two steps: binding of the DNA to the outer surface of the cell, where it is still susceptible to external agents, and entry of the DNA into the cell *(59,68)*. Only double-stranded DNA is efficiently bound *(78)*, but no sequence specificity is required *(77)*. DNA from many different species can bind to and enter cells of *S. pneumoniae (59)*. Binding is an active process, requiring energy *(63)*, and it causes random single-strand breaks averaging ~3 kb apart *(58,64)*. The donor DNA is presumably bound to the cell surface at the site of strand breakage.

The binding step can be separated from the entry step by mutations in the *endA* gene *(68,70)*, which encodes a membrane-located nuclease *(65,113)*, or by a chelating agent *(129)*, which binds divalent cations required for entry. During entry, one strand of donor DNA is degraded by the EndA nuclease *(113)* to oligonucleotides, which remain outside the cell *(63,98)*, and the other strand (previously fragmented on binding) is introduced into the cell in a single-stranded form *(54,67)*. These entering strand segments are coated by the single-stranded DNA binding protein produced during the development of competence *(96)*.

2.3. *Recombination of Donor and Recipient DNA*

Although double-stranded DNA from any source will be taken up and converted to single strands, only if the DNA is sufficiently homologous to that of *S. pneumoniae* will it undergo recombination with the chromosomal DNA. Quantitative analysis of the fate of labeled homologous DNA showed that more than half of the single strands that entered were physically integrated as lengthy segments into the chromosomal DNA (reviewed in ref. *59*). When it is genetically marked, the integration of a single-strand donor segment gives a heteroduplex recombination intermediate that would normally revert to a homoduplex form only on replication of the chromosomal DNA.

Quantitative kinetic experiments on the genetic fate of DNA, in which transforming activity was measured in DNA extracted from cells that had taken up genetically marked DNA, reveal a brief eclipse period in which transforming activity for the marker in test cells is not observed *(27)*. This eclipse corresponds to the time in which the donor DNA is in a single-stranded form after its entry and prior to its integration *(54)*. The eclipse is a consequence of the conversion to single strands and the inability of single strands to transform the test cells. The half-time for recovery from eclipse (for a single-site marker in the absence of mismatch repair) at 30°C is 5 min *(35)*. This is brief compared to the replication time for the chromosome, which is 60 min under these conditions. Except for the case of multisite markers (involving lengthy insertions or deletions; *see* ref. *35*), heteroduplex DNA readily transforms cells.

Little is known about the mechanism of chromosomal integration in *S. pneumoniae*. It has been proposed that the donor strand segment interacts with the double-stranded chromosomal DNA to produce either a three-stranded *(46,56)* or a D-loop *(60,130)* intermediate. The recipient strand segment displaced by the donor segment is somehow eliminated. During the brief interval between replacement of the recipient strand segment and ligation of the free ends of the donor segment to the free ends of the remaining recipient strand, a mismatch repair system can act on the heteroduplex intermediate to eliminate the donor contribution. Otherwise, the donor information is ligated into the chromosome and then converted to a homoduplex by chromosomal replication.

3. MISMATCH REPAIR IN TRANSFORMATION AND MUTATION AVOIDANCE

3.1. *Marker Dependence of Transformation Frequencies*

Analysis of transformation frequencies for a large number of mutations in both the *malM* gene *(56,66)* and the *amiA* locus *(22)* revealed discrete classes of relative transformation frequency (or efficiency) ranging from 0.05 to 1.0, which, on the basis of mutagenic origin, could be correlated to the four pairs of mismatches, A/C and G/T (0.05), A/A and T/T (0.2), G/G and C/C (0.5), and A/G and T/C (1.0), that form in the heteroduplex products of transformation by the complementary strands of donor DNA. Mismatches owing to frameshift mutations corresponding to single nucleotide deletions or insertions give frequencies of 0.1 *(31,32)*, but insertion/deletion mismatches larger than 5 nucleotides are not recognized by this system *(34,82)*. Determination of the base mismatches by DNA sequencing amply confirmed these correlations *(13,14,32,70)*, although some A/G and T/C mismatches can be targeted to a certain extent *(10,70)*, which presumably depends on their DNA context.

The above frequencies represent average values for the mismatches formed by the opposite donor strands. Single mismatch frequencies were determined by constructing donor DNA consisting of one wild-type and one mutant strand *(12)*. The only class in which the pairs always deviated from identity was the one corresponding to a GC to CG transversion: the G/G frequency is 0.05, whereas the C/C frequency is 1.0, so the average is 0.5 *(10)*.

3.2. *Donor DNA Targeting*

Each mismatched base pair has a characteristic recognition level, which leads to repair and proportionately reduced transforming efficiency, but repair is not biased, and reciprocal transformations give identically low or high frequencies *(56)*. For example, whether the G in a G/T mismatch comes from the donor or the recipient, the transformation frequency is low because the donor moiety is eliminated in either case.

Mismatch repair acts very early in the transformation process, as indicated by examining the kinetics of recovery from eclipse for low-efficiency and high-efficiency markers. Although a high-efficiency (1.0) marker recovers with a half-time of 5 min at 30°C, a low-efficiency marker appears to recover faster, with a half-time of 1.5 min *(35)*, but to a lower level. A more careful analysis of low-efficiency marker recovery revealed a spike of recovered activity at early times, followed by a reduction to the final value, indicating that initially recovered marker was subsequently lost *(131)*. The interplay of recovery and repair gives, as a net result, an apparently shorter recovery time. These results suggest that the substrate for mismatch repair is the early heteroduplex transformation product in which donor strands are not yet ligated to the recipient chromosome. The mismatch repair system, therefore, must recognize both the strand breaks and the mismatch itself to correct the donor moiety.

Mismatch repair during transformation involves removal and destruction of the entire donor segment that would otherwise be integrated. Genetic analysis showed that a linked marker, which when tested alone would show high efficiency, is excluded by a nearby low-efficiency marker to an extent proportional to its linkage distance *(56)*. Inasmuch as recombination between linked markers is largely owing to fragmentation of the donor segment, this implies that the entire donor segment is eliminated by the repair process, and biochemical analysis demonstrated such degradation *(92)*.

3.3. *Cellular Mechanism of Repair*

Two elements critical for realizing that the mechanism underlying the variation in marker transformation efficiencies was a repair process were (1) the recognition that the mismatched bases in the recombination product determined the efficiency *(56)* and (2) the suggestion that a mismatch repair system was responsible for eliminating genetic information in one of the two heteroduplex product strands *(21)*. It was proposed that integration of low-efficiency markers involved correction of the recipient component to give a homoduplex donor marker *(21)*, and some evidence was adduced to support this notion *(20)*. However, it is now clear that the repair system removes only the donor contribution to the mismatch along with much or all of the associated donor strand segment *(56,92)*. Surviving low-efficiency markers retain a heteroduplex configuration; they manage to escape the repair process when the fragments containing them are ligated into the chromosome prior to correction *(61,131)*. This will be discussed further in Section 7.2.

Considerable support for a cellular repair mechanism came from the discovery of mutations in *S. pneumoniae* that blocked the effect *(57)*. These cellular mutations were called *hex,* because in such mutant recipients all markers transformed with *h*igh *e*fficiency for (at that time) unknown *(x)* reasons. These mutations were elicited with *N*-methyl-*N'*-nitro-*N*-nitrosoguanidine (MNNG) and appeared at surprisingly high frequency in the treated populations *(57)*; the reason for this was recently shown to be the enhanced resistance to the agent conferred by *hex* mutations *(52)*. The *hex* mutations fell in two distinct genetic loci, *hexA* and *hexB (4,11)*. Both loci were isolated and analyzed *(4,109,110,112)*.

The *hexA* locus consists of a single gene encoding HexA, a polypeptide of 95 kDa *(4)*. Its amino acid sequence contains an ATP binding motif *(107)*. HexA is homologous to MutS *(107)*, a component of mismatch repair systems in enteric bacteria *(41)*, and it can interfere with MutS function when introduced into *E. coli (110)*. MutS was shown to bind to DNA mismatches in vitro *(139)*; presumably, HexA also recognizes mismatches in DNA.

The *hexB* locus encodes a 74-kDa polypeptide *(106,109)* homologous to MutL in *E. coli* and *Salmonella typhimurium (109)*. Sequence analysis of the locus indicates a promoter and terminator that would produce a 2.1-kb monocistronic mRNA *(109)*. A repeated DNA sequence called BOX, which is found at multiple locations in the *S. pneumoniae* chromosome and may be involved in gene regulation *(88)*, is upstream from *hexB (109)*. However, the location of this BOX with respect to the putative promoters in the region *(109,125)* makes it more likely that it regulates the adjacent, oppositely transcribed *orfL* gene rather than *hexB*.

The precise mechanism by which HexA and HexB interact to recognize the DNA mismatch and the strand breaks in the target strand, and to remove the donor moiety, is not known. By analogy with the Mut system of *E. coli,* which has been more completely analyzed in vitro (*71*; *see* Chapter 11), a plausible model for the Hex system can be proposed. First, HexA would bind to a DNA mismatch, and then, in association with HexB, the adjacent DNA would be scanned for strand breaks, possibly by looping the DNA on both sides past a stationary Hex protein complex. Energy for this DNA translocation might come from ATP bound to the HexA moiety. Once a strand break is reached, the Hex complex would degrade the donor strand, and repair synthesis would complete the process.

3.4. Mutation Avoidance

Hex^- mutants have much higher spontaneous mutation rates than wild-type *S. pneumoniae* for both forward and reverse mutations *(142,143)*. Mutations in either the *hexA* or *hexB* genes increase *malM* reversion rates to the same extent, that is, 50- to 500-fold depending on the *malM* mutation *(69)*. Apparently, the same mechanism that recognizes transformation mismatches and removes the donor strand segment also recognizes mismatches in newly (mis)replicated DNA and removes the newly synthesized strand segment, thereby correcting the replication error. It is possible that the Hex mismatch repair system evolved to repair misreplicated DNA, and its action on transforming DNA is only incidental.

3.5. Hex System as a Generalized Mismatch Repair System

The Hex system of *S. pneumoniae* displays a number of features, which it shares with other such systems, that led to its characterization as a generalized mismatch repair system *(61)*. First, it recognizes a variety of DNA mismatches, but varies in its efficiency of correcting them. Base mismatches are corrected with decreasing efficiency as follows: A/C = G/T = G/G > A/A = T/T > C/T > A/G > C/C *(10)*. Insertion/deletion mismatches of 1 or 2 nt are recognized almost as well as the transition mismatches A/C and G/T *(32,70)*; those longer than 5 nt are not recognized at all *(34,56,70,82)*. Second, when the system repairs a mismatch, it removes a large tract of one strand of the DNA, which can be thousands of nucleotides in length *(56,92)*. Third, defects in the repair system greatly increase the spontaneous mutation rate. A generalized mismatch repair system related to the Hex system appears to be almost universally present in living cells.

3.6. Homology Among Generalized Mismatch Repair Systems

That the Hex system of *S. pneumoniae* is evolutionarily related to the Mut system of the enteric bacteria is indicated by the 36% identity of amino acid sequence between HexA and MutS of *S. typhimurium* and *E. coli (39,61,107)*. Similarly, the amino-terminal 400 residues of HexB and MutL of *S. typhimurium* show 40% identity *(85,109)*. It was previously evident that the Hex and Mut systems were very similar *(4,10)*. The spectrum of mismatches that each system recognizes is the same both qualitatively and quantitatively *(10,18,50,61,140)*. Only one strand is targeted for repair, without regard to the direction of the mismatch, and large tracts of that strand are removed *(56,150)*. Although the Mut system can target strands on the basis of methylation status *(111)*, strand breaks can also be used for targeting, both in vivo *(72)* and in vitro *(71)*, as in the Hex system.

In the yeast, *Saccharomyces cerevisiae*, *pms1* mutations increase postmeiotic segregation *(154)* and spontaneous mutation *(156)*, indicating a role of the gene in generalized mismatch repair *(10,61,156)*. Convincing evidence for this conclusion was the homology found between the *PMS1* gene product and HexB and MutL *(51,85,109)*. Subsequently, another homolog of HexB/MutL *(MLH1)* was found to be essential for the system *(108)*. On the basis of conserved sequences in *hexA* and *mutS*, PCR was used to identify two *hexA/mutS* homologs in *S. cerevisiae (117)*. One, *MSH2*, prevents mutations in nuclear genes, and the other, *MSH1*, in mitochondrial DNA *(116)*. A third homolog, *MSH3*, is not required for generalized mismatch repair, but may have a role in repair of larger deletions/insertions *(137)*; it is more similar *(100)* to a mammalian HexA/MutS homolog located near the dihydrofolate dehydrogenase *(DHFR)* gene *(29)*. Two additional homologs of HexA/MutS in *S. cerevisiae, MSH4 (123)* and *MSH5 (44)*, appear to play a role in meiosis, but not in mismatch repair (*see* Chapter 19). These results and those for human cells described below indicate two differences between bacteria and eukaryotes with respect to mismatch repair. One is that eukaryotes require two different HexB/MutL-like proteins, whereas the bacteria require only one type. The other is that eukaryotes have additional, less similar homologs of the bacterial proteins, which perhaps function in other DNA-related reactions. In addition to yeasts, evidence from biochemical activity in extracts or from genetic homology attests to the presence of a related generalized mismatch repair system in various eukaryotes, including *Xenopus laevis (148)*, *Drosophila melanogaster (45)*, and *Homo sapiens (45)*.

Using the same HexA/MutS-based PCR primers that were effective in yeast, a human homolog, *hMSH2*, was found and located on chromosome 2; it corresponded to a gene responsible for a familial predisposition to colorectal cancer *(25,74)*. In addition to *hMSH2* and another homolog, *hMSH6,* three human genes homologous to *hexB/mutL*—*hMLH1, hPMS1,* and *hPMS2*—were found, and defects in all of them have been related to colorectal cancer *(7,101,102)*. It is intriguing how arcane studies of the frequency of transformation in bacteria led to the revelation of an important mechanism for the causation of human cancer.

3.7. Target Strand Recognition

In the generalized mismatch repair system, recognition of the component of the mismatch to be altered depends not on directionality within the mismatch, but rather on properties of the DNA strand containing that component. During transformation, the Hex system recognizes strand breaks at each end of the donor segment, a feature essential for its role in mutation avoidance by correction of errors in newly replicated DNA. Although the state of newly synthesized DNA has not been examined in *S. pneumoniae*, newly replicated DNA in *E. coli* and *Bacillus subtilis* contains breaks on both nascent strands, which are spaced ~1 kb apart *(136,145)*. It is commonly accepted that breaks in the lagging strand result from segmental synthesis, but it is not known what causes breaks in the leading strand, which is synthesized continuously *(49)*. It is possible that the incorporation of aberrant nucleotides and their removal gives rise to such breaks. For example, in *E. coli*, incorporation of uracil from dUTP would result in glycolytic removal of the uracil and cleavage of the strand by an apurinic, apyrimidinic (AP)-endonuclease *(145)*. However, in *E. coli*, the fragmentation of nascent DNA is found in *ung* mutants *(145)*, and *ung* mutants of *S. pneumoniae* still exhibit mutation avoidance by the Hex system *(9)*. Quite possibly, there are multiple, redundant systems for producing the strand breaks used for Hex system targeting.

It has been shown in vivo *(72)* and in vitro *(71)* that single-strand breaks in *E. coli* can direct mismatch repair. In addition, MutH makes strand breaks at unmethylated 5'-GATC-3' sites *(153)* present in newly synthesized DNA. Thus, in *E. coli*, the repair system has a second chance to repair the mismatch prior to methylation of these sites by the Dam methylase *(111*; *see* Chapter 11). In fact, this methylation overlay enhances the extent of mutation avoidance. Whereas the Hex system reduces spontaneous mutations ~100-fold *(70,143)*, the Mut system reduces them ~1000-fold *(36)*. The methylation overlay may be a specialized feature of Gram-negative bacteria. A nonenteric Gram-negative species, *Haemophilus influenzae*, contains homologs of *mutH* and *dam*, as well as *hexA/mutS* and *hexB/mutL (26)*. The Dam methylase, which is homologous to the *Dpn*II.M methylase, could have been recruited from a now defunct *Dpn*II-like restriction system in the progenitor of *E. coli (61)*.

Most organisms do not show the type of DNA methylation (adenine methylation in a four-base recognition sequence) used by *E. coli* to enhance mismatch repair, and they presumably rely on pre-existing strand breaks for targeting the nascent strand in mismatch repair. Some species in which mismatch repair has been demonstrated, such as *S. cerevisiae* and *D. melanogaster*, have no methylated DNA *(41,146)*.

3.8. Specialized or Short-Patch Mismatch Repair

In contrast to generalized mismatch repair, which recognizes a variety of mismatches, specialized mismatch repair recognizes a single type of mismatch, often only in a particular sequence context, and repair is directional and independent of strand location. In the cases studied to date, only a very short segment of DNA, <10 nt, is removed (*see* Chapter 11, this vol. and Chapter 8, vol. 2). One such very short-patch (VSP) repair process in *S. pneumoniae* is the conversion of A to C in mismatches of the sequence 5'-ATT**A**AT/TAA**G**TA-5', which results from a mutation in the *amiA* locus from 5'-ATT**C**AT to 5'-ATT**A**AT and was signaled by the abnormally high recombination frequencies shown by this marker *(30,75)*. The repair is always in the direction of the wild type *(75)*, and occurs only in 5'-ATT**A**AT/TAA**G**TA-5' and not the complementary heteroduplex *(103,132)*.

The A to C repair in *S. pneumoniae* is very similar to the VSP repair in *E. coli* of **T** to **C** in 5'-C**T**AG/G**G**TC-5' heteroduplexes *(79)*, which also occurs in a specific sequence context. This conversion requires the *vsr* gene *(135)*, which encodes a sequence-specific mismatch endonuclease *(42)* and is adjacent to *dcm*, the product of which methylates the second cytosine at 5'-CC(A or T)GG-3'*(93)* in *E. coli* strains that carry this gene pair. The function of this repair, therefore, is to prevent mutations that would result from deamination of the 5-methylcytosine to give T opposite G in that sequence. The pneumococcal repair resembles the *mutY* system *(2,83)* of *E. coli* in that it converts A/G mismatches to C/G. However, MutY acts independently of sequence context, and its action on A/G may be incidental to its intended function of correcting A/8-oxoG to C/G, where 8-oxoG is 8-oxoguanine, a common oxidation product of guanine that can pair with A during replication *(95*; *see* Chapter 6). Unlike the VSP and MutY systems of *E. coli,* no specific gene or protein has yet been implicated in the VSP repair of *S. pneumoniae*. However, all three systems require a pol I-like enzyme *(19,103,114)*. They may share a common mechanism of action in which initial recognition of the mismatch causes a strand break on the 5'-side of the nucleotide to be removed. The 5'-to-3' exonuclease of pol I then removes the incorrect nucleotide and up to 10 additional nucleotides, as it resynthesizes the very short patch.

3.9. Large Deletion/Insertion Mismatches

Although deletion/insertion mismatches larger than 5 nt are not recognized by the Hex repair system, large deletions (or insertions) in the donor or recipient DNA can affect recombination during transformation. In the *amiA* locus, the recombination frequency (to give wild type) between a point mutation in the chromosome and a deletion or insertion in the donor DNA was consistently 20% higher than the reciprocal transformation with the deletion or insertion mutation in the chromosome. The enhanced recombination occurred in Hex$^-$ strains and was evident for deletions or insertions >200 bp. The enhanced recombination frequency could result either from conversion of the donor component of the deletion/insertion mismatch in a transformation intermediate or from its exclusion (together with distal markers) from synaptic pairing *(76,104)*. Three-point genetic crosses can distinguish these alternatives, and initial results indicated that donor deletions were in fact converted in 20% of the crosses *(76)*. However, subsequent data from the same laboratory for both donor deletion and insertion mutations led to the conclusion that the donor DNA corresponding to the deletion or insertion was excluded to this extent from synaptic pairing *(104)*. At present, there is no clear evidence for repair of large deletion/insertion mismatches in *S. pneumoniae*.

4. PHYSICAL AND CHEMICAL DAMAGE TO DNA

Heating of DNA above a critical temperature causes a precipitous drop in transforming activity, because cells cannot take up the denatured DNA *(77)*. Annealing of the complementary strands restores much of the original activity *(86)*. This demonstration of DNA renaturation in the transformation of *S. pneumoniae* paved the way for the widespread use of nucleic acid hybridization in molecular biology.

Heating of DNA at subcritical temperatures causes a slow exponential reduction in transforming activity owing to depurination and consequent strand breaks *(37,66,122)*. Strand breaks in vivo are subject to repair, but in transforming DNA, they reduce the probability of synaptic pairing of incoming strands with recipient cell DNA. The effect of strand breaks on transforming activity is readily demonstrated by treatment of donor DNA with pancreatic DNase, which makes mainly single-strand breaks *(56,78)*. When assessing damage by other agents on transforming activity of DNA, the effects of strand breaks must be taken into account.

4.1. Chemical Alteration of Bases

Transforming DNA is inactivated by treatment in vitro with chemical agents that react with DNA bases. They may be oxidizing agents, such as hydrogen peroxide *(56)*, or alkylating agents, such as nitrogen mustards *(78)* and ethyl methane sulfonate *(56)*. Their kinetics of action indicate one-hit inactivation. Single-site markers show equal sensitivity to a given agent, independent of mismatch repair, and multisite markers show greater sensitivity, depending on the length of the multisite region. These agents produce a variety of altered bases, and the precise products likely to inactivate transformation in *S. pneumoniae* have not been determined; neither have their mechanisms of inactivation, nor the significance of their apparent target sizes been ascertained.

4.2. Nitrous Acid

Inactivation by nitrous acid has been thoroughly investigated. This agent also inactivates markers exponentially, at least initially *(55,81)*. Low-efficiency, single-site markers are inactivated at about twice the rate of high-efficiency markers, and multisite markers are somewhat more sensitive, with the marker corresponding to the largest deletion in the *malM* region (perhaps 10 kb in length) inactivated at only twice the rate of a low-efficiency, single-site marker. Although nitrous acid can deaminate C, A, and G residues in DNA to convert them, respectively, to uracil, hypoxanthine, and xanthine residues, deamination of C and A is relatively slow in duplex DNA *(81)*. Guanine residues are more readily altered, and ~10% of them react as rapidly as free guanine in solution *(81)*. These particularly sensitive guanine residues may occur in GC tracts. Nitrous acid-induced mutations are frequently found in such tracts (*see* Section 5.3.). Experimental data show that when 10% of G had already reacted, no C and only 1% of A residues were deaminated, whereas 99.9% of a single-site marker activity was destroyed *(81)*. Clearly, the G reaction product, presumably xanthine, is deleterious to the marker, possibly by causing a GC to AT mutation at essential sites within the marker gene or neighboring genes.

4.3. UV Light

UV irradiation stands out from other agents in showing a striking difference, more than 10-fold, in rates of inactivation of high-efficiency and low-efficiency single-site markers. This marker difference was noted in the earliest report on UV inactivation of transforming DNA *(78)*, and it was soon related to the marker integration efficiency *(22,55,81)*. The interconnection between Hex mismatch repair and UV inactivation, which gives an inverse relationship between integration efficiency and rate of inactivation *(55)*, is presented in Section 7.2.

Another distinctive feature of UV inactivation is that unlike most other agents, loss of transforming activity is not exponential, but, as first found for *H. influenzae* transformation *(124)*, survival of the marker is inversely proportional to the square of the UV dose *(53)*. A plausible explanation for this behavior is that survival of the marker requires both the absence of a lethal lesion in the target region that is ultimately integrated and avoidance of marker removal by excision or recombination repair of a lesion. As the UV dose increases, the lesions will be closer, and the probability of a recombination break between each of them and the marker will be inversely proportional to the dose; since two such switches are required (one on either side of the marker), the relationship will be to the square of the dose. In the absence of repair processes, marker loss is exponential, as shown for *H. influenzae (128)*. The most common UV lesions are cyclobutane pyrimidine dimers (CPDs) *(149)*, and since wild-type *S. pneumoniae* can excise CPDs (*see* Section 6.1.), the lethal lesions are presumably unrepaired CPDs.

There is no evidence for UV induction of mutations in *S. pneumoniae (33)*. Although mutations were obtained in the *malM* locus after UV irradiation of cells *(56)*, the frequency of mutants was not significantly higher than the spontaneous level *(53)*. The lack of UV mutagenesis was once taken to indicate the absence of an SOS repair system *(33)*, which in other bacteria is induced by DNA damage and enables recombinatory repair and error-prone synthesis of DNA *(15,158*; *see* Chapter 7). However, lysogenic induction by DNA damage, which results from action of SOS systems *(15,158)*, does occur in *S. pneumoniae (5,89)*. Error-prone repair requires particular genes, *umuC* and *umuD*, in the enteric bacteria *(151)*, as well as the SOS system, and the lack of UV mutagenesis in *S. pneumoniae* probably reflects the absence of *umu* homologs *(89)*.

5. MUTAGENESIS

5.1. Chemical Mutagens

Mutants have been isolated after transformation of wild-type *S. pneumoniae* by DNA treated in vitro with various chemical agents that affect its transforming activity, such as hydrogen peroxide and ethyl methane sulfonate, but for the most part, no significant increase over spontaneous mutation levels were reported *(22,56)*. An exception is nitrous acid, discussed in Section 5.3.

Treatment of cells with quinacrine produced a significant level of mutations in vivo *(31)*, which were shown by DNA sequencing to correspond to 1- and 2-nt insertions and deletions *(32)*. These mutations presumably resulted from intercalation of the quinacrine aromatic ring structure between bases in the cellular DNA, as was proposed for other acriflavine derivatives that produce frameshift mutations in *E. coli (138)*. The widely used mutagen MNNG is also highly mutagenic for *S. pneumoniae*; multiple treatments with the reagent gives mutation frequencies as high as 0.001 for various genes *(57)*.

5.2. MNNG and Its Potentiation by Glutathione

A recent investigation of the action of MNNG on cells of S. *pneumoniae* underlined the importance of intracellular activation of this agent *(52)*. Both the lethal and mutagenic actions of MNNG on cells of *S. pneumoniae* are increased >100-fold by addition of yeast extract to the culture medium *(52)*. The effective component in yeast extract is oxidized glutathione (GSSG). Mutations in the gene *glt*, which block transport of either glutathione (GSH) or GSSG into the cell, prevent the potentiation of MNNG action. The mechanism of the potentiating effect apparently depends on GSSG uptake, followed by its reduction in the cell to GSH by glutathione reductase, and then activation of internalized MNNG by GSH to give methyldiazohydroxide, which directly methylates bases in DNA *(52)*. Thus, GSSG, a seemingly innocuous substance in the cellular environment, renders low levels of MNNG highly cytotoxic and genotoxic for *S. pneumoniae*.

5.3. Nitrous Acid

Treatment of *S. pneumoniae* DNA with nitrous acid prior to its use in transformation of cells is an effective way to introduce randomly located mutations into a gene of this species. Both forward mutations *(22,56,81)* and reverse mutations of certain mutant genes *(56)* are readily obtained. Nitrous acid, however, does not provide a broad spectrum of mutations, because it produces almost exclusively GC to AT transitions *(9,22,40,56,70)*. Conversely, only those mutations resulting from AT to GC are reverted by nitrous acid *(9,56,70)*. When the frequencies of reversion are compared to inactivating hits on the DNA, it appears that a large proportion of hits, perhaps all, give rise to mutations *(56)*.

Although the chemical mechanism of mutagenesis by nitrous acid proceeds by deamination of the DNA bases, the precise reactions that inactivate transforming DNA, which is double-stranded, are not those that were initially suspected, that is, the deamination of A and C *(29)*. Deamination of cytosine to give uracil and of adenine to give hypoxanthine would cause, respectively, GC to AT and AT to GC transitions on replication of DNA containing the deaminated base, and both of these transitions are found after nitrous acid treatment of single-stranded bacteriophage DNA *(148)*. However, with double-stranded transforming DNA, nitrous acid produced only G:C to A:T transitions in the *malM* locus *(56)*. These were not reverted by nitrous acid, but other mutations, later shown to correspond to AT to GC forward transitions *(9,70)*, were reverted *(56)*.

A structural basis exists for the unidirectional transitions seen on reaction of transforming DNA. The amino groups of A and C are deaminated much more slowly than G in native DNA, whereas all three bases are rapidly deaminated in denatured DNA *(127)*.A component comprising ~10% of the G residues reacts particularly rapidly *(81)*; by the time this component had reacted, only 1% of A and no C was deaminated, yet 99.9% of the transforming activity was lost *(81)*. It is likely that the deaminative conversion of guanine to xanthine itself is mutagenic, because xanthine can pair with T as well as C *(23)*. Another notable feature of nitrous acid mutagenesis is that it occurs at "hotspots" in DNA containing runs of G and C *(9,40,70)*. Such runs might correspond to the more reactive 10% of G residues, which on conversion to xanthine become mutagenic. Alternatively, but less likely, they might correspond to sites of guanine crosslinking that, by perturbing the DNA structure, could render adjacent C residues susceptible to deamination *(40)*.

5.4. Sequence Repeats and Hotspots

An analysis of spontaneous reverse mutation frequencies for a set of 76 mutations at the *malM* locus *(56)* showed that almost all single-site mutations gave a detectable reversion frequency ($>10^{-10}$). Considerable variation in mutation frequency was found, with some mutants reverting with a frequency as high as 10^{-6}. (These measurements were made in a Hex^+ strain.)

Deletions in *S. pneumoniae*, as in other organisms *(1)*, often occur between directly repeated sequences *(70,76)*. The mechanism creating the deletion is presumably slippage of the DNA template from one repeat to the other during replication so as to eliminate the intervening DNA in the daughter strand *(1,138)*. Curiously, the single-site reversion mechanism creating the hotspot for *malM532* reversion also appears to depend on strand slippage during replication. The forward mutation, an AT to GC transition, occurred at position 2716 in the *malM* region in the sequence, $_{2710}$CAGAAG**A**C-3', where the mutated A is underlined *(9)*. Just 93 bp downstream the 8-bp sequence is repeated: $_{2803}$CAGAAGAC-3' *(70)*. Apparently, when the *malM532* mutant sequence ($_{2710}$CAGAAG**G**C-3') is replicated, the strand being synthesized occasionally slips down to the following sequence, where the template restores the wild-type AT configuration. The nascent strand would have to slip back to the original position upstream to avoid creating a deletion. Incidentally, the mutation *malM564* is just such a deletion *(70)*. The slippage back and forth to produce the *malM532* reversion occurs at a frequency of 5×10^{-8}, but this is 10-fold higher than the typical transition reversion frequency *(56)*, and it is sufficiently frequent to cause the mutational hotspot.

6. CELLULAR MECHANISMS FOR REPAIR OF DNA DAMAGE

6.1. UV Dimer Excision Repair

The first UV-sensitive mutant reported in *S. pneumoniae*, *uvr-1*, increased the sensitivity of cells by a factor of four *(57)*, indicating a repair system that removed 75% of the lethal DNA damage, presumably pyrimidine dimers. Transformation of the *uvr-1* strain with UV-irradiated DNA gave a fourfold lower yield of transformants, as compared to a Uvr^+ recipient *(57)*. This showed that damage to DNA irradiated outside the cell could be repaired after its entry into the cell, presumably after restoration of the donor DNA to a double-stranded form on integration into the chromosome. Such repair of transforming DNA is only observed for high-efficiency markers *(57)*, for reasons discussed in Section 7.2.

Another mutant, *uvr-402*, which was similarly sensitive to UV, was directly shown to be deficient in removal of CPDs *(24,133)*. It also rendered cells sensitive to mitomycin C, 4-nitroquinoline-1-oxide, and *cis*-diaminedichloroplatinum, which indicates that the CPD repair system is directed against bulky adducts in general *(133)*. The *uvr-402* mutation is complemented by a cloned segment of *S. pneumoniae* DNA *(133)* that expresses a gene homologous to *uvrB* of *E. coli (134)*. The cloned gene also complements the *uvr-1* mutation (N. Sicard, personal communication). It appears, therefore, that S. *pneumoniae* has an excision repair system similar to the *uvrA,B,C* system of *E. coli*.

Wild-type strains of *S. pneumoniae* are considerably more sensitive to UV irradiation than is *E. coli*. S_{37} values calculated from data for *S. pneumoniae* range between 4.2 and 8.7 J/m^2 *(17,32,57,89,133)*. Under similar conditions of irradiation, wild-type *E. coli* strains gave S_{37} values between 21.7 and 32.6 J/m^2 *(17,33)*. The relative susceptibility of *S. pneumoniae* has been attributed to the lack of an SOS repair mechanism that, as in *E. coli*, could overcome UV-damage by enabling DNA replication past CPDs *(33)*. The *lex-1* mutant of *E. coli*, which cannot carry out SOS repair, is only slightly more UV-sensitive than wild-type *S. pneumoniae (33)*. Absence of this error-prone feature of SOS repair would also explain the lack of UV mutability in *S. pneumoniae (33)*. Nevertheless, some features of the SOS system, such as lysogenic induction by DNA damage, are exhibited in *S. pneumoniae* (*see* Section 6.5.).

Similarly to *E. coli*, the excision repair of UV damage in *S. pneumoniae* requires a pol I type of DNA polymerase. The gene for the homologous pneumococcal enzyme was isolated *(90)*, analyzed *(83)*, and found to contain 5'-to-3'-exonuclease in addition to DNA polymerase activity *(16)* but no 3'-to-5'-exonuclease activity *(16)*. Its exonuclease activity is essential for cell viability, but that domain can be expressed separately from the polymerase domain *(17)*. A mutant pol I protein defective in polymerase activity resulted in much greater sensitivity to UV, reducing S_{37} to 1.4 J/m^2 *(17)*, which is approximately the resistance level of a *uvrB* mutant of *E. coli*.

6.2. AP-Endonuclease

The *exoA* gene product of *S. pneumoniae* is a DNA phosphatase-exonuclease that removes 3'-terminal nucleotide or phosphate residues from double-stranded DNA *(62)*. Continued exonucleolytic action converts the complementary strand to a single-stranded form *(62)*. This behavior is similar to that of *E. coli* ExoIII *(120)*, and cloning of the *exoA* gene showed the ExoA product and ExoIII to be related with 28% identity in amino acid sequence *(112)*. Like ExoIII *(152)*, ExoA exhibits a strand nicking activity at AP sites in DNA *(112)*. This activity could work in conjunction with glycosylases that remove damaged bases to repair DNA. Null *exoA* mutations in *S. pneumoniae* have not yet been made, and the functions of this enzyme in DNA repair are not known.

6.3. DNA–Uracil Glycosylase

A common base alteration that occurs spontaneously in DNA is the deamination of cytosine to give uracil, producing GC to AT mutations *(80)*. Many organisms, including *E. coli (80*; *see* Chapter 3), therefore, harbor a glycosylase that removes uracil from DNA. Subsequent single-strand cleavage at the apyrimidinic site by an AP endonuclease and removal of the ribose–phosphate residue, followed by repair synthesis, can restore the DNA. *S. pneumoniae* DNA-uracil glycosylase is encoded by its *ung* gene *(9)*. Mutants lacking the glycosylase show a 10-fold higher mutation frequency for GC to AT transitions, but no increase for AT to GC transitions in the *malM* gene *(9)*. In vitro deamination of cytosine in transforming DNA by treatment with bisulfite gives rise to mutations at the *ami* locus in a recipient population only when the recipient carries an *ung* mutation *(94)*. In an Ung$^+$ strain, the glycosylase could remove uracil from the incoming DNA strand before its integration into the chromosome, thereby eliminating the mutagenic effect of bisulfite. No such potentiation occurs for nitrous acid-induced mutations *(94)*, which supports the notion that the induced GC to AT mutations do not arise from deamination of cytosine, but rather from deamination of guanine, as indicated in Section 5.3., above. As indicated in Section 3.7., *ung* mutants show no reduction in mutation avoidance by the Hex system *(9)*.

6.4. 8-Oxo-dGTP Pyrophosphohydrolase

When the *ung* gene of *S. pneumoniae* was isolated, a gene downstream and apparently transcribed together with *ung* was found when blocked, to increase the spontaneous mutation rate of *S. pneumoniae* to streptomycin resistance *(93,94)*. This gene, called *mutX*, is homologous to the *mutT* gene of *E. coli (6*; *see* Chapter 6). The mutations to resistance at the *str* locus, which corresponds to a gene encoding the S12 ribosomal protein *(126)*, were AT to CG transversions *(93)*, the same type of mutation that occurs in *E. coli mutT* strains *(157)*. The *E. coli mutT* gene encodes an 8-oxo-dGTP pyrophosphohydrolase that by hydrolyzing 8-oxo-dGTP, prevents incorporation of 8-oxo-guanine into replicating DNA opposite adenine in the template strand *(95)*. Oxidative processes presumably convert significant amounts of dGTP to 8-oxo-dGTP in the cell *(95)*. Although it can pair with A, the 8-oxo-G residue pairs with A, thereby resulting in an AT to CG mutations *(95)*. The significance of finding the *ung* and *mutX* genes in the same operon remains to be determined.

6.5. Recombination Functions

Three different types of Rec$^-$ mutants, which take up DNA normally, but which are blocked in genetic transformation, have been found in *S. pneumoniae*. Two types were isolated directly by screening for a transformation defect *(99)*; they have been tentatively assigned to hypothetical loci *recP* and *recQ (60)*. In *recP* cells plasmid establishment is normal, but chromosomal transformation is reduced to ~5% of the normal level; in *recQ* cells, both plasmid transfer and chromosomal transformation are reduced to ~1% of normal levels *(99)*. Of the two, only *recP* has been cloned and sequenced *(115,119)*. Its gene product, a 72-kDa protein, is related to a family of transketolases *(99)*, and it is not clear how it affects recombination.

A *recA* gene from *S. pneumoniae* was identified by testing expressed products of cloned pneumococcal DNA for crossreaction with antibody against *E. coli* RecA *(87)*. The RecA protein from *S. pneumoniae is* 60% identical to that of *E. coli (87)*. An insertion in the *recA* gene had no effect on DNA uptake, but it reduced genetic transformation to <0.001% of the normal level. The pneumococcal *recA* gene was also obtained in a search for transformation defects of clones containing inserts in exportable proteins *(105)*. One clone contained *recA* together with a defective upstream gene called *exp10 (105)* (or *cinA [89*]), which is competence inducible. The normal product of this upstream gene is a 46-kDa membrane-associated protein. Defects in *exp10* reduce transformation to ~2% of normal *(108)*, but it was not reported whether uptake or recombination is affected. It appears that *recA* is constitutively transcribed from a promoter near its translation start site, but during the development of competence, an mRNA that starts upstream of *exp10/cinA* is made in greater amount *(89)*, which results in an approximately threefold increase of the RecA product in competent cells *(89)*.

That the pneumococcal RecA acts also in DNA repair is shown by the ~10-fold greater sensitivity to UV irradiation of cells containing an insertion mutation in *recA (89)*. The affected repair process in this case is presumably postreplication recombinational repair of UV damage, as opposed to excision repair by the *uvrB* system. The *recA* mutant is also slightly more sensitive to methyl methane sulfonate and mitomycin C than the wild type *(89)*. As in *E. coli*, the *recA* product of *S. pneumoniae* is required for lysogenic induction *(89)*. Lysogenic induction is an SOS function in *E. coli (121*; *see*

Table 1
S. pneumoniae* Genes with DNA Repair Functions and Their Homologs in *E. coli

Repair function	*S. pneumoniae*	*E. coli*	Ref.
Base mismatch recognition	*hexA*	*mutS*	*107*
Base mismatch removal	*hexB*	*mutL*	*109*
CPD recognition and incision	*uvr-1*	*uvrB*	*134*
Gap-filling replication	*polA*	*polA*	*83*
DNA-uracil removal	*ung*	*ung*	*94*
Recombinational pairing	*recA*	*recA*	*87*
8-oxo-GTP hydrolysis	*mutX*	*mutT*	*93*
Abasic site incision	*exoA*	*xth*	*112*

Chapter 7). *S. pneumoniae* appears to carry out some SOS functions, but not others. It can recognize the signal for DNA damage, but that signal does not lead to a significant increase in *recA* transcription *(89)*. Also missing are functions for error-prone replication (past UV-induced CPDs, for example) that would give rise to UV-induced phage reactivation and UV mutagenesis *(33)*. As pointed out *(89)*, the latter functions require the products of *umu* genes that may be absent in *S. pneumoniae*. Inasmuch as *S. pneumoniae* shows both excision repair and postreplication repair of UV damage, its greater sensitivity to UV irradiation compared to *E. coli* can perhaps be ascribed to its lack of an error-prone repair mechanism for bypassing UV damage. A summary of homologous genes in *S. pneumoniae* and *E. coli* with similar repair functions is given in Table 1.

7. INTERACTION OF CELLULAR REPAIR SYSTEMS

7.1. *Alkylation and Mismatch Repair*

Wild-type (Hex$^+$) cells of *S. pneumoniae* are significantly more sensitive to MNNG (by 2.5-fold) than are Hex$^-$ cells *(52)*. A plausible explanation for this difference is the convergent effect of two repair systems: the Hex mismatch repair system and a glycosylase-based repair of particular methylated bases in DNA *(141)*. An attempt by these two systems to repair DNA simultaneously on opposite strands could produce irreparable double-strand breaks in DNA. Two products of MNNG action on DNA are *O*-6-methylguanine and 3-methyladenine *(73)*. Generalized mismatch repair systems can recognize *O*-6-methylguanine paired with thymine *(48)*. In replicating DNA, the Hex system would remove a long segment of the newly synthesized strand in DNA containing the mismatch. Removal of a 3-methyladenine lesion in the opposite strand could give rise to a strand break by successive action of a 3-methyladenine DNA glycosylase similar to the enzyme reported in *E. coli (141)* and an AP-endonuclease, such as ExoA in *S. pneumoniae (112)*. The resulting double-strand break in the DNA could lead to cell death; the absence of such breaks could explain the enhanced resistance to MNNG of Hex$^-$ strains.

7.2. *UV Damage and Mismatch Repair*

Low-efficiency markers are very sensitive to UV irradiation of the donor DNA *(81)*, and this was amply documented for markers in the *malM* gene *(55)* and the *amiA* locus

(22). For example, a low-efficiency marker is reduced to 1% by a dose that only reduces of a high-efficiency marker to 35% *(55)*. In a Hex$^-$ recipient strain, in which all markers transform with high efficiency, both types of markers show 35% survival *(57)*. In a recipient strain containing *uvr-1*, which renders the cells defective in repair of UV damage (*see* Section 6.1.), the high-efficiency marker in irradiated DNA shows reduced survival compared to the Uvr$^+$ recipient, but the low-efficiency marker shows no difference, as if the UV damage was not repaired in the latter case *(57,61)*.

Why is it that low-efficiency markers do not appear to benefit from the excision repair process? Apparently, breaks resulting from excision of CPDs in the integrated donor segment give the mismatch repair system additional opportunities to correct the mismatch and remove the donor DNA *(61)*. The low efficiency level of marker integration presumably results from the mismatched DNA escaping correction in that fraction of the transformed population in which the breaks at the ends of the donor strand segment are closed prior to recognition of the mismatch. Excision of a CPD within such a donor segment reintroduces a break in the donor strand and subjects the marker once more to elimination by the Hex system. Any excision repair event that would remove lethal damage would eliminate the low-efficiency marker, as well. This interesting interplay of the Uvr excision repair system and the Hex mismatch repair system is apparently responsible for the hypersensitivity of low-efficiency markers to UV irradiation.

ACKNOWLEDGMENTS

I am indebted to Rollin D. Hotchkiss for introducing me to genetic transformation in *S. pneumoniae* and for thoughtful guidance and inspiration ever since. I also thank many other colleagues who have been helpful at various times. This chapter was written at Brookhaven National Laboratory under the auspices of the US Department of Energy Office of Health and Environmental Research and supported in part by US Public Health Service grant AI14885 from the National Institutes of Health.

REFERENCES

1. Albertini, A. M., M. Hofer, M. P. Calos, and J. H. Miller. 1982. On the formation of spontaneous deletions: the importance of short sequence homologies in the generation of large deletions. *Cell* **29:** 319–328.
2. Au, K. G., M. Cabrera, J. H. Miller, and P. Modrich. 1988. *Escherichia coli mutY* gene product is required for specific A-G → C.G mismatch correction. *Proc. Natl. Acad. Sci. USA* **85:** 9163–9166.
3. Avery, O. T., C. M. MacLeod, and M. McCarty. 1944. Studies on the chemical nature of the substance inducing transformation of pneumococcal types. Induction of transformation by a desoxyribonucleic acid fraction isolated from pneumococcus type III. *J. Exp. Med.* **89:** 137–158.
4. Balganesh, T. S. and S. A Lacks. 1985. Heteroduplex DNA mismatch repair system of *Streptococcus pneumoniae:* Cloning and expression of the *hexA* gene. *J. Bacteriol.* **162:** 979–984.
5. Bernheimer, H. P. 1977. Lysogeny in pneumococci freshly isolated from man. *Science* **195:** 66–68.
6. Bhatnagar, S. K. and M. J. Bessmam. 1988. Studies on the mutator gene, *mulT* of *Escherichia coli. J. Biol. Chem.* **263:** 8953–8957.
7. Bronner, C. E., S. M. Baker, P. T. Morrison, G. Warren, L. G. Smith, M. K. Lescoe, M. Kane, C. Earabino, J. Lipford, A. Lindblom, P. Tannergard, R. J. Bollag, A. R. Godwin,

D. C. Ward, M. Nordensjold, R. Fishel, R. Kolodner, and R. M. Liskay. 1994. Mutation in the DNA mismatch repair gene homologue *hMLH1* is associated with hereditary nonpolyposis colon cancer. *Nature* **368:** 258–261.

8. Chandler, M. and D. A. Morrisom. 1987. Competence for genetic transformation in *Streptococcus pneumoniae:* molecular cloning of *com,* a competence control locus. *J. Bacteriol.* **169:** 2005–2011.
9. Chen, J. D. and S. A. Lacks. 1991. Role of uracil-DNA glycosylase in mutation avoidance by *Streptococcus pneumoniae. J. Bacteriol.* **173:** 283–290.
10. Claverys, J. P. and S. A. Lacks. 1986. Heteroduplex deoxyribonucleic acid base mismatch repair in bacteria. *Microbiol. Rev.* **50:** 133–165.
11. Claverys, J. P., H. Prats, H. Vasseghi, and M. Gherardi. 1984. Identification of *Streptococcus pneumoniae* mismatch repair genes by an additive transformation approach. *Mol. Gen. Genet.* **196:** 91–96.
12. Claverys, J. P., M. Roger, and A. M. Sicard. 1980. Excision and repair of mismatched base pairs in transformation of *Streptococcus pneumoniae. Mol. Gen. Genet.* **178:** 191–201.
13. Claverys, J. P., V. Mejean, A. M. Gasc, and A. M. Sicard. 1983. Mismatch repair in *Streptococcus pneumoniae:* relationship between base mismatches and transformation efficiencies. *Proc. Natl. Acad. Sci. USA* **80:** 5956–5960.
14. Claverys, J. P., V. Mejean, A. M. Gasc, F. Galibert, and A. M. Sicard. 1981. Base specificity of mismatch repair in *Streptococcus pneumoniae. Nucleic Acids Res.* **9:** 2267–2280.
15. Defais, M., P. Fauquet, M. Radman, and M. Errera. 1971. Ultraviolet reactivation and ultraviolet mutagenesis of lambda in different genetic systems. *Virology* **43:** 495–503.
16. Diaz, A., M. E. Pons, S. A. Lacks, and P. Lopez. 1992. *Streptococcus pneumoniae* DNA polymerase I lacks 3'-to-5' exonuclease activity: Localization of the 5'-to-3' exonucleolytic domain. *J. Bacteriol.* **174:** 2014–2024.
17. Diaz, A., S. A. Lacks, and P. Lopez. 1992. The 5' to 3' exonuclease activity of DNA polymerase I is essential for *Streptococcus pneumoniae. Mol. Microbiol.* **6:** 3009–3019.
18. Dohet, C., R. Wagner, and M. Radman. 1985. Repair of defined single base-pair mismatches in *Escherichia coli. Proc. Natl. Acad. Sci. USA* **82:** 503–505.
19. Dzidic, S. and M. Radmam 1989. Genetic requirements for hyper-recombination by very short patch mismatch repair: Involvement of *Escherichia coli* DNA polymerase I. *Mol. Gen. Genet.* **217:** 254–256.
20. Ephrussi-Taylor, H. 1966. Genetic recombination in DNA-induced transformation of pneumococcus. IV. The pattern of transmission and phenotypic expression of high and low efficiency donor sites in the *ami-A* locus. *Genetics* **54:** 211–222.
21. Ephrussi-Taylor, H. and T. C. Gray. 1966. Genetic studies of recombining DNA in pneumococcal transformation. *J. Gen. Physiol.* **49, part 2:** 211–231.
22. Ephrussi-Taylor, H., A. M. Sicard, and R. Kamen. 1965. Genetic recombination in DNA-induced transformation of pneumococcus. I. The problem of relative efficiency of transforming factors. *Genetics* **51:** 455–475.
23. Eritja, R., D. M. Horowitz, P. A. Walker, J. P. Ziehler-Martin, M. S. Boosalis, M. F., Good, K. Itakura, and B. E. Kaplan. 1986. Synthesis and properties of oligonucleotides containing 2'-deoxynebularine and 2'-deoxyxanthosine. *Nucleic Acids Res.* **14:** 8135–8153.
24. Estevenon, A. M. and N. Sicard. 1989. Excision-repair capacity of UV-irradiated strains of *Escherichia coli* and *Streptococcus pneumoniae*, estimated by plasmid recovery. *J. Photochem. Photobiol. B Biol.* **3:** 185–192.
25. Fishel, R., M. K. Lescoe, M. R. S. Rao, N. G. Copeland, N. A. Jenkins, J. Garber, M. Kane, and R. Kolodner. 1993. The human mutator gene homolog *MSH2* and its association with hereditary nonpolyposis colon cancer. *Cell* **75:** 1027–1038.
26. Fleischmann, R. D., M. D. Adams, O. White, R. A. Clayton, E. F. Kirkness, A. R. Kerlavage, C. J. Bult, J.-F. Tomb, B. A. Dougherty, J. M. Merrick, K. McKenney, G. Sutton, W. FitzHugh, C. Fields, J. D. Gocayne, J. Scott, R. Shirley, L.-I. Liu, A. Glodek, J. M. Kelley,

J. F. Weidman, C. A. Phillips, T., Spriggs, E. Hedblom, M. D. Cotton, T. R. Utterback, M. C. Hanna, D. T. Nguyen, D. M. Saudek, R. C. Brandon, L. D. Fine, J. L. Fritchman, J. L. Fuhrmann, N. S. M. Geoghagen, C. L. Gnehm, L A. McDonald, K. V., Small C. M. Fraser, H. O. Smith, and J. C. Venter. 1995. Whole-genome random sequencing and assembly of *Haemophilus influenzae* Rd. *Science* **269:** 449–604.

27. Fox, M. S. 1960. Fate of transforming deoxyribonucleate following fixation by transforming bacteria. II. *Nature* **187:** 1004–1001.
28. Freese, E. 1959. On the molecular explanation of spontaneous and induced mutations. *Brookhaven Symp. Biol.* **12:** 63–75.
29. Fujii H. and T. Shimada 1989. Isolation and characterization of cDNA clones derived from the divergently transcribed gene in the region upstream from the human dihydrofolate reductase gene. *J. Biol. Chem.* **264:** 10,057–10,064.
30. Garcia, P., A. M. Gasc, X. Kyriakidis, D. Baty, and A. M. Sicard. 1988. DNA sequences required to induce localized conversion in *Streptococcus pneumoniae* transformation. *Mol. Gen. Genet.* **214:** 509–513.
31. Gasc, A. M. and A. M. Sicard. 1978. Genetic studies of acridine-induced mutants in *Streptococcus pneumoniae. Genetics* **90:** 1–18.
32. Gasc, A. M., and A. M. Sicard. 1986. Frame-shift mutants induced by quinacrine are recognized by the mismatch repair system in *Streptococcus pneumoniae. Mol. Gen. Genet.* **203:** 269–273.
33. Gasc, A. M., N. Sicard, J. P. Claverys, and A. M. Sicard. 1980. Lack of SOS repair in *Streptococcus pneumoniae. Mutat. Res.* **70:** 157–165.
34. Gasc, A. M., P. Garcia D. Baty, and A. M. Sicard. 1987. Mismatch repair during pneumococcal transformation of small deletions produced by site-directed mutagenesis. *Mol. Gen. Genet.* **210:** 369–372.
35. Ghei, O. K. and S. A. Lacks. 1967. Recovery of donor deoxyribonucleic acid marker activity from eclipse in pneumococcal transformation. *J. Bacteriol.* **93:** 816–829.
36. Glickman, B. W. and M. Radman. 1980. *Escherichia coli* mutator mutants deficient in methylation-instructed DNA mismatch correction. *Proc. Natl. Acad. Sci. USA* **77:** 1063–1067.
37. Greer, S. and S. Zamenhof. 1962. Studies on depurination of DNA by heat. *J. Mol. Biol.* **4:** 123–141.
38. Griffith, F. 1928. The significance of pneumococcal types. *J. Hyg.* **27:** 113–159.
39. Haber, L. T., P. P. Pang, D. L. Sobell, J. A. Mankovich, and G. C. Walker. 1988. Nucleotide sequence of the *Salmonella typhimurium mutS* gene required for mismatch repair: homology of MutS and HexA of *Streptococcus pneumoniae. J. Bacteriol.* **170:** 197–202.
40. Hartman, Z., E. N. Henrikson, P. E. Hartman, and T. A. Cebula. 1994. Molecular models that may account for nitrous acid mutagenesis in organisms containing double-stranded DNA. *Environ. Mol. Mutagen.* **24:** 168–175.
41. Hattman, S., C. Kenny, L. Berger, and K. Pratt. 1978. Comparative study of DNA methylation in three unicellular eucaryotes. *J. Bacteriol.* **135:** 1115–1157.
42. Haverstein, L. S., G. Coomaraswamy, and D. A Morrison. 1995. An unmodified heptadecapeptide induces competence for genetic transformation in *Streptococcus pneumoniae. Proc. Natl. Acad. Sci. USA* **92:** 1114–11144.
43. Hennecke, F., H. Kolmar, K Brundl, and H. J. Fritz. 1991. The *vsr* gene product of *E. coli* R12 is a strand- and sequence-specific DNA mismatch endonuclease. *Nature* **353:** 776–778.
44. Hollingsworth, N. M., L. Ponti, and C. Halsey. 1995. *MSH5*, a novel MutS homolog, facilitates meiotic reciprocal recombination between homologs in *Saccharomyces cerevisiae* but not mismatch repair. *Genes Dev.* **9:** 1728–1739.
45. Holmes, J., S. Clark, and P. Modrich. 1990. Strand-specific mismatch correction in nuclear extracts of human and *Drosophila melanogaster* cell lines. *Proc. Natl. Acad. Sci. USA* **87:** 5837–5841.

46. Hsieh, P., C. S. Camerini-Otero, and R. D. Camerini-Otero. 1990. Pairing of homologous DNA sequences by proteins: evidence for three-stranded DNA. *Genes Dev.* **4:** 1951–1963.
47. Hui, F. M. and D. A. Morrison. 1991. Genetic transformation in *Streptococcus pneumoniae:* Nucleotide sequence analysis shows *comA*, a gene required for competence induction, to be a member of the bacterial ATP-dependent transport-protein-family. *J. Bacteriol.* **173:** 372–381.
48. Karran, P. and M. G. Marinus. 1982. Mismatch correction at O^6-methylguanine residues in *E. coli* DNA. *Nature* **296:** 868,869.
49. Kornberg, A. and T. A. Baker. 1992. *DNA Replication.* W. H. Freeman, New York.
50. Kramer, B., W. Kramer, and H. J. Fritz. 1984. Different base/base mismatches are corrected with different efficiencies by the methyl-directed DNA mismatch-repair system of *E. coli. Cell* **38:** 879–887.
51. Kramer, W. B., Kramer, M. S. Williamson, and S. Fogel. 1989. Cloning and nucleotide sequence of DNA mismatch repair gene *PMS1* from *Saccharomyces cerevisiae:* homology of PMS1 to procaryotic MutL and HexB. *J. Bacteriol.* **171:** 5339–5346.
52. Kumaresan, K. R., S. S. Springhorn, and S. A. Lacks. 1995. Lethal and mutagenic action of *N*-methyl-*N'*-nitro-*N*-nitrosoguanidine potentiated by oxidized glutathione, a seemingly innocuous substance in the cellular environment. *J. Bacteriol.* **177:** 3641–3646.
53. Lacks, S. Unpublished data.
54. Lacks, S. 1962. Molecular fate of DNA in genetic transformation of pneumococcus. *J. Mol. Biol.* **5:** 119–131.
55. Lacks, S. 1965. Genetic recombination in pneumococcus, in *The Physiology of Gene and Mutation Expression* (Kohoutova, M. and J. Hubacek, ed.), Academia, Prague, pp. 159–164.
56. Lacks, S. 1966. Integration efficiency and genetic recombination in pneumococcal transformation. *Genetics* **53:** 207–235.
57. Lacks, S. 1970. Mutants of *Diplococcus pneumoniae* that lack deoxyribonucleases and other activities possibly pertinent to genetic transformation. *J. Bacteriol.* **101:** 373–383.
58. Lacks, S. 1979. Uptake of circular deoxyribonucleic acid and mechanism of deoxyribonucleic acid transport in genetic transformation of *Streptococcus pneumoniae. J. Bacteriol.* **138:** 404–409.
59. Lacks, S. A. 1977. Binding and entry of DNA in bacterial transformation, in *Microbial Interactions, Receptors and Recognition* (Reissig, J. L., ed.), Chapman and Hall, London, pp. 179–232.
60. Lacks, S. A. 1988. Mechanisms of genetic recombination in gram-positive bacteria, in *Genetic Recombination* (Kucherlapati, R. and G. Smith, ed.), American Society for Microbiology, Washington, DC, pp. 43–85.
61. Lacks, S. A. 1989. Generalized DNA mismatch repair—its molecular basis in *Streptococcus pneumoniae* and other organisms, in *Genetic Transformation and Expression* (Butler, L. O., C. Harwood, and B. E. B. Moseley, eds.), Intercept, Wimbourne, England, pp. 325–339.
62. Lacks, S. and B. Greenberg. 1967. Deoxyribonucleases of pneumococcus. *J. Biol. Chem.* **242:** 3108–3120.
63. Lacks, S. and B. Greenberg. 1973. Competence for deoxyribonucleic acid uptake and deoxyribonuclease action external to cells in the genetic transformation of *Diplococcus pneumoniae. J. Bacteriol.* **114:** 152–163.
64. Lacks, S. and B. Greenberg. 1976. Single-strand breakage on binding of DNA to cells in the genetic transformation of *Diplococcus pneumoniae. J. Mol. Biol.* **101:** 255–275.
65. Lacks, S. and M. Neuberger. 1975. Membrane location of a deoxyribonuclease implicated in the genetic transformation of *Diplococcus pneumoniae. J. Bacteriol.* **124:** 1321–1329.
66. Lacks, S. and R. D. Hotchkiss. 1960. A study of the genetic material determining an enzyme activity in pneumococcus. *Biochim. Biophys. Acta* **39:** 508–517.
67. Lacks, S., B. Greenberg, and K. Carlson. 1967. Fate of donor DNA in pneumococcal transformation. *J. Mol. Biol.* **29:** 327–347.

68. Lacks, S., B. Greenberg, and M. Neuberger. 1974. Role of a deoxyribonuclease in the genetic transformation of *Diplococcus pneumoniae*. *Proc. Natl. Acad. Sci. USA* **71:** 2305–2309.
69. Lacks, S., B. Greenberg, and M. Neuberger. 1975. Identification of a deoxyribonuclease implicated in genetic transformation of *Diplococcus pneumoniae*. *J. Bacteriol.* **123:** 222–232.
70. Lacks, S. A, J. J. Dunn, and B. Greenberg. 1982. Identification of base mismatches recognized by the heteroduplex-DNA-repair system of *Streptococcus pneumoniae*. *Cell* **31:** 327–336.
71. Lahue, R. S., K. G. Au, and P. Modrich. 1989. DNA mismatch correction in a defined system. *Science* **245:** 160–164.
72. Langle-Rouault, F. M., M. G. Maenhaut, and M. Radman. 1987. GATC sequences, DNA nicks and the MutH function in *Escherichia coli* mismatch repair. *EMBO J.* **6:** 1121–1127.
73. Lawley, P. D. and C. J. Thatcher. 1970. Methylation of deoxyribonucleic acid in cultured mammalian cells by *N*-methyl-*N'*-nitro-*N*-nitrosoguanidine. *Biochem. J.* **116:** 693–707.
74. Leach, F. S., N. C. Nicolaides, N. Papadopoulos, B. Liu, J. Jen, R. Parsons, P. Peltonmaki, P. Sistonen, L A. Aaltonen, M. Nystrom-Lahti, X.-Y. Guan, J. Zhang, P. S. Meltzer, J.-W. Yu, F.-T. Kao, D. J. Chen, K M. Cerosaletti, R. E. K. Fournier, S. Todd, T. Lewis, R. J. Leach, S. L Naylor, J. Weissenbach, J.-P. Mecklin, H. Jarvinen, G. M. Petersen, S. R. Hamilton, J. Green, J. Jass, P. Watson, H. T. Lynch, J. M. Trent, A. de la Chapelle, K W. Kinzler, and B. Vogelstein. 1993. Mutations of a *mutS* homolog in hereditary nonpolyposis colorectal cancer. *Cell* 75:1215–1225.
75. Lefevre, J. C., A. M. Gasc, A. C. Burger, H. Mostachfi, and A. M. Sicard. 1984. Hyperrecombination at a specific DNA sequence in pneumococcal transformation. *Proc. Natl. Acad. Sci. USA* **81:** 5184–5188.
76. Lefevre, J. C., P. Mostachfi, A. M. Gasc, E. Guillot, F. Pasta, and M. Sicard. 1989. Conversion of deletions during recombination in pneumococcal transformation. *Genetics* **123:** 455–464.
77. Lerman, R. S. and L. J. Tolmach 1957. Genetic transformation. I. Cellular incorporation of DNA accompanying transformation in pneumococcus. *Biochim. Biophys. Acta* **28:** 68–82.
78. Lerman, L. S. and L. J. Tolmach. 1959. Genetic transformation. II. The significance of damage to the molecule. *Biochim. Biophys. Acta* **33:** 371–387.
79. Lieb, M. 1985. Recombination in the lambda repressor gene. Evidence that very short patch (VSP) mismatch correction restores a specific sequence. *Mol. Gen. Genet.* **199:** 465–470.
80. Lindahl, T. 1974. An N-glycosidase from *Escherichia coli* that releases free uracil from DNA containing deaminated cytosine residues. *Proc. Natl. Acad. Sci. USA* **71:** 3649–3653.
81. Litman, R. M. 1961. Genetic and chemical alterations in the transforming DNA of pneumococcus caused by ultraviolet light and by nitrous acid. *J. Chim. Phys.* **58:** 997–1004.
82. Lopez, P., M. Espinosa, B. Greenberg, and S. A. Lacks. 1987. Sulfonamide resistance in *Streptococcus pneumoniae:* DNA sequence of the gene encoding dihydropteroate synthase and characterization of the enzyme. *J. Bacteriol.* **169:** 4320–4326.
83. Lopez, P., S. Martinez, A. Diaz, M. Espinosa, and S. A. Lacks. 1989. Characterization of the *polA* gene of *Streptococcus pneumoniae* and comparison of the DNA polymerase I it encodes to homologous enzymes from *Escherichia coli* and phage T7. *J. Biol. Chem.* **264:** 4255–4263.
84. Lu, A. L. and D. Y. Chang. 1988. A novel nucleotide excision repair for the conversion of an A/G mismatch to C/G base pair in *E. coli*. *Cell* **54:** 805–812.
85. Mankovich, J. A., C. A. McIntyre, and G. C. Walker. 1989. Nucleotide sequence of the *Salmonella typhimurium mutL* gene required for mismatch repair: homology of MutL to HexB of *Streptococcus pneumoniae* and to PMS1 of the yeast *Saccharomyces cerevisiae*. *J. Bacteriol.* **171:** 5325–5331.

86. Marmur, J.and D. Lane. 1960. Strand separation and specific recombination in deoxyribonucleic acids: Biological studies. *Proc. Natl. Acad. Sci. USA* **46:** 453–461.
87. Martin, B., J. M. Ruellan, J. F. Angulo, R. Devoret, and J. P. Claverys. 1992. Identification of the *recA* gene of *Streptococcus pneumoniae. Nucleic Acids Res.* **20:** 6412.
88. Martin, B., O. Humbert, M. Camara, E. Guenzi, J. Walker, T. Mitchell, P. Andrew, M. Prudhomme, G. Alloing, R. Hakenbeck, D. A. Morrison, G. J. Boulnois, and J. P. Claverys. 1992. A highly conserved repeated DNA element located in the chromosome of *Streptococcus pneumoniae. Nucleic Acids Res.* **20:** 3479–3483.
89. Martin, B., P. Garcia, M. P. Castanie, and J. P. Claverys. 1995. The *recA* gene of *Streptococcus pneumoniae* is part of a competence-induced operon and controls lysogenic induction. *Mol. Microbiol.* **15:** 367–379.
90. Martinez, S., P. Lopez, M. Espinosa, and S. A. Lacks. 1986. Cloning of a gene encoding a DNA polymerase-exonuclease of *Streptococcus pneumoniae.* Gene **44:** 79–88.
91. May, M. S. and S. Hattman. 1975. Deoxyribonucleic acid-cytosine methylation by host- and plasmid-controlled enzymes. *J. Bacteriol.* **122:** 129–138.
92. Mejean, V. and J. P. Claverys. 1984. Effect of mismatched base pairs on the fate of donor DNA in transformation of *Streptococcus pneumoniae. Mol. Gen. Genet.* **197:** 467–471.
93. Mejean, V., C. Salles, L. C. Bullions, M. J. Bessman, and J. P. Claverys. 1994. Characterization of the *mutX* gene of *Streptococcus pneumoniae* as a homologue of *Escherichia coli mutT,* and tentative definition of a catalytic domain of the dGTP pyrophosphohydrolases. *Mol. Microbiol.* **11:** 323–330.
94. Mejean, V., J. C. Devedjian, I. Rives, G. Alloing, and J. P. Claverys. 1991. Uracil-DNA glycosylase affects mismatch repair efficiency in transformation and bisulphite-induced mutagenesis in *Streptococcus pneumoniae. Nucleic Acids Res.* **19:** 5525–5531.
95. Michaels, M. L and J. H. Miller. 1992. The GO system protects organisms from the mutagenic effeect of the spontaneous lesion 8-hydroxyguanine (7,8-dihydro-8-oxoguanine). *J. Bacteriol.* **174:** 6321-6325.
96. Morrison, D. A. and B. Mannarelli. 1979. Transformation in pneumococcus: nuclease resistance of deoxyribonucleic acid in the eclipse complex. *J. Bacteriol.* **140:** 655–665.
97. Morrison, D. A. and M. F. Baker. 1979. Association of competence for genetic transformation in Pneumococcus with synthesis of a small set of proteins. *Nature* **282:** 215–217.
98. Morrison, D. A. and W. R. Guild. 1973. Breakage prior to entry of donor DNA in *pneumococcus* transformation. *Biochim. Biophys. Acta* **299:** 545–556.
99. Morrison, D. A., S. A. Lacks, W. G. Guild, and J. M. Hageman. 1983. Isolation and characterization of three new classes of transformation-deficient mutants of *Streptococcus pneumoniae* that are defective in DNA transport and genetic recombination. *J. Bacteriol.* **156:** 281–290.
100. New, L., K. Liu, and G. F. Crouse. 1993. The yeast gene *MSH3* defines a new class of eukaryotic MutS homologs. *Mol. Gen. Genet.* **239:** 97–108.
101. Nicolaides, N. C., N. Papadopoulos, B. Liu, Y.-F. Wei, K. C. Carter, S. M. Ruben, C. A. Rosen, W. A. Haseltine, R. D. Fleischmann, C. M. Fraser, M. D. Adams, J. C. Venter, M. G. Dunlop, S. R. Hamilton, G. M. Petersen, A. de la Chapelle, B. Vogelstein, and K. W. Kinzler. 1994. Mutations of two *PMS* homologues in hereditary nonpolyposis colon cancer. *Nature* **371:** 75–80.
102. Papadopoulos, N., N. C. Nicolaides, B. Liu, R. Parsons, C. Lengauer, F. Palombo, A. D'Arrigo, S. Markowitz, J. K. V. Willson, K. W. Kinzler, J. Jiricny, and B. Vogelstein. 1995. Mutations of *GTBP* in genetically unstable cells. *Science* **208:** 1915–1917.
103. Pasta, F. and M. A. Sicard. 1994. Hyperrecombination in pneumococcus: A/G to C.G repair and requirement for DNA polymerase I. *Mut. Res.* **315:** 113–122.
104. Pasta, F. and M. A. Sicard. 1996. Exclusion of long heterologous insertions and deletions from the pairing synapsis in pneumococcal transformation. *Microbiology* **142:** 695–705.

105. Pearce, B. J., A. M. Naughton, E. A. Campbell, and H. R. Masure. 1995. The *rec* locus, a competence-induced operon in *Streptococcus pneumoniae. J. Bacteriol.* **177:** 86–93.
106. Prats, H., B. Martin, and J. P. Claverys. 1985. The *hexB* mismatch repair gene of *Streptococcus pneumoniae:* characterization, cloning and identification of the product. *Mol. Gen. Genet.* **200:** 482–489.
107. Priebe, S. D, S. M. Hadi, B. Greenberg, and S. A. Lacks. 1988. Nucleotide sequence of the *hexA* gene for DNA mismatch repair in *Streptococcus pneumoniae* and homology of *hexA* to *mutS* of *Escherichia coli* and *Salmonella typhimurium. J. Bacteriol.* **170:** 190–196.
108. Prolla, T. A, D. M. Christie, and R. M. Liskay. 1994. Dual requirement in yeast DNA mismatch repair *for MLH1 and PMS1,* two homologs of the bacterial *mutL* gene. *Mol. Cell. Biol.* **14:** 407–415.
109. Prudhomme, M., B. Martin, V. Mejean, and J. P. Claverys. 1989. Nucleotide sequence of the *Streptococcus pneumoniae hexB* mismatch repair gene: homology of HexB to MutL of *Salmonella typhimurium* and to PMS1 of *Saccharomyces cerevisiae. J. Bacteriol.* **171:** 5332–5338.
110. Prudhomme, M., V. Mejean, B. Martin, and J. P. Claverys. 1991. Mismatch repair genes of *Streptococcus pneumoniae:* HexA confers a mutator phenotype in *Escherichia coli* by negative complementation. *J. Bacteriol.* **173:** 7196–7203.
111. Pukkila, P. J., J. Peterson, G. Herman, P. Modrich, P., and M. Meselson. 1983. Effects of high levels of DNA adenine methylation on methyl-directed mismatch repair in *Escherichia coli. Genetics* **104:** 571–582.
112. Puyet, A., B. Greenberg, and S. A. Lacks. 1989. The *exoA* gene of *Streptococcus pneumoniae* and its product, a DNA exonuclease with apurinic endonuclease activity. *J. Bacteriol.* **171:** 2278–2286.
113. Puyet, A., B. Greenberg, and S. A. Jacks. 1990. Genetic and structural characterization of EndA, a membrane-bound nuclease required for transformation of *Streptococcus pneumoniae. J. Mol. Biol.* **213:** 727–738.
114. Radicella, J. P., E. A. Clark, S. Chen, and M. S. Fox. 1993. Patch length of localized repair events: role of DNA polymerase I in *mutY*-dependent mismatch repair. *J. Bacteriol.* **175:**7732–7736.
115. Radnis, B. A., D. K Rhee, and D. A. Morrison. 1990. Genetic transformation in *Streptococcus pneumoniae:* nucleotide sequence and predicted amino acid sequence of *recP. J. Bacteriol.* **172:** 3669–3674.
116. Reenan, R. A. G. and R. D. Kolodner. 1992. Characterization of insertion mutations in the *Saccharomyces cerevisiae MSH1* and *MSH2* genes: evidence for separate mitochondrial and nuclear functions. *Genetics* **132:** 975–985.
117. Reenan, R. A. G. and R. D. Kolodner. 1992. Isolation and characterization of two *Saccharomyces cerevisiae* genes encoding homologs of the bacterial HexA and MutS mismatch repair proteins *Genetics* **132:** 963–973.
118. Reizer, J., A. Reizer, A. Bairoch, and M. H. Saier. 1993. A diverse transketolase family that includes the RecP protein of *Streptococcus pneumoniae,* a protein implicated in genetic recombination. *Res. Microbiol.* **144:** 34; 347.
119. Rhee, D. K. and D. A. Morrison. 1988. Genetic transformation in *Streptococcus pneumoniae:* Molecular cloning and characterization of *recP,* a gene required for genetic recombination. *J. Bacteriol.* **170:** 630–637.
120. Richardson, C. C, I. R. Lehman, and A. Kornberg. 1964. A deoxyribonucleic acid phosphatase-exonuclease from *Escherichia coli.* II. Characterization of the exonuclease activity. *J. Biol. Chem.* **239:** 251–258.
121. Roberts, J. W., C. W. Roberts, and N. L. Craig. 1978. *Escherichia coli recA* product inactivates phage lambda repressor. *Proc. Natl. Acad. Sci. USA* **75:** 4714–4718.
122. Roger, M. and R. D. Hotchkiss. 1961. Selective heat inactivation of pneumococcal transforming deoxyribonudeate. *Proc. Natl. Acad. Sci. USA* **47:** 653–669.

123. Ross-Macdonald, P. and G. S. Roeder. 1994. Mutation of a meiosis-specific MutS homolog decreases crossing over but not mismatch correction. *Cell* **79:** 1069–1080.
124. Rupert, C S. and S. H. Goodgal. 1960. Shape of ultra-violet inactivation curves of transforming deoxyribonucleic acid. *Nature* **185:** 556,557.
125. Sabelnikov, A. G., B. Greenberg, and S. A. Lacks. 1995. An extended -10 promoter alone directs transcription of the *DpnII* operon of *Streptococcus pneumoniae. J. Mol. Biol.* **250:** 144–155.
126. Salles, C., L. Creancier, J. P. Claverys, and V. Mejean. 1994. The high level streptomycin resistance gene from *Streptococcus pneumoniae is* a homologue of the ribosomal protein S12 gene from *Escherichia coli. Nucleic Acids Res.* **20:** 6103.
127. Schuster, H. 1960. The reaction of nitrous acid with deoxyribonucleic acid. *Biochem. Biophys. Res. Commun.* **2:** 320–323.
128. Setlow, J. K. 1977. The shape of the ultraviolet inactivation curve for transforming DNA. *Nature* **268:** 169,170.
129. Seto, H. and A. Tomasz. 1974. Early stages in DNA binding and uptake during genetic transformation of pneumococci. *Proc. Natl. Acad. Sci. USA* **71:** 1493–1498.
130. Shibata, T., C. DasGupta, R. P. Cunningham, and C. M. Radding. 1979. Purified *Escherichia coli recA* protein catalyzes homologous pairing of superhelical DNA and single-stranded fragments. *Proc. Natl. Acad. Sci. USA* **76:** 1638–1642.
131. Shoemaker, N. B. and W. R. Guild. 1974. Destruction of low efficiency markers is a slow process occuring at a heteroduplex stage of transformation. *Mol. Gen. Genet.* **128:** 283–290.
132. Sicard. A. M., J. C. Lefevre, P. Mostachfi, A. M. Gasc, and C. Sarda. 1985. Localized conversion in *Streptococcus pneumoniae* recombination: heteroduplex preference. *Genetics* **110:** 557–568.
133. Sicard, N. and A. M. Estevenon. 1990. Excision-repair capacity in *Streptococcus pneumoniae:* cloning and expression of a uvr-like gene. *Mutat. Res.* **235:** 195–201.
134. Sicard, N., J. Oreglia, and A. M. Estevenon. 1992. Structure of the gene complementing *uvr-402* in *Streptococcus pneumoniae:* Homology with *Escherichia coli uvrB* and the homologous gene in *Micrococcus luteus. J. Bacteriol.* **174:** 2412–2415.
135. Sohail, A., M. Lieb, M. Dar, and A. S. Bhagwat. 1990. A gene required for very short patch repair in *Escherichia coli* is adjacent to the DNA cytosine methylase gene. *J. Bacteriol.* **172:** 4214–4221.
136. Steruglanz, R., H. F. Wang, and J. J. Donegan. 1976. Evidence that both growing DNA chains at a replication fork are synthesized discontinuously. *Biochemistry* **15:** 1838–1843.
137. Strand, M., M. C. Earley, G. F. Crouse, and T. D. Petes. 1995. Mutations in the *MSH3* gene preferentially lead to deletions within tracts of simple repetitive DNA in *Saccharomyces cerevisiae. Proc. Natl. Acad. Sci. USA* **92:** 10,418–10,421.
138. Streisinger, G., Y. Okada, J. Emrich J. Newton, A. Tsugita, E. Terzaghi, and M. Inouye. 1966. Frameshift mutations and the genetic code. *Cold Spring Harbor Symp. Quant. Biol.* **31:** 77–84.
139. Su, S. S. and P. Modrich. 1986. *Escherichia coli* mutS-encoded protein binds to mismatched DNA base pairs. *Proc. Natl. Acad. Sci. USA* **83:** 5057–5061.
140. Su, S. S., R. S. Lahue, K. G. Au, and P. Modrich. 1988. Mispair specificity of methyl-directed DNA mismatch correction *in vitro. J. Biol. Chem.* **263:** 6829–6835.
141. Thomas, L., C. H. Yang, and D. A. Goldthwait. 1982. Two DNA glycosylases in *Escherichia coli* which release primarily 3-methyladenine. *Biochemistry* **21:** 1162–1169.
142. Tiraby, G. and M. A. Sicard. 1973. Integration efficiencies of spontaneous mutant alleles of *amiA* locus in pneumococcal transformation. *J. Bacteriol.* **116:** 1130–1135.
143. Tiraby, G. and M. S. Fox. 1973. Marker discrimination in transformation and mutation of pneumococcus. *Proc. Natl. Acad. Sci. USA* **70:** 3541–3545.
144. Tomasz, A. 1966. Model for the mechanism controlling the expression of competent state in pneumococcus cultures. *J. Bacteriol.* **91:** 1050–1061.

145. Tye, B. K., J. Chien, I. R. Lehman, B. K. Duncan, and H. R. Warner. 1978. Uracil incorporation: A source of pulse-labeled DNA fragments in the replication of the *Escherichia coli* chromosome. *Proc. Natl. Acad. Sci. USA* **75:** 233–237.
146. Urieli-Shoval, S., Y. Gruenbaum, J. Sedat, and A. Razin. 1982. The absence of detectable methylated bases in *Drosophila melanogaster* DNA. *FEBS Lett.* **146:** 148–152.
147. Vanderbilt, A. S. and I. Tessmam. 1970. Identification of the altered bases in mutated single-stranded DNA. IV. Nitrous acid induction of the transitions guanine to adenine and thymine to cytosine. *Genetics* **66:** 1–10.
148. Varlet, L., C. Pallard, M. Radman, I. Moreau, and N. de Wind. 1994. Cloning and expression of the *Xenopus* and mouse *Msh2* DNA mismatch repair genes. *Nucleic Acids Res.* **22:** 5723–5728.
149. Wacker, A. 1963. Molecular mechanisms of radiation effects. *Prog. Nucleic Acid Res.* **1:** 369–399.
150. Wagner, R. and M. Meselson. 1976. Repair tracts in mismatched DNA heteroduplexes. *Proc. Natl. Acad. Sci. USA* **73:** 4135–4139.
151. Walker, G. C. 1984. Mutagenesis and inducible responses to deoxyribonucleic acid damage in *Escherichia coli. Microbiol. Rev.* **48:** 60–93.
152. Weiss, B. 1976. Endonuclease II of *Escherichia coli is* exonuclease III. *J. Biol. Chem.* **251:** 1896–1901.
153. Welsh, K. M., A. L. Lu, S. Clark, and P. Modrich. 1987. Isolation and characterization of the *Escherichia coli mutH* gene product. *J. Biol. Chem.* **262:** 15,624–15,629.
154. White, J. H., K. Lusnak, and S. Fogel. 1985. Mismatch-specific postmeiotic segregation frequency in yeast suggests a heteroduplex recombination intermediate. *Nature* **315:** 350–352.
155. Wildenberg, J. and M. Meselson 1975. Mismatch repair in heteroduplex DNA. *Proc. Natl. Acad. Sci. USA* **72:** 2202–2206.
156. Williamson, M. S., J. C. Game, and S. Fogel. 1985. Meiotic gene conversion in *Saccharomyces cerevisiae.* I. Isolation and characterization of *pms1-1* and *pms1-2. Genetics* **110:** 609–646.
157. Yanofsky, C., E. C. Cox, and V. Horn 1966. The unusual mutagenic specificity of an *E. coli* mutator gene. *Proc. Natl. Acad. Sci. USA* **55:** 274–281.
158. Yasbin, R. E., D. L. Cheo, and K. W. Bayles. 1991. The SOB system of *Bacillus subtilis:* a global regulon involved in DNA repair and differentiation. *Res. Microbiol.* **142:** 885–892.

14

DNA Repair in *Deinococcus radiodurans*

John R. Battista

1. INTRODUCTION

In his introduction to the first review that discussed the radiobiology of *Deinococcus radiodurans*, Moseley *(49)* observed that this bacterium and its relatives were "regarded largely as a scientific curiosity" and that "the research effort put into them has been paltry compared with that devoted to *E. coli*." Thirteen years later, this commentary still holds true. Relative to other well-studied prokaryotes, very little is known of *D. radiodurans* and its unique DNA repair capabilities, even though this organism is the most DNA-damage-resistant ever identified. This is not to say that our understanding of *D. radiodurans* strategies for surviving genetic damage has not advanced since 1983. To the contrary, a number of significant observations have been made during the past decade that have helped to better define DNA repair in *D. radiodurans*; observations that suggest that *D. radiodurans* response to DNA damage is remarkably complex and perhaps different from that of any other prokaryote examined to date. In this chapter is emphasized a discussion of those aspects of *D. radiodurans* DNA damage resistance that seem to set it apart from other prokaryotes, while providing a description of what is known of DNA repair in this organism.

2. SUMMARY DESCRIPTION OF *D. RADIODURANS*

2.1. General Characteristics

The genus *Deinococcus* consists of four characterized species—*D. radiodurans*, *D. proteolyticus*, *D. radiopugnans*, and *D. radiophilus*—with *D. radiodurans* the type species for the genus *(7,57)*. These nonsporing, nonmotile, spherical bacteria are from 1.5–3 μm in diameter. Cells are chemoorganotrophic with respiratory metabolism. The optimal growth temperature is 30°C, with growth remaining strong to 37°C, but ceasing at 45°C.

The *D. radiodurans* chromosome is estimated to be 3×10^6 base pairs *(23)*, and there is believed to be a minimum of four identical copies of the chromosome per stationary-phase cell *(31)*. In exponentially growing cells, the number of chromosome copies increases to as many as 10/cell. The chromosome appears to be a single covalently closed circular molecule. The base composition of all deinococcal species is characterized by a high GC content, ranging from 65 to 71%.

From: DNA Damage and Repair, Vol. 1: DNA Repair in Prokaryotes and Lower Eukaryotes
Edited by: J. A. Nickoloff and M. F. Hoekstra © Humana Press Inc., Totowa, NJ

The natural habitat of the deinococci has not been defined, even though *Deinococcus* strains have been identified in a variety of locations worldwide. The nutritional requirements of the deinococci suggest that they are native to rich organic environments, and the successful isolation of these organisms from soil *(39,58,63)*, animal feces *(35)*, processed meat *(22,39)*, and sewage *(35)* supports this contention. It is, however, possible that in nature the deinococci grow in the presence of other microorganisms that provide them with complex organic nutrients, greatly expanding the number of possible environments that the deinococci could exploit. *D. radiopugnans*, for example, has been identified as part of an endolithic community in the Ross desert of Antarctica, considered to be the most hostile environment on Earth *(59)*, and these bacterial communities are subject to dehydration, severe cold, and intense UV light, receiving between 425 and 1000 h of sunlight annually. For a vegetative organism, *D. radiodurans* is highly resistant to desiccation, exhibiting 10% viability, according to one anecdotal report, after storage in a desiccator for 6 yr *(57)*.

2.2. DNA Damage Tolerance

D. radiodurans displays an extraordinary ability to tolerate the lethal effects of a variety of DNA-damaging agents, with unusually high resistance to ionizing radiation and UV light *(45,49)*. *D. radiodurans* exhibits resistance to killing by crosslinking agents, nitrous acid, hydroxylamine, *N*-methyl-*N'*-nitro-*N*-nitrosoguanidine (MNNG), and 4-nitroquinoline-*N*-oxide *(66,67)*. In contrast, *D. radiodurans* is as sensitive as *Escherichia coli* to other simple alkylating agents such as *N*-methyl-*N*-nitrourethane, ethyl methane sulfonate, and β-propriolactone. There is also a single report that *D. radiodurans* is unusually sensitive to near-UV light (300–400 nm) *(8)*. Treatment of *D. radiodurans* cultures with many of these agents, even at highly lethal doses, fails to increase the frequency of mutation above that of the spontaneous mutation rate *(67)*. Only alkylating agents effectively induce mutagenesis.

2.2.1. UV Light

Wild-type *D. radiodurans* cultures in exponential-phase growth are extremely resistant to UV light (254 nm), surviving doses as high a 1000 J/m^2 *(54)*. As illustrated in Fig. 1, a typical UV survival curve is characterized by a shoulder of resistance that extends to approx 500 J/m^2, clearly distinguishing *D. radiodurans* from other vegetative bacteria. For comparison, Fig. 1 also depicts the survival curve of *E. coli* B/r. *D. radiodurans* cultures irradiated with doses lower than 500 J/m^2 exhibit negligible loss of viability, but beyond this dose, there is an exponential reduction in survival. The D_{37} dose (i.e., the dose that, on average, is required to inactivate a single colony-forming unit [CFU] of *D. radiodurans*) is between 550 and 600 J/m^2 for exponential cultures. The D_{37} value for *E. coli* B/r is 30 J/m^2 *(67)*. Approximately 1% of the irradiated cell's thymine is present as part of a pyrimidine dimer after exposure to 500 J/m^2 UV *(4,70)*. Assuming that the *D. radiodurans* genome is 3 × 10^6 bp with 67% GC content, it is possible that as many as 5000 thymine-containing pyrimidine dimers form per genome when the cell is subjected to a 500 J/m^2 dose, with an average distance between thymine-containing pyrimidine dimers of only 600 bp *(4,64)*.

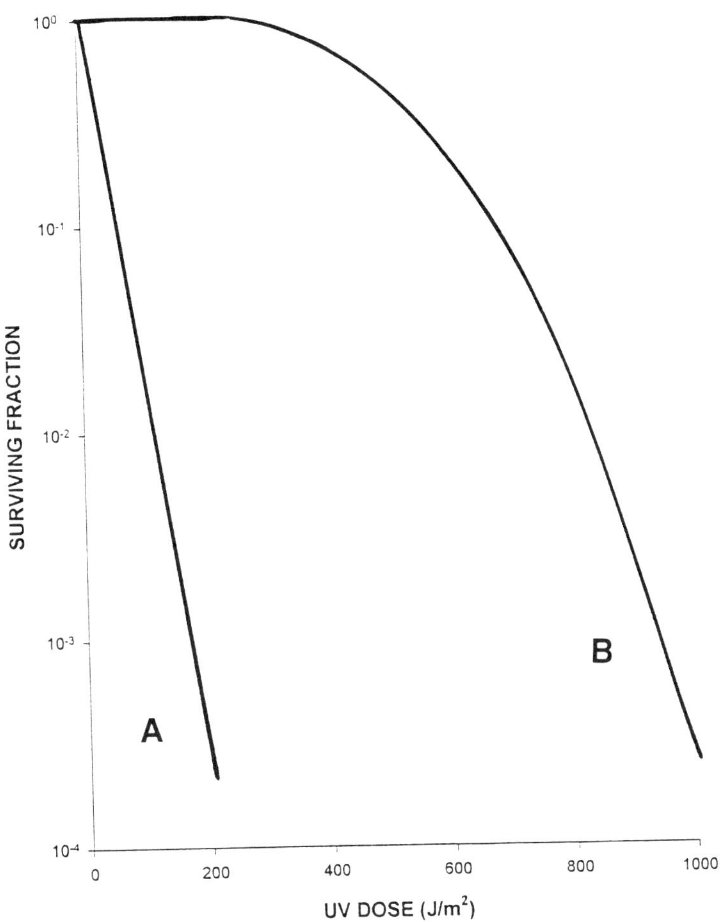

Fig. 1. Representative survival curves for *E. coli* B/r **(A)** and *D. radiodurans* R1 **(B)** following exposure to UV radiation.

2.2.2. *Ionizing Radiation*

The deinococci's most distinguishing characteristic is their ability to tolerate ionizing radiation. In fact, they are the most ionizing-radiation-resistant organisms ever identified. It is this property that led to their initial isolation and that is still used to facilitate their isolation from other natural microflora *(35)*. The typical γ-radiation survival curve (Fig. 2) for *D. radiodurans* R1 exhibits a shoulder of resistance to 5000 Gy, in which there is no loss of viability *(54)*. At doses above 5000 Gy, damage apparently begins to overwhelm the cell's defenses, producing an exponential loss of viability. Exponential-phase cultures routinely survive 15 kGy γ-radiation *(54)*, however, and there are reports of deinococcal strains surviving as much as 50 kGy *(3)*. The D_{37} dose following exposure to ionizing radiation for *D. radiodurans'* R1 in exponential growth is approx 6000 Gy. To put *D. radiodurans* ionizing radiation resistance in perspective, the survival of E *coli* B/r is also plotted in Fig. 2.

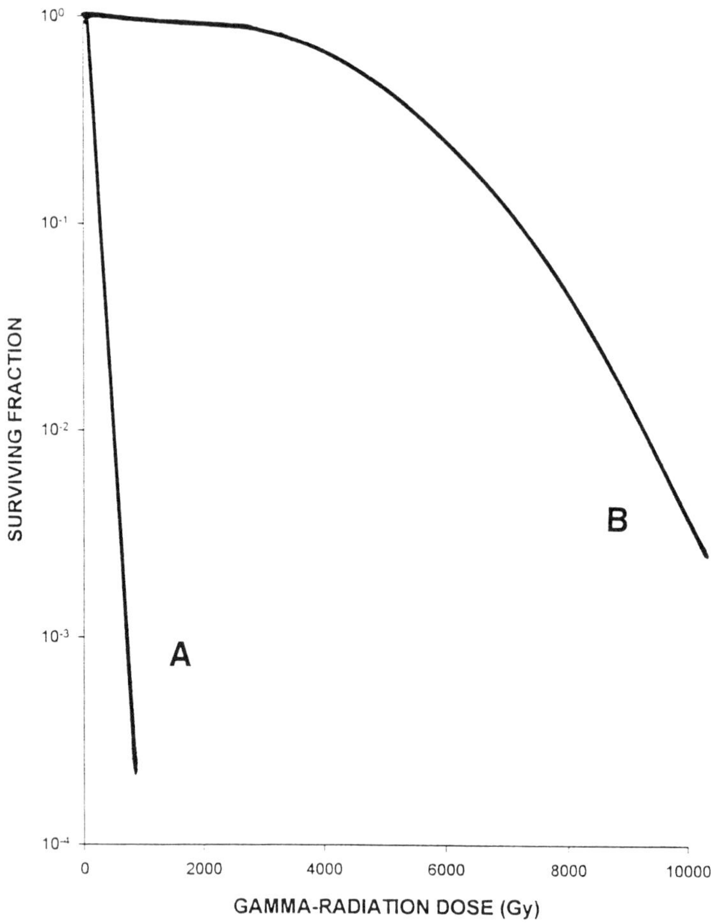

Fig. 2. Representative survival curves for *E. coli* B/r **(A)** and *D. radiodurans* R1 **(B)** following exposure to γ-radiation.

This dose of γ-radiation will induce approx 200 DNA double-strand breaks (DSBs) *(38)*, over 3000 single-strand breaks *(5)*, and greater than 1000 sites of base damage per *D. radiodurans* chromosome. The effect of 5000 Gy γ-radiation on the *D. radiodurans* chromosome is dramatically illustrated by comparing the chromosomal DNA preparations from irradiated and unirradiated cultures by pulsed field gel electrophoresis (Fig. 3). In irradiated cultures, the chromosome has been reduced from a single 3×10^6 bp fragment to a wide band of fragments 50 kbp or smaller in size. At 5200 Gy, there is 100% survival in the irradiated population.

2.2.3. Crosslinking Agents

D. radiodurans is also quite resistant to the crosslinking agent mitomycin C, surviving incubation with high concentrations of this antibiotic. In the presence of 20 μg/mL mitomycin for 10 min at 30°C, there is no loss of viability, and, after 40 min of incubation at this concentration, 1% of the culture still remains viable *(36)*, making *D. radiodurans* four times more resistant to mitomycin C than *E. coli* B/r *(67)*. Although

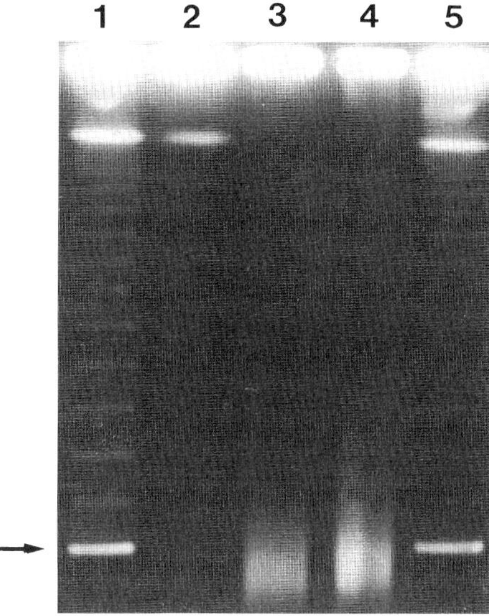

Fig. 3. The accumulation of chromosomal DNA DSBS in *D. radiodurans* R1 cultures subjected to ionizing radiation and desiccation. Lane 1 is a λ ladder size standard. Lane 2 is *D. radiodurans* R1 chromosomal DNA prepared from an untreated culture. Lane 3 is R1 chromosomal DNA prepared from a culture that had been desiccated for 6 wk. Lane 4 is R1 chromosomal DNA prepared from a culture that had been exposed to 5200 Gy γ-radiation.

the number of crosslinks formed are difficult to estimate under these conditions, Kitayama *(36)* has reported that >90% of the chromosomal DNA isolated from cultures treated for 10 min exists as fragments of nondenaturable double-stranded DNA with an average mol wt of 2×10^7 Dalton, indicating that at least 100 mitomycin C-induced crosslinks form/chromosome at this dose.

3. DNA DAMAGE-SENSITIVE STRAINS OF *D. RADIODURANS*

Even though *D. radiodurans* is extremely resistant to the lethal and mutagenic effects of many DNA-damaging agents, cultures are mutable when treated with MNNG. In fact, virtually all of the known base substitution mutations that render *D. radiodurans* sensitive to DNA damage were generated using alkylating agents. Transposon mutagenesis has not been reported in *D. radiodurans*, probably because the drug resistance markers used to identify successful transposition are not expressed. The promoter structure of *D. radiodurans* seems to be quite different from that of other prokaryotes *(65)*. In general, *D. radiodurans* promoters are not recognized by the RNA polymerases of other species, and *D. radiodurans* RNA polymerase does not recognize the promoters of other species. Mutants with increased sensitivity to mitomycin C *(53)*, UV *(48)*, and ionizing radiation *(68)* have been isolated.

4. THE ENZYMOLOGY OF DNA DAMAGE REPAIR

The capacity to survive such massive insults to their genetic integrity suggests that the *Deinococcaceae* have evolved distinctive mechanisms of DNA damage tolerance, and genetic evidence argues that efficient DNA repair is an integral part of this tolerance. However, only a handful of studies in the last 40 yr have attempted to detail the biochemistry of any process associated with deinococcal DNA repair. Thus far, only three types of DNA repair have been described:

1. Nucleotide excision repair;
2. Base excision repair; and
3. Recombinational repair.

Attempts at identifying photoreactivation and error-prone repair systems *(48)* have been unsuccessful. There are no reports of attempts to identify systems for correcting mismatched bases.

4.1. Evidence for Nucleotide Excision Repair

Studies of *D. radiodurans* R1 have shown that resistance to UV-induced DNA damage is, in part, mediated by the activity of a protein identified as endonuclease α *(53)*. Endonuclease α recognizes a broad range of DNA damage, incising the DNA at or near the site of that damage. Strains that express a defective form of endonuclease α are sensitive to mitomycin C *(51)*. Mutational inactivation of either of two loci, designated *mtcA* and *mtcB*, was shown to inactivate endonuclease α. Until very recently, it was believed that *mtcA* and *mtcB* encoded different proteins, but Agostini and Minton *(1,44)* reported that the *mtcAB* region of *D. radiodurans* is a single gene that is 60% homologous with the *uvrA* locus of *E. coli* and *Micrococcus luteus*, suggesting that endonuclease α is a functional homolog of the UvrABC excinuclease of *E. coli.* To date, only this UvrA-like protein has been identified in *D. radiodurans.* There is no evidence that *D. radiodurans* expresses homologs of the *uvrB* and *uvrC* gene products, but a concerted effort to search for these proteins or their coding sequences has not been described. Kitayama and colleagues *(37)* reported that there were as many as five additional independent *mtc* loci, designated *mtcC* through *mtcG,* associated with the repair of mitomycin C-induced DNA damage. These *mtc* loci have not been characterized.

The *uvrA (mtcA,B)* strains are not sensitive to UV light or γ-radiation, and they are recombination proficient. The inactivation of a second locus, designated *uvs*, is required before a *uvrA* strain becomes UV-sensitive *(20,53)*. Three *uvs* loci—*uvsC*, *uvsD*, and *uvsE*—have been identified. Strains that carry only a *uvrA* or *uvs* mutation exhibit near wild-type levels of UV resistance, indicating that the *uvrA* and *uvs* gene products encode functionally redundant proteins. *uvs* mutants exhibit wild-type resistance to mitomycin C and γ-radiation. *uvrA uvs* double mutants are sensitive to mitomycin C, but are as resistant to γ-radiation as wild-type *D. radiodurans.*

Recently, 45,000 MNNG-treated colonies of the *uvrA* strain *D. radiodurans* 302 were screened for ionizing radiation sensitivity, and 49 putative ionizing-radiation-sensitive (IRS) strains were identified *(68)*. Of these, 40 were subsequently shown to be truly IRS, and those were subdivided into 16 linkage groups *(43)*. Three of the IRS strains were identified as *pol* mutants, and three others as *rec* mutants. The remaining loci were designated *irr* (*i*onizing *r*adiation *r*esistance), and each linkage group was assigned a different letter, *irrA–irrL*. Because they were derived from a *uvrA* strain, all 40 of the IRS strains are sensitive to mitomycin C. Udupa et al. *(68)* showed that when IRS18 *(uvrA1, irrB1)* was transformed to mitomycin C resistance with the appropriate *uvrA*$^+$ sequence, the strain's ionizing radiation resistance was largely restored, indicating that the *uvrA* gene product plays a role in both ionizing radiation and UV resistance.

4.2. *Evidence for Base Excision Repair*

Base excision repair in *D. radiodurans* has not been examined in detail, even though this type of repair would be expected in an organism capable of withstanding ionizing radiation-induced DNA damage. There have been reports describing an apurinic/apyrimidinic (AP) endonuclease activity *(41)*, a uracil-*N*-glycosylase activity *(41)*, a DNA deoxyribophophodiesterase (dRPase) activity *(56)*, and a thymine glycol glycosylase activity *(56)* in cell extracts of *D. radiodurans*. The AP endonuclease has been partially purified, and it has a mol wt of 34.5 kDa. It did not exhibit glycosylase activity toward DNA containing uracil, alkylated bases, or UV photoproducts. The AP endonuclease does not require a metal, since it was active in the presence of EDTA. The uracil-*N*-glycosylase was specific for uracil-containing DNA and has no metal requirement. The molecular weight of dRPase and the thymine glycol glycosylase were estimated at between 25 and 30 kDa, respectively. The dRPase requires Mg^{2+} for activity, whereas the thymine glycol glycosylase does not require a metal for activity. The influence of these proteins on cellular DNA repair and viability subsequent to DNA damage has not yet been assessed.

There are also reports that suggest that *D. radiodurans* expresses a pyrimidine-dimer DNA glycosylase (PD DNA glycosylase) analogous to the PD DNA glycosylases isolated from *Micrococcus luteus* and bacteriophage T4. Gutman et al. *(28)* have shown that expression of the *denV* gene of T4 in a *uvrA uvs* strain of *D. radiodurans* partially restores UV resistance to these strains, indicating that the *uvs* gene product, also called endonuclease β, has some degree of specificity for pyrimidine dimers and suggesting that the *denV* gene product and endonuclease β have similar functions. Endonuclease β is a 36-kDa protein that has been partially purified, and it appears to be momomeric *(19)*. Its activity does not require ATP and it exhibits a novel requirement for Mn^{+2}. Three mutations—in loci designated *uvsC, uvsD*, and *uvsE*— have been isolated that inactivate endonuclease β activity *(20)*. It has been suggested that each mutation affects a separate coding sequence, because genomic DNA from each *uvs* mutant will restore wild type UV resistance to the other two *uvs* mutants.

Like the characterized PD DNA glycosylases, endonuclease β catalyzes an incision adjacent to pyrimidine dimers, facilitating their removal, but the action of the *D. radiodurans'* enzyme in vitro is not associated with an AP lyase activity *(18)*. The *M. luteus* and T4 PD DNA glycosylase activities result in the cleavage of the *N*-glycosyl bond of the 5' base in the pyrimidine dimer, producing a structure that will release a free thymine on photoreversal. The action of endonuclease β apparently does not cleave this bond, however, since free thymine is not released when photoreversal is attempted on UV-irradiated DNA treated with endonuclease β. Alternatively, the enzyme may have lost this activity during purification.

4.3. *Evidence for DSB Repair*

When an exponential-phase culture of *D. radiodurans* is exposed to 5000 Gy γ-radiation, >150 DNA DSBs are introduced into the chromosome *(38)*. The DSBs are not only repaired without loss of viability, but the chromosome is also apparently reassembled so that the linear continuity of the genome is unaffected by either the damage or the repair process. Mutational inactivation of the *rec* locus of *D. radiodurans* R1 results in ionizing radiation sensitivity and defective DSB repair *(10)*. *rec* strains are also sensitive to mitomycin C and UV light, however, indicating that the *rec* gene product is also involved in repair pathways that act on other types of DNA damage. The *rec* gene has been cloned, sequenced, and its gene product shown to be 56% identical to *E. coli* RecA protein *(25)*. Despite this similarity, *recA* from *D. radiodurans* will not complement an *E. coli recA* mutant. The expression of the deinococcal RecA protein is toxic to *E. coli*; even low-level expression is lethal, suggesting a fundamental difference in the activity of the two proteins.

Daly and Minton have followed the kinetics of DNA DSB repair in stationary-phase cells exposed to 17.5 kGy, which produces approx 150 DSBs/cell, using pulsed-field gel electrophoresis *(9)*. In the first 1.5 h postirradiation, there is a *recA*-independent increase in the molecular weight of chromosomal DNA of the irradiated population, followed at three hours by a *recA*-dependent increase in molecular weight, which continues over the next 29 h until the chromosome is restored to its normal size. These authors infer that the *recA* independent reassembly of the chromosome is the consequence of a single-strand annealing reaction, such as that catalyzed by the RecE and RecT proteins of *E. coli,* but there is no evidence verifying this prediction.

4.4. Evidence for a Deinococcal DNA Polymerase I Homolog

Gutman et al. *(27)* cloned and sequenced a gene encoding DNA polymerase that is necessary for DNA damage resistance in *D. radiodurans*. This *pol* gene product shares 51.1% homology with DNA polymerase I of *E. coli.* As is observed with *E. coli polA* strains *(24)*, insertional mutagenesis of the *pol* coding sequence generates a strain that is extremely sensitive to UV, ionizing radiation, and mitomycin C. In addition, a *pol* mutant can be restored to DNA damage resistance by expression of *E. coli* DNA polymerase I *(26)*. It appears that the *pol* gene product of *D. radiodurans* and *E. coli* DNA polymerase I have similar, if not identical, activities.

4.5. The IrrI and IrrB Proteins

Two new loci, *irrB* and *irrI,* that are involved in *D. radiodurans* DNA damage resistance have been identified *(68)*. Inactivation of either locus results in a partial loss of resistance to ionizing radiation. The magnitude of this loss is locus specific and differentially affected by inactivation of the *uvrA* gene product. An *irrB urvA* double mutant is more sensitive to ionizing radiation than is an *irrB* mutant. In contrast, the *irrI uvrA* double mutant and the *irrI* mutant are equally sensitive to ionizing radiation. The *irrB* and *irrI* mutations also reduce *D. radiodurans* resistance to UV radiation, this effect being most pronounced in *uvrA*$^+$ backgrounds.

The functions of the IrrI and IrrB gene products are unknown, but there is preliminary evidence that IrrI may be a regulatory protein. A *uvrA irrI1* strain is only slightly more sensitive to UV than strain 302, its *irrI*$^+$ parent. In contrast, a *uvrA*$^+$ *irrI1* strain is extremely sensitive to UV light. This result suggests that the *irrI* gene product regulates either the activity of endonuclease α or an enzymatic activity that arises subsequent to the action of endonuclease α.

In many respects, the effect of the *irrI* mutation in a *uvr*$^+$ background following UV irradiation is similar to the effects observed when wild-type *D. radiodurans* is treated with chloramphenicol prior to UV irradiation *(50)*. Immediately following UV irradiation, there is evidence of extensive DNA degradation, presumably resulting from the enzymatic removal of DNA damage *(50,70)*. This process appears to be regulated by an inducible protein since the administration of chloramphenicol prior to irradiation allows DNA degradation to proceed unchecked with lethal consequences. In *uvrA* mutants, the lethal effect of chloramphenicol addition is not observed, indicating that chloramphenicol stops the synthesis of a protein that is downregulating the process of DNA degradation *(32)*. Presumably, inactivation of endonuclease α prevents lethal degradation, because this enzyme either catalyzes the degradation or initiates the degradative process. Endonuclease α could, by incising the DNA at the site of damage, trigger the action of an exonuclease that degrades DNA until it is inactivated by a second protein. Conceptually, this type of interaction is analogous to that observed when the *gam* protein of phage λ inactivates RecBCD exonuclease of *E. coli (61)*. The similarities between the effects of chloramphenicol and the *irrI* mutation on UV resistance suggest that the *irrI* gene product is a regulatory protein and that it may be the relevant protein inhibited by chloramphenicol pretreatment.

5. CELLULAR RESPONSES TO DNA DAMAGE

Although most of the work discussed in this section was conducted more than 20 years ago, its relevance to the study of *D. radiodurans* DNA repair capability should not be underestimated, since it suggests that the repair processes of this organism are far more complex than might be predicted from what is known of the enzymology of *D. radiodurans* DNA repair.

5.1. *Inhibition of DNA Synthesis*

Exposure to UV or γ-radiation triggers a fairly well-defined sequence of events in *D. radiodurans*. Initially, DNA replication is inhibited as measured by the uptake of ^{3}H-thymidine, which correlates with the introduction of DNA damage. In other organisms, UV irradiation produces bulky lesions which physically block the movement of the replicative polymerase, preventing DNA replication *(29,47)*, and there is no reason to believe that this type of blockage does not occur in *D. radiodurans.* Alternatively, the inhibition of DNA replication may be part of a regulatory mechanism initiated in response to DNA damage. The latter possibility is mentioned, because it has been noted in numerous studies that *D. radiodurans* chromosomal DNA replication is delayed until the completion of DNA repair *(11,40,52)*, suggesting that the cell can assess the extent of DNA damage and control DNA polymerase activity accordingly. The inhibition of DNA replication does not affect DNA repair, however. Studies with inhibitors of DNA replication *(16,17)* and with the temperature-sensitive mutants *(55)* that block DNA replication at the nonpermissive temperature clearly show that DNA damage is repaired even though replication is blocked.

The delay in DNA replication following DNA damage is linearly related to the radiation dose and to the extent of DNA repair as long as the dose administered is sublethal. Following a lethal dose of radiation, there is no correlation between the length of damage repair and the delay in DNA replication. DNA synthesis stops following a lethal dose of radiation, but restarts before the completion of repair *(11)*. This premature reinitiation of replication may, at least in part, be responsible for the observed lethality.

5.2. Induction of Protein Synthesis

One of the most surprising aspects of *D. radiodurans'* response to DNA damage is the limited increase in new protein synthesis. Following exposure to ionizing radiation, Only four *(30)* proteins, designated α, β, γ, and δ are induced as measured by ^{35}S-methionine incorporation. In contrast, induction of the SOS response in *E. coli* results in the expression of over 20 gene products *(72)*. The deinococcal proteins appear to be critical to cell survival subsequent to γ-irradiation, since inhibition of protein synthesis dramatically affects the DNA repair processes. Chloramphenicol pretreatment prevents the restitution of ionizing radiation-induced single- *(5,12,13)* and double-strand DNA breaks *(38)* and the excision of thymine glycol *(33)* from irradiated cells. The lack of significant increases in protein synthesis in *D. radiodurans* cells exposed to ionizing radiation suggests either that little protein synthesis is necessary (i.e., that most of the proteins needed for repair are expressed constitutively) and/or that very few proteins are needed for effective DNA repair in this organism.

5.3. Postirradiation DNA Degradation and the Export of Damaged DNA

Following exposure of *D. radiodurans* to UV or γ-radiation, there is an immediate burst of intracellular DNA degradation that accompanies the inhibition of DNA replication. The rate of degradation is independent of the dose of radiation administered, with approx 0.1% of total genomic DNA being degraded/min *(40,50,70)*. However, the absolute amount of DNA degraded (i.e., the length of time the degradative process continues) is dependent on the amount of radiation axâministered. After sublethal exposure, DNA replication is delayed until the degradation ceases, indicating that the degradative process is a manifestation of DNA repair.

The primary degradation products appear within 5 min of irradiation and are DNA fragments about 2000 bp long *(71)*. These are detected both intracellularly and in the growth medium, indicating that the excised material is exported from the cell. Once extracellular, the oligonucleotides are rapidly reduced to nucleotides and nucleosides, presumably by an exonuclease and a 5'-phosphatidase located in the outer membrane *(21,46)*. Within an hour, the fragments are no longer detected, but nucleotides continue to accumulate in the growth medium *(70)*.

The rapid export of damaged DNA from the irradiated cell is a DNA damage response of *D. radiodurans* that was first noted in 1966 *(4)*, but it has not been studied further. The removal of damaged nucleotides from the intracellular nucleotide pool and their subsequent conversion to nucleosides could represent a strategy to prevent the reincorporation of base damage into the chromosome. The facility with which this occurs in *D. radiodurans* certainly suggests this possibility.

Hariharan and Cerrutti followed the release of ^{3}H-thymine and its oxidized derivatives from the DNA of irradiated *D. radiodurans* cultures, and demonstrated that product release is biphasic *(33,34)*. There is a linear increase in thymine release into the medium and cytoplasm during the first 30 min postirradiation, followed by a 30-min halt in thymine release. After this delay, a second, slower phase of thymine release is initiated. The second phase is of variable duration. On average, one oxidized thymine is removed for every 300 undamaged thymines released, and approx 46% of the thymine glycol is released in the first phase and 54% in the second. Between 70 and 75% of damaged bases appear in the growth medium.

5.4. *Interchromosomal Recombination*

As discussed in Section 2.1., a *D. radiodurans* cell contains at least four copies of its genome, and this number can increase to as many as 10 copies in exponentially growing cells *(31)*. Multiple copies of the genome provide the cell with a reservoir of genetic information. The mere existence of multiple genomes, however, cannot by itself account for the DNA damage resistance of the deinococci, since other organisms with multiple genomes, such as *Azotobacter vinelandii (60)* are not highly resistant to radiation. The deinococci would need a way of utilizing the redundant genetic information.

Daly and Minton *(9)* suggested interchromosomal recombinational repair as one mechanism by which *D. radiodurans* may take advantage of the additional information present in multiple chromosomal copies. They have provided physical evidence that interchromosomal recombination occurs following irradiation and have demonstrated that γ-irradiation induces 175 Holliday junctions/chromosome in stationary-phase cells, some of which could be the result of interchromosomal events *(9)*. This hypothesis, although compelling for its logical appeal, requires additional experimental support, for the necessity of multiple chromosomal copies in DNA damage resistance has not yet been confirmed. It is critical to establish that by reducing the chromosomal copy number, there is a concomitant reduction in DNA damage resistance. Without this correlation, the significance of interchromosomal recombination in DNA damage resistance is unresolved.

6. THE POTENTIAL ROLE OF CHECKPOINTS IN THE DNA DAMAGE TOLERANCE OF *D. RADIODURANS*

A checkpoint is a mechanism that allows the cell to sense the level of DNA damage and to control the cell cycle so that DNA repair can be completed before the cell attempts to replicate a damaged chromosome. Bridges *(6)* has recently argued that DNA damage checkpoints are in operation in *E. coli*, but that they are primitive relative to those in eukaryotic systems. Several observations made in this chapter suggest that DNA damage checkpoints function in *D. radiodurans*. As detailed in Sections 5.1. and 5.3., there is ample evidence to indicate that DNA replication is sensitive to the intracellular level of DNA damage and that there are proteins that sense DNA damage. Whether there are DNA-damage-sensitive proteins that affect the activity of the replicative DNA polymerase remains an open question.

As discussed in Section 4.5., postirradiation degradation is regulated by a protein synthesized following irradiation, possibly IrrI *(68)*. Failure to generate this protein might result in cell death, because the degradative process cannot be turned off. Preliminary evidence indicates that IrrI is part of this regulatory process and that its action either modifies endonuclease α or plays a role in events subsequent to endonuclease α activity *(68)*. Postirradiation DNA degradation is not observed following UV irradiation in a *uvrA* background *(32)*, however. The production of a protein that inhibits the postirradiation DNA degradation is an intriguing aspect of this phenomenon, and it suggests that the cell possesses not only the ability to regulate this DNA repair process, but also the ability to sense either the extent of DNA damage or the completeness of DNA repair. Recall that the length of the degradative process correlates with the dose of radiation administered and, presumably, the degree of DNA damage. The enzymes that mediate postirradiation degradation behave like an exonuclease, degrading DNA until they are stopped or until the DNA is destroyed. The cell produces a protein (IrrI?) that can stop degradation, but this inhibition must itself be regulated, because the exonuclease activity needs to remove an appropriate amount of DNA damage before it is inactivated. It would seem that *D. radiodurans* must have a master control system, analogous to the SOS system, that is sensitive to the level of DNA damage, and that activates and inactivates repair functions as needed.

7. EXTREME RADIORESISTANCE OF *D. RADIODURANS* CORRELATES WITH RESISTANCE TO DESICCATION

Almost since *D. radiodurans* was discovered in 1956 *(2)*, investigators have asked why any organism would have evolved such remarkable DNA repair capabilities, especially toward ionizing radiation. Deinococcal radioresistance cannot be an adaptation (i.e., an evolutionary modification of a character under selection) to ionizing radiation, because there is no selective advantage to being ionizing-radiation-resistant in the natural world. There are no terrestrial environments that generate such a high flux of ionizing radiation *(69)*. It must, therefore, be assumed that deinococcal radioresistance is an incidental use of the cell's DNA repair capability.

It is possible to isolate ionizing-radiation-resistant bacteria, including *D. radiodurans*, from natural microflora without using irradiation by selecting for desiccation resistance *(62)*, suggesting that at least a subset of the cellular functions necessary to survive exposure to ionizing radiation are also necessary to survive desiccation. Since dehydration induces DNA damage in bacteria *(14,15)*, it is likely that the ability to repair DNA damage is one of these functions. To test this possibility, Mattimore and Battista *(42)* evaluated 41 ionizing-radiation-sensitive derivatives of *D. radiodurans* for the ability to survive 6 wk of desiccation. The findings were striking; every ionizing-radiation-sensitive strain was sensitive to desiccation (Fig. 4). In addition, it was established that the process of dehydration induced significant DNA damage that accumulated with time. In Fig. 3, the number of DNA DSBs formed following 6 wk of desiccation are compared with the number generated by 5200 Gy γ-radiation. From these data, it appears that for *D. radiodurans* ionizing radiation resistance and desiccation resistance are functionally interrelated phenomena and that by losing the ability to repair ionizing-radiation-induced DNA damage, *D. radiodurans* is sensitized to the lethal effects of desiccation. It appears that *D. radiodurans* is an organism that has adapted to dehydration and that its DNA repair capability is a manifestation of that evolutionary adaptation.

ACKNOWLEDGMENTS

The author is grateful to Valerie Mattimore and Chris Healy for critically evaluating this manuscript. I also thank Kenneth W. Minton for sharing unpublished results.

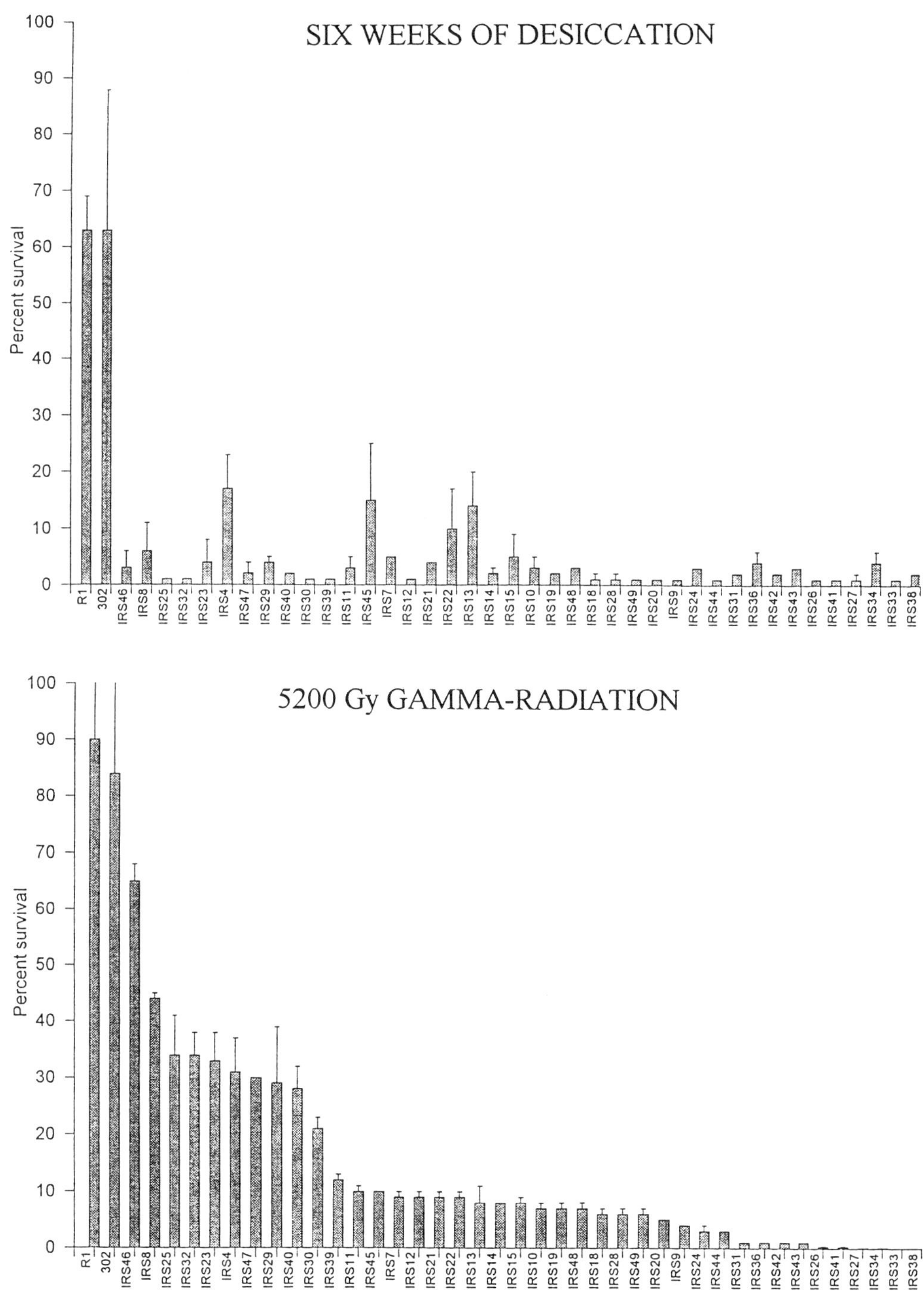

Fig. 4. The effect of 6 wk of desiccation on the survival of ionizing-radiation-sensitive strains of *D. radiodurans* (upper panel). Values represented are the mean percent survivals of two separate trials, three replicates per trial. Each survival reported is relative to that strain's titer immediately prior to desiccation. The lower panel indicates the survival of each strain following exposure to 5200 Gy γ-radiation.

REFERENCES

1. Agostini, H. and Minton, K. 1995. Identification and characterization of a repair endonuclease in *Deinococcus radiodurans. J. Cell. Biochem.* **21A:** 274.
2. Anderson, A., Nordan, H., Cain, R. Parrish, G., and D. Duggan. 1956. Studies on a radioresistant micrococcus. I. Isolation, morphology, cultural characteristics, and resistance to gamma radiation. *Food Technol.* **10:** 575–578.
3. Auda, H. and C. Emborg. 1973. Studies on post-irradiation DNA degradation in *Micrococcus radiodurans*, strain $R_{II}5$. *Radiat. Res.* **53:** 273–280.
4. Boling, M. E. and J. K. Setlow. 1966. The resistance of *Micrococcus radiodurans* to ultraviolet radiation. III. A repair mechanism. *Biochim. Biophys. Acta.* **123:** 26–33.
5. Bonura, T. and A. K. Bruce. 1974. The repair of single-strand breaks in a radiosensitive mutant of *Micrococcus radiodurans. Radiat. Res.* **57:** 260–275.
6. Bridges, B. A. 1995. Are there DNA damage checkpoints in *E. coli*? *Bioessays* **17(1):** 63–70.
7. Brooks, B. W. and R. G. E. Murray. 1981. Nomenclature for *"Micrococcus radiodurans"* and other radiation resistant cocci: *Deinococcus* fam. nov. and *Deinococcus* gen nov., including five species. *Int. J. Syst. Bacteriol.* **31:** 353–360.
8. Caimi, P. and A. Eisenstark. 1986. Sensitivity of *Deinococcus radiodurans* to near ultraviolet radiation. *Mutat. Res.* **162:** 145–151.
9. Daly, M. J. and K. W. Minton. 1995. Interchromosomal recombination in the extremely radioresistant bacterium *Deinococcus radiodurans. J. Bacteriol.* **177(19):** 5495–5505.
10. Daly, M. J., L. Ouyang, P. Fuchs, and K. W. Minton. 1994. In vivo damage and *recA*-dependent repair of plasmid and chromosomal DNA in the radiation-resistant bacterium. *J. Bacteriol.* **176:** 3608–3517.
11 Dean, C. J., P. Feldschreiber, and J. T. Lett. 1966. Repair of X-ray damage to the deoxyribonucleic acid in *Micrococcus radiodurans. Nature* **209:** 49–52.
12. Dean, C. J., J. G. Little, and R. W. Serianni. 1970. The control of post-irradiation DNA breakdown in *Micrococcus radiodurans. Biochem. Biophys. Res. Commun.* **39:** 126–134.
13. Dean, C. J., M. G. Ormerod, R. W. Serianni, and P. Alexander. 1969. DNA strand breakage in cells irradiated with X-rays. *Nature* **222:** 1042–1044.
14. Dose, K., A. Bieger-Dose, M. Labusch, and M. Gill. 1992. Survival in extreme dryness and DNA-single strand breaks. *Adv. Space Res.* **12(4):** 221–229.
15. Dose, K., A. Bieger-Dose, O. Kerz, and M. Gill. 1991. DNA-strand breaks limit survival in extreme dryness. *Origins Life Evol. Biosphere.* **21:** 177–187.
16. Dnedger, A. A. and M. J. Grayston. 1971. The effects of nalidixic acid on X-ray induced DNA degradation and repair in *Micrococcus radiodurans. Can. J. Bacterial.* **17:** 501–505.
17. Driedger, A. A. and M. J. Grayston. 1971. The enhancement of X-ray-induced DNA degradation in *Micrococcus radiodurans* by phenethyl alcohol. *Can. J. Microbiol.* **17:** 487–493.
18. Evans, D. M. and B. E. B. Moseley. 1988. *Deinococcus radiodurans* UV endonuclease β DNA incisions do not generate photoreversible thymine residues. *Mutat. Res.* **207:** 117 119.
19. Evans, D. M. and B. E. B. Moseley. 1985. Identification and initial characterization of a pyrimidine dimer UV endonuclease (UV endonuclease β) from *Deinococcus radiodurans:* a DNA-repair enzyme that requires manganese ions. *Mutat. Res.* **145:** 119–128.
20. Evans, D. M. and B. E. B. Moseley. 1983. Roles of the *uvsC*, *uvsD*, *uvsE*, and *mtcA* genes in the two pyrimidine dimer excision repair pathways of *Deinococcus radiodurans. J. Bacteriol.* **156:** 576–583.
21. Gentner, N. E. and R. E. J. Mitchell 1975. Ionizing radiation-induced release of a cell surface nuclease from *Micrococcus radiodurans. Radiat. Res.* **61:** 204–215.
22. Grant, L. R. and M. F. Patterson. 1989. A novel radiation resistant *Deinobacter* sp. isolated from irradiated pork. *Lett. Appl. Microbiol.* **8:** 21–24.

23. Grimsley, J. K., C. L. Masters, E. P. Clark, and K. W. Minton. 1991. Analysis by pulsed field gel electrophoresis of DNA double-strand breakage and repair in *Deinococcus radiodurans* and a radiosensitive mutant. *Int. J. Radiat. Biol.* **60:** 613–626.
24. Gross, J. and M. Gross. 1969. Genetic analysis of and *E. coli* strain with a mutation affecting DNA polymerase. *Nature* **224:** 1166–1168.
25. Gutman, P. D., J. D. Carroll, C. L. Masters, and K. W. Minton. 1994. Sequencing, targeted mutagenesis and expression of a *recA* gene required for extreme radioresistance of *Deinococcus radiodurans. Gene* **141:** 31–37.
26. Gutman, P. D., P. Fuchs, and K. W. Minton. 1994. Restoration of the DNA damage resistance of *Deinococcus radiodurans* DNA polymerase mutants by *Escherichia coli* DNA polymerase I and Klenow fragment. *Mutat. Res.* **314:** 87–97.
27. Gutman, P. D., P. Fuchs, L. Ouyang, and K. W. Minton. 1993. Identification, sequencing, and targeted mutagenesis of a DNA polymerase gene required for the extreme radioresistance of *Deinococcus radiodurans. J. Bacteriol.* **175:** 3581–3590.
28. Gutman, P. D., H. Yao, and K. W. Minton. 1991. Partial complementation of the UV sensitivity of *Deinococcus radiodurans* excision repair mutants by the cloned *denV* gene of bacteriophage T4. *Mutat. Res.* **254:** 207–215.
29. Hall, J. D., and D. W. Mount. 1981. Mechanisms of DNA replication and mutagenesis in ultraviolet-irradiated bacteria and mammalian cells. *Prog. Nucleic Acid Res. Mol. Biol.* **30:** 53–65.
30. Hansen, M. T. 1980. Four proteins synthesized in response to deoxyribonucleic acid damage in *Micrococcus radiodurans. J. Bacteriol.* **141:** 81–86.
31. Hansen, M. T. 1978. Multiplicity of genome equivalents in the radiation-resistant bacterium *Micrococcus radiodurans. J. Bacteriol.* **134:** 71–75.
32. Hansen, M. T. 1982. Rescue of mitomycin C- or psoralen- inactivated *Micrococcus radiodurans* by additional exposure to radiation or alkylating agents. *J. Bacteriol.* **152:** 976–982.
33. Hariharan, P. V. and P. A. Cerutti. 1972. Formation and repair of γ-ray-induced thymine damage in *Micrococcus radiodurans. J. Mol. Biol.* **66:** 65–81.
34. Hariharan, P. V. and P. A. Cerutti. 1971. Repair of γ-ray-induced thymine damage in *Micrococcus radiodurans. Nature New Biol.* **229:** 247–249.
35. Ito, H., H. Watanabe, M. Takeshia, and H. Iizuka. 1983. Isolation and identification of radiation-resistant cocci belonging to the genus *Deinococcus* from sewage sludges and animal feeds. *Agric. Biol. Chem.* **47:** 1239–1247.
36. Kitayama, S. 1982. Adaptive repair of cross-links in DNA of *Micrococcus radiodurans. Biochim. Biophys. Acta.* **C97:** 381–384.
37. Kitayama, S., S. Asaka, and K. Totsuka. 1983. DNA double-strand breakage and removal of cross-links in *Deinococcus radiodurans. J. Bacteriol.* **155:** 1200–1207.
38. Kitayama, S. and A. Matsuyama. 1971. Double-strand scissions in DNA of gamma-irradiated *Micrococcus radiodurans* and their repair during postirradiation incubation. *Agric. Biol. Chem.* **35:** 644–652.
39. Krabbenhoft, K. L., A. W. Anderson, and P. R. Elliker. 1965. Ecology of *Micrococcus radiodurans. Appl. Microbiol.* **13:** 1030–1037.
40. Lett, J. T., P. Feldschreiber, J. G. Little, K. Steele, and C. J. Dean. 1967. The repair of X-ray damage to the deoxyribonucleic acid in *Micrococcus radiodurans:* a study of the excision process. *Proc. R. Soc. Lond. (Biol.)* **167:** 184–201.
41. Masters, C. L., B. E. B. Moseley, and K. W. Minton. 1991. AP endonuclease and uracil DNA glycosylase activities in *Deinococcus radiodurans. Mutat. Res.* **254:** 263–272.
42. Mattimore, V. and J. R. Battista. 1995. Radioresistance of *Deinococcus radiodurans:* functions necessary to survive ionizing radiation are also necessary to survive prolonged desiccation. *J. Bacteriol.* **178:** 633–637.

43. Mattimore, V., K. S. Udupa, G. A. Berne, and J. R. Battista. 1995. Genetic characterization of forty ionizing radiation-sensitive strains of *Deinococcus radiodurans:* linkage information from transformation. *J. Bacteriol.* **18:** 5232–5237.
44. Minton, K. W. 1995. Personal communication.
45. Minton, K. W. 1994. DNA repair in the extremely radioresistant bacterium *Deinococcus radiodurans. Mol. Microbiol.* **13:** 9–15.
46. Mitchel, R. E. J. 1975. Origin of cell surface proteins released from *Micrococcus radiodurans* by ionizing radiation. *Radiat. Res.* **64:** 380–387.
47. Moore, P. and B. S. Strauss. 1979. Sites of inhibition of in vitro DNA synthesis in cartcinogen- and UV-treated DNA. *Nature* **278:** 664–666.
48. Moseley, B. E. B. 1967. The isolation and some properties of radiation sensitive mutants of *Micrococcus radiodurans. J. Gen. Microbiol.* **49:** 293–300.
49. Moseley, B. E. B. 1983. Photobiology and radiobiology of *Micrococcus (Deinococcus) radiodurans. Photochem. Photobiol. Rev.* **7:** 223–275.
50. Moseley, B. E. B. 1967. The repair of DNA in *Micrococcus radiodurans* following ultraviolet irradiation. *J. Gen. Microbiol.* **48:** 4–24.
51. Moseley, B. E. B. and H. J. R. Copland. 1978. Four mutants of *Micrococcus radiodurans* defective in the ability to repair DNA damaged by mitomycin C, two of which have wild-type resistance to ultraviolet radiation. *Mol. Gen. Genet.* **160:** 331–337.
52. Moseley, B. E. B. and H. J. R Copland. 1976. The rate of recombination repair and its relationship to the radiation-induced delay in DNA synthesis *Micrococcus radiodurans. J. Gen. Microbiol.* **93:** 251–258.
53. Moseley, B. E. B. and D. M: Evans. 1983. Isolation and properties of strains of *Micrococcus(Deinococcus) radiodurans* unable to excise ultraviolet light-induced pyrimidine dimers from DNA: evidence of two excision pathways. *J. Gen. Microbiol.* **129:** 2437–2445.
54. Moseley, B. E. B. and A. Mattingly. 1971. Repair of irradiated transforming deoxyribonucleic acid in wild-type and a radiation-sensitive mutant of *Micrococcus radiodurans. J. Bacteriol.* **105:** 976–983.
55. Moseley, B. E. B., A. Mattingly, and H. J. R. Copland. 1972. Sensitization to radiation by loss of recombination ability in a temperature-sensitive DNA mutant of *Micrococcus radiodurans* held at its restrictive temperature. *J. Gen. Microbiol.* **72:** 329–338.
56. Mun, C., J. Del Rowe, M. Sandigursky, K. W. Minton, and W. A. Franklin. 1994. DNA deoxyribophosphodiesterase and an activity that cleaves DNA containing thymine glycol adducts in *Deinococcus radiodurans. Radiat. Res.* **138:** 282–285.
57. Murray, R. G. E. 1992. The family *Deinococcaceae*, in *The Prokaryotes*, vol. 4 (Balows, A., H. Tupper, G. Dworkin, M. Harder, and K.H. Schleifer, eds.), Springer-Verlag, New York, pp. 3732–3744.
58. Murray, R. G. E. and B. W. Brooks. 1986. Genus 1. *Deinococcus* Brooks and Murray 1981, in *Bergey's Manual of Systematic Bacteriology*, vol. 2. (Sneath, P. H. A., Mair, N. S., Sharpe, M. E., and Holt, J. E., eds.), Williams & Wilkins, Baltimore, pp. 1035–1043.
59. Potts, M. 1994. Dcsiccation tolcrance of prokaryotes. *Microbiol. Rev.* **58:** 755–780.
60. Ponita, S. J. and H. K. Reddy. 1989. Multiple chromosomes of *Azotobacter vinlandii. J. Bacteriol.* **171:** 3133–3138.
61. Sakai, Y., E. Karu, S. Linn, and H. Echols. 1973. Purification and properties of the γ-protein specified by bacteriophage lambda: and inhibitor of the host RecBC recombination enzyme. *Proc. Natl. Acad. Sci. USA* **70:** 2215–2219.
62. Sanders, S. W. and R. B. Maxcy. 1979. Isolation of radiation-resistant bacteria without exposure to irradiation. *Appl. Environ. Microbiol.* **38:** 436–439.
63. Sanders, S. W., and R. B. Maxcy. 1979. Patterns of cell division, DNA base compositions, and fine structures of some radiation resistant vegetative bacteria found on food. *Appl. Environ. Microbiol.* **38:** 436–439.

64. Setlow, J. K. and D. Duggan. 1964. The resistance of *Micrococcus radiodurans* to ultraviolet radiation. I. Ultraviolet-induced lesions in the cell's DNA. *Biochim. Biophys. Acta.* **87:** 664–668.
65. Smith, M. D., C. L. Masters, E. Lennon, L. B. McNeil, and K W. Minton. 1991. Gene expression in *Deinococcus radiodurans. Gene* **98:** 45–52.
66. Sweet, D. M. and B. E. B. Moseley. 1974. Accurate repair of ultraviolet-induced damage in *Micrococcus radiodurans. Mutat. Res.* **23:** 311–318.
67. Sweet, D. M. and B. E. B. Moseley. 1976. The resistance of *Micrococcus radiodurans* to killing and mutation by agents which damage DNA. *Mutat. Res.* **34:** 175–186.
68. Udupa, K., P. A. O'Cain, V. Mattimore, and J. R. Battista. 1994. Novel ionizing radiation-sensitive mutants of *Micrococcus radiodurans. J. Bacteriol.* **176:** 7439–7446.
69. United Nations Scientific Committee on the Effects of Atomic Radiation. 1982. Ionizing radiation: sources and biological effects. United Nations Publication No. E82.IX.8 United Nations, New York.
70. Varghese, A. J. and R. S. Day. 1970. Excision of cytosine-thymine adduct from the DNA of ultraviolet-irradiated *Micrococcus radiodurans. Photochem. Photobiol.* **11:** 511–517.
71. Vukovic-Nagy, B., B. W. Fox, and M. Fox. 1974. The release of DNA fragments after X-irradiation of *Micrococcus radiodurans. Int. J. Radiat. Biol.* **25:** 329–337.
72. Walker, G. C. 1987. The SOS response of *Escherichia coli*, in Escherichia coli *and* Salmonella yphimurium: *Cellular and Molecular Biology*, vol. 2. (Ingraham, J. L., K. B. Low, B. Magasanik, M. Schaechter, and H. E. Umbarger, eds.) American Society for Microbiology, Washington DC, pp. 1346–1357.

Part II

DNA Repair in Lower Eukaryotes

15

The Genetics and Biochemistry of the Repair of UV-Induced DNA Damage in *Saccharomyces cerevisiae*

Wolfram Siede

1. INTRODUCTION

Since the early days of radiation biology, the yeast *Saccharomyces cerevisiae* has been the preferred eukaryotic organism for studies of cellular responses to DNA damage. In recent years, the validity of yeast as a model for higher eukaryotic cells has indeed been impressively confirmed in important areas of DNA repair research, such as nucleotide excision repair and mismatch repair, where the genes involved turned out to be highly conserved. Clues to their molecular function in mammalian cells were frequently deduced from the phenotype of the well-characterized corresponding yeast mutants, and on more than one occasion a mammalian repair gene was isolated by using a repair gene from *S. cerevisiae* or *Schizosaccharomyces pombe* as a probe (but also vice versa). The unique advantages of yeast have been well documented. Examples are the availability of cells in the haploid and diploid stage facilitating strain construction, the preference of homologous recombination enabling sophisticated genetic manipulation, and knowledge of the DNA sequence of the yeast genome.

In the early 1970s, an extensive collection of yeast mutants (termed *rad)* that are sensitive to UV radiation at 254 nm and/or to ionizing radiation was established (e.g., *25,37,40*). The available mutants could be grouped into three phenotypic groups, one group of mutants being predominantly sensitive to UV light, another group of mutants predominantly sensitive to γ-irradiation (with the *rad* gene numbers over 50 set aside for this group *[37]*), and a third group including mutants sensitive to both agents. It soon became clear that these different phenotypic groups represented in fact different epistasis groups, which can be interpreted as nonoverlapping repair pathways. Typically, a double mutant constructed from single mutants of the same phenotypic group was found to be no more sensitive than the most sensitive of the single mutants, whereas a double mutant constructed from single mutants of different phenotypic groups showed synergistically enhanced sensitivity, thus indicating a defect in two pathways competing for the same type of lesion *(18,38,39)*. With some refinements, this scheme of three repair pathways still remains valid.

From: DNA Damage and Repair, Vol. 1: DNA Repair in Prokaryotes and Lower Eukaryotes
Edited by: J. A. Nickoloff and M. F. Hoekstra © Humana Press Inc., Totowa, NJ

This chapter will focus on the two epistasis groups of mutants that are markedly UV-sensitive. The first is termed the *RAD3* epistasis group, and includes mutants that are extremely sensitive to UV radiation and UV-mimetic chemicals. However, their sensitivity toward ionizing radiation is at best marginal. These mutants are involved in nucleotide excision repair (NER; *see* Chapter 2 for a general discussion), and in synchrony with the emerging insights into NER in mammalian cells (*36,60,108*; *see also* Chapter 10, vol. 2), our understanding of the detailed function of the corresponding gene products has advanced dramatically in recent years. The reaction sequence is now amenable to analysis in cell-free extracts *(148,149)*, and the initial steps can be studied in a reconstituted system *(45)*. In *S. cerevisiae*, no data suggest the existence of alternative excision pathways other than the "classic" NER mechanism characterized by a dual incision mode, such as the second excision pathway described for *S. pombe* (*see* Chapter 20).

The second, considerably less-well-defined epistasis group is usually referred to as the *RAD6* epistasis group, and includes mutants showing varying degrees of UV- and γ-ray sensitivity, defects in postreplication repair, and in radiation- or chemical-induced mutagenesis. The detected or predicted activities of several of these gene products seem to suggest specific functions in mutagenic and postreplicative repair, but even a faint outline of a repair mode of this "pathway" cannot be given at the present time. The cloned yeast genes involved in repair of UV-induced DNA damage and the known activities of the encoded proteins are summarized in Table 1.

Three additional topics should be regarded as integral parts of a discussion of UV damage repair in yeast. However, these are discussed in other chapters: photoreactivation (Chapter 2, vol. 2), inducibility of repair genes (Chapter 18) and cell-cycle checkpoint controls (Chapter 17). Regarding checkpoint control, it is important to note that some of the gene products involved most likely have dual functions and should be considered repair proteins as well (*see* Chapter 17). The focus of this chapter is predominantly on the most recent studies on UV repair. Other recent reviews are given in refs. *35* and *102*. Additionally, because of extensive discussion of phenotypic peculiarities of the various mutants (that are often forgotten nowadays), older reviews *(55,73)* still remain a valuable source of information.

2. NUCLEOTIDE EXCISION REPAIR: THE *RAD3* EPISTASIS GROUP

As described in Chapter 2, NER consists of the processes of damage recognition, dual incision, release of an oligonucleotide fragment carrying the damage, resynthesis of the gap, and finally ligation. First, the main players of the initial steps resulting in the incision reaction in yeast will be introduced (Rad14p, Rad1p, Rad10p, Rad2p, the subunits of the transcription factor TFIIH, and replication protein A), and a general scheme will be outlined. Then, postincision events and differential damage removal will be discussed.

Analysis of the highly UV-sensitive mutants of the *RAD3* epistasis group revealed that some of these were defective in single-strand incision following treatment with UV radiation or crosslinking agents. The methodology involved the use of an additional *cdc9* mutation conferring thermoconditional ligase activity and the analysis of DNA profiles in alkaline sucrose gradients with repair-induced single-strand incisions being "frozen" by incubation at restrictive temperature (e.g., *154*). It became clear that

Table 1
Genes Involved in UV Damage Repair in *S. cerevisiae*

Gene	Size of protein, kDa	Activities and other features of the protein	Interacting proteins
1. Photoreactivation			
PHR1	66	Photolyase	
2. Nucleotide excision repair			
RAD1	126	Endonuclease in complex with Rad10p cleaves duplex/single-stranded DNA junctions from unpaired 3' end	Rad10p, Rad14p
RAD2	117	Endonuclease, human homolog cleaves duplex/single-stranded DNA junctions from unpaired 5'-end, UV-inducible	Ssl2p,Tfb1p
RAD3	85	Component of TFIIH, essential, DNA · DNA, DNA · RNA 5' → 3' helicase with single-stranded DNA-dependent ATPase activity	Ss1p, Ss12p
RAD4	87		Rad23p, TFIIH
RAD7	64	Specifically required for repair of transcriptionally silent DNA, UV-inducible	Sir3p
RAD10	24	Endonuclease in complex with Rad10p, cuts unpaired single-stranded DNA from 3'-end	Rad1p
RAD14	43	Binds specifically to UV-damaged DNA	Rad23p, Radlp
RAD16	91	Helicase domains, specifically required for repair of transcriptionally silent DNA, UV and heat-shock-inducible	
RAD23	42	Ubiquitin-like domain, UV-inducible	Rad4p, Radl4p, Tfb1p
RAD26	125	Specifically required for preferential repair of the transcribed strand	
SSL1	52	Core TFIIH component, essential	Rad3p, Tfblp
SSL2/ RAD25)	95	TFIIH component, essential, 3'→5' DNA · DNA helicase, ATPase	Rad3p, Rad2p
TFB1	73	Core TFIIH component,essential	Rad2p, Rad23p, Ssl1p
TFB2	55	Core TFIIH component, role in NER not proven	
TFB3	38	Core TFIIH component, role in NER not proven	
RFA1	70	Subunit of replication factor A	
CDC44	95	Subunit of replication factor C	
POL30	29	PCNA	
CDC9	84	DNA ligase	
3. RAD6 epistasis group and genes associated with DNA damage induced mutagenesis			
RAD6	20	Ubiquitin-conjugating E2 enzyme, UV-inducible	Rad18p
RAD18	55	Binds to single-stranded DNA, UV-inducible	Rad6p
REV1	112		
REV2/ RAD5)	134	Helicase domains, zinc finger	
REV3	173	Rev3p/Rad7p = DNA polymerase ζ	Rev7p
REV7	*29*		Rev3p
SRS2	134	3' → 5' DNA helicase, ATPase, cell-cycle regulated	
CDC7	58	Serine-threonine protein kinase required for initiation of replication	Dbf4p

some of these mutants were completely blocked in single-strand incision and the genes in question had to be regarded as absolutely essential for the initial steps of NER, whereas others showed only a partial defect. It is now known that in certain cases these partially defective phenotypes did not result from the leakiness of the mutants available at that time.

2.1. Damage Recognition: Rad14p

The *rad14* point mutants isolated among the original collection of *rad* mutants were not characterized by extreme UV sensitivity. Combination with a mutation in another repair gene *(RAD5)* synergistically enhancing UV sensitivity enabled the isolation of complementing library plasmids *(11)*. Interestingly, the *RAD14* gene represents so far the only example of a DNA repair gene containing an intron (84 bp) *(44)*. The size of the predicted protein is 43 kDa, and the sequence revealed significant homology (26% residue identity, 54% similarity) with the human xeroderma pigmentosum group A (XP-A) gene *(11)*, thus suggesting a central role in the initial steps of NER. Indeed, the sensitivity of the deletion mutant approaches that of other mutants absolutely essential for the incision step of NER, and biochemical analysis confirmed the absence of incision of UV-damaged DNA in the mutant *(11)*. As its human counterpart, the highly hydrophilic protein contains a zinc finger motif ($CX_2CX_{18}CX_2C$) in addition to a second metal-binding sequence ($HX_3HX_{23}CX_2C$) *(11)*. As estimated by atomic emission spectroscopy, 1 mol zinc is bound by 1 mol Rad14p *(11)*.

Purified Rad14p binds specifically to UV-damaged DNA without a requirement for ATP *(48)*. Removal of cyclobutane pyrimidine dimers (CPDs) by photolyase did not affect its affinity to damaged DNA and thus, it must be concluded that Rad14p binds primarily to pyrimidine(6-4)pyrimidone dimers [(6-4)PDs], with an estimated discrimination factor of 5000 over undamaged sites *(48)*. Thus, Rad14p is clearly involved in the recognition of one major species of DNA damage, but a role in the repair of other kinds of DNA damage must also be assumed since the analysis of the deletion mutant indicated a complete incision defect. Rad14p was subsequently shown to interact with other NER components, such as Rad1p, TFIIH, and Rad23p (*see* Sections 2.2.–2.4. and Fig. 1) and must be considered an integral part of the NER machinery *(44)*. Such an interaction may broaden the substrate specificity of Rad14p. However, damage recognition remains the least understood aspect of NER *(88)*. Another NER component (Rad3p) has properties that are consistent with a role in damage recognition (*see* Section 2.3.1.), and it should also be mentioned that indirect evidence suggests an auxiliary role of the photoreactivating enzyme in NER in the dark, most likely in CPD recognition *(109)* (*see* Chapter 2, vol. 2).

2.2. The Dual-Incision Step: Rad1p/Rad10p and Rad2p

Point and deletion mutants of *RAD1* and *RAD10* are highly UV-sensitive and completely blocked in the single-strand incision step of NER. These mutants share a phenotype that is unique among the NER mutants of yeast: certain types of mitotic intrachromosomal recombination events that involve repeated elements are reduced *(69,111,112)*. Indeed, both gene products were found to interact stably and specifically. This was demonstrated by coimmunoprecipitation of the in vitro transcribed and translated proteins *(17)*, by coimmunoprecipitation from yeast cell extracts *(8)*, and by

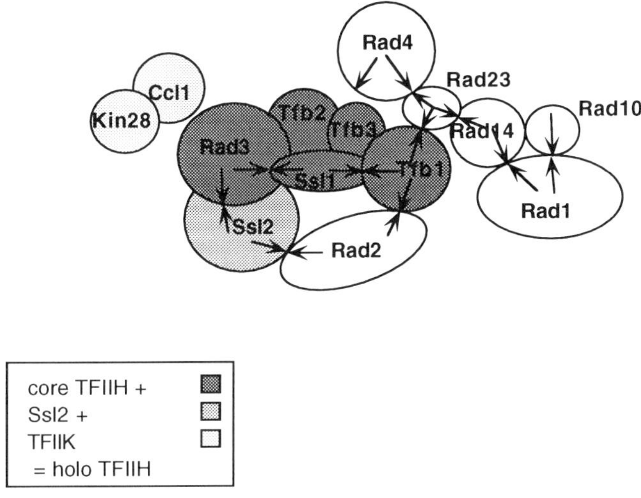

Fig. 1. Known interactions and complex formations between NER proteins. Arrows indicate specific protein–protein interactions demonstrated by coimmunoprecipitation, gel filtration, copurification, or in the yeast two- hybrid system. TFIIK is included here as part of holo TFIIH; it is not involved in NER *(124)*. A different kind of core TFIIH complex termed TFIIHi, lacking both Rad3p and Ssl2p, has also been reported *(124)*. The complexes formed by Rad1p, Rad10p, Rad14p, and by Rad4p, Rad23p have recently been designated NEF1 and NEF2, respectively *(49)*.

the yeast two-hybrid system *(14)*. The interacting domains were mapped *(14)* and indeed, the UV-sensitivity of a highly sensitive *rad1* mutant carrying an amino acid substitution in the Rad10p interacting domain can be specifically complemented by overexpressing Rad10p *(117)*. Thus, the significance of the Rad1p/Rad10p interaction for UV repair was confirmed.

Important clues to the enzymatic function of the Rad1p/Rad10p complex in NER were derived from recombinational studies. In an attempt to delineate the events associated with recombination between duplicated genes on plasmids, the repair of double-strand breaks was studied in vivo following the synchronous induction of site-specific double-strand breaks by regulated expression of the HO endonuclease involved in mating-type switching *(32,33)* (*see also* Chapter 16). *RAD1* and *RAD10,* but no other NER gene *(63)*, were shown to be involved in a recombinational pathway that results in deletion of the intervening sequence separating repeated elements and can best be explained by a single-strand annealing mechanism *(32,33)*. Such a mechanism involves 5' → 3' single-strand resection until homologous pairing between exposed single strands is established. Successful completion also requires the removal of potential non-homologous sequences from the 3'-end, and it is the latter process that appeared to be catalyzed by the Rad1p/Rad10p proteins *(32)*. Indeed, the purified Rad1p/Rad10p complex has nuclease activity *(131,139)* and a closer examination of the activity on Y-shaped, partially duplex substrates proved the specific endonucleolytic cleavage of unpaired regions from the 3'-end (Fig. 2) *(15)*. The identical type of cleavage was found for bubble-type structures containing a centrally unpaired region of 30 nucleotides *(26)*. Since the region around a UV photoproduct can possibly be regarded as locally dena-

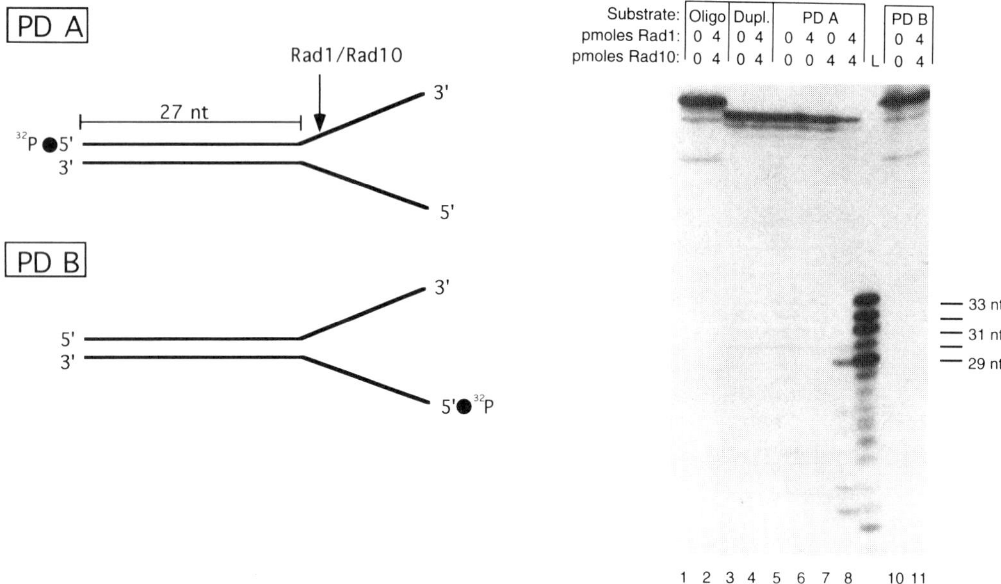

Fig. 2. Substrate specificity of the Rad1p/Rad10p endonuclease. The partial duplex substrates PD A and PD B were 5' ^{32}P end-labeled as shown, incubated with no protein, with Rad1p or Rad10p alone, or a mixture of both proteins. Cleavage products were separated on a denaturing polyacrylamide gel. No cleavage is detected with single-stranded oligonucleotides or duplex substrates. Cleavage occurs only with partial duplex substrates and only when both Rad1p and Rad10p are added. A labeled single-stranded cleavage product of 29 nucleotides is detected (compare lane 8 to size standard lane L) if the label is attached to the 5'-end within the duplex region (PD A). This experiment demonstrates the specificity of the Rad1p/Rad10p endonuclease for 3' single-stranded tails within DNA duplex single-strand junctions. Reproduced from ref. *15* with permission.

tured owing to helicase processing (*see* Section 2.3.), such an activity is ideally suited to introduce the single-strand break at the 5'-site of the lesion. Furthermore, Rad1p/Rad10p forms a complex with Rad14p, and thus, the damage-recognition protein Rad14p seems to direct the endonuclease to the site of the lesion *(49)*.

For Rad10p alone, a renaturing activity for complementary DNA strands has been described *(127)*, but its significance for NER is unknown at the present time. A report indicating Holliday junction cleavage by Rad1p alone *(52)* has not been confirmed *(26,153)*.

The Rad2p protein has Mg^{2+}-dependent endonuclease activity that is specific for single-stranded DNA *(51)*. Similar partial duplex substrates as described above have been used to identify the human Rad2p homolog, XP-G protein, as an endonuclease that can cleave at the 3'-side of the lesion *(26,80,94)* and it was demonstrated that its yeast counterpart plays the same role *(53)*. Rad2p (together with Rad23p) are the only genes indispensable for the initial steps of NER that are UV-inducible on the transcriptional level. However, rendering *RAD2* uninducible has only a moderate effect on UV sensitivity that is primarily detectable in G1-synchronized cells *(118)*. Induced expression does not appear to be an essential precondition for completion of the incision step.

2.3. The Subunits of TFIIH

The demonstration of a dual function of certain NER proteins in repair and transcription as components of TFIIH, an initiation factor essential for most, if not all RNA polymerase II promoters, counts easily among the most startling discoveries made in the DNA repair field in recent years. Before discussing some of the individual components in more detail, it is useful to introduce the terminology introduced primarily by Kornberg's group to distinguish different forms of this initiation factor in yeast (Fig. 1) *(31)*. TFIIH can be isolated as a core complex *(core* TFIIH*)* comprising Rad3p, Ssl1p, Tfb1p, Tfb2p, and Tfb3p. This complex does not support promoter-specific transcription in a reconstituted system, but does so in heat-treated extracts. Other authors have isolated a complex lacking both Rad3p and Ssl2p (termed TFIIHi*)* that lacks NER activity (124). The *holo TFIIH* complex includes additional polypeptides: Rad3p, Ssl2p/Rad25p, as well as Kin28p and polypeptides of 45 (Ccl1p) and 47 kDa. This complex supports transcription in a reconstituted system and phosphorylation of the C-terminal repeat domain of RNA polymerase II. The kinase activity resides in the dissociable subcomplex termed TFIIK consisting of Kin28p and Ccl1p, the latter present in a 45- or 47-kDa form *(133)*. TFIIK, however, is dispensable for NER *(124)*.

2.3.1. Rad3p

RAD3 represents the first cloned yeast DNA repair gene. Deletions were found to be lethal, and for years it remained a mystery why a gene required for NER would also be essential for viability. The identity of Rad3p with a component of the core TFIIH complex finally provided the explanation *(30)*. The analysis of changes in RNA polymerase II transcript levels in a thermoconditionally lethal *rad3* mutant following temperature shift and in vitro transcription studies further contributed to the notion that Rad3p is an essential component of the basal transcriptional machinery *(46)*.

RAD3 encodes a DNA · DNA, DNA · RNA 5' → 3' helicase with single-stranded DNA-dependent ATPase activity *(9,54,91,126,128)*. The results of two studies are consistent with a role of Rad3p in damage recognition. Both helicase and ATPase activities of Rad3p are inhibited by damage in the DNA strand on which the enzyme translocates *(89,90)*. Although Rad3p does not show increased binding to UV-irradiated linear single-stranded DNA, measurements of dissociation rates and competition effects indicated a sequestration of Rad3p at the sites of lesions *(89)*. In a different study, preferred binding of Rad3p to UV- irradiated double-stranded DNA in the presence of ATP was demonstrated, with the degree of binding dependent on the negative superhelicity of the substrate *(132)*. Removal of CPDs by enzymatic photoreactivation did not affect binding, so affinity for (6-4)PDs must account for the majority of preferential binding to UV-irradiated DNA. The mechanism of preferential binding must differ from a helicase stalling-and-sequestration model described previously *(89,90)*, since a lysine-48 → arginine-48 mutation in the "Walker type A" consensus ATP- hydrolysis domain abolishes ATPase and helicase activities of Rad3p *(125)*, but not its preferential binding to UV-treated DNA *(132)*. Also, its essential function remains unaltered *(50,125)*. How all of these properties are affected by the association of Rad3p with other TFIIH components is not known at the present time.

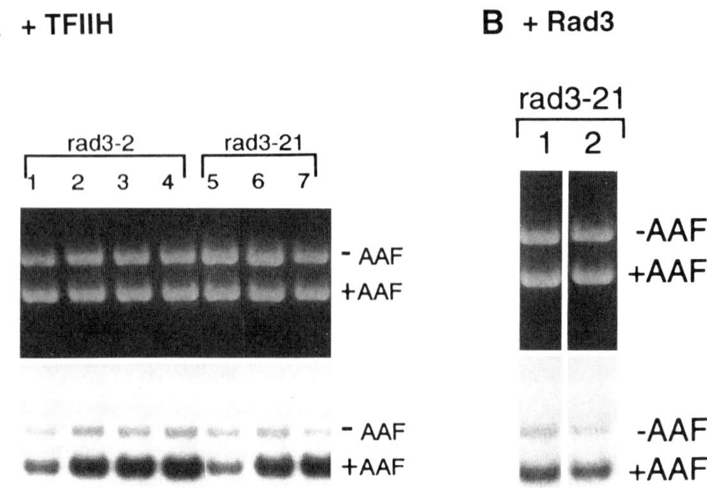

Fig. 3. Rad3p participates in NER as part of TFIIH. NER is measured in vitro by specific incorporation of label into amino-acetoxy-fluorene (AAF) damaged plasmid DNA; an untreated control plasmid (upper band) is always used in the identical reaction. NER deficient Rad3p mutant extracts (lane 1, 5 in **A**, lane 1 in **B**) can be complemented by extracts containing overexpressed Rad3p (lane 2 and 6 in A) or by purified TFIIH (lane 3, 7 = 180 ng, lane 4 = 450 ng, in A), but not by purified Rad3p (lane 2 = 270 ng, in B). Reproduced from ref. *146* with permission.

An essential question concerns the form of Rad3p that is active in DNA repair. Is Rad3p a multifunctional protein that functions independently in repair, but also happens to function as part of a subunit of a transcription factor, or does its association with other components in the form of TFIIH constitute the active repair configuration? Indeed, it could be demonstrated that purified core TFIIH can complement in vitro the defective NER activity of extracts from *rad3* mutants. However, purified Rad3p proved to be inactive (Fig. 3) *(146)*. This is consistent with the notion that Rad3p exerts its repair function as an integral part of TFIIH that participates as a complex in NER and that stable association of Rad3p with TFIIH components can prevent replacement of a mutant Rad3p protein with purified Rad3p wild-type protein in vitro. However, separate functions of Rad3p are not excluded. Rad3p has been implicated in suppression of recombination between short homologous sequences *(6)* and alleles of *RAD3* have been isolated that enhance mitotic recombination and spontaneous mutagenesis without significantly affecting UV sensitivity (*rem* alleles) *(82)*. Studies on a larger collection of *rad3* alleles have shown that the phenotype of enhanced spontaneous mutagenesis is actually quite common among *rad3* alleles, and there is a correlation between the positions of mutations in helicase consensus domains and enhanced mutagenesis phenotypes *(120)*. However, the correlation between alteration in helicase domains and UV sensitivity appears to be stronger *(120,125)*. A role of Rad3p in the repair of spontaneous damage, in preserving fidelity during replication, or in postreplicative mismatch repair could account for such a mutator phenotype. Results of direct measurements of mismatch repair efficiencies in a strain harboring the mutator allele *rad3-1* (which maps to a helicase consensus domain) were inconsistent with a possible involvement of Rad3p in mismatch repair *(156)*.

It remains to be seen whether these "other phenotypes" of *rad3* mutants result from a direct involvement of Rad3p (by itself or in the form of TFIIH) in other processes, such as DNA replication, or whether these are indirect consequences of altered transcription. There are indeed data in the literature that can be explained by the absence of certain transcripts in a *rad3* mutant background *(122)*. This leads to a consideration of the hypothesis of transcriptional syndromes in humans that is discussed in more detail in Chapters 10 and 18, vol. 2. Briefly, a potentially subtle transcription defect may account for the surprising phenotypes associated with alterations in the XP-D gene, the human homolog of *RAD3*, such as a defect in sulfur metabolism resulting in brittle hair in trichothiodystrophy patients. Indeed, the analogous mutation has been constructed in *RAD3* and it was found that the mutant protein cannot rescue the lethality conferred by a *RAD3* mutation inactivating the essential transcription function *(50)*. Thus, the association of trichothiodystrophy phenotypes with altered transcription was further supported by the yeast model.

2.3.2. *Ssl1p and Tfb1p*

The role of both of these core TFIIH subunits as *bona fide* NER proteins has been confirmed. Defective NER was demonstrated in extracts prepared from thermo-conditionally lethal mutant *ssl1* and *tfb1* strains *(145)*. Interestingly, an NER defect together with UV sensitivity is evident at temperatures that are still permissive for growth, and thus, roles for these proteins in transcription that are different from their repair function were suggested *(145)*.

In two-hybrid-assays, the zinc-finger protein Ssl1r *(157)* was found to interact specifically in vivo with Rad3p, Tfb1p and with itself *(16)*. In vitro translated Tfb1p was also found to coimmunoprecipitate with Rad2p and thus, an important link between TFIIH components and other NER components was established *(12)*. The region of Rad2p sufficient for Tfb1p-Rad2p interaction was mapped to residues 642–900, a region that includes acidic and other C-terminal domains that are conserved in Rad2p and its human counterpart, XP-G *(12)*. Figure 1 summarizes the known interactions between TFIIH proteins and other NER proteins.

2.3.3. *Ssl2p/Rad25p*

Ssl2p is part of the holo TFIIH complex, but is less tightly associated than its core components *(124,134)*. The TFIIH interacting region has been mapped to its N-terminal half *(16)*. Specific interaction with the core TFIIH component Rad3p and with Rad2p has been demonstrated (Fig. 1) *(13,16)*. The *SSL2* gene had been previously isolated in two different screens. On the basis of sequence similarity, it has been isolated (as *RAD25)* as the yeast homolog of the human *XP-B/ERCC3* gene *(97)*. Mutations in the genes *SSL2* and *SSL1* were also isolated in a screen for mutations which enable the translation of mRNA species with a stem-loop structure in the 5'-untranslated region that normally prevents efficient translation *(SSL* stands for *s*uppressor of *s*tem-*l*oop formation) *(43)*. This phenotype is not understood; an indirect effect of imbalanced transcription, i.e., enhanced transcription of translational components, has been suggested as an explanation *(103)*. The predicted amino acid structure suggests an ATPase/DNA helicase, and such an activity has indeed been demonstrated for Ssl2p *(47)* and for the human XP-B product *(110)* with which Ssl2p shares 55% residue identity

(97). Analysis of transcript levels in a thermoconditional mutant after temperature shift and the thermosensitivity of RNA polymerase II-dependent transcription activity in extracts from the same mutant confirmed the essential role of Ssl2p in transcription *(47,103)*. In contrast to a similar mutation in Rad3p (*see* Section 2.3.1.), substitution of lysine-392 with arginine in the Walker type-A nucleotide binding sequence inactivates the function required for viability *(97)* as well as in vitro transcription activity *(47)*. However, truncations of Ssl2p at the C-terminus confer UV sensitivity without any effect on viability or in vitro transcription activity, and UV sensitivity is not complemented by the arginine-392 mutant protein *(47,97)*. Additionally, Gal4p-Ssl2p fusions were found to complement the UV sensitivity of *SSL2* point mutants, but not the essential function inactivated by *SSL2* deletions *(16)*. Thus, Ssl2p seems to have domains that are involved both in repair and transcription, but also domains with separable functions.

2.4. Rad4p and Rad23p

Little is known about the function of Rad4p. The cloning of the gene was initially complicated by the fact that Rad4p protein is toxic to *Escherichia coli*. The protein is not essential for viability and it is not part of TFIIH. However, interaction of in vitro translated Rad4p with TFIIH was demonstrated *(12)*. Overexpression of Rad4p and purification from yeast have recently been reported, and it was found that, as is true for the mammalian homologs, Rad4p copurifies with Rad23p *(45)*.

RAD23 encodes a protein that contains a ubiquitin-like region at its N-terminal end. Indeed, the domain is important for its repair function, although this potential degradation signal does not seem to confer a high degree of instability to the Rad23p protein *(151)*. Specific retention of in vitro translated Rad14p and Tfb1p in Rad23p columns was detected, and it was suggested that Rad23p may mediate contacts between TFIIH and the damage recognition factor Rad14p *(44)*.

The fact that even *RAD23* deletion mutants do not approach the UV sensitivity conferred by mutations in other genes essential for NER represents an interesting puzzle. Considering the indirect evidence indicating the functional relationship of Rad23p with Rad7p *(98)* and the similar UV sensitivity of the corresponding mutants, one would have predicted for Rad23p an accessibility function possibly related to Rad7p or Rad16p (*see* Section 2.8.) that may only be required for repair of certain regions of the genome. However, recent data show an absolute deficiency in the overall repair of UV lesions (at least in nonreplicating cells) and not a partial defect *(142)* Thus, the relative UV resistance of *rad23* mutants remains unexplained in this study. However, others have reported some residual repair *(87)*.

2.5. Replication Factor A

Replication factor A (RPA) is a heterotrimeric, single-strand binding protein that is essential for initiation and elongation steps of DNA replication. The UV sensitivity of thermoconditional mutants of *RFA1* encoding the 70-kDa subunit indicates its role in DNA repair in yeast *(77)*. That it is essential for NER has been inferred from mammalian in vitro studies *(24)*, and there is indeed also an absolute requirement of RPA for incision in a reconstituted yeast system *(45)* (*see* Section 2.6.). The interaction of the human counterpart with the XP-A protein *(56,79,85)* and its affinity for (6-4)PDs *(21)* suggest a function in damage recognition.

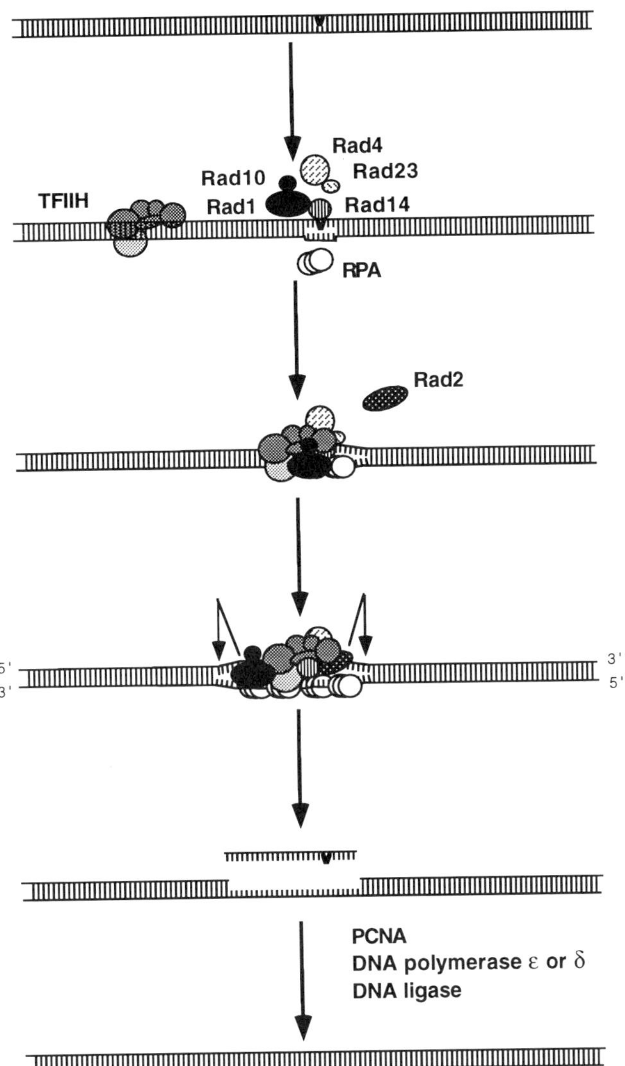

Fig. 4. Model for NER in yeast assuming sequential assembly of repair factors; *see* Section 2.6. for details.

2.6. Synopsis: The Initial Steps of NER

Based on the known properties of the proteins involved and their interactions (Fig. 1), one could imagine the following sequence of events (Fig. 4). Initially, Rad14p recognizes UV damage in a reaction that is enhanced by DNA-bound RPA and attracts Rad4p and Rad23p. Next, both proteins attract TFIIH. However, the possibility exists that TFIIH patrols along DNA even without damage, with the Rad3p component effecting its sequestration to sites of base damage. In any case, it is very likely that damage recognition is not accomplished by a single component, but damage-specific binding is altered in its specificity and enforced by the interaction of several components. The Rad3p and Ssl2p helicase activities associated with TFIIH result in local unwinding of DNA at the site of the lesion and dual incision is accomplished on one side by Rad2p, brought to the site by TFIIH interactions, and on the other side by the Rad1p/Rad10p complex (possibly interacting with Rad14p as inferred from results with the mammalian counterparts *[76]*).

Damage-specific incision activity has now been reconstituted in vitro from purified Rad14p, Rad4p/Rad23p complex, Rad2p and Rad1p/Rad10p nucleases, TFHII and RPA *(45)*. This reaction depends on hydrolysis of ATP. On a UV-damaged template, a fragment of 24–27 nucleotides was released, which is similar in size to that found with human cell extracts *(27–29,62)*. This study also proved the mutual dependency of Rad1p/Rad10p and Rad2p single-strand incision events: not even one-sided incision is found if one of these components is absent. In this respect, the yeast system seems to differ from mammalian NER, where an uncoupling of the incision steps has been observed, i.e., 3'-incision can occur without concomitant 5'-incision *(85)*.

Although the incision machinery can apparently be readily reconstituted from its purified components, there are indications that core TFIIH and Ssl2p are found in a preformed complex together with Rad1p, Rad10p, Rad2p, Rad4p and Rad14p. Following binding of TFIIH to Ni-agarose through a 6-histidine tag on the Tfb1p subunit, immunoblotting and complementation of NER activity of corresponding mutant extracts indicated the coelution of Rad1p, Rad10p, Rad14p, Ssl2p together with core TFIIH components in early fractions *(135)*. Corresponding to a peak in transcriptional activity, a separate peak of coeluting TFIIH core components alone was detected in later fractions *(135)*. It has been suggested that the higher-mol-wt complex corresponds to a core TFIIH containing repairosome that lacks the kinase subunits TFIIK and is inactive in transcription. The suggested balance between different TFIIH complexes with different activities has interesting implications for the possible influence of ongoing NER on transcription, and for indirect consequences of altered repair proteins on transcription. However, it needs to be mentioned that the existence of such a preassembled repairosome was not confirmed in an independent study that indicated a nonspecific interaction of Rad1p, Rad10p, and Rad14p with the Ni-agarose matrix *(49)*.

2.7. *Postincision Steps*

The detailed sequence of events following the bimodal incision step is not known. Helicase activities of TFIIH or of other proteins may be needed to release the single-stranded fragment. The nature of the gap-filling DNA polymerase can be inferred from a recent study *(20)*. Postirradiation DNA synthesis is still possible in single mutants defective in any one of four examined DNA polymerases (*see* ref. *121* for a review). However, the only pairwise combination that results in a repair defect involves polymerase δ and ε indicating that these are essential for repair but can efficiently substitute for each other. Polymerase ε has also been implicated in base excision repair *(147)*.

Following its loading onto DNA, which is mediated by replication factor C, proliferating cell nuclear antigen (PCNA) assembles with polymerase ε and δ into holoenzymes as an essential processivity factor. The large subunit of replication factor C (Cdc44p) was found to play a role in repair of UV radiation damage *(78)*. Studies with mammalian extracts have also identified PCNA as an indispensable component of NER *(93)*. Indeed, in a mutational analysis of yeast PCNA, mutant versions conferring UV sensitivity were readily recovered. However, only a minority of these showed defects in DNA replication, and a role for PCNA in the interaction with NER proteins other than polymerases was suggested *(5)*.

2.8. Determinants of Differential Damage Removal: Rad7p/Rad16p vs Rad26p

Yeast provides convenient systems to study the influence of chromatin structure and transcriptional activity on the rate of removal of UV photoproducts. The yeast mating type is determined by the gene expressed at the *MAT* loci (*MAT*a or *MAT*α), but the same information can also be found at the silent, nontranscribed locus (*HMR* or *HML*), rendered inactive through the activity of the *SIR* gene products. Thus, repair can be studied in identical DNA sequences that are either transcriptionally active or converted into silent chromatin. The rate of CPD removal is considerably slower in the inactive gene, but is accelerated to the level of the active gene in a *sir3* mutant *(138)*. This is clearly an effect of chromatin structure and not of transcription *per se*, since an inactivation of transcription of the MATα locus by deletion of the upstream activating sequence (UAS) does not affect the differential repair kinetics *(19)*.

Two mutants were found that were specifically defective in the repair of the transcriptionally silent *HML*α locus: *rad7* and *rad16 (137)*. Both genes have been cloned *(10,98)*: the Rad16p amino acid sequence contains motifs reminiscent of the seven signature domains found in helicases and also shows homology to the strand-break repair gene *RAD54* (*see* Chapter 16) and to several known transcriptional modulators, such as the yeast gene *SNF2* for which a role in chromatin remodeling has been proposed *(10)*. However, a recent phylogenetic analysis of the Snf2p family of proteins suggests separate subgroups for Rad16p, Rad54p, Rad26p *(see below)*, and for transcriptional activators, such as Snf2p itself *(29)*.

The UV sensitivity of deletion mutants of *RAD7* and *RAD16* is considerably less pronounced than that of mutants of other genes that are indispensable for NER, and indeed, measurement of overall repair suggests that *RAD7* and *RAD16* are only responsible for the removal of a fraction (20–30%) of all CPDs *(141)*. Indeed, the interaction of Rad7p with the transcriptional silencer Sir3p seems to confirm a role in rendering transcriptionally silent regions of the genome accessible for repair *(96)*. However, it became clear that Rad7p and Rad16p are not only involved in the repair of silenced heterochromatin-like regions, but also in the repair of the nontranscribed strand of transcriptionally active genes *(86,141)*. Preferential repair of the transcribed strand of expressed genes has been demonstrated in *E. coli* and mammalian cells (*see* Chapter 9, this vol. and Chapter 10, vol. 2) and there are examples for RNA polymerase II-transcribed yeast genes that follow the same rule (however, this does not appear to be a general feature) *(86,136,141)*. Interestingly, Rad16p has also been implicated in controlling a UV-inducible component of NER that can be detected in split-dose experiments *(150)*.

Rad7p and Rad16p are specifically required for repair of the nontranscribed strand, thus, *rad7* or *rad16* mutants share the same phenotype with human XP-C cells, and Rad7p/Rad16p have been called the functional equivalent of XP-C. However, it is the *S. cerevisiae* Rad4p gene product that shows structural homology with XP-C *(57)*. A *rad4* deletion mutant is highly UV-sensitive and completely blocked for NER. Whether this reflects a different role of Rad4 in yeast compared to XP-C human cells or whether certain, possibly more leaky *rad4* point mutants would indeed show a comparable phenotype, remains to be determined. Recent studies showed that a *rad4* deletion mutant is proficient in the removal of CPDs from the transcribed strand of yeast rDNA and thus indicates a differential defect in the repair of genes transcribed by RNA polymerase I *(143)*.

On the other hand, one could ask whether specific functions are also required for the accelerated rate of dimer removal from the transcribed strand. Indeed, a specific defect in preferential DNA repair is a hallmark of the UV-sensitive cells of Cockayne's syndrome *B (CS-B)* with *ERCC6* being the corresponding gene (*see* Chapter 18, vol. 2). A yeast gene *(RAD26)* displaying a high degree of homology to *ERCC6* was cloned, and again, the consensus domains of the Suf2 subfamily of helicases can be identified in the predicted protein sequence *(140)*. Indeed, preferential repair of the transcribed strand is absent in the *rad26* deletion mutant, and thus, RAD26 can be regarded as a true functional homolog of *CS-B/ERCC6 (140)*. Repair of yeast ribosomal DNA (rDNA) is not affected *(143)*. The deletion mutant shows a somewhat slower recovery of growth following UV treatment that does not translate into enhanced sensitivity of colony formation *(140)*. This is in contrast to the phenotype of the corresponding mammalian cell lines and seems to indicate a diminished role of preferential repair of the transcribed strand in yeast compared to higher eukaryotic cells.

These data have been explained by the a "global" NER pathway that includes Rad7p and Rad16p, and is required for the repair of transcriptionally silent genes and for the repair of the nontranscribed strand of expressed genes. The other subsystem does not require Rad7p and Rad16p, but does require Rad26p and is responsible for preferential repair of transcribed strands. Its activity can be fully compensated for by the global repair system, although at slower repair rates.

Indeed, in the background of a *rad7* or *rad16* deletion, inactivation of *RAD26* results in additional sensitization toward UV radiation. However, the sensitivity of a typical NER mutant (predicted to be blocked in both overlapping pathways) is even more pronounced *(144)*. A *rad7 rad16 rad26* triple mutant exhibits (as expected) a complete failure to repair the nontranscribed strand, but some repair of the transcribed strand is still evident *(144)*. It is difficult to decide whether this indicates the existence of additional unrecognized "preferentiality factors" or some inherent affinity for the transcribed strand of the core NER complex that is part of both pathways.

What constitutes the molecular mechanism for preferential repair? Detailed discussions of the complex interactions between NER and transcription can be found in a recent review *(34)* and in Chapters 10 and 18, vol. 2. Thus far, it is best to avoid using the term "transcription-coupled repair" for yeast, a term coined for *E. coli* (*see* Chapter 9). The mechanism for transcription repair coupling in *S. cerevisiae* and other eukaryotes remains elusive. The best candidate protein for a coupling factor, Rad26p, shows no homology with the *E. coli* transcription repair coupling factor (TRCF, Mfd), although both proteins contain helicase motifs *(29)*. Intuitively, the recruitment of a transcription initiation factor, such as TFIIH for NER (or of an NER component for transcription initiation), seems to suggest explanations for the phenomenon of preferential repair of expressed genes and of transcribed strands within expressed genes. It was hypothesized that TFIIH may actually travel along with the elongating RNA polymerase II and is thus being delivered to lesions that cause stalling of the RNA polymerase. This delivery mechanism would consequently result in a faster repair of lesions in the transcribed strand. However, recent in vitro data using the mammalian components dismiss the possibility that TFIIH is part of the elongation complex: once transcription is initiated, TFIIH stays behind *(158)*. An alternative hypothesis suggests that a blockage of RNA polymerase at a UV lesion may be reminiscent of the situation of transcript initiation, and could quickly attract the binding of TFIIH or of a TFIIH-containing preassembled repairosome. At the present time, there are no biochemical data available to support this idea.

3. POSTREPLICATION REPAIR AND MUTAGENESIS: THE *RAD6* EPISTASIS GROUP

The second epistasis group of mutants that are markedly sensitive to UV radiation is usually referred to as the *RAD6* group. However, the mutants of this group are phenotypically very diverse, and epistatic interactions have not been firmly established for all members regarding all DNA-damaging agents. The group includes mutants that in general are sensitive to both UV and ionizing radiation, and impaired in mutational responses to DNA damage to varying degrees; thus, they superficially resemble SOS repair mutants of *E. coli* (*see* Chapter 7). None of these mutants exhibits a defect in NER; some (but not all) are defective in postreplication repair, presumably owing to a failure to close single-strand gaps left during replication opposite noncoding lesions.

3.1. Ubiquitin Conjugation: Rad6p

The pleiotropic phenotype of *rad6* mutants suggested a multifunctional protein *(100,101,116)*. These mutants are characterized by greatly enhanced sensitivity to UV-radiation, γ-radiation, and various chemical agents. This phenotype is normally associated with the virtually complete absence of enhanced mutability in response to DNA-damaging agents (with the exception of direct acting agents, such as base analogs). However, the functions of Rad6p in repair and induced mutagenesis could be dissociated: suppressors of certain *rad6* point mutants were characterized that suppress UV sensitivity, but not defective mutagenesis. Some, but not all *rad6* point mutants were also found to be defective in sporulation. Spontaneous mutability is enhanced, as are spontaneous and damage-induced recombination. *rad6* deletion mutants are viable, but characterized by a prolonged S phase and low plating efficiencies.

The *RAD6* gene encodes a small protein (19.7 kDa) that contains a remarkably acidic terminus of 13 consecutive Asp residues *(104)*. Most significantly, Rad6p turned out to be a component of the ubiquitin-ligase system *(65)*. Rad6p is one of 12 E2 enzymes that transfer ubiquitin from an E1 enzyme to target molecules *(59,64)*, usually in the form of isopeptide-linked chains. This modification targets the protein for degradation by the 26 S proteasome. Mutations in different E2 enzymes result in distinct phenotypes, thus indicating different substrate specificities, e.g., the E2 enzyme Cdc34p plays an essential role during G1/S transition, however, the sequence similarity with Rad6p notwithstanding, Cdc34p has no function in DNA repair *(42)*. The transfer of ubiquitin to their target proteins can be catalyzed by E2 enzymes directly (as in the case of histones) or may require the participation of an E3 enzyme. The substrate selection by the

E3 enzymes is the basis for the "amino end rule" that defines the structure of protein-destabilizing termini.

The ubiquitin-conjugation activity of Rad6p is indeed essential for all *rad6* mutant phenotypes *(130)*. In vitro, Rad6p can catalyze both the ubiquitination of histones *(65,129)* and, in an E3-dependent reaction, of protein substrates containing destabilizing termini *(123)*. Complex formation of Rad6p with the yeast E3 enzyme Ubr1p was observed and the role of Rad6p in E3-dependent protein degradation was confirmed in vivo by the use of engineered β-galactosidase proteins containing various stabilizing or destabilizing N-termini *(28,152)*.

However, histone ubiquitination by Rad6p cannot be the basis for the DNA repair phenotype of *rad6* mutants. Deletion of the acidic tail abolishes histone ubiquitination in vitro, but this alteration results in a defect in sporulation only *(84,129)*. Indeed, the otherwise highly conserved Rad6p homologs in humans *(HHR6A* and *HHR6B*; *71*) and *Drosophila (Dhr6*; *70*), which lack the acidic terminus, complement the defect of a *rad6* deletion mutant in repair and mutagenesis, but only weakly complement the sporulation defect. Also E3-mediated ubiquitination of nonhistone target proteins is apparently not essential for the function of Rad6p in repair and mutagenesis. Deletion of the highly conserved N-terminus of Rad6p abolishes interaction with Ubr1p, and thus "end-rule"-dependent protein degradation in vivo *(152)*. The mutant is defective in sporulation, but only moderately UV-sensitive and normal with respect to induced mutagenesis. The important clue to the role of Rad6p-mediated ubiquitination in repair and mutagenesis may be its association with Rad18p, another repair protein discussed in the next section.

3.2. Targeting Rad6p to Single-Stranded DNA: Rad18p

Similar to *rad6* mutants, *rad18* mutants display a high degree of sensitivity to UV radiation, X-rays, and various chemical agents, as well as a defect in postreplication repair *(27,99)*. *rad18* mutants were not known to be strong antimutators *(73)* but this phenotype has recently been re-examined, and in *rad18* deletion mutants, a very significant reduction of UV mutability was found with a locus-specific *ochre*-allele reversion system and with a plasmid-borne mutational marker *(4,22)*. The predicted protein structure displays no homology with ubiquitin-conjugating enzymes, but a potential zinc finger (of the RING variety), a nucleotide binding consensus domain, and other structural features are reminiscent of transcriptional activators *(23,67)*. In immunoprecipitation experiments, stable and specific complex formation between Rad18p and Rad6p was demonstrated *(7)*. These complexes are formed separately from those between Rad6p and Ubr1p, the E3 enzyme *(7)*. The biological significance of the Rad6p/Rad18p interaction for DNA repair is indicated by the fact that overexpression of Rad18p can offset the UV sensitivity conferred to a wild-type strain by overexpression of the semidominant Rad6p alanine-88 variant (which lacks any ubiquitin-conjugating activity) *(7)*. Additionally, Rad18p endows the Rad6p/Rad18p complex with affinity to singlestranded DNA *(7)*. Interestingly, *rad18* mutants are not only defective in postreplication repair (mostly likely single-strand gap-filling), but are also impaired in repair of a special type of γ-ray damage that is characterized by locally denatured DNA (clustered base damage resulting in S1 nuclease-sensitive sites; *41*).

3.3. A Helicase with Channeling Function: Srs2p

Suppressors of the UV sensitivity of *rad6* and *rad18* mutants have been mapped to a single genetic locus named *SRS2* (for *s*uppressor of *r*ad *s*ix) or *RADH (2,113)*. Mutations in the same gene have been identified in a screen for mutants with an increased frequency of gene conversion *(107)*. Mutants of *SRS2* are also characterized by defective UV mutability and slightly increased UV sensitivity that seems to be confined to G1 cells *(2)*. The Srs2p structure indicates homology to the *E. coli* helicases UvrD and Rep *(2)*, and indeed, 3' → 5' DNA helicase and ATPase activities of the purified protein were demonstrated *(106)*. The gene is cell-cycle-regulated, with maximal expression at the G1/S boundary, and it is damage-inducible in G2-phase cells *(58)*.

The suppressor effect of *srs2* mutations on mutations in genes of the *RAD6* epistasis group depends on an intact pathway of homologous recombination; it is abolished in a *rad52* mutant background *(113)*. Thus, it has been suggested that Srs2p plays a role in channeling lesions, in particular, single-strand gaps opposite UV damage, into the Rad6p/Rad18p-dependent postreplicative repair pathway by diminishing the accessibility for Rad52p-dependent recombinational processing. Semidominant suppressors of the UV sensitivity of *srs2* mutants map to the *RAD51* gene, a yeast RecA homolog (*see* Chapter 16) *(1)*.

3.4. Activities Specifically Required for Damage-Induced Mutagenesis: The Rev Proteins

Specific selection schemes have been devised to isolate mutants that are defective in DNA-damage-induced mutagenesis. The mutants characterized so far fall into at least seven complementation groups. Mutations in the genes *REV3*, *REV6*, and *REV7* confer a very general defect in induced mutagenesis, i.e., dramatically reduced induced mutation frequencies are found for a number of different mutational systems and various DNA-damaging agents. Mutations in the other genes of this group have effects that are confined to specific mutational systems (frequently locus-specific reversion of highly revertible *ochre*-alleles) or agents. This surprising locus specificity of mutational systems in yeast is not understood. The deletion mutants of the *REV* genes isolated so far were all found to be viable and not significantly impaired in mitotic growth.

Inactivation of the *REV3* gene results in greatly reduced mutability of diverse mutational systems (forward mutation, reversion of nonsense, missense, frameshift alleles, plasmid systems) following treatment with a number of different agents *(73)*. Spontaneous mutability is also reduced, and a *rev3* mutation can diminish the mutator phenotype conferred to a plasmid borne *SUP4*-o locus by deletions of *RAD1*, *RAD6*, *RAD18*, or *RAD52*. However, differential effects on different classes of mutations were observed *(105,114)*. The *REV3* gene encodes a predicted protein of large size (~173 kDa) with a structure that is reminiscent of template-dependent DNA polymerases *(83)*. For example, the protein shares 23.9% amino acid identity and >60% similarity with Epstein-Barr virus DNA polymerase over a stretch of 673 residues and within the same region, six motifs which are highly conserved among DNA polymerases *(155)* can be found in the correct order. Structural similarities between Rev3p and yeast and human polymerase α include two zinc finger domains at the C-terminus *(83)*.

An unstable DNA polymerase activity was indeed found for the purified Rev3p *(92)*. However, this activity was enhanced by complex formation with the gene product of *REV7 (92),* another gene with a very general function in damage-induced mutagenesis *(74,75)*. The Rev3p/Rev7p complex was termed DNA polymerase ζ. If tested with substrates containing a *cis-syn* TT CPD in a defined position, a bypass efficiency of ~10% was found even without applying long incubation periods or forcing conditions *(92)*. By comparison, DNA polymerase α shows only ~1% translesion synthesis.

Rev3p/Rev7p must be regarded a nonessential polymerase that may exclusively be involved in the mutagenic bypass of DNA lesions. Unlike other polymerase transcripts, levels of *REV3* mRNA do not fluctuate during traversal of the cell cycle and are only moderately increased following UV treatment *(119)*. The contribution of Rev3p to the repair of prelethal damage appears to be very minor; the deletion mutant is only slightly UV-sensitive, and overall postreplication repair is not compromised *(99)*. Rev3p does not play a major role in DNA polymerization associated with base excision repair in vitro *(147)* or with UV repair in vivo *(20)*. The latter has been concluded from the fact that the accumulation of single-strand breaks (indicating aborted repair of UV damage) is not different in a quadruple mutant lacking functional polymerases α, δ, ε, and Rev3 from that observed in a polymerase δ and ε double mutant.

Point and deletion mutants of *REV1* show a decrease in UV-induced reversion of certain mutational markers. The only remarkable feature of the predicted protein (112 kDa) is an internal region of 152 residues that shows 25% identity to *E. coli* UmuC protein involved in the SOS response (*see* Chapter 7) *(72)*.

The mutational phenotype conferred by inactivation of *REV2/RAD5* may be even less general. Reduced mutation frequencies following UV treatment are apparently detectable only in certain *ochre* allele reversion systems and confined to *rev2* point mutants; the phenotype has not been confirmed for *REV2* deletion mutants *(66)*. The UV sensitivity of the *REV2* deletion mutant is considerably higher than that of deletion mutants of other *REV* genes, and a primary role in a nonmutagenic subpathway has been suggested *(66)*. Analysis of the structure of the predicted protein (134 kDa) revealed once more the presence of the seven signature domains found in potential helicases, but also a Rad18p-like zinc finger motif and a predicted leucine zipper region *(66)*. Rev2p has been assigned to the Rad16p subfamily of the Snf2p family of proteins (*see* Section 2.8.) *(29)*. The observation that the stability of simple repetitive sequences is enhanced in a *rev2* deletion mutant suggests decreased template slippage and, thus, a role for Rev2p in replication *(66)*.

3.5. Other Proteins Involved in UV Mutagenesis

A *bona fide* replication component, Cdc7p, has been implicated in UV mutagenesis. *CDC7* encodes a protein kinase that is required late during the G1/S transition for initiation of replication *(114)*. Certain (thermoconditional) *cdc7* mutants have been described to be slightly UV-sensitive and impaired in UV-induced reversion of auxotrophic markers at permissive temperature *(68)*. Overexpression of Cdc7p increases mutability *(115)* and a systematic examination of a collection of *cdc7* alleles in an isogenic background identified mutator as well as antimutator alleles. However, no correlation with protein kinase activity was found *(61)*. In general, the highest UV mutability is found if wild-type cells are UV-treated in S phase, and this peak in susceptibility to mutation is absent in an antimutator *cdc7-7* strain *(95)*. There are other mutants (such as *rad8* or *pso4)* known to decrease UV mutability significantly. However, the corresponding genes have not yet been identified. These may be informative, e.g., the phenotype of *pso4* mutants suggests a common function of Pso4p in mutagenesis and recombination *(3,81)*.

3.6. Synopsis

Two of the mutants *(rad6* and *rad18)* that confer the highest degree of UV sensitivity among all mutants of this group have been associated with defective postreplication repair (at least in NER-deficient cells). It seems reasonable to assume a central, initializing role of Rad6p/Rad18p in the bypass of noncoding lesions and the filling of single-strand gaps opposite UV damage. It has been suggested that the affinity of Rad18p for single-strand gaps may target the ubiquitin-conjugation activity of Rad6p to the stalled replication complex *(7)*. The resulting degradation of the replication complex would allow damage-tolerance mechanisms to operate, such as translesion synthesis. The DNA polymerase function of Rev3p/Rev7p seems to be pivotal for translesion synthesis, possibly as part of a specialized replication complex that includes other members of the *REV* gene family as well *(83,92)*. However, it is important to note that this activity is apparently dispensable for tolerating potentially lethal UV damage, and other mechanisms can easily compensate. One of them is most likely gap-filling by homologous recombination, and the *SRS2* helicase seems to play an important role in reducing the probability of gap processing by the *RAD52* pathway. However, it is not possible to explain the UV sensitivity of *rad6*, *rad18*, and, to a lesser extent, *rev2* mutants by the absence of homologous recombination. The existence of another error-free damage tolerance mechanism whose nature is completely unknown at the present time could be postulated. However, whatever model is used to account for mutagenic processing, it will also have to take into account a significant body of circumstantial evidence indicating that a substantial fraction of UV-induced mutations are fixed as two-strand alterations even before DNA replication is initiated *(36)*.

REFERENCES

1. Aboussekhra, A., R. Chanet, A. Adjiri, and F. Fabre. 1992. Semidominant suppressors of Srs2 helicase mutation of *Saccharomyces cerevisiae* map in the *RAD51* gene, whose sequence predicts a protein with similarities to procaryotic RecA proteins. *Mol. Cell. Biol.* **12:** 3224–3234.
2. Aboussekhra, A., R. Chanet, Z. Zgaga, C. Cassier-Chauvat, M. Heude, and F. Fabre. 1989. *RADH*, a gene of *Saccharomyces cerevisiae* encoding a putative DNA helicase involved in DNA repair. Characteristics of *radH* mutants and sequence of the gene. *Nucleic Acids Res.* **17:** 7211–7220.
3. Andrade, H. H., E. K. Marques, A. C. G. Schenberg, and J. A. P. Henriques. 1989. The *PS04* gene is responsible for error-prone recombinational repair in *Saccharomyces cerevisiae. Mol. Gen. Genet.* **217:** 419–426.
4. Armstrong, J. D., D. N. Chadee, and B. A. Kunz. 1994. Roles for the yeast *RAD18* and *RAD52* DNA repair genes in UV mutagenesis. *Mutat. Res.* **315:** 281–293.
5. Ayyagari, R., K. J. Impellizzeri, B. L. Yoder, S. L. Gary, and P. M. J. Burgers. 1995. A mutational analysis of the yeast proliferating cell nuclear antigen indicates distinct roles in DNA replication and DNA repair. *Mol. Cell. Biol.* **15:** 4420–4429.

6. Bailis, A. M., S. Maines, and M. T. Negritto. 1995. The essential helicase gene *RAD3* suppresses short-sequence recombination in *Saccharomyces cerevisiae. Mol. Cell. Biol.* **15:** 3998–4008.
7. Bailly, V., J. Lamb, P. Sung, S. Prakash, and L. Prakash. 1994. Specific complex formation between yeast RAD6 and RAD18 proteins: a potential mechanism for targeting RAD6 ubiquitin-conjugating activity to DNA damage sites. *Genes Dev.* **8:** 811–820.
8. Bailly, V., C. H. Sommers, P. Sung, L. Prakash, and S. Prakash. 1992. Specific complex formation between proteins encoded by the yeast DNA repair and recombination genes RAD1 and RAD10. *Proc. Natl. Acad. Sci. USA* **89:** 8273–8277.
9. Bailly, V., P. Sung, L. Prakash, and S. Prakash. 1991. DNA.RNA helicase activity of Rad3 protein of Saccharomyces cerevisiae. *Proc. Natl. Acad. Sci. USA* **88:** 9712–9725.
10. Bang, D. D., R. Verhage, N. Goosen, J. Bronwer, and P. van de Putte. 1992. Molecular cloning of *RAD16*, a gene involved in differential repair in *Saccharomyces cerevisiae. Nucleic Acids Res.* **20:** 3925–3931.
11. Bankmann, M., L. Prakash, and S. Prakash. 1992. Yeast RAD14 and human xeroderma pigmentosum group A DNA-repair genes encode homologous proteins. *Nature* **355:** 555–558.
12. Bardwell, A. J., L. Bardwell, N. Iyer, J. Q. Svejstrup, W. J. Feaver, R. D. Kornberg, and E. C. Friedberg. 1994. Yeast nucleotide excision repair proteins Rad2 and Rad4 interact with RNA polymerase II basal transcription factor b (TFIIH). *Mol. Cell. Biol.* **14:** 3569–3576.
13. Bardwell, A. J., L. Bardwell, N. Iyer, J. Q. Svejstrup, W. J. Feaver, R. Kornberg, and E. C. Friedberg. 1994. Yeast nucleotide excision repair proteins Rad2 and Rad4 interact with RNA polymerase II basal transcription factor b (TFIIH). *Mol. Cell. Biol.* **14:** 3569–3576.
14. Bardwell, A. J., L. Bardwell, D. K. Johnson, and E. C. Friedberg. 1993. Yeast DNA recombination and repair proteins Rad1 and Rad10 constitute a complex in vivo mediated by localized hydrophobic domains. *Mol. Microbiol.* **8:** 1177–1188.
15. Bardwell, A. J., L. Bardwell, A. E. Tomkinson, and E. C. Friedberg. 1994. Specific cleavage of model recombination and repair intermediates by the yeast Rad1/Rad10 endonuclease. *Science* **265:** 2082–2085.
16. Bardwell, L., A. J. Bardwell, W. J. Feaver, J. Q. Svejstrup, R. D. Kornberg, and E. C. Friedberg. 1994. Yeast RAD3 protein binds directly to both SSL2 and SSL1 proteins: implications for the structure and function of transcription/repair factor b. *Proc. Natl. Acad. Sci. USA* **91:** 3926–3930.
17. Bardwell, L., A. J. Cooper, and E. C. Friedberg. 1992. Stable and specific association between the yeast recombination and DNA repair proteins RAD1 and RAD10 in vitro. *Mol. Cell. Biol.* **12:** 3041–3049.
18. Brendel, M. and R. H. Haynes. 1973. Interactions among genes controlling sensitivity to radiation and alkylation in yeast. *Mol. Gen. Genet.* **125:** 197–216.
19. Bronwer, J., D. d. Bang, R. Verhage, and P. van de Putte. 1992. Preferential repair in *Saccharomyces cerevisiae*, in *DNA Repair Mechanisms* (Bohr, V. A., K. Wassermann, and K. H. Kraemer, eds.), Alfred Benzon Foundation. Munksgaard, Copenhagen, pp. 274–283.
20. Budd, M. E. and J. L. Campbell. 1995. DNA polymerases required for repair of UV-induced damage in *Saccharomyces cerevisiae. Mol. Cell. Biol.* **15:** 2173–2179.
21. Burns, J. L., S. N. Guzder, P. Sung, S. Prakash, and L. Prakash. 1996. An affinity of human replication protein A for ultraviolet-damaged DNA. Implications for damage recognition in nucleotide excision repair. *J. Biol. Chem.* **271:** 11,607–11,610.
22. Cassier-Chauvat, C. and F. Fabre. 1991. A similar defect in UV-induced mutagenesis conferred by the *rad6* and *radl 8* mutations of *Saccharomyces cerevisiae. Mutat. Res.* **254:** 247–253.
23. Chanet, R., N. Magaña-Schwencke, and F. Fabre. 1988. Potential DNA binding domains in the *RAD18* gene product of *Saccharomyces cerevisiae. Gene* **74:** 543–547.

24. Coverley, D., M. K. Kenny, D. P. Lane, and R. D. Wood. 1992. A role for the human single-stranded DNA binding protein HSSB/RPA in an early stage of nucleotide excision repair. *Nucleic Acids Res.* **20:** 3873–3880.
25. Cox, B. S. and J. M. Parry. 1968. The isolation, genetics and survival characteristics of ultraviolet light-sensitive mutants in yeast. *Mutat. Res.* **6:** 37–55.
26. Davies, A. A., E. C. Friedberg, A. E. Tomkinson, R. D. Wood, and S. C. West. 1995. Role of the Rad1 and Rad10 proteins in nucleotide excision repair and recombination. *J. Biol. Chem.* **270:** 24,638–24,641.
27. di Caprio, L. and B. S. Cox. 1981. DNA synthesis in UV-irradiated yeast. *Mutat. Res.* **82:** 69–85.
28. Dohmen, R. J., K. Madura, B. Bartel, and A. Varshavsky. 1991. The N-end rule is mediated by the *UBC2(RAD6)* ubiquitin-conjugating enzyme. *Proc. Natl. Acad. Sci. USA* **88:** 7351–7355.
29. Eisen, J. A., K. Sweder, and P. C. Hanawalt. 1995. Evolution of the SNF2 family of proteins: subfamilies with distinct sequences and function. *Nucleic Acids Res.* **23:** 2715–2723.
30. Feaver, W. J., J. Q. Svejstrup, L. Bardwell, A. J. Bardwell, S. Buratowski, K. D. Gulyas, T. F. Donahue, E. C. Friedberg, and R. D. Kornberg. 1993. Dual roles of a multiprotein complex from *S. cerevisiae* in transcription and DNA repair. *Cell* **75:** 1379–1387.
31. Feaver, W. J., J. Q. Svejstrup, N. L. Henry, and R. D. Kornberg. 1994. Relationship of CDK-activating kinase and RNA polymerase II CTD kinase TFIIH/TFIIK. *Cell* **79:** 1103–1109.
32. Fishman-Lobell, J., and J. Haber. 1992. Removal of nonhomologous DNA ends in double-strand break recombination—the role of the yeast ultraviolet repair gene *RAD1*. *Science* **258:** 480–484.
33. Fishman-Lobell, J., N. Rudin, and J. E. Haber. 1992. Two alternative pathways of double-strand break repair that are kinetically separable and independently modulated. *Mol. Cell. Biol.* **12:** 1292–1303.
34. Friedberg, E. C. 1996. Relationships between DNA repair and transcription. *Annu. Rev. Biochem.* **65:** 1–28.
35. Friedberg, E. C., W. Siede, and A. J. Cooper. 1991. Cellular responses to DNA damage in yeast, in *The Molecular and Cellular Biology of the Yeast* Saccharomyces. *I. Genome Dynamics, Protein Synthesis, and Energetics* (Broach, J. R., J. R. Pringle and E. W. Jones, eds.), Cold Spring Harbor Laboratory, New York, pp. 147–192.
36. Friedberg, E. C., G. C. Walker, and W. Siede. 1995. DNA Repair and *Mutagenesis*. American Society of Microbiology Press, Washington, DC.
37. Game, J. C. and B. S. Cox. 1971. Allelism tests of mutants affecting sensitivity to radiation in yeast and a proposed nomenclature. *Mutat. Res.* **12:** 328–331.
38. Game, J. C. and B. S. Cox. 1972. Epistatic interactions between four *rad* loci in yeast. *Mutat. Res.* **16:** 353–362.
39. Game, J. C. and B. S. Cox. 1973. Synergistic interactions between *rad* mutations in yeast. *Mutat. Res.* **20:** 35–44.
40. Game, J. C. and R. K. Mortimer. 1974. A genetic study of X-ray sensitive mutants in yeast. *Mutat. Res.* **24:** 281–292.
41. Geigl, E.-M. and F. Eckardt-Schupp. 1991. Repair of gamma-ray induced S1 nuclease hypersensitive sites in yeast depends on homologous mitotic recombination and a *RAD18*-dependent function. *Curr. Genet.* **20:** 33–37.
42. Goebl, M. G., J. Yochem, S. Jentsch, J. P. McGrath, A. Varshavsky, and B. Byers. 1988. The yeast cell cycle gene *CDC34* encodes a ubiquitin-conjugating enzyme. *Science* **241:** 1331–1335.
43. Gulyas, K. D. and T. F. Donahue. 1992. SSL2, a suppressor of a stemloop mutation in the HIS4 leader, encodes the yeast homolog of human ERCC-3. *Cell* **69:** 1031–1042.

44. Guzder, S. N., V. Bailly, P. Sung, L. Prakash, and S. Prakash. 1995. Yeast DNA repair protein RAD23 promotes complex formation between transcription factor TFIIH and DNA damage recognition factor RAD14. *J. Biol. Chem.* **270:** 8385–8388.
45. Guzder, S. N., Y. Habraken, P. Sung, L. Prakash, and S. Prakash. 1995. Reconstitution of yeast nucleotide excision repair with purified Rad proteins, replication protein A, and transcription factor TFIIH. *J. Biol. Chem.* **270:** 12,973–12,976.
46. Guzder, S. N., H. Qiu, C. H. Sommers, P. Sung, L. Prakash, and S. Prakash. 1994. DNA repair gene RAD3 of S. cerevisiae is essential for transcription by RNA polymerase II. *Nature* **367:** 91–94.
47. Guzder, S. N., P. Sung, V. Bailly, L. Prakash, and S. Prakash. 1994. RAD25 is a DNA helicase required for DNA repair and RNA polymerase II transcription. *Nature* **369:** 578–581.
48. Guzder, S. N., P. Sung, L. Prakash, and S. Prakash. 1993. Yeast DNA-repair gene RAD14 encodes a zinc metalloprotein with affinity for ultraviolet-damaged DNA. *Proc. Natl. Acad. Sci. USA* **90:** 5433–5437.
49. Guzder, S. N., P. Sung, L. Prakash, and S. Prakash. 1996. Nucleotide excision repair in yeast is mediated by sequential assembly of repair factors and not by a pre-assembled repairosome. *J. Biol. Chem.* **271:** 8903–8910.
50. Guzder, S. N., P. Sung, S. Prakash, and L. Prakash. 1995. Lethality in yeast of trichothiodystrophy (TTD) mutations in the human xeroderma pigmentosum group D gene- implications for transcriptional defect in TTD. *J. Biol. Chem.* **270:** 17,660–17,663.
51. Habraken, Y., P. Sung, L. Prakash, and S. Prakash. 1993. Yeast excision repair gene RAD2 encodes a single-stranded DNA endonuclease. *Nature* **366:** 365–368.
52. Habraken, Y., P. Sung, L. Prakash, and S. Prakash. 1994. Holliday junction cleavage by yeast Rad1 protein. *Nature* **371:** 531–534.
53. Habraken, Y., P. Sung, L. Prakash, and S. Prakash. 1995. Structure-specific nuclease activity in yeast nucleotide excision repair protein Rad2. *J. Biol. Chem.* **270:** 30,194–30,198.
54. Harosh, I., L. Naumovski, and E. C. Friedberg. 1989. Purification and characterization of the Rad3 ATPase/DNA helicase from *Saccharomyces cerevisiae. J. Biol. Chem.* **264:** 20,532–20,539.
55. Haynes, R. H. and B. A. Kunz. 1981. DNA repair and mutagenesis in yeast, in *The Molecular Biology of the Yeast* Saccharomyces, vol. I (Strathern, J. N., E. W. Jones, and J. R. Broach, eds.), Cold Spring Harbor Laboratory, Cold Spring Harbor, NY, pp. 371–414.
56. He, Z., L. A. Henricksen, M. S. Wold, and C. J. Ingles. 1995. RPA involvement in the damage-recognition and incision steps of nucleotide excision repair. *Nature* **374:** 566–569.
57. Henning, K. A., C. Peterson, R. Legerski, and E. C. Friedberg. 1994. Cloning the *Drosophila* homolog of the xeroderma pigmentosum group C gene reveals homology between the predicted human and *Drosophila* polypeptides and that encoded by the yeast RAD4 gene. *Nucleic Acids Res.* **22:** 257–261.
58. Heude, M., R. Chanet, and F. Fabre. 1995. Regulation of the *Saccharomyces cerevisiae* Srs2 helicase during the mitotic cell cycle, meiosis and after irradiation. *Mol. Gen. Genet.* **248:** 59–68.
59. Hochstrasser, M. 1995. Ubiquitin, proteasomes, and the regulation of intracellular protein degradation. *Curr. Opinion Cell Biol.* **7:** 215–223.
60. Hoeijmakers, J. H. J. 1994. Human nucleotide excision repair syndromes: molecular clues to unexpected intricacies. *Eur. J. Cancer* **30A:** 1912–1921.
61. Hollingsworth, R. E., Jr., R. M. Ostroff, M. B. Klein, L. A. Niswander, and R. A. Sclafani. 1992. Molecular genetic studies of the Cdc7 protein kinase and induced mutagenesis in yeast. *Genetics* **132:** 53–62.
62. Huang, J. C., D. L. Svoboda, J. T. Reardon, and A. Sancar. 1992. Human nucleotide excision nuclease removes thymine dimers from DNA by incising the 22nd phosphodiester bond 5' and the 6th phosphodiester bond 3' to the photodimer. *Proc. Natl. Acad. Sci. USA* **89:** 3664–3668.

63. Ivanov, E. L., and J. E. Haber. 1995. *RAD1* and *RAD10*, but not other excision repair genes, are required for double-strand break-induced recombination in *Saccharomyces cerevisiae. Mol. Cell. Biol.* **15:** 2245–2251.
64. Jentsch, S. 1992. The ubiquitin-conjugation system. *Ann. Rev. Genet.* **26:** 179–207.
65. Jentsch, S., J. P. McGrath, and A. Varshavsky. 1987. The yeast DNA repair gene RAD6 encodes a ubiquitin-conjugating enzyme. *Nature* **329:** 131–134.
66. Johnson, R. E., S. T. Henderson, T. D. Petes, S. Prakash, M. Bankmann, and L. Prakash. 1992. *Saccharomyces cerevisiae RAD5*-encoded DNA repair protein contains DNA helicase and zinc-binding sequence motifs and affects the stability of simple repetitive sequences in the genome. *Mol. Cell. Biol.* **12:** 3807–3818.
67. Jones, J. S., S. Weber, and L. Prakash. 1988. The *Saccharomyces cerevisiae RAD18* gene encodes a protein that contains potential zinc finger domains for nucleic acid binding and a putative nucleotide binding sequence. *Nucleic Acids Res.* **16:** 7119–7131.
68. Kilbey, B. J. 1986. *cdc7* alleles and the control of induced mutagenesis in yeast. *Mutagenesis* **1:** 29–31.
69. Klein, H. L. 1988. Different types of recombination events are controlled by the *RAD1* and *RAD52* genes of *Saccharomyces cerevisiae. Genetics* **120:** 367–377.
70. Koken, M., P. Reynolds, D. Bootsma, J. H. J. Hoeijmakers, S. Prakash, and L. Prakash. 1991. *Dhr6*, a *Drosophila* homolog of the yeast DNA repair gene *RAD6. Proc. Natl. Acad. Sci. USA* **88:** 3832–3836.
71. Koken, M. H. M., P. Reynolds, I. Jaspers-Dekker, L. Prakash, S. Prakash, D. Bootsma, and J. H. J. Hoeijmakers. 1991. Structural and functional conservation of two human homologs of the yeast DNA repair gene *RAD6. Proc. Natl. Acad. Sci. USA* **88:** 8865–8869.
72. Larimer, F. W., J. R. Perry, and A. A. Hardigree. 1989. The *REV1* gene of *Saccharomyces cerevisiae:* Isolation, sequence and functional analysis. *J. Bacteriol.* **171:** 230–237.
73. Lawrence, C. W. 1982. Mutagenesis in *Saccharomyces cerevisiae. Adv. Genet.* **21:** 173–254.
74. Lawrence, C. W., G. Das, and R. B. Christensen. 1985. *REV7*, a new gene concerned with UV mutagenesis in yeast. *Mol. Gen. Genet.* **200:** 80–85.
75. Lawrence, C. W., P. E. Nisson, and R. B. Christensen. 1985. UV and chemical mutagenesis in *rev7* mutants of yeast. *Mol. Gen. Genet.* **200:** 86–91.
76. Li, L., S. J. Elledge, C. A. Peterson, E. S. Bales, and R. J. Legerski. 1994. Specific association between the human DNA repair proteins XPA and ERCC1. *Proc. Natl. Acad. Sci. USA* **91:** 5012–5016.
77. Longhese, M. P., P. Plevani, and G. Lucchini. 1994. Replication factor A is required in vivo for DNA replication, repair, and recombination. *Mol. Cell. Biol.* **14:** 7884–7890.
78. MacAlear, M. A., K. M. Tuffo, and C. Holm. 1996. The large subunit of replication factor C (Rfclp/Cdc44p) is required for DNA replication and DNA repair in *Saccharomyces cerevisiae*. *Genetics* **142:** 65–78.
79. Matsuda, T., M. Saijo, I. Kuraoka, T. Kobayashi, Y. Nakatsu, A. Nagai, T. Enjoji, C. Masutani, K. Sugasawa, F. Hanaoka, A. Yasui, and K. Tanaka 1995. DNA repair protein XPA binds replication protein A (RPA). *J. Biol. Chem.* **270:** 4152–4157.
80. Matsunaga, T., D. Mu, C.-H. Park, J. T. Reardon, and A. Sancar. 1995. Human DNA repair excision nuclease. *J. Biol. Chem.* **270:** 20,862–20,869.
81. Meira, L. B., M. B. Fonseca, D. Averbeck, A. C. G. Schenberg, and J. A. P. Henriques. 1992. The *pso4-1* mutation reduces spontaneous mitotic gene conversion and reciprocal recombination in *Saccharomyces cerevisiae. Mol. Gen. Genet.* **235:** 311–316.
82. Montelone, B. A., M. F. Hoekstra, and R. E. Malone. 1988. Spontaneous mitotic recombination in yeast: the hyper-recombinational *rem1* mutations are alleles of the *RAD3* gene. *Genetics* **119:** 289–301.
83. Morrison, A., R. B. Christensen, J. Alley, A. K. Beck, E. G. Bernstine, J. F. Lemontt, and C. W. Lawrence. 1989. *REV3*, a yeast gene whose function is required for induced

mutagenesis, is predicted to encode a nonessential DNA polymerase. *J. Bacteriol.* **171:** 5659–5667.

84. Morrison, A., E. J. Miller, and L. Prakash. 1988. Domain structure and functional analysis of the carboxyl-terminal polyacidic sequence of the *RAD6* protein of *Saccharomyces cerevisiae. Mol. Cell. Biol.* **8:** 1179–1185.
85. Mu, D., D. S. Hsu, and A. Sancar. 1996. Reaction mechanism of human DNA repair excision nuclease. *J. Biol. Chem.* **271:** 8285–8294.
86. Mueller, J. P., and M. J. Smerdon. 1995. Repair of plasmid and genomic DNA in a *rad7Δ* mutant of yeast. *Nucleic Acids Res.* **23:** 3457–3464.
87. Mueller, J. P., and M. J. Smerdon. 1996. Rad23 is required for transcription-coupled repair and efficient overall repair in *Saccharomyces cerevisiae. Mol. Cell. Biol.* **16:** 2361–2368.
88. Naegeli, H. 1995. Mechanisms of DNA damage recognition in mammalian nucleotide excision repair. *FASEB J.* **9:** 1043–1050.
89. Naegeli, H., L. Bardwell, and E. C. Friedberg. 1992. The DNA helicase and adenosine triphosphatase activities of yeast Rad3 protein are inhibited by DNA damage. *J. Biol. Chem.* **267:** 392–398.
90. Naegeli, H., L. Bardwell, and E. C. Friedberg. 1993. Inhibition of Rad3 DNA helicase activity by DNA adducts and abasic sites: implications for the role of a DNA helicase in damage-specific incision of DNA. *Biochemistry* **32:** 613–621.
91. Naegeli, H., L. Bardwell, I. Harosh, and E. C. Friedberg. 1992. Substrate specificity of the Rad3 ATPase/DNA helicase of *Saccharomyces cerevisiae* and binding of Rad3 protein to nucleic acids. *J. Biol. Chem.* **267:** 7839–7844.
92. Nelson, J. R., C. W. Lawrence, and D. C. Hinkle. 1996. Thymine–thymine dimer bypass by yeast DNA polymerase ζ. *Science* **272:** 1646–1649.
93. Nichols, A. F. and A. Sancar. 1992. Purification of PCNA as a nucleotide excision repair protein. *Nucleic Acids Res.* **20:** 2441–2446.
94. O'Donovan, A., A. A. Davies, J. G. Moggs, S. C. West, and R. D. Wood. 1994. XPG endonuclease makes the 3' incision in human DNA nucleotide excision repair. *Nature* **371:** 432–435.
95. Ostroff, R. M. and R. A. Sclafani. 1995. Cell cycle regulation of induced mutagenesis in yeast. *Mutat. Res.* **329:** 143–152.
96. Paetkau, D. W., J. A. Riese, W. S. MacMorran, R. A. Woods, and R. D. Gietz. 1994. Interaction of the yeast *RAD7* and *SIR3* proteins: implications for DNA repair and chromatin structure. *Genes Dev.* **8:** 2035–2045.
97. Park, E., S. N. Guzder, M. H. M. Koken, I. Jaspers-Dekker, G. Weeda, J. H. J. Hoeijmakers, S. Prakash, and L. Prakash. 1992. RAD25 (SSL2), the yeast homolog of the human Xeroderma-pigmentosum group B DNA repair gene, is essential for viability. *Proc. Natl. Acad. Sci. USA* **89:** 11,416–11,420.
98. Perozzi, G. and S. Prakash. 1986. *RAD7* gene of *Saccharomyces cerevisiae:* transcripts, nucleotide sequence analysis and functional relationship between the *RAD7* and *RAD23* gene product. *Mol. Cell. Biol.* **6:** 1497–1507.
99. Prakash, L. 1981. Characterization of postreplication repair in *Saccharomyces cerevisiae* and effects of *rad6, rad18, rev3,* and *rad52* mutations. *Mol. Gen. Genet.* **184:** 471–478.
100. Prakash, L. 1994. The *RAD6* gene and protein of *Saccharomyces cerevisiae. Ann. NY Acad. Sci.* **726:** 267–273.
101. Prakash, L. and S. Prakash. 1980. Genetic analysis of error-prone repair systems in *Saccharomyces cerevisiae,* in *DNA-Repair and Mutagenesis in Eukaryotes* (Generoso,W. M., M. D. Shelby, and F. J. de Serres, eds.), Plenum, New York, pp. 141–158.
102. Prakash, S., P. Sung, and L. Prakash. 1993. DNA repair genes and proteins of *Saccharomyces cerevisiae. Annu. Rev. Genet.* **27:** 33–70.

103. Qiu, H., E. Park, L. Prakash, and S. Prakash. 1993. The *Saccharomyces cerevisiae* DNA repair gene *RAD25* is required for transcription by RNA polymerase II. *Genes Dev.* **7:** 2161–2171.
104. Reynolds, P., S. Weber, and L. Prakash. 1985. *RAD6* gene of *Saccharomyces cerevisiae* encodes a protein containing a tract of 13 consecutive aspartates. *Proc. Natl. Acad. Sci. USA* **82:** 168–172.
105. Roche, H., R. D. Gietz, and B. A. Kunz. 1994. Specificity of the yeast *rev3Δ* antimutator and *REV3* dependency of the mutator resulting from a defect *(radlΔ)* in nucleotide excision repair. *Genetics* **137:** 637–646.
106. Rong, L. and H. Klein. 1993. Purification and characterization of the *SRS2* DNA helicase of the yeast *Saccharomyces cerevisiae*. *J. Biol. Chem.* **268:** 1252–1259.
107. Rong, L., F. Palladino, A. Aguilera, and H. Klein. 1991. The hyper-gene conversion *hpr5-1* mutation of *Saccharomyces cerevisiae* is an allele of the *SRS2/RADH* gene. *Genetics* **127:** 75–85.
108. Sancar, A. 1995. Excision repair in mammalian cells. *J. Biol. Chem.* **270:** 15,915–15,918.
109. Sancar, G. B. and F. W. Smith. 1989. Interactions between yeast photolyase and nucleotide excision repair proteins in *Saccharomyces cerevisiae* and *Escherichia coli*. *Mol. Cell. Biol.* **9:** 4767–4776.
110. Schaeffer, L., R. Roy, S. Humbert, V. Moncollin, W. Vermeulen, J. H. J. Hoeijmakers, P. Chambon, and J.-M. Egly. 1993. DNA repair helicase: a component of BTF2 (TFIIH) basic transcription factor. *Science* **260:** 58–63.
111. Schiestl, R. H. and S. Prakash. 1988. *RAD1*, an excision repair gene of *Saccharomyces cerevisiae*, is also involved in recombination. *Mol. Cell. Biol.* **8:** 3619–3626.
112. Schiestl, R. H. and S. Prakash. 1990. *RAD10*, an excision repair gene of *Saccharomyces cerevisiae*, is also involved in the *RAD1* pathway of mitotic recombination. *Mol. Cell. Biol.* **10:** 2485–2491.
113. Schiestl, R. H., S. Prakash, and L. Prakash. 1990. The *SRS2* suppressor of *rad6* mutations of *Saccharomyces cerevisiae* acts by channeling DNA lesions into the *RAD52* DNA repair pathway. *Genetics* **124:** 817–831.
114. Sclafani, R. A. and A. L. Jackson. 1994. Cdc7 protein kinase for DNA metabolism comes of age. *Mol. Microbiol.* **11:** 805–810.
115. Sclafani, R. A., M. Patterson, J. Rosamond, and W. L. Fangman. 1988. Differential regulation of the yeast *CDC7* gene during mitosis and meiosis. *Mol. Cell. Biol.* **8:** 293–300.
116. Siede, W. 1988. The *RAD6* gene of yeast: a link between DNA repair, chromosome structure and protein degradation. *Radiat. Environ. Biophys.* **27:** 277–286.
117. Siede, W., A. S. Friedberg, and E. C. Friedberg. 1993. Evidence that the Rad1 and Rad10 proteins of *Saccharomyces cerevisiae* participate as a complex in nucleotide excision repair of UV radiation damage. *J. Bacteriol.* **175:** 6345–6347.
118. Siede, W. and E. C. Friedberg. 1992. Regulation of the yeast *RAD2* gene: DNA damage-dependent induction correlates with protein binding to regulatory sequences and their deletion influences survival. *Mol. Gen. Genet.* **232:** 247–256.
119. Singhal, R. K., D. C. Hinkle, and C. W. Lawrence. 1992. The *REV3* gene of *Saccharomyces cerevisiae* is transcriptionally regulated more like a repair gene than one encoding a DNA polymerase. *Mol. Gen. Genet.* **236:** 17–24.
120. Song, J. M., B. A. Montelone, W. Siede, and E. C. Friedberg. 1990. Effects of multiple yeast rad3 mutant alleles on UV sensitivity, mutability, and mitotic recombination. *J. Bacteriol.* **172:** 6620–6630.
121. Sugino, A. 1995. Yeast DNA polymerases and their role at the replication fork. *Tr. Biochem. Sci.* **20:** 319–323.
122. Sugino, A., B. H. Ryu, T. Sugino, L. Naumovski, and E. C. Friedberg. 1986. A new DNA-dependent ATPase which stimulates yeast DNA polymerase I and has DNA-unwinding activity. *J. Biol. Chem.* **261:** 11,744–11,750.

123. Sung, P., E. Berleth, C. Pickart, S. Prakash, and L. Prakash. 1991. Yeast *RAD6* encoded ubiquitin conjugating enzyme mediates protein degradation dependent on the N-end-recognizing E3 enzyme. *EMBO J.* **10:** 2187–2193.
124. Sung, P., S. N. Guzder, L. Prakash, and S. Prakash. 1996. Reconstitution of TFIIH and requirement of its DNA helicase subunits, Rad3 and Rad25, in the incision step of nucleotide excision repair. *J. Biol. Chem.* **271:** 10,821–10,826.
125. Sung, P., D. Higgins, L. Prakash, and S. Prakash. 1988. Mutation of lysine-48 to arginine in the yeast RAD3 protein abolishes its ATPase and DNA helicase activities but not the ability to bind ATP. *EMBO J.* **7:** 3263–3269.
126. Sung, P., L. Prakash, S. W. Matson, and S. Prakash. 1987. *RAD3* protein of *Saccharomyces cerevisiae* is a DNA helicase. *Proc. Natl. Acad. Sci. USA* **84:** 8951–8955.
127. Sung, P., L. Prakash, and S. Prakash. 1992. Renaturation of DNA catalysed by yeast DNA repair and recombination protein RAD10. *Nature* **355:** 743–745.
128. Sung, P., L. Prakash, S. Weber, and S. Prakash. 1987. The *RAD3* gene of *Saccharomyces cerevisiae* encodes a DNA-dependent ATPase. *Proc. Natl. Acad. Sci. USA* **84:** 6045–6049.
129. Sung, P., S. Prakash, and L. Prakash. 1988. The *RAD6* protein of *Saccharomyces cerevisiae* polyubiquitinates histones and its acidic domain mediates this activity. *Genes Dev.* **2:** 1476–1485.
130. Sung, P., S. Prakash, and L. Prakash. 1990. Mutation of cysteine-88 in the *Saccharomyces cerevisiae RAD6* protein abolishes its ubiquitin-conjugating activity and its various biological functions. *Proc. Natl. Acad. Sci. USA* **87:** 2695–2699.
131. Sung, P., P. Reynolds, L. Prakash, and S. Prakash. 1993. Purification and characterization of the *Saccharomyces cerevisiae RAD1/RAD10* endonuclease. *J. Biol. Chem.* **268:** 26,391–26,399.
132. Sung, P., J. F. Watkins, L. Prakash, and S. Prakash. 1994. Negative superhelicity promotes ATP-dependent binding of yeast RAD3 protein to ultraviolet-damaged DNA. *J. Biol. Chem.* **269:** 8303–8308.
133. Svejstrup, J. Q., W. J. Feaver, and R. D. Kornberg. 1996. Subunits of yeast RNA polymerase II transcription factor TFIIH encoded by the CCL1 gene. *J. Biol. Chem.* **271:** 643–645.
134. Svejstrup, J. Q., W. J. Feaver, J. Lapointe, and R. D. Kornberg. 1994. RNA polymerase transcription factor IIH holoenzyme from yeast. *J. Biol. Chem.* **269:** 28,044–28,048.
135. Svejstrup, J. Q., Z. Wang, W. J. Feaver, X. Wu, D. A. Bushnell, T. F. Donahue, E. C. Friedberg, and R. D. Kornberg. 1994. Different forms of TFIIH for transcription and DNA repair: holo-TFIIH and a nucleotide excision repairosome. *Cell* **80:** 21–28.
136. Sweder, K. S. and P. C. Hanawalt. 1992. Preferential repair of cyclobutane pyrimidine dimers in the transcribed strand of a gene in yeast chromosomes and plasmids is dependent on transcription. *Proc. Natl. Acad. Sci. USA* **89:** 10,696–10,700.
137. Terleth, C., P. Schenk, R. Poot, J. Brouwer, and P. van de Putte. 1990. Differential repair of UV damage in *rad* mutants of *Saccharomyces cerevisiae:* a possible function of G2 arrest upon UV irradiation. *Mol. Cell. Biol.* **10:** 4678–4684.
138. Terleth, C., C. A. van Sluis, and P. van de Putte. 1989. Differential repair of UV damage in *Saccharomyces cerevisiae. Nucleic Acids Res.* **17:** 4433–4440.
139. Tomkinson, A., A. J. Bardwell, L. Bardwell, N. J. Tappe, and E. C. Friedberg. 1993. Yeast DNA repair and recombination proteins Rad1 and Rad10 constitute a single-stranded DNA endonuclease. *Nature* **362:** 860–862.
140. van Gool, A. J., R. Verhage, S. M. A. Swagemakers, P. van de Putte, J. Brouwer, C. Troelstra, D. Bootsma, and J. H. J. Hoeijmakers. 1994. *RAD26*, the functional S. *cerevisiae* homolog of the Cockayne syndrome B gene *ERCC6*. *EMBO J.* **13:** 5361–5369.
141. Verhage, R., A.-M. Zeeman, F. Gleig, D. D. Bang, P. van de Putte, and J. Brouwer. 1994. The *RAD7* and *RAD16* genes, which are essential for pyrimidine dimer removal from the silent mating type loci, are also required for repair of the nontranscribed strand of an active gene in *Saccharomyces cerevisiae. Mol. Cell. Biol.* **14:** 6135–6142.

142. Verhage, R., A.-M. Zeeman, M. Lombaerts, P. van de Putte, and J. Brouwer. 1996. Analysis of gene-and strand-specific repair in the moderately UV sensitive *S. cerevisiae rad23* mutant. *Mutat. Res.* **362:** 155–165.
143. Verhage, R. A., P. van de Putte, and J. Brouwer. 1996. Repair of rDNA in *Saccharomyces cerevisiae:* RAD4-independent strand-specific nucleotide excision repair of RNA polymerase I transcribed genes. *Nucleic Acids Res.* **24:** 1020–1025.
144. Verhage, R. A., A. J. van Gool, N. de Groot, J. H. J. Hoeijmakers, P. van de Putte, and J. Brouwer. 1996. Double mutants of *Saccharomyces cerevisiae* with alterations in global genome and transcription-coupled repair. *Mol. Cell. Biol.* **16:** 496–502.
145. Wang, Z., S. Buratowski, J. Q. Svejstrup, W. J. Feaver, X. Wu, R. D. Kornberg, T. F. Donahue, and E. C. Friedberg. 1995. Yeast *TFB1* and *SSL1* genes encoding subunits of transcription factor IIH (TFIIH) are required for nucleotide excision repair. *Mol. Cell. Biol.* **15:** 2288–2293.
146. Wang, Z., J. Q. Svejstrup, W. J. Feaver, X. Wu, R. D. Kornberg, and E. C. Friedberg. 1994. Transcription factor b (TFIIH) is required during nucleotide-excision repair in yeast. *Nature* **368:** 74–76.
147. Wang, Z., X. Wu, and E. C. Friedberg. 1993. DNA repair synthesis during base excision repair *in vitro* is catalyzed by DNA polymerase ε and is influenced by DNA polymerases α and δ in *Saccharomyces cerevisiae. Mol. Cell. Biol.* **13:** 1051–1058.
148. Wang, Z., X. Wu, and E. C. Friedberg. 1993. Nucleotide-excision repair of DNA in cell-free extracts of the yeast *Saccharomyces cerevisiae. Proc. Natl. Acad. Sci. USA* **90:** 4907–4911.
149. Wang, Z., X. Wu, and E. C. Friedberg. 1995. The detection and measurement of base and nucleotide excision repair in cell-free extracts of the yeast *Saccharomyces cerevisiae. Methods: Companion Methods Enzymol.* **7:** 177–186.
150. Waters, R., R. Zhang, and N. J. Jones. 1993. Inducible removal of UV-induced pyrimidine dimers from transcriptionally active and inactive genes of *Saccharomyces cerevisiae. Mol. Gen. Genet.* **239:** 28–32.
151. Watkins, J. F., P. Sung, L. Prakash, and S. Prakash. 1993. The *Saccharomyces cerevisiae* DNA repair gene *RAD23* encodes a nuclear protein containing a ubiquitin-like domain required for biological function. *Mol. Cell. Biol.* **13:** 7757–7773.
152. Watkins, J. F., P. Sung, S. Prakash, and L. Prakash. 1993. The extremely conserved amino terminus of RAD6 ubiquitin-conjugating enzyme is essential for amino-end rule-dependent protein degradation. *Genes Dev.***7:** 250–261.
153. West, S. C. 1995. Holliday junctions cleaved by Rad1? *Nature* **373:** 27,28.
154. Wilcox, D. R. and L. Prakash. 1981. Incision and postincision step of pyrimidine dimer removal in excision-defective mutants of *Saccharomyces cerevisiae. J. Bacteriol.* **148:** 618–623.
155. Wong, S. W., A. F. Wahl, P.-M. Yuan, N. A. Arai, B. E. Pearson, K.-I. Arai, D. Korn, M. W. Hunkapiller, and T. S.-F. Wang. 1988. Human polymerase α gene expression is cell proliferation dependent and its primary structure is similar to both prokaryotic and eukaryotic replicative DNA polymerases. *EMBO J.* **7:** 37–47.
156. Yang, Y., A. L. Johnson, L. H. Johnston, W. Siede, E. C. Friedberg, K. Ramachandran, A. Walichnowski, and B. A. Kunz. A mutation in a *Saccharomyces cerevisiae* gene *(RAD3)* required for nucleotide excision repair and transcription increases the efficiency of mismatch correction. *Genetics* **144:** 459–466.
157. Yoon, H., S. P. Miller, E. K. Pabich, and T. F. Donahue. 1992. *SSL1*, a suppressor of a *HIS4* 5'-UTR stem-loop mutation, is essential for translation initiation and affects UV resistance in yeast. *Genes Dev.* **6:** 2463–2477.
158. Zawel, L., K. P. Kumar, and D. Reinberg. 1995. Recycling of the general transcription factors during RNA polymerase II transcription. *Genes Dev.* **9:** 1479–1490.

16

Double-Strand Break and Recombinational Repair in *Saccharomyces cerevisiae*

Jac A. Nickoloff and Merl F. Hoekstra

1. INTRODUCTION

Double-strand breaks (DSBs) are an important form of DNA damage that if misrepaired can result in mutagenic deletions, rearrangements, or translocations, and unrepaired DSBs can lead to chromosome loss and cell death. Single-strand damage (single-strand breaks [SSBs] or base damage) can be repaired using the immediately accessible complementary strand as a template, but this is not possible for double-strand damage (DSBs, gaps, or opposed base damage). DSBs can be repaired by direct ligation to prevent chromosome loss and increase cell survival, but with a mutagenic cost if imprecise. Alternatively, DSBs can be repaired with high fidelity through homologous recombination. Our understanding of homologous recombination mechanisms has lagged behind that of other DNA metabolic processes, such as replication and transcription. In large part this is because recombination creates or destroys little if any material, but instead rearranges existing material. Thus, radioactive precursors, which have been highly useful in transcription and replication studies, are not generally useful for recombination studies. Also, recombination occurs less frequently than other DNA dynamic processes, which imposes further challenges on the investigator.

DSBs are induced by a variety of DNA-damaging agents, including ionizing radiation, and radiomimetic chemicals, such as methylmethane sulfonate (MMS). DSBs may also arise "spontaneously" from replication past SSBs, processing of endogenous damage, nuclease activity, and torsional strain. Topoisomerases act to reduce torsional strain, and DSBs were found to be more prevalent in temperature-sensitive *top2* mutants at the restrictive temperature *(159)*. It has been suggested that UV killing involves DSB induction; although these DSBs could theoretically arise during excision repair of closely opposed lesions, UV-induced DSBs were also seen in excision repair-defective *rad3* mutants *(80)* and in strains carrying hyper-recombinogenic alleles of *rad3* known as *rem1 (100)*.

Ionizing radiation has been used extensively to induce DSBs. However, DSBs are only a minor component of the radiation damage (*see* vol. 2, Chapter 5), and the complex mixture of lesions makes it difficult to determine if a specific biological effect is caused by a particular lesion. The recombinogenic nature of DSBs was recognized when plasmids with DSBs in regions of shared homology to endogenous loci were found to integrate more efficiently than circular plasmids or plasmids with DSBs in

From: DNA Damage and Repair, Vol. 1: DNA Repair in Prokaryotes and Lower Eukaryotes
Edited by: J. A. Nickoloff and M. F. Hoekstra © Humana Press Inc., Totowa, NJ

nonhomologous regions *(119)*. Yeast mating type interconversion was then shown to be a DSB-induced gene conversion event *(160)*. Chemical agents that create DSBs also stimulate recombination *(44)*. With the advent of nuclease-based systems for producing specific DSBs in vivo, DSBs were shown to stimulate a variety of recombination events *(85,95,108,111,119,123,127,128,139,140)*. Nuclease systems allow studies of DSB repair without interference from other lesions. Although DSBs are only one of several types of damage produced by ionizing radiation, studies in mammalian cells indicate that nuclease-induced DSBs are a reasonable model for radiation damage, since they mimic several effects of ionizing radiation, including the induction of recombination, mutations, chromosomal aberrations, and cell killing *(22,23)*. It has been argued that DSBs are the most relevant lesion with regard to important biological effects of ionizing radiation *(49)*. Because yeast and mammalian mutants hypersensitive to ionizing radiation are cross-sensitive to nuclease-induced DSBs *(10,24,69,111)*, it is likely that radiation- and nuclease-induced DSBs are recognized and processed by the same or overlapping sets of cellular enzymes.

Diploid cells are more resistant to the killing effects of ionizing radiation than haploid cells, and haploid cells in G2 phase are more resistant than G1 cells *(21,129)*; similar findings were obtained for DSBs induced by MMS *(29)*. Thus, increasing genome copy number correlates with increased radioresistance. Mutants unable to perform recombinational repair show increased sensitivity to the killing effects of DSBs. These results emphasize the importance of recombinational repair in conferring resistance to ionizing radiation and other agents that produce DSBs. A single unrepaired DSB in a chromosome is lethal: *MAT* is cleaved by the highly specific HO nuclease, and this is a lethal event in DSB repair-deficient *rad52* mutants *(180)*. This lethality is from a single DSB, since single-base changes in *MAT* ("inconvertible" or *inc* mutations) that prevent HO cleavage *(110,180)* relieve HO-induced lethality in *rad52* mutants *(180)*. Surprisingly, DSBs in a region of a dispensable plasmid lacking homology to genomic sequences have been reported in one study to cause significant cell death *(14)*. This result suggests that yeast sense genome integrity and prevent mitosis when DSB damage is detected, perhaps through signal transduction and/or checkpoint functions (*see* Chapter 17, this vol. and Chapter 21, vol. 2.) Homology is located very efficiently, whether in homologs or at ectopic sites *(60)*. However, G2 diploids preferentially repair ionizing radiation-induced DSBs using sister chromatids over homologs *(77)*. Ionizing radiation also induces recombination competence *(41,50)*, but this remains poorly understood.

This chapter reviews mechanistic aspects of general homologous recombination repair in *Saccharomyces cerevisiae*, including the role of mismatch repair and the genetic control of these processes as revealed in repair-deficient mutants. Only limited discussion of site-specific conversion at *MAT* is presented, since detailed information is available in other reviews *(59,81)*.

2. RECOMBINATION MODELS

2.1. Homologous Recombination Events

Homologous recombination involves interactions between two DNA molecules sharing substantial homology, and includes both reciprocal and nonreciprocal information exchange (referred to as crossing over and gene conversion, respectively). Crossing over is a precise reciprocal exchange that does not alter the content of genome, but does rearrange genetic linkage patterns. In contrast, gene conversion involves information transfer from a donor locus to a recipient locus, and may result in a net gain or loss of functional information.

Early studies focused on meiotic recombination, since spontaneous rates are about 1000-fold higher than in mitotic cells *(122)*. Normally, heteroallelic loci segregate to give 2:2 patterns (or 4:4, if the eight single strands of a single meiosis are considered individually). Aberrant segregation patterns are seen in 2–20% of meioses for typical markers. The majority of aberrant segregations are 3:1 (or 6:2) gene conversions, although several other types are known, notably the 5:3 pattern, in which one of the four spores yields a sectored colony. Sectors result when strand exchange between heterozygous loci creates a mismatch (heteroduplex DNA; hDNA) that escapes repair and segregates in the mitotic division immediately following meiosis; 5:3 patterns therefore reflect postmeiotic segregation (PMS).

2.2. Models for Gene Conversion and Crossing Over

Gene conversion was first explained by a model proposed by Holliday *(70)* in which hDNA was formed by symmetric exchange of single strands, with conversion resulting from mismatch repair of hDNA. However, most hDNA appeared to be asymmetric, since 5:3 segregation patterns were much more frequent than aberrant 4:4 patterns *(122)*. When it became clear that DNA-damaging agents (especially those that create strand breaks) stimulate recombination, Meselson and Radding *(96)* proposed a model in which SSBs initiated recombination by promoting strand exchange, with asymmetric hDNA created by DNA synthesis primed from an invading end. However, no evidence for SSB involvement in spontaneous meiotic or mitotic recombination was forthcoming. SSBs produced in yeast by bacteriophage f1 gene *II* protein were shown to stimulate mitotic recombination, but whether these SSBs were converted to DSBs could not be ruled out *(162)*.

DSBs and double-strand gaps stimulate gene conversion and associated reciprocal exchange *(119)*, which can be explained by the DSB (or gap) repair model *(172)* (Fig. 1A). This model suggests that DSBs are processed to double-stranded gaps, with both 3'-ends invading an undamaged homologous duplex to produce two Holliday junctions (HJs). The 3'-ends prime DNA synthesis using the undamaged duplex as a template to fill the gap. Most conversion is thought to occur in a gap, but conversion is allowed to occur through repair of hDNA adjacent to gaps (not shown in Fig. 1A). The DSB repair model gained support from two lines of evidence. First, double-strand gaps created in vitro were precisely repaired using homologous cellular DNA as a repair template. Second, alleles suffering a DSB were preferentially converted (usually >95%) *(108,111,119,128)*, as would be predicted if conversion occurred in a gap formed by expansion of a DSB. Conversion would not show such bias if it were mediated by unbiased mismatch repair (*see* Section 4.2.). However, the gap repair model was not supported by meiotic studies with mismatch repair mutants. Meiotic conversion normally produces 6:2 segregation, but in mismatch repair-defective *pms1* mutants, 5:3 PMS patterns increase to the same extent that 6:2 patterns decrease, suggesting that 5:3 patterns reflect the failure to repair mismatches, and that 6:2 gene conversion normally reflects

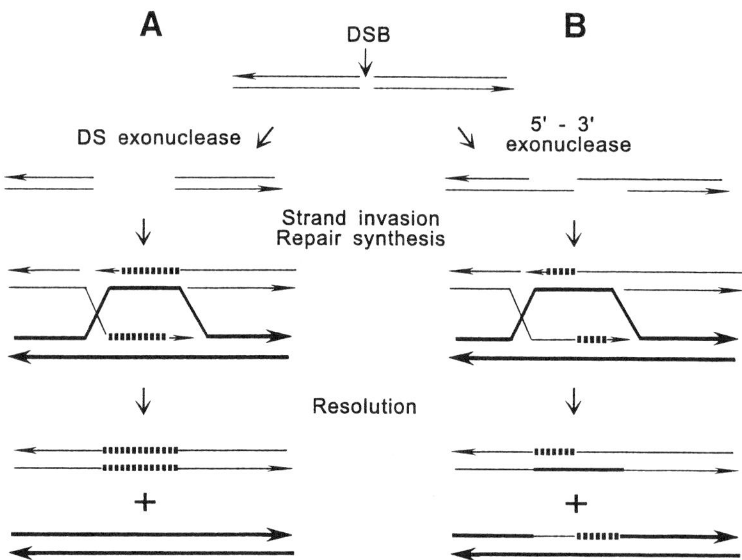

Fig. 1. Double-strand gap and DSB repair models. Thin and thick lines represent the two strands of homologous duplexes, with 3'-ends marked by arrows and repair synthesis by dashed lines. **(A)** Original gap repair model as described by Szostak et al. *(172)*. A DSB is expanded into a double-strand gap. Free ends invade a homologous duplex and prime repair synthesis producing a double HJ structure. Resolution of HJs can produce either crossover or noncrossover products; only the latter are shown. Also not shown is HJ branch migration, which can produce hDNA adjacent to the repaired gap and lead to conversion via mismatch repair (for details, *see* ref. *172*). **(B)** Modified DSB repair model as described by Sun et al. *(166)*. A DSB is processed to two 3'-extensions, which invade and prime repair synthesis, again producing two HJs. This model predicts conversion as a consequence of strand switching between alleles coupled with repair synthesis (thick line paired with dashed line) and mismatch repair of hDNA (thick line paired with thin line). Again, only noncrossover products arising without branch migration are shown. Note that hDNA is present in both alleles, but current evidence suggests that hDNA is not usually present in unbroken alleles.

mismatch repair. Thus, at least for meiotic events, conversion appeared to be mediated predominantly by mismatch repair, not gap repair. Current evidence suggests that little, if any conversion in meiotic or mitotic cells involves gap repair, despite the involvement of DSBs, and the DSB repair model has been modified accordingly *(166)* (Fig. 1B). One problem with the model in Fig. 1B is that it predicts hDNA will frequently occur on both chromatids, and this was not supported in a study of meiotic conversion, leading Gilbertson and Stahl to propose several refinements to this model *(56)*.

It was recognized early that gene conversions are frequently associated with crossovers (up to 50%) *(48)*, and these models incorporate intermediates with at least one HJ that can be resolved in one of two senses to yield crossover or noncrossover products (discussed in ref. *172*). Intrachromosomal recombination between direct repeats can produce deletions that have been accounted for by the single-strand annealing (SSA) model (Fig. 2). Unlike the previous models, SSA is nonconservative, producing only

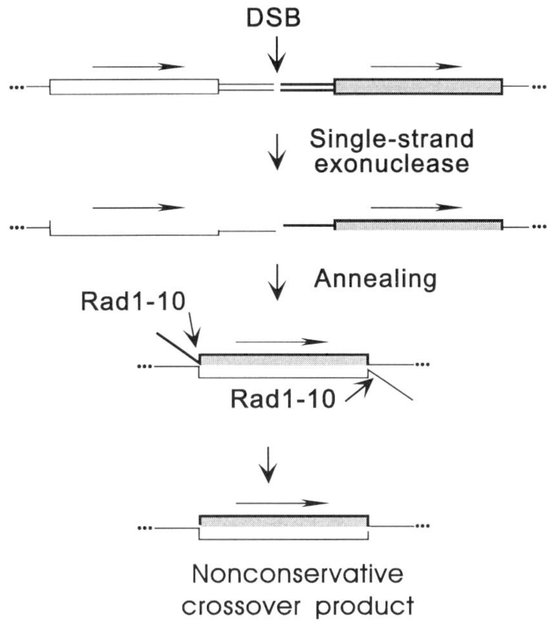

Fig. 2. Single-strand annealing model. Ends produced by a DSB between a nontandem direct repeat (shown by boxes) are processed by a single-strand exonuclease to expose complementary regions in the repeats, which then anneal. Nonhomologous tails are cleaved by Rad1p/10p endonuclease. hDNA is formed on annealing, as shown by half-shaded boxes, and is subject to mismatch repair. This mechanism produces an apparent "crossover" product, but is nonconservative, since only one of two crossover products is formed.

one of two possible crossover products. An alternative to SSA involves sister strand exchange of nascent strands during DNA replication *(92)*, which has the advantage of accounting for both intrachromosomal deletions and triplications *(84)*. For additional discussion of recombination models, *see* Chapter 8.

3. DSBS IN MEIOTIC RECOMBINATION

Crossovers and gene conversion events occur at frequencies about 1000-fold higher in meiosis than mitosis, with 2–20% of tetrads showing aberrant segregation patterns for heterozygous markers. For some loci near very active recombinational hot spots (e.g., *HIS4*), conversion frequencies may reach 30%. Recombination between two mutant alleles (creating either wild-type or double mutant alleles) reflect conversions 10 times more frequently than crossovers. Conversion involves nonreciprocal information transfer from a donor to a recipient; the donor remains unchanged. Conversion is a high-fidelity process (but *see* Section 4.1.), and in events involving multiple markers, conversion tracts are usually continuous. Thus, conversion of two outside markers is almost always associated with conversion of a central marker. Meiotic conversion tract lengths are often several hundred base pairs in length, but have been found to exceed 12 kbp and may be much longer. These aspects of meiotic recombination were reviewed by Petes et al. *(122)*.

3.1. Polarity Gradients and Recombination Initiation Sites

Meiotic gene conversion frequencies often decrease from one end of a gene to the other; the phrase "polarity gradient" is used to describe these differences *(112,122)*. Polarity gradients have been found in at least four genes in *S. cerevisiae* and in several genes from other fungi, including *Ascobolus* and *Neurospora* (reviewed in *112,187*). Commonly, conversion is higher at the 5'-end of genes, but some polarity gradients are reversed, and others have high-frequency conversion at both ends with lower frequencies toward the middle. It was not surprising when *cis*-acting recombination hot spots were identified at the high-frequency conversion ends of several genes. Recombination hot spots are also known that stimulate high levels of crossing over *(122)*.

3.2. DSBs at Meiotic Recombination Hot spots

Since conversion frequencies were often high at 5'-ends of genes, and since transcription stimulates recombination in yeast *(174,177,178)* and mammalian cells *(107,109)*, it seemed possible that hot spots might occur in or near promoters and reflect transcriptional activity. However, strong evidence now implicates meiosis-specific DSBs at hot spots near many genes, including *ARG4 (26,165,190)*, *CYS3 (33)*, *HIS4 (104)*, *THR4*, and *LEU2 (190)*.

Assays to detect DSBs in meiotic cells revealed numerous DSB sites on various chromosomes *(53,190,192)*. Wu and Lichten *(190)* identified 18 DSB sites on chromosome III, all of which were near potential promoter regions of open reading frames. Despite the correspondence of DSB sites with promoters, several studies indicate that DSB induction can be independent of transcription. Partial deletions of the *ARG4* promoter that eliminate transcription failed to reduce gene conversion *(148)*. DSBs were seen at *ARG4* on transcription factor binding, but in the absence of active transcription *(186)*, and these DSBs could be abolished by deletions that did not affect *ARG4* transcription *(58)*. Finally, although the *ARG4* DSB site is within the promoter, it is not transcribed, and modifications that permit transcription through the hot spot region eliminated DSB induction *(32,130)*. Thus, transcription *per se* is not required for elimi latter idea, two studies demonstrated that DSB induction correlates with DNase I and micrococcal nuclease hypersensitivity *(117,190)*, suggesting that an "open" chromatin structure is an important determinant of DSB induction. Similar to the putative meiotic nuclease, cleavage by HO nuclease is influenced by the transcription status (and hence chromatin structure) of its recognition sites in *MAT*, *HML*, and *HMR (66,82)*. However, chromatin structure is not the sole determinant of meiotic DSB induction, since the presence or absence of one DSB site can influence DSB induction at a second site several kilobase pairs away, suggesting that there may be a competition between DSB sites influenced by local concentrations of an enzyme or associated factor *(189)*. Fan and Petes *(43)* showed that DSB formation does not correlate with nuclease hypersensitive sites at the *HIS4* hot spot. At *ARG4*, the "open" structure is apparently controlled to a large degree by a 14-bp poly(dA-dT) tract immediately upstream of the TATA box *(148)*. Substantial modification of the *HIS4* hot spot did not eliminate hot spot activity *(42)*, and hot spots have been generated by chance through genetic engineering *(26)*, indicating a lack of conserved sequence. This idea is further supported by the observation of DSBs at multiple, nonspecific sites at a hot spot *(191)*.

3.3. INDUCTION AND PROCESSING OF MEIOTIC DSBS

There are several lines of evidence indicating that meiotic DSBs are recombination intermediates. DSB levels correlate with recombination frequencies *(26,42,57,113,165)*. Broken ends are processed into long 3'-extensions *(166)* that could invade a homolog and prime synthesis (Fig. 1B). DSBs are transient *(18)*, and appear before joint molecules and recombination products *(58,121,150)*. Finally, mutations in any of several genes block DSB induction and meiotic recombination, including *rad50*, *xrs2*, *mre11*, *spo11*, *mer2*, *rec102*, *rec104*, *rec114*, and *mei4 (131)*. The latter six genes are meiosis-specific and are therefore likely candidates to encode the nuclease.

A protein that binds covalently to the 5'-ends of meiotic DSBs was detected and subsequently identified as the product of *SPO11 (78a)*. However, much remains to be learned about the function of this enzyme in meiosis.

The identification of DSBs at hot spots can account for polarity gradients, with hDNA forming more frequently near DSB sites. However, a different view suggests that hDNA formation is not limiting (at least not for the gene-length distances commonly examined in such studies), with polarity reflecting different patterns of mismatch repair at different locations in a gene *(112)*. In this view, mismatch repair switches from predominantly restoration-type repair at the low end of the polarity gradient to predominantly conversion-type repair at high end of the gradient. Unlike meiotic conversion, polarity gradients are not seen in mitotic conversion *(122)*, and DSBs have not been detected in mitotic cells. Since the "open" chromatin structure at meiotic DSB sites is also present in mitotic cells *(190)*, it is likely that DSB induction is limited to meiosis through regulation of the nuclease or associated factors.

Both *dmc1 (18)* and *rad50S (26)* mutants accumulate DSBs with longer than normal 3'-tails, implicating these gene products in processing steps following DSB induction. Dmc1p is likely to be involved in pairing, since it is related to *Escherichia coli* RecA, *S. cerevisiae* Rad51p *(18,155)* and Rad55p *(91)*, mei-3p from *Neurospora (28)*, ArLIM15 from *Arabidopsis (142)*, and rec2 from *Ustilago maydis (138)*. Several lines of evidence suggest that DSB processing is required for chromosome synapsis (discussed in *132*). Dmc1p and Rad51p (perhaps complexed with Rad52p; ref. *155*) were proposed to be involved in synaptonemal complex formation that depends on DSB processing *(132)*. However, two reports from the Kleckner laboratory suggest that synapsis may proceed independently of DSBs *(181,191)*. DSB induction is independent of the presence of a homolog *(31,55)*.

The original DSB repair model suggested that both ends of a DSB invade a homologous sequence, resulting in the formation of two HJs *(172)* (Fig. 1A). DSB-induced conversion should not require both ends to invade, since repair synthesis primed from one end can produce a region complementary to the opposite 3'-extension (reviewed in reference *12*). However, the identification of double HJs in meiotic recombination intermediates *(149)* provides support for the idea that at least some meiotic events involve two-ended invasions. Two-ended invasions might be expected to yield bidirectional conversion tracts, which were common at *ARG4 (166)*, but not at *HIS4 (124)*. Note that for meiotic events between homologs, both ends produced by a DSB are completely homologous to the duplex that they invade, and this may facilitate two-ended invasions. Factors that may influence one- and two-ended invasion mechanisms are further discussed in Section 4.3. Details of later steps in meiotic recombination were reviewed by Roeder *(132)*, including chromosome synapsis, and the role of recombination in sister chromatid cohesion and chromosome segregation.

3.4. Engineered DSBs in Meiotic Cells

Before the importance of DSBs in meiotic recombination became clear, Kolodkin et al. *(85)* used HO nuclease to induce DSBs in meiosis at *MAT* and at an HO nuclease recognition site inserted into *HIS4*. HO nuclease was regulated by the *GAL10* promoter and induced on sporulation medium containing galactose. DSBs greatly increased conversion at both *MAT* and *HIS4* (up to 400-fold), and as found in mitotic studies, alleles suffering a DSB were preferentially converted (*see* Section 4.2.). However, unlike typical meiotic conversion, which produces 3:1 tetrads, both *MAT* and *HIS4* conversions yielded 4:0 tetrads, suggesting that these events occurred in G1 (i.e., at the four-strand stage) rather than G2 (the eight-strand stage). Recently, the sporulation-specific *SPO13* promoter was fused to the HO nuclease gene, allowing meiosis-specific expression of HO nuclease to mimic natural meiotic DSBs. *rad50Δ* strains fail to produce DSBs in meiosis and are recombination-defective, but expression of HO nuclease during meiosis in *rad50Δ* strains restored meiotic recombination at the site of the HO-induced DSB. Interestingly, not all DSBs were repaired until after meiosis was completed *(93)*.

4. DSB-INDUCED MITOTIC RECOMBINATION

Early recombination studies in yeast focused on meiotic events, because they occur at high frequency and all products of a single meiosis can be analyzed. Because spontaneous mitotic recombination occurs at very low frequencies, selective strategies based on recombination between two inactive heteroalleles are normally employed. Such systems usually restrict analysis to only those molecules that gain a functional, selectable genotype. Although early studies employed radiation or chemicals to increase recombination, the development of systems for delivering DSBs to specific sites in recombination substrates in vitro with restriction enzymes *(119)* or in vivo with HO nuclease *(108)* allowed analysis of the effects of single, targeted DSBs. By inducing recombination in vivo with HO nuclease, recombination frequencies >1% can be attained, allowing recombinants to be identified among nonselected colonies. These systems thus permit more focused and complete analyses of recombination events.

In contrast to meiotic recombination, it is not yet clear whether spontaneous mitotic recombination results from DSBs, other types of endogenous damage, or normal DNA dynamics. Similar gene conversion:crossover ratios were found for spontaneous and DSB-induced events in two studies *(108,128)*. These results are consistent with the idea, but not proof, that spontaneous events are initiated by DSBs. Mitotic conversion does not display polarity gradients, suggesting that mitotic recombination initiates at random sites. If DSBs are responsible for spontaneous mitotic recombination, the putative rare and random DSBs will be difficult to detect in physical assays.

4.1. DSBs Stimulate Gene Conversion, Crossing Over, and Nonconservative Recombination

DSBs in regions of shared homology stimulate interchromosomal *(95)*, plasmid × chromosome *(110,118,119,169)*, extrachromosomal *(47,140)*, and intrachromosomal recombination *(108,111,128,140)*. Intrachromosomal events include gene conversions, which retain the gross structure of the duplication, and deletions, which excise DNA between direct repeats or invert DNA between inverted repeats. DSB-induced deletions were originally classified as conservative crossover events on the basis of experiments by Ray et al. *(128)*, who used an *ade4* duplication that flanked *ARS1*, which allowed crossover products to be recovered as freely replicating circular molecules. In this system, DSBs delivered to one copy of *ade4* stimulated primarily gene conversions without associated crossovers, but 6% were deletion products, and all of these arose by conservative crossovers. At *ura3*, DSBs in homology yielded 30–50% deletion products, but these could not be unequivocally identified as conservative crossovers, since excised regions did not carry an *ARS* element and were not recoverable *(108,111)*. DSB-induced recombination between direct repeats in extrachromosomal substrates yielded 58–83% deletions, all of which arose via a nonconservative SSA mechanism *(47)*.

Deletions predominate when DSBs occur in nonhomologous DNA between repeated regions *(111,139,163)*, and these arise by the nonconservative SSA pathway *(139,163)*. Unlike gene conversion and conservative crossing over, DSB-induced deletions (and triplications) at chromosomal arrays of repeated genes are Rad52p-independent *(84,120)*. Klein *(84)* proposed that the Rad52p-independent deletions might arise by SSA or by the mechanism proposed by Lovett et al. *(92)* involving sister strand exchange of nascent strands during DNA replication. Since SSA cannot produce triplications, the sister strand exchange model is an attractive way to explain both Rad52p-independent DSB-induced deletions and triplications *(84)*. Several factors may influence the ratio of conversions associated and unassociated with crossovers, and whether deletion products are formed by conservative or nonconservative mechanisms, such as lengths and numbers of duplicated regions; length and identity of DNA between duplicated regions; substrate context (chromosomal vs extrachromosomal); and DSB location relative to duplicated regions.

rad1 mutants are UV-sensitive (*see* Chapter 15) and X-ray-resistant *(158)*, yet they show defects in certain types of DSB-induced recombination *(46)*. Rad1p and Rad10p form a complex with endonucleolytic activity *(9)*, and it has been proposed that Rad1p/10p removes terminal nonhomologies during DSB-induced recombination *(46,141)*. DSBs in HO sites inserted into repeated genes produce nonhomologous termini that must be processed by Rad1p/10p to complete gene conversion and SSA-mediated deletion *(46)*. DSBs in nonhomologous DNA between repeated genes also require Rad1p/10p, presumably to process nonhomologous tails produced during SSA *(46)* (Fig. 3); however, the requirement for Rad1p/10p was relaxed for DSBs located a short distance (8–27 bp) from homology *(125)*.

As with meiotic conversion, DSB-induced mitotic conversion generally occurs with high fidelity *(169)*. However, Strathern et al. *(161)* examined mutation rates in a diploid for a homozygous marker located 300 bp from DSBs introduced by HO nuclease and found it reverted 100-fold more often among DSB-induced conversion products (detected at adjacent heteroalleles) than in uninduced cells. Furthermore, when the reverted allele could be assigned unambiguously to one chromosome (in noncrossover products), it always occurred on the chromosome suffering the DSB. These results indicate DNA synthesis associated with DSB repair has a higher rate of mis-

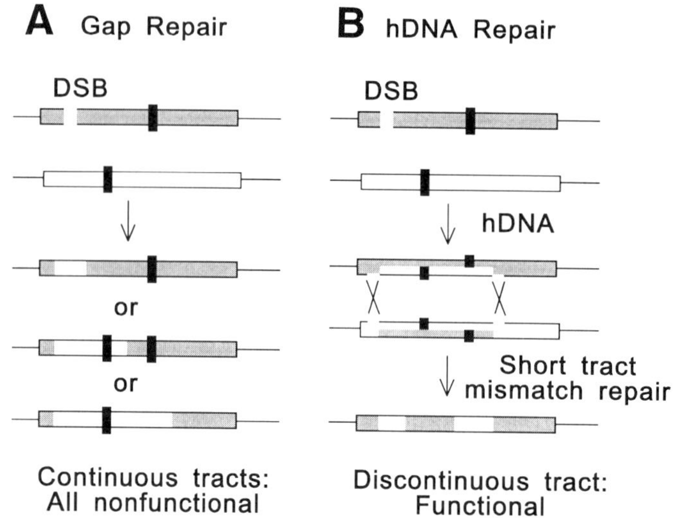

Fig. 3. Crosses restricted for gap or long-tract mismatch repair. Two alleles are shown by the white and shaded boxes, and inactivating markers are shown by black bars. **(A)** Conversion by gap repair cannot produce a functional allele when a marker in the donor locus is between the DSB and the marker in *cis* to the DSB. In the products shown below, converted regions in the shaded allele are indicated by nonshaded regions. **(B)** Strand exchange initiated by a DSB produces hDNA that is subject to mismatch repair. Short-tract mismatch repair produces discontinuous conversion tracts to produce a functional product. This diagram is not intended to show the details of hDNA formation.

incorporation than replicative DNA synthesis, implicating a specific, error-prone DNA polymerase in recombinational repair. Reasonable candidates are DNA polymerases δ and ε, which are thought to be involved in excision repair and may also have roles in recombinational repair *(25)*.

4.2. Mitotic DSB-Induced Gene Conversion: Gap Repair or hDNA Repair?

In both mitotic and meiotic cells, nearly all conversion tracts are continuous *(3,4,19,20,76,169,171,188)*, and DSBs stimulate conversion of broken alleles preferentially *(85,95,106,108,110,111,113,119,128,139,140,165,169)*. These observations, plus the fact that gapped substrates are repaired with information donated by a homologous duplex *(119)*, lended support to the gap-repair model. However, as discussed in Section 2.2., most or all meiotic conversion involves mismatch repair of hDNA, not gap repair, and increasing evidence suggests that mitotic DSB-induced recombination also involves hDNA repair.

Early evidence for hDNA intermediates in spontaneous mitotic conversions was provided by Ronne and Rothstein *(136)*, who found that 15% of products gave sectored colonies. Physical studies of DSB-induced conversion at *MAT* and other loci yielded no evidence of double-strand gap formation; instead, long 3'-extensions were formed at broken ends *(30,139,163,185)*, similar to those seen at meiotic DSBs. Recent genetic evidence also argues against gap repair. Priebe et al. *(126)* studied homeologous conversion between *his4* alleles from *S. cerevisiae* and *Saccharomyces carlsbergensis* stimulated by a double-strand gap produced in vitro. These genes, though related, have numerous closely spaced sequence differences that allowed conversion tracts to be characterized at high resolution. Conversion tracts adjacent to the gap usually extended one direction from the gap. Similar results were seen for conversions of sites within 6 bp of an HO site insertion in *ura3 (106)*. These highly directional tracts are not consistent with DSB-induced conversion involving double-strand gap formation (as in Fig. 1A), since double-strand exonucleolytic activity is unlikely to degrade only one end. Recently, direct evidence was obtained for hDNA repair-mediating mitotic DSB-induced conversion *(182)*. DSBs were induced in an HO site located 22 bp from a 14-bp palindromic frameshift insertion mutation in *ura3*. If present in hDNA, the palindromic insertion forms a stable stem-loop structure, and is expected to escape mismatch repair frequently *(105)* and produce sectored Ura^+/Ura^- colonies. No sectors are expected if conversion involves gap repair. About 85% of converted products were sectored, suggesting that most DSB-induced mitotic conversion involves mismatch repair of hDNA *(182)*. Although the 15% nonsectored colonies might have arisen by gap repair, it seems more likely that these reflect residual mismatch repair.

By definition, gap repair would convert only markers in *cis* to a DSB. Several studies showing that DSBs stimulate conversion of markers in *trans* are therefore better explained as hDNA-mediated events. DSBs stimulate conversion of unbroken alleles, although at low levels *(95,111,133)*. Ray et al. *(127)* found that a DSB stimulated recombination between two unbroken chromosomes at a locus more than 8 kbp from a DSB, and when only one of the recombining partners shared homology with the chromosome suffering the DSB. Similarly, a transforming plasmid with a DSB in an *his3* allele stimulated recombination between chromosomal *his3* heteroalleles to His^+, even though the plasmid-borne allele was unable to donate wild-type information *(156)*. Thus, both physical and genetic evidence now indicate that most, if not all, DSB-induced mitotic conversion involves hDNA repair (two versions of which are shown in Figs. 1B and 4), rather than gap repair (Fig. 1A).

It was difficult to reconcile conversion mediated by mismatch repair in hDNA with preferential conversion of broken alleles in view of early studies showing that most mismatches are repaired efficiently and with parity, or at most with a twofold disparity *(16,17,87)*. These studies were performed by transforming into yeast covalently closed circular plasmids containing mismatches constructed in vitro. In contrast, mismatch repair of natural hDNA (in meiosis) showed marked disparity in favor of conversion-type repair over restoration-type repair, suggesting that donor or recipient strands are "tagged" *(36)*. For DSB-induced events (either meiotic or mitotic), preferential conversion of broken alleles probably reflects differential strand tagging such that invading strands are targeted for repair. The broken end is a likely tag, since this could direct repair to the invading strand, analogous to nick-directed mismatch repair in *E. coli (88)* and mammalian cells *(34,62)*. Nick-directed mismatch repair could involve exonuclease digestion as in *E. coli* or a traveling repair complex that enters at ends. The predominance of continuous conversion tracts would suggest that mismatch repair usually involves long repair tracts, as mediated by the *E. coli mutHLS* system (*99*; Chapter 11). Long-tract mismatch repair has been demonstrated in both meiotic and mitotic yeast cells *(17,35)*. Although short-tract repair systems are known in *E. coli* (*89*; Chapter 11) and mammalian cells (*see* Chapter 8, vol. 2), none have yet been identified in *S. cerevisiae*.

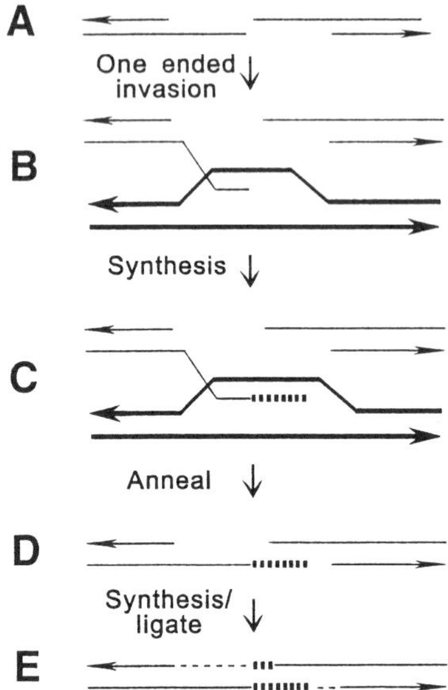

Fig. 4. One-ended invasion model. **(A)** A DSB is processed to expose 3'-ends as in Fig. 1B. **(B,C)** One end invades an unbroken homologous donor locus (thick lines) and primes repair synthesis across the DSB (dashed lines). **(D)** The newly synthesized DNA dissociates from the donor and anneals with the exposed 3'-tail that failed to invade. **(E)** Additional synthesis and ligation completes DSB repair. Conversion can occur when hDNA formed by annealing is subjected to mismatch repair.

By arranging DSB and heteroallelic frameshift mutations appropriately, crosses can be designed in which conversion to wild type of alleles suffering a DSB is not possible by gap or long-tract mismatch repair (Fig. 3A). In these crosses, conversion of alleles suffering a DSB requires at least two repair events: one that repairs the DSB (and that may convert nearby markers) and a second mismatch repair event at the mutation in *cis* to the DSB (Fig. 3B). This is because a single, continuous repair tract (either in a gap or via long-tract mismatch repair) that includes both the DSB and the mutation in *cis* to the DSB would necessarily transfer mutant information from the unbroken allele. Three studies that have used such restricted crosses have shown that this constraint significantly reduces recombination frequencies, and eliminates preferential conversion of alleles suffering a break. Conversion frequencies were shown to decrease as the distance between the DSB and the marker in *cis* increases *(118)*, consistent with products arising by hDNA repair, rather than by multiple crossovers, since crossovers should increase with increasing marker separation. Using interchromosomal recombination substrates, McGill et al. *(95)* found evidence for symmetric hDNA, which is expected to form during branch migration of HJs. Tract structures were analyzed using multiple, phenotypically silent markers in a plasmid × chromosome cross induced by HO nuclease *(183)*. The product spectrum indicated that hDNA formation is generally limiting, but some hDNA regions extended more than half the length of the 1.2-kbp region of shared homology, and markers separated by 20 bp were independently repaired about 40% of the time. Very short mismatch repair tracts were also seen with artificial hDNA processed in vitro *(103)*. These short mismatch repair tracts might result from premature termination of long-tract repair or from an unknown short-tract repair system.

4.3. End Invasion: One End or Two?

Most HO-induced conversion tracts in plasmid × chromosome crosses are unidirectional *(169)*. It is possible that uni- and bidirectional tracts result from one-ended invasions *(12)* and two-ended invasions, respectively. An alternative view that unidirectional tracts reflect differential mismatch repair on either side of the DSB in two-ended invasion intermediates was not supported *(106)*. It is reasonable to suppose that one-ended events would be favored if invasion is rate-limiting, which is suggested by observations that invasion by two ends is temporally distinct *(30,170)*, and by the lack of cooperativity between two invading ends during integration *(64)*. Invasion may also be influenced by the degree of homology at DNA ends. Unidirectional tracts may predominate for HO-induced events, because DSBs in an inserted HO site produce nonhomologous ends that may inhibit strand invasion and promote one-ended events. Unidirectional tracts were also predominant when transforming plasmids with homeologous ends interacted with a chromosomal locus *(106,126)*, a result consistent with the finding that Rad1p/10p does not cleave homology/homeology boundaries *(106)*. Contrary to expectation, further reducing the degree of end homology by flanking nonhomologous ends with 35-bp homeologous regions increased the proportion of bidirectional tracts, suggesting that additional factors, such as mismatch repair, may influence invasion and tract directionality *(106)*.

It is possible that one-ended invasions are favored in plasmid × chromosome crosses because the two ends of the DSB are linked through the plasmid backbone, which might promote intramolecular annealing following repair synthesis (Fig. 4), thus eliminating the need for a second end invasion. Such a model would predict more frequent two-ended invasions when DSBs occur in chromosomes, since these ends would not be topologically linked. Consistent with this model, twofold more bidirectional tracts were observed in an intrachromosomal cross than a related plasmid × chromosome cross *(106)*. *MAT* conversion appears to be a special case, since chromosomal DSBs stimulate largely unidirectional conversion, but this reflects differential 5'- to 3'-strand excision, with one side protected by interactions with silent donor sequences *(185)*; differential strand excision was not seen for DSBs between a *ura3* duplication *(163)*. Thus, although the degree of end homology, topology of the interacting partners, and strand excision from DSBs are likely to influence rates of one- and two-ended invasion pathways, the relative importance of these factors is not known.

4.4. Biochemistry of DSB Repair

Because recombination occurs at low frequencies and involves no or limited biosynthesis, it has been difficult to characterize at the biochemical level. Symington *(170)* demonstrated end-joining activity and DSB-induced recombination in vitro for plasmids incubated with nuclear extracts yielding predominantly crossover products. Consistent with these products arising by SSA, a 5'- to 3'-exonuclease activity has been identified *(71)*. Extracts from mutants defective in recombinational repair retained exonucleolytic activity, which is consistent with the idea that SSA is independent of the major, Rad52p-dependent recombinational repair pathway (*84*; *see* Section 4.1.).

Biochemical studies of general recombination enzymes in yeast, such as strand-exchange proteins, helicases, and HJ resolvases have paralleled studies in *E. coli*. Several strand-exchange proteins with probable roles in recombination have been identified (reviewed in *67*). Rad51p is a RecA homolog, but Rad51p catalyzes exchange initiated at 5'-ends, opposite to RecA *(167)*. Sep1p was first characterized as a strand-exchange protein *(86)*, but it also has exonuclease activity *(73)*, and mutations in *SEP1* have been identified in several genetic screens, suggesting biological roles in other processes in addition to recombination, including RNA metabolism and cytoskeletal architecture. *sep1* mutants exhibit normal mitotic recombination, but meiotic recombination is altered *(175)*. Although Sep1p levels increase twofold during meiosis, Sep1p is located principally in the cytoplasm *(68)*. It has been hypothesized that *sep1* mutants fail to initiate pairing in meiosis *(90)*, but this idea has been questioned *(176)*. Srs2p is a helicase with ATPase activity *(134)*, and *srs2* mutants are hyperrecombinogenic *(135)*. Srs2p is thought to oppose Rad51p and Rad52p during recombinational repair *(97)*.

HJ resolvases from yeast are not as well characterized as in *E. coli* (*see* Chapter 10). Rad1p/10p endonuclease was proposed to be an HJ resolvase, because it cleaved an artificial HJ in vitro and because mutations in these genes influenced certain recombination events *(61)*. However, since Rad1p cleaves at junctions of duplex and single-stranded DNA during UV repair (Chapter 15) and certain recombination intermediates (*see* Section 4.1.), West *(184)* argued that Rad1p may recognize single-strandedness of the particular artificial HJ used, and not natural HJs. Vaccinia virus topoisomerase I cleaves HJs in vitro *(152)*, and a role for topoisomerase in HJ resolution in yeast has been suggested *(63)*, but not demonstrated. Cce1p cleaves cruciforms and HJs in vitro, but *cce1* mutants exhibit normal recombination phenotypes *(83)* and Cce1p is located in mitochondria *(40,83)*.

4.5. Nonhomologous Recombination

S. cerevisiae is extremely efficient at catalyzing homologous recombination. Low-level nonhomologous recombination can be detected by transforming cells with linear DNA carrying selectable markers sharing no homology to genomic DNA *(144)*. Such DNA integrates into chromosomal DNA or ligates to mitochondrial DNA containing a replication origin to produce a freely replicating circular molecule *(143)*. DNA integrates nonrandomly, with nearly half integrating at topoisomerase I cleavage sites *(143)*. Nonhomologous recombination is increased in strains overexpressing topoisomerase I, and *top1*mutants show reduced capacity to integrate nonhomologous DNA into cognate cleavage sites *(193)*. Nonhomologous recombination also can be enhanced by cotransformation of DNA and restriction enzymes *(145)*. A role for the *S. cerevisiae* homolog of Ku (*see* Chapters 16 and 17 in vol. 2) in nonhomologous recombination has been suggested by three studies *(20a,151,176a)*.

5. GENES AND GENE PRODUCTS INVOLVED IN DSB REPAIR

The genes involved in the repair of DSBs in *S. cerevisiae* have been identified and characterized through several approaches:

1. The identification of genes essential for the repair of ionizing radiation-induced DNA damage *(51)*;
2. The study of genes required for meiosis *(5)*;

3. The characterization of recombination-defective mutants *(52)*; and
4. The characterization of yeast mutants affected in the repair of site-specific DNA strand lesions, such as those created by the HO nuclease *(69,101,164,173)*.

These studies have identified about a dozen genes that have been called the *RAD52* epistasis group. Although the members of the *RAD52* epistasis group share the property of conferring resistance to ionizing radiation, they show considerable variation in assays for DSB repair and recombination. The *RAD50*, *XRS2*, and *MRE11* genes form a subgroup with similar properties. Physical interactions among the gene products of these three genes has been detected by the two-hybrid system *(75)*. The persistence of DSBs in mutant strains lacking these proteins suggests the presence of a complex that might associate with chromatin. Mutations in other members of the *RAD52* group (*RAD51*, *RAD52*, *RAD54*, *RAD57*, and *RAD59*) produce DSB repair defects, a reduction in spontaneous and induced mitotic recombination, and defects in mating-type interconversion and sporulation *(8,52)*. The most extensively studied members of the *RAD52* group are *RAD50*, *RAD51*, and *RAD52*. *RAD53* encodes a protein kinase with a checkpoint function (*see* Chapters 17 and 18). *RAD54* encodes a putative ATPase/helicase *(39)* in the *SNF/SWI* family *(27)*. *RAD56* has been identified as *SGE1* and has a predicted role in multidrug resistance *(38,168)*.

5.1. RAD50, MRE11, *and* XRS2 *Subgroup*

The *RAD50*, *MRE11*, and *XRS2* genes appear to be similarly involved in DNA repair and recombination *(6,7,72,75)*. Mutations in these genes confer sensitivity to DNA-damaging agents, including ionizing radiation and MMS. These mutants show elevated mitotic recombination levels, deletion mutants are deficient in meiotic recombination and spore formation, and meiosis-specific DSBs are not observed. A nonnull, recessive allele of *RAD50*, *rad50S*, exhibits unique properties compared to null mutations *(6)*. *rad50S* strains are MMS-resistant, and spontaneous recombination is near wild-type levels. During meiosis, *rad50S* strains are capable of generating meiotic DSBs, but are deficient in DSB processing. Thus, in a *rad50S* mutant, the meiotic recombination pathway is blocked after DSB formation, suggesting a role for Rad50p in DNA end processing.

The primary sequence and protein interactions of Rad50p, Xrs2p, and Mre11p shed some light on biochemical properties of these proteins. Although Xrs2p shows little similarity to proteins in the existing data base, Rad50p is an *SMC* family member that, along with Mre11p, is suggested to show structural and potentially functional similarities to *E. coli* SbcCD proteins *(153)*. Two-hybrid and biochemical assays indicate that the Rad50p, Xrs2p, and Mre11p proteins interact with each other *(75,114)*, suggesting that these molecules are part of a complex that acts in DSB formation and repair. Because of the structural similarity to *E. coli* Sbc proteins and the altered processing of DSBs in mutant strains, it is possible that the biochemical activity of the Rad50p/Mre11p/Xrs2p complex is nucleolytic. Alternatively, the complex may play a role in nonhomologous end filling/rejoining *(101)*, and it is possible that the complex acts in pairing or cohesion between sister chromatids to prevent strand degradation.

5.2. RAD52

Of the entire *RAD52* epistasis group, *rad52* mutants confer the most severe defects in DSB repair and recombination, suggesting a critical role in these processes. *rad52* mutants are defective in the repair of DSBs produced by ionizing radiation *(129)*, are unable to perform mating-type interconversion *(94)*, and are sensitive to HO-created strand breaks *(111)*. Likewise, *rad52* strains are impaired in their commitment to meiotic recombination, and produce few or no viable spores *(54)*. The cloning and sequencing of *RAD52* were reported in the early 1980s *(2,147)*. Rad52p has homologs in a variety of organisms *(13,15,98,154)*, but lacks recognizable sequence motifs that might be indicative of function.

The search for biochemical activity for Rad52p has been elusive. Rad52p has been reported to possess a weak strand-transfer activity *(115)* and DNA strand annealing is promoted by Rad52p *(102)*. Either activity would be consistent with a central role for Rad52p in DSB repair. The homology-dependent DNA strand-annealing activity for Rad52p is puzzling in light of specific defects that have been revealed in mutant analyses. Chromosomal inverted-repeat recombination shows little requirement for the *recA* homologs in yeast (*RAD51*, *RAD55*, and *RAD57*) *(78,91,155)*, and the inferred single-strand annealing component of direct repeat recombination appeared to be *RAD52*-independent in some studies (e.g., *120*). *RAD52* is required for inverted-repeat recombination, which cannot proceed by single-strand annealing. The strand-annealing property of Rad52p is separated into two domains in the protein. The amino-terminus contains the DNA binding domain *(102)*, whereas the carboxy-terminus contains a Rad51p-interacting domain *(98)*.

RAD59 is highly homologous to *RAD52*, and *rad59* mutants are also sensitive to ionizing radiation *(8)*. The *RAD59* gene is required for Rad51p-independent mitotic recombination. A *rad59* mutant was isolated in a search for new recombination-defective mutants by mutagenizing a *rad51* strain (which is proficient for intrachromosomal inverted-repeat recombination). The *rad59* mutation reduced recombination 1200-fold in the presence of *rad51*, but only four- to fivefold in wild-type cells. This synergism suggests that Rad51p and Rad59p participate in separate recombination pathways. The *rad59* mutation reduced spontaneous and DSB-induced recombination between inverted repeats, but elevated interchromosomal recombination. Overexpression of *RAD52* suppressed *rad59* recombination and repair defects, raising the possibility that Rad59p and Rad52p have overlapping functions. Given the suggested biochemical properties for Rad52p, it is tempting to suggest similar strand transfer or strand-annealing properties for Rad59p. In light of the recent observations that yeast does not employ Rad52p in the repair of retrotransposon reverse transcriptase-mediated strand breaks, and that Ku-like proteins in yeast are essential for the repair of DSBs only in the absence of *RAD52 (151)*, it is possible that Rad59p might participate in strand break repair in both homologous and nonhomologous pathways.

5.3. RAD51, RAD55, *and* RAD57

The *RAD51*, *RAD55*, and *RAD57* gene products have sequence homology to prokaryotic RecA proteins *(1,11,78,91,155)*, which are required for homologous recombination and the SOS response. The three proteins contain conserved putative Walker A-type and B-type nucleotide binding motifs. Mutation of the conserved lysine in the A-motif inactivates Rad51p and Rad55p, but not Rad57p for DNA repair and meiosis *(74,155)*. As with RecA, electron microscopic analysis indicated that Rad51p forms helical filaments on single-stranded and double-stranded DNA. However, only the

Rad51p single-stranded DNA nucleoprotein filament complex is functionally relevant for strand exchange *(116,167)*.

5.4. Rad Protein Interactions

A number of experiments indicate that Rad proteins function in a complex that contains a variety of activities required for recognizing, responding to, and repairing strand breaks. The Rad51p and Rad52p proteins interact with each other as measured by affinity chromatography and two-hybrid analyses *(65,98,155)*. The region of interaction has been mapped to the amino-terminal region of Rad51p and the carboxy-terminus of Rad52p *(37,98)*. Likewise, by two-hybrid approaches, interactions between Rad51p and Rad55p, Rad55p and Rad57p, and Rad51p with itself have been reported *(65,74)*, and an interaction between Rad51p and Rad54p has been suggested *(8)*. Genetic suppression experiments further extend the notion that Rad proteins act in a complex. *RAD51* or *RAD52* can suppress the DNA repair and recombination defects of *rad55* and *rad57* mutants *(65,74)*, and overexpression of *RAD51* can suppress certain *rad52* alleles *(98,146)*. The *RFA1* gene, encoding the large subunit of the replication-associated single-stranded DNA binding activity, has been shown to be involved in recombination, and an *rfa* missense mutation is suppressed by *RAD52 (45,157)*. Furthermore, *rad55* and *rad57* mutants are cold-sensitive, and have more severe repair and recombination defects at lower temperatures. The cold-sensitive phenotype is consistent with Rad55p and Rad57p acting to form and stabilize a protein complex. An ATP-requiring protein complex with *SNF/SWI*-like ATPase and DNA helicase activity, DNA strand-annealing and strand-exchange properties, and single-stranded DNA binding activity is entirely consistent with a model by which Rad proteins recognize strand lesions, remodel and reorganize chromatin, recognize a homologous DNA strand, and mediate the recombination reaction through a strand-annealing and exchange reaction.

6. PERSPECTIVES AND FUTURE QUESTIONS

Early studies of mitotic and meiotic recombination revealed several differences between these two processes, including markedly different frequencies and the presence of gene conversion polarity gradients only in meiotic cells. Although some overlap was apparent, especially with respect to similar effects of various *rad* mutations, these two processes have typically been discussed separately. As we learn more about both processes, the features that they share are beginning to outweigh real or perceived differences. For example, we can now explain the high recombination levels in meiosis, and at least some aspects of polarity gradients as consequences of meiosis-specific DSBs. The types of recombination events that occur and at least some mechanistic details of these events (such as the role for mismatch repair in gene conversion) are similar for meiotic recombination and DSB-induced mitotic recombination. Thus, the principal difference between meiotic and mitotic recombination may simply be that meiotic cells express a nuclease that is not expressed in mitotic cells.

Studies in yeast have played a leading role in furthering our understanding of eukaryotic recombination, yet our understanding of recombination mechanisms and their regulation remains incomplete. Despite numerous studies that have detected DSBs at meiotic hot spots, and a fully sequenced genome, the meiosis-specific nuclease has yet to be identified. Many questions remain with regard to mechanistic details of various recombination events. For example, it is not clear whether spontaneous mitotic recombination reflects repair of spontaneous DSBs and/or other endogenous DNA damage. It is not known whether DSB-induced events involve one- and/or two-ended invasions. Broken ends locate homologous regions efficiently, even at ectopic sites, yet the homology search mechanism is still a mystery. It is not known how much hybrid DNA is formed initially on strand invasion, or later by branch migration or by a DNA synthesis/reannealing process (Fig. 4). When an invading end has a region of terminal nonhomology, Rad1p/10p endonuclease cleaves at or near the homology/nonhomology junction. However, the minimum length of the nonhomologous tail that requires Rad1p/10p processing is not well defined. Intrachromosomal deletion events have been found to occur by both conservative and nonconservative (SSA) exchange mechanisms. Several factors might influence the choice between these mechanisms, such as chromosomal context, repeat structure, and the location of the DSB relative to repeated elements, but this remains a topic of speculation. Finally, the functions and interactions of the gene products involved in recombination processing or regulation are just beginning to be uncovered. Information about recombination mechanisms has importance in areas beyond an academic interest in DNA dynamics. Two (of many) examples are gene therapy and genome instability in cancer. Recombination is an essential feature of plasmid integration and gene targeting, and studies of DSB-induced recombination have driven much of the research aimed at optimizing recombination-based gene therapy strategies. Recombination is a highly regulated process, but when regulation fails, genome instability results, as is common in cancer. Instability can lead to greatly increased mutation rates, which in turn may drive tumor progression. Studies of recombination mechanisms and its regulation will continue to play key roles in furthering our understanding of basic cellular processes in normal and disease states.

ACKNOWLEDGMENTS

We thank Andres Aguilera, Paul Berg, Larry Gilbertson, Nancy Kleckner, Hannah Klein, Alain Nicolas, Tomoko Ogawa, Shirleen Roeder, Frank Stahl, and Jeffrey Strathern for sharing information prior to publication and for providing reprints. We also thank the members of our laboratories for their many contributions. Research in JAN's laboratory is supported by grants CA55302, CA62058, and CA54079 from the National Cancer Institute of the NIH.

REFERENCES

1. Aboussekhra, A., R. Chanet, A. Adjiri, and F. Fabre. 1992. Semi-dominant suppressors of *srs2* helicase mutations of *Saccharomyces cerevisiae* map in the *RAD51* gene, whose sequence predicts a protein with similarities to procaryotic RecA proteins. *Mol. Cell. Biol.* **12:** 3224–3234.
2. Adzuma, K., T. Ogawa, and H. Ogawa. 1984. Primary structure of the *RAD52* gene in *Saccharomyces cerevisiae*. *Mol. Cell. Biol.* **4:** 2735–2744.
3. Aguilera, A. and H. L. Klein. 1989. Yeast intrachromosomal recombination: long gene conversion tracts are preferentially associated with reciprocal exchange and require the *RAD1* and *RAD3* gene products. *Genetics* **123:** 683–694.
4. Ahn, B.-Y. and D. M. Livingston. 1986. Mitotic gene conversion lengths, coconversion patterns, and the incidence of reciprocal recombination in a *Saccharomyces cerevisiae* plasmid system. *Mol. Cell. Biol.* **6:** 3685–3693.

5. Ajimura, M., S. H. Leem, and H. Ogawa. 1993. Identification of new genes required for meiotic recombination in *Saccharomyces cerevisiae. Genetics* **133:** 51–66.
6. Alani, E., R. Padmore, and N. Kleckner. 1990. Analysis of wild-type and *rad50* mutants of yeast suggest an intimate relationship between meiotic chromosome synapsis and recombination. *Cell* **61:** 419–436.
7. Alani, E., S. Subbiah, and N. Kleckner. 1989. The yeast *RAD50* gene encodes a predicted 153-kD protein containing a purine nucleotide-binding domain and two large heptad-repeat regions. *Genetics* **122:** 47–57.
8. Bai, Y. and L. S. Symington. 1996. A *RAD52* homolog is required for *RAD51*-independent mitotic recombination in *Saccharomyces cerevisiae. Genes Dev.* **10:** 2025–2037.
9. Bardwell, A. J., L. Bardwell, A. E. Tomkinson, and E. C. Friedberg. 1994. Specific cleavage of model recombination and repair intermediates by the yeast Rad1–Rad10 DNA endonuclease. *Science* **265:** 2082–2085.
10. Barnes, G. and J. Rine. 1985. Regulated expression of endonuclease *Eco*RI in *Saccharomyces cerevisiae*: Nuclear entry and biological consequences. *Proc. Natl. Acad. Sci. USA* **82:** 1354–1358.
11. Basile, G., M. Aker, and R. K. Mortimer. 1992. Nucleotide sequence and transcriptional regulation of yeast recombinational repair gene *RAD51. Mol. Cell. Biol.* **12:** 3235–3246.
12. Belmaaza, A. and P. Chartrand. 1994. One-sided invasion events in homologous recombination at double-strand breaks. *Mutat. Res.* **314:** 199–208.
13. Bendixen, C., I. Sunjevaric, R. Bauchwitz, and R. Rothstein. 1994. Identification of a mouse homologue of the *Saccharomyces cerevisiae* recombination and repair gene, *RAD52. Genomics* **23:** 300–303.
14. Bennett, C. B., A. L. Lewis, K. K. Baldwin, and M. A. Resnick. 1993. Lethality induced by a single site-specific double-strand break in a dispensable yeast plasmid. *Proc. Natl. Acad. Sci. USA* **90:** 5613–5617.
15. Bezzubova, O. Y., H. Schmidt, K. Ostermann, W. D. Heyer, and J. M. Buerstedde. 1993. Identification of a chicken *RAD52* homolog suggests conservation of the *RAD52* recombination pathway throughout evolution of higher eucaryotes. *Nucleic Acids Res.* **21:** 5945–5949.
16. Bishop, D. K., J. Andersen, and R. D. Kolodner. 1989. Specificity of mismatch repair following transformation of *Saccharomyces cerevisiae* with heteroduplex plasmid DNA. *Proc. Natl. Acad. Sci. USA* **86:** 3713–3717.
17. Bishop, D. K. and R. D. Kolodner. 1986. Repair of heteroduplex plasmid DNA after transformation into *Saccharomyces cerevisiae. Mol. Cell. Biol.* **6:** 3401–3409.
18. Bishop, D. K., D. Park, L. Xu, and N. Kleckner. 1992. *DMC1*: a meiosis-specific yeast homolog of *E. coli* recA required for recombination, synaptonemal complex formation, and cell cycle progression. *Cell* **69:** 439–456.
19. Borts, R. H. and J. E. Haber. 1989. Length and distribution of meiotic gene conversion tracts and crossovers in *Saccharomyces cerevisiae. Genetics* **123:** 69–80.
20. Borts, R. H. and J. E. Haber. 1987. Meiotic recombination in yeast: alteration by multiple heterozygosities. *Science* **237:** 1459–1465.

20a. Boulton, S. J. and S. P. Jackson. 1996. *Saccharomyces cerevisiae* Ku70 potentiates illegitimate DNA double-strand break repair and serves as a barrier to error-prone DNA repair pathways. *EMBO J.* **15.**

21. Brunborg, G., M. A. Resnick, and D. H. and Williamson. 1980. Cell-cycle-specific repair of DNA double-strand breaks in *Saccharomyces cerevisiae. Radiat. Res.* **82:** 547–558.
22. Bryant, P. E. 1984. Enzymatic restriction of mammalian cell DNA using *Pvu*I and *Bam*HI: evidence for the double-strand break origin of chromosomal aberrations. *Int. J. Radiat. Biol.* **46:** 57–65.
23. Bryant, P. E. 1985. Enzymatic restriction of mammalian cell DNA: evidence for double-strand breaks as potentially lethal lesions. *Int. J. Radiat. Biol.* **48:** 55–60.

24. Bryant, P. E., D. A. Birch, and P. A. Jeggo. 1987. High chromosomal sensitivity of Chinese hamster *xrs-5* cells to restriction endonuclease-induced double-strand breaks. *Int. J. Radiat. Biol.* **52:** 537–554.
25. Budd, M. E. and J. L. Campbell. 1995. DNA polymerases required for repair of UV-induced damage in *Saccharomyces cerevisiae. Mol. Cell. Biol.* **15:** 2173–2179.
26. Cao, L., E. Alani, and N. Kleckner. 1990. A pathway for generation and processing of double-strand breaks during meiotic recombination in *S. cerevisiae. Cell* **61:** 1089–1101.
27. Carlson, M. and B. C. Laurent. 1994. The SNF/SWI family of global transcriptional activators. *Curr. Opinion Cell Biol.* **6:** 396–402.
28. Cheng, R., T. I. Baker, C. E. Cords, and R. J. Radloff. 1993. mei-3, a recombination and repair gene of *Neurospora crassa*, encodes a RecA-like protein. *Mutat. Res.* **294:** 223–234.
29. Chlebowicz, E. and W. J. Jachymczyk. 1979. Repair of mms-induced DNA double-strand breaks in haploid cells of *Saccharomyces cerevisiae*, which requires the presence of a duplicate genome. *Mol. Gen. Genet.* **167:** 279–286.
30. Connolly, B., C. I. White, and J. E. Haber. 1988. Physical monitoring of mating type switching in *Saccharomyces cerevisiae. Mol. Cell. Biol.* **8:** 2342–2349.
31. de Massy, B., F. Baudat, and A. Nicolas. 1994. Initiation of recombination in *Saccharomyces cerevisiae* haploid meiosis. *Proc. Natl. Acad. Sci. USA* **91:** 11,929–11,933.
32. de Massy, B. and A. Nicolas. 1993. The control in cis of the position and the amount of the ARG4 meiotic double-strand break of *Saccharomyces cerevisiae. EMBO J.* **12:** 1459–1466.
33. de Massy, B., V. Rocco, and A. Nicolas. 1995. The nucleotide mapping of DNA double-strand breaks at the *CYS3* initiation site of meiotic recombination in *Saccharomyces cerevisiae. EMBO J.* **14:** 4589–4598.
34. Deng, W. P. and J. A. Nickoloff. 1994. Mismatch repair of heteroduplex DNA intermediates of extrachromosomal recombination in mammalian cells. *Mol. Cell. Biol.* **14:** 400–406.
35. Detloff, P. and T. D. Petes. 1992. Measurements of excision repair tracts formed during meiotic recombination in *Saccharomyces cerevisiae. Mol. Cell. Biol.* **12:** 1805–1814.
36. Detloff, P., J. Sieber, and T. D. Petes. 1991. Repair of specific base pair mismatches formed during meiotic recombination in the yeast *Saccharomyces cerevisiae. Mol. Cell. Biol.* **11:** 737–745.
37. Donovan, J. W., G. T. Milne, and D. T. Weaver. 1994. Homotypic and heterotypic protein associations control Rad51 function in double-strand break repair. *Genes Dev* **8:** 2552–2562.
38. Ehrenhoffer-Murry, A. E., F. E. Wurgler, and C. Sengstag. 1994. The *Saccharomyces cerevisiae* SGE1 gene product: A novel drug-resistance protein within the major facilitator superfamily. *Mol. Gen. Genet.* **244:** 287–294.
39. Emery, H. S., D. Schild, D. E. Kellogg, and R. K. Mortimer. 1991. Sequence of *RAD54*, a *Saccharomyces cerevisiae* gene involved in recombination and repair. *Gene* **104:** 103–106.
40. Ezekiel, U. R. and H. P. Zassenhaus. 1993. Localization of a cruciform cutting endonuclease to yeast mitochondria. *Mol. Gen. Genet.* **240:** 414–418.
41. Fabre, F. and H. Roman. 1977. Genetic evidence for inducibility of recombination competence in yeast. *Proc. Natl. Acad. Sci. USA* **74:** 1667–1671.
42. Fan Q., F. Xu, and T. D. Petes. 1995. Meiosis-specific double-strand DNA breaks at the HIS4 recombination hot spot in the yeast *Saccharomyces cerevisiae*: control in *cis* and *trans*. *Mol. Cell. Biol.* **15:** 1679–1688.
43. Fan, Q.-Q. and T. D. Petes. 1996. Relationship between nuclease-hypersensitive sites and meiotic recombination hot spot activity at the *HIS4* locus of *Saccharomyces cerevisiae. Mol. Cell. Biol.* **16:** 2037–2043.
44. Fasullo, M., P. Dave, and R. Rothstein. 1994. DNA-damaging agents stimulate the formation of directed reciprocal translocations in *Saccharomyces cerevisiae. Mutat. Res.* **314:** 121–133.

45. Firmenich, A. A., M. Elias-Arnanz, and P. Berg. 1995. A novel allele of *Saccharomyces cerevisiae RFA1* that is deficient in recombination and repair and suppressible by RAD52. *Mol. Cell. Biol.* **15:** 1620–1631.
46. Fishman-Lobell, J. and J. E. Haber. 1992. Removal of nonhomologous DNA ends in double-strand break recombination: the role of the yeast ultraviolet repair gene *RAD1*. *Science* **258:** 480–484.
47. Fishman-Lobell, J., N. Rudin, and J. E. Haber. 1992. Two alternative pathways of double-strand break repair that are kinetically separable and independently modulated. *Mol. Cell. Biol.* **12:** 1292–1303.
48. Fogel, S. and D. D. Hurst. 1967. Meiotic gene conversion in yeast tetrads and the theory of recombination. *Genetics* **57:** 455–481.
49. Frankenberg-Schwager, M. 1990. Induction, repair and biological relevance of radiation-induced DNA lesions in eukaryotic cells. *Radiat. Environ. Biophys.* **29:** 273–292.
50. Frankenberg-Schwager, M. and D. Frankenberg. 1994. Survival curves with shoulders: damage interaction, unsaturated but dose-dependent rejoining kinetics or inducible repair of DNA double-strand breaks? *Radiat. Res.* **138:** S97–S100.
51. Game, J. and R. K. Mortimer. 1974. A genetic study of X-ray sensitive mutants in yeast. *Mutat. Res.* **24:** 281–292.
52. Game, J. C. 1993. DNA double-strand breaks and the *RAD50–RAD57* genes in *Saccharomyces*. *Semin. Cancer Biol.* **4:** 73–83.
53. Game, J. C. 1992. Pulsed-field gel analysis of the pattern of DNA double-strand breaks in the *Saccharomyces* genome during meiosis. *Dev. Genet.* **13:** 485–497.
54. Game, J. C., T. J. Zamb, R. J. Braun, M. Resnick, and R. M. Roth. 1980. The role of radiation *(rad)* genes in meiotic recombination in yeast. *Genetics* **94:** 51–68.
55. Gilbertson, L. A. and F. W. Stahl. 1994. Initiation of meiotic recombination is independent of interhomologue interactions. *Proc. Natl. Acad. Sci. USA* **91:** 11,934–11,937.
56. Gilbertson, L. A. and F. W. Stahl. 1996. A test of the double-strand break repair model for meiotic recombination in *Saccharomyces cerevisiae*. *Genetics* **144:** 27–41.
57. Goldway, M., A. Sherman, D. Zenvirth, T. Arbel, and G. Simchen. 1993. A short chromosomal region with major roles in yeast chromosome III meiotic disjunction, recombination and double strand breaks. *Genetics* **133:** 159–169.
58. Goyon, C. and M. Lichten. 1993. Timing of molecular events in meiosis in *Saccharomyces cerevisiae*: stable heteroduplex DNA is formed late in meiotic prophase. *Mol. Cell. Biol.* **13:** 373–382.
59. Haber, J. E. 1992. Mating-type gene switching in *Saccharomyces cerevisiae*. *Trends Genet.* **8:** 446–452.
60. Haber, J. E., W. Y. Leung, R. H. Borts, and M. Lichten. 1991. The frequency of meiotic recombination in yeast is independent of the number and position of homologous donor sequences: implications for chromosome pairing. *Proc. Natl. Acad. Sci. USA* **88:** 1120–1124.
61. Habraken, Y., P. Sung, L. Prakash, and S. Prakash. 1994. Holliday junction cleavage by yeast Rad1 protein. *Nature* **371:** 531–534.
62. Hare, J. T. and J. H. Taylor. 1985. One role for DNA methylation in vertebrate cells is strand discrimination in mismatch repair. *Proc. Natl. Acad. Sci. USA* **82:** 7350–7354.
63. Hastings, P. J. 1992. Mechanism and control of recombination in fungi. *Mutat. Res.* **284:** 97–110.
64. Hastings, P. J., C. McGill, B. Shafer, and J. N. Strathern. 1993. Ends-in vs. ends-out recombination in yeast. *Genetics* **135:** 973–980.
65. Hays, S. L., A. A. Firmenich, and P. Berg. 1995. Complex formation in yeast double-strand break repair: Participation of Rad51, Rad55, and Rad57 proteins. *Proc. Natl. Acad. Sci. USA* **92:** 6925–6929.
66. Herskowitz, I., J. Rine, and J. N. Strathern. 1992. Mating-type determination and mating-type interconversion in *Saccharomyces cerevisiae*, in *The Molecular and Cellular Biol-*

ogy of the Yeast Saccharomyces, vol. 2 (Jones E. W., J. R. Pringle, and J. R. Broach, eds.), Cold Spring Harbor Laboratory Press, Cold Spring Harbor, NY, pp. 583–656.
67. Heyer, W. D. 1994. The search for the right partner: homologous pairing and DNA strand exchange proteins in eukaryotes. *Experientia* **50:** 223–233.
68. Heyer, W.-D., A. W. Johnson, U. Reinhart, and R. D. Kolodner. 1995. Regulation and intracellular localization of *Saccharomyces cerevisiae* strand exchange protein 1 (Sep1/Xrn1/Kem1), a multifunctional exonuclease. *Mol. Cell. Biol.* **15:** 2728–2736.
69. Hoekstra, M. F., R. M. Liskay, A. C. Ou, A. J. DeMaggio, D. G. Burbee, and F. Heffron. 1991. HRR25, a putative protein kinase from budding yeast: association with repair of damaged DNA. *Science* **253:** 1031–1034.
70. Holliday, R. 1964. A mechanism for gene conversion in fungi. *Genet. Res. Camb.* **5:** 282–304.
71. Huang, K. N. and L. S. Symington. 1993. A 5'-3' exonuclease from *Saccharomyces cerevisiae* is required for in vitro recombination between linear DNA molecules with overlapping homology. *Mol. Cell. Biol.* **13:** 3125–3134.
72. Ivanov, E. L., V. G. Korolev, and F. Fabre. 1992. *XRS2*, a DNA repair gene of *Saccharomyces cerevisiae*, is needed for meiotic recombination. *Genetics* **132:** 651–664.
73. Johnson, A. and R. D. Kolodner. 1991. Strand exchange protein 1 from *Saccharomyces cerevisiae*. A novel multifunctional protein that contains DNA strand exchange and exonuclease activities. *J. Biol. Chem.* **266:** 14046–14054.
74. Johnson, R. D. and L. S. Symington. 1995. Functional differences and interactions among the putative RecA homologues Rad51, Rad55, and Rad57. *Mol. Cell. Biol.* **15:** 4843–4850.
75. Johzuka, K. and H. Ogawa. 1995. Interaction of Mre11 and Rad50: two proteins required for DNA repair and meiosis-specific double-strand break formation in *Saccharomyces cerevisiae. Genetics* **139:** 1521–1532.
76. Judd, S. R. and T. D. Petes. 1988. Physical lengths of meiotic and mitotic gene conversion tracts in *Saccharomyces cerevisiae. Genetics* **118:** 401–410.
77. Kadyk, L. C. and L. H. Hartwell. 1992. Sister chromatids are preferred over homologs as substrates for recombinational repair in *Saccharomyces cerevisiae. Genetics* **132:** 387–402.
78. Kans, J. A. and R. K. Mortimer. 1991. Nucleotide sequence of the *RAD57* gene of *Saccharomyces cerevisiae. Gene* **105:** 139–140.
78a. Keeney, S., C. N. Giroux, and N. Kleckner. 1997. Meiosis-specific DNA double-strand breaks are catalyzed by Spo11, a member of a widely conserved protein family. *Cell* **88:** 375–384.
79. Keeney, S. and N. Kleckner. 1995. Covalent protein-DNA complexes at the 5' strand termini of meiosis-specific double-strand breaks in yeast. *Proc. Natl. Acad. Sci. USA* **92:** 11,274–11,278.
80. Kiefer, J. and M. Feige. 1993. The significance of DNA double-strand breaks in the UV inactivation of yeast cells. *Mutat. Res.* **299:** 219–224.
81. Klar, A. 1993. Lineage-dependent mating-type transposition in fission and budding yeast. *Curr. Opinion Genet. Dev.* **3:** 745–751.
82. Klar, A. J. S., J. N. Strathern, and J. B. Hicks. 1981. A position-effect control for gene transposition: state of the expression of yeast mating-type genes affects their ability to switch. *Cell* **25:** 517–524.
83. Kleff, S., B. Kemper, and R. Sternglanz. 1992. Identification and characterization of yeast mutants and the gene for a cruciform cutting endonuclease. *EMBO J.* **11:** 699–704.
84. Klein, H. L. 1995. Genetic control of intrachromosomal recombination. *Bioessays* **17:** 147–159.
85. Kolodkin, A. L., A. J. S. Klar, and F. W. Stahl. 1986. Double-strand breaks can initiate meiotic recombination in *S. cerevisiae. Cell* **46:** 733–740.

86. Kolodner, R. D., D. H. Evans, and P. T. Morrison. 1987. Purification and characterization of an activity from *Saccharomyces cerevisiae* that catalyzes homologous pairing and strand exchange. *Proc. Natl. Acad. Sci. USA* **84:** 5560–5564.
87. Kramer, B., W. Kramer, M. S. Williamson, and S. Fogel. 1989. Heteroduplex DNA correction in *Saccharomyces cerevisiae* is mismatch specific and requires functional *PMS* genes. *Mol. Cell. Biol.* **9:** 4432–4440.
88. Lahue, R. S., K. G. Au, and P. Modrich. 1989. DNA mismatch correction in a defined system. *Science* **245:** 160–164.
89. Lieb, M. 1991. Spontaneous mutation at a 5-methylcytosine hotspot is prevented by very short patch (VSP) mismatch repair. *Genetics* **128:** 23–27.
90. Liu, Z. and W. Gilbert. 1994. The yeast *KEM1* gene encodes a nuclease specific for G4 tetraplex DNA: implication of in vivo functions for this novel DNA substrate. *Cell* **77:** 1083–1092.
91. Lovett, S. T. 1994. Sequence of the *RAD55* gene of *Saccharomyces cerevisiae*: similarity of *RAD55* to prokaryotic RecA and other RecA-like proteins. *Gene* **142:** 103–106.
92. Lovett, S. T., P. T. Drapkin, V. A. Sutera, Jr., and T. J. Gluckman-Peskind. 1993. A sister-strand exchange mechanism for recA-independent deletion of repeated DNA sequences in *Escherichia coli*. *Genetics* **135:** 631–642.
93. Malkova, A., L. Ross, D. Dawson, M. F. Hoekstra, and J. Haber. 1996. Meiotic recombination initiated by a double-strand break in *rad50Δ* yeast cells otherwise unable to initiate recombination. *Genetics* **143:** 741–754.
94. Malone, R. E. and R. E. Esposito. 1980. The *RAD52* gene is required for homothallic interconversion of mating types and spontaneous recombination in yeast. *Proc. Natl. Acad. Sci. USA* **77:** 503–507.
95. McGill, C. B., B. K. Shafer, L. K. Derr, and J. N. Strathern. 1993. Recombination initiated by double-strand breaks. *Curr. Genet.* **23:** 305–314.
96. Meselson, M. and C. M. Radding. 1975. A general model for genetic recombination. *Proc. Natl. Acad. Sci. USA* **72:** 358–361.
97. Milne, G. T., T. Ho, and D. T. Weaver. 1995. Modulation of *Saccharomyces cerevisiae* DNA double-strand break repair by *SRS2* and *RAD51*. *Genetics* **139:** 1189–1199.
98. Milne, G. T. and D. T. Weaver. 1993. Dominant negative alleles of *RAD52* reveal a DNA repair/recombination complex including Rad51 and Rad52. *Genes Dev.* **7:** 1755–1765.
99. Modrich, P. 1991. Mechanisms and biological effects of mismatch repair. *Ann. Rev. Genet.* **25:** 229–253.
100. Montelone, B. A., M. F. Hoekstra, and R. E. Malone. 1988. Spontaneous mitotic recombination in yeast: the hyper-recombinational *rem1* mutations are alleles of the *RAD3* gene. *Genetics* **119:** 289–301.
101. Moore, J. K. and J. E. Haber. 1996. Cell cycle and genetic requirements of two pathways of nonhomologous end-joining repair of double-strand breaks in *Saccharomyces cerevisiae*. *Mol. Cell. Biol.* **16:** 2164–73.
102. Mortensen, U. H., C. Bendixen, I. Sunjevaric, and R. Rothstein. 1996. DNA strand annealing is promoted by the yeast Rad52 protein. *Proc. Natl. Acad. Sci. USA* **93:** 10,729–10,734.
103. Muster-Nassal, C. and R. Kolodner. 1986. Mismatch correction catalyzed by cell-free extracts of *Saccharomyces cerevisiae*. *Proc. Natl. Acad. Sci. USA* **83:** 7618–7622.
104. Nag, D. K. and T. D. Petes. 1993. Physical detection of heteroduplexes during meiotic recombination in the yeast *Saccharomyces cerevisiae*. *Mol. Cell. Biol.* **13:** 2324–2331.
105. Nag, D. K., M. A. White, and T. D. Petes. 1989. Palindromic sequences in heteroduplex DNA inhibit mismatch repair in yeast. *Nature* **340:** 318–320.
106. Nelson, H. H., D. B. Sweetser, and J. A. Nickoloff. 1996. Effects of terminal nonhomology and homeology on double-strand break-induced gene conversion tract directionality. *Mol. Cell. Biol.* **16:** 2951–2957.

107. Nickoloff, J. A. 1992. Transcription enhances intrachromosomal homologous recombination in mammalian cells. *Mol. Cell. Biol.* **12:** 5311–5318.
108. Nickoloff, J. A., E. Y. C. Chen, and F. Heffron. 1986. A 24-base-pair sequence from the *MAT* locus stimulates intergenic recombination in yeast. *Proc. Natl. Acad. Sci. USA* **83:** 7831–7835.
109. Nickoloff, J. A. and R. J. Reynolds. 1990. Transcription stimulates homologous recombination in mammalian cells. *Mol. Cell. Biol.* **10:** 4837–4845.
110. Nickoloff, J. A., J. D. Singer, and F. Heffron. 1990. In vivo analysis of the *Saccharomyces cerevisiae* HO nuclease recognition site by site-directed mutagenesis. *Mol. Cell. Biol.* **10:** 1174–1179.
111. Nickoloff, J. A., J. D. Singer, M. F. Hoekstra, and F. Heffron. 1989. Double-strand breaks stimulate alternative mechanisms of recombination repair. *J. Mol. Biol.* **207:** 527–541.
112. Nicolas, A. and T. D. Petes. 1994. Polarity of meiotic gene conversion in fungi: contrasting views. *Experientia* **50:** 242–252.
113. Nicolas, A., D. Treco, N. P. Shultes, and J. W. Szostak. 1989. An initiation site for meiotic gene conversion in the yeast *Saccharomyces cerevisiae. Nature* **338:** 35–39.
114. Ogawa, H. K., K. Johzuka, T. Nakagawa, S.-H. Leem, and A. H. Hagihara. 1995. Functions of the yeast meiotic recombination genes *MRE11* and *MRE2. Adv. Biophys.* **31:** 67–76.
115. Ogawa, T., A. Shinohara, A. Nabetani, T. Ikeya, X. Yu, E. H. Egelman, and H. Ogawa. 1993. RecA-like recombination proteins in eukaryotes: Functions and structures of *RAD51* genes. *Cold Spring Harb. Symp. Quant. Biol.* **58:** 567–576.
116. Ogawa, T., X. Yu, A. Shinohara, and E. H. Egelman. 1993. Similarity of the yeast Rad51 filament to the bacterial RecA filament. *Science* **259:** 1896–1899.
117. Ohta, K., T. Shibata, and A. Nicolas. 1994. Changes in chromatin structure at recombination initiation sites during yeast meiosis. *EMBO J.* **13:** 5754–5763.
118. Orr-Weaver, T. L., A. Nicolas, and J. W. Szostak. 1988. Gene conversion adjacent to regions of double-strand break repair. *Mol. Cell. Biol.* **8:** 5292–5298.
119. Orr-Weaver, T. L., J. W. Szostak, and R. J. Rothstein. 1981. Yeast transformation: a model system for the study of recombination. *Proc. Natl. Acad. Sci. USA* **78:** 6354–6358.
120. Ozenberger, B. A. and G. S. Roeder. 1991. A unique pathway of double-strand break repair operates in tandemly repeated genes. *Mol. Cell. Biol.* **11:** 1222–1231.
121. Padmore, R., L. Cao, and N. Kleckner. 1991. Temporal comparison of recombination and synaptonemal complex formation during meiosis in *S. cerevisiae. Cell* **66:** 1239–1256.
122. Petes, T. D., R. E. Malone, and L. S. Symington. 1991. Recombination in yeast, in *The Molecular and Cellular Biology of the Yeast* Saccharomyces: *Genome Dynamics, Protein Synthesis, and Energetics*, vol. I (Broach, J. R., J. R. Pringle, and E. W. Jones, eds.), Cold Spring Harbor Laboratory Press, Cold Spring Harbor, NY, pp. 407–521.
123. Plessis, A., A. Perrin, J. E. Haber, and B. Dujon. 1992. Site-specific recombination determined by I-SceI, a mitochondrial group I intron-encoded endonuclease expressed in the yeast nucleus. *Genetics* **130:** 451–460.
124. Porter, S. E., M. A. White, and T. D. Petes. 1993. Genetic evidence that the meiotic recombination hotspot at the *HIS4* locus of *Saccharomyces cerevisiae* does not represent a site for a symmetrically processed double-strand break. *Genetics* **134:** 5–19.
125. Prado, F. and A. Aguilera. 1995. Role of reciprocal exchange, one-ended invasion crossover and single-strand annealing on inverted and direct repeat recombination in yeast: different requirements for the *RAD1*, *RAD10*, and *RAD52* genes. *Genetics* **139:** 109–123.
126. Priebe, S. D., J. Westmoreland, T. Nilsson-Tillgren, and M. A. Resnick. 1994. Induction of recombination between homologous and diverged DNAs by double-strand gaps and breaks and role of mismatch repair. *Mol. Cell. Biol.* **14:** 4802–4814.
127. Ray, A., N. Machin, and F. W. Stahl. 1989. A DNA double chain break stimulates triparental recombination in *Saccharomyces cerevisiae. Proc. Natl. Acad. Sci. USA* **86:** 6225–6229.

128. Ray, A., I. Siddiqi, A. L. Kolodkin, and F. W. Stahl. 1988. Intrachromosomal gene conversion induced by a DNA double-strand break in *Saccharomyces cerevisiae*. *J. Mol. Biol.* **201:** 247–260.
129. Resnick, M. A. and P. Martin. 1976. The repair of double-strand breaks in the nuclear DNA of *Saccharomyces cerevisiae* and its genetic control. *Mol. Gen. Genet.* **143:** 119–129.
130. Rocco, V., B. de Massy, and A. Nicolas. 1992. The *Saccharomyces cerevisiae ARG4* initiator of meiotic gene conversion and its associated double-strand DNA breaks can be inhibited by transcriptional interference. *Proc. Natl. Acad. Sci. USA* **89:** 12,068–12,072.
131. Rockmill, B., J. A. Engebrecht, H. Scherthan, J. Loidl, and G. S. Roeder. 1995. The yeast *MER2* gene is required for chromosome synapsis and the initiation of meiotic recombination. *Genetics* **141:** 49–59.
132. Roeder, G. S. 1995. Sex and the single cell: meiosis in yeast. *Proc. Natl. Acad. Sci. USA* **92:** 10,450–10,456.
133. Roitgrund, C., R. Steinlauf, and M. Kupiec. 1993. Donation of information to the unbroken chromosome in double-strand break repair. *Curr. Genet.* **23**: 414–422.
134. Rong, L. and H. Klein. 1993. Purification and characterization of the Srs2 DNA helicase of the yeast *Saccharomyces cerevisiae*. *J. Biol. Chem.* **268:** 1252–1259.
135. Rong, L., F. Palladino, A. Aguilera, and H. L. Klein. 1991. The hyper-gene conversion *hpr5-1* mutation of *Saccharomyces cerevisiae* is an allele of the *SRS2/RADH* gene. *Genetics* **127:** 75–85.
136. Ronne, H. and R. Rothstein. 1988. Mitotic sectored colonies: evidence of heteroduplex DNA formation during direct repeat recombination. *Proc. Natl. Acad. Sci. USA* **85:** 2696–2700.
137. Rose, D., W. Thomas, and C. Holm. 1990. Segregation of recombined chromosomes in meiosis I requires DNA topoisomerase II. *Cell* **60:** 1009–1017.
138. Rubin, B. P., D. O. Ferguson, and W. K. Holloman. 1994. Structure of *REC2,* a recombinational repair gene of *Ustilago maydis*, and its function in homologous recombination between plasmid and chromosomal sequences. *Mol. Cell. Biol.* **14:** 6287–6296.
139. Rudin, N. and J. E. Haber. 1988. Efficient repair of HO-induced chromosomal breaks in *Saccharomyces cerevisiae* by recombination between flanking homologous sequences. *Mol. Cell. Biol.* **8:** 3918–3928.
140. Rudin, N., E. Sugarman, and J. E. Haber. 1989. Genetic and physical analysis of double-strand break repair and recombination in *Saccharomyces cerevisiae*. *Genetics* **122:** 519–534.
141. Saffran, W. A., R. B. Greenberg, M. S. Thaler-Scheer, and M. M. Jones. 1994. Single strand and double strand DNA damage-induced reciprocal recombination in yeast. Dependence on nucleotide excision repair and *RAD1* recombination. *Nucleic Acids Res.* **22:** 2823–2829.
142. Sato, S., Y. Hotta, and S. Tabata. 1995. Structural analysis of a recA-like gene in the genome of *Arabidopsis thaliana*. *DNA. Res.* **2:** 89–93.
143. Schiestl, R. H., M. Dominska, and T. D. Petes. 1993. Transformation of *Saccharomyces cerevisiae* with nonhomologous DNA: illegitimate integration of transforming DNA into yeast chromosomes and in vivo ligation of transforming DNA to mitochondrial DNA sequences. *Mol. Cell. Biol.* **13:** 2697–2705.
144. Schiestl, R. H. and T. D. Petes. 1991. Integration of DNA fragments by illegitimate recombination in *Saccharomyces cerevisiae*. *Proc. Natl. Acad. Sci. USA* **88:** 7585–7589.
145. Schiestl, R. H., J. Zhu, and T. D. Petes. 1994. Effect of mutations in genes affecting homologous recombination on restriction enzyme-mediated and illegitimate recombination in *Saccharomyces cerevisiae*. *Mol. Cell. Biol.* **14:** 4493–4500.
146. Schild, D. 1995. Suppression of a new allele of yeast *RAD52* by overexpression of *RAD51*, mutations in *srs2* and *ccr4*, or mating-type heterozygosity. *Genetics* **140:** 115–127.
147. Schild, D., B. Konforti, C. Perez, W. Gish, and R. Mortimer. 1983. Isolation and characterization of yeast DNA repair genes. I. Cloning of the *RAD52* gene. *Curr. Genet.* **7:** 85–92.

148. Schultes, N. P. and J. W. Szostak. 1991. A poly(dA · dT) tract is a component of the recombination initiation site at the *ARG4* locus in *Saccharomyces cerevisiae*. *Mol. Cell. Biol.* **11:** 322–328.
149. Schwacha, A. and N. Kleckner. 1995. Identification of double Holliday junctions as intermediates in meiotic recombination. *Cell* **83:** 783–791.
150. Schwacha, A. and N. Kleckner. 1994. Identification of joint molecules that form frequently between homologs but rarely between sister chromatids during yeast meiosis. *Cell* **76:** 51–63.
151. Seide, W., A. A. Friedl, I. Dianova, F. Eckhardt-Schupp, and E. C. Friedberg. 1996. The *Saccharomyces cerevisiae* Ku autoantigen homologue affects radiosensitivity only in the absence of homologous recombination. *Genetics* **142:** 91–102.
152. Sekiguchi, J., N. C. Seeman, and S. Shuman. 1996. Resolution of Holliday junctions by eukaryotic DNA topoisomerase I. *Proc. Natl. Acad. Sci. USA* **93:** 785–789.
153. Sharples, G. J. and D. R. Leach. 1995. Structural and functional similarities between the SbcCD proteins of *Escherichia coli* and the *RAD50* and *MRE11 (RAD32)* recombination and repair proteins of yeast. *Mol. Microbiol.* **17:** 1215–1217.
154. Shen, Z., K. Denison, R. Lobb, J. M. Gatewood, and D. J. Chen. 1995. The human and mouse homologs of the yeast *RAD52* gene: cDNA cloning, sequence analysis, assignment to human chromosome 12p12.2-p13, and mRNA expression in mouse tissues. *Genomics* **25:** 199–206.
155. Shinohara, A., H. Ogawa, and T. Ogawa. 1992. Rad51 protein involved in repair and recombination in S. cerevisiae is a RecA-like protein [published erratum appears in *Cell* 1992 Oct 2;71(1): following 180]. *Cell* **69:** 457–470.
156. Silberman, R. and M. Kupiec. 1994. Plasmid-mediated induction of recombination in yeast. *Genetics* **137:** 41–48.
157. Smith, J. and R. Rothstein. 1995. A mutation in the gene encoding *Saccharomyces cerevisiae* single-stranded DNA-binding protein Rfa1 stimulates a *RAD52*-independent pathway for direct-repeat recombination. *Mol. Cell. Biol.* **15:** 1632–1641.
158. Snow, R. 1968. Recombination in ultraviolet-sensitive strains of *Saccharomyces cerevisiae*. *Mutat. Res.* **6:** 409–418.
159. Spell, R. M. and C. Holm. 1994. Nature and distribution of chromosomal intertwinings in *Saccharomyces cerevisiae*. *Mol. Cell. Biol.* **14:** 1465–1476.
160. Strathern, J. N., A. J. S. Klar, J. B. Hicks, J. A. Abraham, J. M. Ivy, K. A. Nasmyth, and C. McGill. 1982. Homothallic switching of yeast mating type cassettes is initiated by a double-stranded cut in the *MAT* locus. *Cell* **31:** 183–192.
161. Strathern, J. N., B. K. Shafer, and C. B. McGill. 1995. DNA synthesis errors associated with double-strand break repair. *Genetics* **140:** 965–972.
162. Strathern, J. N., K. G. Weinstock, D. R. Higgins, and C. B. McGill. 1991. A novel recombinator in yeast based on gene *II* protein from bacteriophage f1. *Genetics* **127:** 61–73.
163. Sugawara, N. and J. E. Haber. 1992. Characterization of double-strand break-induced recombination: homology requirements and single-stranded DNA formation. *Mol. Cell. Biol.* **12:** 563–575.
164. Sugawara, N., E. L. Ivanov, J. Fishman-Lobell, B. L. Ray, X. Wu, and J. E. Haber. 1995. DNA structure-dependent requirements for yeast *RAD* genes in gene conversion. *Nature* **373:** 84–86.
165. Sun, H., D. Treco, N. P. Schultes, and J. W. Szostak. 1989. Double-strand breaks at an initiation site for meiotic gene conversion. *Nature* **338:** 87–90.
166. Sun, H., D. Treco, and J. W. Szostak. 1991. Extensive 3'-overhanging, single-stranded DNA associated with meiosis-specific double-strand breaks at the *ARG4* recombination initiation site. *Cell* **64:** 1155–1161.
167. Sung, P. and D. L. Robberson. 1995. DNA strand exchange catalyzed by a RAD51-ssDNA nucleoprotein filament with polarity opposite to that of RecA. *Cell* **82:** 453–461.

168. Suzuki, A. H., M. Nishizawa, and T. Fukasawa. 1993. Isolation and characterization of *SGE1*: a yeast gene that partially suppresses the *gal11* mutation in multiple copies. *Genetics* **134:** 675–683.
169. Sweetser, D. B., H. Hough, J. F. Whelden, M. Arbuckle, and J. A. Nickoloff. 1994. Fine-resolution mapping of spontaneous and double-strand break-induced gene conversion tracts in *Saccharomyces cerevisiae* reveals reversible mitotic conversion polarity. *Mol. Cell. Biol.* **14:** 3863–3875.
170. Symington, L. S. 1991. Double-strand-break repair and recombination catalyzed by a nuclear extract of *Saccharomyces cerevisiae*. *EMBO J.* **10:** 987–996.
171. Symington, L. S. and T. Petes. 1988. Expansions and contractions of the genetic map relative to the physical map of yeast chromosome III. *Mol. Cell. Biol.* **8:** 595–604.
172. Szostak, J. W., T. L. Orr-Weaver, R. J. Rothstein, and F. W. Stahl. 1983. The double-strand break repair model for recombination. *Cell* **33:** 25–35.
173. Teng, S.-C., B. Kim, and A. Gabriel. 1996. Retrotransposon reverse-transcriptase-mediated repair of chromosomal breaks. *Nature* **383:** 641–644.
174. Thomas, B. J. and R. Rothstein. 1989. Elevated recombination rates in transcriptionally active DNA. *Cell* **56:** 619–630.
175. Tishkoff, D., A. W. Johnson, and R. D. Kolodner. 1991. Molecular and genetic analysis of the gene encoding the *Saccharomyces cerevisiae* strand exchange protein SEP1. *Mol. Cell. Biol.* **11:** 2593–2606.
176. Tishkoff, D. X., B. Rockmill, G. S. Roeder, and R. D. Kolodner. 1995. The *sep1* mutant of *Saccharomyces cerevisiae* arrests in pachytene and is deficient in meiotic recombination. *Genetics* **139:** 495–509.
176a. Tsukamoto, Y., J.-I. Kato, and H. Ikeda. 1996. Hdf1, a yeast Ku-protein homologue, is involved in illegitimate recombination but not in homologous recombination. *Nucleic Acids Res.* **24:** 2067–2072.
177. Voelkel-Meiman, K. and G. S. Roeder. 1990. A chromosome containing *HOT1* preferentially receives information during mitotic interchromosomal gene conversion. *Genetics* **124:** 561–572.
178. Voelkel-Meiman, K. and G. S. Roeder. 1990. Gene conversion tracts stimulated by HOT1-promoted transcription are long and continuous. *Genetics* **126:** 851–867.
179. Wahls, W. P. and G. R. Smith. 1994. A heteromeric protein that binds to a meiotic homologous recombination hot spot: correlation of binding and hot spot activity. *Genes Dev.* **8:** 1693–1702.
180. Weiffenbach, B. and J. E. Haber. 1981. Homothallic mating type switching generates lethal chromosome breaks in *rad52* strains of *Saccharomyces cerevisiae*. *Mol. Cell. Biol.* **1:** 522–534.
181. Weiner, B. M. and N. Kleckner. 1994. Chromosome pairing via multiple interstitial interactions before and during meiosis in yeast. *Cell* **77:** 977–991.
182. Weng, Y.-S. and J. A. Nickoloff. Evidence for independent mismatch repair processing on opposite sites of a double-strand break. *Mol. Cell. Biol.* Submitted.
183. Weng, Y.-S., J. Whelden, L. Gunn, and J. A. Nickoloff. 1996. Double-strand break-induced gene conversion: examination of tract polarity and products of multiple recombinational repair events. *Curr. Genet.* **29:** 335–343.
184. West, S. C. 1995. Holliday junctions cleaved by Rad1? *Nature* **373:** 27–28.
185. White, C. I. and J. E. Haber. 1990. Intermediates of recombination during mating type switching in *Saccharomyces cerevisiae*. *EMBO J.* **9:** 663–673.
186. White, M. A., P. Detloff, M. Strand, and T. D. Petes. 1992. A promoter deletion reduces the rate of mitotic, but not meiotic, recombination at the *HIS4* locus in yeast. *Curr. Genet.* **21:** 109–116.
187. Whitehouse, H. 1982. *Genetic Recombination*. John Wiley, New York.

188. Willis, K. K. and H. L. Klein. 1987. Intrachromosomal recombination in *Saccharomyces cerevisiae*: reciprocal exchange in an inverted repeat and associated gene conversion. *Genetics* **117:** 633–643.
189. Wu, T. C. and M. Lichten. 1995. Factors that affect the location and frequency of meiosis-induced double-strand breaks in *Saccharomyces cerevisiae*. *Genetics* **140:** 55–66.
190. Wu, T. C. and M. Lichten. 1994. Meiosis-induced double-strand break sites determined by yeast chromatin structure. *Science* **263:** 515–518.
191. Xu, L. and N. Kleckner. 1995. Sequence non-specific double-strand breaks and interhomolog interactions prior to double-strand break formation at a meiotic recombination hot spot in yeast. *EMBO J.* **14:** 5115–5128.
192. Zenvirth, D., T. Arbel, A. Sherman, M. Goldway, S. Klein, and G. Simchen. 1992. Multiple sites for double-strand breaks in whole meiotic chromosomes of *Saccharomyces cerevisiae*. *EMBO J.* **11:** 3441–3447.
193. Zhu, J. and R. H. Schiestl. 1996. Topoisomerase I involvement in illegitimate recombination in *Saccharomyces cerevisiae*. *Mol. Cell. Biol.* **16:** 1805–1812.

17

Pathways and Puzzles in DNA Replication and Damage Checkpoints in Yeast

Ted Weinert and David Lydall

1. INTRODUCTION

1.1. Fidelity and Order of Cell-Cycle Events

The events of cell division occur in a precise order; mitosis always follows DNA replication, and cell septation follows mitosis. Several types of cell-cycle controls ensure the order of events, even in stressed cells. Defects in cell cycle controls leading to an imprecise order of events can result in cell death or mutation. This chapter focuses on a type of control that orders events called checkpoint control.

Checkpoints respond to cell damage and delay progression until the damage is repaired *(11,13,32,61,68,94)*. One type of checkpoint senses DNA damage or delays in DNA replication, whereas a second type senses defects in the spindle and its assembly, called the spindle assembly checkpoint. The spindle assembly checkpoint *(28,56,62,68)* and checkpoints that detect other types of cellular damage (chromosome condensation, budding pattern in budding yeast) are discussed elsewhere *(15,53)*.

The checkpoints that respond to damage are typically not essential for viability and act only in cells with damage. The spontaneous damage in cells probably necessitates the occasional cell-cycle delay, however, because checkpoint mutant cells do lose chromosomes and suffer other forms of genome instability at increased rates compared to wild-type cells *(32,55,57,112,118)*.

1.2. Relevance to Cancer

The relationship between checkpoints and cancer was established through studies of two human genes, *p53* and *ATM*. Mutant cells defective for either gene have checkpoint defects *(7,41,49,69,85)* and suffer genomic instability (*57,118*; discussed in ref. *22*), and these cellular phenotypes probably contribute to the cancer phenotype *(32)*. Both genes have additional roles in other cellular responses to DNA damage, such as apoptosis or transcriptional regulation *(20,76)*. The *p53* gene is a tumor suppressor found mutated in half of all human cancers *(30)*. The *ATM* gene is defective in individuals with ataxia telangiectasia (*90*; reviewed in ref. *52*). In homozygotes, the *ATM* mutation leads to complex phenotypes (discussed in ref. *22*) culminating in death in the second decade of life. Most individuals do not live long enough to get cancer. In het-

From: DNA Damage and Repair, Vol. 1: DNA Repair in Prokaryotes and Lower Eukaryotes
Edited by: J. A. Nickoloff and M. F. Hoekstra © Humana Press Inc., Totowa, NJ

erozygotes, which consist of about 1% of the human population, there is a predisposition to cancer, particularly breast cancer *(102,103)*.

The checkpoint defects in cancer cells may be useful in designing new cancer therapy strategies. Checkpoint defects in yeast cells lead to increased sensitivity to DNA-damaging agents *(111)*, suggesting such defects may be manipulated in mammalian cells as well. The checkpoint defects in *p53* mutant cells do not lead directly to increased damage sensitivity *(81)*, but when combined with a defect in the G2 checkpoint do lead to increased damage-sensitivity *(21,32,77,81)*. It is possible that components of the mammalian G2 checkpoint, or some other aspect of damage responses, could be inactivated selectively by drugs to improve therapeutic outcome.

2. WHAT IS A CHECKPOINT?

The term checkpoint and its context in the cell cycle deserves a brief introduction. Conceptually, the order of events in any process is regulated either intrinsically by the assembly of structures or extrinsically by regulators that monitor those structures. The extrinsic regulators were termed checkpoints *(33)*. If order is regulated by assembly, a structure from an early event is a prerequisite for the next event. To make the distinction clear, an analogy to construction of a building is useful. If the second floor of a building is constructed only after the first is completed because you need the structure of the first floor to build the second, then order is based on assembly. In contrast, if the second floor is built only when the regulators allow it to be built, regardless of whether the first floor exists, then order is based on extrinsic regulators (termed checkpoints in cell division). Remarkably, many events in cell division are regulated by extrinsic regulators and not by assembly.

Whether the order between two events is ordered by assembly or by extrinsic regulators can be distinguished by an empirical genetic criterion; a mutation that allows a late event to occur without completion of an earlier event argues against regulation by assembly and for regulation by a checkpoint. Bacteriophage assembly appears regulated by assembly; progress in one step of assembly requires intrinsically the structure assembled in the previous step *(115)*. In bacteriophage assembly, a defect in an early step leads to obligatory arrest, and late events do not occur. In cell division, in contrast, order is largely achieved by extrinsic regulatory mechanisms. The proper cell structures are surprisingly not usually intrinsic requirements for cell-cycle progression, though progression without the proper structure has disastrous consequences. Rather, in cell division, a defect in an early step prevents a later step only because regulators identify the defect and block progression.

2.1. *What Is a Checkpoint in Cell Division?*

In the context of cell division, two general types of controls regulate the order of events. One type of control is essential, acts during each cell cycle, and initiates events *(54)*. The second ensures the completion of an event before the next event begins. The first type of control involves genes that act in the beginning of cell division in G1 and in key events of the cell cycle, DNA replication, and mitosis. These genes include $p34^{CDK}$- cyclin, the protein kinase that drives many events, its many regulators (*Schizosaccharomyces pombe cdc25^{+}, wee1^{+}, rum1^{+} [51,63,70]*; *Saccharomyces* *cerevisiae FAR1* and *SIC1 [79,92]*). The essential regulators also include genes involved in initiation of DNA replication (*S. pombe cdc18$^+$*, *cut5$^+$/rad4$^+$*, *[43,87]*;*S. cerevisiae CDC6*, *[80]*; *CDC7*, *DBF4*, *[106]*). How p34^{CDK}-cyclin, its regulators, and regulators of DNA replication control the order of events is a key and unsolved problem.

The essential controls regulate the initiation of events and their alternation (e.g., alternation of mitosis and DNA replication). The damage-responsive checkpoints have a more restricted role to ensure that early events are complete before late events. Termed the alternation and completion problems by Murray and Hunt *(66)*, the two problems are perhaps surprisingly distinct. Apparently, the essential controls provide enough time between events to complete each event. Only when there is damage does an event take longer to complete, necessitating a delay of the late event imposed by checkpoints.

Certain mutations in essential genes also show checkpoint defects; in the mutants, late events no longer require completion of early events. Remarkably, in these mutants, the early event does not occur at all! For example, in null mutants of *S. pombe cdc18$^+$* cells progress from G1 directly to mitosis without any DNA replication *(39,43)*. Mutations in other essential regulators, like *cdc2$^+$* and cyclin, can also disrupt the order of events such that events are skipped entirely *(9,34)*. In contrast, mutants in the damage-responsive checkpoint genes typically do not skip events but rather early events do occur. However, they are not completed owing to damage before late events begin. Damage-responsive checkpoint genes may act through the essential genes, p34^{CDK} and its regulators, a hypothesis discussed in Section 5.

2.2. DNA Replication and Damage Checkpoints in Budding Yeast

The DNA replication and damage checkpoints affect several physiological responses in budding yeast (Fig. 1). The responses include delays in G1 (the G1 checkpoint; *95*), delay of progression of DNA replication (intra-S-phase control; *78*), delay of mitosis when DNA replication is blocked (the S/M checkpoint or S-phase checkpoint; *3,114*), delay of mitosis in G2 cells with DNA damage (the G2/M checkpoint or G2 checkpoint; *111*), and delay of meiosis I when recombination is blocked (meiotic prophase checkpoint; *8,58*). Checkpoint genes also regulate transcriptional induction of genes involved in DNA repair (*3,45,71*; Chapter 18) and may have additional functions in DNA repair as well (*see* Fig.1).

Clues to gene functions are emerging from studies of budding and fission yeast (reviewed in *12,16,61,67,68,94*), frog embryos *(47,50)*, mammalian cells *(36,74)*, and *Drosophila (29)*. Since all cell types possess checkpoints, there are many sources of clues to mechanism. Comparative studies of controls in different cell types show that the control pathways do appear conserved, even to the molecular level. For example, the human *ATM* and budding yeast *MEC1p* gene products are similar in sequence, and each is required for checkpoint responses *(90)*. The budding and fission yeast checkpoint genes include four pairs of homologous genes (Table 1), even though the two yeasts diverged about 1 billion years ago *(96)*. The conservation between yeast genes may lead to identification of human counterparts. The details of the pathways may well differ, and comparison of gene functions in the two yeast cell types does provide clues as well as puzzles regarding mechanism. Reviews on fission yeast checkpoints are recommended for certain contrasting views to those presented here (*11,94*; *see* Chapter 20).

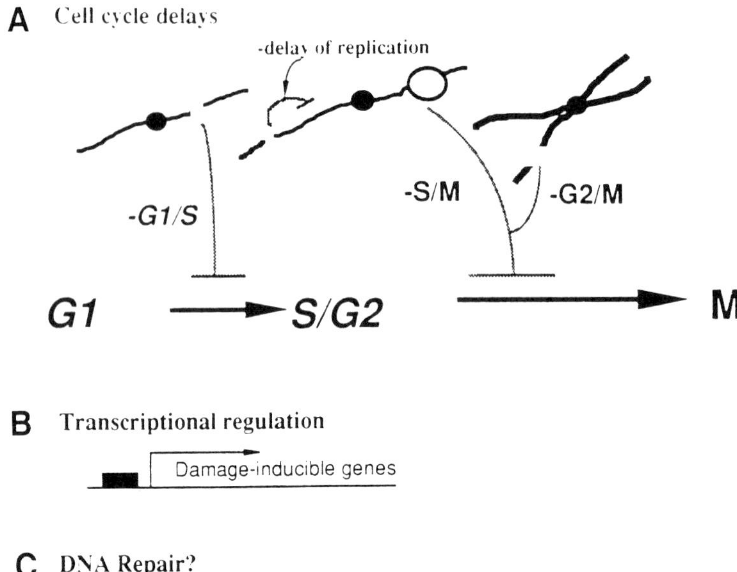

Fig. 1. Cell responses to be mediated by checkpoint genes. **(A)** A chromosome in G1, S, and G2, and progression through stages of the cell cycle. Four checkpoints are shown; in G1/S, S/M, G2/M, and a delay of progression of DNA replication. The responses are induced either by a DNA break or a stalled replication fork (bubble). **(B)** Checkpoint genes also have roles in transcriptional regulation of damage-inducible genes. **(C)** Checkpoint genes probably have roles in DNA repair that are as yet to be defined.

2.3. *Possible Roles for Budding Yeast Checkpoint Genes*

Budding yeast checkpoint genes were identified in screens for mutant cells that failed to arrest after DNA damage or after a delay in DNA replication *(3,71,111,113)*. Characterization of mutants and corresponding gene products leads to the hypothesis that arrest is mediated by two types of gene products: sensors and signal transducers (Fig. 2). The sensors may interact directly with DNA damage. Sensor genes act either in the S/M or in the G2/M checkpoint pathways. The basis for this specificity is key to understanding checkpoint gene functions and is discussed in Section 3. The two pathways and requirements for checkpoint genes are suggested to be distinct, because the types of primary damage, a stalled replication fork and DNA breaks, are different. Each type of damage is converted to an intermediate that signals arrest and conversion, or processing of the damage requires different pathways. The signal transducers, discussed in Section 5., may detect sensor proteins bound to damaged DNA, become activated, and mediate downstream events leading to arrest. The current distinction between genes that are sensors and signal transducers should be considered a preliminary hypothesis since many details remain to be determined.

3. SENSORS

Early clues to the functions of this class of checkpoint genes came from analysis of gene sequence. The S/M phase-specific genes identified thus far include the

Table 1
Checkpoint Genes of *S. cerevisiae*[a]

Class	*S. cerevisiae* checkpoint gene	Checkpoint position: G_1/S	Intra-S	S/M	G_2M	Meiosis 1	Essential?	Homologs	Sequence motifs	Refs.
I	*MEC1*	Yes	Yes	Yes	Yes	Yes	Yes	*rad3*[+] (*S. pombe*)	Lipid/protein kinase	*40,42,58, 93,101,114*
								ATM (Human)		*90*
								MEI41 (Drosophila)		*29*
	RAD53	Yes	Yes	Yes	Yes	?	Yes	*cds1*[+] (*S. pombe*)	Protein kinase	*3,65,98,114*
II	*POL2*	No	?	Yes	No	?	Yes	—	DNA polymerase ε	*71*
	DBP11			Yes	?		Yes	—		*5*
	RFC5			Yes	?		Yes			*99*
III	*RAD9*	Yes	Partial	No	Yes	No	No	—	—	*911,111,113,122*
IV	*RAD17*	Yes	Partial	No	Yes	Yes	No	*rad1*[+] (*S. pombe*)	3' to 5' DNA	*27,58,60,75, 113,121,122*
								rec1 (U. maydis)	exonuclease	*See* Chapter 23
	RAD24	Yes	Partial	No	Yes	Yes	No	*rad17*[+] (*S. pombe*)	Weak homology to RFC	*27,58,59, 60,114,122*
	MEC3	(Yes?)	Partial	No	Yes	(Yes?)	No	—	—	*59,60,120*
V	*PDS1*	?	?	No	Yes	?	No	—	—	*116,117*

[a]Modified from ref. *61*.

The budding yeast mitotic cell cycle is delayed at the G1/S, S/M and G2/M transitions after DNA is damaged. In addition, progression of DNA replication in S phase is slowed, called the intra-S control, and meiosis I is delayed by recombination events. "Yes" indicates the gene is required for that control by analysis of single mutants, and "No" indicates the gene is not required. *See text* for details and references. The G1/S delay is transient, lasting on the order of an hour or longer in some strain backgrounds *(95)*. The intra-S delay slows S phase by 1–2 h when DNA is chronically damaged with the alkylating agent methyl methanesulfonate *(78)*. The intra-S delay is significantly reduced by *mec1* and *rad53* mutations and to a lesser extent by other checkpoint mutations. The S/M checkpoint does not respond to DNA damage *per se*, but to the inhibition of DNA synthesis by the ribonucleotide reductase poison, hydroxyurea *(3,114)*. The G2/M delay responds to DNA damage and is the major point at which *S. cerevisiae* cells delay their cell cycle after DNA damage. *MEC1* is also known as *ESR1 (42)*, *RAD53* was identified as *MEC2 (114)*, *SAD1 (3)*, and *SPK1 (98)*. Question marks indicate that the results are not yet known. The brackets and question marks for *MEC3* indicate that the requirements for this gene have not yet been tested, but on the basis of similarity of mutant phenotypes with *RAD17* and *RAD24*, it is expected *MEC3* will have similar functions *(59,60,114)*. The degree of similarity between checkpoint genes in different species is not high (generally 20–25% identical with 40–50% similarity), and as yet there is no evidence for cross-species complementation. The *cds1*[+] gene, an *S. pombe* homologue of the *S. cerevisiae RAD53* gene, has a subtle checkpoint defect; it is not required for arrest of cell division when S phase is inhibited by hydroxyurea, but it seems to help stop precocious mitosis as cells recover from hydroxyurea.

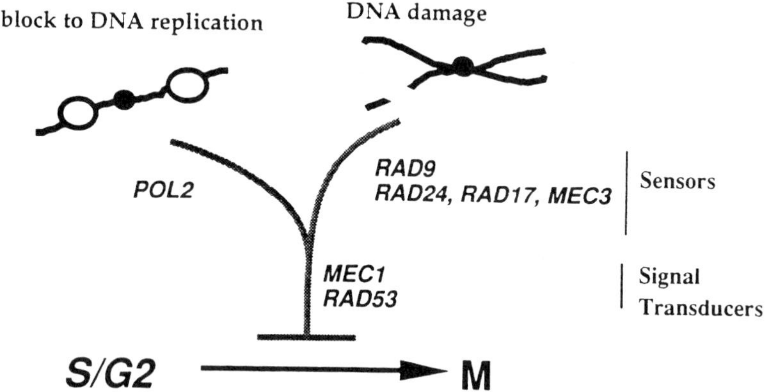

Fig 2. Sensors and signal transducers in S/M and G2/M checkpoints in budding yeast. Sensor proteins interact directly with DNA, and signal transducer proteins are activated by the sensors to prevent entry into mitosis.

POL2 gene *(71,99)*, which encodes DNA polymerase ε, essential for some feature of DNA replication, and *RFC5* and *DPB11*, which encode proteins that interact with DNA polymerases *(5,46,99)*. Mutants in *POL2* have been most extensively characterized. DNA polymerase ε appears to act only during DNA replication in the S/M checkpoint, and not in cells in with damage in the G2/M checkpoint *(71)*. What does DNA polymerase ε sense? It may act continuously during replication providing a signal that prevents mitosis. Another idea is that DNA polymerase ε senses a stalled replication fork as a form of damage and converts it to a signal that leads to arrest.

Analysis of the G2/M class of checkpoint genes predicts that they too interact with DNA directly. One gene, *RAD17*, encodes a putative 3'-5' DNA exonuclease *(60)* because of its homology with a 3'-5' DNA exonuclease, *REC1*, from *Ustilago maydis* (*105*; *see* Chapter 23). The *S. cerevisiae* Rad17p appears to degrade DNA in vivo when cells arrest at the G2/M checkpoint *(60)* with damage resulting from a defect in replication of sequences near telomeres (in *cdc13* mutants; *23*). *cdc13* mutant cells complete DNA replication at the restrictive temperature, but chromosome ends accumulate single-stranded DNA (ssDNA; *23*). The details of how Rad17p mediates degradation in *cdc13* mutant cells is unclear. One hypothesis is that a nick during lagging strand synthesis remains in *cdc13*-defective cells *(60)*. The nick is converted to a gap by degradation mediated by the products of *RAD17*, *RAD24*, and *MEC3*, the G2/M checkpoint genes that simultaneously elicit the arrest response. *RAD9* may also act on processing DNA damage, but in a different way. Degradation of DNA in *cdc13* mutants was inhibited in *RAD9* cells compared to *rad9* mutants, and double mutant analysis suggested inhibition by *RAD9* occurs through the *RAD17*-mediated degradation *(60)*. How *RAD9* might act as an inhibitor of degradation, and how that activity is linked to cell-cycle arrest and repair is also unexplained. In the model presented in this section, it is suggested that *RAD9* may slow the protein complex degrading DNA to allow detection of ssDNA-protein complex by signal transducers.

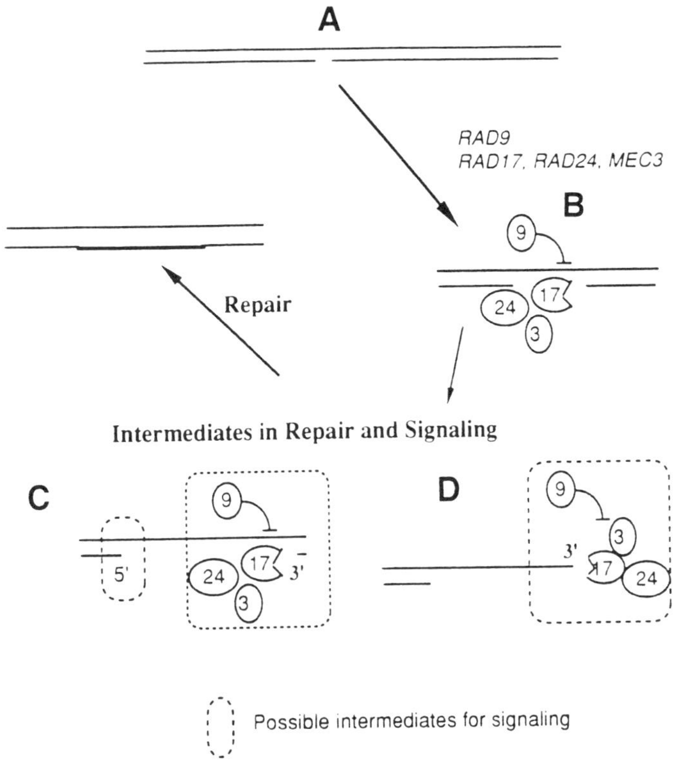

Fig. 3. Processing DNA damage to a damaged intermediate that acts in DNA repair and in signaling. An initial DNA break **(A)** is not sufficient for arrest. It is processed by checkpoint proteins (encoded by *RAD9*, *RAD24*, *RAD17*, and *MEC3* genes) by DNA degradation **(B)** to form an intermediate that signals arrest (**C** and/or **D**). (In structure **D**, only half of a DSB is shown. Rad9p may sometimes be involved in this structure; arrest after a DSB break requires Rad9p in mitotic cells, but not in meiotic prophase.) The intermediates also contribute to DNA repair. The damage signal consists of either a structure generated by degradation (small box in **C**) or the complex of proteins assembled on DNA (large box in **C** and **D**). We now favor the model of ssDNA with associated checkpoint proteins.

3.1. A Model of Checkpoint Sensor Gene Function

How might DNA degradation by the G2/M genes be linked to cell-cycle arrest and DNA repair? The following model is suggested (Fig. 3). An initial DNA break (Fig. 3A) is not sufficient for arrest; rather it must first be converted to ssDNA by degradation (Fig. 3B). The intermediate structure (Fig. 3C and/or D) contains the damage signal that leads to arrest and repair. Each of these two roles are considered in turn.

3.1.1. Checkpoint Proteins in Formation of a Damage Signal

There are at least two possible ways to explain how DNA degradation generates a signal that leads to arrest. In one view, the damage signal may be some aspect of the DNA structure, for example, double-strand/single-strand junctions formed by degradation (small box in Fig. 3C). Some DNA binding proteins do recognize specific DNA damage structures, such as Rad1p and Rad10p, in excision repair *(6)* and DNA-PK

(64). DNA-PK is a protein kinase in the PI3 kinase enzyme class to which *Mec1p* belongs; *(31,38)*. DNA-PK is activated by DNA strand junctions and not by ssDNA *per se (24,64)*. Alternatively, the damage signal may consist of a DNA structure together with associated proteins. For example, sensor proteins may assemble on ssDNA or on a dsDNA-ssDNA junction (large box, Fig. 3C,D). We now favor the hypothesis that the damage signal consists of checkpoint proteins assembled on ssDNA for reasons discussed in Section 3.2.

3.1.2. Checkpoint Proteins in DNA Repair

Evidence in budding yeast suggests checkpoint genes have a role in DNA repair as well as in cell-cycle arrest. A direct role in repair is suggested by analysis of double mutants of the G2/M genes *(60)*. Specific combinations of double mutants (e.g., *rad9 rad24*) are more sensitive to DNA damage than are the corresponding single mutants. Since each single mutant is completely defective for arrest, the increased sensitivity to DNA damage in double mutants indicates additional roles for checkpoint genes, probably in DNA repair itself. The identity of the repair pathway(s) is unknown, but it may involve recombinational repair, since ssDNA can be generated by DNA degradation, presumably by the *RAD17*-encoded exonuclease. Checkpoint gene-mediated repair may constitute a secondary pathway, because checkpoint mutants have relatively modest radiation sensitivity phenotypes *(35)*.

3.2. Different Lesions Require Different Pathways of Processing

To provide a framework for understanding the roles of putative sensor genes, a scheme is presented that correlates the genes of budding yeast required for arrest with the types of DNA damage (Fig. 4). This scheme implies that the function of many genes is specific for different types of damage, though certainly details of the damage and checkpoint protein association are unknown.

Multiple pathways (solid lines) mediate arrest from different types of damage, labeled as Lesions 1–6. Three pathways defined genetically connect specific lesions to cell-cycle arrest. The genes initially identified as checkpoint genes are grouped in distinct classes because mutants have similar phenotypes *(3,58,59,60,71,110,114)*. Along the main trunk of the diagram are the *RAD/MEC* checkpoint genes, including the *RAD9* class, the *RAD24* class (including *RAD17* and *MEC3*), and the *MEC1* class (including *RAD53*). Along the side branches are additional genes required for arrest, including *RAD1* and *PHR1* after UV damage (Lesion 4; *44*), *RAD50* during meiotic recombination (Lesion 5; *2,8,58*), and *POL2* when DNA replication is stalled (Lesion 6; *71*). For example, arrest in cells with Lesion 1 arising in *cdc13* mutants requires all three classes of checkpoint genes: the *RAD9* class, the *RAD24* class, and the *MEC1* class. Arrest in cells with Lesion 6 requires the *POL2* and *MEC1* gene classes, and does not require, for example, the *RAD9* nor *RAD24* gene classes *(3,71,114)*.

It is tempting, but incorrect, to interpret this diagram as a biochemical pathway where the genes higher in the pathway act first. Indeed, the genes above the intermediate, the putative sensors, may act before the proteins below the intermediate, the putative signal transducers. However, the relative order of function of the sensor proteins is unknown.

This model suggests that the roles of checkpoint proteins are best understood by considering the specific types of damage. The checkpoint proteins may interact with

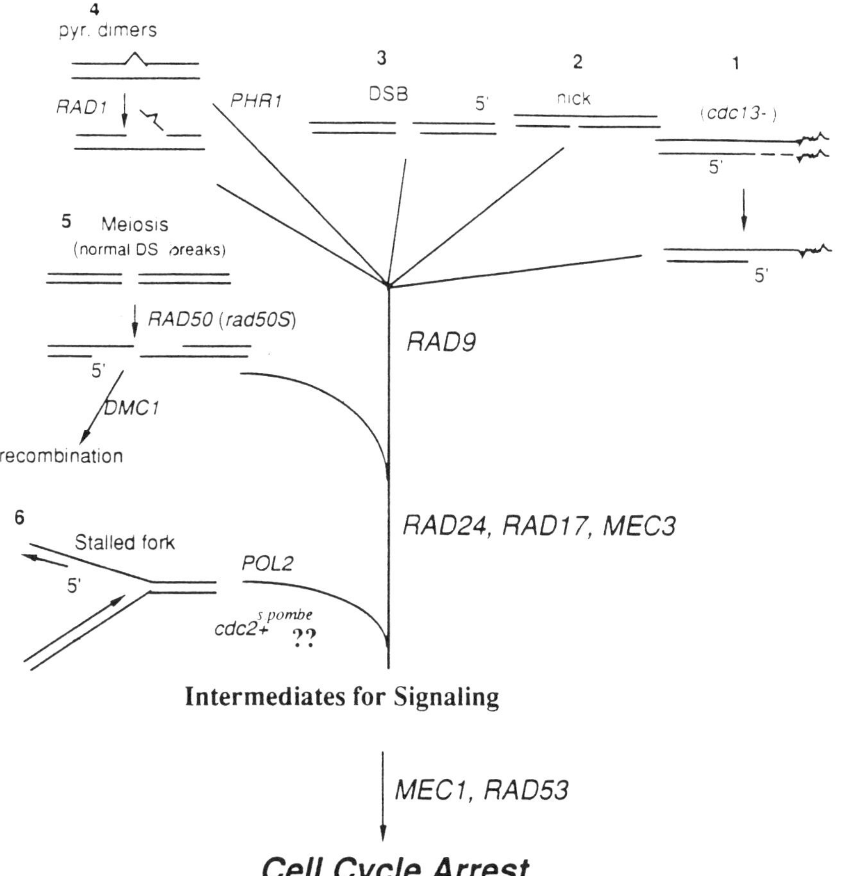

Fig. 4. The Lesion tree—processing pathways to intermediates for repair and signaling. The scheme correlates genes, DNA damage structures, and cell-cycle arrest from studies in budding yeast. Lesion 1 occurs in *cdc13*-defective cells; Lesion 2 is a nick in DNA ligase-defective cells; Lesion 3 is a DSB from HO endonuclease or X-rays; Lesion 4 is pyrimidine dimer after UV irradiation; Lesion 5 is a DSB during meiotic recombination. In *rad50S* mutants, the DSB forms, but is not degraded to ssDNA *(2,8)*. Lesion 6 is a stalled replication fork. *See text* for details. The structures of the lesions are speculative. All checkpoint genes have not yet been tested for roles in arrest after all types of DNA lesions; current data are consistent with this scheme. The role of *S. pombe cdc2*$^+$ in the S/M checkpoint is also speculative *(see text)*.

specific types of damage, to convert them into intermediates that signal arrest. By this hypothesis, *RAD9* is required for arrest after some types of damage because Rad9p is involved in processing and formation of the intermediate. *RAD9* is not required for arrest in S phase, because it does not process a stalled replication fork, a function instead carried out by DNA polymerase ε.

3.3. Evidence on Lesion Processing and Arrest

There is evidence that for three types of lesions, the damage must be processed before arrest. First, for Lesion 1 formed in *cdc13* mutant cells, dsDNA is converted to ssDNA by checkpoint proteins encoded by *RAD24*, *RAD17*, and *MEC3*, as well as by *RAD9* as discussed in Section 3.1. Second, after UV irradiation, pyrimidine dimers themselves are insufficient to cause arrest at the G2/M checkpoint *(44)*. Arrest requires either incision of dimers by *RAD1*-encoded endonuclease or repair by photolyase encoded by *PHR1* (*22*; *see* Fig. 4). How incision and photoreactivation contribute to the damage signal is unknown. *RAD1* and *PHR1*, the photolyase gene, do not encode general checkpoint proteins, because they are not required for arrest after other types of damage (e.g., Lesions 1 and 3 in Fig. 4).

A third example of processing is in a meiotic recombination checkpoint. During meiotic recombination, double strand breaks (DSBs) are normally created and repaired before chromosome segregation in meiosis I *(8)*. When recombination is blocked (in the meiotic mutant *dmc1*), DSBs accumulate, and cells arrest in meiotic prophase. The link between arrest and processing of DSBs is provided by study of two meiotic mutants. In *dmc1* mutants, DSBs are degraded to ssDNA, leaving a 3'-extension, and cells arrest. In *rad50s* mutants, the DSBs remain undegraded and cells do not arrest *(8,58)*. This implies that processing of DSBs to ssDNA may be required for arrest.

The study of the meiotic recombination checkpoint has further defined the roles of sensor checkpoint genes. The meiotic recombination checkpoint requires many of the mitotic checkpoint genes, including *RAD24*, *RAD17*, and *MEC1* (*see* Fig. 4; *58*). Formation of the ssDNA required for meiotic arrest probably requires a 5'-3' exonuclease because a 3' extension is formed from dsDNA, so in contrast to *cdc13*, damage degradation probably does not involve the 3' → 5' exonuclease encoded by *RAD17*. In fact, checkpoint genes do not have an essential role in generating the ssDNA formed during meiotic arrest *(58)*, yet the checkpoint genes, including *RAD17*, are required for arrest. These results suggest DNA degradation linked to arrest can occur independent of checkpoint gene functions. Therefore, checkpoint proteins during meiotic recombination may recognize ssDNA to form a complex needed for arrest. An economical model that accounts for observations in mitotic and meiotic cells hypothesizes that ssDNA and associated checkpoint proteins form the signal. In some cases, the formation of ssDNA involves the checkpoint proteins themselves, and in other cases, formation of ssDNA occurs by other mechanisms.

3.4. Relationship Among Degradation, Arrest, and Repair

A key to understanding the checkpoint sensor proteins that may bind directly DNA is the relationship between degradation, arrest, and repair. Analysis of the meiotic recombination checkpoint suggests that degradation may be required, yet need not be mediated by checkpoint proteins themselves. It is possible that degradation may be involved in the putative repair function of checkpoint proteins. Degradation may be linked to repair, whereas binding to ssDNA is linked to cell-cycle arrest. This hypothesis predicts that alleles of *S. cerevisiae RAD17* that are defective for exonuclease activity, but still bind ssDNA will be defective for repair and proficient for at least some types of arrest (e.g., during meiotic recombination). Alleles of certain fission yeast checkpoint genes may indeed separate repair and arrest, though the relationship to degradation in those alleles is unclear *(1,27)*.

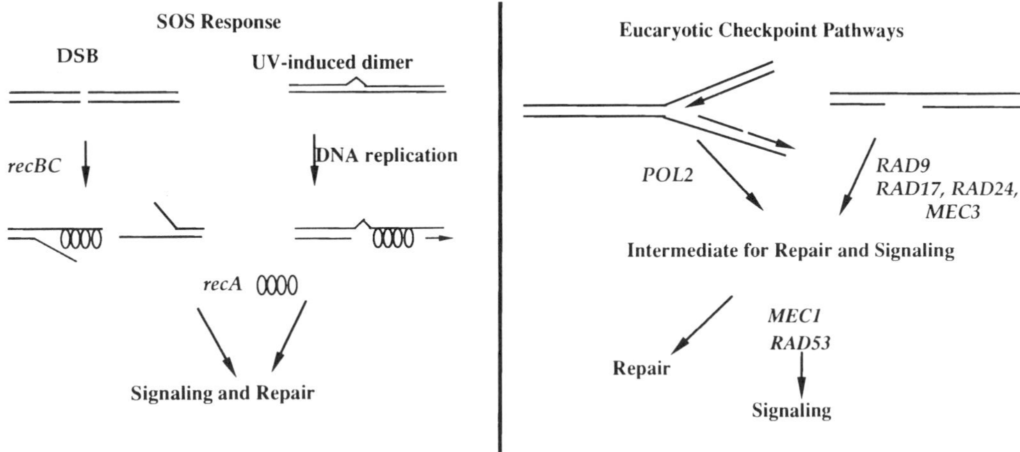

Fig. 5. Parallels in the SOS response in bacteria and DNA replication and damage checkpoints in budding yeast. SOS response *(left)*. A DSB is converted to ssDNA by RecBC. A UV dimer is converted to ssDNA by DNA replication past the dimer. RecA protein is activated by binding to ssDNA and signals downstream events.

Eukaryotic response *(right)*. A stalled replication fork or a DNA break is processed by distinct pathways to form intermediate, the damage signals, that lead to DNA repair and activation of signal transducers Mec1p and Rad53p that mediate downstream events.

3.5. Lesion Processing in the Bacterial SOS Response

Studies of the SOS response in bacteria provide precedent for understanding the roles of eukaryotic checkpoint sensor proteins in processing damage. In bacteria, DNA damage leads to changes in cell-cycle progression and in DNA repair, both mediated by RecA protein (Fig. 5; reviewed in ref. *48,108*; *see* Chapter 7). After a DSB is formed, ssDNA is generated by RecBC (either through its 5'-3' exonuclease or helicase activities; discussed in ref. *89*). ssDNA activates RecA protein on binding. After UV damage, ssDNA is formed by a different mechanism involving replication past dimers (ref. *89* and references therein). Activation of RecA leads directly to cleavage and inactivation of the LexA transcriptional repressor protein, and derepression of LexA-regulated genes, which include those involved in DNA repair and cell-cycle septation. After damage is repaired, LexA protein levels increase, cells resume septation, and DNA repair functions return to normal levels.

The SOS and the proposed eukaryotic checkpoint responses share several features (Fig. 5). First, the repair and signaling intermediates are formed by distinct processing pathways depending on the type of initial damage. In the SOS response, ssDNA is generated from a DSB by RecBC or from UV dimers by DNA replication past the dimer. In the eukaryotic response different lesions may also require different pathways. Both a DSB and UV dimer require Rad9p for arrest, whereas a stalled replication fork requires Pol2p and not Rad9p. Second, in both responses, signal transduction occurs from an intermediate formed by processing; ssDNA in the SOS response and perhaps complexes of ssDNA and checkpoint proteins in the eukaryotic response. Third, the intermediate also functions as a substrate in DNA repair. In the SOS response, RecA mediates recombinational repair, and in the eukaryotic response, the intermediate may also be repaired by recombination.

Clearly, there are differences in the two responses. It appears that in the SOS response, RecBC has no direct role in signaling after processing damage. In the eukaryotic response, checkpoint proteins may both form ssDNA and be needed to associate with ssDNA to signal.

4. PUZZLES IN UNDERSTANDING CHECKPOINT GENES CONSERVED IN BUDDING AND FISSION YEASTS

Two sets of checkpoint genes in budding and fission yeasts are homologous yet have distinctive roles in checkpoint controls. The include *S. pombe rad1⁺/S. cerevisiae RAD17 (60)* and *S. pombe cdc2⁺/S. cerevisiae CDC28*. The *S. pombe rad1⁺* and *cdc2⁺* have roles in the S/M checkpoint in fission yeast *(17–19,83)*, whereas their counterparts in budding yeast, *RAD17* and *CDC28*, do not *(4,97,113)*.

There are two plausible explanations for these differences. First, the two yeasts may have different mechanisms of cell-cycle progression into mitosis and therefore have different mechanisms of arrest. Budding and fission yeasts do have distinctive cell cycles, the subject of a long-standing, entertaining, and unresolved debate *(72,86)*. The similarity in cell-cycle controls with higher eukaryotes has even won fission yeast acclaim as a "Micro-mammal" by some (V. Simanis, personal communication). Meanwhile, budding yeast is good-naturedly scorned as somehow different and bizarre. However, the two yeasts may have different checkpoint controls because they have different mechanisms of processing DNA damage, and therefore different requirements for checkpoint proteins.

4.1. *The* S. pombe rad1⁺ *and* S. cerevisiae RAD17 *Genes*

The argument suggesting that *rad1⁺* and *RAD17* have roles in damage processing in each yeast cell type is simple: both encode putative 3' → 5' DNA exonucleases (*60*; see Chapter 23). Both are required at the G2/M checkpoint, a function that may involve their interaction with DNA as discussed for *RAD17* in Section 3. However, only *S. pombe rad1⁺* is required at the S/M checkpoint, whereas *RAD17* is not. An economical explanation is that rad1p in *S. pombe* has a role in processing (or recognition) of a stalled replication fork in fission yeast, whereas Rad17p in budding yeast does not. Perhaps in fission yeast, the structures of a stalled replication fork and DNA break are similar, and rad1p binds to and/or processes each. In budding yeast, a stalled replication fork and a DNA break may differ in some fundamental way. How stalled replication forks might differ in the two cell types is unknown.

4.2. *The* S. pombe cdc2⁺ *and* S. cerevisiae CDC28 *Genes*

A related though more enigmatic checkpoint puzzle concerns the roles of *cdc2⁺* and *CDC28*. These genes encode the catalytic subunits of the central regulatory protein kinase, $p34^{CDC2}$, that mediates both G1-S and G2-M transitions *(73)*. At issue is their additional roles in checkpoint controls. *cdc2⁺* has a role in the S/M checkpoint; activation of *cdc2⁺* through removal of inhibitory phosphorylation sites or overexpression of the activating phosphatase cdc25p *(18,19)* fails to arrest in interphase when DNA replication is blocked (an S/M checkpoint defect). Similar results were obtained when $p34^{CDC2}$ activity was manipulated in frog embryos and in mammalian cells *(14,36,107)*. Budding yeast does not fit this paradigm; *S. cerevisiae CDC28* does not appear to have a role in the S/M checkpoint, because activated alleles do not inactivate the checkpoint responses *(4,97)*.

Why is *cdc2* required for the S/M checkpoint in one cell type and not in another? A favored explanation is that *cdc2*$^+$ is the downstream target of the S/M checkpoint in all cell types, except budding yeast. This may be true, but it is notable that *cdc2*$^+$ has not been reported to play a role in the G2/M checkpoint in fission yeast (at least the G2/M checkpoint is intact in mutant cells with certain *cdc2*$^+$ alleles defective for the S/M checkpoint *[82])*.

There is a second explanation for this enigma: *cdc2*$^+$ and *CDC28* may have roles in processing initial damage much like sensor proteins (*see* Fig. 4, Lesion 6). These roles may be species-specific, depending on the nature of the lesion. For example, a stalled replication fork in fission yeast may require *cdc2*$^+$ function to form the appropriate damage signal, whereas a stalled replication fork in budding yeast forms the damage signal by a different mechanism not requiring *CDC28*. (For *cdc2*$^+$ to have a role in processing damage, either the active form of the kinase inhibits processing or the phosphorylated form is essential for processing. p34^{CDC2} has been found associated with replication complexes *in vivo* and its functions in those complexes are unknown *[10]*. Cyclin A also seems to have a role in the S/M checkpoint in frog embryos *[107]*. Perhaps cyclin A may have some role in forming the damage signal for arrest as well.) If *cdc2* $^+$ is involved in processing of damage, it may be through phosphorylation of proteins associated with damage. There is a precedent for roles for protein kinases in DNA repair *(37)* and association of a CDK protein kinase with DNA damage as well *(84)*.

5. SIGNAL TRANSDUCERS AND PROTEIN KINASE CASCADES

The damage signal, perhaps consisting of ssDNA and associated checkpoint sensor proteins, may be detected by signal transducers that then mediate downstream responses (e.g., arrest, transcriptional regulation). In budding and fission yeast, the putative signal transducers for many responses are *S. cerevisiae MEC1/S. pombe rad3*$^+$ and *S. cervisiae RAD53/S. pombe cds1*$^+$ (reviewed in refs. *12* and *61*). Rad53p and Cds1p+ are classical protein kinases *(65,98)*. Mec1p and rad3p are putative phosphotidylinositol kinases (PI3 kinase), a class of enzyme that phosphoryates either phosphotidylinositol or protein *(38)*. Other PI3 kinases in this class have roles in DNA metabolism as well, including the budding yeast *TEL1* gene involved in telomere metabolism *(26)*; *ATM*, a human disease and checkpoint control gene *(90)*, and DNA-PK, a DNA damage-activated transcriptional regulator *(25)*. *MEC1/rad3*$^+$ and *RAD53/cds1*$^+$ protein kinases may act in a kinase cascade that is activated by DNA damage. Rad53p becomes phosphorylated in vivo after activation of the S/M checkpoint. Phosphorylation requires *MEC1 (88,100)*, suggesting that *MEC1* acts upstream of *RAD53*. In fission yeast a protein kinase cascade activated by DNA damage may also exist. The chk1p kinase becomes phosphorylated after DNA damage *(109)*. Phosphorylation requires *rad3*$^+$, suggesting that *rad3*$^+$ acts upstream of *chk1*$^+$.

Inferring order of function from phosphorylation of proteins has the caveat that downstream-acting protein kinases may feed back and regulate upstream proteins. For *chk1*$^+$, the *rad3*-dependent phosphorylation changes on *chk1*$^+$ protein occurred within minutes of the damage, suggesting that feedback is not involved *(109)*, yet phosphorylation of *chk1*$^+$ required the *chk1*$^+$ protein kinase itself, suggesting phosphorylation may occur by a feedback response. Alternative roles as sensors in forming the damage signal for any of these protein kinases remain to be determined.

6. HOW CELLS ARREST IS UNKNOWN

Once the checkpoint controls are activated, how do they alter cell physiology? Little is known about the regulation by checkpoint controls of downstream events. p34^{CDK4} cyclin is involved in the G1 checkpoint in mammalian cells *(104)* (*see* Chapter 21, vol. 2) and the S/M checkpoint in some cell types (*14,18,19,36,107*; *see* discussion in Section 4.2.). One gene in budding yeast involved in transcriptional regulation after DNA damage, *DUN1*, has been identified *(119)* and may interact with *RAD53*. Recently, a gene was identified called *PDS1* that regulates the separation of sister chromatids and may be the target of the G2/M control *(116,117)*. How a cell converts DNA damage and alters cellular responses remains a black box; our current understanding of this aspect has not progressed much beyond the observation of arrest.

7. SUMMARY

Checkpoint controls in eukaryotic cells recognize damage and arrest the cell cycle, as well as regulate transcriptional induction and progression of DNA replication. Genes required for arrest have been identified in budding and fission yeasts, and consist of at least two functions, the sensors that associate with DNA and the signal transducers that mediate downstream events by phosphorylation. ssDNA may be a part of the signal. Specific sensor proteins may interact with damage to convert them into ssDNA as well as associate with the ssDNA to form an intermediate. The intermediate may be involved in both repair and in signaling arrest. A protein kinase signal transducer may recognize the ssDNA and associated proteins, become active, and mediate cellular responses through phosphorylation of substrates yet to be identified. Little is known about the downstream target proteins that directly affect cellular responses.

REFERENCES

1. Al-Khodairy, G., E. Fotou, K. S. Sheldrick, D. J. F. Griffiths, A. R. Lehmann, and A. M. Carr. 1994. Identification and characterization of new elements involved in checkpoint and feedback controls in fission yeast. *Mol. Biol. Cell* **5:** 147–160.
2. Alani, E., R. Padmore, and N. Kleckner. 1990. Analysis of wild-type and rad50 mutants of yeast suggests an intimate relationship between meiotic chromosome synapsis and recombination. *Cell* **61:** 419–436.
3. Allen, J. B., Z. Zhou, W. Siede, E. C. Friedberg, and S. J. Elledge. 1994. The SAD1/RAD53 protein kinase controls multiple checkpoints and DNA damage-induced transcription in yeast. *Genes Dev.* **8:** 2416–2428.
4. Amon, A., U. Surana, I. Muroff, and K. Nasmyth. 1992. Regulation of p34^{cdc28} tyrosine phosphorylation is not required for entry into mitosis in *S. cerevisiae. Nature* **355:** 368–371.
5. Araki, H., S. Leem, A. Phongdara, and A. Sugino. 1995. Dpb11, which interacts with DNA polymerase II(ε) in *S. cerevisiae*, has a dual role in S-phase progression and at a cell cycle checkpoint. *Proc. Natl. Acad. Sci. USA* **92:** 11791–795.
6. Bardwell, A. J., L. Bardwell, A. E. Tomkinson, and E. C. Friedberg. 1994. Specific cleavage of model recombination and repair intermediates by the yeast Rad1-Rad10 endonuclease. *Science* **265:** 2082–2085.

7. Beamish, H. and M. F. Lavin. 1994. Radiosensitivity in ataxia-telangiectasia: anomalies in radiation-induced cell cycle delay. *Int. J. Radiat. Biol.* **65:** 175–184.
8. Bishop, D. K., D. Park, L. Xu, and N. Kleckner. 1992. DMC1: a meiosis-specific yeast homolog of E. coli recA required for recombination, synaptonemal complex formation, and cell cycle progression. *Cell* **69:** 439–456.
9. Broek, D., R. Bartlett, K. Crawford, and P. Nurse. 1991. Involvement of $p34^{CDC2}$ in establishing the dependency of S phase on mitosis. *Nature* **349:** 388–393.
10. Cardoso, M. C., H. Leonhardt, and B. Nadal-Ginard. 1993. Reversal of terminal differentiation and control of DNA replication: cyclin A and Cdk2 specifically localize at subnuclear sites of DNA replication. *Cell* **74:** 979–992.
11. Carr, A. M. 1995. DNA structure checkpoints in fission yeast. *Semin. Cell. Biol.* **6:** 65–72.
12. Carr, A. M. and M. F. Hoekstra. 1995. The cellular responses to DNA damage. *Trends Cell Biol.* **5:** 32–40.
13. D'Urso, G. and P. Nurse. 1995. Checkpoints in the cell cycle of fission yeast. *Curr. Opinion Genet. Dev.* **5:** 12–16.
14. Dasso, M. and J. W. Newport. 1990. Completion of DNA replication is monitored by a feedback system that controls the initiation of mitosis in vitro: studies in *Xenopus*. *Cell* **61:** 811–823.
15. Downes, C. S., D. J. Clarke, A. M. Mullinger, J. F. Giminez-Abian, A. M. Creighton, and R. T. Johnson. 1994. A topoisomerase II-dependent G2 cycle chckpoint in mammalian cells. *Nature* **372:** 467–470.
16. Dunphy, W. G. 1994. The decision to enter mitosis. *Trends Biochem. Sci.* **4:** 202–207.
17. Enoch, T., A. M. Carr, and P. Nurse. 1992. Fission yeast genes involved in coupling mitosis to completion of DNA replication. *Genes Dev.* **6:** 2035–2046.
18. Enoch, T., K. L. Gould, and P. Nurse. 1991. Mitotic checkpoint control in fission yeast. *Cold Spring Harbor Sym. Quant. Biol.* **56:** 409–416.
19. Enoch, T. and P. Nurse. 1990. Mutation of fission yeast cell cycle control genes abolishes dependence of mitosis on DNA replication. *Cell* **60:** 665–673.
20. Evan, G. I., L. Brown, M. Whyte, and E. Harrington. 1995. Apoptosis and the cell cycle. *Curr. Biol.* **7:** 825–834.
21. Fan, S., M. L. Smith, D. J. Rivet, D. Duba, Q. Zhan, K. W. Kohn, A. J. Fornace, and P. M. O'Connor. 1995. Disruption of p53 function sensitizes breast cancer MCF-7 cells to cisplatin and pentoxifylline. *Cancer Res.* **55:** 1649–1654.
22. Friedberg, E. C., G. C. Walker and W. Siede. 1995. *DNA Repair and Mutagenesis*, 2nd ed., ASM, Washington, DC.
23. Garvik, B., M. Carson, and L. Hartwell. 1995. Single-stranded DNA arising at telomeres in cdc13 mutants may consistitue a specific signal for the RAD9 checkpoint. *Mol. Cell. Biol.* **15:** 6128–6138.
24. Gottlieb, T. M. and S. P. Jackson. 1993. The DNA-dependent protein kinase: requirement for DNA ends and association with Ku antigen. *Cell* **72:** 131–142.
25. Gottlieb, T. M. and S. P. Jackson. 1994. Protein kinases and DNA damage. *TIBS* **19:** 500–503
26. Greenwell, P. W., S. L. Kronmal, S. E. Porter, J. Gassenhuber, B. Obermaier, and T. D. Petes. 1995. *TEL1*, a gene involved in controlling telomere length in *S. cerevisiae*, is homologous to the human ataxia telangiectasia gene. *Cell* **82:** 823–829.
27. Griffiths, D. J. G., N. C. Barbet, S. McCready, A. R. Lehmann, and A. M. Carr. 1995. Fission yeast rad17: a homolog of budding yeast RAD24 that shares regions of seqeunce similarity with DNA polymerase accessory proteins. *EMBO J.* **14:** 101–112.
28. Hardwick, K. G. and A. W. Murray. 1995. Mad1p, a phosphoprotein component of the spindle assembly checkpoint in budding yeast. *J. Cell Biol.* **131:** 709–720.
29. Hari, L. K., A. Santerre, J. J. Sekelsky, K. S. McKim, J. B. Boyd and R. S. Hawley. 1995. The mei-41 gene of D. melanogaster is a structural and functional homolog of the human ataxia telangiectasia gene. *Cell* **82:** 815–821.

30. Harris, C. C. and M. Hollstein. 1993. Clinical implication of the p53 tumor-suppressor gene. *New Engl. J. Med.* **329:** 1318–1327.
31. Hartley, K. O., D. Gell, G. C. M. Smith, H. Zhang, N. Divecha, M. A. Connelly, A. Admon, S. P. Lees-Miller, C. W. Anderson and S. P. Jackson. 1995. DNA-dependent protein kinase catalytic subunit: a relative of phosphatidylinositol 3-kinase and the ataxia telangiectasia gene product. *Cell* **82:** 849–856.
32. Hartwell, L. H. and M. B. Kastan. 1994. Cell cycle control and cancer. *Science* **266:** 1821–1828.
33. Hartwell, L. H. and T. A. Weinert. 1989. Checkpoints: controls that ensure the order of cell cycle events. *Science* **246:** 629–634.
34. Hayles, J., D. Fisher, A. Woollard, and P. Nurse. 1994. Temporal order of S phase and mitosis in fission yeast is determined by the state of the $p34^{CDC2}$-mitotic B cyclin complex. *Cell* **78:** 813–822.
35. Haynes, R. H. and B. A. Kunz. 1981. DNA repair and mutagenesis in yeast. *The Molecular Biology of Yeast, Saccharomyces, Life Cycle and Inheritance.* Cold Spring Harbor Laboratory, Cold Spring Harbor, NY, pp. 371–414.
36. Heald, R., M. MeLoughlin, and F. McKeon. 1993. Human Wee1 maintains mitotic timing by protecting the nucleus from cytoplasmically activated cdc2 kinase. *Cell* **74:** 463–474.
37. Hoekstra, M. F., A. J. Demaggio, and N. Dhillon. 1991. Genetically identified protein kinases in yeast. *TIG* **7:** 256–261.
38. Hunter, T. 1995. When is a lipid kinase not a lipid kinase? When it is a protein kinase. *Cell* **83:** 1–4.
39. Hunter, T. and J. Pines. 1994. Cyclins and cancer II: cyclin D and CDK inhibitors come of age. *Cell* **79:** 573–582.
40. Jimenez, G., J. Yucel, R. Rowley, and S. Subramani. 1992. The $rad3^+$ gene of Schizosaccharomyces pombe is involved in multiple checkpoint function and in DNA repair. *Proc. Natl. Acad. Sci. USA* **89:** 4952–4956.
41. Kastan, M. B., O. Onyekwere, D. Sidransky, B. Vogelstein, and R. W. Craig. 1991. Participation of p53 protein in the cellular response to DNA damage. *Cancer Res.* **51:** 6304–6311.
42. Kato, R. and H. Ogawa. 1994. An essential gene, ESR1, is required for mitotic cell growth, DNA repair and meiotic recombination in *Saccharomyces cerevisiae. Nucleic Acids Res.* **22:** 3104–3112.
43. Kelley, T. J., G. S. Martin, S. L. Forsburg, R. J. Stephensen, A. Russo, and P. Nurse. 1993. The fission yeast $cdc18^+$ gene product couples S phase to START and mitosis. *Cell* **74:** 371–382.
44. Kiser, G., A. Admire, S. Kim, and T. Weinert. Unpublished.
45. Kiser, G. L. and T. A. Weinert. 1996. Distinct roles of yeast *MEC* and *RAD* checkpoint genes in transcriptional induction after DNA damage and implications for function. *Mol. Biol. Cell* **7:** 703–718.
46. Kornberg, A. and T. Baker. 1992. *DNA Replication.* Freeman, New York. p. 931.
47. Kornbluth, S., C. Smythe, and J. W. Newport. 1992. In vitro cell cycle arrest induced by using artificial DNA templates. *Mol. Cell. Biol.* **12:** 3216–3223.
48. Kowalczykoski, S. C., D. A. Dixon, A. K. Eggleston, S. D. Lauder, and W. M. Rehrauer. 1994. Biochemistry of homologous recombination in *Escherichia coli. Micro. Rev.* **58**: 401–465.
49. Kuerbitz, S. J., B. S. Plunkett, W. V. Walsh, and M. B. Kastan. 1992. Wild-type p53 is a cell cycle checkpoint determinant following irradiation. *Proc. Natl. Acad. Sci. USA* **89:** 7491–7495.
50. Kumagai, A. and W. G. Dunphy. 1995. Control of the Cdc2/cyclin B complex in *Xenopus* egg extracts arrested at a G2/M checkpoint with DNA synthesis inhibitors. *Mol. Biol. Cell* **6:** 199–213.

51. Lees, E. 1995. Cyclin dependent kinase regulation. *Curr. Opinion Cell. Biol.* **7:** 773–780.
52. Lehmann, A. R. and A. M. Carr. 1995. The ataxia-telangiectasia gene: a link between checkpoint controls, neurodegeneration and cancer. *TIG* **11:** 375–377.
53. Lew, D. J. and S. I. Reed. 1995. A cell cycle checkpoint monitors cell morphogenesis in budding yeast. *J. Cell. Biol.* **129:** 739–749.
54. Li, J. J. and R. J. Deshaies. 1993. Exercising self-restraint: discouraging illicit acts of S and M in eukaryotes. *Cell* **74:** 223–226.
55. Li, R. and A. W. Murray. 1991. Feedback control of mitosis in budding yeast. *Cell* **66:** 519–531.
56. Li, X. and R. B. Nicklas. 1995. Mitotic forces control a cell-cycle checkpoint. *Nature* **373:** 630–632.
57. Livingston, L. R., A. White, J. Sprouse, E. Livanos, T. Jacks, and T. D. Tisty. 1992. Altered cell cycle arrest and gene amplification potential accompany loss of wild-type p53. *Cell* **70:** 923–935.
58. Lydall, D., Y. Nikolsky, D. Bishop, and T. Weinert. 1996. A meiotic recombination checkpoint controlled by mitotic checkpoint genes. *Nature* **383:** 840–843.
59. Lydall, D. and T. Weinert. Submitted.
60. Lydall, D. and T. Weinert. 1995. Yeast checkpoint genes in DNA damage processing: implications for repair and arrest. *Science* **270:** 1488–1491.
61. Lydall, D. and T. Weinert. 1996. From DNA damage to cell cycle arrest and suicide: a budding yeast perspective. *Curr. Opinion Gen. Dev.* **6:** 4–11.
62. Minshull, J., H. Sun, N. K. Tonks, and A. W. Murray. 1994. A MAP kinase-dependent spindle assembly checkpoint in *Xenopus* egg extracts. *Cell* **79:** 475–486.
63. Moreno, S. and P. Nurse. 1994. Regulation of progression through G1 phase of the cell cycle by the rum1^{+} gene. *Nature* **367:** 236–242.
64. Morozov, V. E., M. Falzon, C. W. Anderson, and E. L. Kuff. 1994. DNA-dependent protein kinase is activated by nicks and larger single-stranded gaps. *J. Biol. Chem.* **269:** 16,684–16,688.
65. Murakami, H. and H. Okayama. 1995. A kinase from fission yeast responsible for blocking mitosis in S phase. *Nature* **374:** 817–819.
66. Murray, A. and T. Hunt. 1993. *The Cell Cycle.* W. H. Freeman, New York. p. 251.
67. Murray, A. W. 1992. Creative blocks: cell-cycle checkpoints and feedback controls. *Nature* **359:** 599–604.
68. Murray, A. W. 1995. The genetics of cell cycle checkpoints. *Curr. Opinion Genet. Dev.* **5:** 5–11.
69. Nagasawa, H., S. A. Latt, M. E. Lalande, and J. B. Little. 1985. Effects of x-irradiation on cell-cycle progression, induction of chromosomal aberrations and cell killing in ataxia-telangiectasia (AT) fibroblasts. *Mutat. Res.* **148:** 71–82.
70. Nasmyth, K. 1993. Control of the yeast cell cycle by the Cdc28 protein kinase. *Curr. Opinion Cell. Biol.* **5:** 166–179.
71. Navas, T. A., Z. Zhou, and S. J. Elledge. 1995. DNA polymerase epsilon links the DNA replication machinery to the S phase checkpoint. *Cell* **80:** 29–39.
72. Nurse, P. 1985. Cell cycle control genes in yeast. TIGS. **2:** 51–55.
73. Nurse, P. 1990. Universal control mechanism regulating onset of M-phase. *Nature* **344:** 503–508.
74. O'Connor, P. M., D. K. Ferris, I. Hoffmann, J. Jackman, G. Draetta, and G. Kohn. 1994. Role of the CDC25C phosphatase in G2 arrest induced by nitrogen mustard in human lymphoma cells. *Proc. Natl. Acad. Sci. USA* **91:** 9480–9484.
75. Onel, K., A. Koff, R. L. Bennett, P. Unrau, and W. K. Holloman. 1996. The *REC1* gene of *Ustilago maydis*, which encodes a 3' to 5' exonuclease, couples DNA repair and completion of DNA synthesis to a mitotic checkpoint. *Genetics* **143:** 165–174.

76. Papathanasiou, M. A., N. C. Kerr, J. H. Robbins, O. W. McBride, I. Alamo, S. F. Barret, I. D. Hickson, A. J. Fornace, et al. 1991. Induction by ionizing radiation of the GADD45 gene in cultured human cells: lack of mediation by protein kinase C. *Mol. Cell. Biol.* **11:** 1009–1016.
77. Paules, R. S., E. N. Levedakou, S. J. Wilson, C. L. Innes, N. Rhodes, T. D. Tlsty, D. A. Galloway, L. A. Donehower, M. A. Tainsky, W. K. Kaufmann, et al. 1995. Defective G2 checkpoint function in cells from individuals with familial cancer syndromes. *Cancer Res.* **55:** 1763–1773.
78. Paulovich, A. G. and L. H. Hartwell. 1995. A checkpoint regulates the rate of progression through S phase in *S. cerevisiae* in response to DNA damage. *Cell* **82:** 841–847.
79. Peter, M., A. Gartner, J. Horecka, G. Ammerer, and I. Herskowitz. 1993. FAR1 links the signal transduction pathway to the cell cycle machinery in yeast. *Cell* **73:** 747–760.
80. Piatti, S., C. Lengauer, and K. Nasmyth. 1995. Cdc6 is an unstable protein whose de novo synthesis in G1 is important for the onset of S phase and for preventing a "reductional" anaphase in the budding yeast *S. cerevisiae. EMBO J.* **14:** 3788–3799.
81. Powell, S. N., J. S. DeFrank, P. Connell, M. Edgan, F. Preffer, D. Dombkowski, W. Tang, S. Friend, et al. 1995. Defferential sensitivity of $p53^-$ and $p53^+$ cells to caffeine-induced radiosensitization of G2 delay. *Cancer Res.* **55:** 1643–1648.
82. Rowley, R., J. Hudson, and P. G. Young. 1992. The *wee1* protein kinase is required for radiation-induced mitotic delay. *Nature* **356:** 353–355.
83. Rowley, R., S. Subramani, and P. G. Young. 1992. Checkpoint controls in Schizosaccharomyces pombe: *rad1*. *EMBO J.* **11:** 1335–1342.
84. Roy, R., J. P. Adamczewski, T. Seroz, W. Vermeulen, J. P. Tassan, L. Schaeffer, E. A. Nigg, J. J. J. Hoeijmakers, and J. M. Egly. 1994. The MO15 cell cycle kinase is associated with the TFIIH transcription-DNA repair factor. *Cell* **79:** 1093–1101.
85. Rudolph, N. S. and S. A. Latt. 1989. Flow cytometric analysis of X-ray sensitivity in ataxia-telangiectasia. Mut.Res. **211:** 31–41.
86. Russell, P. and P. Nurse. 1986. *Schizosaccharomyces pombe* and *Saccharomyces cerevisiae*: A look at yeasts divided. *Cell* **45:** 781,782.
87. Saka, Y. and M. Yanagida. 1993. Fission yeast $cut5^+$, required for S phase onset and M phase restraint, is identical to the radiation-damage repair gene $rad4^+$. *Cell* **74:** 383–393.
88. Sanchez, Y., B. A. Desany, W. J. Jones, Q. Liu, B. Wang, and S. Elledge. 1996. Regulation of *RAD53* by the *ATM*-like kinases *MEC1* and *TEL1* in yeast cell cycle checkpoint pathways. *Science* **271:** 357–360.
89. Sassanfar, M. and J. W. Roberts. 1990. Nature of the SOS-inducing signal in *Eschericia coli. J. Mol. Biol.* **212:** 79–96.
90. Savitsky, K., A. Bar-Shira, S. Gilad, G. Rotman, Y. Ziv, L. Vanagaite, D. A. Tagle, S. Smith, T. Uziel, S. Sfez, M. Ashkenazi, and Y. Shilon. 1995. A single ataxia telangiectasia gene with a product similar to PI-3 kinase. *Science* **268:** 1749–1752.
91. Schiestl R. H, P. Reynolds, S. Prakash, and L. Prakash. 1989. Cloning and sequence analysis of the Saccharomyces cerevisiae RAD9 gene and further evidence that its product is required for cell cycle arrest induced by DNA damage. *Mol. Cell Biol.* **9**: 1882–1896.
92. Schwob, E., T. Bohm, M. D. Mendenhall, and K. Nasmyth. 1994. The B-type cyclin kinase inhibitor p40 (SIC1) controls the G1 to S transition in *S. cerevisiae. Cell* **79:** 233–44.
93. Seaton, B. L., J. Yucel, P. Sunnerhagen, and S. Subramani. 1992. Isolation and characterization of the Schizosaccharomyces pombe rad3 gene, involved in DNA damage and DNA synthesis checkpoints. *Gene* **119:** 83–89.
94. Sheldrick, K. S. and A. M. Carr. 1993. Feedback controls and G2 checkpoints: fission yeast as a model system. *BioEssays* **15:** 775–782.
95. Siede, W., A. S. Friedberg, and E. C. Freidberg. 1993. RAD9-dependent G1 arrest defines a second checkpoint for damaged DNA in the cell cycle of *Saccharomyces cerevisiae. Proc. Natl. Acad. Sci. USA* **90:** 7985–7989.

96. Sipiczki, M. 1989. Taxonomy and phylogenesis, in *Molecular Biology of the Fission Yeast*, pp. 431–452.
97. Sorger, P. K. and A. W. Murray. 1992. S-phase feedback control in budding yeast independent of tyrosine phophorylation of p34^{cdc28}. *Nature* **355:** 365–368.
98. Stern, D. F., P. Zheng, D. R. Beidler, and C. Zerillo. 1991. Spk1, a new kinase from *Saccharomyces cerevisiae*, phosphorylates proteins on serine, threonine, and tyrosine. *Mol. Cell. Biol.* **11:** 987–1001.
99. Sugimoto, K., T. Shinomura, K. Hashimoto, H. Araki, and A. Sugino. 1996. Rfc5, a subunit required for DNA replication, is involved in coupling DNA replication to mitosis. *Proc. Natl. Acad. Sci. USA* **93:** 7048–7052.
100. Sun, Z., D. S. Fay, F. Marini, M. Foiani, and D. F. Stern. 1996. Spk1/Rad53 is regulated by Mec1-dependent protein phosphorylation in DNA replication and damage checkpoint pathways. *Genes Dev.* **10:** 395–406.
101. Sunnerhagen, P., B. L. Seaton, A. Nasim, and S. Subramani. 1990. Cloning and analysis of a gene involved in DNA repair and recombination, the rad1 gene of *Schizosaccharomyces pombe*. *Mol. Cell. Biol.* **10:** 3750–3760.
102 Swift, M., D. Morrell, R. B. Massey, and C. L. Chase. 1991. Incidence of cancer in 161 families affected by ataxia telangiectasia. *New Engl. J. Med.* **325:** 1831–1836.
103. Swift, M., P. J. Reitnauer, D. Morrel, and C. L. Chase. 1987. Breast and other cancers in families with ataxia-telangiectasia. *New Engl. J. Med.* **316:** 1289–1294.
104. Terada, Y., M. Tatsuka, S. Jinno, and H. Okayama. 1995. Requirement for tyrosine phosphorylation of Cdk4 in G1 arrest induced by ultraviolet irradiation. *Nature* **376:** 358–362.
105. Thelen, M. P., K. Onel, and W. K. Holloman. 1994. The REC1 gene of Ustilago maydis involved in the cellular response to DNA damage encodes an exonuclease. *J. Biol. Chem.* **269:** 747–754.
106. Toyn, J. H., A. L. Johnson, and L. H. Johnston. 1995. Segregation of unreplicated chromosomes in *Saccharomyces cerevisiae* reveals a novel G1/M phase checkpoint. *Mol.Cell. Biol.* **15:** 5312–5321.
107. Walker, D. H. and J. L. Maller. 1991. Role of cyclin A in the dependence of mitosis on completion of DNA replication. *Nature* **354:** 314–317.
108. Walker, G. C. 1985. Inducible DNA repair systems. *Ann. Rev. Biochem.* **54:** 425–457.
109. Walworth, N. C. and R. Bernards. 1996. rad-Dependent response of the chk1-encoded protein kinase at the DNA damage checkpoint. *Science* **271:** 353–356.
110. Weinert, T. and D. Lydall. 1993. Cell cycle checkpoints, genetic instability and cancer. *Semin. Cancer Biol.* **4:** 129–140.
111. Weinert, T. A. and L. H. Hartwell. 1988. The *RAD9* gene controls the cell cycle response to DNA damage in *Saccharomyces cerevisiae*. *Science* **241:** 317–322.
112. Weinert, T. A. and L. H. Hartwell. 1990. Characterization of *RAD9* of *Saccharomyces cerevisiae* and evidence that its function acts posttranslationally in cell cycle arrest after DNA damage. *Mol. Cell. Biol.* **10:** 6554–6564.
113. Weinert, T. A. and L. H. Hartwell. 1993. Cell cycle arrest of *cdc* mutants and specificity of the *RAD9* checkpoint. *Genetics* **134:** 63–80.
114. Weinert, T. A., G. L. Kiser, and L. H. Hartwell. 1994. Mitotic checkpoint genes in budding yeast and the dependence of mitosis on DNA replication and repair. *Genes Dev.* **8:** 652–665.
115. Wood, W. B. 1992. Assembly of a complex bacteriophage in vitro. *Bioessays* **14:** 635–640.
116. Yamamoto, A., V. Guacci, and D. Koshland. 1996. Pds1p is required for faithful execution of anaphase in yeast, *Saccharomyces cerevisiae*. *J. Cell Biol.* **133:** 85–97.
117. Yamamoto, A., V. Guacci, and D. Koshland. 1996. Pds1p, an inhibitor of anaphase in budding yeast, plays a critical role in the APC and checkpoint pathways. *J. Cell. Biol.* **133:** 99–110.

118. Yin, Y., M. A. Tainsky, F. Z. Bischoff, L. C. Strong, and G. M. Wahl. 1992. Wild-type p53 restores cell cycle control and inhibits gene amplification in cells with mutant p53 alleles. *Cell* **70:** 937–948.
119. Zhou, Z. and S. J. Elledge. 1993. DUN1 encodes a protein kinase that controls the DNA damage response in yeast. *Cell* **75:** 1119–1127.
120. Longhese, M. P., R. Fraschini, P. Plevani, and G. Lucchini. 1996. Yeast *pip3/mec3* mutants fail to delay entry into S phase and to slow DNA replication in response to DNA damage, and they define a functional link between Mec1 and DNA DNnase. *Mol. Cell. Biol.* **16:** 3235–3244.
121. Siede, W., G. Nusspaumer, V. Portillo, R. Rodriguez, and E. C. Friedberg. 1996. Cloning and characterization of *Rad17*, a gene controlling cell cycle responses to DNA damage in *Saccharomyces cerevisiae. Nucleic Acids Res.* **24:** 1669–1675.
122. Paulovich, A. G., R. U. Margulies, B. Gorvik, and L. H. Hartwell. *RAD9*, *RAD17*, and *RAD24* are required for S phase regulation in *S. cerevisiae* in response to DNA damage. *Genetics*. (in press).

18

Regulatory Networks That Control DNA Damage-Inducible Genes in *Saccharomyces cerevisiae*

Jeffrey B. Bachant and Stephen J. Elledge

1. INTRODUCTION

All organisms are exposed to physical and chemical agents that cause genomic injury, either by modification of DNA or by interfering with DNA synthesis. In response to DNA damage, both prokaryotic and eukaryotic cells activate stress responses that result in specific alterations in patterns of gene expression and an active inhibition of cell division. In mammalian cells, the response to DNA damage appears to contribute to overall genomic stability, since mutations in genes controlling this response, such as the transcription factor *p53* and ataxia telangiectasia mutated *(ATM)*, promote genomic instability and cause an increased susceptibility to cancer *(41,59,93)*. These genes are likely to be components of larger genetic programs that can detect DNA damage and control appropriate cellular responses. However, the nature of this molecular circuitry is largely undefined.

The SOS regulatory network of *Escherichia coli* is a stress response that is understood in considerable mechanistic detail and provides a useful framework for considering the eukaryotic response to DNA damage (*111*; Chapter 7). In this organism, exposure to a variety of DNA-damaging agents induces the expression of approx 20 genes through a common mode of transcriptional derepression controlled by the *recA* and *lexA* gene products. As a consequence of DNA damage or interference with DNA synthesis, single-stranded DNA accumulates and activates RecA molecules. Activated RecA potentiates the autoproteolytic cleavage of the LexA repressor that binds within the operators of SOS-inducible genes; as a result, these genes are expressed at higher levels. The SOS-inducible genes mediate a number of functions related to DNA metabolism, DNA repair, and cell septation. Mutations that block the inducibility of these genes make cells sensitive to DNA-damaging treatments. Thus, in the SOS stress response, diverse forms of DNA damage result in the generation of a common damage signal that activates a concerted program of gene expression. Accumulation of DNA damage-inducible gene products then promotes cell viability by creating a state that facilitates DNA repair processes.

The focus of this chapter is the transcriptional response to DNA damage in the budding yeast *Saccharomyces cerevisiae*. The powerful genetic, molecular, and biochemi-

From: DNA Damage and Repair, Vol.1: DNA Repair in Prokaryotes and Lower Eukaryotes
Edited by: J. A. Nickoloff and M. F. Hoekstra © *Humana Press Inc., Totowa, NJ*

cal techniques available in yeast have already provided insights into the eukaryotic response to DNA damage, and it may be anticipated that these technical advantages will ultimately allow an understanding of the yeast DNA damage response comparable to that of *E. coli*. Progress in understanding the yeast transcriptional response to DNA damage has come principally from three areas. First, the transcripts of a number of yeast genes have been demonstrated to accumulate after exposure to DNA-damaging agents, and at least some of these function directly in DNA repair. Second, the biochemistry of DNA repair in yeast is becoming increasingly well defined *(37)*, and this information allows the physiological significance of the transcriptional response to be examined in greater detail. Finally, regulatory genes that control the transcriptional response to DNA damage have recently been identified. Interestingly, some of these genes also function in cell-cycle checkpoint controls that couple cell-cycle progression to the status of genome duplication and integrity. Several checkpoints that respond to DNA damage have been defined in *S. cerevisiae* (*24,42*; Chapter 17). Interfering with DNA replication activates or prolongs the activation of an S-phase checkpoint that prevents entry into mitosis with unreplicated DNA. DNA lesions encountered outside of S phase activate additional checkpoint pathways that delay cell-cycle progression prior to the G1/S and G2/M transitions, presumably to prevent the replication or segregation of damaged chromosomes. The first part of this chapter surveys the known DNA damage-inducible genes in *S. cerevisiae*. Subsequent sections are concerned with the organization of the signal transduction systems that control transcriptional induction and the regulatory *cis*-acting promotor sequences and *trans*-acting factors that mediate this response. The coordinate regulation of DNA damage checkpoints and the transcriptional response to DNA damage is also emphasized.

2. DNA DAMAGE INDUCIBLE GENES IN *S. CEREVISIAE*

Several studies have demonstrated that some DNA repair capacities in *S. cerevisiae* have inducible components that require protein synthesis, suggesting that DNA damage activates the expression of genes involved in DNA repair *(8,25,34,95,96)*. Consistent with this idea, 15 genes with known or putative functions in DNA repair processes have been shown to be transcriptionally induced in response to DNA-damaging agents (Table 1). These genes can be divided into two classes, one that functions directly in the repair of damaged DNA, and another that functions primarily in nucleotide metabolism and DNA synthesis. DNA damage-inducible genes have also been identified using genetic approaches, revealing a different set of genes whose role, if any, in processing DNA lesions is unclear *(69,83)*. This section surveys the known DNA damage-inducible genes in *S. cerevisiae* and considers whether the induction of these genes is coordinately regulated by a system that promotes the cellular capacity for DNA repair.

2.1. DNA Damage Inducible Genes that Function in DNA Repair

A large number of yeast radiation-sensitive *(rad)* mutants have been isolated that are hypersensitive to UV or ionizing radiation. Based on the sensitivity of double mutant combinations, the *rad* mutants have been divided into radiation complementation groups that are thought to define three largely nonoverlapping pathways for repair of DNA damage (*37*; Chapters 15 and 16). These complementation groups include *RAD* genes that function in nucleotide excision repair (NER; *RAD3* epistasis group), double-strand break (DSB) or recombinational repair (*RAD52* epistasis group), and less-well-defined processes associated with postreplication repair and spontaneous and damage-induced mutagenesis (*RAD6* epistasis group). Analysis of the transcriptional regulation of many of these genes has shown that some members of all three pathways are induced by DNA-damaging agents.

Within the NER genes, the expression of *RAD2*, *RAD7*, and *RAD23* is inducible approximately fivefold by UV irradiation (Table 1; *54,64,66,82*). In contrast, no induction of the NER genes *RAD1*, *RAD3*, *RAD4*, or *RAD10* was observed under these conditions *(36,82)*. *RAD2* encodes an endonuclease required for the incision step of NER, whereas *RAD7* and *RAD23* appear to fulfill accessory or regulatory roles in NER biochemistry (*39,79*; Chapter 15). The response of *RAD2* to DNA damage has been studied in some detail. In addition to UV irradiation, *RAD2* is induced in response to other DNA-damaging agents, many of which result in lesions that are not processed by the NER machinery *(64,82,97)*. The inducibility of *RAD2* may be increased at certain points in the cell cycle, since *RAD2* induction was reduced in both G1-arrested and stationary-phase cells *(97)*. Furthermore, *RAD2* transcript accumulation apparently does not require *de novo* protein synthesis; in fact, cycloheximide itself can elicit transcript accumulation *(97)*.

Within the *RAD52* epistasis group, *RAD51*, which encodes a *recA* homolog, and *RAD54*, are DNA damage-inducible. Both these genes function in general recombination and recombinational repair (*7,33,94*; Chapter 16). The expression of these genes is induced approx 5- to 10-fold by agents that cause DSBs, including X-irradiation, the alkylating agent methyl methane sulfonate (MMS), and in vivo activation of the *Eco*RI restriction endonuclease (Table 1; *7,15,94*). Like *RAD2*, *RAD54* is also induced by agents to which *RAD54* mutants are not sensitive, such as UV radiation. It has been postulated that the promiscuous inducibility of these genes occurs either because the sensory/signaling system(s) for inducible NER and recombinational repair genes recognize several forms of DNA damage, or that different forms of DNA damage are converted into a discrete form or intermediate that acts as a damage signal (*15*; *see* Section 3.). However, the response of *RAD2* and the inducible recombinational repair genes does not appear to be completely equivalent, since both *RAD51* and *RAD54* are still inducible in G1-arrested cells, whereas *RAD2* induction is substantially reduced under these conditions (*7,15,97*). DNA damage inducibility seems to be restricted to a subset of genes involved in recombinational repair, since other genes within this epistasis group, such as *RAD52*, do not appear to be inducible *(15)*. Among the genes within the *RAD6* epistasis group, *RAD6* and *RAD18* have been shown to be induced several fold in response to UV irradiation (Table 1; *52 67*). *RAD6* encodes an E2 ubiquitin conjugating enzyme that is required for postreplication repair in response to UV irradiation *(49)*. *RAD18* encodes a protein with DNA binding domains that is capable of binding single-stranded DNA in vitro *(5,11,53)*. Coimmunoprecipitation experiments suggest that *RAD6* and *RAD18* physically interact, indicating that these genes may be coordinately regulated in response to DNA damage *(5)*. The damage inducibility of other genes in the *RAD6* epistasis group has not been examined directly.

In addition to the genes described, two other genes that function directly in DNA repair, *PHR1* and *MAG*, are induced in response to DNA damage (Table 1). *PHR1* encodes the yeast photoreactivating enzyme that catalyzes the light-dependent repair

Table 1
Genes Induced by DNA-Damaging Agents in *Saccharomyces cerevisiae*

Gene	Function	Inducing agents[a]	Refs.
DNA damage-inducible genes required for DNA repair			
RAD2	Encodes a single-stranded endonuclease required for the incision step of nucleotide excison repair	UV irradiation (5X); NQO (3.4X); γ-irradiation (4.9X); nalidixic acid (3.6X); bleomycin (2.5X); nitrogen mustard (2.4X); mitomycin C (2.3X); MMS (1.9X); methotrexate (2.0X); Cyclohexamide; also induced duringsporulation (8X)	*39,64,65, 82,97*
RAD7	Accessory role in NER; required for repair of transcriptionally inactive DNA	UV irradiation (6X); also induced during sporulation (15X)	*54,79*
RAD23	Accessory role in NER; regions of homology to ubiquitin; can functionally substitute for RAD7 in some circumstances	UV irradiation (5X); also induced during sporulation (6X)	*66,79*
RAD6	Required for postreplication repair, induced mutagenesis, and sporulation; encodes a ubiquitin-conjugating enzyme (E2)	UV irradiation; also induced during sporulation	*49,67*
RAD18	Required for postreplication repair; Rad18p physically associates with Rad6p and binds single-stranded DNA in vitro	UV irradiation; also induced during sporulation (4X)	*5,11, 52,53*
RAD51	Functions in general recombination and recombinational repair; encodes yeast RecA homolog	X-ray irradiation (3X); MMS (4X); also induced during sporulation (30X)	*7,94*
RAD54	Encodes a putative helicase required for recombinational repair and in meiotic recombination	X-ray irradiation (5.8X); UV irradation (4.2X); MMS (12X); *Eco*RI K(3.1X); also induced during sporulation (10–15X)	*15,16,32*
PHR1	Encodes yeast photolysase; required for light-dependent repair of pyrimidine dimers; stimulates NER in dark	UV irradiation (13.7X); NQO (9X); MMS (7.2X); MNNG (10.8X); bleomycin (2.3X)	*86,90*
MAG	Encodes a 3-methyladenine DNA glycosylase specifically required for repair of alkylated DNA lesions	MMS; MNNG; UV irradiation; NQO	*12–14*
DNA damage-inducible genes required for DNA synthesis			
CDC8	Thymidylate kinase	NQO (2.5X)	*26*
CDC9	DNA ligase	UV irradiation (8X)	*50,80*
CDC17/ POL1	DNA polymerase I catalytic subunit	UV irradiation (20X); also induced during sporulation (20X)	*51*

(continued)

Table 1 ***(continued)***

Gene	Function	Inducing agents[a]	Refs.
RNR1	Encodes large subunit of ribonucleotide reductase	NQO (3-5X); MMS; HU	*29*
RNR3/ DIN1	Encodes alternative large subunit of ribonucleotide reductase	NQO (50–100X); MMS (50–100X); UV irradiation (50–200X); HU; methotrexate; γ-irradiation	*29,83,117*
Genes induced by both DNA-damaging agents and thermal stress			
UBI4	Polyubiquitin; functions in protein degradation	NQO (20X); MNNG; HU; heat shock; also increased during sporulation (3–5X)	*35,108*
DDRA2	Unknown	UV irradiation; NQO; MMS; heat shock	*69,70*
DDR49	Function unknown; encodes a hydrophilic protein with ATP/GTP hydrolysis activity	UV irradiation; NQO; heat shock	*69,70 92,109*
Other genes induced by DNA-damaging agents			
DIN2,3,4,6	Unknown	UV irradiation; NQO; MMS; MNNG; γ-irradiation; methotrexate	*83*
DIN5	Unknown	UV irradiation; NQO; methotrexate independent of DNA damage signal	*83*
HIS3	Histidine metabolism	UV irradiation, induction thought to be independent of DNA damage signal	*33*
HIS4	Histidine metabolism	UV irradiation; induction thought to be independent of DNA damage signal	*33*

[a]Where available, maximum reported values for fold induction over basal levels in asynchronous populations are provided. In most cases, these data have been obtained either from Northern blot analysis or from studies with *lacZ* reporter constructs. The time of induction and concentration or intensity of inducing agents have not been standardized between experiments with different damage-inducible genes.

of pyrimidine dimers in DNA and stimulates NER activity in the dark *(86)*. *PHR1* is specific for the repair of pyrimidine dimers, but is induced by a variety of DNA-damaging agents, including agents that chemically modify DNA and agents that cause DSBs *(90)*. *MAG* encodes a 3'-methyladenosine DNA glycosylase specifically involved in the repair of alkylated bases *(12)*. Like *PHR1*, *MAG* is induced in response to damaging agents, such as UV irradiation, that create lesions not directly repaired by glycosylase activity *(13,14)*. *MAG* is inducible in mating pheromone arrested cells, and transcript accumulation in response to UV irradiation requires ongoing translation *(14)*.

With the exception of *RAD51*, which exhibits increased transcript abundance during late G1/early S phase *(7,94)*, all the damage-inducible genes in this class are transcribed at relatively low constitutive levels throughout the mitotic cell cycle. During meiosis, however, all these genes show elevated transcript levels (Table 1; *16,52,65,66 67*). The available evidence indicates that the induction of these genes in response to DNA-damaging agents and during meiosis is likely to be mechanistically distinct. Many yeast genes also display a similar pattern of transcript accumulation during meiosis, suggesting this increase is a general phenomenon that is not specific for damage-inducible genes *(55)*. Furthermore, deletion analysis of the *RAD2* and *RAD54* promotors indicates that transcriptional induction during meiosis and in response to DNA damage are controlled by different promotor elements (*see* Section 5.;*17,97*). It therefore appears that induction of these genes in response to DNA-damaging agents is a distinct regulatory response.

2.2. DNA Damage-Inducible Genes that Function in DNA Synthesis and Nucleotide Biosynthesis

A second class of DNA damage-inducible genes functions in DNA replication and in the production of deoxyribonucleotides required for DNA synthesis (Table 1). Since some DNA repair mechanisms require components of the DNA replication machinery, it seems likely that some of these genes may also function in damage repair processes. This class includes *CDC17/POL1*, which encodes the catalytic subunit of DNA polymerase α, *CDC8*, which encodes thymidylate kinase, *CDC9*, which encodes DNA ligase, and *RNR1*, *RNR2*, and *RNR3*, which encode subunits of ribonucleotide reductase *(6,26,29,47,51,80,117)*.

Within this class, the damage inducibility of the *RNR* genes has received the most attention. Ribonucleotide reductase catalyzes the first and rate-limiting step in the production of deoxyribonucleotides, and the genes that encode the subunits of the enzyme are inducible by DNA-damaging agents in all organisms examined *(30,31)*. In *S. cerevisiae, RNR2* encodes the small catalytic subunit, whereas *RNR1* and *RNR3* encode alternative forms of the large subunit necessary for both allosteric regulation of the enzyme and for catalytic activity *(26,29,47,117)*. The amount of *RNR1* and *RNR2* transcripts is increased in response to a variety of DNA-damaging agents, including UV irradiation, UV mimetic compounds, bleomycin, and MMS, which is an especially potent inducer *(26,29,47,48)*. Both genes are also induced by hydroxyurea (HU), an inhibitor of ribonucleotide reductase. Treatment of cells with HU results in depletion of deoxyribonucleotide pools and an accumulation of cells in S phase. *RNR2* mutants show hypersensitivity not only to HU, but also to MMS, indicating ribonucleotide reductase activity may also be required for some process associated with DNA repair *(26)*.

RNR1 and *RNR2* are essential genes required for mitotic growth *(26,29)*. In contrast, *RNR3* is not essential, but it does encode a functional protein since it can substitute for *RNR1* when overexpressed *(29,117)*. *RNR3* is normally expressed at very low levels. However, in response to DNA damage or HU-mediated replication blocks, *RNR3* transcripts increase approx 50–100-fold. The *RNR3* gene presumably provides a selective advantage to yeast, since both its ability to function and its regulatory mechanisms have been conserved; however, the nature of this advantage is not understood. It is possible that ribonucleotide reductase containing Rnr3p exhibits regulatory functions or biosynthetic capacities that allow cells to withstand certain environmental conditions or periodic episodes of DNA damage. For example, fungicidal compounds frequently target key regulatory enzymes, such as ribonucleotide reductase, and the evolution of isozyme forms that provide resistance to these compounds may have resulted from an amplification of the *RNR* gene family *(30)*.

With the possible exception of *RNR3*, the expression of each gene in this class is cell-cycle-regulated, with members displaying an increase in transcript accumulation as cells traverse the G1/S boundary *(6,26,29,51,80,114)*. In some cases, it has been shown that the damage inducibility of these genes is a direct response to DNA-damaging agents *per se* and does not result from cell-cycle synchronization caused by DNA damage or replication blocks. For example, *CDC9* is still inducible by UV radiation when cells are arrested in stationary phase by starvation *(50)*. Furthermore, the amount of transcript accumulation observed for *RAD51* in response to X-irradiation is comparable between G1-arrested cells and asynchronous populations *(7)*. With the exception of *RNR2*, all these cell-cycle-regulated, DNA damage-inducible genes contain *Mlu*I cell cycle boxes in their promotor regions, suggesting that cell-cycle-regulated expression is mediated by the *SWI4/MBP1* pathway that controls the expression of other genes involved in DNA synthesis that are not induced by DNA damage *(58)*. The role of this pathway in mediating damage-inducible gene expression has not been examined directly. However, in at least one case, the promotor elements that are responsible for regulated cell-cycle expression of *CDC9* appear to be distinct from those that control the response to DNA damage *(115)*.

2.3. *Additional DNA Damage-Inducible Genes*

Two types of genetic screens have been employed to identify DNA damage-inducible genes directly. Six *DIN*, or *d*amage *in*ducible, genes were identified by screening libraries of random genomic/*lacZ* fusions for those that displayed β-galactosidase activity specifically in response to DNA damage (Table 1; *83*). With the exception of *DIN5*, all the *DIN* genes are induced by UV irradiation, γ-irradiation, methotrexate, and the alkylating agents MMS and *N*-methyl-*N*'-nitro-*N*-nitrosoguanidine (MNNG); *DIN5* appears to be differentially regulated, since it is induced by UV and thymidine starvation, but not by other damaging agents. *DIN1* is identical to *RNR3*, but the identities of the other *DIN* genes and their involvement in DNA repair are unknown *(117)*. Based on the frequency with which damage-inducible fusions were obtained in this screen, it was estimated that approx 80 genes in *S. cerevisiae* should be damage-inducible. About 20 genes are currently known to be induced in response to DNA damage (Table 1).

DNA damage-inducible genes have also been identified using differential hybridization techniques to isolate DNA damage regulation *(DDR)* genes that displayed increased transcript levels after treatment with UV irradiation or the UV mimetic agent 4-nitroquinoine-1-oxide (NQO) *(69)*. This screen resulted in the identification of two unique transcripts, *DDRA2* and *DDR48*, whose functions are currently unknown (Table 1). An open reading frame in the *DDR48* transcript has been identified that is predicted to encode an extremely hydrophilic approx 45 kDa protein *(109)*. Ddr48p is glycosylated and capable of hydrolyzing both ATP and GTP *(92)*. Disruption of *DDR48* increases spontaneous mutation rates 6- to 14-fold, but does not markedly increase sensitivity to DNA-damaging agents *(109)*.

Unlike the DNA damage-inducible genes discussed so far, *DDRA2* and *DDR48* are inducible by both DNA-damaging agents and heat shock, as is the polyubiquitin gene *UBI4* (Table 1; *35,70,108*). *UBI4* is required for protein degradation and resistance to starvation conditions, but is not known to play a role in the repair of DNA damage *(35)*. As discussed in Section 4.2., *DDR48* and *UBI4* are not regulated by the same genetic circuit known to control the DNA damage inducibility of the *RNR* genes *(3,119)*. Thus, *DDRA2*, *DDR48*, and *UBI4* may define a distinct subset of inducible genes. Whether these genes are dually regulated by thermal stress and DNA damage or whether they are coordinately regulated in response to common damage signals produced by both heat shock and DNA-damaging agents is currently unclear. The expression of the histidine metabolism genes *HIS3* and *HIS4* has recently been shown to be induced by UV irradiation, and in this case, the transcriptional response is mediated by a pathway that probably does not respond to a direct DNA damage signal (Table 1; *33*; *see* Section 3.1.).

2.4. Physiological Relevance of DNA Damage Induced Gene Expression

Although there are probably additional DNA damage-inducible genes that have yet to be described, it is clear that only a limited number of DNA repair genes are actually induced by DNA-damaging treatments. What is the significance of inducing a particular set of genes involved in DNA metabolism and repair? One function of the damage response may be to provide an inducible source of gene products necessary for DNA replication. Since the expression of many DNA damage-inducible genes involved in DNA replication is normally confined to a narrow temporal window at the beginning of S phase, damage inducibility may facilitate DNA repair synthesis during other stages in the cell cycle. In the case of the *RNR* genes, transcriptional induction by damage may provide a source of deoxyribonucleotides for repair synthesis *(27)*. In addition, DNA-damaging treatments can also modify unincorporated nucleotides, and it has been proposed that one function of *RNR* induction may be to prevent the incorporation of damaged bases into DNA by greatly increasing the availability of newly synthesized dNTPs *(27)*.

Other genes induced by DNA damage may encode functions that are rate-limiting for repair, as hypothesized for *RAD2 (98)*, or regulate the accessibility of damaged DNA to repair enzymes. In this regard, the induction of *RAD6* and *RAD18* is of particular interest. Based on the finding that Rad6p and Rad18p associate in vivo, it has been proposed that the DNA binding activity of Rad18p may direct Rad6p to single-stranded regions of DNA that accumulate when DNA polymerases are stalled on UV-damaged chromosomal templates *(5)*. Once targeted to these lesions, the ubiquitin conjugation activity of Rad6p may initiate the degradation or modification of polymerase subunits at these sites, permitting greater access to the postreplicational repair machinery. A similar mechanism may underlie the damage inducibility of *RAD7*. It has been shown that *rad7* and *rad16* mutants are defective in the removal of UV-induced pyrimidine dimers from the transcriptionally inactive *HML*α mating type cassette, whereas the removal of dimers from the transcriptionally active *MAT* locus is unaffected *(106,107)*. Deletion of *SIR3* was able to partially restore *HML*α dimer removal in *rad7* mutants *(106)*; furthermore, Rad7p and Sir3p have recently been demonstrated to interact physically in vitro *(77)*. *SIR3* is required for the transcriptional silencing of the cryptic mating type loci, and it is thought to mediate this function by packaging DNA into a heterochromatic conformation *(60)*. Based on these findings, *rad7* may function to alter the structure of transcriptionally inactive chromatin to allow access and/or assembly of the "repairosome" complex, which contains many of the NER components required for damage recognition and excision repair *(77)*. This putative function of *RAD7* may not be limited to transcriptionally silenced regions in the genome, since *RAD7* and *RAD16* mutants are also defective in the repair of DNA damage in the nontranscribed strand of a transcriptionally active gene *(110)*.

In a few cases, the physiological significance of the damage induciblity of particular genes has been examined through the use of promotor mutations that eliminate damage inducibility. Promotor deletions that render *RAD54* damage uninducible had no effect on the short-term growth or survival relative to wild-type strains in response to MMS or X-rays *(17)*. Similarly, asynchronous cell populations harboring a damage-uninducible *RAD2* gene did not show an enhanced sensitivity to UV irradiation *(97)*. However, the situation for damage-uniducible *RAD2* alleles is more complex. Deletion of promotor elements that eliminated inducibility, but did not affect constitutive expression, resulted in an increased sensitivity to UV irradiation during either G1/S or G2/M *(98)*. However, deletions that simultaneously eliminated inducibility and reduced constitutive expression only resulted in sensitivity during G1/S. These observations are important, since they represent the first example of a cellular phenotype associated with defective inducibility of a yeast DNA damage-inducible gene. However, interpretation of this phenotype is complicated by the fact that no simple correlation between UV sensitivity and *RAD2* expression levels was observed in these experiments. One possible explanation for this apparent paradox may be that reducing constitutive *RAD2* expression levels (and possibly also the overall efficacy of NER) may activate alternative repair pathways or checkpoint mechanisms that facilitate repair specifically in G2 *(98)*. These mechanisms may not be activated by *RAD2* promotor deletions that only affect inducibility. If this type of regulatory redundancy exists, it may be difficult to assess the significance of the damage inducibility of a single damage-inducible gene.

3. REGULATORY PATHWAYS THAT CONTROL DNA DAMAGE-INDUCIBLE GENES: SIGNALS AND SENSORS

The transcriptional response to DNA damage implies the existence of signal transduction pathways that detect DNA damage and convert damage signals into specific increases in the level of gene expression. In general, pathways controlling the response to DNA damage are likely to consist of three types of components: sensors capable of detecting DNA damage signals, proteins that interact with sensor elements to transduce the damage signal, and effectors, which directly influence levels of gene expression. The remainder of the chapter is organized around this theme, and this section discusses the nature of the signal or signals generated by DNA-damaging agents and their detection. At least two distinct sensory pathways are known to control transcriptional responses to DNA-damaging agents, one that responds directly to DNA lesions and/or replication blocks and a second pathway induced by UV irradiation that is independent of a nuclear DNA damage signal.

3.1. DNA-Damaging Agents Can Generate Stress Signals that Are Unrelated to DNA Damage

As discussed in the previous section, a large number of DNA-damaging agents elicit transcription induction, and these cause a variety of different DNA lesions or blocks to DNA replication. In addition to causing DNA damage, many of these damaging agents can modify other cellular components, including proteins, lipids, RNA, nucleotides, and their precursors. Therefore, it is important to determine whether cells sense DNA damage directly or whether some other effect of damaging agents constitutes the actual signal for the transcriptional response. Recent evidence indicates that UV irradiation activates multiple pathways, including a pathway known as the UV response that is independent of a direct DNA damage signal *(21)*. In mammalian cells, the UV response consists of an early component that is independent of protein synthesis and a late component that requires ongoing translation (*44*; Chapter 15, vol. 2). The early phase of the UV response is executed, at least in part, by phosphorylation of the AP-1 and NF-κB transcription factors on sites that potentiate transcriptional activation *(19,102)*. Phosphorylation of AP-1 in response to UV irradiation is mediated through a pathway involving growth factor receptors, src-family tyrosine kinases, Ras, and other factors involved in cytoplasmic signal transduction pathways initiated at the plasma membrane *(20,81,84)*. Activation of NF-κB in response to UV can occur in enucleated cells *(21)*, indicating that the signal for induction of the UV response in mammalian cells is not dependent on a DNA damage signal generated within the nucleus *(20,21)*. Pretreatment of HeLa cells with the antioxidant *N*-acetylcysteine prevents src kinase activation and increased transcription after irradiation *(20)*. This result suggests oxidative stress at the plasma membrane may be the signal that elicits activation of this pathway.

The overall features of this response to UV irradiation are conserved in *S. cerevisiae*. Transcription of *HIS3* and *HIS4* is increased following UV irradiation, and it has been shown that this response is dependent on a signaling pathway that includes Ras and the transcription factor *GCN4*, which is analogous to AP-1 *(33)*. *GCN4* is also responsible for the transcriptional induction of these genes in response to amino acid or purine starvation *(45)*. Activation of *GCN4* during the starvation response is mediated by phosphorylation of translation initiation factor 2α by the Gcn2p protein kinase, resulting in ribosomal readthrough of upstream start codons and increased *GCN4* translation *(22)*. Induction of *HIS3* and *HIS4* after UV irradiation is also mediated by increased *GCN4* translation, but the mechanism of translational regulation in this case appears to be distinct, since it does not depend upon Gcn2p phosphorylation *(33)*. It is not known whether other damage-inducible genes are controlled by this Ras-dependent pathway. However, this pathway does appear to have a protective function in yeast, since mutants that constitutively activate Ras show an approx 3.5-fold increase in UV resistance compared to wild-type cells, whereas *gcn4* mutant strains are approx fivefold more sensitive *(33)*.

The induction of *UBI4*, *DDRA2*, and *DDR48* by both DNA-damaging agents and heat shock suggests these genes may also be regulated by a pathway that does not respond directly to DNA damage. In particular, the role of *UBI4* in protein degradation and resistance to heat shock suggests that induction of this set of genes may be elicited by a signal generated by aberrant proteins rather than lesions in DNA *(108)*. *UBI4* induction would provide a logical genetic end point for a pathway activated by damaged proteins. Furthermore, recent experiments have shown that the response of *UBI4* and the *DDR* genes to DNA-damaging agents is controlled by a different set of regulatory components than other DNA damage-inducible genes (*3,119*; *see* Section 4.2.). However, the induction of *DDRA2* and *DDR48* in response to NQO and MNNG is altered or eliminated in *rad3*, *rad6*, and *rad52* mutants, whereas induction after heat shock is unaffected (68). These results indicate that the cellular capacity for DNA repair can differentially influence the induction of the *DDRA2* and *DDR48* genes. Therefore, it is also possible that the response of this subset of damage-inducible genes to DNA-damaging agents is mediated by potentially overlapping signaling pathways that are activated by both thermal stress, and by DNA damage or damage repair processes. It remains to be seen whether the induction of the *UBI4* or *DDR* genes is regulated by the Ras-dependent UV response pathway.

3.2. The Role of Replication Blocks and DNA Polymerases in the Damage Inducibility of the RNR Genes

In contrast to the examples cited above, studies of the damage inducibility of the *RNR* genes have provided evidence that some damage-inducible genes do respond directly to signals generated by DNA damage. Analysis of *RNR* inducibility by DNA damage is complicated by the fact that ribonucleotide reductase is also regulated in response to nucleotide depletion. Nucleotide depletion by DNA repair processes is therefore a plausible mechanism to account for *RNR* inducibility *(27)*. In the case of *RNR2*, some evidence indicates that transcriptional induction is not activated solely in response to nucleotide availability. *rad4* mutants are defective in the incision step of NER, and are thus unable to deplete nucleotide pools through repair synthesis. The finding that *RNR2* shows increased inducibility at low concentrations of NQO in a *rad4* mutant background therefore suggests that DNA damage itself, and not nucleotide depletion, acts as a primary induction signal *(27)*.

If cells can sense DNA damage directly, are there multiple sensors that detect specific DNA lesions, or do different damages lead to a common repair intermediate that acts as a damage signal? One possibility is that many DNA lesions may interfere with DNA and RNA polymerases, resulting in some change in chromatin structure that activates the signaling pathway. For example, DNA polymerases stalled by replication blocks may generate regions of single-stranded DNA owing to gaps created in the newly synthesized strand or by the activity of helicases in front of the replication fork. Alternatively, some change in the conformation or activity of the stalled polymerase may constitute the damage signal. These models are prompted in large measure by the requirement for single stranded DNA in the *E. coli* SOS response. In response to UV irradiation, it has been shown that activation of the SOS response requires ongoing DNA replication to generate these single stranded regions (*88*; Chapter 7).

Analysis of *crt5* and *dun2* mutants has provided evidence that DNA polymerases are also involved in DNA damage signaling in *S. cerevisiae*, at least for some damage-inducible genes. These mutants were isolated using two complementary genetic screens designed to identify components involved in controlling the damage inducibility of the *RNR* genes *(118,119)*. Mutations that either constitutively activate *RNR3* transcription (Crt$^-$ phenotype, 9 complementation groups) or render *RNR3* uninducible in response to DNA damage (Dun$^-$phenotype, 5 complementation groups) were identified. It was anticipated that Crt$^-$ mutants would define negative regulators of *RNR3* transcription or components that inappropriately activated the DNA damage pathway when mutated. Dun$^-$ mutants, on the other hand, would likely define positive-acting sensors and transducers of DNA damage signals. Analysis of the *crt5* mutant showed it defined a new allele of *CDC17/POL1*, indicating that defects in a DNA polymerase can directly result in a constitutive, endogenous DNA damage signal. Mutations in *POL3* have been identified that also display a Crt$^-$phenotype *(4)*. The *crt5/pol1* mutant caused an increase in the transcript levels of all three *RNR* genes, suggesting that these genes comprise a regulon that is coordinately regulated in response to DNA damage. Whether the expression of other damage-inducible genes is affected by the *crt5* mutation is not known.

dun2 mutants are defective in the transcriptional induction of *RNR1*, *RNR2*, and *RNR3* in response to HU and MMS *(119)*. In addition, *dun2* mutants display reduced survival when treated with HU and inappropriately enter mitosis instead of activating cell-cycle arrest; *dun2* mutants are thus also defective in the S-phase checkpoint that maintains the dependence of mitosis on prior completion of DNA replication (*75*; Chapter 17). *DUN2* is identical to *POL2*, the gene encoding the catalytic subunit of DNA polymerase ε (pol ε). Pol ε is organized with an amino-terminal catalytic domain and a large carboxyl-terminal region that is unique to the pol ε family and is conserved between yeast and humans *(57,71)*. A subset of *pol2* alleles also shows Dun⁻ and checkpoint deficient phenotypes, and these mutations map to the extreme carboxyl-terminus *(75)*. This set of *pol2* mutants is also temperature sensitive for growth and unable to synthesize DNA at the nonpermissive temperature; thus, mutations that inactivate the DNA damage response apparently also alter polymerase function *(9)*. The identification of mutations in pol ε that display both damage-uniducible and checkpoint deficient phenotypes provides a link among DNA replication, S-phase checkpoint control, and induction of DNA damage-inducible genes. As a central component of DNA replication complexes, pol ε could directly monitor the functionality of the DNA replication machinery. As such, it seems likely that pol ε is a sensor that communicates a DNA damage signal to both checkpoint and transcription regulatory elements, possibly via an activity of its carboxyl-terminal domain *(75)*.

The existence of a DNA replication component that can function as a sensor of S-phase progression is also predicted by a class of S-phase checkpoint defective mutants that fail to initiate DNA replication properly and, as a result, inappropriately enter mitosis (*46*; Chapter 17). To explain the checkpoint deficient phenotype of these mutants, it has been proposed that progression into S-phase results in the assembly of a critical replication structure that activates the S phase checkpoint and ensures the dependency of events during a normal mitotic cell cycle *(61)*. Following the completion of DNA synthesis, these complexes or intermediates would no longer be present, permitting entry into mitosis. Based on the results discussed above, it is possible that pol ε is a component of the DNA replication machinery that transmits the status of ongoing DNA replication. If so, the actual signal or activity communicated to the checkpoint machinery by pol ε may normally be present during S phase, and not exclusively associated with stalled polymerases or DNA damage. In this context it is important to emphasize that *RNR3* is not normally transcribed during S phase. Thus, if *POL2* functions to coordinate S phase and mitosis in an unperturbed cell cycle, the presence of DNA damage or replication blocks must convey additional information to the *POL2*-dependent transcriptional pathway. This information may be distinct from the information that mediates checkpoint signaling during a normal S phase; alternatively, the transcriptional pathway may have a higher signaling threshold to activate damage-induced *RNR* transcription. This elevated level of checkpoint signaling would only be attained when polymerases were stalled on the DNA owing to damage or nucleotide depletion.

4. *MEC1, RAD53,* AND *DUN1* ARE TRANSDUCERS OF DNA DAMAGE SIGNALS

The observation that damage-uniducible alleles of pol ε are defective in replication checkpoint control indicates that both cell-cycle checkpoints and the transcriptional response to DNA damage may be controlled by a common set of regulatory components *(75)*. Support for this idea has emerged through the characterization of additional genes involved in DNA damage signal transduction pathways. In particular, the *S. cerevisiae MEC1/SAD3/ESR1* and *RAD53/SAD1/MEC2/SPK1* genes appear to play central roles in these complex pathways, and are necessary for both the transcriptional and cell-cycle arrest responses to DNA damage *(3,99,113)*. *DUN1*, on the other hand, functions specifically in the control of transcriptional induction *(119)*. *MEC1*, *RAD53*, and *DUN1* encode proteins with homology to protein serine/threonine (*RAD53*, *DUN1*) and lipid kinases (*MEC1*), consistent with the idea that these genes are transducers of DNA damage signals. With the exception of *DUN1*, homologs of these genes have been identified in other organisms, indicating that DNA damage signal transduction mechanisms are evolutionarily conserved.

4.1. MEC1 *and* RAD53 *Control Multiple Cell-Cycle Checkpoints and DNA Damage Induced Transcription*

Alleles of *MEC1* have been isolated in genetic screens to identify mutants defective in either S-phase checkpoint regulation (*SAD3*) or in genes required to maintain the G2 arrest of temperature sensitive *cdc13* mutants at the nonpermissive temperature *(MEC1) (3,113)*. Interestingly, *mec1* strains were also isolated in a screen for mutants displaying meiotic recombination defects (*ESR1*; *56*). In addition to checkpoint-deficient phenotypes, *mec1* mutants do not induce *RNR3* transcription in response to UV irradiation *(76)*. *MEC1* is therefore also required for the inducibility of some DNA damage-responsive genes. *MEC1* is also involved in a distinct form of DNA damage regulation that actively slows the rate of ongoing S phase in response to alkylating agents *(78)*.

Both structurally and functionally, *MEC1* is related to the recently identified human *ATM* gene defective in the autosomal hereditary disease ataxia telangiectasia (*72,87,89*; Chapter 19, vol. 2). At the cellular level, defects in the *ATM* gene are associated with a number of abnormalities, including short telomeres, chromosomal breaks and rearrangements, sensitivity to X-irradiation, and defective G1/S and G2/M checkpoints *(93)*. Both *MEC1* and the *ATM* gene are members of a growing family of *ATM*-related kinases that display homology with the catalytic domain of phosphatidylinositol 3'-kinases. The *S. pombe rad3* gene and the *Drosophila mei-41* gene are members of this family, and are also required for checkpoint control in response to DNA damage and replication blocks (*1,40*; Chapter 20). It is currently unclear if any of these kinases function as lipid kinases in vitro or inside the cell. However, at least two members of this family phosphorylate protein substrates in vitro, suggesting that these kinases may function as protein kinases *(23,101)*.

Based on sequence comparisons, the *ATM* gene is more similar to *TEL1*, another *S. cerevisiae ATM*-related gene, than it is to *MEC1 (38,72,87)*. Cells with a disrupted *TEL1* gene have shortened telomeres, but no other obvious growth or checkpoint defects *(38)*. However, compared to *mec1* mutants, *tel1 mec1* double mutants display an enhanced sensitivity to DNA damage and replication blocks *(38,72,87)*. *tel1 mec1* double mutants also exhibit additional synergistic phenotypes, including growth defects and sensitivity to DNA-damaging agents that have no effect on the viability of either single mutant. Furthermore, an extra copy of the *TEL1* gene can suppress the lethality associated with a truncation allele of *MEC1 (72)*. Taken together, these results demonstrate that *TEL1* and *MEC1* mediate overlapping functions, and that in some circumstances, *TEL1* can substitute for the essential functions of *MEC1*.

Although *RAD53* was originally assigned to the *RAD52* X-ray-sensitive recombinational repair epistasis group, *rad53* mutants are sensitive to a number of different forms of DNA damage and replication blocks. Mutations in *RAD53* were isolated in the same genetic screens that identified *MEC1* (as *SAD1* and *MEC2*), and it appears that the damage sensitivity of *rad53* mutants reflects a requirement for *RAD53* in DNA damage checkpoint control *(3,113)*. *rad53* mutants display the same range of defects exhibited by *mec1* mutants, including defects in *RNR* gene induction after DNA-damaging treatments and DNA replication inhibition in response to MMS *(3,78,113)*. In addition, *RAD53* is required for execution of the G1 DNA damage checkpoint *(99)*, indicating that *RAD53* regulates a wide array of checkpoint functions (*MEC1* is also thought to be required for activation of the G1 DNA damage checkpoint, but this has yet to be directly demonstrated). *RAD53* encodes an essential serine/threonine protein kinase that was cloned *(SPK1)* by virtue of its ability to phosphorylate tyrosine residues in vitro; *RAD53* is thus a member of the so-called dual-specificity family of kinases *(103)*. Rad53p shows extensive amino acid sequence similarity with the *S. pombe* Cds1p protein kinase *(73)*. Like *RAD53*, *cds1* functions in maintaining viability in the presence of replication blocks.

Two sets of genetic and biochemical observations suggest that *RAD53* functions downstream of *MEC1* in the DNA damage signaling pathway *(87,105)*. First, overproduction of Rad53p can suppress the lethality associated with an *mec1* deletion, and can partially suppress the HU sensitivity of both *mec1* mutants and *tel1 mec1* double mutants. Second, Rad53p shows altered electrophoretic mobility and increased in vitro autophosphorylation activity after HU and MMS treatment. Both the mobility shift and increase in Rad53p kinase activity are eliminated in *mec1* and *rad53* mutants. These observations are consistent with the idea that *RAD53* and *MEC1* function together as central transducer component(s) in a number of pathways controlling different checkpoints and the transcriptional response of the *RNR* gene family. According to this scenario, in response to different forms of DNA damage and replication blocks, the products of *MEC1* and possibly also *TEL1* either directly or indirectly communicate damage signals to the Rad53p kinase, resulting in specific changes in the posttranslational modification state of Rad53p. These changes presumably modulate the activity of Rad53p or alter its access to critical downstream substrates. It therefore appears that the signaling pathways for multiple cell-cycle checkpoints and the transcriptional response to DNA damage (at least the pathway controlling inducibility of the *RNR* genes) are, to a large degree, coordinately regulated. Furthermore, the fact that both *MEC1* and *RAD53* are essential genes indicates either that some aspect of the DNA damage response is necessary for normal mitotic growth or that these genes fulfill additional, uncharacterized functions.

4.2. DUN1 Controls the Transcriptional Response to DNA Damage

How do *RAD53* and *MEC1* control damage-inducible gene expression? Insight into regulation of this DNA damage response has come from analysis of the *DUN1* gene. Like *rad53* and *mec1* mutants, *dun1* cells are defective in *RNR2* and *RNR3* induction in response to HU and MMS (119). Although *dun1* mutants are sensitive to DNA-damaging agents, *DUN1* is not required for activation of S-phase or G2 DNA damage checkpoints. *DUN1* thus appears to function specifically in damage-inducible gene expression, although it is formally possible that *DUN1* has a role in checkpoint activation that is redundant with other genes.

DUN1 encodes a nonessential serine/threonine protein kinase, suggesting that *DUN1* is involved in DNA damage signal transduction *(119)*. Kinase activity is necessary for *RNR* transcriptional induction, since point mutations within the kinase catalytic domain engender damage-uninducible phenotypes. In vivo, Dun1p is a phosphoprotein that has a basal level of phosphorylation in the absence of DNA damage. Treatment with DNA damaging agents increases the phosphorylation state of Dun1p. Kinase-deficient *dun1* mutants maintain the basal level of Dun1p phosphorylation, but are unable to increase phosphorylation in response to damage. These results suggest that Dun1p is activated in response to DNA damage, and the increase in Dun1p phosphorylation results either from autophosphorylation or the activity of a distinct kinase activated in response to Dun1p kinase activity.

Several observations suggest *DUN1* functions in the same signal transduction pathway as *POL2*, *RAD53*, and *MEC1* to regulate *RNR* transcription in response to DNA damage. Damage-induced Dun1p kinase activation is eliminated in both Dun$^-$*pol2* and *rad53* mutants, suggesting these components of the damage response are upstream of *DUN1* and regulate *DUN1* activity *(3,75)*. Dun1p may be a direct substrate of the Rad53p kinase, although basal levels of Dun1p phosphorylation appear unaltered either in the presence or absence of DNA damage in *rad53* mutants, arguing against this possibility *(3)*. However, incorporation of single phosphate changes may have escaped detection in Dun1p phosphorylation assays. It was recently demonstrated that overproduction of *DUN1* could restore viability in the complete absence of *MEC1* or *RAD53* gene function, raising the intriguing possibility that deregulation of a *DUN1* target gene may be able to compensate for the essential functions of these genes *(74,87)*. It should be noted, however, that overproduction of *DUN1* is not sufficient either to autoactivate Dun1p kinase activity or to activate constitutive *RNR3* transcription *(119)*. Thus, further work will be required to decipher the nature of *DUN1*-mediated suppression.

The extent to which other DNA damage-inducible genes besides the *RNR* genes are under the control of *DUN1* is currently unknown. *DUN1* is apparently also required for the damage inducibility of *MAG (85)*; however, *DDR48* and *UBI4* damage induction is *DUN1*-independent, revealing the existence of at least one other pathway for transcriptional induction. In this light, the sensitivity of *dun1* mutants to replication blocks and DNA damage is of interest with respect to the physiological significance of the transcriptional response to DNA damage. Since DNA damage checkpoints are still functional in *dun1* cells, sensitivity to damaging agents may be caused by a failure to induce *DUN1* controlled damage-inducible genes. A requirement for *DUN1* controlled genes may be especially pronounced when cells experience replication blocks due to nucleotide depletion, since it seems unlikely that alternative mechanisms could compensate for a critical threshold of *RNR* gene activity. The damage sensitivity of *dun1* strains may therefore indicate that the transcriptional response controlled by *DUN1* is essential for surviving episodic exposure to some forms of DNA damage or replication blocks. An alternative possibility is that *DUN1* may have an unappreciated role in DNA repair processes. These issues will only be resolved by identifying the relevant substrates of the Dun1 kinase.

A model that depicts the general features of the response to DNA damage in *S. cerevisiae* is shown in Fig. 1. In this model, *RAD53* is positioned in the damage response pathway as a central transducer of signals to both the transcriptional apparatus and the cell-cycle arrest machinery. As discussed above, the activation of Rad53p kinase activity in response to damage signals is normally dependent on *MEC1*, although *TEL1* appears to be able to substitute functionally for *MEC1* in some experimental situations. The nature of this dependency is not understood, but if the *ATM*-related kinases actually function as protein kinases, it is possible that Rad53p may be a direct Mec1p substrate. Upstream of *MEC1*, *POL2* functions as a sensor that communicates the presence of replication blocks and DNA damage during S phase of the cell cycle. Dun$^-$ *pol2* mutants are proficient for the G1 and G2 DNA damage checkpoints, revealing the existence of additional, *POL2*-independent, checkpoint sensory/signaling mechanisms (75). Candidates for these sensors include the *RAD9*, *RAD17*, *RAD24,* and *MEC3* gene products, which function in executing the G1 and G2 DNA damage checkpoints (*99,112,113*; Chapter 17). It has recently been demonstrated that the *RAD9* group of checkpoint genes are involved in processing some forms of DNA damage, and it was proposed that this processing results in checkpoint activation *(63)*. These results suggest that the *RAD9* group of genes functions upstream of *RAD53* and *MEC1* in DNA damage checkpoint signaling pathways. Downstream of *RAD53*, *DUN1* functions to activate specifically the transcriptional response to DNA damage, at least for the *RNR* genes. A separate pathway(s) controlled by *RAD53* activates checkpoint mechanisms that execute cell-cycle arrest at both the G1/S and G2/M transitions. In *S. cerevisiae*, the identity of the critical cell-cycle effectors for these checkpoints are unknown. Finally, in response to treatment of cells with MMS, the functions of both *MEC1* and *RAD53* are also required to inhibit progression through S phase *(78)*.

5. EFFECTORS OF THE TRANSCRIPTIONAL RESPONSE TO DNA DAMAGE

Although DNA damage-induced gene expression could occur by a number of mechanisms, current knowledge suggests that regulation of transcription initiation is the main strategy employed during the transcriptional response to DNA damage in *S. cerevisiae.* The ultimate effectors of DNA damage signal transduction pathways are therefore likely to be *cis*-acting elements within the promotors of damage-inducible genes and the transcription factors that interact with these sequences. The promotors of several damage-inducible genes have been analyzed to define these elements, and these studies have shown that a variety of mechanisms appear to be responsible for damage-induced transcription.

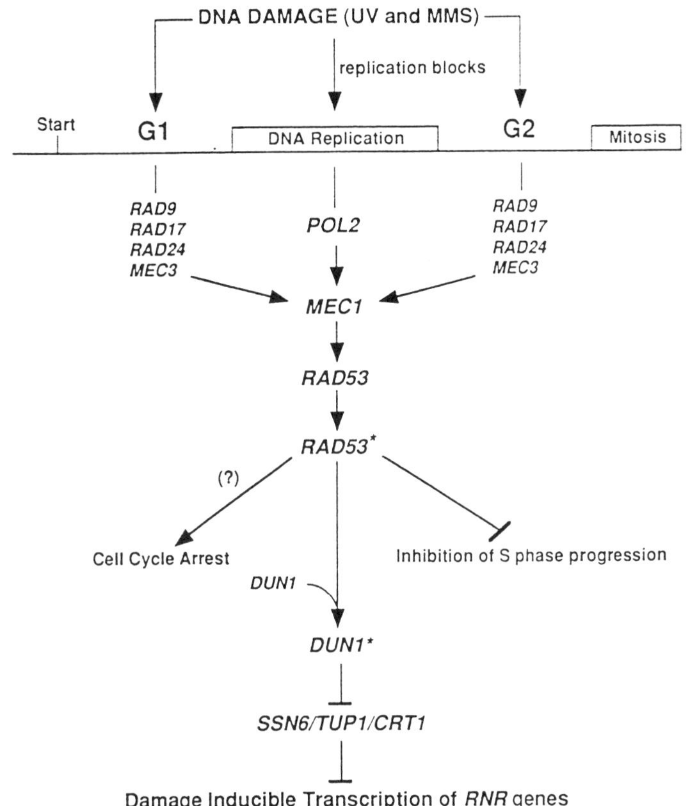

Fig. 1. A model of the cellular response to DNA damage in *S. cerevisiae*. The key features of this model are outlined in the text. In response to DNA damage or replication blocks, Rad53p protein kinase is activated (*) via an *MEC1*-dependent pathway. *RAD53* mediates a number of damage-induced functions, including activating cell-cycle arrest at the G1/S and G2/M transitions by an unknown mechanism (?), and actively inhibiting S-phase progression. *RAD53* is also required for activation of the Dun1p protein kinase and the transcriptional induction of some DNA damage-inducible genes. In the case of the *RNR* genes, transcriptional induction appears to be accomplished by relief of repression mediated by the *SSN6*, *TUP1*, and *CRT1* genes. It is proposed that *POL2* and the *RAD9* group of checkpoint genes function as sensors or damage signal generating components upstream of *RAD53* and *MEC1* to communicate the presence of replication blocks or DNA damage.

5.1. Transcriptional Regulation of DNA Damage-Inducible Genes

DNA damage inducibility of the *RAD2* gene appears to be accomplished by a positive acting mechanism. Deletions within the *RAD2* promotor defined two related sequences, called damage response elements (DRE1 and DRE2), that were necessary for induction of a *RAD2-lacZ* fusion in response to UV irradiation *(97)*. DRE1 also appeared to act as an upstream activating sequence (UAS) necessary for constitutive expression, since deletion of this region resulted in both uninducibility and reduced basal transcription levels. No deletions that resulted in enhanced constitutive expression were identified in these studies, suggesting relief of repression is not involved in

RAD2 inducibility. Gel-retardation studies revealed that one or more proteins specifically bind to DRE1 and DRE2 *(98)*. The protein/DNA complex was present in the uninduced state, but increased following exposure to DNA damage. Treatment of the cells with cycloheximide inhibited increased complex formation. Taken together, these data suggest that the response of *RAD2* to DNA damage is controlled by distinct regulatory elements, and that some step required for full induction requires the *de novo* synthesis of a limiting factor. The DNA damage inducibility of *RAD54* is apparently also under positive control. Deletion analysis defined a 29-bp element in the 5'-regulatory region required for efficient induction in response to DNA-damaging agents, including X-rays, UV, and MMS *(17)*. Deletion of this element did not affect constitutive expression, implicating inducibility by positive regulation. The involvement of a single promotor region in the response to a variety of damaging agents is consistent with a coordinated transcriptional response of *RAD54* to a number of different forms of DNA damage.

In contrast to *RAD2* and *RAD54*, relief of repression appears to underlie inducibility of *PHR1*. Electromobility shift assays were used to detect a binding activity, termed photolyase regulatory protein (PRP), that interacted with the *PHR1* 5'-regulatory region *(91)*. PRP activity was present in undamaged yeast extracts and disappeared within 30 minutes after UV irradiation. Footprint analysis of PRP–DNA complexes revealed that PRP-protected a 39-bp sequence within the *PHR1* promotor. Deletion of this sequence decreased inducibility of *PHR1* in response to UV and MMS, and increased the basal level of *PHR1* expression in vivo. Insertion of the PRP protected sequence into the *CYC1* promotor was sufficient to repress basal levels of transcription and confer damage inducibility, indicating the PRP binding domain functions as an upstream repressing sequence (URS). It thus appears that *PHR1* induction is mediated through inactivation of the PRP repressor complex.

Analysis of the *RNR2* promotor has defined both positive and negative regulatory elements that control the expression of this gene *(28,48)*. A 42-bp DRE was identified that was necessary for correct regulation in response to treatment with HU and MMS *(28)*. This region could confer a low level of inducibility on a heterologous promotor, indicating that additional sequences outside the DRE were also involved. The DRE contains both URS and UAS sequences, suggesting that damage inducibility of *RNR2* could be accomplished either by relief of repression transduced through the URS element or by the activation of transcription over a basal level of repression mediated by the specific interaction of a factor with the UAS. Four DNA binding factors were identified that interacted with the DRE. One of these factors was identified as Rap1p, a DNA binding protein that can act as both a positive and negative regulator of gene expression depending on promotor context. The presence of DNA damage did not alter the pattern or abundance of protein/DNA complexes detected in these studies *(121)*.

Regulation of the *MAG* promotor also appears to be complex. Promotor deletion analysis identified one 46-bp UAS, the deletion of which resulted in a substantial decrease in the MMS inducibility of a *MAG* promotor/*lacZ* fusion construct *(116)*. This region contained a putative Rap1p binding site; however, Rap1p binding to this sequence was not detected. Deletion and heterologous promotor analysis also indicated the presence of two URSs in the *MAG* promotor. One of these URS sequences was capable of reducing expression levels when incorporated into a heterologous promotor, but this construct was not sufficient to confer MMS inducibility. Two proteins of 39

Table 2
A 10-bp Homologous Sequence Found Within the Promoters of 12 Genes Involved in DNA Repair and Nucleotide Biosynthesis[a]

Gene	Position[b]	Sequence	Refs.
RAD2 (DRE1)[c]	–169	C G T G G A G G C A	*64,90,97,98,116*
RAD51[c]	–157	C G T G G T G G G A	*7,94,116*
MAG1 (URS1)[c]	–215[f]	G T A G G T C G A A	*116*
PHR1[c]	–103	C G A G G A A G C A	*90,116*
	–109[f]	*C G A G G A A G A A*	
RNR2[c]	–424	C G A G G T C G C A	*28,48,90,16*
RNR3[c]	–467	C T A G G T A G C A	*29,116,117*
DDR48[c]	–271	C G A G G A T G A C	*109,116*
	–322	C G T G G T T G A T	
RAD1[d]	–201	T G A G G T G G A A	*116*
RAD4[d]	–364	C G T G G A T G A A	*90,116*
RAD10[d]	–312[f]	C G A G G A A G A A	*90,116*
RAD16[e]	–309	C A T G G T T G C C	*116*
MGT1[d]	–210	G G T G G A G G C C	*116*

[a]This table is adapted from references *90*, *116*, and *120*.
[b]Relative to the first nucleotide in the translation initiation ATG of each gene.
[c]DNA damage inducible.
[d]Not inducible by DNA damaging agents that have been examined.
[e]Damage inducibility has not been examined directly.
[f]Indicated sequence is present on the complementary strand.

and 26 kDa were shown to bind the *MAG* URS, and were present in extracts of both damaged and undamaged cells.

5.2. *A Putative Replication Protein A Binding Site May Define a Common Damage-Inducible Gene Regulatory Element*

Comparisons of sequences within the 5'-regulatory regions of damage-inducible genes have led two groups of investigators to note the presence of a consensus 10-bp sequence (5'-C G T/A G G T/A N G C/A C/A-3') within the promotors of a number of damage-inducible genes, including the DRE1 of *RAD2*, *RAD51*, *MAG* URS sequences, *PHR1*, *DDR48*, *RNR2*, and *RNR3* (Table 2; *90,116*). Sequences homologous to this decamer element are also present in the promotors of *RAD* genes that are not induced by DNA damage (for examples, *see* ref. *116* and Table 2). Nonetheless, the presence of this sequence suggests that all these damage-inducible genes may be regulated by a common *trans*-acting factor. If this is the case, the interaction of this factor with the decamer sequence would be required to mediate both positive and negative influences on gene expression, since the decamer sequence is present in UAS sequences in *RAD2*, and URS sequences in *MAG* and *RNR2*.

Other homologous promotor elements within the family of DNA damage-inducible genes have also been described. The nucleotide sequences of the *RAD2* DRE1 and DRE2 are quite similar, and a comparison of these sequences with other DNA damage-inducible genes has suggested that similar sequence elements are found within the promotors of *RAD7*, *RAD23*, and *PHR1* (Table 3; *98*). In the case of the *RAD2* DRE1

Table 3
Alignment of Homologous Sequences Found Within the Promotors of *RAD2, PHR1, RAD7*, and *RAD23*[a]

Gene	Postition	Sequence[b]
RAD2 (DRE1)	–93	GTGGAGGCATTAAAA
RAD2 (DRE2)	–168	TTAAAGGGATTGAAA
PHR1	–102	GAGGAAGCAGTCAAA
RAD7	–122	ATGGAAGCAAAAATG
RAD23	–296	GTGGCGAAATTGAAA

[a]Alignment is as described in Fig. 6 of ref. *98*.
[b]Underlined sequences also occur within the decamer concensus sequence depicted in Table 2.

and *PHR1*, this sequence element overlaps the decamer consensus sequence described in Section 5.2. Oligonucleotides bearing the *RAD2* DRE1 and DRE2, *RAD7*, *RAD23*, and *PHR1* versions of this element all form specific protein complexes when incubated with undamaged yeast extracts *(98)*. However, these complexes do not appear to be completely equivalent, since the different oligonucleotides do not all compete in the same manner for complex binding.

Recent evidence suggests that one factor capable of interacting with the promotors of a number of DNA damage-inducible genes is replication protein A (RPA). RPA is a multiprotein complex that has single-stranded DNA binding activity and functions in DNA replication, homologous recombination, and in both the incision and DNA synthesis steps of NER *(2,18,43,104)*. Purified RPA can bind to oligonucleotides containing the decamer sequence within the URS element of the *MAG* promotor, and the regulatory regions of *MGT1*, *RAD1*, *RAD2* (DRE1), *RAD4*, *RAD10*, *RAD16*, *RAD51*, *RNR2*, *RNR3*, *DDR48*, and *PHR1 (100)*. Furthermore, the *MAG* URS/protein complexes formed in undamaged yeast extracts can be supershifted in electrophoresis mobility experiments using antisera specific for RPA. The role of RPA in the regulation of these genes has not yet been clarified. However, human RPA has been shown to be phosphorylated when cells are exposed to DNA damage *(10,62)*, indicating that phosphorylation of RPA in response to damaging agents may redistribute RPA to sites of DNA repair *(100)*. This redistribution would allow RPA to function in recombination and NER, while simultaneously effecting a block to replication fork progression and a derepression of damage-inducible genes.

5.3. TUP1, SSN6, *and* CRT1 *Control Damage Inducibility of RNR Gene Transcription*

In the case of the *RNR* genes, analysis of Crt⁻ mutants that constitutively activate *RNR3* expression has yielded insights into the components responsible for transcriptional induction in response to DNA damage. Genetic analysis of *crt1*, *crt4*, and *crt8* alleles showed that the Crt⁻ phenotype of these mutants was epistatic in *crt dun1* double mutants *(121)*. This same group of *crt* mutations was also capable of relieving repression by the *RNR2* DRE element on the heterologous *CYC1* promotor. These results provide a strong genetic argument that *CRT1*, *CRT4*, and *CRT8* function downstream of the Dun1p protein kinase to repress *RNR2* and *RNR3* transcription, and this regulation is directed through the same promotor elements that have been shown to control *RNR2* damage inducibility (Fig. 1). Consistent with this interpretation, *CRT4* and *CRT8* are identical to the *TUP1* and *SSN6* genes, which encode transcription factors that negatively regulate several unrelated genes *(118)*.

Since *TUP1* and *SSN6* are general repressors, these factors are unlikely to be the targets that are directly regulated in response to DNA damage. How then is transcriptional derepression of the *RNR* gene family accomplished? Molecular characterization of the *CRT1* gene has shed light on this question. Mutant alleles of this gene formed the largest complementation group isolated in the Crt$^-$ screen, suggesting *CRT1* is the most sensitive target for constitutive derepression of *RNR3* transcription *(118)*. *CRT1* was cloned and shown to encode a novel protein that displays homology to the DNA binding domain of the mammalian RFX family of transcriptional activators *(121)*. Crt1p binds to specific sequences within the promotors of *RNR2* and *RNR3* in vitro, and overproduction of Crt1p reduces the damage inducibility of *RNR3* in a *TUP1/SSN6*-dependent manner. Repression of the *RNR* genes is therefore likely to be controlled by Crt1p binding within the promotors of these genes and by recruiting *TUP1/SSN6* complexes. It will be interesting to define the exact mechanism by which *CRT1/TUP1/SSN6*-mediated repression is abolished or overcome in response to DNA damage and to determine if this mode of regulation is employed in the expression of other damage-inducible genes.

6. CONCLUDING REMARKS

A survey of DNA damage induced gene expression in *S. cerevisiae* indicates that the transcriptional response to DNA damage in this organism is multifaceted and complex. A number of DNA damage-inducible genes have been described, and a subset of these function directly in DNA repair or DNA repair synthesis. Not all DNA repair genes, however, are induced by DNA-damaging treatments and some of the genes that are induced have no known role in DNA repair. Two remaining challenges, then, are to understand how elevated expression of particular DNA repair genes enhances specific repair pathways, and to determine the function of uncharacterized DNA damage-inducible genes. Recent studies on the role of *RAD6*, *RAD7*, and *RAD18* suggest that one function of inducible genes may be to modulate the structure of damaged chromosomes to permit greater access of the repair machinery to sites of DNA damage. As the biochemistry and genetics of DNA repair in *S. cerevisiae* continue to advance, it will be interesting to see to what extent regulation of template accessibility contributes to the overall efficiency of DNA repair and to genomic stability.

One central focus of this chapter has been whether transcriptional activation of yeast DNA damage-inducible genes occurs by a single induction pathway that responds directly to signals generated by damaged DNA. With the exception of the *DDR/UBI4* genes, the majority of DNA damage-inducible genes appear to be induced in response to a variety of DNA-damaging agents. Characterization of DNA damage-uninducible mutations in DNA polymerase ε has shown that cells possess at least one sensory apparatus that permits them to monitor the presence of DNA damage and replication blocks. However, analysis of the Ras dependent UV response has demonstrated that DNA-damaging agents can activate damage signaling pathways that appear to be independent of lesions in DNA. In addition, studies of the transcriptional regulation of damage-inducible genes indicate that a variety of both positive and negative acting influences underlie damage inducibility. Therefore, unlike the SOS response of *E. coli*, there is currently no direct evidence to suggest that the induction of yeast DNA damage-inducible genes is controlled by a single mode of regulation activated by lesions in DNA. However, it is also important to emphasize that the mechanisms of transcriptional induction for many damage-inducible genes are largely undefined; thus, it is entirely possible that these genes are regulated by components that contribute to a coordinated transcriptional response.

Ultimately, a detailed understanding of the transcriptional response to DNA damage in *S. cerevisiae* will require elaboration of the pathways that control damage induced gene expression. In the case of the *RNR* genes, this type of analysis has shown that many aspects of the cellular response to DNA damage and replication blocks, including activation of cell-cycle checkpoints, inhibition of DNA synthesis, and the regulation of some damage-induced genes, are controlled by overlapping signal transduction pathways. The *MEC1* and *RAD53* genes are central signal transducers in these responses, and it will be of considerable interest to determine the mechanisms by which DNA damage regulates the activity of these genes and to identify critical downstream targets. The concerted regulation of several different responses to DNA damage mediated by *MEC1* and *RAD53* lends support to the idea that there may exist coordinated transcriptional regulatory pathways for different damage-inducible genes. The *DUN1* gene plays a pivotal regulatory role in the inducibility of the *RNR* genes, and it is possible that *DUN1* may control other damage-inducible genes as well. Thus, at this time, it should be possible to initiate studies to compare directly the genetic requirements for the inducibility of different DNA damage-inducible genes. Further analysis of factors controlling the response of damage-inducible genes should greatly enrich our understanding of the regulatory complexity underlying the transcriptional response to DNA damage.

ACKNOWLEDGMENTS

The authors wish to thank our colleagues who generously communicated their results prior to publication. We also wish to thank Yoli Sanchez and Connie Nugent for critical readings of the manuscript, and to members of the Elledge research group for enthusiastic scientific discussions. J. B. B. is supported by postdoctoral fellowship 1F32CA68673 from the National Cancer Institute. S. J. E. is a Pew Scholar in the Biomedical Sciences and an Investigator of the Howard Hughes Medical Institute. This work was supported by NIH grant GM44664.

REFERENCES

1. Al-Khodairy, F., E. Fotou, K. S. Sheldrake, D. J. F. Grittiths, A. R. Lehmann, and A. M. Carr. 1994. Identification and characterization of new elements involved in checkpoints and feedback controls in fission yeast. *Mol. Biol. Cell* **5:** 147–160.
2. Alani, E., R. Thresheer, J. D. Grittith, and R. D. Kolodner. 1992. Characterization of DNA binding and strand exchange stimulation properties of y-RPA, a yeast single stranded DNA binding protein. *J. Mol. Biol.* **227:** 54–70.

3. Allen, J. B., Z. Zhou, W. Siede, E. C. Friedberg, and S. J. Elledge. 1994. The *SAD1/RAD53* protein kinase controls multiple checkpoints and DNA damage induced transcription in yeast. *Genes Dev.* **8:** 2416–2428.
4. Allen, J. B. and S. J. Elledge. Unpublished observations.
5. Bailly, V., J. Lamb, P. Sung, S. Prakash, and L. Prakash. 1994. Specific complex formation between yeast *RAD6* and *RAD18* proteins: a potential mechanism for targeting *RAD6* ubiquitin-conjugating activity to DNA damage sites. *Genes Dev.* **8:** 801–810.
6. Barker, D. G., J. M. White, and L. H. Johnston. 1985. The nucleotide sequence of the DNA ligase gene (*CDC9*) from *Saccharomyces cerevisiae*: a gene which is cell cycle regulated and induced in response to DNA damage. *Nucleic Acid Res.* **13:** 8223–8237.
7. Basile, G., M. Aker, and R. K. Mortimer. 1992. Nucleotide sequence and transcriptional regulation of the yeast recombinational repair gene *RAD51*. *Mol. Cell. Biol.* **12:** 3235–3246.
8. Budd, M. and R. K. Mortimer. 1984. The effect of cyclohexamide on repair in a termperature conditional radiation-sensitive mutant of *Saccharomyces cerevisiae. Radiat. Res.* **98:** 581–589.
9. Budd, M. E. and J. L. Campbell. 1993. DNA polymerases delta and epsilon are required for chromosomal replication in *Saccharomyces cerevisiae. Mol. Cell. Biol.* **13:** 496–505.
10. Carty, M. P., M. Zernick-Kobak, S. McGrath, and K. Dixon. 1994. UV light induced DNA synthesis arrest in HeLa cells is associated with changes in phosphorylation of human single stranded DNA binding protein. *EMBO J.* **13:** 2114–2123.
11. Chanet, R. N., N. Magaña-Schwenke, and F. Fabre. 1988. Potential DNA binding domain in the *RAD18* gene product of *Saccharomyces cerevisiae* . *Gene* **73:** 543–547.
12. Chen, J. B., B. Derfler, A. Maskati, and L. Samson. 1989. Cloning a eucaryotic DNA glycosylase repair gene by the suppression of a DNA repair defect in *Escherichia coli. Proc. Natl. Acad. Sci. USA* **85:** 7851–7854.
13. Chen, J., B. Derfler, and L. Samson. 1990. *Saccharomyces cerevisiae* 3-methyladenine DNA glycosylase has homology to the AlkA glycosylase of *E. coli* and is induced in response to DNA alkylation damage. *EMBO J.* **9:** 4568–4574.
14. Chen, J. and L. Samson. 1991. Induction of *S. cerevisiae* mag 3-methyladenine DNA glycosylase transcript levels in response to DNA damage. *Nucleic Acid Res.* **19:** 6327–6332.
15. Cole, G. M., D. Schild, S. T. Lovett, and R. K. Mortimer. 1987. Regulation of *RAD54* and *RAD52* lacZ gene fusions in *Saccharomyces cerevisiae* in response to DNA damage. *Mol. Cell. Biol.* **7:** 1077–1083.
16. Cole, G. M., D. Schild, and R. K. Mortimer. 1989. Two DNA repair and recombination genes in *Saccharomyces cerevisiae*, *RAD52* and *RAD54*, are induced during meiosis. *Mol. Cell. Biol.* **9:** 3101–3104.
17. Cole, G. M. and R. K. Mortimer. 1989. Failure to induce a DNA repair gene, *RAD54*, in *Saccharomyces cerevisiae* does not affect DNA repair of recombination phenotypes. *Mol. Cell. Biol.* **9:** 3314–3322.
18. Coverly, D., M. D. Kenny, M. Munn, W. D. Rupp, D. P. Lane, and R. D. Wood. 1992. Requirement for the replication protein SSB in human DNA excision repair. *Nature (Lond.)* **349:** 538–541.
19. Devary, Y., R. A. Gottlieb, L. F. Lau, and M. Karin. 1991. Rapid and preferential activation of the c-jun gene during the mammalian UV response. *Mol. Cell. Biol.* **11:** 2794–2801.
20. Devary, Y., R. A. Gottlieb, T. Smeal, and M. Karin. 1992. The mammalian ultraviolet response is triggered by activation of Src tyrosine kinases. *Cell* **70:** 1080–1090.
21. Devary, Y., C. Rosette, J. A. DiDonato, and M. Karin. 1993. NF-kB activation by ultraviolet light not dependent on a nuclear signal. *Science* **261:** 1442–1445.
22. Dever, T. E., L. Feng, R. C. Wek, A. M. Cigan, T. R. Donahue, and A. G. Hinnebusch. 1992. Phosphorylation of initiation factor 2α by protein kinase *GCN2* mediates gene specific translational control of *GNC4* in yeast. *Cell* **67:** 584–596.

23. Dhand, R., I. Hiles, G. Panayotou, S. Roche, M. J. Fry, I. Gout, N. F. Totty, O. Truong, P. Vicendo, and K. Yorezawa. 1994. PI 3 kinase is a dual specificity enzyme: autoregulation by an intrinsic protein serine kinase activity. *EMBO J.* **13:** 522–533.
24. Downes, C. S. and A. S. Wilkens. 1994. Cell cycle checkpoints, DNA repair and DNA replication strategies. *BioEssays* **16:** 74–78.
25. Eckardt, F., E. Moustacchi, and R. H. Haynes. 1978. On the inducibility of error prone repair in yeast, in *DNA Repair Mechansims* (Hanawalt, P., E. Friedberg, and C. Fox, eds.), Academic, New York, pp. 421–423.
26. Elledge, S. J. and R. W. Davis. 1987. Identification and isolation of the gene encoding the small subunit of ribonucleotide reductase from *Saccharomyces cerevisiae*: DNA damage inducible gene required for mitotic viability. *Mol. Cell. Biol.* **7:** 2773–2782.
27. Elledge, S. J. and R. W. Davis. 1989. DNA damage induction of ribonucleotide reductase. *Mol. Cell. Biol.* **9:** 4932–4940.
28. Elledge, S. J. and R. W. Davis. 1989. Identification of the DNA damage responsive element of *RNR2* and evidence that four distinct cellular factors bind it. *Mol. Cell. Biol.* **9:** 5372–5385.
29. Elledge, S. J. and R. W. Davis. 1990. Two genes differentially regulated in the cell cycle and by DNA damaging agents encode alternative regulatory subunits of ribonucleotide reductase. *Genes Dev.* **4:** 730–741.
30. Elledge, S. J., Z. Zhou, and J. B. Allen. 1992. Ribonucleotide reductase: regulation, regulation, regulation. *Trends Biochem. Sci.* **17:** 119–123.
31. Elledge, S. J., Z. Zhou, J. B. Allen, and T. A. Navas. 1993. DNA damage and cell cycle regulation of ribonucleotide reductase. *BioEssays* **15:** 333–339.
32. Emery, H. S, D. Schild, D. E. Kellogg, and R. K. Mortimer. 1991. Sequence analysis of *RAD54*, a *Saccharomyces cerevisiae* gene involved in recombination and repair. *Gene* **104:** 103–106.
33. Engelberg, D., C. Klein, H. Martinetto, K. Struhl, and M. Karin. 1994. The UV response involving the Ras signalling pathway and AP-1 transcription factors is conserved between yeast and mammals. *Cell* **76:** 380–389.
34. Fabre, F., and F. Roman. 1977. *Gene*tic evidence for inducibility of recombination competence in yeast. *Proc. Natl. Acad. Sci. USA* **73:** 1657–1661.
35. Finley, D., E. Ozkaynak, and A. Varshavsky. 1987. The yeast polyubiquitin gene is essential for resistance to high temperatures, starvation, and other stresses. *Cell* **48:** 1035–1046.
36. Friedberg, E. C. 1988. Deoxyribonucleic acid repair in the yeast *Saccharomyces cerevisiae*. *Microbiol. Rev.* **52:** 69–102.
37. Friedberg, E. C., W. Siede, and A. J. Cooper. 1991. Cellular responses to DNA damage in yeast, in *The Molecular Biology of the Yeast Saccharomyces: Genome Dynamics, Protein Synthesis, and Energetics*, vol. I (Jones, E., J. R. Pringle, and J. Broach, eds.), Cold Spring Harbor Laboratory, Cold Spring Harbor, NY, pp. 147–192.
38. Greenwell, P. W., S. L. Kronmal, S. E. Porter, J. Gassenhuber, B. Obermaier, and T. D. Petes. 1995. *TEL1*, a gene involved in controlling telomere length in *S. cerevisiae*, is homologous to the human ataxia telangiectasia gene. *Cell* **84:** 813–819.
39. Habraken, Y., P. Sung, L. Prakash, and S. Prakash. 1993. Yeast excision repair gene *RAD2* encodes a single stranded DNA endonuclease. *Nature (Lond.)* **365:** 364–367.
40. Hari, K. L., A. Santerre, J. J. Sekelsky, K. S. McKim, J. B. Boyd, R. S. Hawley. 1995. The *mei-41* gene of *D. melanogaster* is a structural and functional homologue of the human ataxia telangiectasia gene. *Cell* **81:** 805–811.
41. Hartwell, L. 1992. Defects in a cell cycle checkpoint may be responsible for the genomic instability of cancer cells. *Cell* **70:** 543–546.
42. Hartwell, L. and T. A. Weinert. 1989. Checkpoints: controls that ensure the order of cell cycle events. *Science* **246:** 619–624.
43. He, Z., L. A. Henricksen, M. S. Wold, and C. J. Ingles. 1995. RPA involvement in the damage recognition and incision steps of nucleotide excision repair. *Nature (Lond.)* **373:** 565–568.

44. Herrlich, P., H. Ponta, and H. J. Rahmsdorf. 1992. DNA damage induced gene expression: signal transduction and relation to growth factor signalling. *Rev. Physiol. Biochem. Pharmacol.* **119:** 186–223.
45. Hinnebusch, A. G. 1988. Mechanisms of gene regulation in the general control of amino acid biosynthesis in *Saccharomyces cerevisiae* . *Microbiol. Rev.* **52:** 248–272.
46. Humphrey, T. and T. Enoch. 1995. Keeping mitosis in check. *Curr. Biol.* **5:** 375–378.
47. Hurd, H. K., C. W. Roberts, and J. W. Roberts. 1987. Identification of the gene for the yeast ribonucleotide reductase small subunit and its inducibility by methyl methanesulfonate. *Mol. Cell. Biol.* **7:** 3663–3667.
48. Hurd, H. K. and J. W. Roberts. 1989. Upstream regulatory sequences of the yeast *RNR2* gene include a repression sequence and an activation site that bind the *RAP1* protein. *Mol. Cell. Biol.* **9:** 5359–5371.
49. Jentsch, S., J. P. McGrath, and A. Varshavsky. 1987. The yeast DNA repair gene *RAD6* encodes a ubiquitin conjugating enzyme. *Nature (Lond.)* **329:** 131–134.
50. Johnson, A. L., D. G. Barker, and L. H. Jonhston. 1986. Induction of yeast DNA ligase genes in exponential and stationary phase cultures in response to DNA damaging agent. *Curr. Genet.* **11:** 107–112.
51. Johnston, L. H., J. H. M. White, A. L. Johnson, G. Lucchini, and P. Plevani. 1987. The yeast DNA polymerase I transcript is regulated in both the mitotic cell cycle and in meiosis and is also induced after DNA damage. *Nucleic Acid Res.* **15:** 5017–5030.
52. Jones, J. S., and L. Prakash. 1991. Transcript levels of the *Saccharomyces cerevisiae* DNA repair gene *RAD18* increase in UV irradiated cells and during meiosis but not during the mitotic cell cycle. *Nucleic Acid Res.* **19:** 883–888.
53. Jones, J. S., S. Weber, and L. Prakash. 1988. The *Saccharomyces cerevisiae RAD18* gene encodes a protein that contains potential zinc finger domains for nucleic acid binding and a putative nucleotide binding sequence. *Nucleic Acid Res.* **16:** 7019–7031.
54. Jones, J. S., L. Prakash, and S. Prakash. 1990. Regulated expression of the *Saccharomyces cerevisiae* DNA repair gene *RAD7* in response to DNA damage and during sporulation. *Nucleic Acid Res.* **18:** 3280–3284.
55. Kaback, D. B. and L. R. Feldberg. 1985. *Saccharomyces cerevisiae* exhibits a sporulation specific temporal pattern of transcript accumulation. *Mol. Cell. Biol.* **5:** 741–751.
56. Kato, R. and H. Ogawa. 1994. An essential gene, *ESR1*, is required for mitotic cell growth, DNA repair and meiotic recombination in *Saccharomyces cerevisiae. Nucleic Acid Res.* **22:** 3104–3112.
57. Kesti, T., H. Frantti, and J. E. Syraoja. 1993. Molecular cloning of the cDNA for the catalytic subunit of human DNA polymerase epsilon. *J. Biol. Chem.* **267:** 10,287–10,245.
58. Koch, C. and K. Nasmyth. 1994. Cell cycle regulated transcription in yeast. *Curr. Opinion Cell Biol.* **6:** 451–459.
59. Lane, D. P. 1992. p53, guardian of the genome. *Nature (Lond.)* **358:** 15,16.
60. Laurenson, P. and J. Rine. 1992. Silencers, silencing, and heritable transcriptional states. *Microbiol. Rev.* **56:** 543–560.
61. Li, J. J. and R. Deshaies. 1993. Excercising self restraint: discouraging illicit acts of S and M in eucaryotes. *Cell* **73:** 223–226.
62. Liu, V. F. and D. T. Weaver. 1993. The inonizing radiation induced replication protein A phosphorylation response differs between ataxia telangiectasia and normal human cells. *Mol. Cell. Biol.* **13:** 7122–7131.
63. Lydall, D. and T. Weinert. 1995. Yeast checkpoint genes in DNA damage processing: implications for repair and arrest. *Science* **270:** 1488–1491.
64. Madura, K. and S. Prakash. 1986. Nucleotide sequence, transcript mapping, and regulation of the *RAD2* gene of *Saccharomyces cerevisiae*. *J. Bacteriol.* **165:** 904–923.
65. Madura, K. and S. Prakash. 1990. The *Saccharomyces cerevisiae* DNA repair gene *RAD2* is regulated in meiosis but not during the mitotic cell cycle. *Mol. Cell. Biol.* **10:** 3256–3257.

66. Madura, K. and S. Prakash. 1990. Transcript levels of the *Saccharomyces cerevisiae* DNA repair gene *RAD23* increase in response to UV light and in meiosis but remain constant in the mitotic cell cycle. *Nucleic Acid Res.* **18:** 4727–4732.
67. Madura, K., S. Prakash, and L. Prakash. 1990. Expression of the *Saccharomyces cerevisiae* DNA repair gene *RAD6* that encodes a ubiquitin conjugating enzyme, increases in response to DNA damage and in meiosis but remains constant during the mitotic cell cycle. *Nucleic Acid Res.* **18:** 760–768.
68. Maga, J. A., T. A. McClanahan, and K. McEntee. 1986. Transcriptional regulation of DNA damage responsive (*DDR*) genes in different rad mutant strains of *Saccharomyces cerevisiae* . *Mol. Gen. Genet.* **205:** 275–283.
69. McClanahan, T., and K. McEntee. 1984. Specific transcripts are elevated in *Saccharomyces cerevisiae* in response to DNA damage. *Mol. Cell. Biol.* **4:** 2356–2362.
70. McClanahan, T. and K. McEntee. 1986. DNA damage and heat shock dually regulate genes in *Saccharomyces cerevisiae*. *Mol. Cell. Biol.* **6:** 89–95.
71. Morrison, A. L., H. Araki, A. B. Clark, R. K. Hamatake, and A. Sugino. 1990. A third essential DNA polymerase in *Saccharomyces cerevisiae* . *Cell* **61:** 1142–1151.
72. Morrow, D. M., D. A. Tagle, Y. Shiloh, F. S. Collins, and P. Hieter. 1995. *TEL1*, an *S. cerevisiae* homologue of the human gene mutated in ataxia telangiectasia, is functionally related to the yeast checkpoint gene *MEC1*. *Cell* **81:** 821–830.
73. Murakami, H. and H. Okayama. 1995. A kinase from fission yeast responsible for blocking mitosis in S phase. *Nature (Lond.)* **373:** 807–809.
74. Nasr, F., A. M. Becam, P. P. Slonimski, and C. J. Herbert. 1994. *YBR1013* an essential gene from *S. cerevisiae*: construction of an RNA antisense conditional allele and isolation of a multicopy suppressor. *C. R. Acad. Sci. III* **317:** 607–613.
75. Navas, T. A., Z. Zhou, and S. J. Elledge. 1995. DNA polymerase links the DNA replication machinery to the S phase checkpoint. *Cell* **79:** 29–39.
76. Navas, T. A., Y. Sanchez, and S. J. Elledge. 1996. RAD9 and DNA polymerase epsilon form parallel sensory branches for transducing the DNA damage checkpoint signal in Saccharomyces cerevisiae. *Genes Dev.* **10:** 2632–2643.
77. Paetkau, D. W., J. A. Riese, W. S. MacMorran, R. A. Woods, and R. D. Gietz. 1994. Interaction of the yeast *RAD7* and *SIR3* proteins: implications for DNA repair and chromatin structure. *Genes Dev.* **8:** 2035–2045.
78. Paulovich, A. G. and L. H. Hartwell. 1995. A checkpoint regulates the rate of progression through S phase in *S. cerevisiae* in response to DNA damage. *Cell* **81:** 831–837.
79. Perozzi, G. and S. Prakash. 1986. *RAD7* gene of *Saccharomyces cerevisiae*: transcripts, nucleotide sequence analysis, and functional relationship between *RAD7* and *RAD23* gene products. *Mol. Cell. Biol.* **6:** 1497–1507.
80. Peterson, T. A., L. Prakash, S. Prakash, M. A. Osley, and S. I. Reed. 1985. Regulation of *CDC9*, the *Saccharomyces cerevisiae* gene that encodes DNA ligase. *Mol. Cell. Biol.* **5:** 226–235.
81. Radler-Pohl, A., C. Sachsenmaier, S. Gebel, H.-P. Auer, J. T. Bruder, U. Rapp, P. Angel, H. J. Rahmsdorf, and P. Herrlich. 1993. UV induced activation of AP-1 involves obligatory extranuclear steps including Raf-1 kinase. *EMBO J.* **12:** 1005–1012.
82. Robinson, G. W., C. M. Nicolet, D. Kalainov and E. C. Friedberg. 1986. A yeast excision repair gene is inducible by DNA damaging agents. *Proc. Natl. Acad. Sci. USA* **82:** 1832–1836.
83. Ruby, S. W. and J. W. Szostak. 1985. Specific *Saccharomyces cerevisiae* genes are expressed in response to DNA damaging agents. *Mol. Cell. Biol.* **5:** 74–83.
84. Sachsenmaier, C., A. Radler-Pohl, R. Zinck, A. Nordheim, P. Herrlich, and H. J. Rahmsdorf. 1994. Involvement of growth factor receptors in the mammalian UVC response. *Cell* **77:** 962–971.
85. Samson, L. and K. Singh. Personal communication.

86. Sancar, G. B. 1990. DNA photolyases: physical properties, action mechanism, and roles in dark repair. *Mutat. Res.* **236:** 147–160.
87. Sanchez, Y., B. A. Desany, W. J. Jones, Q. Liu, B. Wang, and S. J. Elledge. 1996. *RAD53* functions downstream of the ATM-like lipid kinases *MEC1* and *TEL1* in *S. cerevisiae* cell cycle checkpoint pathways. *Science* **271:** 357–360.
88. Sassanfar, M. and J. W. Roberts. 1990. Nature of the SOS inducing signal in *Escherichia coli*: the involvement of DNA replication. *J. Mol. Biol.* **212:** 78–95.
89. Savitsky, K., A. Bar-Shira, S. Gilad, G. Rotman, Y. Ziv, L. Vanagaite, D. A. Tagle, S. Smith, T. Uziel, S. Sfez, M. Ashkenazi, I. Pecker, M. Frydman, R. Harnik, S. R. Patanjali, A. Simmons, G. A. Clines, A. Sartiel, R. A. Gatti, L. Chessa, O. Sanal, M. F. Lavin, N. G. J. Jaspers, A. Malcolm, R. Taylor, C. F. Arlett, T. Miki, S. M. Weissman, M. Lovett, F. S. Collins, and Y. Shiloh. 1995. A single ataxia telangiectasia gene with a product similar to PI-3 kinases. *Science* **267:** 1739–1743.
90. Sebastian, J., B. Kraus, and G. B. Sancar. 1990. Expression of the yeast *PHR1* gene is induced by DNA damaging agents. *Mol. Cell. Biol.* **10:** 4620–4627.
91. Sebastian, J. and G. B. Sancar. 1991. A damage responsive DNA binding protein regulates transcription of the yeast DNA repair gene *PHR1*. *Proc. Natl. Acad. Sci. USA* **87:** 11,251–11,255.
92. Sheng, S. and S. M. Schuster. 1993. Purification and characterization of the *Saccharomyces cerevisiae* DNA damage responsive protein 48 (DDRP 48). *J. Biol. Chem.* **267:** 4742–4748.
93. Shiloh, Y. 1995. Ataxia-telangiectasia: closer to unraveling the mystery. *Eur. J. Hum. Genet.* **3:** 116–138.
94. Shinohara, A., H. Ogawa, and T. Ogawa. 1992. *RAD51* protein involved in repair and recombination in *Saccharomyces cerevisiae* is a *RecA* like protein. *Cell* **68:** 457–469.
95. Siede, W., F. Eckardt, and M. Brendel. 1983. Analysis of mutagenic DNA repair in a thermoconditional mutant of *Saccharomyces cerevisiae*. I. Influence of cycloheximide on UV-irradiated stationary phase rev2^{ts} cells. *Mol. Gen. Genet.* **189:** 406–412.
96. Siede, W., F. Eckardt and M. Brendel. 1983. Analysis of mutagenic DNA repair in a thermoconditional mutant of *Saccharomyces cerevisiae*. I. Influence of cycloheximide on UV-irradiated exponentially growing rev2^{ts} cells. *Mol. Gen. Genet.* **189:** 413–416.
97. Siede, W., G. W. Robinson, D. Kalainov, T. Malley and E. C. Friedberg 1989. Regulation of the *RAD2* gene of *Saccharomyces cerevisiae*. *Mol. Microbiol.* **3:** 1686–1697.
98. Siede, W. and E. C. Friedberg. 1992. Regulation of the yeast *RAD2* gene: DNA damage dependent induction correlates with protein binding to regulatory sequences andtheir deletion influences survival. *Mol. Gen. Genet.* **232:** 247–256.
99. Siede, W., A. S. Friedberg, and E. C. Friedberg. 1993. *RAD9* dependent G1 arrest defines a second checkpoint for damaged DNA in the cell cycle of *Saccharomyces cerevisiae*. *Proc. Natl. Acad. Sci. USA* **89:** 7874–7879.
100. Singh, K. K. and L. Samson. 1995. Replication protein A binds to regulatory elements in yeast DNA repair and DNA metabolism genes. *Proc. Natl. Acad. Sci. USA* **91:** 4897–4901.
101. Stack, J. H. and S. D. Emr. 1994. Vps34p required for yeast vacuolar protein sorting is a multiple specificity kinase that exhibits both protein kinase and phosphatidylinositol specific PI 3 kinase activities. *J. Biol. Chem.* **268:** 31,552–31,561.
102. Stein, B., H. J. Rahmsdorf, A. Steffen, M. Litfin, and P. Herrlich. 1989. UV induced DNA damage is an intermediate step in UV induced expression of human immunodeficiency virus type I, collagenase, c-fos, and metallothionein. *Mol. Cell. Biol.* **9:** 5168–5180.
103. Stern, D. F., P. Zheng, D. R. Beidler, and C. Zerillo. 1991. Spk1, a new kinase from *Sacchromyces cerevisiae*, phosphorylates proteins on serine, threonine, and tyrosine. *Mol. Cell. Biol.* **11:** 986–1001.
104. Stillman, B. 1989. Initiation of eucaryotic DNA replication *in vitro*. *Ann. Rev. Cell Biol.* **5:** 197–245.

105. Sun, Z., D. S. Fay, F. Marini, M. Fioani, and D. F. Stern. 1996. Spk1/Rad53 is regulated by Mec1-dependent protein phosphorylation in DNA replication and damage checkpoint pathways. *Genes Dev.* **10:** 395–406.
106. Terleth, C., C. A. van Sluis, and P. van de Putte. 1989. Differential repair of UV damage in *Saccharomyces cerevisiae. Nucleic Acid Res.* **17:** 4433–4439.
107. Terleth, C., P. Schenk, R. Poot, J. Brouwer, and P. van de Putte. 1990. Differential repair of UV damage in *rad* mutants of *Saccharomyces cerevisiae*: a possible function of G2 arrest upon UV irradiation. *Mol. Cell. Biol.* **10:** 4667–4673.
108. Treger, J. M., K. A. Heichman, and K. McEntee. 1988. Expression of the yeast *UBI4* gene increases in response to DNA damaging agents in meiosis. *Mol. Cell. Biol.* **8:** 1132–1136.
109. Treger, J. M. and K. McEntee. 1990. Structure of the DNA damage inducible gene *DDR48* and evidence for its role in mutagenesis in *Saccharomyces cerevisiae. Mol. Cell. Biol.* **10:** 3173–3183.
110. Verhage, R., A.-M. Zeeman, N. de Groot, F. Gleig, D. D. Bang, P. van de Putte, and J. Brouwer. 1994. The *RAD7* and *RAD16* genes, which are essential for pyrimidine dimer removal from the silent mating type loci, are also required for repair of the nontranscribed strand of an active gene in *Saccharomyces cerevisiae. Mol. Cell. Biol.* **14:** 6135–6142.
111. Walker, G. C. 1986. The SOS response of *Escherichia coli*, in Escherichia coli *and* Salmonella typhmurium: *cellular and molecular biology* (J. L. Ingraham, K. B. Low, B. Magasanik, F. C. Neidhardt, M. Schaechter, and H. E. Umbarger, eds.), American Society for Microbiology, Washington, D. C, pp. 1346–1357.
112. Weinert, T. A. and L. H. Hartwell. 1988. The *RAD9* gene controls the cell cycle response to DNA damage in *Saccharomyces cerevisiae. Science* **241:** 317–322.
113. Weinert, T. A., G. L. Kiser, and L. H. Hartwell. 1994. Mitotic checkpoint genes in budding yeast and the dependence of mitosis on DNA replication repair. *Genes Dev.* **8:** 642–654.
114. White, J. M. H., S. R. Green, D. G. Barker, L. B. Dumas, and L. H. Johnston. 1988. The *CDC8* transcript is cell cycle regulated in yeast and is expressed coordinately with *CDC9* and *CDC21* at a point preceding histone transcription. *Exp. Cell Res.* **170:** 223–231.
115. White, J. M. H., A. L. Johnson, N. F. Lowndes, and L. H. Johnston. 1991. The yeast DNA ligase gene *CDC9* is controlled by six orientation specific upstream activating sequences that respond to cellular proliferation but which alone cannot mediate cell cycle regulation. *Nucleic Acid Res.* **19:** 359–363.
116. Xiao, W., K. K. Singh, B. Chen, and L. Samson. 1993. A common element involved in transcriptional regulation of two DNA alkylation genes (*mag* and *MGT1*) of *Saccharomyces cerevisiae. Mol. Cell. Biol.* **13:** 7113–7121.
117. Yagle, K. and K. McEntee. 1990. The DNA damage inducible gene *DIN1* of *Saccharomyces cerevisiae* encodes a regulatory subunit of ribonulceotide reductase and is identical to *RNR3*. *Mol. Cell. Biol.* **10:** 5553–5557.
118. Zhou, Z. and S. J. Elledge. 1992. Isolation of *crt* mutants constituitive for transcription of the DNA damage inducible gene *RNR3* in *Saccharomyces cerevisiae. Genetics* **131:** 841–855.
119. Zhou, Z. and S. J. Elledge. 1993. *DUN1* encodes a protein kinase that controls the DNA damage response in yeast. *Cell* **74:** 1119–1127.
120. Zhou, Z. 1994. Genetic and biochemical analysis of the DNA damage response pathway in yeast *Saccharomyces cerevisiae.* Ph. D. thesis, Baylor College of Medicine, Houston, TX.
121. Zhou, Z. and S. J. Elledge. Unpublished observations.

19

Mismatch Repair Systems in *Saccharomyces cerevisiae*

Gray F. Crouse

1. INTRODUCTION

Mismatch repair has been best defined in bacteria, primarily *Escherichia coli*, *Salmonella typhimurium*, and *Streptococcus pneumoniae* (*see* Chapters 11 and 13). The existence and characteristics of mismatch repair in fungi have been determined from a variety of genetic data obtained over more than two decades. However, 1989 marked the start of a rapid growth in knowledge about mismatch repair in *Saccharomyces cerevisiae*, as cloned mismatch repair genes became available. As will be discussed below, we now know that the mismatch repair system in *S. cerevisiae* shows strong conservation with prokaryotic mismatch repair, but is much more complex. Mismatch repair in vertebrates is similarly complex, and yeast serves as an ideal model system to study the genetics and biochemistry of eukaryotic mismatch repair. Yeast is the first eukaryotic organism whose sequence is completely known, and this has been, and will continue to be, a great aid in identifying genes involved in mismatch repair.

Intense interest in mismatch repair has been sparked by the finding that defects in mismatch repair genes in humans can lead to cancer (*see* Chapter 20, vol. 2). Studies in yeast have shown that not only is the mismatch repair system involved in repair of replication errors, but that parts of the system are important components of recombination pathways. These products can either block or stimulate recombination, playing important roles in genome stabilization, and in both mitotic and meiotic recombination. This chapter will focus on repair and recombination pathways that involve genes related to the bacterial MutS and MutL proteins, since little information is known about other mismatch repair pathways in yeast.

2. EVIDENCE FOR MISMATCH REPAIR IN YEAST AND OTHER EUKARYOTES

2.1. *Genetic Evidence for Mismatch Repair in Yeast*

Among ascomycetes, meiotic gene conversion was first observed in forms yielding eight spores in an ascus. Whereas heterozygous alleles *A* and *a* usually segregate 4*A*:4*a*, gene conversion events were observed as 6*A*:2*a* and 2*A*:6*a* events. As long as 25 years ago, favored models explained gene conversion as the result of repair of heteroduplex

From: DNA Damage and Repair, Vol. 1: DNA Repair in Prokaryotes and Lower Eukaryotes
Edited by: J. A. Nickoloff and M. F. Hoekstra © Humana Press Inc., Totowa, NJ

DNA that was formed as an intermediate during recombination. Another type of aberrant segregation referred to as postmeiotic segregation (PMS) led to 5*A*:3*a* and 3*A*:5*a* events in which the first mitotic division of the haploid cell after meiosis resulted in two daughter cells differing in genotype. These events were explained as being the result of failure to repair the heteroduplex region in a recombination intermediate *(26)*. Thus, gene conversion was an indication of successful mismatch repair, and PMS of the failure of mismatch repair. In four-spored asci, such as those produced by *S. cerevisiae*, gene conversion events are observed as 3*A*:1a and 1A:3*a* segregation. Esposito demonstrated that PMS events in *S. cerevisiae* could be detected as events in which a sectored colony is produced *(27)*. In order to have a nomenclature consistent among all ascomycetes, meiotic events in *S. cerevisiae* are counted as if eight spores were produced such that a sectored colony would yield 1*A* and 1*a*, and the total for a PMS event in *S. cerevisiae* would be 5*A*:3*a* or 3*A*:5*a*. A more complete discussion of recombination and PMS is given in various reviews *(34,73,77)*. A model illustrating the relationship of gene conversion, PMS, and mismatch repair to the process of recombination is shown in Fig. 1. Note that in each case, whether a crossover occurs is dependent on which way the Holliday junction is resolved. If resolution is unbiased, approximately one-half of the events would be accompanied by a crossover (if the crossover product is viable).

In addition to gene conversion and PMS, the existence of a mismatch repair system in yeast similar to that in prokaryotes was indicated by the finding that meiotic ratios of PMS vs gene conversion were similar to what would be expected for mismatch repair in prokaryotes. For example, heteroduplexes with a postulated single base pair insertion never gave rise to PMS in wild-type strains, and heteroduplexes with an A/A, T/T, A/C, or G/T mismatch gave low levels of PMS *(110)*. All of these mismatches are repaired efficiently in prokaryotes. However, heteroduplexes that would give a C/C mismatch or a large loop mismatch gave high levels of PMS; these same mismatches are repaired poorly in prokaryotes *(110)*. Similarly, meiotic recombination in a slightly different system found that C/C mispairs were poorly repaired, whereas all other mispairs (including G/G mispairs) were corrected at approximately the same efficiency *(21)*.

The hypothesized association of mismatch repair, gene conversion, and PMS was considerably strengthened by the finding of mutants in *S. cerevisiae* that increased PMS, decreased gene conversion, and increased mitotic mutation rates *(111)*. Mutations in four separate *PMS* loci were obtained (originally named *COR* for correction-deficient) *(34)*, but only *PMS1* was further characterized *(111)*. It was not until 1989 that *PMS2* and *PMS3* were shown to be necessary for mismatch repair *(50)*. In 1994, additional *PMS* alleles were obtained with the same mutant screen as before, and it was also demonstrated that *PMS2* and *PMS4* were allelic *(46)*. The relation between the mismatch repair mutants obtained in meiotic screens and mitotic mismatch repair was indicated by the fact that the *pms* mutants were mitotic mutators *(111)*. This result indicated that the same genes involved in repair of heteroduplex DNA formed during meiotic recombination were also involved in replication repair.

Mismatch repair in mitotic cells was demonstrated in transformation studies with plasmids containing specific mismatches *(8)*. Based on corepair frequencies of mis-

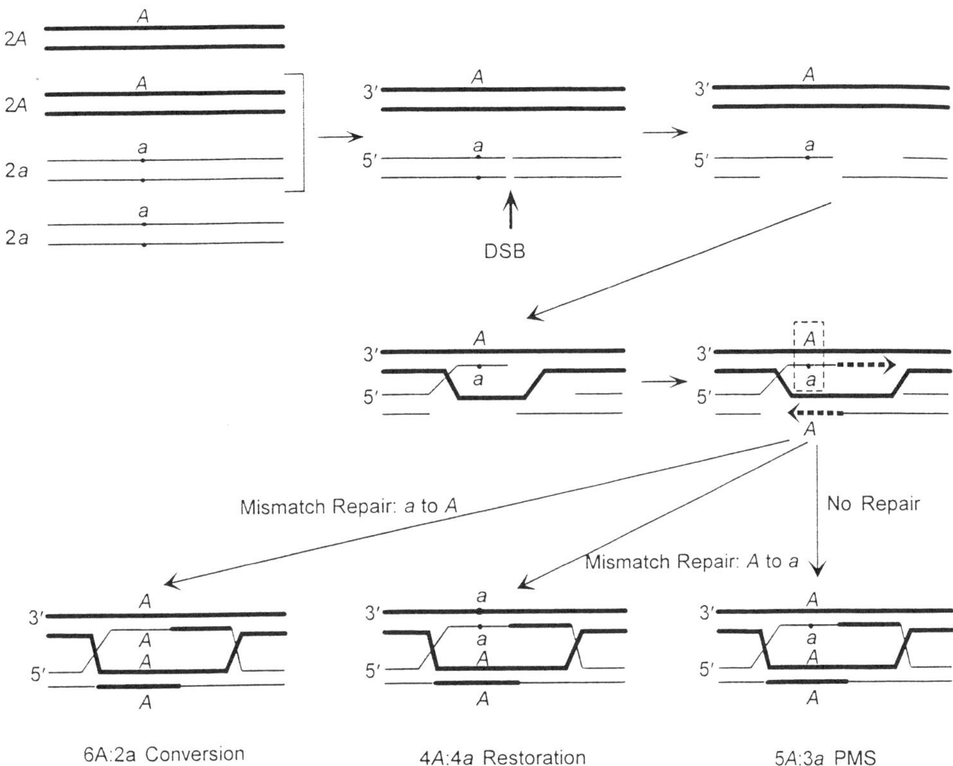

Fig. 1. DSB model of recombination leading to gene conversion or PMS. Each pair of lines represents a chromatid present after DNA replication; thick lines represent one homolog, and the thin lines represent the other homolog. In this diagram, it is assumed that the two homologs are identical with the exception of a mutation in gene *A*, represented by •, to give allele *a*. A DSB is made, followed by resection of the 5'-ends of the break, and an invasion of one of the 3'-ends of the DNA into the homolog. Subsequent DNA synthesis and branch migration lead to the formation of a structure with an *A*/*a* mispair. The mispair could be corrected in one of two ways, one direction leading to restoration, and the other to gene conversion. When the DSB is close to the mutation, it appears from various lines of evidence that there is a strong bias for conversion events, presumably using the end of the invading strand as a marker for biased mismatch repair. If the mispair is not corrected, the result is 3*A* and 1*a* strands from the recombination, which combined with the 2*A* and 2*a* alleles on the other chromatids results in a 5*A*:3*a* segregation and PMS. Note that in all cases, the Holliday junctions could be resolved in one of two ways, resulting in crossover or noncrossover products.

matches, excision-resynthesis tracts were shown often to be <1 kb. Similar experiments indicated that in wild-type strains, single base pair loops were repaired very efficiently (94–97%), and that C/C mispairs and a 38-bp loop were repaired very inefficiently (24–40%), with the rest of the mispairs being repaired with intermediate efficiencies *(50)*. This general repair profile is comparable to that observed in prokaryotes, and further strengthened the notion that there were similar mismatch repair systems in yeast and prokaryotes.

2.2. Biochemical Evidence for Mismatch Repair in Yeast

In vitro assays using mitotic yeast cell extracts and artificial heteroduplex molecules containing different mismatches indicated that certain substrates, such as a 4-bp insertion/deletion and A/C or G/T mismatches, were repaired efficiently *(68)*. However, other mismatches were repaired poorly, if at all. In addition, the repair tracts were only 10–20 nucleotides in length. Although repair was observed, it is not known what repair systems were contributing to the results seen; at the time the experiments were done, it was not possible to demonstrate the dependence of the observed repair on known mismatch repair genes.

An endonuclease activity was found in yeast extracts that nicked specifically at base mismatches with varying efficiencies *(15)*. This activity has not been further characterized, but it was suggested that the activity could be part of a mismatch repair pathway. A similar type of nicking activity in mammalian cells was later identified as an activity of topoisomerase I *(113)*. The relationship of the yeast nicking activity to any type of mismatch repair is at this time unclear.

3. GENETIC ANALYSIS OF MISMATCH REPAIR USING *PMS1* MUTANTS

3.1. Analysis of PMS1 in Meiosis

From 1985 until the isolation of MutS and MutL homologs in *S. cerevisiae* by PCR, the only available characterized mismatch repair mutant was *PMS1*. Although it is now known that *PMS1* is a homolog of MutL (*see* Section 4.1.), at the time of the early experiments, the nature of the *PMS1* gene was unknown. *pms1* mutations are recessive, reduce spore viability, and at most of 13 loci examined, PMS was significantly increased *(111)*. The percentage of aberrant segregation events (the sum of 6:2 and 2:6 gene conversion events and 5:3 and 3:5 PMS events plus a few much rarer events) that were owing to PMS ranged from 20 to 80%. In wild-type cells, no PMS was observed at nine loci, and 5–60% PMS was seen at four loci. The increase in PMS was interpreted as the failure to repair mismatches in heteroduplex DNA formed during recombination. At some loci, total aberrant segregation increased in *pms1* strains along with the increase in PMS. It should be noted also that there were gene conversion events found in *pms1* strains at all loci, indicating either that these did not require mismatch repair, or that even in *pms1* strains, some mismatch repair still takes place, either owing to partial *PMS1* activity or repair pathways not requiring *PMS1*. The double-strand break (DSB) or gap repair model *(106)* is one mechanism for gene conversion that does not require mismatch repair (Fig. 2). As discussed in Section 7.2., and more extensively elsewhere (*34,73,77*; *see* Chapter 16), a variety of physical and genetic evidence suggests that gap repair is unlikely to be the predominant mechanism for gene conversion in wild-type cells.

The length of heteroduplexes formed in recombination intermediates has been estimated from co-PMS events using markers at known positions. Such analysis showed that for two *arg4* mutations, *arg4-16* and *arg4-17*, heteroduplex could frequently cover 214-bp *(34,110)*, and a later analysis using two *HIS4* alleles demonstrated that heteroduplexes could be >1.8-kbp *(20)*. The length of repair tracts during mismatch repair can be estimated from the effects of low PMS alleles on high PMS alleles. High PMS

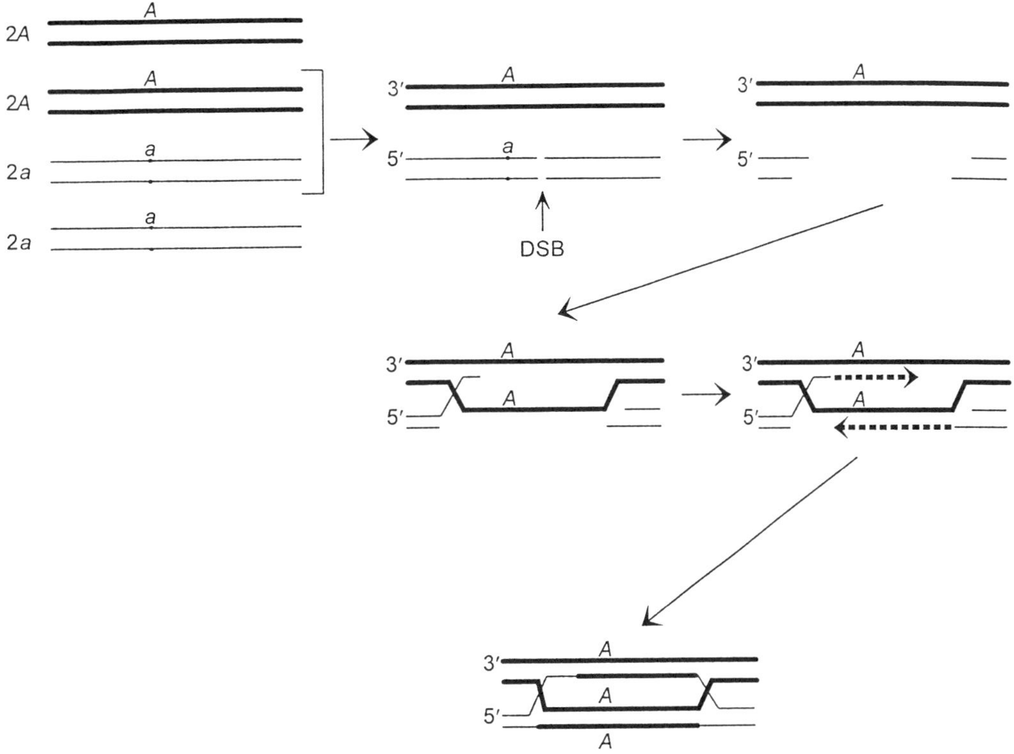

Fig. 2. Double-strand gap model of repair. The lines and alleles are as in the previous figure. Recombination is initiated by a DSB that is enlarged by resection of both 5'- and 3'-ends through the mismatch in allele *a*. Strand invasion, DNA synthesis, and branch migration are as in Fig. 1, but in this case because the invading strand has been resected past the *a* allele, there is no mismatch in the heteroduplex to be repaired. Note that because both strands have been resected, it is possible that there could be a shorter amount of DNA in the invading strand to form a heteroduplex, which might have implications for the stability of the recombination complex compared to the DSB model *(see text)*. As in Fig. 1, the Holliday junctions can be resolved to give either a crossover or noncrossover product.

alleles are presumably poorly recognized and repaired by the mismatch repair system, whereas low PMS alleles are readily recognized. If the alleles are spaced closely enough, an excision tract originating from the low PMS allele may include the high PMS allele. The effect of the low PMS allele *arg4-17* on the high PMS allele *arg4-16* suggested that repair tracts would span 200-bp about half the time *(20,33,110)*. The effect of the low PMS allele *his4-712* on the high PMS allele *his4-IR9* showed that repair tracts could extend 900-bp about one-third of the time *(20)*, and other markers in this locus indicated that almost all repair tracts were longer than 26-bp. Repair tracts >1-kbp are also seen in the MutHLS mismatch repair pathway in *E. coli* (Chapter 11).

3.2. *Analysis of* PMS1 *in Mitotic Cells*

In *S. cerevisiae*, mitotic mutation rates are frequently measured as forward mutation rates to canavanine (CAN) resistance at the *CAN1* locus *(109)*. The *CAN1* locus encodes an arginine permease that renders cells susceptible to the toxic drug canavanine; mutation in the *CAN1* gene thus blocks transport of canavanine into the cells and yields canavanine resistance. The gene is a particularly good target for mutation because of its large size (1770-bp) *(39)* and simple selection. The disadvantage of this assay is that until very recently, no one had attempted to analyze the nature of the mutants; the tacit assumption was that most mutations in mismatch repair defective backgrounds were point mutations. The actual nature of these mutations will be discussed in Section 5.2. By this assay, mutation rates in *pms1* strains are up to 45-fold higher than in wild-type strains, as would be expected for strains deficient in mismatch repair *(111)*. However, *pms1* strains are not hypersensitive to UV or ionizing radiation *(111)*.

Transformation of plasmids with artificial heteroduplexes into *PMS1* and *pms1* strains indicated that repair was reduced in *pms1 (9)*. Also, in the absence of Pms1p, much of the repair that was observed resulted from deletion of the insertion loops, whereas in wild-type cells, there was not a strong bias toward an insertion or deletion. Later work showed that wild-type yeast could repair any of the single base pair mismatches, but with variable efficiencies *(7)*. Some of the relative repair efficiencies were expected, such as G/T being among the best repaired species and C/C being among the poorest, but other results were unexpected. For example, a T/T mismatch was repaired less efficiently than a C/C, which contrasts with meiotic PMS frequencies *(110)* or repair seen in prokaryotes. Repair efficiency of certain mismatches was reduced in *pms1* strains, but surprisingly some mismatches were unaffected and in even the most severely affected mismatches over 25% of the molecules showed repair. These unexpected results might reflect sequence context effects, that mitotic repair is different than meiotic repair, or that repair in transformed plasmids does not perfectly reflect chromosomal repair. Another possibility is that some of the observed repair could have been owing to a second mismatch repair system that does not involve *PMS1*; a minor mismatch repair system that recognizes C/C mismatches has been observed in *Schizosaccharomyces pombe* (*see* Section 14.).

The role of Pms1p in repair of replication errors was indicated by analysis of double mutants that were defective in the proofreading exonuclease activity of DNA polymerase III (pol III; encoded by *POL3*) combined with *PMS1* mutations *(66)*. Each mutation alone led to elevated mutation rates, and diploids deficient in both proofreading and *PMS1* (haploids were not viable) showed a multiplicative increase in mutation. These experiments suggest that *PMS1*-mismatch repair can act in a common pathway with the pol III 3' $\rightarrow$ 5' exonuclease for replication error correction.

An analysis of mating type switching in a strain with a mutation in the *MAT***a** locus that reduces switching, the *MAT***a**-*stk* mutation, was consistent with the above view of mismatch repair and recombination *(84)*. The *MAT***a**-*stk* mutation is a single base pair mutation 8-bp from the HO cleavage site, and its presence caused 23% of the switches in wild-type cells to exhibit postswitching segregation (PSS) in which one of the two progeny had retained the *stk* mutation. PSS is thus a mitotic equivalent to PMS and increased to 59% in *pms1* strains, indicating that heteroduplex DNA involving a sequence just 8 bp from the DSB is normally part of the recombination complex in mating type switching, and that there is usually little degradation of the 3'-end of the DSB. Gene conversion in mating type switching is normally strongly biased in favor of converting the invading strand, suggesting that the end of the invading strand can serve as a signal for mismatch correction. In *pms1* strains, there was still gene conversion seen in 15% of the switches, indicating either mismatch correction via a *PMS1*-independent pathway or conversion by gap repair. Mismatch correction of the *stk* mutation is very rapid, occurring with a half-life of 6–10 min *(36)*. Thus, mismatch correction is one of the first events in recombination and occurs faster than the 60 min required to complete repair of the DSB *(36)*.

4. *MUTS* AND *MUTL* HOMOLOGS

4.1. PMS1

The cloning of *PMS1 (52)* made possible construction of strains that were known to be nulls, alleviating any concerns about partial activity in *pms1* strains. *PMS1* is a homolog of the bacterial *mut*L and *hex*B genes, confirming its involvement in DNA mismatch repair, and indicating a conservation of mismatch repair functions between prokaryotes and eukaryotes.

4.2. Cloning Homologs by PCR

An important development in the study of DNA mismatch repair was the finding in 1989 of mammalian MutS homologs *(35,59)*. These homologs were discovered by virtue of their tight linkage to the mouse *(59)* and human *(35) DHFR* genes. There was no information on the function of these genes, but their homology to MutS and the existence of a yeast MutL homolog further suggested that eukaryotic mismatch repair shared the same fundamental mechanism with prokaryotes. Conserved regions were used to design degenerate primers for PCR, which yielded three yeast MutS homologs: *MSH1*, *MSH2 (85,86)*, and *MSH3 (70)*. In spite of the obvious success of this approach, the limitations of this method can be seen in the fact that the primers used to isolate *MSH1* and *MSH2* did not detect *MSH3*, and the primers used for *MSH3* did not detect *MSH1* and *MSH2*. None of these primers, as well as numerous other primer pairs, have detected other MutS homologs in yeast (Kolodner, personal communication), although three other MutS homologs are present in the yeast genome (*see* Section 4.3.). The unexpected finding of multiple MutS homologs in yeast suggested that there might be additional MutL homologs, and a PCR approach similar to that used with the MutS homologs was undertaken. This approach was also successful, leading to the isolation of *MLH1* and *MLH2 (82)*. All of the newly found genes have been named *MSH* for MutS homolog, or *MLH* for MutL homolog. *PMS1* is the only exception to this scheme and is considered to be one of the *MLH* genes.

4.3. Other MutS and MutL Homologs

Two other yeast MutS homologs were identified in screens designed to find meiotic specific genes. *MSH4* was obtained in a screen for random *lacZ* fusions expressed during meiosis *(14,90)*. *MSH5* was obtained in a screen for mutants defective in recombination between homologous chromosomes in meiosis *(40)*. The sequencing of the yeast genome, completed in early 1996, was instrumental in revealing homologs that had not been found by other approaches. The first of these homologs to be identified was *MSH6*, which is now known to be a homolog of the human p160/GTBP protein (*24,74,75*; Chapters 7 and 8, vol. 2). The yeast sequence also revealed a previously

Table 1
Yeast MLH and MSH Genes

Gene	Function	Other eukaryotic homologs
MSH1	Mitochondrial repair	?
MSH2	Base mismatch and loop repair; recombination; forms heterodimers with Msh3p and Msh6p	Homologs found in *S. pombe*, *Drosophila*, *Xenopus*, mouse, rat, and human
MSH3	Loop repair and recombination; forms heterodimer with Msh2p	Homologs in *S. pombe*, mouse, and human
MSH4	Meiotic recombination; heterodimer with Msh5p (?)	?
MSH5	Meiotic recombination; heterodimer with Msh4p (?)	?
MSH6	Loop and base mismatch repair; forms heterodimer with Msh2p	Homolog in human (GTBP/p160) and mouse
PMS1	Base mismatch and loop repair; forms heterodimer with Mlh1p	Homologs in *S. pombe*, mouse, and human
MLH1	Base mismatch and loop repair; forms heterodimer with Pms1p	Homologs in mouse and human
MLH2	?	?
MLH3	?	?

unknown *MLH* gene, *MLH3*, making a total of six MutS homologs and four MutL homologs in the yeast genome.

Of the six *pms* mutants isolated *(46,111)*, genes have been isolated from all but one, and found to be one of the known *MLH* and *MSH* genes. The first cloned was *PMS1*, which is an MutL homolog *(52)*, *PMS2* and *PMS4* are *MLH1* (*45*; W. Kramer, personal communication), *PMS3* is *MSH6 (61)*, and *PMS5* is *MSH2 (46)* (*see* Table 1). Only *PMS6* remains unidentified.

4.4. Sequence Relationships of Yeast MutS and MutL Homologs

Alignment of MutS homologs reveals regions of similarity throughout the proteins *(70)*. Similarity is greatest in the nucleotide binding domain at the C-terminal end of the proteins. This domain is currently the only region of the protein whose function has been demonstrated experimentally, since a point mutation affected the ATPase activity of the *S. typhimurium* MutS *(37)*. A dendrogram of the sequence relationships of the MutS homologs is shown in Fig. 3. It is particularly striking that the yeast *MSH2*, *MSH3*, and *MSH6* genes show a greater degree of similarity with their mammalian counterparts than they do with each other, suggesting that understanding the function of the yeast genes should help in determining the role of the homologous genes in other eukaryotic organisms.

Alignment of the known MutL homologs is shown in Fig. 4, and a dendrogram illustrating the relationships is shown in Fig. 5. Not included are the large number of human genes that have been found to be related to the human *PMS2* gene *(41,71)*. Other than the human *PMS2* gene itself, the function of these other family members is

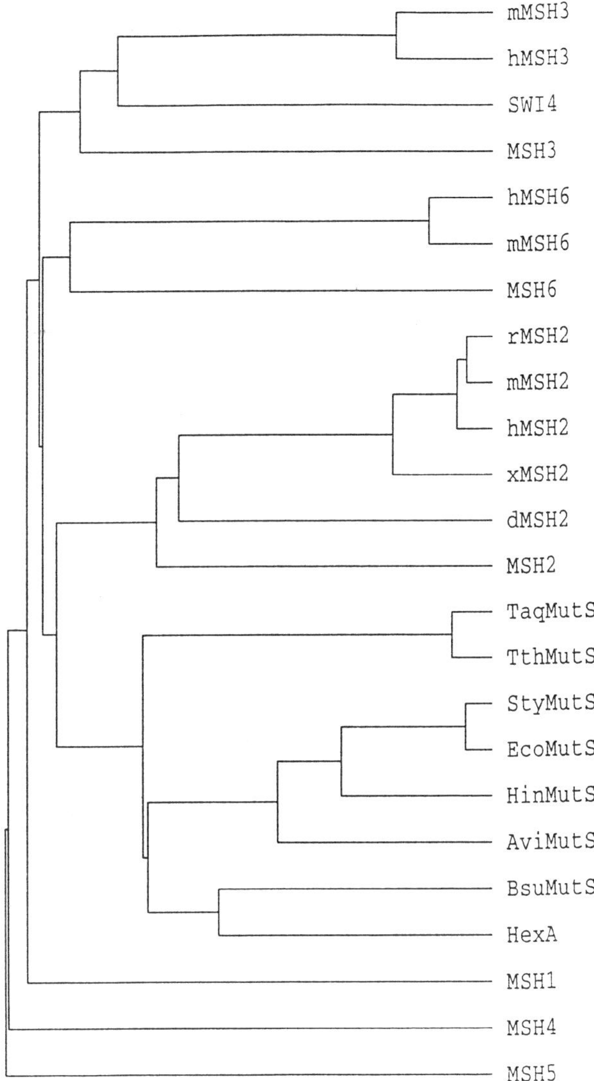

Fig. 3. Dendrogram of MutS-related proteins generated with the PILEUP program of GCG *(23)*. Although the dendrogram is not a phylogenetic reconstruction, it does show the relative similarity of sequences. The distance along the horizontal branch is proportional to the differences between sequences. The sequences and accession numbers are as follows: mMSH3 = mouse homolog of *MSH3* (formerly *Rep3*), M80360; hMSH3 = human homolog of *MSH3* (formerly *Dup1*), J04810; SWI4 = *S. pombe* Swi4p, X61306; MSH3 = Msh3p, M96250; hMSH6 = human homolog of *MSH6*, also known as p160/GTBP, U28946; mMSH6 = mouse homolog of *MSH6*, U42190; MSH6 = Msh6p, Z47746; rMSH2, mMSH2, hMSH2, xMSH2, and dMSH2 = rat, mouse, human, *Xenopus*, and *Drosophila* homologs of *MSH2*, respectively: X93591, X81143, U03911, L26599, U17893, and M84170 for Msh2p; TaqMutS = *Thermus aquaticus* MutS, U33117; TthMutS = *Thermus thermophilus* MutS, D63810; StyMutS = *S. typhimurium* MutS, M18965; EcoMutS = *E. coli* MutS, M64730; HinMutS = *Haemophilus influenzae* MutS, U32699; AviMutS = *Azotobacter vinelandii* MutS, M63007; BsuMutS = *Bacillus subtilis* MutS, U27343; HexA = *S. pneumoniae* HexA, M18729; MSH1 = Msh1p, M84169; MSH4 = Msh4p, U13999; MSH5 = Msh5p, L42517.

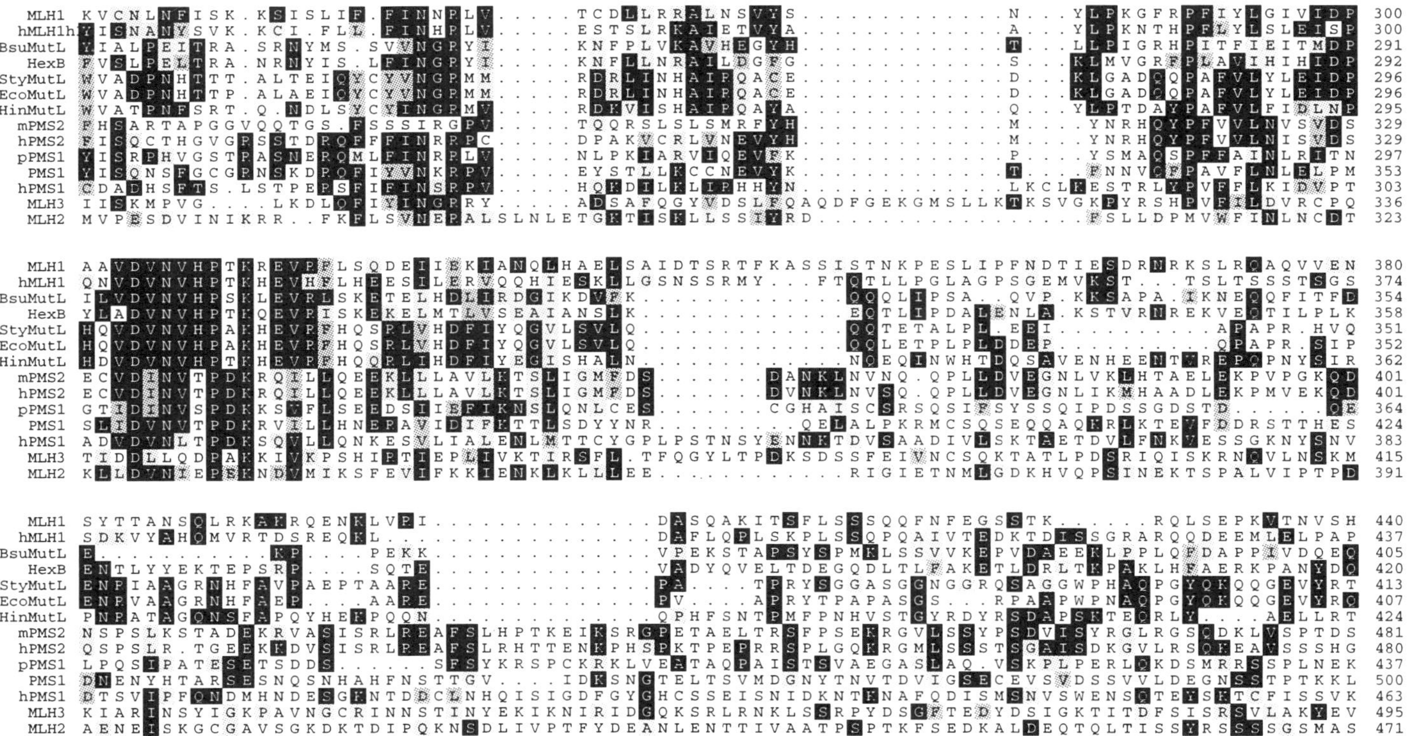

Fig. 4. Comparison of the protein sequences of all known MutL homologs (with the exception of human homologs related to *hPMS2*). Sequences were aligned with the PILEUP program of GCG *(23)* and PrettyBox by Rick Westerman. Amino acids in a solid box are identical in at least three sequences. The values of nonidentical comparisons are determined by a scaled log odds form of the amino acid similarity matrix of Schwartz and Dayhoff *(97)*, such that perfect matches are set to 1.5 and other values range from 1.4 to –1.2 with a mean of 0. Amino acids in darkly shaded boxes have a similarity value of 1.0–1.5 and, in lightly shaded boxes, have a similarity value of 0.5–1.0. The sequences and accession numbers are as follows: MLH1 = Mlh1p, U07187; hMLH1 = human homolog of *MLH1*, U07418; BsuMutL = *B. subtilis* MutL, U27343; HexB = *S. pneumoniae* HexB, M29686; StyMutL = *S. typhimurium* MutL, M29687; EcoMutL = *E. coli* MutL, Z11831; HinMutL = *H. influenzae* MutL, U32801; mPMS2 and hPMS2 = mouse and human MutL homologs, U28724 and U14658; pPMS1 = *S. pombe* homolog of PMS1, X96581; *PMS1* = Pms1p, M29688; hPMS1 = human MutL homolog, U13695; MLH3 = Mlh3p, Z73520; MLH2 = Mlh2p, Z73207 *(continued on next page)*.

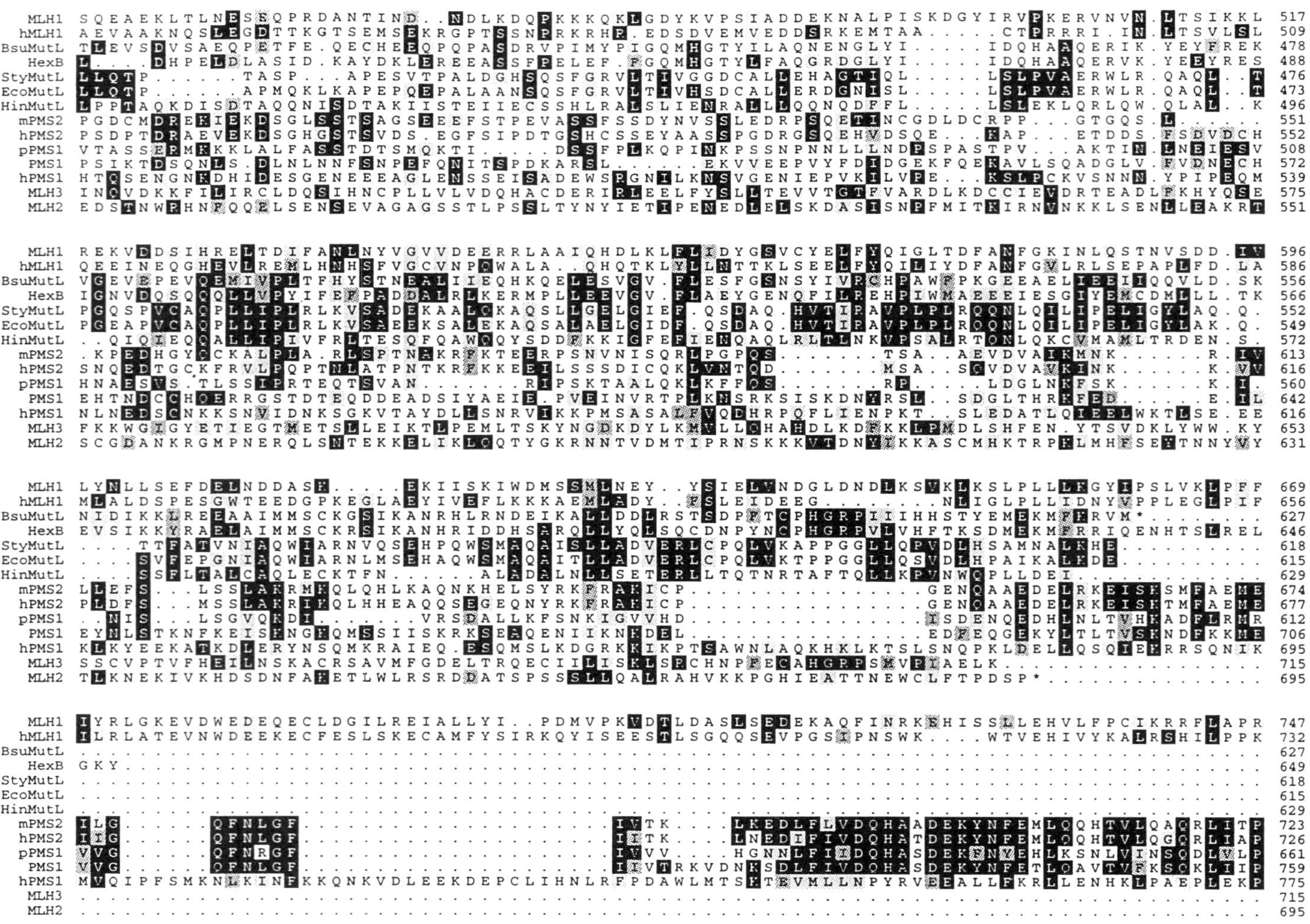

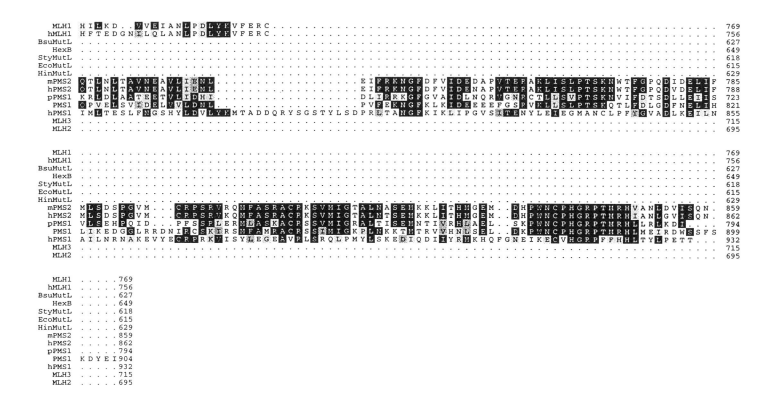

Fig. 4 (*continued*)

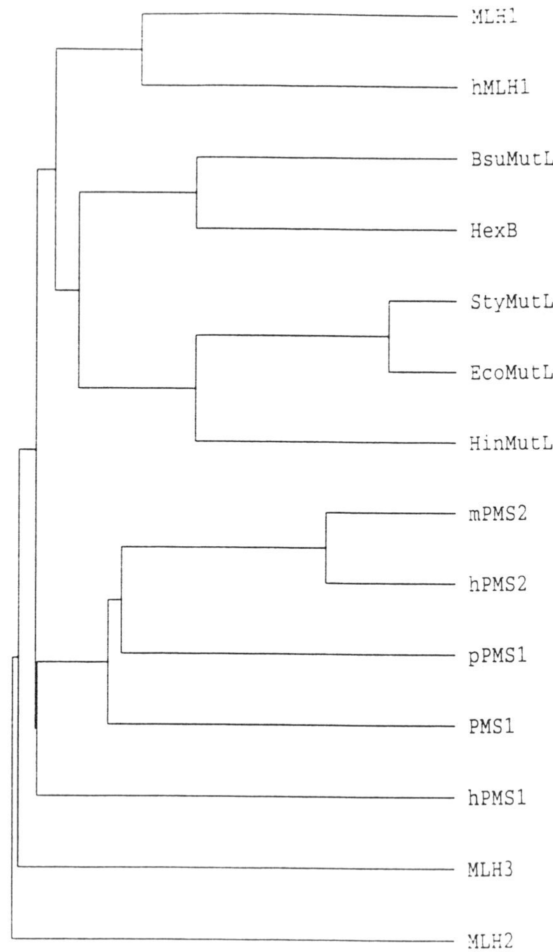

Fig. 5. Dendrogram of MutL-related proteins.

not known, and the various *PMS2*-related genes show substantial sequence variation outside a core region corresponding to the first five exons of the human *PMS2* gene. Even excluding these genes, it is interesting to note that the MutL homologs as a whole show much less sequence conservation with each other than do the MutS homologs. This lack of conservation has made finding new homologs more difficult, and suggests that the functions and interactions of MutL homologs may vary more than that of the MutS homologs.

5. INVOLVEMENT OF *MSH* AND *MLH* GENES IN PMS, GENE CONVERSION, AND NUCLEAR MUTATION RATES

5.1. Postmeiotic Segregation and Gene Conversion

Many of the *MSH* and *MLH* genes have been tested for their effect on PMS and gene conversion. The null phenotype of *pms1* mutants is indistinguishable from that of the earlier *pms1* strains *(52)*. The effects of *MSH1* on PMS have not been examined, but none would be expected, since *MSH1* appears to function solely in mitochondria *(16,86)*. The effect of *MSH2* on gene conversion was studied using two different *his*4 alleles. For one allele, total aberrant segregation was similar in *MSH2* and *msh2* strains, whereas PMS increased from 0 to 46.4%. At the other allele (at the low end of a gene conversion polarity gradient; *see* Section 7.2.), total aberrant segregation rose from 0.8 to 6.5%, and PMS increased from 0 to 30% *(86)*. For single base pair mismatches examined in a variety of heteroalleles, total aberrant segregation was low in *msh2* and wild-type strains, whereas PMS increased to >70% in *msh2* strains *(3)*. For 1-, 2-, and 4-bp insertions, aberrant segregation generally increased in *msh2* strains compared to wild-type. There was always a significant increase in PMS in the *msh2* strains, ranging from a total of 30% to 90%, but with about half of the alleles examined, there was no significant decrease in gene conversion *(3)*. In a limited study, comparable PMS and conversion were found in *msh2* and *pms1* strains, and in *msh2 pms1* double mutants *(3)*.

The effect of *MSH3* on gene conversion and PMS was tested using heteroalleles at four loci. There was no change in total aberrant segregation, but two of the four alleles showed an increase in PMS in *msh3* strains, with one showing a statistically significant increase, from 2.9 to 22% *(70)*. These results cannot be compared directly to those above, but it seems clear that *MSH3* has a much smaller effect on PMS than either *MSH2* or *PMS1*.

The effect of *MSH4* on gene conversion and PMS was examined at five loci *(90)*. Gene conversion increased slightly in *msh4* strains, with only one locus showing a statistically significant increase. There was no increase in PMS at any locus in *msh4* strains. *MSH5* behaves similarly to *MSH4*; using the same background as for *MSH2 (86)*, gene conversion increased slightly in *msh5*, whereas there was no change in PMS *(40)*.

mlh1 and *pms1* mutants showed only slight increases in total aberrant segregation at four loci, whereas PMS increased from 0% to between 33 and 88% at all loci *(82)*. There was no difference between single mutant and the *mlh1 pms1* double mutant strain.

One of the difficulties in analyzing data on gene conversion and PMS is that even in mismatch repair-defective strains, the number of meiotic products that show a PMS event is rarely more than a few percent, and so hundreds of tetrads must be examined. If most gene conversions take place through a heteroduplex intermediate as illustrated in Fig. 1, then one would expect that a defective mismatch repair machinery would decrease the number of gene conversion events and increase the number of PMS events. The second part of this prediction is true; PMS is clearly increased at many different loci in mutant strains. However, the case with gene conversion events is not so clear. Although some loci do show a decrease in gene conversion events in mutant strains, in many cases, gene conversion stays essentially the same, resulting in an increase in total aberrant segregation. For a further discussion of this phenomenon, *see* Section 7.2.

5.2. Mutation Rates

Another expectation of defects in mismatch repair is that mutation rates should increase. The increase in forward mutation rate in *pms1*Δ strains was 20- to 30-fold at *CAN1* as measured by CAN resistance, comparable to the 50-fold increase observed in the original *pms1* mutants *(52)*. In contrast, the reversion rate of the *hom3-10* frameshift allele was elevated up to several thousand-fold *(52)*. *mlh1* mutants have similarly increased mutation rates, but *pms1 mlh1* double mutants have the same phenotype as either single mutant *(82)*.

msh2 strains showed a similar increase in mutation rates to CAN resistance and a slightly lower rate of *hom3-10* frameshift reversion than *pms1* strains *(61,86)*. In contrast, *msh3* strains had less than a two-fold increase in CAN resistance, and a 4- to 12-fold increase in reversion of *hom3-10 (61,70)*. *msh6* mutants had increases in CAN resistance approximately equal to that for *msh2* mutants, and reversion rates of the *hom3-10* allele similar to that in *msh3* strains *(61)*. Mutation rates in *msh3 msh6* strains were similar to those in *msh2* strains, and the triple *msh2 msh3 msh6* strains showed essentially the same rates of mutation as the *msh3 msh6* double mutants *(48,61)*. The implications of this finding are considered in Section 10. Mutation rates in *msh1*, *msh4*, and *msh5* strains are normal *(40,86,90)*. In summary, *MSH2*, *PMS1*, and *MLH1* have similar effects on nuclear mutation rates; *MSH1*, *MSH4*, and *MSH5* have no effect on nuclear mutation rates; and *MSH3* and *MSH*6 appear to have partially overlapping function.

Why the large difference in increase of mutation rates as measured by CAN resistance compared to frameshift reversion? The tacit assumption has been that most mutations in the *CAN1* gene generated in mismatch repair-defective strains are owing to point mutations. It is very likely that most of the spontaneously occurring mutations in *CAN1* are point mutations. There are no gross alterations in *CAN1* structure in most spontaneous mutations *(78)*, and a similar forward mutation assay in the *URA3* gene found that almost 90% of the mutations were point mutations *(58)*. However, in contrast to expectations, a recent analysis of 20 *msh2 can1* mutants revealed that 85% were the result of frameshift mutations, as were 68% of *msh3 msh6 can1* mutants *(61)*. These data suggest that the mismatch repair system has relatively little effect on repair of mismatched bases compared to frameshift errors. If most of the spontaneous mutations in *CAN1* are owing to point mutations and not frameshifts, the increase in overall mutation rate would be less than that of *hom3-10*, since it would be the minor class of mutations that would be greatly increased. In agreement with this hypothesis, the rate of mutation in *CAN1* was measured to be approx 100-fold higher than the rate of *hom3-10* reversion in wild-type cells, whereas the two rates were similar in *pms1*Δ strains *(52)*. Almost all of the identified frameshift mutations in the *CAN1* gene in mismatch repair-defective cells were owing to slippage events in short homopolymeric runs of nucleotides. These are the type of frameshift mutations that Streisinger et al. predicted would be most common *(102)*. A more extensive study using a plasmid-based *SUP4-o* assay found that the vast majority of mutations in *msh2* or *pms1* strains were frameshifts and not point mutations (Kunz, personal communication). An extensive spectrum of frameshift mutations in a region of the *LYS2* gene has found that almost all of the frameshift mutations (-1 frameshifts are selected) in both wild-type and mismatch repair-defective cells occur in homopolymeric runs, as predicted (Jinks-Robertson, personal communication). However, the spectra are quite different in various mismatch repair-defective cells, an effect that is not yet understood. There is evidence, at least with prokaryotic polymerases, that proofreading of slippage events in homopolymeric runs becomes less efficient as the length of the run increases, presumably because looped bases could form that would be relatively far from the proofreading exonuclease and thus less likely to undergo correction *(53)*. This would suggest a greater dependence on the mismatch repair system for frameshift correction than for correction of point mutations. Together, these results suggest that the major role of the mismatch repair system in replication repair is not in correction of base mismatches, but rather in correction of frameshift mutations, principally owing to DNA polymerase slippage. This suggestion would also explain the large effect of the mismatch repair system on dinucleotide repeat stability (*see* Section 6.). The mismatch repair system also repairs point mutations that escape proofreading, and a plasmid-based *SUP4-o* assay found that transition and transversion mutations were repaired at the same efficiency (Kunz, personal communication).

The action of the mismatch repair system apparently depends in part on the source of the mutations. For example, 1-bp deletions that occur in *rad52* cells, presumably because of error-prone repair, were not a substrate for the mismatch repair system, since the increase in reversion rate in a *rad52 pms1* strain was additive rather than synergistic *(107)*.

Although transcription-coupled repair is defective in both bacterial and mammalian cells that are defective in mismatch repair (*see* Chapter 9), mutations in *MSH2*, *MSH3*, *PMS1*, or *MLH1* have no effect on transcription-coupled repair in yeast *(104)*. This difference between yeast and the other systems is not yet understood.

6. DINUCLEOTIDE REPEAT STABILITY

6.1. Rates of DNA Polymerase Slippage

The effect of mismatch repair functions on stability of simple repeats can be measured using selectable markers containing GT repeats *(101)*. Defects in *MSH2*, *PMS1*, or *MLH1* increase GT tract instability by several hundred-fold *(101)*. This suggested that the microsatellite instability typical of some hereditary nonpolyposis colorectal cancer (HNPCC) cells might be the result of defects in mismatch repair genes, a finding subsequently confirmed (*12,29,57,72,76*; Chapter 20, vol. 2). *MSH3* also has a significant effect on simple repeat instability, but the magnitude is 10–30% of the level of the other mismatch repair genes *(100)*. No human diseases have been found to be associated with defects in the human *MSH3* gene; given the weaker phenotype of the yeast *MSH3* gene compared to the other genes, defects in the human *MSH3* gene might result in a weaker phenotype. The effect of *MSH6* is 5- to 15-fold less than that of *MSH3*, but *msh3 msh6* mutants show the same rate of repeat instability as that of *msh2* mutants *(48)*.

Another system has been used to select for deletions in a mutated *LYS2* gene that would give reversion to a wild-type phenotype *(107)*. In this assay, deletions in *PMS1*, *MSH2*, or *MSH3* increased reversion rates by 3- to 10-fold, essentially all 1-bp deletions in homopolymeric runs, and in contrast to the above results with dinucleotide repeats, the effects of *MSH2* and *MSH3* were similar. With the possible additivity of *MSH2* and *MSH3* reversion rates, there was no difference in reversion rates of single or double mutants of *PMS1*, *MSH2*, or *MSH3*. Based on double mutants with a *pol3-t* proofreading mutant, it appears that the mismatch repair system can recognize and correct loops of 1 and 7 bases, but has no effect on loops as large as 31 bases.

6.2. Products of DNA Polymerase Slippage

The products of simple repeat instability are deletions or insertions that are multiples of the repeat length. The distribution of these products is nonrandom. In wild-type and mutant strains, most tract length changes in dinucleotide repeats are either +2 or –2 bp, with a few +4 and –4 changes. In one case, 4 out of 24 changes in wild-type cells showed much larger deletions ranging from –10 to –16 *(100)*. These larger deletions are not seen in the mutant strains, which suggests that the mismatch repair genes investigated do not recognize such large deletions, and that the frequency of large deletions is thus not increased in mismatch-deficient strains.

Insertions are caused by a failure to repair slippage loops in the primer strand during replication, whereas deletions are caused by a failure to repair loops in the template strand. In wild-type strains, the ratio of deletions to insertions was measured to be 0.37 *(100)*. In *pms1* and *mlh1* strains, there was a slight change in this ratio to 0.71 and 0.67, respectively *(101)*. The ratio in *msh6* strains is similar to that in wild-type cells *(48)*. A dramatic shift was seen in *msh2* strains, with a bias in favor of deletions, yielding a deletion: insertion ratio of 2.1 *(100)*. Although *msh3* strains do not show as great a rate of instability as *msh2* strains, the shift in product bias was even more striking, with a deletion: insertion ratio of 5.5 *(100)*. Therefore, the rate of instability is similar between *msh2* and either *pms1* or *mlh1* strains, but product spectra are different; *msh3* strains have lower rates of instability than *msh2* strains, but have a stronger bias for deletions. There may be different repair complexes associated with the leading vs the lagging strand of replication, different types of slippage events that occur on the two strands, or differing specificities of repair.

It was reported that a single variant dinucleotide repeat inserted into the middle of a 51-bp repeat tract could stabilize repeat slippage by over 50-fold and that this stabilization was dependent on the mismatch repair system *(38)*. It now appears that this effect is substantially less than first reported (6- to 10-fold) (Petes, personal communication), but these results still have important implications for human disease, since a number of disease alleles are known to arise by repeat slippage, and the genes are normally protected against slippage by the presence of variants within each of the repeat tracts *(38)*.

7. EFFECTS OF *MLH* AND *MSH* GENES ON RECOMBINATION

7.1. Mitotic Recombination

The first indication of any effects of the mismatch repair system on mitotic recombination was that the rate of prototroph formation in heteroallelic diploids increased between three- and sixfold in *pms1* strains *(111)*, a result likely owing to the failure of the *pms1* mutant to suppress recombination of the heteroalleles (Kolodner, personal communication). These were gene conversion events; the nature of the assay precluded any observation of the effect of *PMS1* on completely homologous sequences. The absence of *PMS1* function also increased ectopic recombination between *SAM1* and *SAM2* alleles by 4.5-fold *(4)*. However, the same magnitude of increase in *pms1* strains was also seen with homologous substrates *(4)*. This result will be discussed further below. *PMS1* had no effect on spontaneous recombination between *his4* alleles on homeologous chromosomes, but its absence did increase the number of His$^+$ prototrophs in response to ionizing radiation *(87)*. *MSH2* had no effect on homeologous recombination, as measured by a gapped plasmid assay *(3)*.

In plasmid transformations, using a plasmid with either a DSB or a gap, *PMS1* had essentially no effect on gap repair or plasmid integration frequencies with homologous DNA *(80)*. However, the types of events produced in *PMS1* and *pms1* strains were significantly different. A homologous gapped plasmid most frequently showed gene conversion on both ends of the plasmid in a wild-type strain, but such conversion genotypes were absent, and the overall amount of gene conversion was decreased in *pms1* strains. In addition, *pms1* increased mixed populations of plasmid, reflecting a failure to repair recombination heteroduplexes. It is important to note that in this case, small heterologies were introduced near the ends of the integrated sequences in order to monitor gene conversion. There was a 4:1 bias in favor of gene conversion over restoration in a *PMS1* strain, and essentially no bias in a *pms1* strain. These results provide further support for the hypothesis that recombination intermediates involve heteroduplex DNA and that this DNA is a substrate for the mismatch repair system. In addition, the bias in favor of gene conversion over restoration suggests that the mismatch repair system is biased in favor of repair of the invading DNA strand; this is consistent with the end of the DNA strand serving as a signal for strand discrimination in mismatch repair. It is also important to note that gene conversion was observed in the absence of *PMS1*. This fact suggests that those conversions were mediated by a *PMS1*-independent mismatch repair system or by extension of the original gap, as illustrated in Fig. 2. There was a substantial difference seen in the effect of *PMS1* when homeologous instead of homologous DNA was used. The rate of homeologous recombination as measured by either gap repair or plasmid integration with a substrate approx 15% diverged was about 30-fold lower than homologous recombination in a wild-type strain; this rate was increased only slightly, if at all, in a *pms1* strain. In contrast to the results seen with homologous DNA, there was no difference seen in gap repair conversion patterns of homeologous DNA in *PMS1* vs *pms1* strains. In particular, there were no mixed populations of plasmids recovered, suggesting that for the homeologous DNA, little heteroduplex DNA was formed or was stable in the recombination complex. Reduced heteroduplex DNA formation during homeologous recombination might reflect reduced strand invasion. Homeologous recombination with sequences 14% diverged that could occur only by gene conversion was also examined between a chromosome and a single-copy plasmid *(79)*. There was no effect of either *PMS1* or *MSH2* on either homologous or homeologous recombination.

The effects of *PMS1*, *MSH2*, and *MSH3* on intrachromosomal recombination were examined between homologous or 25% mismatched regions of DNA consisting of an *S. cerevisiae SPT15* gene and, either upstream or downstream, its *S. pombe* homolog present as a direct repeat *(98)*. In wild-type strains, recombination between the homeologous genes was depressed 150- to 180-fold compared to homologous controls containing only one mismatch. When the *S. pombe* gene was placed downstream, *msh2* mutations increased recombination by 17-fold, *msh3* increased recombination by 9.6-fold, and together *msh2 msh3* mutations increased recombination by 43-fold. In the opposite orientation, only *msh2* mutations showed a modest effect on recombination. *PMS1* did not show any effect on homeologous recombination rates. Recombination in this system could occur either by crossovers or gene conversion, and mutations in *MSH2* and *MSH3* increased each recombinant class (Lahue, personal communication).

The *SPT15* system described above has the disadvantage that it cannot be used to study the effects of mismatch repair genes on recombination between regions with differing levels of homeology. A different assay system solves this problem, but can only be used to study the effects of mismatch repair on crossovers *(19)*. Recombination is monitored between duplicated introns inserted into *HIS3*, and any desired sequences less than about 400-bp in length can be inserted into the intron. In this system, *MSH2* and *MSH3* had only modest effects (about fivefold) on recombination rates between sequences diverged by 23%, and *PMS1* had an insignificant effect. In contrast, when sequences differing by only 9% were used, recombination rates increased 7-fold in *msh2* strains, 9-fold in *msh3* strains, and 11-fold in *pms1* strains relative to wild-type levels. An *msh2 msh3* strain was not significantly different from an *msh2* strain. Recombination of the 9% diverged sequences was 88% of the wild-type level in *msh2 msh3* strains. These results suggest that below a certain level of similarity, recombination is blocked by an inability to form stable heteroduplexes, whereas at higher levels of similarity, the heteroduplexes can form, but are disrupted by the mismatch repair system if they contain mismatches. Later work *(18)* showed that one mismatch is sufficient to impact recombination in the presence of the mismatch repair system, and this might explain in part the increase in "homologous" recombination observed previously in *pms1* strains *(4)*, since these assays use alleles with at least one heterology. One important feature of the intron system is that recombination can be examined between completely homologous sequences as well as between homeologous sequences. It has been consistently found that recombination of completely homologous sequences is increased severalfold in mismatch repair-defective cells *(18,19)*. One explanation for this result that could also apply to the previous work *(4)* is that DNA in the process of forming heteroduplex in recombination intermediates could have transient secondary structures that would be recognized by the mismatch repair system.

It is not clear why some studies have found effects of *MSH* and *MLH* genes on homeologous recombination, and others have not. The assay systems that have shown an effect of *MSH* and *MLH* genes on homeologous recombination have examined chromosome/chromosome recombination rather than plasmid/chromosome recombination. Whether this difference is important is still to be determined.

The involvement of *MSH2* and *MSH3* in mitotic recombination may not be solely owing to mismatches in the recombining DNA sequences. A recent paper indicates that both *MSH2* and *MSH3* belong to the *RAD1-RAD10* epistasis group *(91)*. In addition to their roles in excision repair and repair of DSBs via single-strand annealing, *RAD1* and *RAD10* are required for intrachromosomal recombination between direct repeats and affect homologous integration of linear DNA into the chromosome *(94,95)*. Homologous recombination in *msh3* and *rad1* strains was similarly reduced for direct repeats and integrating plasmids, and recombination rates in the *msh3 rad1* double mutant were the same as either single mutant. Direct repeat recombination in *msh2*, *msh3*, and *rad1* strains was similarly reduced, but an *msh2* strain had a threefold less effect on homologous integration of linear DNA than did either *msh3* or *rad1*. Neither *PMS1* nor *MLH1* had any effect on either type of recombination.

In summary, gene conversion both requires and is inhibited by mismatch repair activity, including both MutS and MutL homologs and the direction of repair appears to be targeted by the end of the invading strand of DNA. In the case of homeologous recombination, greater effects of the mismatch repair system are seen with relatively low levels of homeology, presumably because high divergence produces unstable heteroduplexes. In terms of antirecombination activity, *MSH2* has a strong effect, *MSH3* is weaker (presumably because it has overlapping specificities with *MSH6*), and *PMS1* has relatively weak effects. The binding of the MutS homologs appears to be sufficient to block recombination, without having to recruit MutL homologs. The role of the MutL homologs may be to stabilize the interactions of the MutS homologs with the mismatched DNA; thus, in the absence of MutL homologs, the MutS homologs are not as effective in blocking recombination.

Both *MSH2* and *MSH3* appear to act in the *RAD1-RAD10* recombination pathway, and their absence leads to a decrease in recombination that occurs via this pathway. It is not clear how *MSH2*, *MSH3*, *RAD1*, and *RAD10* function in recombination, but it may be that *MSH2* and *MSH3* have a role in recognizing the Holliday junction and then recruit *RAD1* and *RAD10* to cleave the junction *(91)*. A role for mismatch repair proteins in junction resolution is attractive, because it places the proteins in an ideal position to monitor DNA sequences and terminate branch migration if mismatches are encountered. If it is true that the absence of *MSH3* has a greater effect on the integration of linear DNA than the absence of *MSH2*, then *MSH3* must be able to act independently of *MSH2*. This would be a significant finding, because it would indicate that Msh3p does not always act as a heterodimer with Msh2p. Neither *PMS1* nor *MLH1* has any effect on this recombination pathway, indicating either that other MutL homologs are involved, or that *MSH2* and *MSH3* do not require them for this activity.

7.2. Meiotic Recombination

Several loci in yeast have higher rates of gene conversion at one end of the gene than at the other, and these gene conversion polarity gradients have been postulated to reflect differential mismatch repair activity *(3,22,86)*. Low PMS alleles near the 5'-end of the *HIS4* gene had much higher levels of gene conversion that low PMS alleles at the 3'-end of the gene (typical of the *HIS4* polarity gradient). However, high PMS alleles did not display a conversion gradient *(22)*. This was explained as resulting from a difference in the amount of restoration-type repair compared to conversion-type repair at the two ends of the gene; high PMS alleles are repaired very inefficiently at both ends of the gene and therefore do not exhibit a polarity gradient. Since most, if not all, genes displaying polarity gradients have a meiotic-specific DSB at the high conversion end of the gradient, the end of the invading strand might serve as the loading site for the mismatch repair system. The distance from this end to the site of a given allele could then determine the probability of conversion vs restoration repair.

An alternative model to explain polarity gradients was developed in a study of *pms1* and *msh2* mutants (3). *msh2* strains, just like *pms1* strains, have a large decrease in gene conversions and a concomitant increase in PMS, with most of the polarity gradients at *ARG4* and *HIS4* abolished. These results were explained in terms of a heteroduplex rejection model. In this model, heteroduplex DNA is initiated at the point of an invading DNA end and extended by branch migration from asymmetric heteroduplex to symmetric heteroduplex. Mismatch repair of a marker in asymmetric heteroduplex would result in a gene conversion, but repair of a marker in symmetric heteroduplex would result in an apparent wild-type segregation. The model further proposes that the mismatch repair system regulates branch migration and that when a mispair is detected, branch migration is either reversed or the recombination process is terminated by resolution of the Holliday junction followed by symmetric repair of the recombinants. Thus, markers more distal to the initiation site would be less likely to be in asymmetric heteroduplex and less likely to undergo gene conversion. Most of the data obtained in these studies are consistent with either model. In many ways, the first model is more appealing. It is clear that there is a high degree of gene conversion near the initiating DSB, and there is agreement that this is likely owing to the end of the invading strand serving as a signal for the direction of mismatch repair. It is known that in *E. coli*, the nick formed by MutH can serve as a strand direction signal for up to 1-kb, but that beyond this distance, directionality resulting from the nick is very low (*13,56*; Chapter 11). A similar process may operate in yeast, in which the directionality of mismatch repair decreases as a function of the distance from the DSB. In mitotic cells, as discussed in Section 7.1., *MSH2* appears to provide a stronger block to homeologous recombination than *PMS1*. Similar results are found in meiotic cells *(43) (see below)*. These results suggest that recognition of a mismatch by Msh2p is sufficient to block recombination without involvement of an MutL homolog. If Msh2p were similarly effective in blocking branch migration in meiotic cells, the heteroduplex rejection model would predict different effects on the polarity gradient in *msh2* and *pms1* mutants, but in fact they behave identically.

If meiotic recombination normally proceeds via a DSB and heteroduplex mechanism as illustrated in Fig. 1, and there is much evidence to suggest that this is true, how does one explain the persistence of gene conversion events? It is particularly striking that for certain alleles, gene conversion does not decrease, although PMS increases in *msh2* or *pms1* mutants *(3,111)*. It cannot be ruled out that there is another mismatch repair system operating, although the system would have to be quite active, since some alleles are converted in up to 5–10% of all aberrant segregations in *msh2* and *pms1* mutants *(3,111)*. A more likely explanation is that these conversions arise through gap repair as illustrated in Fig. 2. This possibility would suggest a larger role for gap repair than is usually thought to occur. One resolution of this paradox would be that recombination intermediates formed through gap repair that contained certain types of mismatched DNA would be more sensitive to the blocking effects of the mismatch repair system than would intermediates formed without the excision of both ends of a DSB. Thus, in wild-type cells, most gap repair intermediates would be blocked by the mismatch repair system, whereas in mismatch repair-deficient strains, these intermediates could result in gene conversions. These pathways might differ in the stability of the intermediate, controlled by the length of the invading strand. If this length depends on the amount of single-stranded DNA available, it is possible that in double strand gap repair, which requires excision of both 3'- and 5'-ends, less free single-strand DNA would be available. The resulting heteroduplex DNA would be shorter and less stable than a longer duplex, and thus more susceptible to disruption by the mismatch repair system. Another possibility is that some of the conversions observed in mismatch repair-defective strains could represent a PMS event in which one of the cells formed by the first mitotic division of the spore fails to grow (Petes, personal communication), but there is as of yet no experimental evidence for this.

The initial characterization of the *pms* mutants showed the involvement of the mismatch repair system in meiotic recombination between heteroalleles. A later study found that the introduction of certain heterologies within a given region reduced the frequency of crossovers and increased the frequency of gene conversions and ectopic recombination between flanking homologous DNA *(10)*. Recombination was restored to a nearly wild-type pattern in the absence of *PMS1 (11)*. The authors postulated that this change in recombination pattern was owing to additional recombination events stimulated by the effect of the mismatch repair system on mismatched recombination intermediates, though alternative explanations are possible, such as the channeling of recombination into other pathways because of the mismatch repair system blocking certain types of events.

It was recently found that placing high PMS alleles very close (from 8–20 bp away) to normal alleles induced hyperrecombination between the markers *(60)*. This result was explained by the formation of a partial mismatch repair complex on the high PMS allele that would sterically hinder the formation of a complete mismatch repair complex on the normal allele and thus prevent corepair of the markers. This is just one example of how mismatch distributions and types influence recombination.

The effect of *PMS1* and *MSH2* on meiotic recombination in an interspecific hybrid of *S. cerevisiae* and *Saccharomyces paradoxus* was determined *(43)*. In the absence of either of these genes, meiotic recombination between the homeologous chromosomes was increased, chromosome nondisjunction was decreased, and spore viability was improved. This work is one of the most direct demonstrations of the antirecombination activity of mismatch repair in yeast serving as a barrier to genetic exchange between different organisms. In all cases, the effect of a mutation in *MSH2* was greater than that of a mutation in *PMS1*. In a related study, a diploid of *S. cerevisiae* was constructed with one copy of chromosome III replaced by *S. paradoxus* chromosome III *(114a)*. Meiotic recombination between the homeologous chromosomes was reduced, as was spore viability. As before, homeologous recombination was increased to a greater extent in *msh2* strains than in *pms1* strains. Recombination was further enhanced in an *msh2 pms1* strain, suggesting that *MSH2* and *PMS1* may act independently in meiosis. The effect of low levels of heterology on meiotic recombination within a short 5.5-kbp interval was also studied (Borts, personal communication). Surprisingly, mutations in *MSH3* appeared to reduce total levels of recombination, suggesting that *MSH3* might play a role in stimulating meiotic recombination when heterologies are present. These studies show that components of the mismatch repair system are also meiotically active in blocking homeologous recombination. However, the recombination enhancement seen in the *msh2 pms1* strain and the possibility of a stimulatory effect of *MSH3* on recombination indicate that there may be differences in the action of these proteins in mitotic vs meiotic recombination.

8. *MLH* AND *MSH* GENES IN MEIOSIS

MSH4 and *MSH5* appear to function only in meiosis and not in mitosis *(40,90)*. Neither gene has any effect on mismatch repair or gene conversion. However, mutations in either gene reduce meiotic reciprocal exchange, increase meiosis I chromosome nondisjunction, and decrease spore viability. *MSH4* is not transcribed in mitotic cells *(90)*; data on mitotic expression are not available for *MSH5*. The genes are in the same epistasis group *(40)*. How is it that homologs of genes that are involved in DNA mismatch repair and that block recombination between mismatched sequences appear themselves to play no role in mismatch repair and seem to increase recombination between identical sequences? It is likely that *MSH2*, *MSH3*, and *MSH6* recognize particular structures of DNA that contain distortions owing to mismatched base pairs and small loops. If such structures are found during replication, the rest of the mismatch repair machinery may be recruited to repair the mispaired region. If such structures are found during recombination, the recombination intermediate is destroyed. The homology of *MSH4* and *MSH5* to the other *MSH* genes suggests that they might recognize some types of DNA distortions. It is possible that recombination intermediates contain structures similar to those recognized by *MSH2*, *MSH3*, and *MSH6*; when these structures are bound by *MSH4* and *MSH5*, the result would be an enhancement of recombination rather than a destruction of the intermediate.

Spore viability is also decreased by mutations in *MSH2*, *PMS1*, or *MLH1 (82,86)*. Although this effect has been ascribed to mutations in essential genes uncovered in the haploid state, it is likely that these genes play additional roles in meiosis and that the loss in spore viability is not owing solely to increased mutations. There are indications, for example, that *MSH2* and *MSH3* are induced during meiosis *(51)*. It is also perhaps relevant that *Pms2*$^{-/-}$ mice are male sterile and *Mlh1*$^{-/-}$ mice are both male and female sterile (*5,6,25*; Chapter 20, vol. 2).

9. MITOCHONDRIAL MISMATCH REPAIR

Mutation rates are much greater for mitochondrial DNA than for nuclear DNA, and one attractive explanation for this finding has been that mitochondria lack mismatch repair. Thus, it was a surprise when one of the first yeast MutS homologs to be found, *MSH1*, appeared to be involved in mitochondrial mismatch repair *(85,86)*. *MSH1* has a mitochondrial targeting sequence *(85)*, and *msh1* mutants rapidly develop a petite phenotype *(86)*. For example, in fewer than 20 generations, <0.01% of the cells in a colony still had functioning mitochondria. Many of the petites appeared to be hypersuppressive petites, which contain mitochondrial DNA that displaces wild-type mitochondrial DNA and which result from large-scale rearrangements and deletions. Msh1p was localized to mitochondria by using an epitope tag *(16)*. The levels of Msh1p are limiting and point mutation rates of mitochondrial DNA are elevated sevenfold in *msh1/MSH1* heterozygotes, further indicating a role for Msh1p in mismatch repair. However, the extreme rapidity of petite formation and the common hypersuppressive phenotype suggests that an even more important role of *MSH1* in mitochondria may be the suppression of homeologous recombination, thereby preventing the rearrangements and deletions seen in *msh1* strains.

One question that is still unanswered is whether there is an MutL homolog that functions with *MSH1*. Given what is known about nuclear mismatch repair in all organisms, one would expect such an MutL homolog. One would further expect that there would be a petite phenotype observed in its absence. However, there is no report of a petite phenotype with either *PMS1* or *MLH1*; disruption of *MLH2* or *MLH3* also does not lead to a petite phenotype (Kolodner, personal communication).

10. INTERACTION OF *MLH* AND *MSH* PROTEINS

As described in Chapter 11, MutS in *E. coli*, probably as a homodimer or larger oligomer, recognizes small loops or mismatched base pairs. A homodimer of MutL then adds to this complex. A central question for yeast, with its multiple MutS and MutL homologs, is whether the proteins function as homodimers or as heterodimers. There is both genetic and biochemical evidence suggesting that most Mlh and Msh proteins may function primarily, if not exclusively, as heterodimers.

The phenotypes of disruption of either single or double mutants of *MLH1* and *PMS1* were found to be similar *(82)*. An analysis with the yeast two-hybrid system showed an association between Mlh1p and Pms1p, but neither protein associated with itself or with Msh2p *(83)*. Each of the three proteins was made as a fusion protein with the *E. coli* maltose binding protein and their affinity judged by binding an amylose resin. Again, only Mlh1p and Pms1p showed substantial interaction with each other. A complex formed with Msh2p and heteroduplex DNA is supershifted by the addition of both Mlh1p and Pms1p, indicating that a complex of all three proteins can form in the presence of a mismatch. It should be noted that a complex also forms in the presence of completely homologous DNA, but to a lesser extent, owing presumably to the affinity of Msh2p for DNA.

As mentioned previously, there is strong genetic evidence for an interaction between *MSH2* and either *MSH3* or *MSH6*, and this interaction has been directly demonstrated *(61)*. These experiments suggest several important features of these MutS homologs, at least in terms of their interactions with single base pair mismatches and small loops. First, Msh2p is unlikely to function as a homodimer, since *msh2* mutants have essentially the same phenotype as an *msh3 msh6* strain. Second, it appears that Msh2p acts as a heterodimer with either Msh3p or Msh6p; the Msh2p/Msh3p heterodimer apparently has little affinity for single basepair mismatches, but does recognize various small loops, whereas the Msh2p/Msh6p heterodimer recognizes both single basepair mismatches and loops. It is not yet known whether Msh2p might interact with yet other MutS homologs or if there are other homologs that might recognize the same type of substrates, but the evidence suggests that if there are such additional combinations, they would have a relatively minor effect on overall mismatch repair.

Although there is no direct evidence on interaction of the encoded proteins of *MSH4* and *MSH5*, it is likely that they act as a heterodimer. As discussed in Section 8., single and double *msh4 msh5* mutants have identical phenotypes. The likelihood that other MutS homologs act as heterodimers also supports the hypothesis that Msh4p and Msh5p directly interact. It will be interesting to determine if their action requires an MutL homolog.

11. IN VITRO PROPERTIES OF *MLH* AND *MSH* PROTEINS

The substrate properties of some of the yeast MutL and MutS homologs have been inferred by the types of mutations created in their absence. However, there is obviously a desire to observe the properties of these proteins in a purified state, as has been done with *E. coli* MutS and to a lesser extent MutL. There are several major problems in trying to study the properties of the purified proteins. Although the *E. coli* mismatch repair system has been reconstituted in vitro *(55)*, the exact nature of the MutL function is not known, although it presumably acts as a matchmaker with MutS and MutH. Thus, most attention has been focused on the MutS homologs, since in vitro experiments with MutS have been more revealing. However, MutS in *E. coli* presumably acts as a homodimer (Chapter 11), and as mentioned above, most data suggest that many of the MutS (and MutL) homologs in yeast act as heterodimers and not as homodimers. It was shown that yeast extracts had a mismatch binding activity that specifically recognized various mismatches, including T/G, G/G, G/A, A/C, and T/C, but only weakly bound to C/C mismatches *(64)*. Furthermore, this activity was abolished in extracts from *msh2* strains, but not from *msh3* strains. The recognized mismatches are repaired in vivo, implying that Msh2p is responsible for recognition of base-pair mismatches (although the experiments did not rule out the involvement of other homologs, such as Msh6p).

Msh1p has been purified and characterized in vitro *(16,17)*. Msh1p was found to have both DNA binding and ATPase activity. It recognizes various mismatches with a relative affinity similar to *E. coli* MutS, with G/T mismatches being the best recognized and C/C the least well recognized. Small loops did not appear to be recognized as well as with MutS. The affinity shown by Msh1p for heteroduplex DNA is less sensitive to NaCl concentration compared to the affinity for homoduplex DNA, and Msh1p is more stably bound to heteroduplex DNA. It appears that Msh1p binds to mismatches and ATP cooperatively, and this cooperativity may add to the preferential stability of complexes with heteroduplex DNA compared to homoduplex-containing DNA *(17)*.

Msh2p has been overexpressed and purified *(2)*. The protein shows modest specificity for binding of mispaired bases over homoduplexes, but a much greater specificity for binding to loops. One of the best binding substrates was a 14-bp palindromic loop that is known to be poorly repaired in vivo.

All of the data obtained on purified Msh proteins will have to be re-evaluated in light of the finding that many, if not most, of the yeast MutS homologs appear to function mainly as heterodimers. In the case of Msh2p, it is now fairly clear that its function in recognition of mismatches is done primarily in conjunction with Msh3p and Msh6p. Thus, the binding affinities of Msh2p alone, which do not match in vivo repair efficiencies, are presumably modified in heterodimer formation. It is more likely that Msh1p functions independently of any other MutS homolog, and so the data obtained for Msh1p may be an accurate reflection of the in vivo specificities of this protein.

Recently, yeast MutSα has been purified and shown to consist of two peptides, Msh2p and Msh6p (44). This complex exhibits specific mismatch binding that is sensitive to ATP, as is human MutSα *(44)*. MutSα also binds to +1 insertion mismatches and to palindromic loops, but not to +2 or +4 insertion mismatches *(1a)*. However, the binding to palindromic loops, which are not repaired well in vivo, is not sensitive to ATP.

12. REGULATION OF *MLH* AND *MSH* GENES

PMS1 mRNA is cell-cycle-regulated in the same manner as ribonucleotide reductase, which peaks at the G1/S boundary *(66)*. This is consistent with Pms1p functioning in replication repair. Several of the *MLH* and *MSH* genes have been recently examined by RT-PCR for cell-cycle regulation. *MSH2* and *MSH6* mRNAs have also been found to be cell-cycle-regulated with a peak expression at the G1/S boundary *(51)*. Putative MCB elements, a sequence implicated in cell-cycle regulation, are found in all three genes *(51,62,66)*. When the MCB elements in *PMS1* were mutated such that the gene was transcribed only at a basal level, expression appeared insufficient, resulting in a mutator phenotype *(51)*. Neither *MSH3* nor *MLH1* mRNA showed cell-cycle regulation. One caveat to this work is that levels of mRNA and not protein were being measured. It is possible, though unlikely, that the levels of Msh3p and/or Mlh1p could also show cell-cycle regulation owing to regulated protein stability, for example. Assuming that the levels of protein in the cell correspond to mRNA levels, it is striking that *PMS1* and *MLH1* are differentially regulated. Although at present there is no evidence for any other associations of Pms1p and Mlh1p than with each other and Msh2p, the differing regulation of the two genes suggests that there might be some functions that require only one and not both of the proteins. The difference in regulation of *MSH3* compared to *MSH2* and *MSH*6 is also noteworthy. It appears that *MSH3* may have a greater effect on the *RAD1* recombination pathway than does *MSH2 (91)*, suggesting that *MSH3* may exert some or all of its effects in this pathway in an *MSH2*-independent manner. *MSH3* may also make a smaller contribution to loop recognition during S phase than at other times in the cell cycle when levels of Msh6p are lower.

The efficiency of repair of a heteroduplex *CEN* plasmid transfected into a cell is increased by 25% in a *rad3* or *pol3-4* (polymerase δ proofreading mutation) strain *(112)*. This phenomenon is not associated with a change in transcription of *PMS1*, *MLH1*, or *MSH2*, and suggests that yeast may be able to post-transcriptionally modulate mismatch correction in response to mismatch load.

13. OTHER PROTEINS INVOLVED IN MISMATCH REPAIR

The *E. coli* proteins known to be necessary for mismatch repair are MutS, MutL, MutH, the MutU/UvrD helicase II, RecJ or ExoVII as a 5'-exonuclease, ExoI as a 3'-exonuclease, DNA polymerase III, single-strand binding protein, and DNA ligase (Chapter 11). One would thus expect homologs to most of these proteins to be required for yeast mismatch repair. Since yeast DNA is not methylated *(81)*, one exception would be MutH, which recognizes hemimethylated DNA to allow strand discrimination. There could be an analogous protein that recognizes a different signal for strand selectivity in repair. However, MutH is not required to give proper strand bias in vitro as long as there is a nick present in the newly replicated strand *(55)*. In addition, there is a suggestion based on patterns of codon usage that MutH may have been acquired by *E. coli* by horizontal gene transfer *(63)*, implying that MutH was a late addition to the mechanism of mismatch repair. It is an attractive idea that in yeast, the mismatch repair system tracks very closely along with the replication fork and uses nothing more than the newly replicated end of DNA to provide a strand discrimination signal. This idea is also consistent with the data mentioned in Section 7.1. showing a bias for repair of the invading strand in a heteroduplex recombination intermediate, as well as the data that indicate a possible interaction of PCNA with the mismatch repair system *(see below)*.

A 5' to 3' exonuclease is another likely component of the mismatch repair system, and two different candidates have been proposed for this function. The *RTH1 (RAD27)* gene is a 5' to 3' exonuclease, and appears to function in DNA replication *(99)*. Strains deficient in *RTH1* show dinucleotide repeat instability only severalfold lower than that exhibited by *msh2* strains *(47)*; strains deficient in *RTH1* and either *msh2*, *mlh1*, or *pms1* show a level of instability severalfold greater than that of strains deficient in any of the mismatch repair genes. When the products of slippage were examined in *rth1* strains, it was found that in 51 out of 52 cases, there was an insertion of one or two repeat units, which is consistent with a role for *RTH1* in removal of insertion loops from the newly replicated strand. *rth1* strains also show about the same forward mutation rate at *CAN1* as *msh2* strains, and *msh2 rth1* double mutants show a several fold higher mutation rate. The synergistic effects of *rth1* mutations with mutations in known mismatch repair genes suggests that *RTH1* gene acts in a separate pathway from *MSH2 PMS1 MLH1*, or perhaps in two pathways, one of which is distinct from the *MSH2 PMS1 MLH1* pathway *(49)*.

The *EXO1* gene was identified in a two-hybrid interaction screen with *MSH2* and is therefore likely to be in the same pathway as *MSH2* (*49*; Kolodner, personal communication). The *EXO1* gene appears to be a homolog of the *S. pombe exo1* gene, which is a 5' to 3' exonuclease involved in mismatch correction in *S. pombe* (*see* Section 14.). Strains deficient in *EXO1* have increased mutation rates and dinucleotide repeat instability, but not as high as *msh2* strains (*49*; Kolodner, personal communication). An *exo1 msh2* strain showed mutation rates comparable to an *msh2* strain. *exo1* mutants lack a previously purified 5'-3' exonuclease required for in vitro recombination (*42*; Kolodner, personal communication). *EXO1* was found independently in a two-hybrid screen with *MLH1* and *PMS1* (Simon and Liskay, personal communication), with a much stronger binding to Mlh1p than with Pms1p. The interaction of *EXO1* with *MSH2* and *MLH1*, combined with the phenotypic analysis, strongly suggests that *EXO1* is a component of the mismatch repair pathway. This of course does not rule out the involvement of *RTH1* in mismatch repair as well, though the case for *RTH1* involvement is weaker.

There are at least three helicases in yeast, but there is no compelling evidence that any of them are involved in mismatch repair. *PIF1* encodes a 5' to 3' helicase that was thought originally to be present only in mitochondria, but has also been shown to be involved in telomere formation *(54,96)*. *RAD3* also encodes a 5' to 3' helicase that is a component of the repair/transcription factor b *(28,103,108)*. Of the three known helicases, the most likely to be involved in mismatch repair is that encoded by *SRS2/RADH/HPR5 (1,88,89)*. This gene has extensive homology to the *E. coli* helicase *UvrD* that is involved in mismatch repair, and alleles of *SRS2* have a hyper-gene conversion phenotype *(1,89)*.

A potentially interesting interaction is that of *MLH1* with yeast PCNA *(107a)*. That human PCNA was obtained in a screen with human *PMS2* strengthens the suggestion that this interaction is biologically relevant. In addition, a single amino acid substitution in yeast PCNA has a large destabilizing effect on dinucleotide repeats *(107a)*. This is the first direct evidence that the mismatch repair system may track with the replisome, and indicates at least one of the ways in which the two systems may interact.

One question that has yet to be answered is whether there are additional mismatch repair activities in yeast. A search of the complete yeast genome indicates that there are no additional homologs of MutS. However, there are indications that there is a repair pathway for insertion loops and that this may be independent of *MSH2*. For example, a 26-bp palindromic loop is resistant to mismatch repair, but the introduction of three mismatches into the loop makes it a substrate for mismatch repair *(69)*; *MSH2* is not known to repair loops larger than 4 bases. A candidate for such a repair activity has been recently identified *(65)*. The IMR protein binds specifically to loop mismatches larger than 4 bases, but not to a perfect palindrome, and is independent of *MSH2*, *MSH3*, and *MSH4*. Further work will be necessary to determine if this activity is associated with loop repair.

14. MISMATCH REPAIR IN *S. POMBE*

The *S. pombe swi4* gene is an MutS homolog *(30)*. This was surprising, since the only previously identified phenotype of *swi4* strains was a defect in mating type switching. Since *swi4* strains show no PMS or mitotic mutator phenotypes, it was assumed that *swi4* played no major role in mismatch repair. The closest *S. cerevisiae* homolog to *swi4* is *MSH3*. The phenotype of *swi4* strains prompted an examination of the role of *MSH3* in mating type switching in *S. cerevisiae*, and no effect was found *(70)*, which is not necessarily surprising, since mating type switching is quite different in the two organisms.

Genetic analysis in *S. pombe* suggests that there are two mismatch repair systems: a major pathway that corrects all mismatches except for C/C mismatches, and a minor, inefficient pathway that corrects all mismatches, including C/C mismatches *(92,93)*. Repair tracts appear to average 100 bases in length in the major system and 10 bases in the minor system. The existence of two separate pathways was further supported by the identification of two different mismatch binding activities in *S. pombe* extracts *(32)*. Gel-retardation experiments demonstrated one activity that bound to most single base pair mismatches and single loops, but not to C/C mismatches, and a separate, lower-mol-wt activity that bound to C/C, T/C, C/T, C/A, A/C, and C/– mismatches, weakly to T/T, and not to other mismatches. These activities were present in extracts devoid of the *swi4* gene product, indicating that the *swi4* gene was not involved in single base pair mismatch binding.

The *swi8* gene, also involved in mating type switching, was found to be a mutator *(31)* and appears to be allelic to *mut3*, a previously identified mutator *(67)*. *swi8* is a homolog of *MSH2* (Rudolph and Kohli, personal communication). *swi4* and *swi8* are the only MutS homologs known in *S. pombe*, but several hypotheses can be made based on what is known about mismatch repair in *S. cerevisiae*. Because *MSH3* appears to be involved in loop repair, it will be important to examine the role of the *swi4* gene in correction of loop mismatches. If *swi4* and *swi8* are homologous in function to *MSH3* and *MSH2*, one would predict that a swi4p/swi8p heterodimer would be involved in loop repair, and a heterodimer with swi8p and an as yet unknown protein (a homolog of Msh6p?) would be involved in repair of both single base pair mismatches and loops. The fact that both *swi4* and *swi8* are involved in the same aspect of mating type switching suggests that the swi4p/swi8p heterodimer may be recognizing a certain type of heteroduplex structure that would normally be a termination signal for the switching event, but that is not recognized in its absence. This effect may be related to the observation that in *S. cerevisiae* meiosis, *MSH3* stimulates recombination between sequences with low levels of mismatches (Borts, personal communication). This role of *MSH3* appears to be unrelated to mismatch repair *per se*, but would be related to its function in recognition of certain types of DNA structures.

An *S. pombe* MutL homolog, *pms1*, has been recently identified *(91a)*. Disruption of the *pms1* gene gives a strong mitotic mutator phenotype and increased *PMS*, but has no effect on mating type switching, suggesting that MutL homologs might not be involved in the mating type switching mechanism. If true, this would be another indication that there might be functions of the MutS homologs in *S. cerevisiae* that would also be independent of Mlh proteins. As is also observed in *S. cerevisiae*, *S. pombe* strains deficient in *pms1* still show some level of gene conversion, suggesting that some recombination events do not involve mismatch repair, or that there are alternate mismatch repair pathways for gene conversion.

The only other *S. pombe* gene that has been implicated in mismatch repair is the *exo1* gene *(105)*, which appears to be homologous to the *S. cerevisiae EXO1* gene (*see* Section 13.). *exo1* mutants show a mutator phenotype similar in magnitude to that of the *swi8/msh2* mutants *(31,67,105)*. *exo1* mutants also exhibit altered recombination between closely spaced markers, as expected for a mismatch repair mutant *(92,93,105)*. The *exo1* gene maps on the same chromosomal fragment as the *mut2* gene *(67)* and so may be allelic with this gene *(105)*.

At present, it appears that the *S. pombe* mismatch repair system has close similarities to that in *S. cerevisiae*. The involvement of some components of the mismatch repair system in mating type switching in *S. pombe* and not in *S. cerevisiae* may reflect the difference in the mechanism of the two mating type systems, and does not necessarily reflect a difference in the activities of the proteins.

15. SUMMARY

It is misleading to refer to all *MSH* and *MLH* genes as mismatch repair genes. Some of the genes, such as *MSH4* and *MSH5*, do not appear to be involved in any aspect of mismatch repair. Most, if not all, of the genes have some functions that are unrelated to mismatch repair. Even in their repair functions, the recognition of small loops in order to prevent frameshift mutations is probably more important than recognition of single base pair mismatches.

There are many questions yet to be answered about this system. For the replication repair functions, how is strand discrimination made so that the proper strand can be repaired? In many ways, the most attractive model is that the repair system tracks along with the replisome and uses the primer end of the DNA to distinguish strands. The finding of an association between some of the proteins and PCNA is supportive of this model, but much work remains to be done. If the repair complexes do track with the replisome, does each replisome have a complement of all repair complexes, or only a subset, and is there any distinction between the leading and lagging strand?

What Msh and Mlh components are involved in repair? The current model is that nuclear repair complexes consist of heterodimers of either Msh2p and Msh3p or Msh2p and Msh6p, complexed with a heterodimer of Pms1p and Mlh1p. This is almost surely an oversimplification. Some mutation rates and spectra are not compatible with this simple model, nor is the apparent difference in cell-cycle regulation of the genes. Although no phenotypes have yet been observed for them, it is possible that Mlh2p or Mlh3p may also be involved in repair. What other proteins are involved in mismatch repair, and how do they interact with the Msh and Mlh proteins?

What is the role of the mismatch repair system in mitochondria? Mlh1p is certainly involved in mitochondrial repair, although its primary role may be in antirecombination rather than repair of loops or single base mismatches. Does it function as a homodimer? Are there any *MLH* genes involved in mitochondrial repair? For replication repair functions, one would expect that an MutL homolog would be present, perhaps as a homodimer, but none has yet been found to be involved in mitochondrial repair. It may be that the mitochondrial phenotypes of any *MLH* gene are more subtle than expected, or less likely, that no Mlh protein functions in mitochondria.

What is the role of the *MSH* and *MLH* genes in mitotic recombination? Msh2p and Msh3p seem to be involved in the *RAD1-RAD10* recombination pathway. It is unclear how exactly they function, and it is interesting that *MSH3* seems to have some phenotypes that are stronger than those of *MSH2*, suggesting that *MSH3* may function independently of *MSH2* in this pathway. No involvement of the *MLH* genes in the *RAD1-RAD10* pathway has been found. A number of *MSH* and *MLH* genes are involved in blocking recombination between diverged sequences, but there is much that is unknown. Not all assay systems show an effect of these genes on homologous recombination, and in general, the *MSH* genes appear to have a stronger phenotype than do the *MLH* genes.

What role do the *MSH* and *MLH* genes play in meiosis? There are two *MSH* genes, *MSH4* and *MSH5*, that appear to be expressed only in meiosis. It is suspected that they function as a heterodimer; their function is not understood, but it appears that they are important for proper chromosome segregation. It is also not known which *MLH* genes, if any, interact with *MSH4* and *MSH5*. Many of the other *MSH* and *MLH* genes are active in meiosis, as can be seen from the increase in PMS in their absence, but other roles that they may play in meiosis are not known.

How do the Msh and Mlh proteins function? There are many questions remaining about the interactions of these proteins. Do all Mshp functions involve Mlhp proteins? Do all Mshp and Mlhp interact solely as heterodimers, or do they ever interact as homodimers? It appears that most of the known proteins may function solely as heterodimers, though Msh1p may be an exception. Finally, does a given Mshp heterodimer always interact with the same Mlhp heterodimer? There is very little information about the functional domains of any Msh or Mlh protein. The regions that bind DNA, that are responsible for dimerization, that are responsible for Msh–Mlh interaction, or that interact with other proteins, such as PCNA, are unknown. The overall function of Msh proteins to recognize DNA is understood, but the function of the Mlh proteins remains a mystery.

The study of eukaryotic mismatch repair is most advanced in *S. cerevisiae*, yet there is much to be learned about the system. Yeast has already served as a useful model for finding and beginning to understand mammalian genes involved in mismatch repair. The accessibility of both genetics and biochemistry should keep yeast in the forefront of eukaryotic mismatch repair research.

ACKNOWLEDGMENTS

I am very grateful to Eric Alani, Rhona Borts, Sue Jinks-Robertson, Jürg Kohli, Bernie Kunz, Richard Kolodner, Willfried Kramer, Bob Lahue, Mike Liskay, Tom Petes, and Louise Prakash for supplying data prior to publication and for comments on this manuscript. This manuscript has been greatly helped by suggestions from Sue Jinks-Robertson, Abhijit Datta, Gregg Orloff, and members of my lab. My research has been supported in part by grants from the National Cancer Institute.

REFERENCES

1. Aboussekhra, A., R. Chanet, Z. Zgaga, C. Cassier-Chauvat, M. Heude, and F. Fabre. 1989. RADH, a gene of *Saccharomyces cerevisiae* encoding a putative DNA helicase involved in DNA repair. Characteristics of *radH* mutants and sequence of the gene. *Nucleic Acids Res.* **17:** 7211–7219.

1a. Alani, E. 1996. The *Saccharomyces cerevisiae* Msh2 and Msh6 proteins form a complex that specifically binds to duplex oligonucleotides containing mismatched DNA base pairs. *Mol. Cell Biol.* **16:** 5604–5615.

2. Alani, E., N.-W. Chi, and R. Kolodner. 1995. The *Saccharomyces cerevisiae* Msh2 protein specifically binds to duplex oligonucleotides containing mismatched DNA base pairs and insertions. *Genes Dev.* **9:** 234–247.

3. Alani, E., R. A. G. Reenan, and R. D. Kolodner. 1994. Interaction between mismatch repair and genetic recombination in *Saccharomyces cerevisiae*. *Genetics* **137:** 19–39.

4. Bailis, A. M. and R. Rothstein. 1990. A defect in mismatch repair in *Saccharomyces cerevisiae* stimulates ectopic recombination between homeologous genes by an excision repair dependent process. *Genetics* **126:** 535–547.

5. Baker, S. M., C. E. Bronner, L. Zhang, A. W. Plug, M. Robatzek, G. Warren, E. A. Elliott, J. Yu, T. Ashley, N. Arnheim, R. A. Flavell, and R. M. Liskay. 1995. Male mice defective in the DNA mismatch repair gene PMS2 exhibit abnormal chromosome synapsis in meiosis. *Cell* **82:** 309–319.

6. Baker, S. M., A. W. Plug, T. A. Prolla, C. E. Bronner, A. C. Harris, X. Yao, D.-M. Christie, C. Monell, N. Arnheim, A. Bradley, T. Ashley, and R. M. Liskay. 1996. Involvement of mouse *Mlh1* in DNA mismatch repair and meiotic crossing over. *Nature Genet.* **13:** 336–342.

7. Bishop, D. K., J. Andersen, and R. D. Kolodner. 1989. Specificity of mismatch repair following transformation of *Saccharomyces cerevisiae* with heteroduplex plasmid DNA. *Proc. Natl. Acad. Sci. USA* **86:** 3713–3717.

8. Bishop, D. K. and R. D. Kolodner. 1986. Repair of heteroduplex plasmid DNA after transformation into *Saccharomyces cerevisiae*. *Mol. Cell. Biol.* **6:** 3401–3409.

9. Bishop, D. K., M. S. Williamson, S. Fogel, and R. D. Kolodner. 1987. The role of heteroduplex correction in gene conversion in *Saccharomyces cerevisiae*. *Nature* **328:** 362–364.

10. Borts, R. H. and J. E. Haber. 1987. Meiotic recombination in yeast: alteration by multiple heterozygosities. *Science* **237:** 1459–1465.

11. Borts, R. H., W.-Y. Leung, W. Kramer, B. Kramer, M. Williamson, S. Fogel, and J. E. Haber. 1990. Mismatch repair-induced meiotic recombination requires the *PMS1* gene product. *Genetics* **124:** 573–584.

12. Bronner, C. E., S. M. Baker, P. T. Morrison, G. Warren, L. G. Smith, M. K. Lescoe, M. Kane, C. Earabino, J. Lipford, A. Lindblom, P. Tannergård, R. J. Bollag, A. R. Godwin, D. C. Ward, M. Nordenskjold, R. Fishel, R. Kolodner, and R. M. Liskay. 1994. Mutation in the DNA mismatch repair gene homologue *hMLH1* is associated with hereditary non-polyposis colon cancer. *Nature* **368:** 258–261.

13. Bruni, R., D. Martin, and J. Jiricny. 1988. d(GATC) sequences influence *Escherichia coli* mismatch repair in a distance-dependent manner from positions both upstream and downstream of the mismatch. *Nucleic Acids Res.* **16:** 4875–4890.
14. Burns, N., B. Grimwade, P. B. Ross-Macdonald, E.-Y. Choi, K. Finberg, G. S. Roeder, and M. Snyder. 1994. Large-scale analysis of gene expression, protein localization, and gene disruption in *Saccharomyces cerevisiae. Genes Dev.* **8:** 1087–1105.
14a. Chambers, S. R., N. Hunter, E. J. Louis, and R. H. Borts. 1996. The mismatch repair system reduces meiotic homeologous recombination and stimulates recombination-dependent chromosome loss. *Mol. Cell Biol.* **16:** 6110–6120.
15. Chang, D.-Y. and A.-L. Lu. 1991. Base mismatch-specific endonuclease activity in extracts from *Saccharomyces cerevisiae. Nucleic Acids Res.* **19:** 4761–4766.
16. Chi, N.-W. and R. D. Kolodner. 1994. Purification and characterization of MSH1, a yeast mitochondrial protein that binds to DNA mismatches. *J. Biol. Chem.* **269:** 29,984–29,992.
17. Chi, N.-W. and R. D. Kolodner. 1994. The effect of DNA mismatches on the ATPase activity of MSH1, a protein in yeast mitochondria that recognizes DNA mismatches. *J. Biol. Chem.* **269:** 29,993–29,997.
18. Datta A. 1995. Genome instability in *Saccharomyces cerevisiae*: studies on spontaneous mutation and homeologous recombination. Ph.D. dissertation. Emory University, Atlanta, GA.
19. Datta, A., A. Adjiri, L. New, G. F. Crouse, and S. Jinks-Robertson. 1996. Mitotic crossovers between diverged sequences are regulated by mismatch repair proteins in yeast. *Mol. Cell. Biol.* **16:** 1085–1093.
20. Detloff, P. and T. D. Petes. 1992. Measurements of excision repair tracts formed during meiotic recombination in *Saccharomyces cerevisiae. Mol. Cell. Biol.* **12:** 1805–1814.
21. Detloff, P., J. Sieber, and T. D. Petes. 1991. Repair of specific base pair mismatches formed during meiotic recombination in the yeast *Saccharomyces cerevisiae. Mol. Cell. Biol.* **11:** 737–745.
22. Detloff, P., M. A. White, and T. D. Petes. 1992. Analysis of a gene conversion gradient at the *HIS4* locus in *Saccharomyces cerevisiae. Genetics* **132:** 113–123.
23. Devereux, J., P. Haeberli, and O. Smithies. 1984. A comprehensive set of sequence analysis programs for the VAX. *Nucleic Acids Res.* **12:** 387–395.
24. Drummond, J. T., G.-M. Li, M. J. Longley, and P. Modrich. 1995. Isolation of an hMSH2-p160 heterodimer that restores DNA mismatch repair to tumor cells. *Science* **268:** 1909–1912.
25. Edelmann, W., P. E. Cohen, M. Kane, K. Lau, B. Morrow, S. Bennett, A. Umar, T. A. Kunkel, G. Cattoretti, R. Chaganti, J. W. Pollard, R. D. Kolodner, and R. Kucherlapati. 1996. Meiotic pachytene arrest in MLH1-deficient mice. *Cell* **85:** 1125–1134.
26. Emerson, S. 1969. Linkage and recombination at the chromosome level, in *Genetic Organization* (Caspari, E. W. and A. W. Ravin, eds.), Academic, New York, pp. 267–360.
27. Esposito, M. S. 1971. Postmeiotic segregation in *Saccharomyces. Mol. Gen. Genet.* **111:** 297–299.
28. Feaver, W. J., J. Q. Svejstrup, L. Bardwell, A. J. Bardwell, S. Buratowski, K. D. Gulyas, T. F. Donahue, E. C. Friedberg, and R. D. Kornberg. 1993. Dual roles of a multiprotein complex from *S. cerevisiae* in transcription and DNA repair. *Cell* **75:** 1379–1387.
29. Fishel, R., M. K. Lescoe, M. R. S. Rao, N. G. Copeland, N. A. Jenkins, J. Garber, M. Kane, and R. Kolodner. 1993. The human mutator gene homolog *MSH2* and its association with hereditary nonpolyposis colon cancer. *Cell* **75:** 1027–1038.
30. Fleck, O., H. Michael, and L. Heim. 1992. The *swi4*$^+$ gene of *Schizosaccharomyces pombe* encodes a homologue of mismatch repair enzymes. *Nucleic Acids Res.* **20:** 2271–2278.
31. Fleck, O., C. Rudolph, A. Albrecht, A. Lorentz, P. Schär, and H. Schmidt. 1994. The mutator gene *swi8* effects specific mutations in the mating-type region of *Schizosaccharomyces pombe. Genetics* **138:** 621–632.

32. Fleck, O., P. Schär, and J. Kohli. 1994. Identification of two mismatch-binding activities in protein extracts of *Schizosaccharomyces pombe. Nucleic Acids Res.* **22:** 5289–5295.
33. Fogel, S., R. Mortimer, K. Lusnak, and F. Tavares. 1979. Meiotic gene conversion: a signal of the basic recombination event in yeast. *Cold Spring Harbor Symp. Quant. Biol.* **43:** 1325–1341.
34. Fogel, S., R. K. Mortimer, and K. Lusnak. 1981. Mechanisms of meiotic gene conversion, or "wanderings on a foreign strand," in *The Molecular Biology of the Yeast* Saccharomyces (Strathern, J. N., E. W. Jones, and J. R. Broach, eds.), Cold Spring Harbor Laboratory, Cold Spring Harbor, NY, pp. 289–339.
35. Fujii, H. and T. Shimada. 1989. Isolation and characterization of cDNA clones derived from the divergently transcribed gene in the region upstream from the human dihydrofolate reductase gene. *J. Biol. Chem.* **264:** 10,057–10,064.
36. Haber, J. E., B. L. Ray, J. M. Kolb, and C. I. White. 1993. Rapid kinetics of mismatch repair of heteroduplex DNA that is formed during recombination in yeast. *Proc. Natl. Acad. Sci. USA* **90:** 3363–3367.
37. Haber, L. T. and G. C. Walker. 1991. Altering the conserved nucleotide binding motif in the *Salmonella typhimurium* MutS mismatch repair protein affects both its ATPase and mismatch binding activities. *EMBO J.* **10:** 2707–2715.
38. Heale, S. M. and T. D. Petes. 1995. The stabilization of repetitive tracts of DNA by variant repeats requires a functional DNA mismatch repair system. *Cell* **83:** 539–545.
39. Hoffmann, W. 1985. Molecular characterization of the *CAN1* locus in *Saccharomyces cerevisiae.* A transmembrane protein without N-terminal hydrophobic signal sequence. *J. Biol. Chem.* **260:** 11,831–11,837.
40. Hollingsworth, N. M., L. Ponte, and C. Halsey. 1995. *MSH5*, a novel MutS homolog, facilitates meiotic reciprocal recombination between homologs in *Saccharomyces cerevisiae* but not mismatch repair. *Genes Dev.* **9:** 1728–1739.
41. Horii, A., H.-J. Han, S. Sasaki, M. Shimada, and Y. Nakamura. 1994. Cloning, characterization and chromosomal assignment of the human genes homologous to yeast *PMS1*, a member of mismatch repair genes. *Biochem. Biophys. Res. Commun.* **204:** 1257–1264.
42. Huang, K. N. and L. S. Symington. 1993. A 5'-3' exonuclease from *Saccharomyces cerevisiae* is required for *in vitro* recombination between linear DNA molecules with overlapping homology. *Mol. Cell. Biol.* **13:** 3125–3134.
43. Hunter, N., S. R. Chambers, E. J. Louis, and R. H. Borts. 1996. The mismatch repair system contributes to meiotic sterility in an interspecific yeast hybrid. *EMBO J.* **15:** 1726–1733.
44. Iaccarino, I., F. Palombo, J. Drummond, N. F. Totty, J. J. Hsuan, P. Modrich, and J. Jiricny. 1996. MSH6, a *Saccharomyces cerevisiae* protein that binds to mismatches as a heterodimer with MSH2. *Curr. Biol.* **6:** 484–486.
45. Jeyaprakash, A., R. Das Gupta, and R. Kolodner. 1996. *Saccharomyces cerevisiae pms2* mutations are alleles of *MLH1*, and *pms2-2* corresponds to a hereditary nonpolyposis colorectal carcinoma-causing missense mutation. *Mol. Cell. Biol.* **16:** 3008–3011.
46. Jeyaprakash, A., J. W. Welch, and S. Fogel. 1994. Mutagenesis of yeast MW104-1B strain has identified the uncharacterized *PMS6* DNA mismatch repair gene locus and additional alleles of existing *PMS1*, *PMS2* and *MSH2* genes. *Mutat. Res. Lett.* **325:** 21–29.
47. Johnson, R. E., G. K. Kovvali, L. Prakash, and S. Prakash. 1995. Requirement of the yeast *RTH1* 5' to 3' exonuclease for the stability of simple repetitive DNA. *Science* **269:** 238–240.
48. Johnson, R. E., G. K. Kovvali, L. Prakash, and S. Prakash. 1996. Requirement of the yeast *MSH3* and *MSH6* genes for *MSH2*-dependent genomic stability. *J. Biol. Chem.* **271:** 7285–7288.

49. Kolodner, R. 1996. Biochemistry and genetics of eukaryotic mismatch repair. *Genes Dev.* **10:** 1433–1442.
50. Kramer, B., W. Kramer, M. S. Williamson, and S. Fogel. 1989. Heteroduplex DNA correction in *Saccharomyces cerevisiae* is mismatch specific and requires functional *PMS* genes. *Mol. Cell. Biol.* **9:** 4432–4440.
51. Kramer, W., B. Fartmann, and E. C. Ringbeck. 1996. Transcription of *mutS*- and *mutL*-homologous genes in *Saccharomyces cerevisiae* during the cell cycle. *Mol. Gen. Genet.* **252:** 275–283.
52. Kramer, W., B. Kramer, M. S. Williamson, and S. Fogel. 1989. Cloning and nucleotide sequence of DNA mismatch repair gene *PMS1* from *Saccharomyces cerevisiae*: homology to procaryotic MutL and HexB. *J. Bacteriol.* **171:** 5339–5346.
53. Kroutil, L. C., K. Register, K. Bebenek, and T. A. Kunkel. 1996. Exonucleolytic proofreading during replication of repetitive DNA. *Biochemistry* **35:** 1046–1053.
54. Lahaye, A., H. Stahl, D. Thines-Sempoux, and F. Foury. 1991. PIF**1:** a DNA helicase in yeast mitochondria. *EMBO J.* **10:** 997–1007.
55. Lahue, R. S., K. G. Au, and P. Modrich. 1989. DNA mismatch correction in a defined system. *Science* **245:** 160–164.
56. Lahue, R. S., S. S. Su, and P. Modrich. 1987. Requirement for d(GATC) sequences in *Escherichia coli mutHLS* mismatch correction. *Proc. Natl. Acad. Sci. USA* **84:** 1482–1486.
57. Leach, F. S., N. C. Nicolaides, N. Papadopoulos, B. Liu, J. Jen, R. Parsons, P. Peltomäki, P. Sistonen, L. A. Aaltonen, M. Nyström-Lahti, X.-Y. Guan, J. Zhang, P. S. Meltzer, J.-W. Yu, F.-T. Kao, D. J. Chen, K. M. Cerosaletti, R. E. K. Fournier, S. Todd, T. Lewis, R. J. Leach, S. L. Naylor, J. Weissenbach, J.-P. Mecklin, H. Järvinen, G. M. Petersen, S. R. Hamilton, J. Green, J. Jass, P. Watson, H. T. Lynch, J. M. Trent, A. De la Chapelle, K. W. Kinzler, and B. Vogelstein. 1993. Mutations of a *mutS* homolog in hereditary nonpolyposis colorectal cancer. *Cell* **75:** 1215–1225.
58. Lee, G. S., E. A. Savage, R. G. Ritzel, and R. C. von Borstel. 1988. The base-alteration spectrum of spontaneous and ultraviolet radiation-induced forward mutations in the *URA3* locus of *Saccharomyces cerevisiae. Mol. Gen. Genetics* **214:** 396–404.
59. Linton, J. P., J.-Y. J. Yen, E. Selby, Z. Chen, J. M. Chinsky, K. Liu, R. E. Kellems, and G. F. Crouse. 1989. Dual bidirectional promoters at the mouse *dhfr* locus: cloning and characterization of two mRNA classes of the divergently transcribed *Rep-1* gene. *Mol. Cell. Biol.* **9:** 3058–3072.
60. Manivasakam, P., S. M. Rosenberg, and P. J. Hastings. 1996. Poorly repaired mismatches in heteroduplex DNA are hyper-recombinagenic in *Saccharomyces cerevisiae. Genetics* **142:** 407–416.
61. Marsischky, G. T., N. Filosi, M. F. Kane, and R. Kolodner. 1996. Redundancy of *Saccharomyces cerevisiae MSH3* and *MSH6* in *MSH2*-dependent mismatch repair. *Genes Dev.* **10:** 407–420.
62. McIntosh, E. M. 1993. MCB elements and the regulation of DNA replication genes in yeast. *Curr. Genet.* **24:** 185–192.
63. Médigue, C., T. Rouxel, P. Vigier, A. Hénaut, and A. Danchin. 1991. Evidence for horizontal gene transfer in *Escherichia coli* speciation. *J. Mol. Biol.* **222:** 851–856.
64. Miret, J. J., M. G. Milla, and R. S. Lahue. 1993. Characterization of a DNA mismatch-binding activity in yeast extracts. *J. Biol. Chem.* **268:** 3507–3513.
65. Miret, J. J., B. O. Parker, and R. S. Lahue. 1996. Recognition of DNA insertion deletion mismatches by an activity in *Saccharomyces cerevisiae. Nucleic Acids Res.* **24:** 721–729.
66. Morrison, A., A. L. Johnson, L. H. Johnston, and A. Sugino. 1993. Pathway correcting DNA replication errors in *Saccharomyces cerevisiae. EMBO J.* **12:** 1467–1473.
67. Munz, P. 1975. On some properties of five mutator alleles in *Schizosaccharomyces pombe. Mutat. Res.* **29:** 155–157.

68. Muster-Nassal, C. and R. Kolodner. 1986. Mismatch correction catalyzed by cell-free extracts of *Saccharomyces cerevisiae*. *Proc. Natl. Acad. Sci. USA* **83:** 7618–7622.
69. Nag, D. K., M. A. White, and T. D. Petes. 1989. Palindromic sequences in heteroduplex DNA inhibit mismatch repair in yeast. *Nature* **340:** 318–320.
70. New, L., K. Liu, and G. F. Crouse. 1993. The yeast gene *MSH3* defines a new class of eukaryotic MutS homologues. *Mol. Gen. Genet.* **239:** 97–108.
71. Nicolaides, N. C., K. C. Carter, B. K. Shell, N. Papadopoulos, B. Vogelstein, and K. W. Kinzler. 1995. Genomic organization of the human PMS2 gene family. *Genomics* **30:** 195–206.
72. Nicolaides, N. C., N. Papadopoulos, B. Liu, Y.-F. Wei, K. C. Carter, S. M. Ruben, C. A. Rosen, W. A. Haseltine, R. D. Fleischmann, C. M. Fraser, M. D. Adams, J. C. Venter, M. G. Dunlop, S. R. Hamilton, G. M. Petersen, A. De la Chapelle, B. Vogelstein, and K. W. Kinzler. 1994. Mutations of two PMS homologues in hereditary nonpolyposis colon cancer. *Nature* **371:** 75–80.
73. Nicolas, A. and T. D. Petes. 1994. Polarity of meiotic gene conversion in fungi: Contrasting views. *Experientia* **50:** 242–252.
74. Palombo, F., P. Gallinari, I. Iaccarino, T. Lettieri, M. Hughes, A. D'Arrigo, O. Truong, J. J. Hsuan, and J. Jiricny. 1995. GTBP, a 160-kilodalton protein essential for mismatch-binding activity in human cells. *Science* **268:** 1912–1914.
75. Papadopoulos, N., N. C. Nicolaides, B. Liu, R. Parsons, C. Lengauer, F. Palombo, A. D'Arrigo, S. Markowitz, J. K. V. Willson, K. W. Kinzler, J. Jiricny, and B. Vogelstein. 1995. Mutations of *GTBP* in genetically unstable cells. *Science* **268:** 1915–1917.
76. Papadopoulos, N., N. C. Nicolaides, Y.-F. Wei, S. M. Ruben, K. C. Carter, C. A. Rosen, W. A. Haseltine, R. D. Fleischmann, C. M. Fraser, M. D. Adams, J. C. Venter, S. R. Hamilton, G. M. Petersen, P. Watson, H. T. Lynch, P. Peltomäki, J.-P. Mecklin, A. De la Chapelle, K. W. Kinzler, and B. Vogelstein. 1994. Mutation of a *mutL* homolog in hereditary colon cancer. *Science* **263:** 1625–1629.
77. Petes, T. D., R. E. Malone, and L. S. Symington. 1991. Recombination in yeast, in *The Molecular and Cellular Biology of the Yeast* Saccharomyces (Broach, J. R., J. R. Pringle, and E. W. Jones, eds.), Cold Spring Harbor Laboratory, Cold Spring Harbor, NY, pp. 407–521.
78. Picologlou, S., N. Brown, and S. W. Liebman. 1990. Mutations in *RAD6*, a yeast gene encoding a ubiquitin-conjugating enzyme, stimulate retrotransposition. *Mol. Cell. Biol.* **10:** 1017–1022.
79. Porter, G., J. Westmoreland, S. Priebe, and M. A. Resnick. 1996. Homologous and homeologous intermolecular gene conversion are not differentially affected by mutations in the DNA damage or the mismatch repair genes *RAD1*, *RAD50*, *RAD51*, *RAD52*, *RAD54*, *PMS1* and *MSH2*. *Genetics* **143:** 755–767.
80. Priebe, S. D., J. Westmoreland, T. Nilsson-Tillgren, and M. A. Resnick. 1994. Induction of recombination between homologous and diverged DNAs by double-strand gaps and breaks and role of mismatch repair. *Mol. Cell. Biol.* **14:** 4802–4814.
81. Proffitt, J. H., J. R. Davie, D. Swinton, and S. Hattman. 1984. 5-Methylcytosine is not detectable in *Saccharomyces cerevisiae* DNA. *Mol. Cell. Biol.* **4:** 985–988.
82. Prolla, T. A., D.-M. Christie, and R. M. Liskay. 1994. Dual requirement in yeast DNA mismatch repair for *MLH1* and *PMS1*, two homologs of the bacterial *mutL* gene. *Mol. Cell. Biol.* **14:** 407–415.
83. Prolla, T. A., Q. Pang, E. Alani, R. D. Kolodner, and R. M. Liskay. 1994. MLH1, PMS1, and MSH2 interactions during the initiation of DNA mismatch repair in yeast. *Science* **265:** 1091–1093.
84. Ray, B. L., C. I. White, and J. E. Haber. 1991. Heteroduplex formation and mismatch repair of the "stuck" mutation during mating-type switching in *Saccharomyces cerevisiae*. *Mol. Cell. Biol.* **11:** 5372–5380.

85. Reenan, R. A. G. and R. D. Kolodner. 1992. Isolation and characterization of two *Saccharomyces cerevisiae* genes encoding homologs of the bacterial HexA and MutS mismatch repair proteins. *Genetics* **132:** 963–973.
86. Reenan, R. A. G. and R. D. Kolodner. 1992. Characterization of insertion mutations in the *Saccharomyces cerevisiae MSH1* and *MSH2* genes: evidence for separate mitochondrial and nuclear functions. *Genetics* **132:** 975–985.
87. Resnick, M. A., Z. Zgaga, P. Hieter, J. Westmoreland, S. Fogel, and T. Nilsson-Tillgren. 1992. Recombinational repair of diverged DNAs: A study of homoeologous chromosomes and mammalian YACs in yeast. *Mol. Gen. Genet.* **234:** 65–73.
88. Rong, L. and H. L. Klein. 1993. Purification and characterization of the *SRS2* DNA helicase of the yeast *Saccharomyces cerevisiae. J. Biol. Chem.* **268:** 1252–1259.
89. Rong, L., F. Palladino, A. Aguilera, and H. L. Klein. 1991. The hyper-gene conversion *hpr5-1* mutation of *Saccharomyces cerevisiae* is an allele of the *SRS2/RADH* gene. *Genetics* **127:** 75–85.
90. Ross-Macdonald, P. and G. S. Roeder. 1994. Mutation of a meiosis-specific MutS homolog decreases crossing over but not mismatch correction. *Cell* **79:** 1069–1080.
91. Saparbaev, M., L. Prakash, and S. Prakash. 1996. Requirement of mismatch repair genes *MSH2* and *MSH3* in the *RAD1-RAD10* pathway of mitotic recombination in *Saccharomyces cerevisiae. Genetics* **142:** 727–736.
91a. Schär, P., M. Baur, C. Schneider, and J. Kohli. 1997. Mismatch repair in *Saccharomyces pombe* requires the MutL homologous gene *pms1*: molecular cloning and functional analysis. *Genetics* **146:** 1275–1286.
92. Schär, P. and J. Kohli. 1993. Marker effects of G to C transversions on intragenic recombination and mismatch repair in *Schizosaccharomyces pombe. Genetics* **133:** 825–835.
93. Schär, P., P. Munz, and J. Kohli. 1993. Meiotic mismatch repair quantified on the basis of segregation patterns in *Schizosaccharomyces pombe. Genetics* **133:** 815–824.
94. Schiestl, R. H. and S. Prakash. 1988. *RAD1*, an excision repair gene of *Saccharomyces cerevisiae*, is also involved in recombination. *Mol. Cell. Biol.* **8:** 3619–3626.
95. Schiestl, R. H. and S. Prakash. 1990. *RAD10*, an excision repair gene of *Saccharomyces cerevisiae*, is involved in the *RAD1* pathway of mitotic recombination. *Mol. Cell. Biol.* **10:** 2485–2491.
96. Schulz, V. P. and V. A. Zakian. 1994. The *Saccharomyces PIF1* DNA helicase inhibits telomere elongation and de novo telomere formation. *Cell* **76:** 145–155.
97. Schwartz, R. M. and M. O. Dayhoff. 1979. Matrices for detecting distant relationships, in *Atlas of Protein Sequence and Structure* (Dayhoff, M. O., ed.), National Biomedical Research Foundation, Washington, DC, pp. 353–358.
98. Selva, E. M., L. New, G. F. Crouse, and R. S. Lahue. 1995. Mismatch correction acts as a barrier to homeologous recombination in *Saccharomyces cerevisiae. Genetics* **139:** 1175–1188.
99. Sommers, C. H., E. J. Miller, B. Dujon, S. Prakash, and L. Prakash. 1995. Conditional lethality of null mutations in *RTH1* that encodes the yeast counterpart of a mammalian 5'- to 3'-exonuclease required for lagging strand DNA synthesis in reconstituted systems. *J. Biol. Chem.* **270:** 4193–4196.
100. Strand, M., M. C. Earley, G. F. Crouse, and T. D. Petes. 1995. Mutations in the *MSH3* gene preferentially lead to deletions within tracts of simple repetitive DNA in *Saccharomyces cerevisiae. Proc. Natl. Acad. Sci. USA* **92:** 10,418–10,421.
101. Strand, M., T. A. Prolla, R. M. Liskay, and T. D. Petes. 1993. Destabilization of tracts of simple repetitive DNA in yeast by mutations affecting DNA mismatch repair. *Nature* **365:** 274–276.
102. Streisinger, G., Y. Okada, J. Emrich, J. Newton, A. Tsugita, E. Terzaghi, and M. Inouye. 1966. Frameshift mutations and the genetic code. *Cold Spring Harbor Symp. Quant. Biol.* **31:** 77–84.

103. Sung, P., L. Prakash, S. W. Matson, and S. Prakash. 1987. RAD3 protein of *Saccharomyces cerevisiae* is a DNA helicase. *Proc. Natl. Acad. Sci. USA* **84:** 8951–8955.
104. Sweder, K. S., R. A. Verhage, D. J. Crowley, G. F. Crouse, J. Brouwer, and P. C. Hanawalt. 1996. Mismatch repair mutants in yeast are not defective in transcription-coupled DNA repair of UV-induced DNA damage. *Genetics* **143:** 1127–1135.
105. Szankasi, P. and G. R. Smith. 1995. A role for exonuclease I from *S. pombe* in mutation avoidance and mismatch correction. *Science* **267:** 1166–1169.
106. Szostak, J. W., T. L. Orr-Weaver, R. J. Rothstein, and F. W. Stahl. 1983. The double-strand-break repair model for recombination. *Cell* **33:** 25–35.
107. Tran, H. T., D. A. Gordenin, and M. A. Resnick. 1996. The prevention of repeat-associated deletions in *Saccharomyces cerevisiae* by mismatch repair depends on size and origin of deletions. *Genetics* **143:** 1579–1587.
107a. Umar, A., B. Buermeyer, J. A. Simon, D. C. Thomas, A. B. Clark, R. M. Liskay, and T. A. Kunkel. 1996. Requirement for PCNA in DNA mismatch repair at a step preceding DNA resynthesis. *Cell* **87:** 65–73.
108. Wang, Z., J. Q. Svejstrup, W. J. Feaver, X. Wu, R. D. Kornberg, and E. C. Friedberg. 1994. Transcription factor b (TFIIH) is required during nucleotide-excision repair in yeast. *Nature* **368:** 74–76.
109. Whelan, W. L., E. Gocke, and T. R. Manney. 1979. The CAN1 locus of *Saccharomyces cerevisiae*: fine-structure analysis and forward mutation rates. *Genetics* **91:** 35–51.
110. White, J. H., K. Lusnak, and S. Fogel. 1985. Mismatch-specific post-meiotic segregation frequency in yeast suggests a heteroduplex recombination intermediate. *Nature* **315:** 350–352.
111. Williamson, M. S., J. C. Game, and S. Fogel. 1985. Meiotic gene conversion mutants in *Saccharomyces cerevisiae.* I Isolation and characterization of *pms1-1* and *pms1-2. Genetics* **110:** 609–646.
112. Yang, Y., A. L. Johnson, L. H. Johnston, W. Siede, E. C. Friedberg, K. Ramachandran, and B. A. Kunz. 1996. A mutation in a *Saccharomyces cerevisiae* gene *(RAD3)* required for nucleotide excision repair and transcription increases the efficiency of mismatch correction. *Genetics* **144:** 459–466.
113. Yeh, Y. C., H. F. Liu, C. A. Ellis, and A. L. Lu. 1994. Mammalian topoisomerase I has base mismatch nicking activity. *J. Biol. Chem.* **269:** 15,498–15,504.

20

DNA Repair in *Schizosaccharomyces pombe*

Dominic J. F. Griffiths and Antony M. Carr

1. INTRODUCTION

All living cells have evolved complex DNA repair mechanisms that remove the lesions inflicted on their genetic material by a variety of environmental and chemical mutagens. Such lesions, if left unrepaired, cause mutations to become fixed during DNA replication, and these can often prove detrimental to the survival of the cell. The investigation of the DNA repair pathways in higher eukaryotes has concentrated on the analysis of DNA repair-deficient genetic disorders, such as xeroderma pigmentosum and Cockayne's syndrome, as well as mutant rodent cell lines. However, a more comprehensive picture of DNA damage response pathways and their relationship to each other has been provided by the study of lower, unicellular eukaryotes. Early work focused on the budding yeast *Saccharomyces cerevisiae*, where mutants defective in the pathways involved in processing DNA damage have been identified. The genes corresponding to these mutants have subsequently been isolated and many of the protein products have now been biochemically characterized.

The fission yeast *Schizosaccharomyces pombe*, despite being phylogenetically classified alongside the budding yeast, is, in fact, quite different. These species diverged between 600 and 330 million years ago *(13)*, and they diverged from humans about 1000 million years ago *(96,104)*. The identification of DNA repair genes that possess significant amino acid sequence similarity between these two yeasts has demonstrated that DNA repair pathways are substantially conserved throughout evolution. More importantly, this conservation has permitted the extrapolation of genetic and biochemical analysis carried out in the yeasts to the analogous processes in higher eukaryotes, including humans.

Both yeasts, especially when studied in parallel, have provided excellent model systems for the dissection of a number of fundamental processes common to all eukaryotic cells. The similarities (i.e., conserved proteins and protein domains) and the differences (i.e., contrasting control mechanisms between budding and fission yeast) have proven informative in a wide variety of biological systems, from cell cycle to protein transport mechanisms. This chapter outlines the current understanding of the cellular responses that occur following exposure of *S. pombe* cells to DNA-damaging agents. For information on mismatch repair in *S. pombe*, *see* Chapter 19.

From: DNA Damage and Repair, Vol. 1: DNA Repair in Prokaryotes and Lower Eukaryotes
Edited by: J. A. Nickoloff and M. F. Hoekstra © Humana Press Inc., Totowa, NJ

1.1. *The Fission Yeast* S. pombe

First isolated from an East African millet beer, called *Pombe (54)*, the fission yeast is now among the best-characterized of eukaryotic organisms. The cell cycles of budding and fission yeasts are quite different: *S. pombe* divides by medial fission, whereas *S. cerevisiae* divides by budding daughter cells. This is reflected in the fact that approx 70–75% of the *S. pombe* cell-cycle is spent in the G2 phase of the cycle, where bulk DNA replication has been completed. In rapidly proliferating cells, the major control point exerted on the cell cycle is therefore the decision of when to enter mitosis (G2/M boundary). In contrast, budding yeast has a very short G2 phase, and G1 is protracted. In this yeast, the G1/S transition—"START" (the "restriction point" in mammalian cells) is the key regulatory event in rapidly growing cells, and marks the decision to enter a complete round of mitotic cell division. The fact that fission yeast spends the majority of its cell cycle in G2, with a replicated genome (2C DNA content) has important implications for both DNA repair and cell-cycle checkpoint processes.

The main attraction of the use of yeast in DNA repair studies is the ability to perform genetic analysis. Unlike mammalian cells, yeast exists predominantly in a haploid state, which allows the phenotypic analysis of recessive mutations that would be masked by the wild-type allele if present in a diploid cell. Additionally, classical genetic complementation studies and epistasis analysis can be easily performed. Genetic complementation analysis is a rapid and easy way of defining individual genes involved in a particular process. Epistasis analysis using DNA damage-sensitive mutants has allowed these different repair genes to be assigned to distinct pathways. To perform epistasis analysis, a double mutant strain is phenotypically compared to the respective single mutant strains. If this double mutant is no more sensitive to the agent being examined (e.g., UV irradiation) than either of the single mutants, this is taken as evidence that the two mutated genes act in the same response pathway. However, if the double mutant is more sensitive than the respective single mutants, this is often interpreted to indicate that the two mutated genes act in distinct pathways or subpathways *(37)*.

A further advantage of yeast is that, in comparison to mammalian cells, it is a relatively straightforward procedure to isolate the gene that is defective in a mutant strain of interest. This is most commonly achieved by functional complementation, whereby a library representing total genomic DNA (in an appropriate yeast vector) is transformed into the mutant strain, and plasmids are isolated that are capable of rescuing the mutant phenotype. The DNA fragments responsible for such correction usually contain the wild-type gene of interest, which can then be isolated and analyzed.

1.2. *DNA Damage*

There are a wide range of lesions inflicted on DNA by a variety of chemical and physical mutagens (*32*; *see* Chapter 5, vol. 2). In addition to induced damage, a significant level of spontaneous DNA damage can occur as a result of hydrolytic decay, DNA methylation and oxidative damage. In fission yeast, the majority of research has focused on the pathways that respond to exposure to short-wave UV (UV-C) and γ-irradiation. UV-C light inflicts covalent joining of adjacent bases, forming mainly cyclobutane pyrimidine dimers (CPDs) and pyrimidine(6-4)pyrimidone photodimers [(6-4)PDs]. Such alterations induce distortional changes in the DNA helix, which can result in mispairing and misincorporation during DNA synthesis. Exposure to γ-irradiation also results in a variety of base alterations, but mainly it produces single-and double-strand breaks (DSBs) in the phosphodiester backbone. These arise largely as a result of the formation of highly reactive radical species within the cells. DSBs represent particularly toxic lesions, which, if left unrepaired, can result in chromosome breakage and subsequent loss during cell division.

2. NUCLEOTIDE EXCISION REPAIR OF UV-INDUCED DNA DAMAGE

By far the most extensively researched pathway that removes UV-induced DNA damage is the complex nucleotide excision repair pathway (NER). This process, which is capable of removing a wide variety of lesions that distort the shape of the DNA helix, is mechanistically conserved among *S. cerevisiae*, *S. pombe*, and mammalian cells. NER is estimated to require the products of at least 30 genes *(32)*. NER results in the excision of an oligomer of approx 30 bases that includes the damage site, and the remaining single-stranded patch then serves as a template for resynthesis of the excised region. Because the recognition event occurs at the level of helix distortion rather than detecting the structure of specific altered bases, it is suitable for the removal of a wide variety of bulky DNA adducts. This is in contrast to the process of base excision repair, in which individual enzymes each recognize and cleave a specific altered base structure. Once such cleavage has occurred, the abasic sites generated by these enzymes are processed by an AP endonuclease, which removes the deoxyribose monophosphate moiety from the DNA.

2.1. Nucleotide Excision Repair in S. pombe

Although NER in *S. cerevisiae* is reasonably well-characterized (*see* Chapter 15), in fission yeast this is certainly not the case, predominantly because the number of identified mutants is less extensive, as is the genetic and epistatic analysis that has been performed on the available radiation-sensitive *"rad"* mutants. However, the isolation of several of the genes corresponding to UV-sensitive mutants has revealed that several predicted gene products are homologs of the *S. cerevisiae* NER genes (*RAD3* epistasis group) *(20)*. The conservation of these genes is not confined to the yeasts: many of the genes from the excision repair defective human disorder xeroderma pigmentosum (*XP-A* through *XP-G*) and rodent *E*xcision *R*epair *C*ross *C*omplementing *(ERCC1-6)* genes have been cloned from UV-sensitive cell lines of higher eukaryotes (*see* Chapter 18, vol. 2). These have also been found to be homologs of the *S. cerevisiae* NER genes (Table 1). The conserved nature of the NER pathway indicates that much of the biochemical analysis of the proteins in *S. cerevisiae* and human cells can be related directly to the corresponding *S. pombe* proteins.

Before detailing each of the mutants and corresponding genes involved in *S. pombe* NER, it is necessary to explain the nomenclature. Both the *S. cerevisiae* and *S. pombe* mutants are presented in lower-case italics, whereas the genes corresponding to these mutants are in either upper case, i.e., *RAD (S. cerevisiae)* or lower case, i.e., *rad (S. pombe)*. For protein designation, in both organisms, the first letter is capitalized, but the letters are not italicized and names end with "p" (Radp). In no case are numerically corresponding mutants structural or functional homologs: *rad16* is not the fission yeast homolog of *S. cerevisiae RAD16*. This is predominantly owing to the fact that the

Table 1
Proteins Involved in the Responses to DNA Damage

S. pombe	*S. cerevisiae*	Mammalian	Comment[a]
Excision repair proteins			
$Rad13p^{Sp}$	$Rad2p^{Sc}$	$ERCC5^{Hs}$	Nuclease
$Rad15p^{Sp}/rhp3p^{Sp}$	$Rad3p^{Sc}$	$ERCC2^{Hs}$	Helicase, TFIIH
$ERCC3p^{Sp}$	$Rad25p^{Sc}/Ssl2p^{Sc}$	$ERCC3^{Hs}$	Helicase, TFIIH
$Rad16p^{Sp}$	$Rad1p^{Sc}$	$ERCC4^{Hs}$	Nuclease subunit
$Swi10p^{Sp}$	$Rad10p^{Sc}$	$ERCC1^{Hs}$	Nuclease subunit
$Rad2p^{Sp}$	$Rad27p^{Sc}$	$MF\text{-}1/FEN1^{Hs}$	Flap endonuclease
$Rad18p^{Sp}$	$Rhc18p^{Sc}$	—	Essential
$Rhp51p^{Sp}$	$Rad51p^{Sc}$	$hRAD51^{Hs}$	RecA homology
$SPDEp^{Sp}$	—	—	Damage nuclease
Recombination repair proteins			
$Rhp54p^{Sp}$	$Rad54p^{Sc}$	—	Helicase
$Rad22p^{Sp}$	$Rad52p^{Sc}$	—	Associates with $Rad51p^{Sc}$
$Rad32p^{Sp}$	$Mre11p^{Sc}$	—	—
$Rad21p^{Sp}$	—	—	Essential
$Hhp1p/Hhp2p^{Sp}$	$Hhr1p^{Sc}$	CKI^{Hs}	Casein kinase
Miscellaneous repair proteins			
$Rad8p^{Sp}$	$(Rad5p^{Sc}/Rev2p^{Sc})b$	—	Helicase
$Rhp6p^{Sp}$	$Rad6p^{Sc}$	$HHR6^{Hs}$	Ubiqitin conjugation
Yeast checkpoint proteins			
$Rad1p^{Sp}$	$Rad17p^{Sc}$	—	Nuclease
$Rad3p^{Sp}$	$Mec1p^{Sc}$	—	Lipid and protein kinase domains
$Rad4p^{Sp}/Cut5p^{Sp}$	$Dbp11p^{Sc}$	—	Replication
$Rad9p^{Sp}$	—	—	—
$Rad17p^{Sp}$	$Rad24p^{Sc}$	—	Limited RFC homology
$Rad26p^{Sp}$	—	—	—
$Chk1p^{Sp}$	—	—	Protein kinase
$Rad24p^{Sp}$ and $Rad25p^{Sp}$	$Bmh1p^{Sc}$ and $Bmh2p^{Sc}$	14-3-3	Binds phosphoserine
$Hus5p^{Sp}$	$Ubc9p^{Sc}$	—	Essential
$Cds1p^{Sp}$	$Rad53p^{Sc}$	—	Kinase

[a]Putative activity often based on amino acid homology.
[b]Best match, may not be true homolog.
^{Sp}, *S. pombe*; ^{Sc}, *S. cerevisiae*; ^{Hs}, *Homo sapiens*.

mutants were discovered in the two yeasts and designated before such homology was established. In recent years, corresponding genes between the two yeasts, which have been newly isolated, have been named either *rhp* (*rad h*omolog in *p*ombe) or *RHC* (*R*ad *H*omolog in *C*erevisiae) in a vain attempt to minimize confusion. Further efforts to standardize corresponding mutants between yeasts have generally failed. For the purpose of this chapter, *S. cerevisiae* mutants, their genes, and the corresponding proteins will be superscripted "Sc", whereas *S. pombe* will be superscripted "Sp." It should be noted, however, that this is not a universally used nomenclature.

2.2. S. pombe *Genes Involved in NER*

2.2.1. rad13^{Sp}

The *rad13^{Sp}* gene is not essential for cell growth, and was isolated by complementation of the UV sensitivity of the *rad13.A* mutant *(23)*, which is sensitive to UV but not γ-irradiation. It had previously been reported that the *RAD2^{Sc}* gene was capable of partially rescuing the UV sensitivity of the *rad13.A* mutant *(61)*. It was therefore not surprising that the 1113 amino acid protein encoded by the *rad13^{Sp}* gene was found to be homologous to Rad2p^{Sc}. The sequence similarity in Rad13p^{Sp} is predominantly confined to three regions, which display 74, 74, and 61% identity over 66, 35, and 115 amino acid stretches, respectively. Such conserved regions are likely to delineate evolutionarily conserved functional domains. The Rad2p^{Sc} protein possesses single-strand DNA endonuclease activity *(35)*, and it is postulated that this protein induces a 3'-nick in the excision repair patch. Although no biochemical evidence has been presented for Rad13p^{Sp}, it is believed that it will possess the same activity.

Two of the domains within the Rad13p^{Sp}/Rad2p^{Sc} proteins also share significant homology to another pair of conserved gene products: the *S. pombe* Rad2p^{Sp} and *S. cerevisiae* Rad27p^{Sc} proteins *(23)*. Although the corresponding *rad2^{Sp}* and *rad27^{Sc}* mutants are UV sensitive, neither are apparently involved in NER. Rad2p^{Sp} and Rad27p^{Sc} are homologs of the mammalian FEN-1 protein, which is a structure specific endonuclease involved in DNA replication (*see* Section 2.3.1.). Therefore, the structural conservation seen between the Rad13p^{Sp}/Rad2p^{Sc} and Rad2p^{Sp}/Rad27p^{Sc} family of proteins probably reflects a nuclease activity. This is supported by the fact that a similar conserved motif is seen in the *S. pombe* exonuclease I protein *(94)*.

2.2.2. rad15^{Sp}

The *rad15^{Sp}* gene was isolated by complementation of the UV sensitivity of the *rad15.P m*utant, which is highly sensitive to UV, but not γ-irradiation *(67)*, and also by cross-species hybridization with an*RAD3^{Sc}* probe *(67,78)*. This identifies the 772 amino acid Rad15^{Sp} protein as the fission yeast homolog of Rad3p^{Sc}. The overall sequence similarity between the two proteins is approx 65%, and is most pronounced in the seven domains that identify Rad15p^{Sp} and Rad3p^{Sc} as helicases. Such helicase activity in the Rad3p^{Sc} protein is capable of unwinding both DNA–DNA and DNA–RNA duplexes in an ATP-dependent reaction, and may facilitate the release of the excised patch during NER. Both the Rad15p^{Sp} and Rad3p^{Sc} are essential for cell growth, and this has been explained by the identification of Rad3p^{Sc} as a component of the basal transcription factor TFIIH *(34)*. Rad15p^{Sp} and Rad3p^{Sc} are therefore essential components of both the RNA polymerase II transcription machinery, and the "repairosome" complex that is proposed to be responsible for repair of UV induced damage (*92*; *see* Chapter 15). The overlap between these two processes is currently an area of intense research, and may help to resolve paradigms, such as transcription-coupled repair, a process that preferentially repairs the transcribed strand of actively transcribed genes, that has been observed from bacteria to humans (*93*; *see* Chapter 9; this vol., and Chapter 10, vol. 2).

2.2.3. ERCC3p^{Sp}

As the name implies, the ERCC3p^{Sp} protein is the fission yeast homolog of the human ERCC3 protein, which exhibits DNA helicase activity in vitro *(50,57)*. This, in turn, is the homolog of the Rad25p^{Sc} (Ssl2p) protein, which, like Rad15p^{Sp}/Rad3p^{Sc}, also encodes a DNA helicase *(34)*. In addition, the *ERCC3^{Sp}* gene is essential for cell growth, probably reflecting the fact that Rad25p^{Sc} has been demonstrated to be part of TFIIH *(34)*. However, the directionality of the Rad25p^{Sc} helicase activity is opposite to that of the Rad3p^{Sc} protein, and such bidirectional unwinding of the double helix suggests that Rad15p^{Sp} and ERCC3p^{Sp} may act in concert to provide localized melting of damaged areas, thereby permitting access of the 5'- and 3'-endonucleases to the sugar-phosphate backbone of the nucleotides flanking the damage.

2.2.4. rad16^{Sp}

The *rad16^{Sp}*, *rad10^{Sp}*, *rad20^{Sp}*, and *swi9* mutants were all demonstrated to be allelic *(86)*. The three *rad* mutants were identified by their UV sensitivity, whereas *swi9* mutants are defective in the ability to switch mating types. The *rad16^{Sp}* gene has been isolated by genomic complementation *(22)*. Sequencing analysis revealed an open reading frame (ORF) of 2676 bases, interrupted by the presence of seven introns. The predicted protein sequence of this ORF shares approx 30% overall identity with Rad1p^{Sc}, which is absolutely required for NER.

Rad1p^{Sc} has been shown biochemically and genetically to interact with Rad10p^{Sc} *(11,12)*. This association constitutes a structure-specific endonuclease, that provides the 5' incision in the initial step of excision repair *(99,100)*. Using the yeast two-hybrid system, this interaction has been shown to be evolutionarily conserved: Rad16p^{Sp} interacts with Swi10p, the *S. pombe* homolog of Rad10p^{Sc} *(22)*. It is likely, although not biochemically proven, that the Rad16p^{Sp}/Swi10p complex will have the same activity as the corresponding *S. cerevisiae* complex.

2.2.5. swi10

swi10, like *swi9*, is one of a collection of mutants that are deficient in the ability to switch mating types. *S. pombe* is a haploid unicellular organism, and cells exist as either a Minus or Plus mating type. Homothallic strains (h^{90}) possess the ability to switch between these two mating types, via unidirectional transpositions of specific DNA fragments into the *mat1* locus (reviewed in ref. *47*). Mutants defective in mating type switching are thus unable to perform one of the many steps required. For example, *swi1*, *swi3*, and *swi7* are defective in the ability to generate the DSB at the *mat1* locus; *swi2*, *swi5*, and *swi6* are defective in the utilization of this break to instigate DNA transposition; and *swi4*, *swi8*, *swi9 (rad16^{Sp})*, *swi10*, and *rad22^{Sp}* are unable to perform the recombination event. Several of these mutants are also sensitive to the effects of both UV and ionizing radiation, implying that mating type switching and DNA repair share common elements.

The *swi10* gene was isolated by complementation of the switching defect of the *swi10-154* mutant *(80)*. The *swi10* ORF is 252 amino acids, with a predicted protein mol wt of 29 kDa. Swi10^{Sp} is not essential for mitotic growth and displays considerable sequence similarity at the amino-terminus to both the human Ercc1p and Rad10p^{Sc}, particularly in the DNA binding domain and nucleotide binding site. The *ERCC1* cDNA fails to complement any of the *swi10* defects, demonstrating that this conservation is not reflected at the functional level. However, the proven interaction between Rad1p^{Sc} and Rad10p^{Sc} is conserved between the two yeasts: Swi10p has been shown to interact with Rad16p^{Sp}, the fission yeast homolog of Rad1p^{Sc} *(22)*. The lack of functional complementation may represent an inability to form the correct protein–protein interactions in different species rather than different roles for these proteins in each organism.

2.3. Evidence for a Second Repair Pathway for Damage Caused by Exposure to UV

S. cerevisiae null mutants at the *RAD1^{Sc}* locus are highly sensitive to the killing effects of UV. The *rad1^{Sc}* null mutant cells are totally defective in the ability to repair CPDs or (6-4)PDs *(58,60)*. In contrast, *S. pombe* cells deleted for the *rad16^{Sp}* gene display a reduction in, but not the complete loss of the ability to repair CPDs and (6-4)PDs. This feature is in common with other *S. pombe* classical excision repair mutants *(59)* and demonstrates the existence of a second repair pathway in *S. pombe* that removes UV lesions. The extended G2 phase observed in rapidly growing fission yeast allows additional opportunity for recombination-based repair pathways (a process that requires a 2C DNA content) and epistasis analysis and dimer removal kinetics has indicated that the *rad18^{Sp}*, *rad2^{Sp}*, and *rhp51* genes may be involved in this second repair pathway, which removes UV-induced DNA damage *(51,64,68)*. Double mutants of either of these three mutants with *rad13^{Sp}* (a classical excision repair mutant—*see* Section 2.2.1.) result in a marked increase in sensitivity to UV irradiation. This is in contrast to the possible double mutant combinations of the three mutants (*rad18^{Sp} rhp51* and *rad18^{Sp} rad2^{Sp}*) which display no such increase in sensitivity (*rad2^{Sp} rhp51* is inviable and cannot be tested). These epistatic interactions suggest that they are all defective in the same repair pathway. The current understanding of this novel repair pathway is discussed below.

2.3.1. rad2^{Sp}

The *rad2^{Sp}* mutant is sensitive to UV but not γ-irradiation. The 380 amino acid predicted protein product of the *rad2^{Sp}* gene, isolated by complementation of UV sensitivity *(68)*, displays homology to the Rad13p^{Sp}/Rad2p^{Sc} family of proteins (*see* Section 2.2.1.), and itself has a budding yeast homolog, Rad27p^{Sc} *(75)*. In addition, Rad2^{Sp} has a human homolog, *FEN1 (38,68)*, which encodes an endonuclease (a property implied by the budding yeast Rad2p^{Sc} homology—*see* Section 2.2.1.). This endonuclease displays substrate specificity for "flap" structures (hence *F*lap *EN*donuclease 1—*FEN1*), and has previously been identified as the enzyme activity DNase IV *(80)*. Finally, *FEN1* is identical to the nuclease MF-1, which is essential for lagging strand synthesis during in vitro SV40 DNA replication with purified proteins *(103)*. This suggests that Rad2p^{Sp} may have a role in both DNA repair and DNA replication. The viability of *rad2^{Sp}* and *rad27^{Sc}* null mutants is therefore surprising and may be explained if lesions processed by MF-1 during DNA replication can also be repaired by the recombination repair pathway. This possibility is supported by the apparent inviability of fission yeast *rad2^{Sp} rhp51 (68)* and *rad2^{Sp} rhp54 (65)* double mutants (*rhp51* and *rhp54* are homologs of *RAD51^{Sc}* and *RAD54^{Sc}*; *see* Sections 2.3.3. and 3.1.). It has also been reported that the equivalent double mutants between the budding yeast *rad27^{Sc}* and the *rad50^{Sc}* mutant series cannot be isolated *(75)*.

2.3.2. rad18^{Sp}

Unlike the *rad2^{Sp}* mutant, the *rad18.X* mutant is sensitive to both UV and γ-irradiation. The role of Rad18p^{Sp} in γ-radiation repair may reflect a role in recombination repair, borne out by epistasis analysis with *rhp51* mutants. Deletion of the *rad18*^{Sp} gene is lethal, and suggests that Rad18p^{Sp} is involved in normal cell growth *(51)*. The epistatic interaction between *rad18*^{Sp} and *rad2*^{Sp}, and the role in recombination repair imply that the essential nature may again be concerned with an aspect of DNA replication. The Rad18p^{Sp} has a budding yeast homolog, Rhc18p, identified by sequence similarity with a genomic fragment of the *S. cerevisiae* genome sequencing project, and deletion of the *RHC18* gene is also lethal *(51)*. Both proteins contain regions of protein sequence that place them in the structural maintenance of chromosomes (SMC) superfamily of proteins. Members of this family show structural features reminiscent of myosin and kinesin, and are concerned with chromosome structure, condensation, and/or segregation during mitosis. This homology, and the known roles of related proteins suggest that Rad18p^{Sp} and Rhc18p may be motor proteins involved in structural alterations to the chromatin that are required during repair processes and possibly during DNA replication (*see* Chapter 13, vol. 2).

2.3.3. rhp51

The *rhp51* gene was isolated by heterologous hybridization using an *S. cerevisiae RAD51*^{Sc} probe *(44,64)*. Rad51^{Sc} is apparently not involved in the repair of UV-induced DNA damage, since the *rad51*^{Sc} null mutant exhibits very little sensitivity to this agent. However, the *rhp51* null mutant is very sensitive to UV damage, and Rhp51p has been implicated in the second repair pathway for the removal of UV-induced damage. Rhp51p is 365 amino acids, with an estimated mol wt of 40 kDa, which displays 69% identical amino acids to the Rad51p protein. A null allele of *rhp51* is viable and renders the cells extremely sensitive to ionizing radiation, which is consistent with the postulated role for Rhp51p in DSB repair. Like *RAD51*^{Sc}, the levels of *rhp51* expression are likely to be cell-cycle-regulated, as predicted by the presence of two *MluI* cell-cycle boxes (MCBs) in the *rhp51* promoter, suggesting increased expression at the G1/S transition. Additionally, as with *RAD51*^{Sc}, there is also evidence for transcriptional increase following treatment with methyl methanesulfonate *(44)*.

Both Rad51p^{Sc} and Rhp51p display moderate homology to the bacterial strand exchange protein RecA, which is involved in many cellular processes, including recombination (*see* Chapter 8) and the SOS response to DNA damage (*see* Chapter 7). The recent identification of *rhp51/RAD51*^{Sc} homologs in humans, mouse, and chicken implies that the mechanism for the repair of DSBs may be conserved in higher eukaryotes *(14,90)*.

2.3.4. SPDE

An alternative approach toward identifying components involved in the repair of damaged DNA is to identify components of cell-free extracts that contain a specific activity involved in DNA repair. This approach has yielded an activity that is capable of repairing plasmid DNA damaged by UV irradiation *(91)*. Protein fractionation has subsequently been used in order to characterize further components of this extract that are specifically involved in DNA repair, and has identified SPDE—an S. *pombe* DNA Endonuclease *(17)*. This enzyme recognizes a DNA template containing either CPDs or (6-4)PDs, and catalyzes an incision immediately 5' to the damage site. This reaction is ATP-dependent, and produces cleavage products that contain 5'-phosphoryl- and 3'-hydroxl-termini. The characteristics of the reaction catalyzed by this 120-kDa protein implicate it in a mechanism that is distinct from both NER and base excision repair. Despite this, further analysis is required before SPDE can be placed alongside *rad2^{Sp}*, *rad18^{Sp}*, and *rhp51* in the repair of UV-induced DNA damage.

Recent work has demonstrated that SPDE activity is not present in *rad12^{Sp}* mutants *(31)*. *rad12^{Sp} rad13^{Sp}* double mutants are more sensitive to UV than either corresponding single mutant, implicating *rad12^{Sp}* in a repair process that is distinct from NER. This is in contrast to the *rad12^{Sp} rad2^{Sp}* double mutant, which is no more sensitive than *rad12^{Sp}*. However, *rad12^{Sp}* does not encode the SPDE nuclease, and thus the epistasis analysis with *rad12^{Sp}* mutants reported by Freyer et al. *(31)* cannot be interpreted until the nature of the *rad12^{Sp}* gene and the relationship between SPDE and *rad12^{Sp}* are understood.

3. DSB REPAIR IN *S. POMBE*

In comparison to NER, little is known about the repair of DSBs induced by physical (e.g., γ-irradiation) and chemical (e.g., methyl methanesulfonate) agents in fission yeast. Some of the mutants defective in the repair of DSBs have been identified from the collection of available mutants, whereas others have been generated by deletion of genes sharing homology with *S. cerevisiae* DSB repair genes (Table 1). *rhp51* is described in Section 2.3.3. owing to its involvement in UV induced DNA repair. Other DSB repair genes appear to be involved in UV repair, but this has not been well characterized.

3.1. rhp54

The *rhp54* gene, like *rhp51*, was isolated by low stringency hybridization *(65)*. The 852 amino acid predicted product of the *rhp54* gene is 67% similar and 51% identical to Rad54p^{Sc}, which is required for recombination repair, and probably represents its true homolog. Rad54p^{Sc}/Rhp54p proteins display significant similarity to the Snf2p subfamily of potential helicases. The *rhp54* deletion mutant has a phenotype indistinguishable from the *rhp51* deletion mutant. It is extremely sensitive to ionizing radiation and is also sensitive to UV radiation. Both *rhp51* and *rhp54* mutant cells are elongated compared to wild-type cells, suggesting a delay to mitosis *(65)*. This delay is dependent on an intact S-phase mitosis checkpoint pathway, and checkpoint-defective (e.g., *rad17*) *rhp54* null double mutants form microcolonies, but are essentially inviable. This inviability correlates with abortive attempts at mitosis. These data suggest that *rhp54* cells are defective in DNA replication. Further evidence for this comes from the observation that both *rhp51* and *rhp54* null cells display a 500-fold increase in the levels of chromosome loss in a minichromosome maintenance assay. Examination of DNA content by FACS analysis shows that *rhp54* single mutant cells mainly have a 2C DNA content, indicating that any S-phase defect must be manifested at the later stages of S phase *(65)*. Together, these data are consistent with a model whereby recombination repair proteins play a role in late S phase, perhaps to process the lesions generated by lagging strand DNA synthesis that can alternatively by processed by the Rad2p/MF-1/FEN1 nuclease.

3.2. rad22^{Sp}

In addition to the γ-ray and UV sensitivity, *rad22^{Sp}* mutants are also defective in mating type switching. As previously described, several other mutants that are radiation-sensitive are unable to interconvert their mating type, inferring that these two processes share many common aspects. *rad22*^{Sp} along with *swi4*, *swi8*, *swi9 (rad16*^{Sp}*)*, and *swi10* are involved in the resolution of the recombination structure that arises at the termination of the process, which requires the ability to repair a DSB *(47)*. The *rad22*^{Sp} gene was cloned by genomic complementation of this switching type defect, and encodes a 469 amino acid, 52-kDa protein *(72)*. This protein displays considerable sequence similarity to the product of *RAD52*^{Sc}, particularly in the N-terminal half (56% identity). *rad52*^{Sc} mutants are also defective in mating type switching, mitotic recombination, and meiosis, and are sensitive to ionizing radiation. Rad52p^{Sc} has also been demonstrated to interact with Rad51p^{Sc}, although the corresponding interaction of Rad22p^{Sp}/Rhpp51 has not yet been reported. *rad22*^{Sp} is not essential for viability in a heterothallic strain (h^+ or h^-), but homothallic (h^{90}) *rad22*^{Sp} cells are inviable. This has been attributed to the increased level of mating type switching events (and initiating DSBs) that occur in homothallic strains. A homolog of Rad22p^{Sp}/Rad52p^{Sc} proteins has been identified in chicken and humans *(14,63)*, demonstrating that this gene is evolutionarily conserved.

3.3. rad32^{Sp}

The *rad32-1* mutant was identified in a mutant screen designed specifically to isolate genes involved in the repair of DSBs *(95)*. In common with the other mutants identified on this basis, the *rad32-1* mutant is sensitive to both γ- and UV irradiation. The *rad32*^{Sp} mutant is defective in the ability to repair DSBs, as determined by pulse-field gel electrophoresis. Additionally, *rad32*^{Sp} null mutant cells are defective in meiotic recombination, (having a 15-fold reduction in meiotic recombination at the *ade6* locus) and display a 300-fold increase in minichromosome loss during mitotic cell division. The *rad32*^{Sp} gene was isolated by complementation of UV sensitivity, and encodes a protein of 648 amino acids (73.5 kDa), with homology to the *S. cerevisiae* meiotic recombination protein Mre11p. There is also a homolog in human cells *(3,74)*. Although deletion of the *rad32*^{Sp} gene is not lethal, *rad32.d* spores have a very poor viability, suggesting that Rad32p^{Sp} plays an important role in meiosis and/or spore development.

3.4. rad21^{Sp}

The inability of the *rad21-45* mutant to repair DSBs was determined by pulse field gel electrophoresis (*see* Chapter 24, vol. 2). Unlike wild-type cells, the chromosomes from γ-irradiated *rad21-45* cells were not resolved, even after a 4-h recovery period, and the DNA appeared to run as a smear of fragmented DNA. This indicates the continued presence of DNA containing DSBs *(15)*. It has also been reported that the *rad21*^{Sp} mutant displays a DNA damage checkpoint defect *(4)*, although this was not seen in a separate study *(15)*. *rad21*^{Sp} mutant cells certainly delay mitosis after irradiation, but then enter mitosis with unrepaired damage and exhibit evidence of mitotic segregation defects after DNA damage *(15)*. The *rad21*^{Sp} gene was isolated by correction of the extreme γ-ray sensitivity of the *rad21-45* mutant, and encodes a protein of 628 amino acids (mol wt 68 kDa) with no sequence homology to any known proteins. Deletion of *rad21*^{Sp} produces inviable cells, indicating that in addition to the role in DSB repair, Rad21p^{Sp} is essential for mitotic growth.

Rad21p^{Sp} is localized to the nucleus and exists in phosphorylated forms in vivo *(16)*. The protein is phosphorylated at multiple serine and threonine sites. Phosphorylation levels vary throughout the cell cycle, with the least phosphorylated form existing at the peak of septation. This stage also corresponds to the highest levels of the *rad21^{Sp}* transcript, and suggests that Rad21p^{Sp} mRNA and protein levels peak during the G1/S transition. In contrast, the most phosphorylated form peaks in late G2, when mRNA levels are at their lowest, and this form is stable through mitosis. The defect in the *rad21-45* mutant has been associated with the inability to phosphorylate fully Rad21p^{Sp}, although exposure to γ-irradiation does not appear to induce phosphorylation of the protein. The exact role of these phosphorylation events are not currently understood.

3.5. hhp1, hhp2

A genetic screen to isolate mutants that are defective in DSB repair in *S. cerevisiae* led to the identification of the *hrr25* mutant *(40)*. *HRR25* encodes a serine/threonine protein kinase, which is the homolog of mammalian casein kinase I. A subsequent PCR-based approach was employed to identify *S. pombe* homologs of this subfamily of proteins, and two such homologs have been discovered *(24)* named *hhp1* and *hhp2* (*HRR25 h*omolog in *S. pombe*). The 42.5- and 45.8-kDa protein products of these genes display 72 and 65% sequence identity to Hrr25p^{Sc}, respectively. Both Hhp1p and Hhp2p display dual-specificity protein kinase activity (serine/threonine and tyrosine) in vitro *(39)*. Expression of the *hhp1*, and to a lesser extent *hhp2* cDNA in an *hrr25* null mutant, rescued the morphological and growth defects as well as the MMS sensitivity. It is therefore likely that these genes encode functional homologs of *HRR25*, although *hhp1* may be closer to a true homolog. Neither Hhp1p nor Hhp2p is essential for viability, but a strain bearing both deletions is severely impaired in its growth rate, and tends to divide at a cell length that is approximately double that of wild-type cells. This suggests that these proteins may have a role in regulating normal cell growth. Single deletion strains of *hhp1* and *hhp2* are resistant to γ-irradiation, but deletion of both renders the cells sensitive. This phenotype is mirrored by the moderate UV sensitivity of these strains. Pulse-field gel electrophoresis was used to demonstrate that the sensitivity associated with the double mutant was attributable to the inability to repair strand breaks. Such a phenotype suggests a degree of functional redundancy between the Hhp1p and Hhp2p protein kinases.

4. MISCELLANEOUS REPAIR GENES

4.1. rad8^{Sp}

The *rad8^{Sp}* mutant is sensitive to both UV and γ-irradiation, and has been implicated in a recombination and/or error prone DNA repair process. The *rad8^{Sp}* gene was cloned by complementation of UV sensitivity, and is nonessential *(25)*. Rad8p^{Sp} is 1133 amino acids, with a predicted mol wt of 129 kDa. This protein displays homology to the *Snf2* subfamily of putative DNA helicases, and also contains a RING finger motif. Four other proteins involved in DNA repair processes, Rad54p^{Sc}(Rhp54p), Rad16p^{Sc}, Rad5p^{Sc}(Rev2p) and human ERCC6p, are also members of this subfamily. Rad5p^{Sc} shows the greatest homology, although predominantly within the helicase domain, but *RAD5^{Sc}* cannot functionally complement the *rad8^{Sp}* null allele. The *rad8^{Sp}* null mutant is not epistatic with *rad13*^{Sp} (NER), *rad9*^{Sp} (damage checkpoint) or *rad21*^{Sp} (DSB repair), implicating *rad8*^{Sp} in an as yet undefined repair mechanism. The killing effects of UV and γ irradiation are most severe in *rad8*^{Sp} null mutant cells at the G1/S period of the cell cycle. This is unlike the sensitivities associated with other repair mutants, suggesting Rad8p^{Sp} may function to repair or tolerate DNA damage at this period of the cell cycle *(25)*.

4.2. rhp6^{Sp}

rhp6^{Sp} was cloned by low-stringency hybridization using *RAD6*^{Sc} *(77)* as a probe, and encodes a protein of 17 kDa, which displays 77% sequence identity to Rad6^{Sc} *(76)*. Deletion of *rhp6*^{Sp} renders cells sensitive to both UV and γ-irradiation, and cells are defective in DNA damage-induced mutagenesis and postreplication repair. In addition, *rhp6*^{Sp} null mutant cells display retarded growth rates and are defective in sporulation, inferring a role for Rhp6p^{Sp} in normal growth. Rhp6p^{Sp} can crosscomplement the UV sensitivity (but not the sporulation defect) of the *rad6*^{Sc} null mutant, an ability that is rare among homologous *rad* genes of the two yeasts, and both proteins encode E2 ubiquitin-conjugating enzymes. The role of Rad6p^{Sc} in postreplication repair may be facilitated by association with the Rad18^{Sc} protein, which is capable of binding single-stranded DNA *(8)*. This potentially provides a mechanism for targeting the Rad6p^{Sc} ubiquitin conjugating enzyme to regions of damaged DNA. However, as yet, no homolog of Rad18p^{Sc} has been reported in fission yeast.

5. CELL-CYCLE CHECKPOINTS

Following exposure to any agent that causes damage to the genetic material, all eukaryotic cells are capable of inducing a reversible cell-cycle arrest. The duration of this arrest is dose-dependent, and provides the cell with sufficient time to repair the inflicted damage before DNA replication or chromosome segregation is attempted. There have been two major cell cycle transitions identified through which such checkpoints operate: at the G1/S transition, and at the G2/M transition (reviewed in ref. *36*; *see* Chapter 17, this vol.; Chapter 21, vol. 2). The G1/S transition represents the commitment of a cell to enter a complete round of mitotic cell division. Following exposure to DNA-damaging agents in the G1 phase, eukaryotic cells are prevented from entering S phase. Failure to prevent entry into S phase would otherwise lead to DNA of a damaged template, which could lead to heritable mutation and genomic instability. In fission yeast, this G1/S transition is awkward to study, since exponentially growing wild-type *S. pombe* have a barely detectable G1 period, and cells progress rapidly from mitosis into the subsequent S phase. Consequently, little is known about this damage checkpoint in *S. pombe*, and studies have been largely confined to *S. cerevisiae* and mammalian cells, which have longer G1 phases.

The second cell-cycle damage checkpoint acts at the G2/M transition, which controls the timing of entry into mitosis. At this stage, the replicated chromosomes are separated, prior to cell division. Attempting to separate damaged chromosomes can prove detrimental to the genetic integrity of the daughter cells, especially in the presence of DSBs, which, if left unrepaired, would lead to chromosome fragmentation and loss of genetic material. Consequently, cells that are exposed to DNA-damaging agents in the S and G2 phases of the cell cycle are prevented from undergoing mitosis until the genome has been repaired.

Several checkpoint mutants have been identified that are defective in the transient arrest of mitosis following exposure to both γ- and UV irradiation *(4,6,26,82)*. No checkpoint-defective mutant has been reported to be defective in G2/M arrest following treatment with one, but not the other type of DNA-damaging agent (but *see* Section 5.1.3.), and it is generally accepted that it is the presence of damage (or its subsequent repair) *per se*, rather than the recognition of specific lesions, that triggers the checkpoint response. Interest in this field arose following the identification of a *S. cerevisiae* mutant, *rad9*^{Sc}, that is sensitive to irradiation, but not as a consequence of a repair deficiency *(107)*. The defect in this strain was found to be in the ability to prevent mitosis prior to completion of DNA repair: cells consequently attempted mitosis with unrepaired chromosomes, with subsequent loss of viability. *rad9*^{Sc} strains also demonstrate slightly increased levels of chromosome loss, presumably as a result of a deficiency in the checkpoint that acts following spontaneous errors (*see* Chapter 17).

Of the 20 available radiation-sensitive mutants (*rad* mutants), four were found to be unable to prevent mitotic entry following exposure to 50 J/m^2 of UV *(4)*. In brief, *rad cdc25* double mutants were constructed (*cdc25* is a temperature-sensitive mutant that arrests in late G2) and synchronized in G2 by means of a temperature shift to the restrictive temperature. Cells were then returned to the permissive temperature and irradiated. At regular time intervals following irradiation, a sample of cells was removed and scored for the percentage of cells that had entered mitosis. In the majority of mutants, a delay to mitotic entry was observed following irradiation, indicating that in these *rad* mutants, the damage checkpoint was functional. However, the response of four of these mutants, *rad1*^{Sp}, *rad3*^{Sp}, *rad9*^{Sp}, and *rad17*^{Sp}, was identical to the control, unirradiated sample, and cells proceeded into mitosis with normal kinetics. Similar results were seen with 50 Gy of γ-irradiation *(82)*. This clearly demonstrated that these four mutants were unable to induce a cell-cycle delay following irradiation, and are therefore defective in a component of this checkpoint response.

Two further loci with equivalent mutant phenotypes have now been identified, *rad26*^{Sp} *(6)* and *hus1 (26)*. All six of these "checkpoint *rad*" mutants are highly sensitive to UV and γ-irradiation but, unlike the *rad9*^{Sc} mutant, this sensitivity is only partially rescued by an artificial G2 delay *(4)*, which implicates the proteins defective in these mutants in additional cellular responses to DNA damage. In addition, all six of these mutants were profoundly defective in the ability to recognize incomplete DNA synthesis following treatment with a chemical inhibitor of DNA synthesis, hydroxyurea (HU). In a similar fashion to the damage checkpoint, any cell that is perturbed during S phase induces a mitotic entry delay until the block has been removed, and prevents cells from undergoing chromosome separation with a partially replicated genome. When the single "checkpoint *rad*" mutants were incubated in the presence of HU, which inhibits the enzyme ribonucleotide reductase and consequently blocks DNA replication through deprivation of deoxyribonucleotide pools, cells entered mitosis with concomitant rapid loss of viability *(4,6,26,82)*. In a similar fashion to the radiation phenotype, HU sensitivity contains a component that is not owing to the inability to impose a cell-cycle delay *(26)*, implicating the "checkpoint Radp" gene products in multiple pathways following blocks to DNA replication. A representative of these six "checkpoint *rad*" mutants was also unable to arrest mitosis in a variety of genetic backgrounds which interfere with DNA synthesis *(21)*. This demonstrates that these six genes are required for all of the DNA structure dependency checkpoints identified in *S. pombe*.

In summary, the six "checkpoint *rad*" mutants (*rad1*^{Sp}, *rad3*^{Sp}, *rad9*^{Sp}, *rad17*^{Sp}, *rad26*^{Sp}, and *hus1*) are unable to delay mitosis following DNA damage, or following an interruption to DNA replication. This indicates that there must be considerable overlap between these two checkpoint processes. Subsequent mutant screens have identified mutants that are defective primarily in the DNA damage checkpoint, but not the DNA replication checkpoint (*chk1*, *rad24*^{Sp}, *rad25*^{Sp}). Together with the *cdc2.3w* mutant and the *hus6* mutant, which are unable to respond to unreplicated DNA, but have an intact damage checkpoint *(6,89)*, these mutants have been used to define genetically two closely related, but distinct checkpoints in fission yeast: one that responds to DNA damage, and one that responds to inhibition of bulk DNA synthesis (Fig. 1). The following section outlines the genes that have been identified, their mutant phenotypes, and what is currently known about their protein products.

5.1. The Checkpoint-Defective Mutants

5.1.1. *rad1*^{Sp}

The *rad1-1* allele was the first radiation-sensitive mutant reported in fission yeast *(69)*. The complementing gene consists of three exons, and encodes a polypeptide of approx 37 kDa. This ORF displays partial homology (48% homology, 24% identity) to the Rec1p DNA exonuclease of *Ustilago maydis (19,42,55)*, and similar levels of homology to the Rad17p^{Sc} checkpoint protein *(56)*. Rec1p is involved in the control of recombination *(72,101)*, and the *rec1* mutant is also sensitive to irradiation and HU exposure. This homology supports the current opinion that the checkpoint Rad proteins possess other roles within the cell. Although *S. pombe* Rad1p^{Sp} is homologous to Rad17p^{Sc}, the *RAD17*^{Sc} cDNA, under control of an inducible *S. pombe* promoter, is not capable of rescuing any of the *rad1*^{Sp} null mutant phenotypes *(33)*. However, functional crosscomplementation between budding and fission yeast checkpoint proteins is rare, and it is still likely that Rad1p^{Sp} and Rad17p^{Sc} possess related functions within the respective yeasts. Although the *rad17*^{Sc} mutant is defective for the G2/M damage checkpoint, unlike its equivalent from *S. pombe* (*rad1*^{Sp}), it maintains an intact DNA replication checkpoint, similar to the original *rad9*^{Sc} mutant. Since this is also true for another pair of homologs (*rad17*^{Sp} and *RAD24*^{Sc}), it is therefore likely that there may be some subtle mechanistic differences in the S-phase checkpoint apparatus between the two yeasts.

5.1.2. *rad3*^{Sp}

The *rad3*^{Sp} gene was originally reported to encode a protein of 1070 amino acids, with no homology to known proteins *(87)*. However, the *rad3*^{Sp} gene actually contains coding regions beyond the 3'- and 5'-limits of the reported sequence *(9)*. The *rad3*^{Sp} gene is, in fact, 7158 bp and encodes a protein of 2386 kDa. This protein displays extensive homology to *S. cerevisiae* Esr1p/Mec1p *(45,108)*. The *mec1* mutant is, like *rad3*^{Sp}, defective in both the S phase and damage checkpoints, although deletion of *MEC1* is lethal *(108)*.

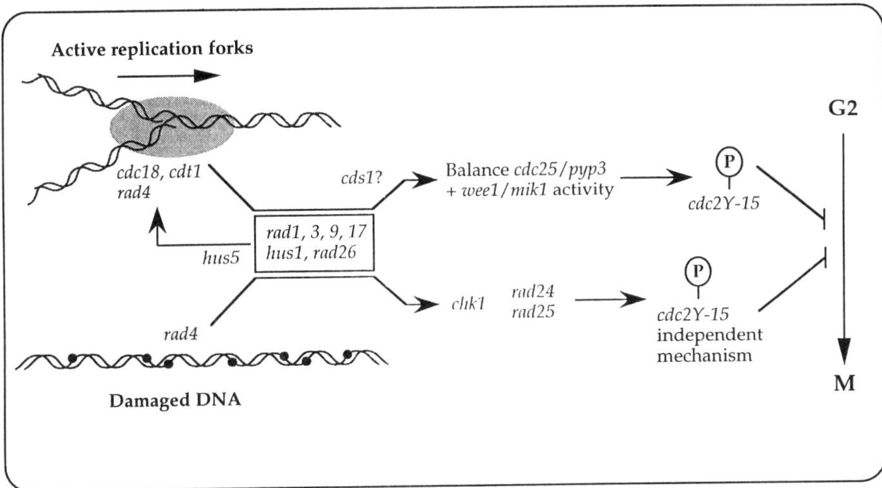

Fig. 1. The checkpoint pathways in *S. pombe*. Two distinct pathways are evident: 1. The S-phase mitosis pathway. Activation of this pathway requires the initiation of S phase (*cdc18 [46], cdt1 [41], rad4 [85]*), and utilizes the "checkpoint *rad*" genes ultimately to signal Cdc2p via tyrosine phosphorylation. 2. The DNA damage pathway. DNA damage, or its repair, activates a "checkpoint *rad*"-dependent signal pathway, which ultimately prevents cell-cycle progression in a manner independent of tyrosine phosphorylation of Cdc2p. Both pathways also impact on DNA replication itself and may also signal to other points in the cell cycle, such as start.

In *S. cerevisiae*, certain aspects of mitosis, such as spindle pole body duplication and spindle formation, are initiated before the completion of S phase, and it is probably necessary to prevent spindle elongation actively during S phase. This is not the case in fission yeast, where a significant G2 period separates the S and M phases, and relative timing (dictated by, for example, cell growth) is sufficient for unperturbed cells to maintain the correct order of cell-cycle events. This difference may also explain why in budding yeast, several of the checkpoint genes are essential for cell growth: loss of the restraint over chromosome separation might initiate mitosis during S phase, with catastrophic consequences.

The sequence similarity between these two proteins is highest in domains that identify both Rad3p^{Sp} and Mec1p as members of the phosphoinositol 3 (PI-3)- kinase subfamily of protein kinases *(109)*. In the case of Rad3p^{Sp}, this protein kinase domain appears essential for function. Mutagenesis of this domain results in a totally defective protein *(9)*. The founding members of this family are responsible for phosphorylation events on phosphotidylinositol, a phospholipid involved in a variety of signal transduction processes. However, recent work has identified several large proteins with sequence similarity in the PI-3 kinase domain, none of which have been shown to phosphorylate phospholipids, and at least one of which is known to encode a protein kinase *(109)*. Interest in this subfamily of large PI-3 like kinases has increased dramatically since the discovery that the protein defective in the genetic disorder ataxia telangiectasia (AT) is also a member (*85*; *see* Chapter 19, vol. 2). Mutations in the *ATM* gene may also be the largest hereditary cause of breast cancer. AT is a multisystem disorder, but the cells from AT patients are sensitive to ionizing radiation, and defective in

S-phase and G2 arrest following such DNA damage. The complicated phenotype of AT cells suggests that these cells have a deficiency in both DNA repair itself and the damage checkpoint. The identification of *ATM* as a member of the same PI-3-like kinase subfamily as the Rad3p^{Sp}/Mec1p homologs provides the first structural homology between yeast checkpoint proteins and a higher eukaryotic checkpoint protein. However, ATM and Rad3p^{Sp} may not be true homologs, since ATM is more closely related to the *S. cerevisiae* Tel1 protein, and a direct human homolog of Rad3p^{Sp} has been characterized that is more related to fission yeast Rad3p^{Sp} than is ATM *(9)*. These data suggest that more than one Rad3p^{Sp}-related protein may control the DNA damage checkpoints in metazoan cells.

5.1.3. rad4^{Sp}

The *rad4.116* mutant is temperature-sensitive for growth, and moderately sensitive to both UV and γ-irradiation. The *rad4^{Sp}* gene was isolated by genomic complementation of the thermosensitivity of the *rad4.116* mutant, and isolated fragments were also capable of rescuing the UV sensitivity of this mutant *(28)*. The *rad4^{Sp}* gene encodes a novel protein of 648 amino acids with an approximate mol wt of 74 kDa *(28,49,84)*. The amino acid sequence contains two regions in the N-terminal half of the protein, which displays strong sequence homology to the product of the human *XRCC1* gene. *XRCC1* was cloned by the ability to complement the DNA repair defect of the Chinese hamster mutant EM9, which is slightly sensitive to UV and γ-irradiation *(98)*. The two conserved regions are proposed to reflect important functional domains in the two proteins, such as an active site or region of protein–protein interactions.

cut5 mutants are allelic to *rad4^{Sp} (84)*. *cut* mutants have been named by virtue of their inappropriate mitotic commitment (*c*ell *u*ntimely *t*orn), since they enter mitosis in the absence of nuclear division. Characterization of the *cut5* mutant has identified Rad4p^{Sp}/Cut5p protein as being essential for correct initiation of S phase: *cut5* mutants at the restrictive temperature are blocked at early S phase, and cells undergo inappropriate catastrophic mitosis with a 1C DNA content. The *cut5* gene product is required to establish and maintain the cell-cycle block induced by HU exposure: cells incubated in HU at the permissive temperature arrest normally, and this arrest is rapidly lost on shift to the nonpermissive temperature. γ-irradiation of synchronous *rad4^{Sp}* and *cut5* mutants has shown that this is not the case for the γ-radiation-induced DNA damage checkpoint. If cells are first irradiated and then Cut5p is inactivated by a temperature shift, the DNA damage checkpoint is intact, and cells arrest normally. However, if the cells are first shifted to the nonpermissive temperature and then irradiated, they display a complete checkpoint defect *(18)*. This therefore indicates that Rad4p^{Sp}/Cut5p is required to initiate, but not maintain, the G2/M cell-cycle delay induced by γ-radiation. This defective checkpoint response to γ-radiation contrasts with the reported checkpoint proficiency of *cut5* mutants at the restrictive temperature following exposure to UV irradiation *(83)* and may represent the first difference seen in the checkpoint responses to different DNA-damaging agents in fission yeast.

5.1.4. rad9^{Sp}

The *rad9^{Sp}* gene encodes a polypeptide of 47.5 kDa that displays no homology to any known protein *(66)*. Rad9p^{Sp} is evolutionarily conserved among fission yeasts, with homologs having been identified in *Schizosaccharomyces octosporus* and

Schizosaccharomyces malidevorans (53). Recently, three extragenic suppressors of the radiation sensitivity of *rad9^{Sp}* mutants, named *srr1*, *srr2*, and *srr3*, have been isolated *(52)*. All three of these suppressors moderately rescue the radiation (UV and ionizing) and HU sensitivity of both the *rad9^{Sp}* null mutant and the loss-of-function *rad9-192* point mutant. However, in no case is this partial rescue owing to a restoration of the G2 DNA damage or DNA replication checkpoints. Such results are consistent with the theory that the checkpoint Radp proteins have other as yet undefined roles within the cell that are required for survival following exposure to radiation or HU, and are separable from the ability to enforce mitotic arrest.

5.1.5. rad17^{Sp}

The *rad17^{Sp}* gene consists of three exons, and encodes a polypeptide of approx 69 kDa *(33)*. This protein displays limited sequence similarity to the subunits of human replication factor C (RF-C). RF-C is a five-subunit processivity factor (i.e., non-catalytic) involved in DNA replication, and all five subunits share sequence similarity to each other *(48,70,103)*. RF-C functions to recognize the primer—template junctions at origins of replication, whereon it binds to these sites and facilitates the loading of proliferating cell nuclear antigen (PCNA) and a DNA polymerase. In addition, RF-C is required in vitro for the gap-resynthesis step in NER *(1)*. The product of the *rad17^{Sp}* gene is, however, unlikely to be a true fission yeast homolog of an RF-C, since the sequence similarity is limited and confined to specific domains, and deletion of *rad17^{Sp}* is not lethal, as would be expected for an RF-C subunit *(43)*. It is tempting to speculate that this limited homology reflects a role of Rad17p^{Sp} in recognizing and binding to DNA that contains strand breaks.

Work from Lydall and Weinert *(56)* has indicated a role for the *S. cerevisiae* homologs of Rad1p^{Sp} and Rad17p^{Sp} (Rad17p^{Sc} and Rad24p^{Sc}, respectively; note the unfortunate confusion in the nomenclature) in processing the strand breaks that are generated in *cdc13^{Sc}* mutants into single-stranded regions. This suggests that these two checkpoint proteins have a direct role in DNA metabolism, although the consequence of such processing is not known. This is consistent with the observation that Rad1p^{Sp} may be a nuclease (by homology to Rec1p from *U. maydis*; *97*) and that Rad17p^{Sp} may be a DNA structure binding protein. It is not yet clear if the processing function reported by Lydall and Weinert corresponds to the generation of the checkpoint signal, which prevents mitosis, or whether it is a separate "repair" event, which occurs in parallel. Mutagenesis of Rad17p^{Sp} at sites conserved between Rad17p^{Sp} and RF-C, created several mutants that are sensitive to γ-irradiation. This sensitivity does not arise from the inability to prevent mitosis, since these mutants all have a normal G2 DNA damage checkpoint and arrest proficiently after γ-irradiation *(33)*. If the RF-C homology corresponds to the potential processing event, then this might imply that processing and checkpoint signaling are not mutually dependent and can be separated under some circumstances.

5.1.6. rad26^{Sp}

The *rad26^{Sp}* gene encodes a polypeptide of 69 kDa, which shows no homology to any known proteins. The *rad26^{Sp}* mutant was identified in a mutant screen to identify additional elements of the DNA damage checkpoint pathway *(6)*. The *rad26^{Sp}* null allele is a classic checkpoint *rad* mutant, displaying both S-phase and damage checkpoint deficiencies. However, the original allele of *rad26^{Sp}*, *rad26.T12*, has been particularly informative in understanding the additional role(s) of Rad26p^{Sp}, and by extrapolation, other checkpoint Radp proteins. *rad26.T12* mutant cells possess a wild-type DNA damage checkpoint, but remain significantly sensitive to radiation. Cell biology and genetic analysis has demonstrated that this sensitivity defines a component of the radiation sensitivity seen in all the checkpoint *rad* mutants that is owing to defects in a mitotic arrest-independent process or processes. Unlike the *rad17*^{Sp} mutants described above, this sensitivity has been specifically mapped to the G1/S phase of the cell cycle, and is not seen during G2, which infers that the checkpoint *rad* mutants may be deficient in regulating DNA replication following DNA damage during S phase *(6)*. This is similar to the phenotype recently reported for *mec1* mutants (the *S. cerevisiae* homolog of *rad3*^{Sp}), where the regulation of S phase after exposure to DNA-damaging agents is defective *(73)*.

5.1.7. chk1/rad27^{Sp}

The *chk1/rad27*^{Sp} gene encodes a nonessential 56-kDa protein kinase, and was isolated independently in two studies. In one, it was identified by complementation of the *rad27.T15* mutant, which is defective in the damage checkpoint *(6)*. In a separate study *(105)*, it was identified as a multicopy suppressor of a cold-sensitive allele of the *cdc2*^{Sp} gene, *cdc2.r4*. The *cdc2*^{Sp} gene encodes the evolutionary ubiquitous protein kinase p34^{Cdc2}, that is required for both the G1/S and G2/M transitions. Such genetic suppression implies a possible interaction between the Chk1p/Rad27p^{Sp} proteins and p34^{Cdc2}, that may prevent cell cycle transitions under certain circumstances. However, it is not clear why a protein that causes mitotic arrest following DNA damage should be capable of relieving the cell-cycle block imposed by the *cdc2.r4* allele at the restrictive temperature. This is further confused by the observation *(30)* that overexpressing Chk1p causes cell-cycle arrest in both wild-type and *cdc2.r4* cells.

Deletion of the *chk1/rad27*^{Sp} gene causes loss of the radiation checkpoint, but results in a sensitivity to radiation much less profound than that seen in the checkpoint *rad* mutants. This reduced sensitivity is consistent with the observation that several checkpoint *rad* mutants are defective in additional processes unrelated to the ability to prevent mitosis following DNA damage *(6,33)*. These additional responses are therefore presumed to be unaffected in the *chk1/rad27*^{Sp} null mutants. Analysis of radiation sensitivity through the cell cycle in *chk1* cells indicates that they are specifically very sensitive to irradiation in late G2, immediately preceding mitosis *(6)*.

The observation that overexpression of the *chk1/rad27*^{Sp} cDNA (from the inducible *nmt1* promoter of pREP41) induces cell-cycle arrest suggests that the amount of active Chk1p/Rad27p^{Sp} protein present may be a rate-limiting step in cell-cycle arrest or may indicate that Chk1p titrates some important cell-cycle control proteins. Such mitotic arrest occurs even in a checkpoint *rad* genetic background, and hence, places Chk1p/Rad27p^{Sp} function downstream of the checkpoint Rad proteins in the damage response. Recent work *(106)* demonstrates that Chk1p/Rad27p^{Sp} is phosphorylated after irradiation, and that this phosphorylation event is dependent on functional checkpoint *rad* genes. This confirms that Chk1p/Rad27p^{Sp} functions downstream of the checkpoint Radp proteins, presumably to regulate directly cell-cycle control proteins. Although no functional homologs of Chk1p/Rad27p^{Sp} have been identified from metazoan cells, it is possible that Chk1p has been conserved through evolution as an *S. cerevisiae* structural homolog has been identified by the sequencing project.

5.1.8. rad24^{Sp}, rad25^{Sp}

The *rad24.T1* mutant was originally identified as being defective in the mitotic arrest following exposure to UV and γ-irradiation *(6)*. In addition, the *rad24.T1* mutant, unlike any of the previously discussed mutants, enters mitosis prematurely during normal growth with a "semiwee" phenotype, indicating an involvement in the timing of mitosis. A genomic fragment was isolated that complemented the UV sensitivity of the *rad24.T1* mutant *(30)*. However, this complementing fragment was found to contain a multicopy suppressor of *rad24.T1*, designated *rad25^{Sp}*. The protein encoded by the *rad25^{Sp}* gene displays 58% identity to the 14-3-3 family of highly conserved eukaryotic proteins. These proteins have been identified in most eukaryotic cells, and possess diverse properties that implicate them in signal transduction pathways (reviewed in ref. *2*). The existence of seven isoforms of the 14-3-3 proteins in mammalian brain tissue suggested that additional members of this family might exist in fission yeast. A single additional homolog was identified by PCR that was 71% identical to Rad25p^{Sp}. Deletion of the gene encoding this homolog produced a strain with identical phenotypes to the *rad24.T1* mutant, and was subsequently mapped to the *rad24.T1* locus.

Strains bearing null mutants of either *rad24^{Sp}* or *rad25^{Sp}* are viable, yet deletion of both genes in the same strain is lethal. This is indicative of the 14-3-3 homologs possessing an essential role in *S. pombe*, and is analogous to simultaneous deletion of the *BMH1* and *BMH2* 14-3-3 homologs in *S. cerevisiae (102)*. Additionally, the *rad24^{Sp}* null mutant displays only 50% loss of the DNA damage checkpoint, whereas deletion of *rad25^{Sp}* inflicts only a marginal defect. Such phenotypes suggest that the Rad24p^{Sp} and Rad25p^{Sp} are functionally redundant. The nature of the role(s) played by these two proteins is currently not understood. However, overexpression of the Chk1p/Rad27p^{Sp} protein kinase (*see* Section 5.1.7.) does not induce cell-cycle arrest in a *rad24^{Sp}* deletion strain *(30)*. This indicates that these proteins are required to mediate the Chk1p/Rad27p^{Sp}-induced arrest, and is consistent with the hypothesis that *rad24^{Sp}* (and *rad25^{Sp}*) are either required for Chk1p/Rad27p^{Sp} function or act downstream in the DNA damage checkpoint (Fig. 1).

5.1.9. hus5

Alleles of *hus5* have been identified in two independent screens for checkpoint defective mutants *(6,26)*. Both of these mutants, *hus5.62* and *hus5.17* are sensitive to HU and radiation, and exhibit a reduced growth rate. However, neither allele displays a clear checkpoint defect. Epistasis analysis of *hus5.62* has implicated this mutant as being defective in the "checkpoint *rad*" specific recovery processes that are required following exposure to irradiation or HU, but that are not directly related to mitotic arrest *(5)*. *hus5* mutants exhibit a wild-type G2 arrest after irradiation. As expected, a double mutant of *hus5.62* with a *chk1/rad27^{Sp}* null mutant (which is defective only in DNA damage-induced mitotic arrest), produces a strain with a UV sensitivity equivalent to the *rad17^{Sp}* null mutant *(5)*. In contrast, a *rad17^{Sp} hus5.62* double mutant is no more sensitive than a *rad17^{Sp}* single mutant, suggesting that Hus5p acts in a Rad17p^{Sp}-dependent DNA damage response mechanism unrelated to G2 arrest. Deletion of *hus5* severely impairs cell growth, and renders them virtually inviable. Cells display a pleiotrophic phenotype, with many displaying abortive mitosis and "cut" nuclei. This is consistent with Hus5p having an essential role during normal mitosis.

The *hus5* gene contains five introns, and encodes a 157 amino acid protein of 21 kDa. This protein displays 38% homology to *S. pombe* Rhp6p, which is the fission yeast homolog of Rad6p^{Sc}. These proteins are both members of the E2 family of ubiquitin-conjugating enzymes, which mediate attachment of ubiquitin molecules to a variety of substrates. Such ubiquitination commonly targets the recipient protein for degradation, although it may serve other purposes (for review, *see* ref. *29*) and the *hus5* mutant implicates this process in the DNA damage (and S-phase) checkpoint related responses. Recently, a budding yeast ubiquitin-conjugating protein has been reported that is homologous to Hus5p. Ubc9p is also essential and is required for the degradation of both S- and M-phase cyclins, a key event in the transition of cell-cycle stages *(88)*.

5.1.10. cds1

The *cds1* cDNA was isolated as a multicopy suppressor of a DNA polymerase α mutant, *swi7-H4*, and it encodes a 460 amino acid protein of predicted molecular weight 52 kDa *(62)*. This protein contains consensus serine/threonine protein kinase domains, and displays 35% identity to the Sad1p/Rad53p^{Sc}/Spk1p protein from *S. cerevisiae*. This latter protein is essential, a key regulator of multiple checkpoints in budding yeast, and controls the expression of a number of DNA damage-inducible genes *(7)*. Deletion of the *cds1* gene produces viable cells that have lost the DNA replication checkpoint, but maintain the ability to delay mitosis following exposure to UV irradiation. Examination of the data of Murakami and Okayama *(62)* indicates that the *cds1* null mutant is capable of inducing a significant mitotic delay in response to HU exposure, and suggests that *cds1* gene product may be involved in the recovery of S phase following perturbations by either HU or irradiation. The relationship between Cds1p and the DNA replication and DNA damage checkpoints is currently a matter of speculation. However, *chk1 cds1* double mutants exhibit an S-phase mitosis checkpoint deficiency in response to HU which is comparable to the checkpoint *rad* mutants *(18)*. This is consistent with a model whereby in the absence of Cds1p lesions are produced (possibly as a consequence of unstable replication complexes) that trigger a Chk1p/Rad27p^{Sp}-dependent mitotic arrest mechanism. Cds1p may therefore be directly responsible for the stabilization of replication complexes when they become stalled by blocks to S-phase progression, such as in the presence of HU.

A total of six checkpoint *rad* mutants have now been identified: *rad1*^{Sp}, *rad3*^{Sp}, *rad9*^{Sp}, *rad17*^{Sp}, *rad26*^{Sp}, and *hus1*. They are all profoundly defective in mitotic arrest following exposure to either HU or radiation, which highlights the apparent overlap between these control mechanisms. Neither sensitivity can be fully rescued by imposition of an artificial G2 delay, thereby indicating that in addition to the checkpoint deficiencies, they have other undefined functions within the cell. Some of these functions have been identified by mutational analysis *(5,6,33)*, confirming that the responses to damage/HU mediated by the checkpoint pathways are extremely complex. Work in *S. cerevisiae* indicates that these include regulation of the G1/S transition, S-phase regulation, and transcriptional induction of DNA damage-inducible genes *(7,73)*. However, there appears to be considerable differences in the details of the organization of the checkpoints in the two yeasts. For example, the budding yeast homologs of Rad1p^{Sp} and Rad17^{Sp} are not involved in arrest following HU exposure. However, there are also considerable similarities: both *rad53^{Sc}/sad1* and *mec1* mutants have phenotypes similar to the checkpoint *rad* mutants.

None of the checkpoint *rad* genes analyzed so far is essential for viability, which suggests that the intrinsic timing mechanisms in the *S. pombe* cell cycle are sufficient to maintain the cell-cycle dependency in unperturbed cells. It has been proposed that these six proteins (and possibly more) interact within the cell, forming a "guardian complex" whose role it is to assimilate signals of the S-phase (replication) and DNA damage checkpoints, and to participate in some aspects of DNA metabolism following DNA damage and blocks to DNA synthesis. In this model, not all of the proteins need be directly responsible for both of the checkpoint pathways, but loss of any one of the six proteins compromises the integrity of the complex, indirectly abolishing both control mechanisms. Subtle differences in the respective protein complexes between the two organisms could thus explain the apparently dramatically different requirements for Rad1p^{Sp} and Rad17p^{Sc} in the S-phase mitosis checkpoint of *S. pombe* and *S. cerevisiae*.

Three genes, *chk1/rad27^{Sp}*, *rad24^{Sp}*, and *rad25^{Sp}* separate the DNA damage and S-phase checkpoint mechanisms. Null mutants of these genes are sensitive to radiation, but not HU exposure, and genetic analysis has tentatively placed the functionally overlapping Rad24p^{Sp} and Rad25p^{Sp} proteins downstream of Chk1p/Rad27p^{Sp} in the damage checkpoint response. These proteins are believed to receive a signal that has been generated, or relayed, by the checkpoint Radp proteins. This signal is then transduced to the mitotic machinery where mitosis is inhibited in a dose-dependent manner. The nature of this signal is currently unknown, but mitosis appears to be inhibited in a manner that is independent of the tyrosine-15 phosphorylation state of the mitotic regulator p34^{Cdc2}, (*10,89*; but *see also* ref. *81*) which appears to be the target of the DNA replication checkpoint *(26,27)*.

6. FUTURE PERSPECTIVES

Much of the work on direct DNA repair mechanisms in *S. pombe* has confirmed the evolutionary conservation of the NER and recombination repair mechanisms. The continued analysis of genes and proteins involved in these DNA repair pathways may often be redundant, since it is almost without question that the major gene products will perform very similar roles in budding yeast, fission yeasts, and humans, and both NER and recombination repair are extensively studied in other systems.

The major advantage of research in fission yeast has been the identification and subsequent characterization of processes that are either unknown or poorly understood in other model organisms and higher eukaryotes. A second mechanism that removes UV damage from DNA has only been characterized in fission yeast, despite the fact that many of the gene products are conserved. Studies on this mechanism promise to reveal information concerning the links among repair, replication, and recombination. The genetic analysis of cell-cycle checkpoints in *S. pombe* has also been informative. Analysis of the G2/M DNA damage checkpoint has complemented work performed in *S. cerevisiae*. The recent identification of a related human gene product strongly suggests that the mechanism by which mitosis is delayed following DNA damage is likely to be conserved in higher, multicellular eukaryotes. Future work in this field will undoubtedly focus predominantly on the genetic and biochemical relationships that exist between the identified components, and how these proteins co-operate to interact with the mitotic machinery, a particularly strong aspect of the *S. pombe* model system.

REFERENCES

1. Aboussekhra, A., M. Biggerstaff, M. K. K. Shivji, J. A. Vilpo, V. Moncollin, V. N. Podust, M. Protic, U. Hubscher, J.-M. Egly, and R. D. Wood. 1995. Mammalian DNA nucleotide excision repair reconstituted with purified components. *Cell* **80:** 859–868.
2. Aitken, A., D. Jones, Y. Soneji, and S. Howell. 1995. 14-3-3 proteins: Biological function and domain structure. *Biochem. Soc. Trans.* **23:** 605–611.
3. Ajimura, M., S. H. Leem, and H. Ogawa. 1993. Identification of new genes required for meiotic recombination in *Saccharomyces cerevisiae*. *Genetics* **133:** 51–66.
4. Al-Khodairy, F. and A. M. Carr. 1992. DNA repair mutants defining G2 checkpoint pathways in *Schizosaccharomyces pombe*. *EMBO J.*, **11:** 1343–1350.
5. Al-Khodairy, F., T. Enoch, I. M. Hagan, and A. M. Carr. 1995. The schizosaccharomyces-pombe hus5 gene encodes a ubiquitin- conjugating enzyme required for normal mitosis. *J. Cell Sci.* **108:** 475–486.
6. Al-Khodairy, F., E. Fotou, K. S. Sheldrick, D. J. F. Griffiths, A. R. Lehmann, and A. M. Carr. 1994. Identification and characterisation of new elements involved in checkpoints and feedback controls in fission yeast. *Mol. Biol. Cell* **5:** 147–160.
7. Allen, J. B., Z. Zhou, W. Siede, E. C. Friedberg, and S. J. Elledge. 1994. The SAD1/RAD53 protein kinase controls multiple checkpoints and DNA damage-induced transcription in yeast. *Genes Dev.* **8:** 2401–2415.
8. Bailly, V., J. Lamb, P. Sung, S. Prakash, and L. Prakash. 1994. Specific complex formation between yeast RAD6 and RAD18 proteins: A potential mechanism for targeting RAD6 ubiquitin-conjugating activity to DNA damage sites. *Genes Dev.* **8:** 811–820.
9. Bentley, N. J., D. A. Holtzman, G. Flaggs, K. S. Keegan, A. DeMaggio, J. C. Ford, M. F. Hoekstra, and Carr, A. M. 1996. The *Schizosaccharomyces pombe rad 3* checkpoint gene. *EMBO J.* **14(23):** 6641–6651.
10. Barbet, N. C. and A. M. Carr. 1993. Fission yeast wee1 protein kinase is not required for DNA damage-dependent mitotic arrest. *Nature* **364:** 824–827.
11. Bardwell, A. J., L. Bardwell, D. K. Johnson, and E. C. Friedberg. 1993. Yeast DNA recombination and repair proteins Rad1 and Rad10 constitute a complex in vivo mediated by localised hydrophobic domains. *Mol. Microbiol.* **8:** 1177–1188.
12. Bardwell, L., A. J. Cooper, and E. C. Friedberg. 1992. Stable and specific association between the yeast recombination and DNA repair proteins RAD1 and RAD10 in vitro. *Mol. Cell. Biol.* **12:** 3041–3049.
13. Berbee, M. L. and J. W. Taylor. 1993. Dating the evolutionary radiations of the true funghi. *Can. J. Botany* **71:** 1114–1127.
14. Bezzubova, O., A. Shinohara, R. G. Mueller, H. Ogawa, and J.-M. Buerstedde. 1993. A chicken RAD51 homologue is expressed at high levels in lymphoid and reproductive organs. *Nucleic Acids Res.* **21:** 1577–1580.
15. Birkenbihl, R. P. and S. Subramani. 1992. Cloning and characterisation of *rad21* an essential gene of *Schizosaccaromyces pombe* involved in DNA double-strand-break repair. *Nucleic Acids Res.* **20:** 6605–6611.
16. Birkenbihl, R. P. and S. Subramani. 1995. The rad21 gene product of *Schizosaccharomyces pombe* is a nuclear, cell cycle-regulated phosphoprotein. *J. Biol. Chem.* **270:** 7703–7711.
17. Bowman, K. K., K. Sidik, C. A. Smith, J.-S. Taylor, P. W. Doetsch, and G. A. Freyer. 1994. A new ATP-independent DNA endonuclease from *Schizosaccharomyces pombe*

that recognizes cyclobutane pyrimidine dimers and 6-4 photoproducts. *Nucleic Acids Res.* **22:** 3026–3032.
18. Carr, A. M. Unpublished data.
19. Carr, A. M. 1994. Radiation checkpoints in model systems. *Int. J. Radiat. Biol.* **66:** S133–S139.
20. Carr, A. M. and M. F. Hoekstra. 1995. The cellular responses to DNA damage. *Trends Cell Biol.* **5:** 32–40.
21. Carr, A. M., M. Moudjou, N. J. Bentley, and I. M. Hagan. 1995. The *chk1* pathway is required to prevent mitosis following cell-cycle arrest at "start." *Curr. Biol.* **5:** 1179–1190.
22. Carr, A. M., H. Schmidt, S. Kirchhoff, W. J. Muriel, K. S. Sheldrick, D. J. Griffiths, C. N. Basmacioglu, S. Subramani, M. Clegg, A. Nasim, and A. R. Lehmann. 1994. The *rad16* gene of *Schizosaccharomyces pombe*: a homolog of the *RAD1* gene of *Saccharomyces cerevisiae*. *Mol. Cell. Biol.* **14:** 2029–2040.
23. Carr, A. M., K. S. Sheldrick, J. M. Murray, R. Al-Harithy, F. Z. Watts, and A. R. Lehmann. 1993. Evolutionary conservation of excision repair in *Schizosaccharomyces pombe*: evidence for a family of sequences related to the *Saccharomyces cerevisiae RAD2* gene. *Nucleic Acids Res.* **21:** 1345–1349.
24. Dhillon, N. and M. Hoekstra. 1994. Characterization of two protein kinases from *Schizosaccharomyces pombe* involved in the regulation of DNA repair. *EMBO J.* **12:** 2777–2788.
25. Doe, C. L., J. M. Murray, M. Shayeghi, M. Hoskins, A. R. Lehmann, A. M. Carr, and F. Z. Watts. 1993. Cloning and characterisation of the *Schizosaccharomyces pombe rad8* gene, a member of the SNF2 helicase family. *Nucleic Acids Res.* **21:** 5964–5971.
26. Enoch, T., A. M. Carr, and P. Nurse. 1992. Fission yeast genes involved in coupling mitosis to completion of DNA replication. *Genes Dev.* **6:** 2035–2046.
27. Enoch, T. and P. Nurse. 1990. Mutation of fission yeast cell cycle control genes abolishes dependece of mitosis on DNA replication. *Cell* **60:** 665–673.
28. Fenech, M., A. M. Carr, J. M. Murray, F. Z. Watts, and A. R. Lehmann. 1991. Cloning and characterisation of the *rad4* gene of *Schizosaccharomyces pombe*. *Nucleic Acids Res.* **19:** 6737–6741.
29. Finley, D. and V. Chau. 1991. Ubiquitination. *Ann. Rev. Cell Biol.* **7:** 25–69.
30. Ford, J. C., F. Al-Khodairy, E. Fotou, K. S. Sheldrick, D. Griffiths, J. F., and A. M. Carr. 1994. 14-3-3 protein homologs required for the DNA damage checkpoint in fission yeast. *Science* **265:** 533–535.
31. Freyer, G. A., S. Davey, J. V. Ferrer, A. M. Martin, D. Beach, and P. W. Doetsch. 1995. An alternative eukaryotic DNA excision repair pathway. *Mol. Cell. Biol.* **15:** 4572–4577.
32. Friedberg, E. C., G. C. Walker, and W. Siede. 1995. *DNA repair and Mutagenesis.* ASM, Washington, DC.
33. Griffiths, D. J. F., N. C. Barbet, S. McCready, A. R. Lehmann, and A. M. Carr. 1995. Fission yeast *rad17*: a homologue of budding yeast *RAD24* that shares regions of sequence similarity with DNA polymerase accessory proteins. *EMBO J.* **14(23):** 5812–5823.
34. Guzdcr, S. N., P. Sung, V. Bailly, L. Prakash, and S. Prakash. 1994. Rad25 is a DNA hclicasc required for RNA repair and RNA-polymerase-II transcription. *Nature* **369:** 578–581.
35. Habraken, Y., P. Sung, L. Prakash, and S. Prakash. 1993. Yeast excision repair gene RAD2 encodes a single-stranded DNA endonuclease. *Nature* **366:** 365–368.
36. Hartwell, L. H. and M. B. Kastan. 1994. Cell cycle control and cancer. *Science* **266:** 1821–1828.
37. Haynes, R. H. and B. A. Kunz. 1981. DNA repair and mutagenesis in yeast, in *The Molecular Biology of the Yeast* Saccharomyces. *Life Cycle and Inheritance* (Strathern, J., E. W. Jones, and J. R. Broach, eds.), Cold Spring Harbor Laboratory, Cold Spring Harbor, NY, pp. 371–414.

38. Hiraoka, L. R., J. J. Harrington, D. S. Gerhard, M. R. Lieber, and C. L. Hsieh. 1995. Sequence of human FEN-1, a structure-specific endonuclease, and chromosomal localization of the gene (FEN1) in mouse and human. *Genomics* **25:** 220–225.
39. Hoekstra, M. F., N. Dhillon, G. Carmel, A. J. DeMaggio, R. A. Lindberg, T. Hunter, and J. Kuret. 1994. Budding and fission yeast casein kinase I isoforms have dual-specificity protein kinase activity. *Mol. Biol. Cell* **5:** 877–886.
40. Hoekstra, M. F., R. M. Liskay, A. C. Ou, A. J. DeMaggio, D. G. Burbee, and F. Heffron. 1991. HRR25, a putative protein-kinase from budding yeast—association with repair of damaged DNA. *Science* **253:** 1031–1034.
41. Hofmann, J. and D. Beach. 1994. Cdt1 is an essential target of the Cdc10/Sct1 transcription factor: Requirement for DNA replication and inhibition of mitosis. *EMBO J.* **13:** 425–434.
42. Holden, D. W., A. Spanos, N. Kanuga, and G. R. Banks. 1991. Cloning the REC1 gene of Ustilago maydis. *Curr. Genet.* **20:** 145–150.
43. Howell, E. A., M. A. McAlear, D. Rose, and C. Holm. 1994. CDC44: A putative nucleotide-binding protein required for cell cycle progression that has homology to subunits of replication factor C. *Mol. Cell. Biol.* **14:** 255–267.
44. Jang, Y. K., Y. H. Jin, E. M. Kim, F. Fabre, S. H. Hong, and S. D. Park. 1994. Cloning and sequence analysis of *rhp51*$^+$, a *Schizosaccharomyces pombe* homolog of the *Saccharomyces cerevisiae RAD51* gene. *Gene* **142:** 207–221.
45. Kato, R. and H. Ogawa. 1994. An essential gene, ESR1, is required for mitotic cell growth, DNA repair and meiotic recombination in *Saccharomyces cerevisiae. Nucleic Acids Res.* **22:** 3104–3112.
46. Kelly, T. J., G. S. Martin, S. L. Forsburg, R. J. Stephen, A. Russo, and P. Nurse. 1993. The fission yeast cdc18+ gene product couples S phase to start and mitosis. *Cell* **74:** 371–382.
47. Klar, A. J. S. 1992. Developmental choices in mating-type interconversion in fission yeast. *Trends Genet.* **8:** 208–213.
48. Lee, S. H., A. D. Kwong, Z. Q. Pan, and J. Hurwitz. 1991. Studies on the activator 1 protein complex, an accessory factor for proliferating cell nuclear antigen-dependent DNA polymerase delta. *J. Biol. Chem.* **266:** 594–602.
49. Lehmann, A. R. 1993. Duplicated region of sequence similarity to the human XRCC1 DNA repair gene in the *Schizosaccharomyces pombe* rad4/cut5 gene. *Nucleic Acids Res.* **21:** 5274.
50. Lehmann, A. R., J. H. J. Hoeijmakers, A. A. van Zeeland, C. M. P. Backendorf, B. A. Bridges, A. Collins, R. P. D. Fuchs, G. P. Margison, R. Montesano, E. Moustacchi, A. T. Natarajan, M. Radman, A. Sarasin, E. Seeberg, C. A. Smith, M. Stefanini, L. H. Thompson, G. P. van der Schans, C. A. Weber, and M. Z. Zdzienicka. 1992. Workshop on DNA repair. *Mutat. Res.* **273(1):** 1–28.
51. Lehmann, A. R., M. Walicka, D. J. F. Griffiths, J. M. Murray, F. Z. Watts, S. McCready and A. M. Carr. 1995. The *rad18* gene of *Schizosaccharomyces pombe* defines a new subgroup of the SMC superfamily involved in DNA repair, *Mol. Cell. Biol.* **15:** 7067–7080.
52. Lieberman, H. B. 1995. Extragenic suppressors of *Schizosaccharomyces pombe* rad9 mutations uncouple radioresistance and hydroxyurea sensitivity from cell cycle checkpoint control. *Genetics* **141:** 107–117.
53. Lieberman, H. B. and K. M. Hopkins. 1994. *Schizosaccharomyces malidevorans* and *Sz. octosporus* homologues of *Sz. pombe* rad9, a gene that mediates radioresistance and cell-cycle progression. *Gene* **150:** 281–286.
54. Lindner, P. (1893) *Schizosaccharomyces Pombe* n. sp., ein neuer Garungserreger. *Wochenschr. Brau* **10:** 1298–1300.
55. Long, K. E., P. Sunnerhagen, and S. Subramani. 1994. The *Schizosaccharomyces pombe* rad1 gene consists of three exons and the cDNA sequence is partially homologous to the Ustilago maydis REC1 cDNA. *Gene* **148:** 155–159.

56. Lydall, D. and Weinert, T. A.. 1995. Yeast checkpoint genes in DNA damage processing: implications for repair and arrest. *Science* **270:** 1488–1491.
57. Ma, L., E. D. Siemssen, M. Noteborn, and A. J. Van der Eb. 1994. The xeroderma pigmentosum group B protein ERCC3 produced in the baculovirus system exhibits DNA helicase activity. *Nucleic Acids Res.* **22:** 4095–4102.
58. McCready, S. 1994. Repair of 6-4 photoproducts and cyclobutane pyrimidine dimers in rad mutants of *Saccharomyces cerevisiae. Mutation Res. DNA Repair* **315:** 261–273.
59. McCready, S., A. M. Carr, and A. R. Lehmann. 1993. Repair of cyclobutane pyrimidine dimers and 6-4 photoproducts in the fission yeast, *Schizosaccharomyces pombe*, *Mol. Microbiol.* **10:** 885–890.
60. McCready, S. and B. Cox. 1993. Repair of 6-4 photoproducts in *Saccharomyces cerevisiae*. *Mutat. Res.* **293:** 233–240.
61. McCready, S. J., H. Burkill, S. Evans, and B. S. Cox (1989) The *Saccharomyces cerevisiae* RAD2 gene complements a *Schizosaccharomyces pombe* repair mutation. *Curr. Genet.* **15:** 27–30.
62. Murakami, H. and H. Okayama. 1995. A kinase from fission yeast responsible for blocking mitosis in S phase. *Nature* **374:** 817–819.
63. Muris, D. F. R., O. Bezzubova, J. M. Buerstedde, K. Vreeken, A. S. Balajee, C. J. Osgood, C. Troelstra, J. H. J. Hoeijmakers, K. Ostermann, H. Schmidt, A. T. Natarajan, J. Eeken, P. Lohman, and A. Pastink. 1994. Cloning of human and mouse genes homologous to *RAD52*, a yeast gene involved in DNA repair and recombination. *Mutat. Res.* **315:** 295–305.
64. Muris, D. F. R., K. Vreeden, A. M. Carr, B. C. Broughton, A. R. Lehmann, P. H. M. Lohman, and A. Pastink. 1993. Cloning the RAD51 homologue of *Schizosaccharomyces pombe*. *Nucleic Acids Res.* **21:** 4586–4591.
65. Muris, D. F. R., K. Vreeken, A. M. Carr, C. Smidt, P. H. M. Lohman, and A. Pastink. 1996. Isolation of the *Schizzosaccharomyces pombe RAD54* homolog, *rhp54⁺*, a gene involved in the repair of radiation damage and replication fidelity, *J. Cell Sci.*, in press.
66. Murray, J. M., A. M. Carr, A. R. Lehmann, and F. Z. Watts. 1991. Cloning and characterization of the DNA repair gene, rad9, from *Schizosaccharomyces pombe*. *Nucleic Acids Res.* **19:** 3525–3531.
67. Murray, J. M., C. Doe, P. Schenk, A. M. Carr, A. R. Lehmann, and F. Z. Watts. 1992. Cloning and characterisation of the *S. pombe* rad15 gene, a homologue to the *S. cerevisiae RAD3* and human *ERCC2* genes. *Nucleic Acids Res.* **20:** 2673–2678.
68. Murray, J. M., M. Tavassoli, R. Al-Harithy, K. S. Sheldrick, A. R. Lehmann, A. M. Carr, and F. Z. Watts. 1994. Structural and functional conservation of the human homolog of the *Schizosaccharomyces pombe rad2* gene, which is required for chromosome segregation and recovery from DNA damage. *Mol. Cell. Biol.* **14:** 4878–4888.
69. Nasim, A. 1968. Repair mechanisms and radiation induced mutation in fission yeast. *Genetics* **59:** 327–333.
70. O'Donnell, M., R. Onrust, F. B. Dean, M. Chen and J. Hurwitz. 1993. Homology in accessory proteins of replicative polymerases–*E. coli* to humans. *Nucleic Acids Res.* **21:** 1–3.
71. Onel, K., M. P. Thelen, D. O. Ferguson, R. L. Bennett, and W. K. Holloman. 1995. Mutation avoidance and DNA repair proficiency in Ustilago maydis are differentially lost with progressive truncation of the *REC1* gene product. *Mol. Cell. Biol.* **15:** 5329–5338.
72. Ostermann, K., A. Lorentz, and H. Schmidt. 1993. The fission yeast *rad22* gene, having a function in mating-type switching and repair of DNA damages, encodes a protein homolog to Rad52 of *Saccharomyces cerevisiae. Nucleic Acids Res.* **21:** 5940–5944.
73. Paulovich, A. G. and L. H. Hartwell. 1995. A checkpoint regulates the rate of progression through S phase in *S. cerevisiae* in response to DNA damage. *Cell* **82:** 841–847.
74. Petrini, J., M. E. Walsh, C. Dimare, X. N. Chen, J. R. Korenberg, and D. T. Weaver. 1995. Isolation and characterization of the human MRE11 homologue. *Genomics* **29:** 80–86.

75. Reagan, M. S., C. Pittenger, W. Siede and E. C. Friedberg. 1995. Characterization of a mutant strain of *Saccharomyces cerevisiae* with a deletion of the *RAD27* gene, a structure homolog of the *RAD2* nucleotide excision repair gene. *J. Bacteriol.* **177:** 364–371.
76. Reynolds, P., M. H. M. Koken, J. H. J. Hoeijmakers, S. Prakash and L. Prakash. 1990. The rhp6+ gene of *Schizosaccharomyces pombe*: a structural and functional homolog of the RAD6 gene from the distantly related yeast *Saccharomyces cerevisiae. EMBO J.* **9:** 1423–1430.
77. Reynolds, P., S. Weber, and L. Prakash. 1985. Rad6 gene of *Saccharomyces cerevisiae* encodes a protein containing a tract of 13 consecutive aspartates. *Proc. Natl. Acad. Sci. USA* **82:** 168–172.
78. Reynolds, P. R., S. Biggar, L. Prakash, and S. Prakash. 1992. The *Schizosaccharomyces pombe* rhp3+ gene required for DNA repair and cell viability is functionally interchangeable with the RAD3 gene of *Saccharomyces cerevisiae. Nucleic Acids Res.* **20:** 2327–2334.
79. Robins, P., D. J. C. Pappin, R. D. Wood and T. Lindahl. 1994. Structural and functional homology between mammalian DNase IV and the 5' nuclease domain of *Escherichia coli* DNA polymerase I. *J. Biol. Chem.* **269:** 28,535–28,538.
80. Rodel, C., S. Kirchhoff, and H. Schmidt. 1992. The protein sequence and some intron positions are conserved between the switching gene *swi10* of *Schizosaccharomyces pombe* and the human excision repair gene *ERCC1. Nucleic Acids Res.* **20:** 6347–6353.
81. Rowley, R., J. Hudson, and P. G. Young. 1992. The wee1 protein kinase is required for radiation induced mitotic delay. *Nature* **356:** 353–355.
82. Rowley, R., S. Subramani, and P. G. Young. 1992. Checkpoint controls in *Schizosaccharomyces pombe*: *rad1*. *EMBO J.* **11:** 1335–1342.
83. Saka, Y., P. Fantes, T. Sutani, C. McInerny, J. Creanor, and M. Yanagida. 1994. Fission yeast cut5 links nuclear chromatin and M phase regulator in the replication checkpoint control. *EMBO J.* **13:** 5319–5329.
84. Saka, Y. and M. Yanagida. 1993. Fission yeast cut5, required for S phase onset and M phase restraint , is identical to the radiation-damage repair gene rad4$^+$. *Cell* **74:** 383–393.
85. Savitsky, K., A. Bar-Shira, S. Gilad, G. Rotman, Y. Ziv, L. Vanagaite, D. A. Tagle, S. Smith, T. Uziel, S. Sfez, M. Ashkenazi, I. Pecker, M. Frydman, R. Harnik, S. R. Patanjali, A. Simmons, G. A. Clines, A. Sartiel, R. A. Gatti, L. Chessa, O. Sanal, M. F. Lavin, N. G. J. Jaspers, M. R. Taylor, C. F. Arlett, T. Miki, S. M. Weissman, M. Lovett, F. S. Collins and Y. Shiloh. 1995. A single ataxia telangiectasia gene with a product similar to PI 3-kinase. *Science* **268:** 1749–1753.
86. Schmidt, H., P. Kapitza-Fecke, E. R. Stephen, and H. Gutz. 1989. Some of the swi genes of *Schizosaccharomyces pombe* also have a function in the repair of radiation damage. *Curr. Genet.* **16:** 89–94.
87. Seaton, B. L., J. Yucel, P. Sunnerhagen and S. Subramani. 1992. Isolation and characterisation of the *Schizosaccharomyces pombe rad3*$^+$ gene which is involved in the DNA damage and DNA synthesis checkpoints. *Gene* **119:** 83–89.
88. Seufert, W., B. Futcher, and S. Jentsch. 1995. Role of a ubiquitin-conjugating enzyme in degradation of S and M- phase cyclins. *Nature* **373:** 78–81.
89. Sheldrick, K. S. and A. M. Carr. 1993. Feedback controls and G2 checkpoints: fission yeast as a model system, *BioEssays* **15:** 775–782.
90. Shinohara, A., H. Ogawa, Y. Matsuda, N. Ushio, K. Ikeo, and T. Ogawa. 1993. Cloning of human, mouse and fission yeast recombination genes homologous to RAD51 and recA, *Nature Genet.* **4:** 239–243.
91. Sidik, K., H. B. Lieberman, and G. A. Freyer. 1992. Repair of DNA damaged by UV light and ionizing radiation by cell-free extracts prepared from *Schizosaccharomyces pombe. Proc. Natl. Acad. Sci. USA* **89:** 12,112–12,116.
92. Svejstrup, J. Q., Z. Wang, W. J. Feaver, X. Wu, D. A. Bushnell, T. F. Donahue, E. C. Friedberg, and R. D. Kornberg. 1995. Different forms of TFIIH for transcription and DNA repair: holo-TFIIH and a nucleotide excision repairosome. *Cell* **80:** 21–28.

93. Sweder, K. S., P. C. Hanawalt, and S. Buratowski. 1993. Transcription-coupled DNA repair. *Science* **262:** 439,440.
94. Szankasi, P. and G. R. Smith. 1995. A role for exonuclease I from *S. pombe* in mutation avoidance and mismatch correction. *Science* **267:** 1166–1169.
95. Tavassoli, M., M. Shayegi, A. Nasim, and F. Z. Watts. 1995. Cloning and characterisation of the *Schizosaccharomyces pombe* rad32 gene: a gene required for repair of double strand breaks and recombination. *Nucleic Acids Res.* **23:** 383–388.
96. Taylor, J. W., B. Bowman, M. L. Berbee, and T. J. White. 1993. Fungal model organisms: phylogenetics of Saccharomyces, Aspergillus and Neurospora. *Systematic Biol.* **42:** 440–457.
97. Thelen, M. P., K. Onel, and W. K. Holloman. 1994. The REC1 gene of Ustilago maydis involved in the cellular response to DNA damage encodes an exonuclease, *J. Biol. Chem.* **269:** 747–754.
98. Thompson, L. H., K. W. Brookman, N. J. Jones, S. A. Allen, and A. V. Carrano. 1990. Molecular cloning of the human XRCC1 gene, which corrects defective DNA strand break repair and sister chromatid exchange. *Mol. Cell. Biol.* **10:** 6160–6171.
99. Tomkinson, A. E., A. J. Bardwell, L. Bardwell, N. J. Tappe, and E. C. Friedberg. 1993. Yeast DNA repair and recombination proteins Rad1 and Rad10 constitute a single-stranded-DNA endonuclease. *Nature* **362:** 860–862.
100. Tomkinson, A. E., A. J. Bardwell, N. Tappe, W. Ramos, and E. C. Friedberg. 1994. Purification of Rad1 Protein from *Saccharomyces cerevisiae* and further characterization of the Rad1/Rad10 endonuclease complex. *Biochemistry* **33:** 5305–5311.
101. Tsukuda, T., R. Bauchwitz, and W. K. Holloman (1989) Isolation of the REC1 gene controlling recombination in *Ustilago maydis*. *Gene* **85:** 335–341.
102. Van Heusden, G. P. H., D. J. F. Griffiths, J. C. Ford, T. F. C. Chinawoeng, P. A. T. Schrader, A. M. Carr and H. Y. Steensma. 1995. The 14-3-3-proteins encoded by the bmh1 and bmh2 genes are essential in the yeast *Saccharomyces cerevisiae* and can be replaced by a plant homolog. *Eur. J. Biochem.* **229:** 45–53.
103. Waga, S. and B. Stillman. 1994. Anatomy of a DNA replication fork revealed by reconstitution of SV40 DNA replication in vitro. *Nature* **369:** 207–212.
104. Wainright, P. O., G. Hinkle, M. L. Sogin, and S. K. Stickel. 1993. Monophyletic origins of the Metazoa: an evolutionary link with funghi. *Science* **260:** 340–342.
105. Walworth, N., S. Davey, and D. Beach. 1993. Fission yeast chk1 protein kinase links the rad checkpoint pathway to cdc2. *Nature* **363:** 368–371.
106. Walworth, N. C. and Bernards, R. 1996. rad-Dependent responses of the chk1-encoded kinase at the DNA damage checkpoint. *Science* **271:** 353–356.
107. Weinert, T. A. and L. H. Hartwell. 1988. The RAD9 gene controls the cell cycle response to DNA damage in *Saccharomyces cerevisiae*. *Science* **241:** 317–322.
108. Weinert, T. A., G. L. Kiser, and L. H. Hartwell. 1994. Mitotic checkpoint genes in budding yeast and the dependence of mitosis on DNA replication and repair. *Genes Dev.* **8:** 652–665.
109. Zakian, V. A.. 1995. ATM-related genes: What do they tell us about functions of the human gene? *Cell* **82:** 685–687.

21

Toward Repair Pathways in *Aspergillus nidulans*

Etta Kafer and Gregory S. May

1. INTRODUCTION

Interest in DNA repair of *Aspergillus nidulans* has grown out of studies of at least three different biological processes:

1. Recombination and mitotic segregation, an important field of investigation in early *Aspergillus* genetics, which aimed to elucidate basic gene structures and functions *(84–86)*; this resulted in the isolation of a large number of mutant loci and the development of methods for mapping and of test systems for recombination and malsegregation, especially useful for genetic analysis of recombinational DNA repair.
2. Inducible responses and enzymes, either in development *(75)* or as a result of environmental changes, as found, e.g., for alkylation repair *(3,46,97)*; this identified processes and enzymes in *Aspergillus*, analogous to intriguing findings in prokaryotes, that so far have barely been recognized in eukaryotic microorganisms, e.g., adaptive responses.
3. Genetic control of the cell cycle, which in *Aspergillus* starts from metabolically inactive conidia (in G_0) and progresses with initially synchronous nuclear divisions during germination, thus providing excellent material for investigations by molecular approaches *(25,71)*. Such investigations in *Aspergillus* often complement the more extensive analyses in the yeasts, *Saccharomyces cerevisiae* and *Schizosaccharomyces pombe*, e.g., by identifying the function of novel genes or of new interactions for known genes not previously shown to function in DNA repair processes *(22)*.

A. nidulans is a haploid homothallic ascomycete, and the currently used strains are derived from a single nucleus *(85)*. Such strains are therefore all interfertile, heterokaryon-compatible, and isogenic, barring induced mutations. Asexual spores (conidia) are uninucleate and quiescent, but produce metabolizing "germlings" with dividing nuclei, which develop into multinucleate mycelia. Mitosis in *A. nidulans* shares many features with that of higher eukaryotes *(72)*. *Aspergillus* germlings, like *S. pombe* and mammalian cells, spend a substantial part of their nuclear division cycle in G_2, and this may be related to an unexpected predominance of postreplication UV repair.

For the sexual cycle, hybrid as well as "selfed" cleistothecia (fruiting bodies) are obtained in profusion from "forced" heterokaryons under suitable conditions. A single hybrid cleistothecium produces $>10^5$ ascospores for meiotic random spore analysis. For the "parasexual cycle," which shares many features with the genetic analysis of somatic mammalian cells, diploids of desired genotypes are selected from heterokaryons as a result of accidental fusion of vegetative nuclei. Such diploids do not progress

From: DNA Damage and Repair, Vol. 1: DNA Repair in Prokaryotes and Lower Eukaryotes
Edited by: J. A. Nickoloff and M. F. Hoekstra © Humana Press Inc., Totowa, NJ

into meiosis, but at low frequency produce recombinants, mainly by mitotic crossing over. However, a fraction are nondisjunctional segregants, initially unstable aneuploids and subsequently haploids, thus completing the "parasexual" cycle *(84)*. When "haploidization" is induced by spindle poisons, haploid segregants with a random assortment of chromosomes are obtained for mitotic mapping of mutations, chromosomal aberrations, and transfected genes *(55)*.

Assignment of genes to chromosomes, followed by meiotic linkage tests, produced the current reliable genetic map of *A. nidulans (19)* and provided the basis for mapping genes to chromosomes by pulsed-field gel electrophoresis *(8,28)*. The use of accumulated mutants and genetic information, combined with an increasing number of available molecular approaches, e.g., the opportunity to clone mapped genes by complementation with cosmids from ordered chromosome-specific libraries *(9,103)*, makes *A. nidulans* an excellent model eukaryote *(98)*.

2. *ASPERGILLUS* MUTANTS DEFICIENT IN DNA REPAIR: MUTAGEN SENSITIVITIES AND EPISTATIC GROUPING

2.1. Introduction

Evidence for DNA repair was first obtained when UV-sensitive mutants were isolated and analyzed in *Escherichia coli*. Results from genetic and biochemical investigations identified three major processes for the repair of radiation damage, namely, excision repair in nondividing cells, and two types of postreplication repair, error-free recombinational repair and error-prone mutagenic repair.

Similar basic mechanisms, but generally involving many more genes and larger repair–protein complexes, are presumed to be active in eukaryotes. This seems especially likely for radiation repair in *S. cerevisiae*; three epistatic groups of radiation repair defective, *rad*, mutants showed patterns of UV and X-ray sensitivities correlated to functions roughly corresponding to the three basic repair types of *E. coli (35)*, namely:

1. UV sensitive mutants of the first group identified genes that function in nucleotide excision repair (NER);
2. Mutants sensitive to X-rays, and also to methylmethane sulfonate (MMS), showed defects in meiosis, recombination, and also in double-strand break repair; and
3. The third group of UV- and X-ray-sensitive mutants showed defects in postreplication repair of several types, with some genes clearly being required for induced mutation, i.e., error-prone repair.

However, when sensitivities to other chemical mutagens were investigated, no similar correlations were obtained *(20)*. Furthermore, many of the gene products identified in yeast, as well as the homologous human proteins active in NER, are now known to function in additional cellular processes, often as components of large complexes, not only in other DNA transactions like recombination, but even in transcription (*36,96,104*; summary in *44*). Thus, a web of interactions rather than separate pathways are being identified, and in spite of striking homologies, significant differences are becoming evident between species. Investigations of DNA repair in different organisms therefore show that varied adaptive use of highly conserved protein domains and of alternate basic mechanisms may result in different repair mutant phenotypes, perhaps indicative of different prevalent repair modes.

2.2. Isolation of Mutants with Increased Sensitivities to UV and Chemical Mutagens

In *A. nidulans*, as in bacteria and yeast, the first DNA repair-deficient types isolated were UV-sensitive mutants, but few of these were cross-sensitive to X-rays (contrastic with yeasts, especially *S. pombe*, where most of the UV-sensitive *rad* mutants are sensitive to ionizing radiation; *66*). In general, the UV sensitivities found in *A. nidulans* were very low and tests were cumbersome until it was found that all significantly UV-sensitive mutants showed cross-sensitivity to MMS. Plate tests and replica transfers made analysis efficient, especially after initial problems were recognized as the result of unexpected MMS sensitivity of all amino acid-requiring mutants in *Aspergillus (58)*. Subsequently, new DNA repair-deficient mutants were therefore mostly "selected" for hypersensitivity to chemical mutagens, especially to various alkylating agents (Section 2.4.).

Four sets of *Aspergillus* mutants have been isolated, two aiming at analysis of radiation repair (*63*; Section 2.3.), and two for investigation of alkylation and recombinational repair (*79*; Section 2.4.). Analysis of epistasis was attempted in three of the four sets as follows:

1. UV- and generally MMS-sensitive mutants, representing nine *uvs* genes were tested in all pairwise combinations (Table 1, details in Section 2.3.3.);
2. MMS sensitive mutants of eight *mus* genes were checked for epistasis with representative members of the Uvs groups (Table 2; Section 2.3.4.);
3. *sag* mutants, specifically sensitive to alkylating agents, defining five genes, were tested in pairwise combinations (Section 2.4.1.);
4. Of many *nuv* mutants, hypersensitive to nitrosoguanidine and some cross-sensitive to UV, 24 are being analyzed; many novel genes, but so far no epistatic groups have been identified (Section 2.4.2.).

In most cases, only a few alleles of each gene were obtained, with the exception of *uvsB* for which 12 alleles are known (isolated in six laboratories). Generally, alleles of each gene showed similar characteristics and clearly differed from mutants in most other genes; this was found even for alleles of *uvsB (64)*, possibly with the exception of one mutant that has not yet become available for comparative tests *(2)*. It was therefore tacitly assumed that most mutants had gene-specific characteristics, as expected from complete loss of gene function. However, only a very few cases are certain to be null mutations. The use of null mutants is important not only for an assessment of primary function, but also for the analysis of epistatic vs synergistic interactions. Some of the current ambiguities will be resolved when gene disruptions are available. Except when two genes were located on the same chromosome, double mutant strains were obtained as haploid segregants from heterozygous diploids and were isogenic for up to six chromosomes.

2.3. uvs and mus Mutants for Analysis of Radiation Repair: Genes in Four Epistatic Groups

2.3.1. Mutagen Sensitivities of uvs and mus Mutants

The mutants of the first set were obtained as UV-sensitive types. Most of these *uvs* mutants showed reduced UV survival when quiescent conidia were irradiated ("Q" in

Table 1
Properties of uvs Mutants: Mutagen Sensitivities, Effects on Mutation, and Recombination *(14,63)*

Epistatic group, genes and mutant alleles	UvsF *uvsF201*	UvsF *uvsH* 77	UvsF *uvsH* 304	UvsF *uvsJ1*	UvsB *uvsB110* *uvsD153*	UvsC *uvsC114* *uvsE182*	UvsC *uvsA101*	UvsI *uvsI501*
Radiation survival								
UV, quiescent (Q)	---	---	--	--	--	0	0	---
pregerminated (P)	–	---	NT	--	--	–	–	–[b]
Ionizing radiation (Q)	0	---	--	–	0	0	NT	NT
pregerminated (P)						--[d]		
Survival after treatment with chemical mutagens								
4-NQO	--	---	NT	---	--	--	±[e]	--[c,e]
MMS (growth on CM + MMS)	--	---	--	–	---	--	±[e]	0[c]
EMS (Q) (at 37°)[a]	--[a]	--[a]	NT	NT	0[a]	NT	NT	
at low temp (20° or 30°)	±[a]	–[a]		--[e]	0[e]	--[e]		0[e]
MNNG (Q)	–	NT	NT	–	NT	NT	NT	0[c]
NA (nitrous acid)	–	NT	–	--	-	NT	NT	NT
Effects on mutation (selenate-resistant forward mutation)								
Spontaneous	+	0	0	+[e]	+	+++	0[e]	0[e]
UV-induced	++	++	+	++[e]	±	--[e]	0[e]	0[e]
Effects on meiosis (homozygous crosses)								
Fertility	0	--	±	--	–	--	0[e]	0[e]
Recombination	0	(NT)	0	(NT)	±	(NT)	NT	0[e]
Spontaneous mitotic recombination								
Allelic	+++[f]	NT	NT	NT	++[d,f]	---[d,f]	NT	0[e]
Intergenic	+++[f]	++	+	+	++[f]	---[f]	NT	0[e]

0, like wild-type.
–, --, --- : slightly to highly reduced compared to *uvs*$^+$.
+, ++, +++: slightly to highly increased.
NT, not tested; (NT), not testable.
[a]*(2)*, [b]*(38)*, [c]*(39)*, [d]*(33,52)*, [e]*(13,15)*, [f]*(64,108)*.

Table 2
Phenotypes of *Aspergillus mus* Mutants: Mutagen Sensitivities, Growth in Diploids, Fertility, and Mitotic Recombination, Relative to mus^+, and Interactions with *uvs* of Four Epistatic Groups (Table 1; *60,61,108*)

Genes, alleles	Linkage group, arm[b]	Survival after mutagen treatment[a]				Mitotic recombination		Mitosis in diploids		Effects on meiosis	Interaction with *uvs*
						Compared to wild-type, =1×					
		UV	γ-rays	MMS	4-NQO	Intergenic[c]	Intragenic	Survival	Abnormals	Fertility	"E, R, or L"[d]
*mus*P234	II L or VII R[b]	--	+	–	--	~1.5 ×	~1.5 ×	100%	1 ×	Sterile	*uvsF*(L),*uvsB* (R)[e]
*mus*R223	II L	+	+	--	±	0.8 ×	~1 ×	97%	1 ×	Normal	*uvsB* (R)
*mus*S224	III R or VII R[b]	+	+	---	±	0.9 ×	~1 ×	94%	1 ×	Normal	Any?
Hyper-rec *mus*											
*mus*B221	IV C	---	±	---	---	3.5 ×	~4 ×	~50%	~2.5 ×	Poor	(=*uvsB*)
*mus*O226	III L or VII R[b]	–	+	--	--	1.5 ×[c]	2.5 ×	80%	4.5 ×	Sterile	*uvsC* (L)
*mus*N227	VII R	+	+	--	–	2.5 ×	3 ×	~100%	1 ×	Sterile	*uvsF* (L), *uvsB* (R), *uvsC* (E)[e]
*mus*Q230	II R	–	–	--	--	~4 ×	9 ×	68%	3.5 ×	Sterile	*uvsF* (L)
Hypo-rec types											
*mus*L222	I R	–	+	---	---	<0.2 ×	~0.2 ×	80%	1.3 ×	Sterile	*uvsF*(L), *uvsC*(E)[e]
*mus*K228	VIII R	+	+	–	±	0.5 ×	~0.5 ×	71%	7 ×	Normal	*uvsF* (L)

[a]+ = like wild type; ±, –, --, --- = increasingly sensitive or deficient

[b]Three *mus* mutations are associated with translocations and the three pairs of translocation breaks have been mapped, but the location of the three active *mus* loci, each at one of the two corresponding breaks, has not yet been determined.

[c]*musO226* is the only mutant which shows a difference between effects on intergenic vs allelic recombination.

[d]Interactions in double mutants resulted either in lethality ("L"), or showed evidence for epistasis ("E") or rescue ("R," i.e., increased growth and survival when combined with *uvsB*).

[e]For each *mus* mutant, four double *mus uvs*, mutant strains were either viable and were tested, or were identified as lethal (L). However, only unambiguous and noteworthy results are shown; i.e., in few cases are more than one type of interaction indicated: e.g., three types of interactions for *musN* (as indicated) and two types each for *musL*, and *musP*.

Table 1). However, a few were sought that are UV-sensitive only after pregermination ("P" in Table 1), i.e., only when germlings with dividing nuclei are exposed *(32,33,51)*.

The second set are prototrophic *mus* mutants sensitive to MMS (Table 2; *60,63*) isolated to obtain mutants affecting recombination, with some presumably being X-ray-sensitive. However, none of the *mus* mutants representing new genes were noticeably radiation-sensitive (exceptions were two new *uvsB* alleles; not listed in Table 2).

All *uvs* and *mus* mutants of both sets showed cross-sensitivity to the quasi-UV-mimetic mutagen 4-nitro-quinoline-1-oxide (4-NQO), and all but one, *uvsI*, also to MMS. Tests for epistasis therefore used treatments with these two DNA-damaging agents *(14,60)*. For double mutants of *uvs* pairs, survival after UV irradiation of quiescent condida was also analyzed. Similarly, pregerminated conidia were exposed to UV, when one of the *uvs* mutations was UV sensitive only in dividing cells *(15)*.

2.3.2. uvs; uvs *and* uvs; mus *Double Mutants: Viability and Growth Patterns, Lethality or Rescue*

When double mutants were isolated between pairs of *uvs* alleles, they were obtained more readily from pairs with similar phenotypes than from pairs of very different *uvs* mutants *(63)*. In the latter case, double mutants often showed reduced viability or were synthetically lethal. Such lethal combinations were taken to indicate different epistatic groups, since lethality is expected for genes that function in alternate essential pathways.

Similarly, when *mus* mutations were combined with *uvs* mutations, viability of double mutant strains varied, and several lethal pairs were found (*61*; Table 2). Surprisingly, however, for certain *uvsB mus* pairs, the opposite effect was found; namely, the resulting double-mutant strains showed apparent suppressor effects of the *mus* mutations or "rescue" from the poor growth typical of single-mutant *uvsB* strains. This was clearly apparent when *uvsB* was combined with any one of three different *mus* mutations (*musP*, *R*, and *N*; Table 2; *13*). In addition, MMS and 4-NQO survival was improved, but only at low-doses of treatments with UV or MMS *(60)*.

2.3.3. The uvs *Genes,* uvsA-uvsJ*, Define Four Epistatic Groups*

When mutants of the nine *uvs* genes were compared for effects on growth, viability, and fertility, as well as for recombination and mutation (Sections 3. and 4.), three pairs of mutants were found, each with very similar pleiotropic phenotypes, namely, *uvsH* and *J*, *uvsC* and *E*, and *uvsB* and *D* (Table1). Tests of double mutants involving the two members of each pair revealed epistatic relationships, whereas interpair double mutants showed synergistic interactions or were nonviable, indicating three different epistatic and possibly functional groups (*63*; details in Section 6.). Mutants of the other three genes, *uvsF*, *uvsI*, and *uvsA*, differed in various ways, not fitting well into this pattern.

uvsF consistently showed epistasis with *uvsH* and *uvsJ (14,63)* and also shared certain phenotypic characteristics, namely, *uvsF* shared with *uvsH* features usually found in excision-defective types, especially increased UV-induced mutation and spontaneous mitotic recombination. However, it differed from *uvsH* and *J*, which are X-ray-sensitive and defective in meiosis (effects not found for *uvsF*; Table 1). *uvsF* is therefore considered to be an unusual member of the epistatic group containing *uvsH* and *J*.

The other two mutations, *uvsI501* and *uvsA1*, differed from the preceding ones, in being able to form viable double mutants with all other *uvs* (and *mus*) mutations *(13,14)*. *uvsA1* is epistatic with *uvsC* and *E*, and shares certain special features, namely UV sensitivity only in dividing cells, and synergism for MMS sensitivity with *uvsI*, which alone is not MMS-sensitive. However, *uvsA1* does not affect recombination or mutation, differing greatly from *uvsC*; it may either be a partial mutant or a mutation in a gene with minor repair function (therefore, *uvsA1* will not be considered further). In contrast *uvsI*, is not epistatic with any of the members of the three postulated Uvs groups and has a unique phenotype for *Aspergillus*. Namely, it shows no defects in meiosis, mitosis, or recombination, but has specific effects on mutation (Section 4.). *uvsI* has recently been confirmed as clearly representing a new, fourth, epistatic Uvs group of *Aspergillus* (*15*; Section 6.4.1.).

Therefore, consistent results identified four epistatic groups of *uvs* genes, each with one to three members. These have been named after the most frequently used member, namely, "UvsT" (also referred to more appropriately as UvsH/F; Section 6.2.1.), "UvsC," "UvsI," and "UvsB" (Table 1). The correlated differences in phenotypes between members of different groups suggest that these epistatic groups may also represent functional groups, i.e., groups of genes active mainly in different repair processes.

2.3.4. mus *Genes: Interaction with* uvs *Mutations and Membership in Uvs Epistatic Groups?*

mus mutants were induced by γ-rays to obtain null mutations for tests of epistasis and gene cloning. Among a dozen *mus* mutants, nine genes, *musK–musS*, were identified, and in three cases, the *mus* mutation was associated with a translocation breakpoint (Table 2; *60*). None were significantly cross-sensitive to radiation, but over half of them were defective for meiosis and/or affected recombination (Section 3.). Assignment to Uvs epistatic groups was difficult, because phenotypic overlaps were slight, and only a few epistatic relationships are indicated (Table 2; Section 6.3.).

For the *mus N*, *L*, and *K* genes, the results suggest assignment to the UvsC epistatic group, which has been confirmed for *musN (13)*, whereas results were less consistent for *musL* and *K*. Of special interest is also *musO*, which was lethal in combination with *uvsC* and showed synergism in tests of all viable *musO*;*uvs* double-mutant strains, suggesting an additional epistatic group.

2.4. sag *and* nuv *Mutants and Genes: Analysis of Alkylation Repair and Recombination*

Alkylation repair has been extensively analyzed in bacteria, but little, to our knowledge, has been published on pathway analysis and epistatic grouping of alkylation-specific mutants in *S. cerevisiae*. Only for very few such mutations have genes been cloned, e.g., by transformation of suitable *E. coli* mutants with yeast DNA, which led to the identification of the *S. cerevisiae mms5* gene as a homolog of *E. coli alkA (16)*. Furthermore, attempts to group or identify epistatic relationships among *rad* mutants involved in alkylation repair revealed a complex pattern of cross-sensitivities to simple monofunctional alkylation agents *(20)*. In *Aspergillus*, some interesting findings were obtained for *sag* and *nuv* mutations. However, systematic analysis of epistasis was carried out only for *sag* mutants *(97)* and no clear picture of epistatic or functional groups has emerged.

2.4.1. *Alkylation-Specific Mutants and Genes* sagA–sagE

In a search for genes involved in alkylation repair, especially genes for various constitutive or inducible enzymes identified biochemically *(3,97)*, *sag* mutants, hypersensitive to *N*-methyl-*N'*-nitro-*N*-nitrosoguanidine (MNNG), but not to UV, were isolated *(79)*. They defined five new loci, *sagA–sagE*. However, all mutants showed normal levels of activity in assays for alkylation repair enzymes.

Tests of epistasis among *sag* mutations clearly showed at least additive levels of sensitivity, when double-mutant strains combining *sagA* and *sagB*, or *sagA* with *sagC*, were tested with two different agents *(97)*. These results and those of further tests suggest that *sag* mutants represent genes in three epistatic groups, and possibly are involved in three different repair types or pathways of alkylation damage repair *(79)*. The latter conclusion is supported by the finding of different patterns of cross-sensitivities to other mutagens. Although all *sag* mutants were sensitive to MMS, but not to X-rays, some, but not all *sag* mutants are cross-sensitive to 4-NQO, and only *sagC* is sensitive to UV (only when germlings are irradiated). However, the repair processes involved presumably are not alkylation-specific, as also suggested by information from the cloning of *sagA*, which showed homology to a membrane protein (Section 5.2.4.).

2.4.2. nuv, *Mutants Hypersensitive to Nitrosoguanidine and Possibly UV*

In a large-scale screen, many mutants (>200) sensitive to MNNG, were isolated for a general analysis of DNA repair and recombination in *Aspergillus* (*50,78,97*; ref. *79* includes partially characterized mutants).

Several streamlined assays were developed for their efficient characterization, e.g., sensitivities to toxic effects of mutagens were roughly assessed by relative growth inhibition of mutants on media with mutagens compared to controls. Analysis of 24 *nuv* mutants with pronounced MNNG sensitivities showed several features similar to those observed for *mus* mutants. Cross-sensitivity to X-rays or to UV was rare, although all were sensitive to 4-NQO. About half of them were sterile in homozygous crosses, and only a few cases of allelic pairs were found *(50)*. In addition, several mutations (6 of 24 tested) caused mitotic instability when homozygous in diploids, and these *nuv* mutations affected recombination in various ways (Section 3.3.).

3. TESTING DNA REPAIR MUTANTS FOR EFFECTS ON RECOMBINATION

3.1. Introduction: Genes for Recombination and Operational Definitions of Tests

Mutants with reduced recombination ("hyporec" or "rec$^-$" types) are expected to identify genes required for recombinational repair, but occasionally increases are the result of (unusual) mutations in such genes. In general, however, increased recombination ("hyperrec" effects) are more likely the result of unrepaired recombinogenic lesions, caused by other types of repair defects. In *Aspergillus*, as in other fungi, mutants with defects in recombinational repair frequently are sterile in homozygous crosses: so far very few mutants with good fertility but abnormal meiotic recombination have been found (for the first exception, see *50,77*; Sections 3.3. and 5.2.2.). Thus, mitotic recombination in vegetative diploids was the assay of choice for comparative tests of putative DNA repair-defective mutants of *Aspergillus*.

Traditionally, effects on recombination are assayed in *A. nidulans* by measuring mitotic recombination frequencies in vegetative diploids. Such recombination may result from reciprocal crossing over between homologs, or from nonreciprocal conversion events. The tests used are of two types:

1. Tests for intergenic recombination that monitor mitotic crossing over between genes, which, although much rarer in mitotic cells than meiosis, can be used to identify the order and relative distances of markers from their centromere(s) *(84)*; crossovers may either appear as color changes in conidia (e.g., as yellow *[yA/yA]* sectors in green *[yA/+]* colonies) or be selected as homozgotes with a growth advantage (e.g., selenate resistant, homozygous *sC* recombinants from *sC/+* diploids; comparable to tests using canavanine resistance in budding yeast).
2. Alternatively, intragenic (=allelic) recombinants are selected from diploids, usually as prototrophs when diploids are auxotrophs, heteroallelic, e.g., for noncomplementing alleles of *pabaA (33,34,51)*, *adE (86,108)* or *niaD (78)*.

3.2. *Tests for Inter- and Intragenic Mitotic Recombination: Effects of* uvs *and* mus *Mutations*

3.2.1. *Nonselective Tests for Spontaneous, Intergenic Recombination*

Tests using recognizable sectors in "parental" colonies have the advantage, that different sectors, being rare, are sure to result from independent recombination events, and hence are easy to quantify (e.g., as colonies with or without sectors). For all *uvs* and *mus* mutants, the frequencies of color sectors were therefore compared to those of controls to assess frequencies of intergenic recombination (Tables 1 and 2). Mutants in about half of the genes tested (7/16) showed increases (hyperrec types), whereas decreases (hyporec mutants) were rarer (4/16 cases; *63,108*). In repair-proficient strains, by far the majority (90%) of such sectors result from mitotic crossing over. However, in some of the mutant diploids, nondisjunction and malsegregation was increased, especially when recombination was reduced. Color segregants needed therefore to be purified and tested for informative linked markers to estimate the relative frequencies of the different types of mitotic segregants.

In contrast, for *nuv* mutants, spontaneous mitotic crossovers were identified among induced haploids *(77,79)*. This test is easier, because after induction by benomyl, secondary haploids form many prominent sectors. However, for hyperrec cases quantitative assessment is complicated by the clonal distribution of events during growth, requiring analysis of several independent samples (as illustrated in Table 3). In fact, even when several samples are analyzed by one of the methods which correct for clonal distribution, such tests may overestimate the increases in mutant diploids, e.g., 50-fold increases are deduced for *uvsH77/4* (Table 3) compared to 5- to 10-fold increases obtained in the same diploid for crossover color sectors *(63)*.

3.2.2. *Intragenic Mitotic Recombination*

Mitotic allelic recombination is relatively frequent in *Aspergillus* (about one-tenth of meiotic values) and good selective systems make measurements convenient. When *uvs* and *mus* mutants were tested using the *adE8/20* system, results confirmed the hyper- and hyporec effects seen in tests of intergenic recombination with one exception (*musO* showed significant increases only for intragenic recombination; Table 2; *60,108*).

For some mutants, a fraction of the selected recombinants was haploidized to identify effects on reciprocal vs conversion events using information from such "half-tetrads" *(34,86)*. Results showed that some hyperrec mutants merely increased the overall

Table 3
Assessing Intergenic Mitotic Recombination in Diploids from the Frequency of Recombinants Among Benomyl-Induced Haploid Segregants

Diploid types	Control, repair proficient		Homozygous *nuvA* *nuvsA11/nuvA11*		Homozygous *uvsH* *uvsH77/uvsH4*	
Percent recombination in each sample	A[a], %	B[b], %	A, %	B, %	A, %	B, %
Sample no.						
1	0	0			2	2
2	0	0	74	45	0	23
3	0.7	1			28	42
4	0	0	42	45	38	39
5	0	3			0	2
6					5	27
Average sample size (no.)	35	130	150	150	45	35
Crossovers/total	2/630	5/760	188/323	173/382	25/221	47/221
Average frequency	0.3%	0.7%	58%	45%	11%	21%
Median frequency	0	0	—	—	2–5%	23–27%

[a]A: intergenic distance of heterozygous markers on chromosome I: 60–75 c*M*.
[b]B: intergenic distance of second heterozygous marker pair: 150–200 c*M*.

frequency of types found for controls (e.g., *uvsF*; 59), whereas others mainly increased reciprocal types *(musN)*, or completely changed the patterns of coincidence for reciprocal and conversion events (e.g., *musQ*; *108*).

3.3. *The Selective* niaD *Test System: Recombination in* nuv *Mutant Strains*

A more efficient test was developed to check recombination frequencies in the many *nuv* mutant strains *(78)*. Homozygous *nuv* diploids, heteroallelic for two nonoverlapping *niaD* deletions, are grown on nitrate plates and produce thin spreading growth. On such plates many niaD$^+$ recombinants formed small colonies in controls, and reduction of such niaD$^+$ recombinants in some of the *nuv/nuv* strains was easily detectable. In contrast to the preceding results for *uvs* and *mus* mutants, hyporec types were more frequent (10/24) than hyperrec ones (4/24) among the tested *nuv* mutants *(78)*.

Interestingly, a first case of a fertile mutant, *nuv-11*, with decreased meiotic and increased mitotic recombination has recently been identified among the highly MNNG-sensitive *nuv* mutant strains *(50)*. *nuv-11*, which is a complementing allele of *uvsH* (Section 5.2.2.; *77*), shares some phenotypic features with other *uvsH* mutants (Table 1; *63*); e.g., it is highly sensitive to MMS as well as UV and X-rays, and shows increases in UV-induced mutation as well as spontaneous mitotic recombination. However, effects on intergenic mitotic recombination appeared to differ in being 5- to 10-fold higher in *nuv-11* diploids than reported for *uvsH77/4*. Considering the different methods of assessment used (Section 3.2.1.) the significance of this difference is not clear. The method of measuring spontaneous intergenic recombination frequencies in diploids by testing derived haploid segregants was therefore used for the *uvsH77/4* diploid *(59)*.

Results show that in this more legitimate comparison differences are smaller, with the *nuv-11* frequency being about twice that found for the *uvsH77/4* diploid (Table 3). In contrast, *nuv-11* shows qualitative differences for meiotic recombination from these *uvsH* alleles, since *nuv-11* is fertile in homozygous crosses, whereas fertility of the *uvsH77* and *uvsH4* mutants was too low to analyze meiotic recombination. In addition, *nuv-11* shows reduced growth in haploids, and higher instability of diploids than *uvsH77/4*, even though the latter diploid also produced 10% abnormal, probably aneuploid types *(63)*. For these latter features, as well as allelic recombination, which has not been tested for the *uvsH* alleles, comparisons are therefore not possible. It will be of interest to compare all of these different alleles as well as different disruption mutants of the *uvsH* gene using virtually isogenic strains and identical assays.

4. TESTING FOR EFFECTS ON SPONTANEOUS AND INDUCED MUTATION

In *Aspergillus*, few putative DNA repair-defective mutants have been shown to reduce mutation frequencies, although several were found to show increases (*uvsF*, *H*, and *J*, Section 2.3.3.). However, two mutants, *uvsI* and *uvsC*, clearly show defects in different types of UV mutagenesis (*13,15*; Section 4.2.).

4.1. Mutation Assays, Forward vs Reverse Mutation Systems

For measurements of mutation frequencies, the most frequently used assays are forward mutation systems. These select all types of mutations when analog resistance (e.g., to selenate; section 4.1.1) is the result of inactivation of specific genes. However, more subtle effects of mutations affecting proofreading or mismatch repair would not be uncovered in such tests. Reversion assays are therefore used as complementary assays for more specific effects (Section 4.1.2.).

4.1.1. Selection for Selenate Resistance, Testing Forward Mutation Frequencies

This preferred forward mutation assay selects for mutants that are sulfite-requiring, mainly alleles of *sB* (sulfate permease) or *sC* (ATP sulfurulase). It was used for the first comparative tests of spontaneous mutation in *uvs* mutants *(52)* and also was one of the three forward mutation assays used to measure UV-induced frequencies in *uvs* strains (Table 1; *(13,63)*. This test system has since been used generally for analysis of mutation frequencies in *mus*, *nuv*, and *sag* mutants, but for mutants in these latter genes, no significant effects on mutagenesis have been reported *(60,79)*.

4.1.2. Reverse Mutation Systems

Some of the selected *sC* mutants, as well as other nutritional mutants (e.g., *pabaA1* or *choA1*; Table 4), served for reversion tests to identify allele-specific patterns of spontaneous and induced mutation in *uvs* strains *(13)*. For *sC* mutations, actual base-pair changes can now be identified, since the *sC* gene region has been cloned and sequenced *(7)*.

4.2. Different Defects of uvsI *and* uvsC *in UV Mutagenesis*

Two mutants known to be deficient in UV mutagenesis, *uvsI* and *uvsC*, have been tested in detail and compared for effects on mutation in *uvs;I uvsC* double-mutant strains and controls *(15)*. It has been known for some time, that *uvsC* (and *E*) are mutators, i.e., mutations in these genes lead to increases in spontaneous mutation *(52)*. In addition, *uvsC* and also *uvsI* have been found to reduce UV-induced mutation in certain tests (Table 1; *14,38,63*).

To investigate the possibility that these two genes act in the same process of UV mutagenesis, their effects were compared in forward mutation systems (Section 4.1.1.) and also in reversion tests (Section 4.1.2.) using different-acting mutagens for induction, namely EMS, bleomycin, and UV (for UV mutagenesis, *see* Table 4). These and additional results from reversion tests of EMS-induced *sC* strains demonstrated a very specific defect in *uvsI* strains, possibly for AT → GC transitions *(13)*. In contrast, UV mutagenesis appears to be affected in a more basic or general way by the *uvsC* gene, which is now known to be a homolog of *E. coli recA* (*100*; Section 5.2.3.1.), since in *uvsC* mutant strains, all mutation assay systems used revealed similar decreases of UV-induced mutation. However, *uvsC* is able to produce chromosomal mutations, e.g., translocations, when these are induced by clastogens like bleomycin *(15)*.

5. CLONING OF ASPERGILLUS DNA REPAIR GENES AND HOMOLOGS IN OTHER ORGANISMS

5.1. Introduction

The molecular analysis of genes that apparently function in DNA repair, based on mutant phenotypes in *A. nidulans*, lags far behind the work in the budding or fission yeasts. A major reason is that methods for the efficient cloning of genes by complementation of recessive mutations have only recently been perfected. Thus far, four genes for DNA repair have been cloned, *uvsH*, *uvsF*, *uvsC*, and *sagA* genes, and the complete sequences have been reported for the first three *(62,100,107)*.

5.2. Cloned Aspergillus Repair Genes

5.2.1. The uvsF Gene

uvsF was the first of the mutagen-sensitive genes of *A. nidulans* to be cloned *(83)*. Sequencing of the gene indicates that it encodes a putative homolog of the large subunit of the human DNA replication factor C *(RFC1; 10,62)*. RFC was shown to be required not only in DNA replication, but also for efficient excision repair *(91)*. In yeast, all five RFC subunits were isolated, and each was found to be essential for cell proliferation *(21)*. A cold-sensitive mutant of *S. cerevisiae RFC1*, *cdc44-1*, shows cell-cycle arrest with G2/M DNA content at the restrictive temperature *(47)*, but at permissive temperature it has repair defects similar to those found for *uvsF201*, e.g., MMS sensitivity (Section 6.2.2.). Since UvsFp is thought to function in UV damage removal by excision and/or postreplication repair, these observations may provide some insight into its primary function, possibly as an essential component in error-free repair replication.

5.2.2. The uvsH Gene

uvsH was independently cloned by two groups. One group cloned *uvsH* by complementing the *uvsH77* mutation sensitivity to MMS *(107)* and, the other by complementing the *nuv-11* mutation, subsequently shown to be a unique allele of *uvsH* (*50*; Section 3.3.). The deduced amino acid sequence for UvsHp showed significant similarity to the *RAD18* gene product from *S. cerevisiae* and even closer overall correspondence to the

Table 4
Frequencies of Spontaneous and UV, or Bleomycin-Induced Mutations in *uvsC, uvsI*, and *uvsI;uvsC* Double Mutants[a]

	Spontaneous			Mutagen-induced			
				Bleomycin,	UV		
	Forward	Reversion of		forward	Forward	Reversion of	
Type of mutation	se[r]	*choA1*	*pabaA1*	se[r]	se[r]	*ChoA1*	*pabaA1*
uvs[+] Frequency	$\times 10^{-6}$ 1.4	$\times 10^{-8}$ 1.0	$\times 10^{-8}$ 0.6	$\times 10^{-6}$ 10	$\times 10^{-6}$ 120	$\times 10^{-8}$ 100	$\times 10^{-8}$ 3
Factor of induction in uvs[+]	—	—	—	(7×)	(80×)	(80×)	(5×)
uvs type							
uvsC	100× increased ↑ mutator			Like +	Much reduced, ↓ ± no induction		
uvsI	Like +	Reduced, ~ 1/3 ↓	1/10	Like +	Like +	Reduced Like +	> 0
uvsI;uvsC	20× ↑ increased	↓ Both with 0 spontaneous reversion		Like +	1/10 ↓ reduced	↓ Both with 0 UV-induced reversion	

[a]Results from forward and reverse mutation assays, compared to those of the uvs+ strain.

uvs-2 gene product of *Neurospora crassa (99)*. *RAD18* is a member of the *RAD6* group of error-prone postreplication repair genes in budding yeast, and *RAD18* is thought to be involved in *RAD6*-dependent, mutagenic repair of UV-induced DNA damage *(1,35)*.

5.2.3. *Two Homologs of* E. coli recA

5.2.3.1. UVSC

The *uvsC* gene was cloned by sib selection after transformation of an *A. nidulans uvsC114* strain with a cosmid library. Selection on MMS media identified two clones that showed complementation of the *uvsC114* mutation. Sequencing of a complementing subclone has revealed strong sequence homology to all known *S. cerevisiae RAD51* homologs and moderate homology to *E. coli recA (100)*. Gene disruption produced a *uvsC*[-] strain with somewhat more extreme defects than found for *uvsC114*, e.g., MMS sensitivity was higher than found for *uvsC114*, and meiosis in homozygous crosses was blocked at an earlier stage.

5.2.3.2. A *RECA*-LIKE GENE WITH STRONG HOMOLOGY TO *MEI*-3 OF *NEUROSPORA*

Oligonucleotide primers were constructed based on conserved *recA*-like domains of the *S. cerevisiae RAD51* and *N. crassa mei-3* genes, and were used to amplify a homologous DNA fragment from *A. nidulans*. This fragment identified a genomic DNA clone that shows significant sequence similarity to fungal *recA*-like genes, especially *mei-3 (65)*. Preliminary results suggest that this new *recA* homolog codes for a gene similar to *uvsC* of *A. nidulans*.

It is evident by now that considerable evolution of *recA* occurred in eukaryotes. Several species have more than one *recA*-like gene, e.g., in *S. cerevisiae*, unique functions have been demonstrated for four genes of this family *(54)*.

5.2.4. *The* sagA *Gene*

sagA was one of five genes identified in a screen for mutations that conferred an increased sensitivity to the alkylating agent MMNG, but not UV *(97)*. *sagA* has been cloned by complementation of a *sagA sagC* double-mutant strain. The product of the *sagA* gene shows extensive homology along its entire length to the *END3* gene of *S. cerevisiae (53)*. End3p is a membrane protein that functions in endocytosis and is required for the proper organization of the actin cytoskeleton. The similarity of the *Aspergillus* SagAp to End3p suggests an indirect role for the *sagA* gene in the repair of DNA damage by alkylating agents. It is possible that *sagA* mutants have increased cell permeability to such chemicals.

5.2.5. *A Possible* S. cerevisiae RAD2 *Homolog*

An apparent *A. nidulans* homolog of the *S. cerevisiae RAD2* and *S. pombe rad13* excision repair genes was tentatively identified using PCR to amplify a genomic DNA segment. Substantial sequence similarity to the corresponding fission and budding yeast genes was found, but no additional information has been reported for this putative *Aspergillus* homolog to excision repair genes *(12)*.

6. POSSIBLE FUNCTION OF UVS EPISTATIC GROUPS IN *ASPERGILLUS*, COMPARED TO YEASTS AND *NEUROSPORA*

Although knowledge of the molecular genetics and cell biology of the cell cycle in *Aspergillus* has increased rapidly in recent years, the analysis of mutagen-sensitive, putative DNA repair defective mutants has used mainly genetic approaches. However, with the emerging overlap of the two fields, it is expected that more direct analysis of function, and increased information from protein homologies will lead to faster progress in DNA repair research in the near future.

At this time, however, discussion of the function of DNA repair genes or groups of genes in *Aspergillus* is clearly speculative. Furthermore, only one of the four sets of mutants and genes has been analyzed sufficiently by tests for epistasis, recombination, and mutation to justify such speculations. Even in this case, the number of genes in each epistatic group is small, as are the number of mutant alleles known for each gene. In addition, in most cases, little is known about the types of mutations that happen to be available, so that it is uncertain how typical and how similar to null mutations the available mutants are. However, a summary of present findings and their patterns, based on unusually thorough and extensive tests, is of interest not only in comparison to the results obtained in other fungi, but also will be helpful for the planning of further investigations.

The description of the epistatic groups among *uvs* mutants demonstrates a clear pattern of correlated phenotypes in *A. nidulans*, at least to the extent that mutants within a group share some characteristics that differ from those of other groups. It is also clear that, as found for fission yeast, comparison with *S. cerevisiae* reveals "a great deal of diversity," and classification of the DNA repair genes will be more complex than originally thought *(74)*. This means that little direct correspondence of the *Aspergillus* Uvs epistatic groups can be found to any of the three epistatic and functional groups of *rad* genes of budding yeast, whereas some, but not all, of the deviating patterns show similarity to recent findings in *Neurospora*.

6.1. Four Epistatic Uvs Groups of **Aspergillus***: No Clear Correlation with Basic Repair Types*

Analysis of *uvsI* led to the proposal of a new epistatic group in *Aspergillus* based on thorough testing *(13,14)*, bringing the total number of Uvs groups to four. Tests using additional DNA-damaging agents and treatment at different stages of the cell cycle expanded and confirmed synergistic interaction of *uvsI* with all members of the three previously proposed groups *(15)*. These four groups each include one to three genes that share a number of features (Table 1).

1. Mutants of three genes in the *uvsF/H* group, namely *uvsF*, *H*, and *J*, show increased UV induced mutation and spontaneous mitotic recombination, features typical for excision repair-defective mutants, with *uvsF* most closely resembling the latter (Section 6.2.).
2. Many *uvsC* group mutants are defective in meiosis and mitotic recombination, similar to other *recA* homologs, but different from *S. cerevisiae rad51* mutants in that they show defects also in UV mutagenesis (Sections 6.3. and 6.4.).
3. *uvsI* shows normal recombination, but in some cases, reduces spontaneous and induced reversion; it presumably is defective for some type of error-prone repair.
4. *uvsB* group mutants show signs of unrepaired double-strand breaks, i.e., high levels of spontaneous deletions and aberrations, and nonreciprocal recombination is increased.

Similarly in *Neurospora*, at least four epistatic groups have been identified for genes of radiation repair (*48*; discussed further in Chapter 22). However, comparing the four or five groups identified in *Neurospora* to those of *Aspergillus*, similar differences to the three groups of budding yeast are apparent for some groups, whereas others differ also between these two species of filamentous fungi.

6.2. The **uvsF/H** *Group: Function in Error-Free Postreplication Repair of UV Damage*

Thus far, none of the highly UV-sensitive mutants of *Aspergillus* correspond to mutants deficient in NER. However, properties of the *uvsF* group mutants, i.e., high UV sensitivity, increased UV-induced mutation and spontaneous mitotic recombination suggest function in error-free repair.

6.2.1. RAD18 *Homologs in* Aspergillus *and* Neurospora

Although NER seemed the most likely process affected in the UvsH/F, group mutants, the recent sequencing of *uvsH*, which showed homology to *RAD18 (107)*, virtually rules out defects in excision function. *RAD18* is a member of the *RAD6* group of error-prone postreplication repair in yeast, but *S. cerevisiae rad18* mutants are highly sensitive to UV and most have only minor changes in mutation *(35)*. *RAD18* therefore is thought to be mainly involved in a poorly defined process of error-free postreplication repair *(87)*. If this is the case, *rad18* and similarly *uvsH* mutants may accumulate unrepaired recombinogenic and mutagenic lesions, as is found for mutants defective in NER.

In *Neurospora*, almost identical results were obtained for *UVS-2* group genes. *uvs-2* originally was thought to function in NER *(24)* and, like *uvsH*, is homologous to *RAD18 (99)*. However, in *Neurospora* even stronger support for implication in postreplication repair was obtained. A second member of the *UVS-2* group, the highly UV-sensitive mutant *mus-8*, which shows nearly normal mutagenesis *(56,57)* was recently found to be homologous to the yeast *RAD6* gene *(92)*; furthermore, all members of the *UVS-2* group were shown to have normal NER function *(4)*.

6.2.2. No NER Genes in Neurospora *or* Aspergillus?

No genes involved in NER have yet been identified in these filamentous fungi, and most likely NER is a less important process in these species than in budding yeast. In *Aspergillus*, however, one member of the UvsF group, *uvsF201*, has a pleiotropic phenotype more like NER-type mutants than *rad18* and its fungal homologs (Sections 2.3.3. and 5.2.1.). The answer to this puzzle was the recent finding that *UVSF* is a homolog of human Replication Factor C (RFC; *62*). As an essential gene active in DNA replication as well as repair, disruption of the gene probably is lethal. However, viable minor mutants, like *uvsF201*, may have various deleterious effects depending on which other proteins of the replication protein complex might be unable to bind or function normally.

In *Neurospora*, on the other hand, the *mus-18* mutant which is moderately, but specifically sensitive to UV light, was recently found to lack excision of dipyrimidine UV damage *(48)*. This gene was shown to encode a new type of UV-endonuclease rather than an NER-specific protein *(106)*, and a similar enzyme was found also in fission yeast (Chapter 20).

6.3. Recombinational Repair: A Variety of Genes in the UvsC Group

Mutants of most of the genes that show epistasis with *uvsC* affect recombination, usually reducing its frequency. Mutants of the following genes have been analyzed: *uvsC* and *uvsE*, and *musK*, *L*, and *N* (Tables 1 and 2). Of these, *uvsE* most closely resembles *uvsC* which is by far the most extensively analyzed mutant (e.g., Section 4.2.; they may also show similar X-ray sensitivity *(52)*, although there are no definite reports in the literature). Both are UV-sensitive only when germlings are irradiated, whereas the *mus* mutants are not UV sensitive. Mutants of all these genes (except *musK*) have defects in meiosis (*uvsC* is blocked in meiotic prophase; *100*); *uvsC*, *E*, and *musL* greatly reduce mitotic recombination and show increased nondisjunctional segregation, as found for *rad52* mutants of *S. cerevisiae* (Section 3.2.1.). However, recombination is only slightly reduced in *musK* and considerably increased in *musN*. All *mus* mutants are normal for spontaneous and UV-induced mutation, in contrast to *uvsC*, which is deficient in UV mutagenesis, and recently has been identified as a *recA* homolog (Section 6.4.; *100*).

6.4. Error-Prone Repair: Interaction of uvsI*, Defective for Reversion, with* uvsC*, a* recA *Homolog*

The next question concerns error-prone repair, and especially the *Aspergillus* epistatic group(s), which might be expected to function in mutagenic repair and mutation.

6.4.1. Aspergillus uvsI*, a Gene Functioning in Basepair-Specific Error-Prone Repair*

uvsI shows defects only for very specific basepair changes *(13)*. It therefore resembles certain *rev* mutations in yeast that cause allele-specific effects. *REV* genes are one of several subtypes of the yeast *RAD6* epistatic group, whereas *rad18* represents another *(35,36,42)*. However, *uvsI* clearly does not represent a UvsC subgroup, since tests using treatments of pregerminated conidia have conclusively established synergistic interactions between *uvsI* and *uvsC*, not only for UV survival but also for effects on certain types of UV-induced mutation *(15)*.

6.4.2. Aspergillus uvsC *with Dual* recA-*Like-Functions: Recombination and Mutagenesis?*

A general defect in mutagenesis was recently demonstrated for *uvsC* (Section 4.2.). Consistent results brought the realization that in *Aspergillus*, the *uvsC* mutant, in addition to causing the most pronounced reduction of recombination among all *uvs* types, also shows the most pronounced defects in UV-induced mutation.

With defects both in mutagenesis and recombination well documented, it becomes impossible to assign *uvsC* to a group of recombination repair like that of *RAD51* in yeast, or to a subtype of the *RAD6* group involved in error-prone repair. The conclusion is that *uvsC* is a true equivalent of the *E. coli recA* gene, and does not correspond to genes in either of these yeast groups. As far as can be judged, it differs also from *mei-3*, the *recA*-related gene of *Neurospora (17,41)*, mutants of which apparently are normal for UV mutagenesis *(49)*. At this time, it is unknown whether any other eukaryotic *recA* homologs resemble *uvsC* and are required for UV mutagenesis. In addition, there is no information regarding whether the differences in gene structure of the four *recA*-type genes, *recA*, *uvsC*, *Rad51*, and *mei-3*, are correlated with these observed differences in function.

6.5. A Fourth Type of Function: UvsB Group Mutants with Accumulated Double-Strand Breaks

Mutants of the UvsB group are moderately UV- and MMS-, but not X-ray sensitive (Table 1). *uvsB* and *uvsD* mutants have similar defects in growth, and in ascospore and conidial viability. In diploids, they show increased recombination mainly of a nonreciprocal type and show high spontaneous frequencies of irregularly sectoring colonies *(63,64)*. This phenotype resembles that of survivors after treatments with X-rays or bleomycin, which result from accumulation of spontaneous chromosome breaks, aberrations, and semidominant deletions that can be eliminated by mitotic crossing over *(55)*. *uvsB* and *uvsD* mutants may therefore show defects in the spontaneous joining of DNA strands required for normal DNA transactions, and may cause increased lethality after UV or MMS treatments in a similar way. The UvsB epistatic group does not correspond to any of the epistatic groups of *rad* mutants in budding yeast, but may either be directly involved in a process of double-strand break repair (perhaps by end joining, as identified in mammals; *76,105)* or indirectly cause defects in the regulation of repair, which leads to accumulation of unrepaired breaks.

7. EFFECTS ON MITOSIS AND THE CELL CYCLE

7.1. Introduction

DNA damage and repair genes have two roles. First, there are genes that ensure that the various cell-cycle stages are completed in the correct order and others that block further cell-cycle progression in response to insults that produce DNA damage. These genes have recently been referred to as checkpoint genes (*23,40,73*; Chapter 17). Considerable effort has gone into the study of cell-cycle genes in *A. nidulans* that function in the G_2/M transition and in mitosis. A collection of conditionally lethal mutations were isolated that were "blocked in mitosis" (the *bim* genes) or "never were in mitosis" (the *nim* genes) and therefore arrested in interphase *(22,26,29,30,70,80,81)*. These genes are discussed here because they have been demonstrated to perform cell cycle checkpoint regulatory functions or have been shown to display alterations in DNA damage responses.

7.2. Specific Gene Functions

7.2.1. nimA *and* nimX *are Regulators of Cell-Cycle Progression*

Among the cell cycle *nim* and *bim* genes studied in *A. nidulans*, four are most relevant to the discussion of DNA repair and damage, *nimA*, *nimX*, *bimD*, and *bimE*. *nimA* codes for a protein kinase required for the G_2/M transition *(80)*. The *nimX* gene codes for the fission yeast $p34^{cdc2}$ homolog in *A. nidulans (82)*. Although it is not known whether *nimA* and *nimX* are involved in DNA damage responses, the products of these genes function to regulate cell-cycle progression at entry into mitosis, and it is reasonable to assume that they will play some role in regulating cell-cycle progression in response to DNA damage. nimAp kinase activity is cell-cycle-regulated, and overexpression of the kinase causes cells to enter prematurely a mitotic-like state, including chromosome condensation and reorganization of the microtubule array *(80,81)*. Expression of *nimA* in vertebrate cells in culture can induce premature mitotic events, and in *Xenopus* oocytes, *nimA* can induce germinal vesicle breakdown without activating Mos, CDC2, or MAP kinases *(68)*. Additional structure–function studies of the nimAp kinase have shown that the noncatalytic carboxyl-terminal domain of a full-length mutant lacking kinase activity will arrest cells in G2 without activating endogenous nimAp or CDC2 kinases when overexpressed *(67,88)*. Expression of carboxyl-terminal truncated versions of nimAp that still have catalytic activity were shown to be toxic to *A. nidulans*. The truncated forms were shown to be more stable, indicating the presence in the carboxyl-terminus of sequences necessary for nimAp degradation *(88)*. One of the fundamental characteristics of cell cycle research has been the demonstration of a high degree of conservation in the regulatory pathways. The recent finding that a nimAp kinase regulatory pathway also exists in vertebrate systems underscores the fundamental nature of these regulatory systems. For this reason, it is likely that *nimX* and possibly *nimA*, like *cdc2* of fission yeast and *CDC28* of budding yeast, will be involved in regulating cell-cycle progression in response to DNA damage *(5,102)*.

7.2.2. *Negative Regulation of Entry into Mitosis by* bimE

The *bimE* gene of *A. nidulans* is among the most interesting of the cell-cycle genes to be discovered in this fungus. *bimE* codes for a high-mol-wt putative membrane-spanning protein that negatively regulates entry into mitosis *(29,81)*. Inactivation of *bimE* function either by gene disruption or use of a heat-sensitive conditionally lethal *bimE7* mutation results in premature entry into mitosis. *A. nidulans bimE7* germlings arrested in S phase by treatment with hydroxyurea and then shifted to restrictive temperature enter mitosis with condensed chromosomes, form spindles, and remain blocked. *bimE* thus acts to regulate entry into mitosis negatively. An interesting feature of the block is that loss of *bimE* function also prevents exit from mitosis. What role, if any, *bimE* may play in regulating entry into mitosis in response to DNA damage has not yet been investigated.

7.2.3. bimD *and Interacting Genes (*sudA, sudB, *and* sudC*) in Cell-Cycle Control and DNA Repair*

7.2.3.1. *BIMD*

The *bimD* gene of *A. nidulans* codes for a large nuclear protein that is a putative DNA binding protein *(22)*. Two different mutant alleles exist for this gene, *bimD5* and *bimD6*, and both mutations have the same heat-sensitive mitotic defect, characterized by a failure of chromosome attachment to the spindles and anaphase arrest. Both mutant alleles also display a sensitivity to DNA-damaging agents at temperatures that are permissive for mitosis, and interact synergistically with several mutants of Uvs group genes. Overexpression of *bimD* in germinating conidia blocks the nuclear division cycle in G_1 or early S phase of the cell cycle and is fully reversible. These observations led to an investigation of whether *bimD* may function as a check point that negatively regulates cell cycle progression in response to DNA damage *(23)*. These studies showed that *bimD* did not have a checkpoint function and that *bimD6* mutant strains displayed the expected delay in nuclear division in response to DNA damage.

7.2.3.2. Extragenic Suppressors of *BIMD6*

A screen for extragenic suppressors of the *bimD6* mutation was undertaken to identify additional genes that function with *bimD* in mitosis and/or DNA repair. This screen identified seven extragenic suppressors among approx 1500 revertants of heat-sensitive mitotic lethality. The seven suppressor mutations conferred a cold-sensitive phenotype of their own and defined four genes, designated *sudA*, *sudB*, *sudC*, and *sudD*, respectively, for suppressors of *bimD*. An interesting feature of these suppressors was that they suppressed only the heat-sensitive mitotic defect and not the sensitivity to MMS *(45)*.

7.2.3.3. Definition of a Possible Protein Complex

The genetic analysis of the suppressors also suggests that they form a complex with *bimD*, thus providing additional insight into the function of these genes. The classical inference of the complementation test is that two mutations that fail to complement in *trans* must both perform the same function and are therefore assumed to lie within one gene. Exceptional behavior in genetic complementation tests may then expose underlying structure and function relationships between products of different genes whose distinct proteins perform redundant or highly similar activities and/or associate in a complex *(6,27,95)*. Screens for unlinked noncomplementing mutations have been used to isolate genes whose protein products participate in the same structure or pathway (*31*; reviewed in *37,101*).

Allele specificity of noncomplementation may allow an understanding of the actual physical means by which the wild-type polypeptide does not function normally in the presence of two different mutant proteins. Specific noncomplementing mutant proteins may compete with wild-type proteins for binding to each other, so that the relative amount of active complexes falls below a critical threshold. On the other hand, the combination of both mutant proteins may produce a complex that dominantly inhibits or "poisons" the process to be complemented *(93)*. The *sudC4* and *sudA3* mutations displayed unlinked noncomplementation which may define a complex with *sudA*, *sudC*, and *bimD* gene products and/or association of SudA and SudC. This putative *sudA/*

sudC protein complex would function in the same pathway or process as *bimD*, either up- or downstream of the *bimD* execution point.

7.2.3.4. *SUDA*

In addition to the noncomplementation seen for the *sudA3* and *sudC4* mutations, the different *sudA* mutant alleles exhibited intragenic complementation. Cloning of *sudA* provided additional insight into *bimD* function. *sudA* mutations can confer a heat-sensitive mitotic lethality and a sensitivity to UV and MMS. *sudA* codes for a member of a family of proteins referred to as the SMC or DA-box proteins *(18,43,89,90,94)*. SMC proteins function in higher-order chromosome structure and segregation at mitosis *(43,89)*, and one member of the family functions as a global transcriptional repressor in dosage compensation in *Caenorhabditis elegans (18)*. Like other SMC proteins, *sudAp* also functions in chromosome segregation *(45)*. A simple model to explain the mitotic and mutagen-sensitive phenotypes displayed by the *bimD5* and *bimD6* mutations is that *bimDp*, in conjunction with other proteins like *sudAp*, has a more general function in chromatin structure that is required to support DNA repair as well as higher-order chromosome structure needed for faithful chromosome segregation during mitosis. The complex genetic interactions displayed by the *sud* genes is certainly consistent with their products cooperatively functioning with *bimD* in chromosome segregation at mitosis, and may also play a role in DNA repair or responses to DNA damage.

8. CONCLUSIONS

The continued molecular analysis of genes involved in DNA repair and mutagen sensitivity in *A. nidulans* will provide an excellent foundation for comparative protein structure and biochemical studies with similar genes in other systems. Such comparative studies will help to identify the functionally important domains in these proteins. Similarly, the study of the role of the cell-cycle genes in *A. nidulans* will provide further insight into the role of these genes in checkpoint functions. In this regard, it is notable that several genes that function in cell-cycle regulation and mitosis have been first identified in this organism. Elucidation of the role that these genes play in DNA repair processes and their control will lead to new insights into regulation of the eukaryotic cell cycle.

Although gene cloning and sequence homologies give impressive evidence of highly conserved domains required for such essential functions, genetic analysis has revealed surprising differences in the organization of pathways. This suggests alternate use and evolutionary shuffling of basic functional units.

A striking finding for *Aspergillus* DNA repair mutations was the high frequency of lethal interactions between pairs of random mutations from genes of different epistatic groups. These provide interesting genetic backgrounds for isolation of additional components of these repair processes. Two types are found that either give information about alternate pathways when two mutations are combined that disrupt different repair types, resulting in unrepairable lesions, as found, e.g., for *rem1* (a *rad3* allele) and *rad52* in yeast *(69)*, or *recA* and *polA* in *E. coli (11)*. Alternatively, if double mutants include specific malfunctioning alleles of proteins that normally interact in complexes, disturbed protein interactions may result. Both of these types of interactions may well be found among the many lethal interactions identified for *uvsF* proteins, which, as components of a DNA replication factor, work in a complex that may be active in several DNA transactions, including one or more types of repair.

ACKNOWLEDGMENTS

The authors are indebted to the laboratories of H.-S. Kang, and P. Strike, for sharing unpublished data from their laboratories. Their contributions have helped to make this chapter as up to date as possible. The work from the laboratory of E. K. was supported by a grant from the Nature Science and Engineering Research Council of Canada, and the work from the laboratory of G. S. M. was supported by grants from the National Institutes of Health and the National Science Foundation.

REFERENCES

1. Armstrong, J. D., D. N. Chadee, and B. A. Kunz. 1995. Roles for the yeast *RAD18* and *RAD52* DNA repair genes in UV mutagenesis. *Mutat. Res.* **315:** 281–293.
2. Babudri, N. and G. Morpurgo. 1988. An *uvsB* mutant of *Aspergillus nidulans* with high variable spontaneous mutation and intergenic recombination frequencies. *Mutat. Res.* **199:** 167–173.
3. Baker, S. M., G. P. Margison, and P. Strike. 1992. Inducible alkyltransferase DNA repair proteins in the filamentous fungus *Aspergillus nidulans*. *Nucleic Acids Res.* **20:** 645–651.
4. Baker, T. I., C. E. Cords, C. A. Howard, and R. J. Radloff. 1990. The nucleotide excision repair group in *Neurospora crassa*. *Curr. Genet.* **18:** 207–209.
5. Barbet, N. C. and A. M. Carr. 1993. Fission yeast *wee1* protein kinase is not required for DNA damage-dependent mitotic arrest. *Nature* **364:** 824–827.
6. Bender, A. and J. R. Pringle. 1991. Use of a screen for synthetic lethal and multicopy suppressor mutants to identify two new genes involved in morphogenesis in *Saccharomyces cerevisiae*. *Mol. Cell. Biol.* **11:** 1295–1305.
7. Borges-Walmsley, M. I., G. Turner, A. M. Bailey, J. Brown, J. Lehmbeck, and I. G. Clausen. 1995. Isolation and characterization of genes for sulphate activation and reduction in *Aspergillus nidulans*: implications for evolution of an allosteric control region by gene duplication. *Mol. Gen. Genet.* **247:** 423–429.
8. Brody, H. and J. Carbon. 1989. Electrophoretic karyotype of *Aspergillus nidulans*. *Proc. Natl. Acad. Sci. USA* **86:** 6260–6263.
9. Brody, H., J. Griffith, A. J. Cuticchia, J. Arnold and W. E. Timberlake. 1991. Chromosome-specific recombinant DNA libraries from the fungus *Aspergillus nidulans*. *Nucleic Acids Res.* **19:** 3105–3109.
10. Bunz, F., R. Kobayashi, and B. Stillman. 1993. cDNAs encoding the large subunit of human replication factor C, RFC, a multisubunit DNA polymerase accessory protein, required for coordinated synthesis of both DNA strands during SV40 DNA replication *in vitro*. *Proc. Natl. Acad. Sci. USA* **90:** 11,014–11,018.
11. Cao, Y. and T. Kogoma. 1995. The mechanism of *recA polA* lethality: Suppression by *recA*-independent recombination repair activated by the *lexA* (Def) mutation in *Escherichia coli*. *Genetics* **139:** 1483–1494.
12. Carr, A. M., K. S. Sheldrick, J. M. Murray, R. al-Harithy, F. Z. Watts, and A. R. Lehmann. 1993. Evolutionary conservation of excision repair in *Schizosaccharomyces pombe*: evidence for a family of sequences related to the *Saccharomyces cerevisiae* RAD2 gene. *Nucleic Acids Res.* **21:** 1345–1349.
13. Chae, S.-K. 1993. DNA repair and mutagenesis in the UV-sensitive mutant *uvsI* of *Aspergillus nidulans*. Ph.D. thesis, McGill University, Montreal, Canada.
14. Chae, S.-K. and E. Kafer. 1993. *uvsI* mutants defective in UV mutagenesis define a fourth epistatic group of *uvs* genes in *Aspergillus*. *Curr. Genet.* **24:** 67–74.

15. Chae, S.-K. and E. Kafer. 1997. Two *Aspergillus* genes with different function in error-prone repair: *uvsI* active in mutagenic specific reversion, and *uvsC*, a *recA* homolog, required for all UV mutagenesis. *Mop. Gen. Genet.*, in press.
16. Chen, J., B. Derfler, and L. Samson. 1990. *Saccharomyces cerevisiae* 3-methyladenine DNA glycosylase has homology to the AlkA glycosylase of *E. coli* and is induced in response to DNA alkylation damage. *EMBO J.* **9:** 4569–4575.
17. Cheng, R., T. I. Baker, C. E. Cords, and R. J. Radloff. 1993. *mei-3*, a recombination and repair gene of *Neurospora crassa*, encodes a RecA-like protein. *Mutat. Res.* **294:** 223–234.
18. Chuang, P.-T., D. G. Albertson, and B. J. Meyer. 1994. DPY–27: a chromosome condensation protein homolog that regulates *C. elegans* dosage compensation through association with the X chromosome. *Cell* **79:** 459–474.
19. Clutterbuck, A. J. 1993. *Aspergillus nidulans*; nuclear genes, in *Genetic Maps*, vol. 3, 6th ed. (O'Brien, S. J., ed.), Cold Spring Harbor Laboratory, Cold Spring Harbor, NY, pp. 3.71–3.84.
20. Cooper, A. J. and R. Waters. 1987. A complex pattern of sensitivity to simple monofunctional alkylating agents exists amongst the *rad* mutants of *Saccharomyces cerevisiae. Mol. Gen. Genet.* **209:** 142–148.
21. Cullman, G., K. Fien, R. Kobayashi, and B. Stillman. 1995. Characterization of the five replication factor C genes of *Saccharomyces cerevisiae. Mol. Cell. Biol.* **15:** 4661–4671.
22. Denison, S. H., E. Kafer, and G. S. May. 1992. Mutation in the *bimD* gene of *Aspergillus nidulans* confers a conditional mitotic block and sensitivity to DNA damaging agents. *Genetics* **134:** 1085–1096.
23. Denison, S. H. and G. S. May. 1994. Mitotic catastrophe is the mechanism of lethality for mutations that confer mutagen sensitivity in *Aspergillus nidulans. Mutat. Res.* **304:** 193–202.
24. de Serres, F. J., H. Inoue, and M. Schüpbach. 1983. Mutagenesis at the *ad-3A* and *ad-3B* loci in haploid UV-sensitive strains of *Neurospora crassa*. VI. Genetic characterization of *ad-3* mutants provides evidence for quantitative differences in the spectrum of genetic alterations between wild type and nucleotide excision-repair-deficient strains. *Mutat. Res.* **108:** 93–108.
25. Doonan, J. H. 1992. Cell division in *Aspergillus. J. Cell Sci.* **103:** 599–611.
26. Doonan, J. H. and N. R. Morris. 1989. The *bimG* gene of *Aspergillus nidulans*, required for completion of anaphase, encodes a homolog of mammalian phosphoprotein phosphatase 1. *Cell* **57:** 987–996.
27. Drubin, D. 1991. Development of cell polarity in budding yeast. *Cell* **65:** 1093–1096.
28. Ehninger, A., S. H. Denison, and G. S. May. 1990. Sequence, organization and expression of the core histone genes of *Aspergillus nidulans. Mol. Gen. Genet.* **222:** 416–424.
29. Engle, D. B., S. A. Osmani, A. H. Osmani, S. Rosborough, X. Xiang, and N. R. Morris. 1990. A negative regulator of mitosis in *Aspergillus* is a putative membrane-spanning protein. *J. Biol. Chem.* **265:** 16,132–16,137.
30. Enos, A. P. and N. R. Morris. 1990. Mutation of a gene that encodes a kinesin-like protein blocks nuclear division in *Aspergillus nidulans. Cell* **60:** 1019–1027.
31. Erickson, J. R. and M. Johnston. 1993. Genetic and molecular characterization of *GAL83*: its interaction and similarities with other genes involved in glucose repression in *Saccharomyces cerevisiae. Genetics* **135:** 655–664.
32. Fortuin, J. J. H. 1971. Another two genes controlling mitotic intragenic recombination and recovery from UV damage in *Aspergillus nidulans*. I. UV sensitivity, complementation and location of six mutants. *Mutat. Res.* **11:** 149–162.
33. Fortuin, J. J. H. 1971. Another two genes controlling mitotic intragenic recombination and recovery from UV damage in *Aspergillus nidulans*. II. Recombination behaviour and X-ray-sensitivity of *uvsD* and *uvsE* mutants. *Mutat. Res.* **11:** 265–277.
34. Fortuin, J. J. H. 1971. Another two genes controlling mitotic intragenic recombination and recovery from UV damage in *Aspergillus nidulans*. IV. Genetic analysis of mitotic

intragenic recombinants from uvs+/uvs+, *uvsD/uvsD* and *uvsE/uvsE* diploids. *Mutat. Res.* **13:** 137–148.
35. Friedberg, E. C. 1988. Deoxyribonucleic acid repair in the yeast *Saccharomyces cerevisiae*. *Microbiol. Rev.* **52:** 70–102.
36. Friedberg, E. C., G. C. Walker, and W. Siede. 1995. *DNA Repair and Mutagenesis*. ASM Press, Washington, DC.
37. Fuller, M. T., L. L. Regan, B. Green, R. Robertson, R. Deuring, and T. S. Hays. 1989. Interacting genes identify interacting proteins involved in microtubule function in *Drosophila*. *Cell Motil. Cytoskeleton* **14:** 128–135.
38. Han, D.-M. and H.-S. Kang. 1985. A study on the error prone DNA repair in *Aspergillus nidulans*: cell lethality and mutation induced by UV radiation in germinating asexual spores. *Korean J. Environ. Mutat. Carcin.* **5(2):** 53–60.
39. Han, D.-M., H.-Y. Suh, K.-H. Choi, and H.-S. Kang. 1983. Isolation and characterization of UV sensitive mutants in *Aspergillus nidulans*. *Korean J. Environ. Mutat. Carcin.* **3(1):** 21–33.
40. Hartwell, L. H. and T. A. Weinert. 1989. Checkpoints: controls that ensure the order of cell cycle events. *Science* **246:** 629–634.
41. Hatakeyama, S., C. Ishii, and H. Inoue. 1995. Identification and expression of the *Neurospora crassa mei–3* gene which encodes a protein homologous to Rad51 of *Saccharomyces cerevisiae*. *Mol. Gen. Genet.* **249:** 439–446.
42. Haynes, R. H. and B. A. Kunz. 1981. DNA repair and mutagenesis in yeast, in *The Molecular Biology of the Yeast* Saccharomyces. *Life Cycle and Inheritance*, vol. 1 (Strathern, J. N., E. W. Jones, and J. R. Broach, eds.), Cold Spring Harbor Laboratory, Cold Spring Harbor, NY, pp. 371–414.
43. Hirano, T. and T. Mitchison. 1994. A heterodimeric coiled-coil protein required for mitotic chromosome condensation *in vitro*. *Cell* **79:** 449–458.
44. Hoeijmakers, J. H. J., G. Weeda, W. Vermeulen, J. de Wit, N. G. J. Jaspers, D. Bootsma, and J.-M. Egly. 1996. Nucleotide excision repair: Molecular mechanisms and relevance to cancer. *Experientia* **52:** A1.
45. Holt, C. L. and G. S. May. 1996. An extragenic suppressor of the mitosis-defective *bimD6* mutation of *Aspergillus nidulans* codes for a chromosomal scaffold protein. *Genetics* **142:** 777–787.
46. Hooley, P., S. G. Shawcross, and P. Strike. 1988. An adaptive response to alkylating agents in *Aspergillus nidulans*. *Curr. Genet.* **14:** 445–449.
47. Howell, E. A., M. A, McAlear, D. Rose, and C. Holm. 1994. *CDC44*: a putative nucleotide binding protein required for cell cycle progression, that has homology to subunits of Replication Factor C. *Mol. Cell. Biol.* **9:** 255–267.
48. Ishii, C., K. Nakamura, and H. Inoue. 1991. A novel phenotype of an excision-repair mutant in *Neurospora crassa*: Mutagen sensitivity of the mus-18 mutant is specific to UV. *Mol. Gen. Genet.* **228:** 33–39.
49. Ishii, C. and H. Inoue. 1994. Mutagenesis and epistatic grouping of the *Neurospora* meiotic mutants, *mei-2* and *mei-3*, which are sensitive to mutagens. *Mutat. Res.* **315:** 249–259.
50. Iwanejko, L., C. Cotton, G. Jones, B. Tomsett, and P. Strike. 1996. Cloning and characterization of *nuvA*, an *Aspergillus nidulans* gene involved in DNA repair and recombination. *Microbiology* **142:** 505–515.
51. Jansen, G. J. O. 1970. Abnormal frequencies of spontaneous mitotic recombinaton in *uvsB* and *uvsC* mutants of *Aspergillus nidulans*. *Mutat. Res.* **10:** 33–41.
52. Jansen, G. J. O. 1972. Mutator activity in *uvs* mutants of *Aspergillus nidulans*. *Mol. Gen. Genet.* **116:** 47–50.
53. Jones, G. W., P. Hooley, S. M. Farrington, and P. Strike. Personal communication.
54. Johnson, R. D. and L. S. Symington. 1995. Functional differences and interactions among the putative RecA homologs Rad51, Rad55, and Rad57. *Mol. Cell. Biol.* **15:** 4843–4850.

55. Kafer, E. 1977. Meiotic and mitotic recombination in *Aspergillus* and its chromosomal aberrations. *Adv. Genet.* **19:** 33–131.
56. Kafer, E. 1981. Mutagen sensitivities and mutator effects of MMS-sensitive mutants in *Neurospora*. *Mutat. Res.* **80:** 43–64.
57. Kafer, E. 1983. Epistatic grouping of repair-deficient mutants in *Neurospora*: comparative analysis of two *uvs-3* alleles, *uvs-6* and their *mus* double mutant strains. *Genetics* **105:** 19–33.
58. Kafer, E. 1987. MMS sensitivity of all amino acid-requiring mutants in *Aspergillus* and its suppression by mutations in a single gene. *Genetics* **115:** 671–676.
59. Kafer, E. Unpublished results.
60. Kafer, E. and S.-K. Chae. 1994. Phenotypic and epistatic grouping of hypo- and hyper-rec *mus* mutants in *Aspergillus*. *Curr. Genet.* **25:** 223–232.
61. Kafer, E. and S.-K. Chae. 1994. Lethality or improved growth as a result of interaction in DNA repair double mutants of *Aspergillus*. *Fungal Genet. Newslett.* **41:** 54–59.
62. Kafer, E. and G. S. May. 1997. The *uvsF* gene region in *Aspergillus nidulans* reveals that *uvsF* codes for a protein with homology to DNA replication factor C. *Gene*, in press
63. Kafer, E. and O. Mayor. 1986. Genetic analysis of DNA repair in *Aspergillus*: evidence for different types of MMS-sensitive hyperrec mutants. *Mutat. Res.* **161:** 119–134.
64. Kafer, E. and P. Zhao. Unpublished results.
65. Kang, H. S. Personal communication.
66. Lehmann, A. R., A. M. Carr, F. Z. Watts, and J. M. Murray. 1991. DNA repair in the fission yeast, *Schizosaccharomyces pombe*. *Mutat. Res.* **250:** 205–210.
67. Lu, K. P. and A. R. Means. 1994. Expression of the noncatalytic domain of the NIMA kinase causes a G2 arrest in *Aspergillus nidulans*. *EMBO J.* **13:** 2103–2113.
68. Lu, K. P. and T. Hunter. 1995. Evidence for a *nimA*-like mitotic pathway in vertebrate cells. *Cell* **81:** 413–424.
69. Malone, R. E. and M. F. Hoekstra. 1984. Relationships between a hyper-rec mutation *(rem1)* and other recombination and repair genes in yeast. *Genetics* **107:** 33–48.
70. May, G. S., C. A. McGoldrick, C. L. Holt, and S. H. Denison. 1992. The *bimB3* mutation of *Aspergillus nidulans* uncouples DNA replication from the completion of mitosis. *J. Biol. Chem.* **267:** 15,737–15,743.
71. Morris, N. R. 1976. Mitotic mutants of *Aspergillus nidulans*. *Genet. Res.* **26:** 237–254.
72. Morris, N. R. and Enos, A. P. 1992. Mitotic gold in a mold: *Aspergillus* genetics and the biology of mitosis. *Trends. Genet.* **8:** 32–37.
73. Murray, A. W. 1992. Creative blocks: cell-cycle checkpoints and feedback controls. *Nature* **359:** 599–604.
74. Nasim, A. and M. A. Hannan. 1993. Cellular recovery, DNA repair and mutagenesis—a tale of two yeasts. *Mutat. Res.* **289:** 55–60.
75. Navarro, R. E., M. A. Stringer, W. Hansberg, W. E. Timberlake, and J. Aguirre. 1996. *catA*, a new *Aspergillus nidulans* gene encoding a developmentally regulated catalase. *Curr. Genet.* **29:** 352–359.
76. Nicolas, A. L., P. L. Munz, and S. C. Young. 1995. A modified single-strand annealing model best explains the joining of DNA double-strand breaks in mammalian cells and extracts. *Nucleic Acids Res.* **23:** 1036–1043.
77. Osman, F., C. Cotton, B. Tomsett, and P. Strike. 1991. Isolation and characterization of *nuv11*, a mutation affecting meiotic and mitotic recombination in *Aspergillus nidulans*. *Biochimie* **73:** 321–327.
78. Osman, F., B. Tomsett, and P. Strike. 1993. The isolation of mutagen-sensitive *nuv* mutants of *Aspergillus nidulans* and their effects on mitotic recombination. *Genetics* **134:** 445–454.
79. Osman, F., B. Tomsett, and P. Strike. 1994. Homologous recombination, in *Aspergillus: 50 Years On* (Martinelli, S. D. and J. R. Kinghorn, eds.), Progress Industrial Microbiology, vol. 29, Elsevier, Amsterdam, pp. 687–732.

80. Osmani, S. A., G. S. May, and N. R. Morris. 1987. Regulation of the mRNA levels of *nimA*, a gene required for the G_2-M transition in *Aspergillus nidulans*. *J. Cell Biol.* **104:** 1495–1504.
81. Osmani, S. A., D. B. Engle, J. H. Doonan, and N. R. Morris. 1988. Spindle formation and chromosome condensation in cells blocked at interphase by mutation of a negative cell cycle control gene. *Cell* **52:** 241–251.
82. Osmani, A. H., N. van Peij, M. Mischke, M. J. O'Connell, and S. A. Osmani. 1994. A single $p34^{cdc2}$ protein kinase (encoded by $nimX^{cdc2}$) is required at G_1 and G_2 in *Aspergillus nidulans*. *J. Cell Sci.* **107:** 1519–1528.
83. Oza, K. and E. Kafer. 1990. Cloning of the DNA repair gene, *uvsF*, by transformation of *Aspergillus nidulans*. *Genetics* **125:** 341–349.
84. Pontecorvo, G. and E. Kafer. 1958. Genetic analysis based on mitotic recombination. *Adv. Genet.* **9:** 71–104.
85. Pontecorvo, G., J. A. Roper, L. M. Hemmons, K. D. Macdonald, and A. W. J. Bufton. 1953. The genetic analysis of *Aspergillus nidulans*. *Adv. Genet.* **5:** 141–238.
86. Pritchard, R. H. 1955. The linear arrangement of a series of alleles of *Aspergillus nidulans*. *Heredity* **9:** 343–371.
87. Prakash, S., P. Sung, and L. Prakash. 1993. DNA repair genes and proteins of *Saccharomyces cerevisiae*. *Ann. Rev. Genet.* **27:** 33–70.
88. Pu R. T. and S. A. Osmani. 1995. Mitotic destruction of the cell cycle regulated NIMA protein kinase of *Aspergillus nidulans* is required for mitotic exit. *EMBO J.* **14:** 995–1003.
89. Saitoh, N., I. G. Goldberg, E. R. Wood, and W. C. Earnshaw. 1994. ScII: an abundant chromosome scaffold protein is a member of a family of putative ATPases with an unusual predicted tertiary structure. *J. Cell Biol.* **127:** 303–318.
90. Saka, Y., T. Sutani, Y. Yamashita, S. Saitoh, M. Takeuchi, Y. Nakaseko, and M. Yanagida. 1994. Fission yeast *cut3* and *cut14*, members of a ubiquitous protein family, are required for chromosome condensation and segregation in mitosis. *EMBO J.* **13:** 4938–4952.
91. Shivji, M. K. K., V. N. Podust, U. Hübscher, and R. D. Wood. 1995. Nucleotide excision repair DNA synthesis by DNA polymerase ε in the presence of PCNA, RFC, and RPA. *Biochemistry* **34:** 5011–5017.
92. Soshi, T., Y. Sakuraba, E. Kafer, and H. Inoue. 1995. The *mus-8* gene of *Neurospora crassa* encodes a structural and functional homolog of the Rad6 protein of *Saccharomyces cerevisiae*. *Curr. Genet.* **30:** 224–231.
93. Stearns, T. and D. Botstein. 1988. Unlinked noncomplementation: isolation of new conditional-lethal mutations in each of the tubulin genes of *Saccharomyces cerevisiae*. *Genetics* **119:** 249–260.
94. Strunnikov, A. V., V. L. Larionov, and D. Koshland. 1993. *SMC1*: an essential yeast gene encoding a putative head-rod-tail protein is required for nuclear division and defines a new ubiquitous protein family. *J. Cell. Biol.* **123:** 1635–1648.
95. Swanson, M. S. and F. Winston. 1992. *SPT4*, *SPT5* and *SPT6* interactions: effects on transcription and viability in *Saccharomyces cerevisiae*. *Genetics* **132:** 325–336.
96. Sweder, K. S., R. Chun, T. Mori, and P. C. Hanawalt. 1996. DNA repair deficiencies associated with mutations in genes encoding subunits of transcription initiation factor TFIIH in yeast. *Nucleic Acids Res.* **24:** 1540–1546.
97. Swirski, R. A., S. G. Shawcross, B. M. Faulkner, and P. Strike. 1988. Repair of alkylation damage in the fungus *Aspergillus nidulans*. *Mutat. Res.* **193:** 255–268.
98. Timberlake, W. E. and M. A. Marshall. 1992. Genetic engineering in filamentous fungi. *Science* **244:** 1313–1317.
99. Tomita, H., T. Soshi, and H. Inoue. 1993. The *Neurospora uvs-2* gene encodes a protein which has homology to yeast Rad18, with unique zinc finger motifs. *Mol. Gen. Genet.* **238:** 225–233.

100. van Heemst, D., K. Swart, E. F. Holub, R. van Dijk, H. H. Offenberg, T. Goosen, H. W. J. van den Broek and C. Heyting. 1996. Cloning, sequence, gene disruption, and mitoic and meiotic phenotypes of *uvsC*, on *Aspergillus nidulaus* homolog of yeast *RAD51*. *Mol. Gen. Genet.*, in press.
101. Vinh., D. B. N., M. D. Welch, A. K. Corsi, K. F. Wertman, and D. G. Drubin. 1993. Genetic evidence for functional interaction between actin noncomplementing (Anc) gene products and actin cytoskeletal proteins in *Sacharomyces cerevisiae*. *Genetics* **135:** 275–286.
102. Walworth, N., S. Davey, and D. Beach. 1993. Fission yeast chk1 protein kinase links the *rad* checkpoint pathway to *cdc2*. *Nature* **363:** 368–371.
103. Wang, Y., R. A. Prade, J. Griffith, W. E. Timberlake, and J. Arnold. 1994. A fast random cost algorithm for physical mapping. *Proc. Natl. Acad. Sci. USA* **91:** 11,095–11,098.
104. Wang, Z., J. Q. Svejstrup, W. J. Feaver, X. Wu, R. D. Kornberg, and E. C. Friedberg. 1994. Transcription factor b (TFIIH) is required during nucleotide-excision repair in yeast. *Nature* **368:** 74–76.
105. Weaver, D. T. 1995. What to do at an end: DNA double-strand-break repair. *Trends Genet.* **11:** 388–392.
106. Yajima, H., M. Takao, S. Yasuhira, J. H. Zhao, C. Ishii, H. Inoue, and A. Yasui. 1995. A eukaryotic gene encoding an endonuclease that specifically repairs DNA damaged by ultraviolet light. *EMBO J.* **14:** 2393–2399.
107. Yoon, J. H., B. J. Lee, and H. S. Kang. 1995. The *Aspergillus uvsH* gene encodes a homolog to yeast *RAD18* and *Neurospora uvs-2*. *Mol. Gen. Genet.* **248:** 174–181.
108. Zhao, P. and E. Kafer. 1992. Effects of mutagen-sensitive *mus* mutations on spontaneous mitotic recombination in *Aspergillus*. *Genetics* **130:** 717–728.

22

DNA Repair in *Neurospora*

Alice L. Schroeder, Hirokazu Inoue, and Matthew S. Sachs

1. INTRODUCTION

Repair of damaged DNA is necessary to maintain a functional, replicating genome. As detailed in this volume, an understanding of how organisms cope with DNA damage has arisen from work in many systems. In this chapter, we describe current understanding of processes important for DNA damage and DNA repair in *Neurospora crassa*. These studies reveal important similarities and differences with other systems. Several older reviews that cover different aspects of the literature related to *N. crassa* DNA damage and repair are available *(33,133,135)*. In addition, compilations of genetic and biochemical studies using *N. crassa*, although not specifically focused on DNA damage and repair, are indispensable resources for obtaining a comprehensive view of the earlier literature *(4,116)*.

The phenomenon of DNA damage in *N. crassa* has been of interest starting with the first successful attempts to induce mutations that led to the formulation of the one gene–one enzyme theory *(10)*. The discovery that gene function could be inactivated without loss of protein led to the concept of point mutations *(175)*, and much effort was put into analyzing the spectrum of mutations induced by different mutagenic conditions *(32,33)*.

The first reports of *Escherichia coli* repair-defective mutants and the excision of cyclobutane pyrimidine dimers (CPDs) in *E. coli (11,12)* spurred efforts to identify analogous mutants affecting these processes in *N. crassa*, *Saccharomyces cerevisiae*, and other fungi. The first mutagen-sensitive mutants in a fungus were obtained in the smut *Ustilago (70)*. A driving force in these studies was also the promise that, like the *recA* mutation in bacteria *(28)*, mutants affecting DNA repair in fungi would also affect recombination *(71)*.

2. EXPERIMENTAL APPROACHES TO DNA REPAIR IN NEUROSPORA

N. crassa is the most extensively studied multinuclear, filamentous ascomycete *(29,113)*. It has both sexual and asexual phases in its life cycle (Fig. 1). It produces multinuclear macroconidia and uninucleate microconidia (10–30% *[121]*) during asexual development and ascospores during heterothallic sexual development. Except for a brief phase preceding meiosis, it is a haploid organism. Its approx 43-Mbp genome *(44,108,110)* contains seven genetic linkage groups *(116)* corresponding to seven cyto-

From: DNA Damage and Repair, Vol. 1: DNA Repair in Prokaryotes and Lower Eukaryotes
Edited by: J. A. Nickoloff and M. F. Hoekstra © Humana Press Inc., Totowa, NJ

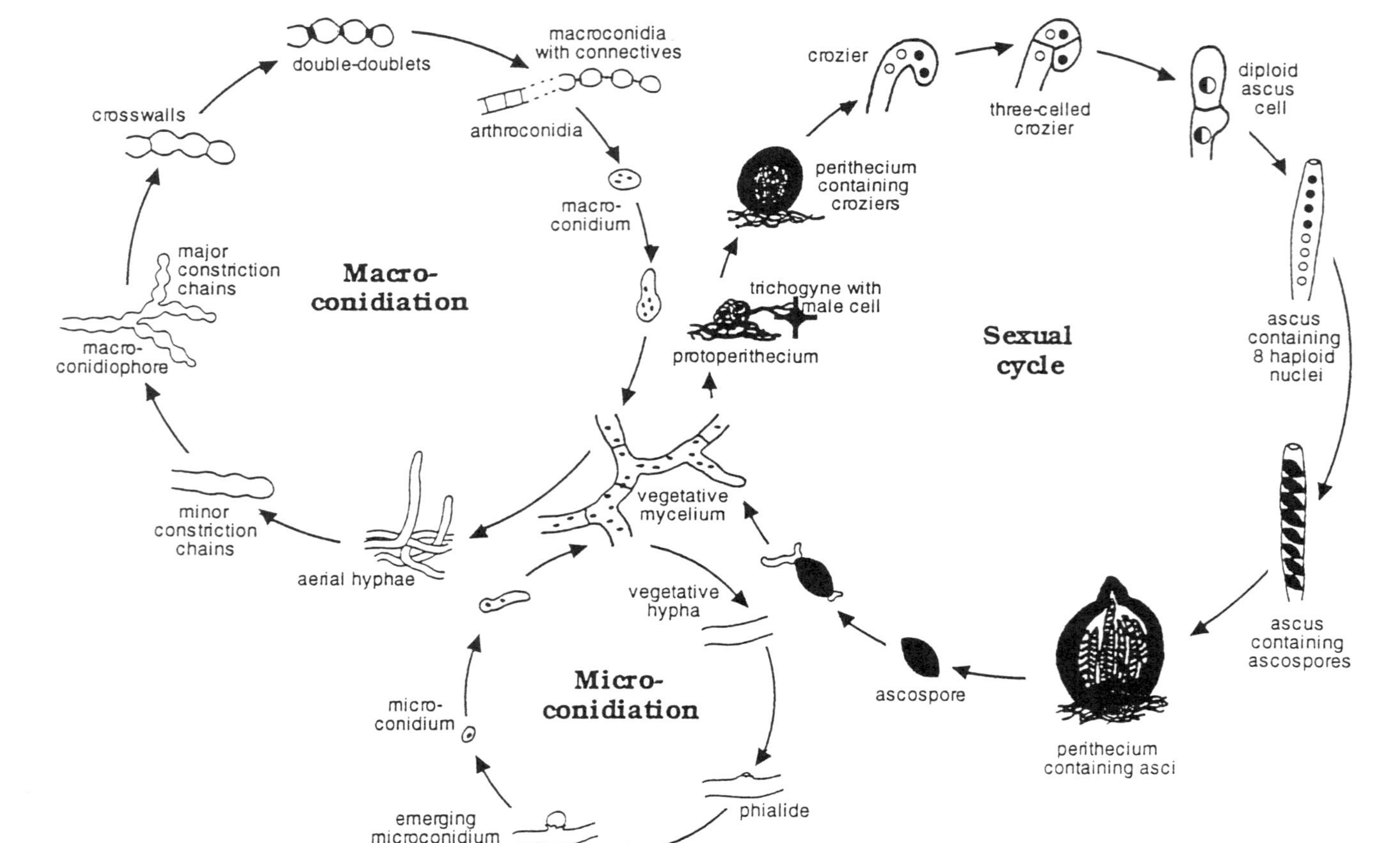

Fig. 1. The *N. crassa* life cycle. The multinuclear, vegetative mycelium can produce three kinds of spores. The asexual process of macroconidiation results in the production of multinuclear spores called macroconidia. The asexual process of microconidiation, which requires special environmental conditions or mutant strains to be the preponderant process, produces uninucleate spores called microconidia. The sexual cycle, which leads to ascospore production, requires the interaction of male and female cells of different mating types. Macroconidia, microconidia and ascospores all germinate to form vegetative mycelium. Reprinted from ref. *151* with permission.

logically observable chromosomes *(99,115)*. Although nuclei in this organism share a common cytoplasm, nuclear division appears to be asynchronous *(144)*, except during the production of ascospores *(120)*.

2.1. *Heterokaryon Formation*

A major advantage of *N. crassa* is its ability to form heterokaryons. This has been particularly useful in mapping and mutation studies. *N. crassa* strains of the same mating type and carrying the same heterokaryon compatibility alleles can fuse. Since the organism grows as a syncytium, the nuclei of the two fusing strains are mixed in the same cytoplasm. If two strains require different nutritional supplements, neither will grow on minimal medium, but the heterokaryon can grow without the nutritional supplements. Nuclear numbers in conidia usually vary between 1 and 5, and up to 30% of the conidia may be uninuclear *(121)*. Thus, both heterokaryotic conidia and homokaryotic conidia containing each of the nuclei present in the heterokaryon will be produced from heterokaryons, allowing selection, separation, and testing of the various nuclear types.

2.2. *Meiotic Recombination*

Meiotic defects are easy to detect in *N. crassa* crosses because homozygous crosses of mutants either fail to produce ascospores or produce a large fraction of inviable white ascospores instead of normal black ascospores. However, auxotrophic markers can sometimes create similar problems, and it important to examine whether specific meiotic blocks are caused by mutations that affect mutagen sensitivity. This must be determined for each mutant. Fortunately, *N. crassa* is suitable for cytological studies of meiosis *(99,119)*.

2.3. *Mitotic Recombination*

Testing of mitotic recombination is difficult in *N. crassa*, since disomics and diploids are unstable, and no autonomously replicating nuclear plasmids have been isolated. Two types of tests have been developed, one based on duplications and the other on the integration of exogenous DNA. Duplications are easily generated by crosses of inversion or translocation strains to normal sequence strains *(104,115)*. If the duplication strain carries both the *A* and *a* ideomorphs *(57)* of the mating type locus or different alleles of a heterokaryon incompatibility locus, the duplication-carrying ascospores will show inhibited growth. Eventually, a small sector will begin to grow normally. In the area of normal growth, the mating type or heterokaryon incompatibility locus will have become hemizygous or homozygous via deletion of one of the duplicated arms or mitotic recombination between the arms. In the case of the most commonly used test, duplications generated by the chromosome I inversion, (IL $\rightarrow$ IR) H4250, both have been shown to occur. However, tests to distinguish between the two are rarely performed *(104)*. Thus, chemicals such as histidine and hydroxyurea (HU), or mutations that increase the speed with which the colony escapes from inhibited growth, are usually referred to simply as increasing mitotic chromosomal instability.

Another assay that is considered to reflect an organism's capacity for recombination repair is the efficiency of integration of exogenous DNA into homologous sites in the chromosome. In *S. cerevisiae*, homologous recombination is very efficient (99%), and almost any homology will serve as a site of integration *(127)*. In *Neurospora*, homologous integration of plasmid DNA is inefficient (~3.5%) and clearly dependent on the length of homology *(48)*. Using exogenous DNA with an *mtr* gene disrupted with a selective marker, homologous integration can be recognized by the inactivation of the endogenous *mtr* gene *(140)*. The *mtr* gene specifies an amino acid transporter and when defective leads to recessive resistance to toxic amino acids *(111)*. Homologous integration can then be verified both by molecular methods and by the linkage of the selective marker and the *mtr* phenotype.

2.4. Mutagenesis

N. crassa has been particularly useful for studies of mutagenesis. As in other microorganisms, very large numbers of cells (conidia) can be easily and inexpensively tested and a large number of nutritional and drug resistance markers are available *(116)*. It is straightforward to control conditions that induce DNA damage, and many mutagens have been tested *(32)*. The uninucleate conidia produced by macroconidiating strains allow forward mutation induction for recessive mutants. Conidial platings can be used to purify dominant mutants. However, accurate rates of mutation/genome are difficult to calculate, and results are usually given as mutations/survivor. The ease with which heterokaryons can be formed allows relatively easy analysis of the recessive lethal mutation rate and also of the molecular type of the mutation by complementation *(33,154)*. Heterokaryons can also be used to examine the ability of various genes and treatments to influence the mutation process during or after damage *(154,158)*. Systems based on the *mtr* gene *(154)* and the *ad-3A* and *ad-3B* genes *(33)* allow selection of both forward and reverse mutations, and have been particularly useful in the analysis of mutation. Both have been used to examine mutation induction in mutagen-sensitive strains.

Strains with an *ad-3* mutation accumulate a purple pigment. By using sorbose to force colonial growth in liquid, millions of colonies can be screened for gain or loss of color *(32)*. Heterokaryons that include mutations at the *ad-2*, *ad-3A*, and *ad-3B* loci in one component expand the system to allow not only mutations in either of the *ad-3* genes in the other nucleus to be detected, but also deletions that include the *ad-3* genes and essential genetic regions lying between and to either side of the *ad-3* loci, or in other essential loci. Complementation in two- and three-component heterokaryons can be used to determine the extent of the mutations and, in the *ad-3B* gene, where intragenic complementation occurs, polarity. Polarity and reversion tests with chemicals thought to cause specific types of DNA damage can pinpoint the most likely type of change at the nucleotide level, including differentiating between point, frameshift, and small deletion mutations *(33,36)*. This system has given the highest-resolution analyses of mutations, without sequencing, that have been obtained in any of the filamentous fungi *(32)*.

Recessive forward mutations in the *mtr* gene can be selected by their resistance to the toxic amino acid analogs, methyltryptophan (MTR) and *p*-fluorophenylalanine (FPA), whereas reversions to normal function allow *trp* auxotrophs to grow on tryptophan (which requires the *mtr* gene product for uptake) as well as anthranilic acid (whose uptake is independent of the *mtr* gene product). In heterokaryons carrying mutations that cause colonial growth with abundant conidiation, recessive lethal mutations can be assayed by replica plating to score homokaryotic and heterokaryotic conidia. Heterokaryons can also be made between germinating conidia to examine rescue phenomena *(157,158)*.

2.5. DNA Biochemistry

As with other microorganisms, release of large DNA requires permeabilization or elimination of the cell wall. Both the enzyme zymolyase, which permeabilizes cell walls *(16,18)*, and the enzyme preparation novozym, which removes cell walls, can be used *(135,168)*. Most fungi, including *N. crassa*, do not have a thymidine kinase *(61)*. Therefore, DNA labeling requires a radioactive ribonucleotide, usually uridine, followed by destruction of the labeled RNA with RNase or alkali *(135,171)*.

2.6. Gene Isolation

Although mitochondrial plasmids are common in *N. crassa (59,60)*, no DNA sequences have been recovered that allow *N. crassa* to maintain plasmids that enable the facile recovery of genes that complement a nuclear mutant phenotype. Thus, different strategies must be used to identify *N. crassa* genes by their function. The initial approach used was sibling (sib) selection *(3)*. Ordered cosmid libraries with dominant selectable markers have been constructed for this purpose *(109,168)*. Cosmid DNAs are pooled and transformed into *N. crassa* of the desired phenotype. The pool that rescues the phenotype is ascertained, and smaller pools of those cosmid DNAs constructed for transformation. The procedure is repeated until a single cosmid containing the rescuing gene is identified. The gene can be further characterized by isolation and characterization of smaller genomic DNA fragments that rescue the phenotype and by the isolation of corresponding cDNA clones. Recent efforts to produce a physical map ordering the cosmids along the chromosomes (J. Arnold, University of Georgia) and correlating them with clones for expressed cDNAs will greatly streamline this procedure by allowing chromosome walks near a gene to be identified based on the genetic position of the gene. RFLP mapping will then allow identification of the cosmid containing the gene in situations where complementation is difficult. Gene isolation by complementation of *E. coli (174)* or *S. cerevisiae (15)* mutants with *N. crassa* cDNA expression libraries and by PCR strategies *(173)* has also been successful.

3. NEUROSPORA MUTAGEN-SENSITIVE MUTANTS

3.1. Overview of the Mutants

The first ultraviolet light (UV)-sensitive mutants in *N. crassa* were reported in 1967 *(20)*, but in subsequent work it was not possible to isolate single genetic loci responsible for the sensitive phenotypes (the original strains are available in the Fungal Genetics Stock Center, University of Kansas). Thus, work on *N. crassa* mutagen-sensitive genes really began with the reports in 1968 of several other mutations affecting UV sensitivity *(128,159)*. At present, more than 50 mutagen-sensitive mutants have been isolated, which map at more than 40 different nuclear genes (Table 1; *79,90,135*). Where more than one allele has been tested, they have, in general, had very similar properties, *(135)*. Thus, allele designations have generally not been used in this chapter. The presence of cytoplasmic factors affecting radiation sensitivity *(21)* has been suggested, but never confirmed *(138)*. The first mutagen-sensitive genes discovered

Table 1
***N. crassa* Genes with Roles in the Repair of DNA Damage**[a,b]

Gene	Epistasis group	Characteristics of mutants	Refs.
upr-1	Upr-1 ?[c]	Sensitive to UV and X-rays, not sensitive to MMS; partially defective in photoreactivation	*8,9,30,34,35,72–74,76,80,82,134,142,162,167,172*
mei-2	Uvs-3, Uvs-6	Not sensitive to UV; sensitive to MMS, γ-rays; no meiotic recombination	50,73,*74,83,134,139–141,152,153*
mei-3	Uvs-6	Sensitive to MMS, MTC, histidine, γ-rays, not sensitive to UV at 25°C, some alleles sensitive at 38°C; high spontaneous mutation frequency; gene resembles *S. cerevisiae RAD51*	*25,66,72–74,83,139,140,152,153*
uvs-2	Uvs-2	Extremely sensitive to UV, 4NQO, MTC, MMS; sensitive to γ-rays, MNNG, ICR170; gene resembles *S. cerevisiae RAD18*	*8,9,16–18,30,34,35,72–74,76,80,86,98,132,138, 140,142,152,153,157–159,162,164,172,176*
uvs-3	Uvs-3	Sensitive to UV, MMS, MTC, HU, histidine, X-rays; high spontaneous mutation frequency; low induced mutation frequency; partial defect in photoreactivation	*8,9,30,34,72–74,76,80,83,86,87,129,130,140, 142,152,153,157,158,162,172*
uvs-4	Uvs-2	Sensitive to UV, slight sensitivity to MMS, not sensitive to X-rays; mitochondrial DNA defects	*30,34,35,67,114,129,130,142*
uvs-5	—	Sensitive to UV, not sensitive to γ-rays; mitochondrial DNA defects	*30,34,35,67,80,114,129,130,142,172*
uvs-6	Uvs-6	Sensitive to UV, γ-rays, MMS, histidine; partially dominant; fails to activate replicon initiation checkpoint	*9,30,34,35,72–74,76,80,83,86,87,94,106140, 142,152,153,157,158,162,172*
mus-7	Uvs-6	Not sensitive to UV; sensitive to MMS, slight sensitivity to X-rays	*77,85,86,90,134,152,153,162*
mus-8	Uvs-2	Extremely sensitive to UV, MMS, MTC, sensitive to X-rays; gene resembles *S. cerevisiae RAD6*	*8,9,85,86,92,118,134,150*
mus-9	Uvs-3	Sensitive to UV, MMS, histidine, X-rays; high spontaneous mutation frequency, low UV and X-ray induced mutation frequency; fails to activate replicon initiation checkpoint	*85,86,90,92,94,134,140,152,153*

mus-10	Uvs-6	Sensitive to UV, MMS, not sensitive to X-rays	*85,86,92,94,134*
mus-11	Uvs-3?	Sensitive to UV, X-rays; extremely sensitive to MMS, histidine; high spontaneous mutation frequency, low UV- and X-ray induced mutation frequency	*85,86,90,92,94,134,140,152,153*
mus-12, 13, 14	—	Mycelia but not conidia are MMS-sensitive	*38,39*
mus-16	—	Not sensitive to UV or γ-rays; sensitive to MMS and nitrogen mustard	*6,81,152,153,162*
mus-18	Mus-18	Sensitive to UV; not sensitive to other mutagens; UV-specific excision repair; gene has unique characteristics	*84,164,174*
mus-25	Uvs-6	Not sensitive to UV; sensitive to MMS; cloned gene resembles *S. cerevisiae RAD54*	*63,75,77,90*
mus-26	Upr-1	Resembles *upr-1*	*8,78,82*
phr-1	Phr-1	Mutant lacks photorepair; gene specifies photolyase enzyme	*14,45,58,173*

[a]Additional *mus* genes include *mus-15, 19, 20 (77)*; *mus-21, 27–30 (90)*; *mus 21, 31–39 (79)*; *mus-17, 22–24 (75)*.

[b]For additional references, *see (116,134,135)*.

[c]The Upr-1 group may be a subgroup of the Uvs-2 group (Section 3.2.2).

were designated *uvs-1–uvs-6* (UV-sensitive) with one exception, *upr-1* (UV-sensitive, photoreactivation-defective), whereas more recently isolated genes have been designated *mus* (mutagen-sensitive), starting with *mus-7*. Additional genes leading to mutagen sensitivity were originally isolated based on their meiotic defects and are thus designated as *mei* genes.

A variety of inducing agents have been used to induce mutagen-sensitive mutants, including UV, *N*-methyl-*N'*-nitro-*N*-nitrosoguanidine (MNNG) and 4-nitroquinoline 1-oxide (4NQO). Because of early interest in the repair of UV-induced damage, the first *N. crassa* mutagen-sensitive mutants were isolated on the basis of their sensitivity to UV (*upr-1*, *uvs-1*, *3-6*; *[20,129,167]*). Subsequent mutants were mainly isolated on the basis of methyl methane sulfonate (MMS) sensitivity, since it is often correlated with γ-ray sensitivity and defects in recombination *(92)*, including *mus-7-15, 19-20, 22-26*, and *31-37 (38,77,79,90,92)*. Notable exceptions are the *uvs-2* mutant, found serendipitously in a wild-type strain on the basis of its increased UV sensitivity *(159)*; the *mei* mutants *(104,149)*; and the *mus-16* mutant, which was selected for sensitivity to nitrogen mustard *(6)*. Attempts to isolate mutagen-sensitive mutants by their inability to secrete nuclease have led to reports of several mutants *(49,101)*; one of these mutations is an allele of *uvs-3 (89)*. Three mutants induced by UV irradiation and selected on the basis of γ-ray sensitivity have been lost *(100)*. Further selection of mutants for sensitivity to γ-irradiation might reveal classes of mutations not yet seen. There also exist a large number of meiotic mutants isolated from wild-collected *Neurospora (97)*, which have not been tested for mutagen sensitivity, but could prove to be an important source of mutagen-sensitive strains.

In general, only mutants with clear-cut properties and reasonable growth rates were isolated and studied further, partially for convenience and partially on the assumption that these most likely would be null mutations. Most work has centered on *upr-1*, *mei-2* and *3*, *uvs-2*, *3* and *6*, and *mus-7-11*, *16*, *18*, and *26*. Both *uvs-4* and *uvs-5* are difficult to work with, because *uvs-4* dies when stored on silica gel or as conidia for more than a few weeks, and *uvs-5* grows poorly and is difficult to recover from crosses. Recently, it has been shown that both mutants accumulate mitochondrial defects on repeated subculturing *(67)*. To alleviate growth problems, Perkins *(114)* has developed *uvs-4* and *uvs-5* stocks sheltered in heterokaryons, which are available from the Fungal Genetics Stock Center.

3.2. Epistasis Groups

In both *E. coli* and *S. cerevisiae*, it was possible to discern mutations affecting genes in the same repair pathway by their sensitivity to UV, γ-rays, or both, and effects on recombination (Section 4.5.) and mutation (Section 6.2.2.). Such groupings were not clear in *N. crassa* or other filamentous fungi (Chapter 21; *134,135*). Almost all of the mutants were sensitive to a wide range of DNA-damaging agents. None showed the pattern of sensitivity expected for mutants defective in nucleotide excision repair (NER): sensitivity only to UV and agents causing bulky additions to DNA. A few mutants, like those in the *S. cerevisiae* RAD50 epistasis group, were sensitive to γ-rays and most chemicals, but not to UV. Thus, epistasis tests of double mutants were undertaken based on the rationale that if a double mutant has no stronger a phenotype than either of the single mutants that comprise it, the mutants belong to the same epistasis

Table 2
Known Epistatic Relationships Among DNA Damage Repair Genes in *N. crassa*[a]

Uvs-2	Uvs-3	Uvs-6	Mus-18	Upr-1?[b]	Phr-1
uvs-2[c]	*uvs-3*	*uvs-6*	*mus-18*[c]	*upr-1*	*phr*[1-c]
uvs-4	*mus-9*	*mus-7*		*mus-26*	
mus-8[c]	*mus-11?*	*mus-10*			
	mei-2[d]	*mus-25*[c]			
		mei-2[d]			
		mei-3[c]			

[a]The names of each of the six epistasis groups are given in the top line.
[b]The Upr-1 group may be a subgroup of the Uvs-2 group (Section 3.2.2.).
[c]Genes that have been isolated.
[d]Shows epistatic interactions with more than one group.

group and are in the same repair pathway. If the mutants show a synergistic interaction, i.e., in the case of mutagen sensitivity, the double mutant is much more sensitive than either single mutant, then the mutant genes are considered to be in different epistasis groups, which presumably represent independent repair pathways *(68)*. Since the molecular defects in the mutant alleles of the *N. crassa* genes are not known, an underlying assumption is that the mutations are in fact null mutations, and mutants showing additive and lethal interactions are considered to be in different epistasis groups. Where it could be tested, different alleles at the same gene showed the same type of interaction, supporting this assumption *(86,92)*. However, it should be noted when two leaky alleles in an essential pathway are combined, the activity of the pathway might be reduced sufficiently to cause lethality. Thus, epistasis grouping based on additive or lethal interactions is always tentative.

On the basis of epistasis tests, three groups were initially defined in *N. crassa*, typified by the *uvs-2*, *uvs-3*, and *uvs-6* genes, respectively *(76,86,92,166)*. In more recent work, it has become clear that *upr-1*, which was originally considered part of the Uvs-2 group, appears to form either a separate or subgroup with *mus-26 (82)*. A single mutant, *mus-18*, forms a separate epistatic group *(84)*, and one mutant, *mei-2*, appears to be epistatic to both the Uvs-3 and Uvs-6 groups *(83)*. Photoreactivation may also be considered a separate epistasis group, Phr-1 *(45)*. The six *N. crassa* epistasis groups affecting mutagen sensitivity, with all known members, are summarized in Table 2. These epistasis groups and their possible relationships to DNA repair pathways in *S. cerevisiae* and *Schizosaccharomyces pombe* are shown in Table 3.

3.2.1. Uvs-2

The Uvs-2 group was thought to be involved with a multiprotein NER pathway on the basis of early data showing a lack of CPD excision *(172)*. The *uvs-2* and *upr-1* mutants also show a normal spontaneous mutation frequency, but high induced mutation frequency *(34,76)* similar to that of other NER-defective mutants *(68)*. However, *uvs-4* and *mus-8* show low to normal spontaneous and induced mutation frequencies *(85)*. Later data showed that mutants in this group could still excise dimers *(9,98)*. Furthermore, the demonstration that *uvs-2* and *mus-8* genes are homologs of the *S. cerevisiae RAD18* and *RAD6* genes, respectively, suggests that the Uvs-2 group is simi-

Table 3
Relation of Repair Systems and Epistasis Groups in *N. crassa*, *S. cerevisiae*, and *S. pombe*

	Repair system					
Organism	Photoreactivation	UV endonuclease	NER[a]	PRR	Recombination repair	Undetermined function
S. cerevisiae	P*hr*-1[b]	NF[c]	R*ad*3	R*ad*6	R*ad*52	—
N. crassa	Phr-1[b]	Mus-18[b]	NF	Uvs-2 Upr-1?[d]	Uvs-6	Uvs-3
S. pombe[e]	NF	present[b,f]	*rad2*	*rhp6*	*rad21*	*rad1*

[a]NER, nucleotide excision repair.
[b]A single protein appears to be responsible for this activity (phr-1 *[45]*; mus-18 *[174]*); *S. pombe (54,160)*.
[c]NF, not found.
[d]The Upr-1 group may be a subgroup of the Uvs-2 group (Section 3.2.2.).
[e]Many members of the *S. pombe* groups have been placed in these groups on the basis of phenotype rather than direct epistasis testing *(96)*.
[f]The protein has been partially purified and the probable gene has been cloned, but not named *(160)*.

lar to the yeast RAD6 group *(150,164)*, which is involved in postreplication repair. The biochemical role of the Uvs-2 group in DNA repair is not understood.

3.2.2. *Upr-1*

The original epistasis tests of *upr-1* showed an epistatic relationship to *uvs-2* with UV, but a synergistic relationship for X-irradiation *(76)*. Because the *upr-1* mutant showed many properties similar to *uvs-2*, including normal fertility, normal spontaneous mutation *(34)*, increased UV- and X-ray-induced mutation in an *ad-3* forward mutation test (Section 2.4. *[76]*), and no excision of pyrimidine dimers *(172)*, it was considered to be in the Uvs-2 epistasis group. The discovery of a mutant, *mus-26*, which has almost identical properties to those of *upr-1*, but is not allelic *(78)*, and data showing that both *uvs-2* and *upr-1* are proficient in dimer excision *(8,9)* led to a re-evaluation of epistasis and mutagenesis in *upr-1 (82)*. The *mus-26* and *upr-1* strains show a clear epistatic relationship with both UV and γ-rays. In these authors' hands, both also showed an additive relationship to *uvs-2* when tested with either radiation. Both *mus-26* and *upr-1* are quite sensitive to 4NQO, but only slightly sensitive to mitomycin C (MTC), MMS, and γ-rays; both are fertile. Both show normal levels of spontaneous mutation, but in an *ad-8* reversion test, both show lower than normal UV-induced mutation, whereas *uvs-2* shows increased mutation induction (Section 6.2.2. *[82]*). Thus, it appears that *upr-1* and *mus-26* form a distinct epistatic group, which, given the additivity of the epistatic relations to *uvs-2*, is possibly a subgroup of the Uvs-2 group (Table 2).

3.2.3. *Uvs-3*

The biochemical role of the Uvs-3 epistasis group is unknown. With the exception of *mei-2*, which appears epistatic to both *uvs-3* and *uvs-6 (83)*, the remaining three members of this group, *uvs-3*, *mus-9*, and *mus-11*, show:

1. High spontaneous mutation;
2. Low or no UV-induced mutation; and
3. Probably low γ-ray-induced mutation *(34,76,85,135)*.

However, induction of recessive lethal mutations by UV appears to be normal in *uvs-3*. Kafer has hypothesized that this may be owing to the different nature of recessive lethals, which may mainly result from chromosomal aberrations *(87)*. The low induced mutation frequencies suggest that the Uvs-3 epistasis group is involved in a pathway that normally is responsible for inducing the majority of the mutations seen with UV and γ-irradiation, i.e., an "error-prone" repair pathway. Both *E. coli recA* and *S. cerevisiae rad6* mutants show similar phenotypes. Several mutants in the *S. cerevisiae* RAD52 epistasis group have also been reported to display similar phenotypes *(68)*. It should be noted that the evidence that *mus-11*, which is difficult to work with, is in the Uvs-3 group is only based on the MMS sensitivity of *mus-11 uvs-3* double mutants. They are slightly more MMS sensitive than single mutants.

3.2.4. Uvs-6

The possible function of the Uvs-6 group is revealed by the characterization of *mei-3*, a gene which is epistatic to *uvs-6* but lethal in combination with *uvs-3* mutants. The *mei-3* gene is a homolog of the *E. coli recA* and *S. cerevisiae RAD51 (25,66)* genes. Given that *RAD51* is in the RAD52 epistasis group, which is involved in recombinational repair, it would seem likely that the Uvs-6 group is involved in recombinational repair. However, although most members of the RAD52 group in yeast show an increase in spontaneous mutation and normal or decreased induced mutation, *mei-3* is unusual in showing considerably increased induced mutation with UV, as well as a strong increase in spontaneous mutation *(83)*. Other members of this group (*uvs-6*, *mus-7*, and *mus-10*) show normal spontaneous and induced mutation frequencies *(34,76,85,87)*.

3.2.5. Mus-18

The *mus-18* gene encodes a unique endonuclease, which recognizes both CPDs and pyrimidine-pyrimidone (6-4) photoproducts ([6-4]PDs), and cleaves at sites immediately 5' to the damage. Thus far, it appears to be the only NER activity in *N. crassa*, since no evidence exists for a multicomponent NER activity *(84,174)* (*see* Note Added in Proof). A protein with a similar activity is found in *S. pombe (42,54)*. The *mus-18* mutant also shows a normal rate of spontaneous mutation, but increased mutation in response to UV, as would be expected for a mutant defective in incision of UV damage.

3.2.6. Phr

Neurospora also has a photolyase, specified by the *phr* gene, which is involved in photoreactivation repair. The gene has been cloned and is very similar to *S. cerevisiae* photolyase *(45,173)*.

4. REPAIR SYSTEMS

4.1. Protection from DNA Damage

Exposure to UV, ionizing radiation, oxidizing agents, and mutagens all result in DNA damage, as do naturally occurring cellular processes. In addition to the DNA repair systems and protective enzymes discussed below, *N. crassa* produces a variety of compounds that act as primary defenses against environmental damaging agents. The sexually produced ascospores (Fig. 1) are heavily melanized and appear black when ripe. The multinuclear asexual spores, called macroconidia (Fig. 1), are rich in carotenoids, giving them an orange color; furthermore, the production of carotenoids in mycelia is induced by blue light. Photoinactivation of conidia by light occurs in the presence of methylene blue, toluidine blue, or acridine orange *(163)*. Strains lacking carotenoids are more sensitive to methlyene blue and toluidine blue, but not acridine orange, consistent with a role of carotenoids in protecting against membrane damage but not nuclear damage *(163)*. Conidia are also rich in the compound ergothionein, which may act as a free radical scavenger and prevent damage by mutagens *(64)*. Ergothionein, which is also ubiquitously found in mammals, remains one of the more mysterious compounds found in cells *(2)*.

N. crassa also contains three forms of catalase *(24)* and superoxide dismutase to scavenge free radicals generated by oxidative processes. Inactivation of the *sod-1* gene, which specifies a CuZn superoxide dismutase *(23)*, results in greater sensitivity to oxidative agents, such as paraquat, but not to UV or γ-irradiation *(22)*. The *sod-1* mutant also has a higher spontaneous mutation frequency *(22)*. The wild-type γ-ray sensitivity of the *sod-1* mutant is somewhat surprising, since the majority of the damage from γ-irradiation is thought to be caused by free radicals generated from the interaction of γ-rays and water.

One might also expect γ-ray-sensitive mutants to be sensitive to hydrogen peroxide (H_2O_2) which produces the same free radicals. However, when the γ-ray sensitive mutants, *uvs-2* and *6*, and *mus-7-11, 21, 25, 28*, and *29* were tested, only *mus-9* and *11* showed significant sensitivity in either buffer or growth medium *(91)*. Although a number of factors may be involved, it seems likely that the protective actions of the catalases and superoxide dismutases are very effective in inactivating most exogenous peroxides and their products before they can reach the DNA, whereas free radicals generated in close proximity to DNA by γ-rays may escape. This possibility is supported by the clear sensitivity of the *N. crassa* γ-ray-sensitive mutants to the drug bleomycin, which is thought to act, in part, by generation of free radicals upon its intercalation in DNA, and by the substantial increase in H_2O_2-sensitivity seen in *Aspergillus nidulans* conidia on disruption of the conidial catalase *(102)*. However, the increased sensitivity of *mus-9* and *11* to H_2O_2 must be more complicated than loss of one of the protective enzymes, since these mutants are sensitive to UV and alkylating agents as well.

4.2. Nucleotide Excision Repair

Among the most toxic UV photoproducts are CPDs and (6-4)PDs. These can be photochemically repaired by photoreactivation using photolyase, or can be repaired following excision of the base or nucleotide moieties. In most organisms, a complex multiprotein excision system exists that can remove a number of types of bulky adducts and structure-distorting lesions, including CPDs and (6-4) PDs *(55)*. Early studies using thin-layer chromatographic detection of radioactive dimers in acid-hydrolyzed *N. crassa* DNA showed that wild-type strains removed dimers quite rapidly, with 80% of the dimers removed in the first 60 min of growth after irradiation. Neither *uvs-2* nor *upr-1* mutants removed UV-induced dimers by this assay. The *uvs-3* mutant showed a delay in removing dimers *(172)*. These data were consistent with the observed extreme UV sensitivity of *uvs-2* and the increase in survival in *uvs-2* seen when cells were held for 2 h in nonnutritive liquid and then exposed to photoreactivating light (*[132]* note that the symbols for immediate photoreactivation and liquid holding in Fig. 3 of ref. *132* are switched). The *uvs-2* mutant also showed other properties typical of excision-defective mutants, including normal fertility, increased UV-induced mutation *(76)*, and inability to rescue a UV-irradiated component of a heterokaryon *(158)*.

It was puzzling that the *uvs-2* and *upr-1* mutants, unlike NER mutants in *E. coli* and yeast, were clearly sensitive to ionizing radiation and MNNG, and showed increased γ-ray-induced mutation *(76)*. Subsequent assays of dimer removal, using (1) loss of *Micrococcus luteus* endonuclease-sensitive sites detected by alkaline sucrose gradient centrifugation and (2) loss of binding of dimer-specific antibodies, indicated that all members of the Uvs-2 and Upr-1 epistasis groups that were tested (*uvs-2*, *upr-1*, *mus-8*, and *mus-26*) removed both CPDs and (6-4)PDs as rapidly as wild-type strains *(8,9,98)*. CPDs and (6-4)PDs were also removed from *uvs-3* and *uvs-6* strains. These lesions seemed to be removed somewhat faster in these tests than in the chromatographic assay, with at least 80% of the CPDs removed within 30 min. Convincing evidence that the Uvs-2 epistasis group does not contain genes involved in multicomponent NER came with the isolation of two members of this group. The *uvs-2* gene is homologous to *S. cerevisiae RAD18 (164)* and *A. nidulans uvsH (176)*, and *mus-8* is homologous to *S. cerevisiae RAD6 (150)*. The products of *S. cerevisiae RAD18* and *RAD6* are involved in postreplication repair and appear to interact with each other *(5)*.

Thus far, there is no evidence for an *N. crassa* pathway comparable to the NER of *E. coli*, yeast, or mammals, which involves multiple protein components *(84,174)*. Only one mutant that is sensitive to UV, but not to MMS or γ-rays has been isolated: *mus-18 (84)*. The *mus-18* mutant is less than twofold more sensitive to UV than wild-type, and not more sensitive to MMS, γ-rays, 4NQO, MNNG, or MTC than wild-type. However, like NER mutants, it shows much higher frequencies of mutation at *ad-3* than wild-type, and it is unable to remove UV-induced *M. luteus* endonuclease-sensitive sites *(84)*.

The *mus-18* gene has been isolated and a single gene product, Mus-18p, purified. Mus-18p is a unique protein that makes single-strand nicks immediately 5' to both CPDs and (6-4)PDs *(174)*. Presumably the dimers are then removed with the aid of other enzymes in the cell, but which enzymes are used is not known. Thus, *Neurospora* has a different NER system than in many other organisms whose repair processes have been examined.

Considerable evidence suggests that *mus-18*-like repair systems may be common among fungi. An enzymatic activity very similar to that coded for by *mus-18* has been found in *S. pombe (42,54)* and a recently isolated gene shows considerable similarity to the *mus-18* gene *(160)*. The *radC* mutant of *Dictyostelium discoideum* is sensitive only to UV light and, in an assay using loss of antibody binding sites, shows a slower removal of UV photoproducts, especially CPDs *(107)*. Thus, this phenotype resembles that of the *N. crassa mus-18* mutant. Furthermore, extensive mutant hunts in *A. nidulans* (*see* Chapter 21), *Dictyostelium*, and *Ustilago* have not given rise to typical NER mutants with sensitivity to UV and bulky photoproducts, but no accompanying sensitivity to γ-rays or MMS. This is particularly impressive in *Dictyostelium*, where an extensive hunt for mutants sensitive to 4NQO produced no mutants with this combination of properties *(13)* (*see* Note Added in Proof).

4.2.1. *What Is the Role of* uvs-2 *in Repair of UV-Induced Damage?*

It is particularly intriguing that *mus-18*, despite preventing the great majority of, if not all dimer incision, shows less than twofold greater sensitivity to UV than wild-type. On the other hand, *uvs-2* is 15 times more sensitive to UV, even when the *mus-18* gene is intact *(84,132)*. The two mutants may have additive or synergistic interactions *(84)*. The *uvs-2* mutants appear normal with respect to vegetative growth, sexual fertility, and meiotic recombination *(159)*. They are the most UV-sensitive *N. crassa* strains known *(85,138)* and are sensitive to a variety of mutagens *(80,116)*. Both UV and γ-rays show increased mutation induction in *uvs-2* compared to wild-type *(76)*.

Analyses of genomic and cDNA sequences *(164)* revealed that *uvs-2* specifies a 501 residue polypeptide that resembles *S. cerevisiae RAD18* and the recently identified product of the *A. nidulans uvsH* gene *(176)*. Both Uvs-2p and Rad18p contain zinc finger domains; Rad18p has a nucleotide binding domain whose conserved features are lacking in Uvs2p.

As discussed above, data on the ability of the *uvs-2* mutant to remove dimers are conflicting. Either the tests using thin-layer chromatographic detection of radioactive dimers in acid-digested DNA were misleading, or in *uvs-2* mutants, dimers are converted into forms not recognized by *M. luteus* endonuclease or dimer-specific antibodies. If the chromatographic data were incorrect, then an obvious hypothesis is that Uvs-2p is necessary for accurate removal of the dimers and perhaps other damage, with a less accurate and often lethal process following Mus-18p incision when Uvs-2p is absent. Conversely, in the absence of Mus-18p, Uvs-2p might allow toleration of dimers. Alternatively, Uvs-2p might be necessary for removal or toleration of a small number of UV photoproducts and other damage in key regions of the genome *(9)*. An example of this is seen in Cockayne's syndrome in humans, where the products of the affected genes are necessary for removal of damage in actively transcribing regions, but overall excision repair is normal (*see* Chapter 18, vol. 2; *55)*.

If in fact the chromatographic data are correct (and there seems little reason to doubt it), the observations must be reconciled differently. *Neurospora* may have an endonuclease that attacks the internal phosphodiester bond of both types of dimers; such an endonuclease activity has been described in human cells *(56)*. If the endonuclease action is inhibited in the presence of Uvs-2p, then when Mus18p is present and Uvs-2p is absent, this endonuclease would rapidly cut the interdimer backbone, enabling rapid religation of the 5'-cut made by Mus-18p to produce a DNA structure that chromatographs as a dimer after acid hydrolysis *(56)*. However, this structure would probably not be recognized by either the very specific antibodies to the photoproducts or by *M. luteus* endonuclease [since *M. luteus* endonuclease can recognize CPDs but not (6-4)PDs], such specificity is reasonable. In the absence of Mus-18p and presence of Uvs-2p, unaltered dimers would remain in the DNA, but would be tolerated. Examination of retention of dimers in *Neurospora* DNA using the HPLC system of Galloway et al. *(56)* could provide some interesting clues to the function of pathways requiring Uvs-2p for activation.

4.3. Base Excision and Mismatch Repair

Base excision via glycosylases and mismatch repair have not been studied in *Neurospora*. Several mutants have phenotypes consistent with roles in these processes, but their roles in DNA damage repair remain to be established. The *mus-16* mutant is sensitive to the alkylating agents MMS and MNNG, but not to UV or γ-rays *(81)* and is also unable to remove nitrogen mustard-induced crosslinks *(6)*. However, no epistatic or other biochemical data are available concerning the role of this mutant, nor are there biochemical studies on two other mutants with similar patterns of mutagen sensitivity, *mus-7* and *mus-20*. All three mutants are homozygous sterile and show normal spontaneous mutation rates. The two that were tested, *mus-7* and *mus-16*, show increased mitotic chromosomal instability *(135)*.

4.4. Postreplication Repair

Pyrimidine dimers inhibit DNA synthesis; in their presence, new DNA is initially synthesized as small fragments, often, although not always, less than or equal to the interdimer distance *(55)*. Eventually, this DNA is extended into normal-length DNA, even if dimers remain in the DNA. In some organisms, new DNA is synthesized in short fragments for a considerable time after irradiation if dimers remain in the template *(117)*. In other organisms, the ability to recover long DNA synthesis occurs fairly quickly *(62)*, although dimers remain in the template. The ability to bypass dimers via bypass synthesis or recombination is called postreplication repair (PRR), although it is actually a tolerance mechanism rather than a repair mechanism *(55)*.

Calza and Schroeder *(16,17)* examined the size of DNA made after UV irradiation using the *uvs-2* mutant, then thought to be excision-defective. In a 45-min labeling period, unirradiated cells synthesized very long DNA. However, in UV-irradiated, germinating conidia, DNA was synthesized in fragments approximating the interdimer distance during the first hour after irradiation. The induction of dimers and the length of new DNA were dose-dependent at least to doses of 80 J/m^2. At a dose of 22 J/m^2, which allowed 37% survival, the small DNA fragments were extended into high-mol-wt DNA over a 2–3 h incubation period after label was removed. If cells were incubated in growth medium for 2 h and then labeled for 45 min, new DNA was nearly as long as DNA synthesized in unirradiated cells. Unlike the situation in *E. coli*, there was no evidence that dimers were bypassed through a recombination mechanism. Thus, it appeared at that time that *Neurospora* had a postreplication tolerance mechanism resembling that in many mammalian cells. However, it now seems highly likely that dimers were being removed from this DNA during the experiments, although tests for endonuclease-sensitive sites at the time indicated that most dimers remained in the DNA. Therefore, it is probable that the delay in new DNA synthesis seen in these experiments was indicative of a rather slow dimer excision process in the *uvs-2* mutant, rather than dimer bypass. In this light, it is not surprising that normal rates of DNA synthesis were seen within 3 h of UV irradiation. Based on the hypotheses concerning the action of *uvs-2* presented above, and the possible role of Uvs-2p in postreplication repair based on similarities with its *S. cerevisiae* homolog, *RAD18*, it would now be very interesting to examine new DNA synthesis in *mus-18* mutants and *mus-18 uvs-2* double mutants.

4.5. Recombination Repair

Recombination repair is necessary for the repair of double-strand breaks (DSBs) in most organisms and is thought to be involved in repair of crosslinked DNA *(55)*. There has been no biochemical demonstration of DSB repair in *Neurospora*. However, it is quite likely that *Neurospora* has a recombinational repair system resembling that controlled by the RAD52 epistasis group of budding yeast. As described later in this section, attempts to study directly meiotic and mitotic recombination have rarely been able to provide convincing evidence of a recombinational defect in any mutagen-sensitive mutant. Strong support for a gene involved in a recombinational repair system comes from the structure of the *mei-3* gene. This gene has extensive similarity to both *S. cerevisiae RAD51* and *E. coli recA*. The *mei*-3 gene was isolated by sib selection based on its ability to confer resistance to 4NQO. Recently Hatakeyama et al. *(66)* have expanded on an earlier report *(25)* to show that the reading frame specifying Mei-3p is 1059 bp. It is interrupted by two introns and would code for a protein of 353 amino acids. The predicted mass of Mei-3p, 38 kDa, agrees well with the predicted size and the size of Mei-3p detected by immunoblotting of perithecial (fruiting body) extracts with antiserum directed against Mei-3p. Mei-3p has both the core I and II regions found in *S. cerevisiae* Rad51p protein homologs in eukaryotes, whereas *E. coli* RecA has only the core II domain. It is quite similar to all the Rad51p homologs, especially that of *S. pombe*, where the core domains I and II are about 67 and 84% homologous at the amino acid level. Both the *S. cerevisiae* and *N. crassa* polypeptides show approx 30% amino acid identity when compared to the core region of RecA, which is responsible for oligomer formation and recombination, and includes the two ATP binding domains *(55)*. The *mei-3* transcript is probably present at low levels throughout the life cycle, but the protein concentrations are considerably higher in perithecia. The *mei-3* mRNA is induced by UV and MMS exposure; induction of *RAD51* transcripts after X-ray and MMS exposure is seen in *S. cerevisiae*.

Similarities between *mei-3* and *RAD51* are further supported by phenotypic similarities of the mutants *(55)*. Mutations in both genes cause sensitivity to γ-rays and MMS, although at least one allele of *mei-3* (N289) is also sensitive to UV at high temperatures *(105,139)*. Each gene is induced by DNA damage *(1,66,147)* and meiosis is defective in each mutant. In *mei-3* mutants, most asci are arrested in zygotene, and there is very little chromosome pairing *(119)*. Mitotic recombination may be reduced in *mei-3* mutants, as it is in *S. cerevisiae rad51* mutants (*see* Chapter 16; *55,105,140)*.

A second gene almost surely involved in recombinational repair is *mei-2*. This mutant was originally isolated by its dominant effect on meiosis *(149)*. However, on repeated backcrossing, the mutant was no longer completely dominant. The *mei-2* mutant is sensitive to γ-rays and MMS, but not to UV *(83,139)*. When both parents contain *mei-2* mutations, little or no chromosome pairing is seen in zygotene. Some viable ascospores are produced in these crosses because the small number of chromosomes in *Neurospora* (seven) provides a reasonable probability that some ascospores will receive at least one copy of each chromosome. There is little or no meiotic recombination seen in these spores, but many are disomic and the mitotic recombination typical of disomic ascospores does occur *(141)*. In budding yeast, *rad50* mutants also show sensitivity to γ-rays and MMS, but not to UV. The *rad50* mutants exhibit no meiotic recombination and little or no chromosome pairing, yet have little, if any, effects on mitotic recombination. Thus, it would seem likely that *mei-2* functions will be related to those of *RAD50*, which is a member of the RAD52 epistasis group and is known to be involved in recombinational repair of DSBs *(55)*.

In addition to *mei-2* and *mei-3*, many other mutagen-sensitive strains are female infertile or homozygous sterile *(135)*. In at least two mutants, *uvs-3* and *uvs-6*, the block to meiosis appears to occur before the haploid nuclei of the parents fuse to form the diploid nucleus that will undergo meiosis *(119)*. A mutation in the *mus-8* gene (a homolog of *S. cerevisiae RAD6 [150]*) has an unusual defect in which meiosis is normal but the postmeiotic mitoses occur without chromosome replication *(118)*. In *uvs-5*, meiosis is normal, until pachytene, when normal ascus development stops and the asci degenerate *(119)*. Combined with recent data on the accumulation of deletions in mitochondrial DNA in this mutant *(67)*, it is possible to speculate that *uvs-5* may be defective in the repair of DNA strand breaks.

In the duplication test for mitotic chromosomal instability (Section 2.3.), all of the mutagen-sensitive mutants in the Uvs-3 and Uvs-6 epistasis groups, except *mus-10 (105,106,129,134)*, show increased escape from inhibited growth. All of the Uvs-2 and Upr-1 group mutants tested (all but *mus-26 [134]*) show an escape pattern identical to wild-type. These data support the idea that genes in both the Uvs-3 and Uvs-6 epistasis groups are involved with the generation or repair of DNA strand breaks. Unfortunately, it gives little information on which process is affected, and whether mitotic recombination or deletion is affected.

All of the mutants that show chromosomal instability in this test are also unusually sensitive to histidine and HU *(134)*. HU and histidine are compounds that inhibit DNA synthesis in *N. crassa* (Section 5.3. *[153]*). HU (and possibly histidine) interacts with the free radical in ribonucleotide reductase, the enzyme responsible for the formation of deoxyribonucleotides. Inhibition of ribonucleotide reductase leads to lowered dNTP pools, DNA strand breaks, and chromosomal instability in a wide variety of organisms *(95)*. Histidine has been shown to produce DNA strand breaks in *N. crassa*, and the steady-state number of nicks is correlated with histidine sensitivity *(73)*. Again, this suggests that the persistence of single-strand breaks might be the basis of the histidine and HU sensitivity, and the mitotic instability of these mutants.

When tested for integration of exogenous *mtr* sequences (Section 2.3.), four mutagen-sensitive strains, *uvs-2*, *uvs-3*, *uvs-6*, and *mei-3*, showed reduced homologous integration, whereas three others, *mei-2*, *mus-9*, and *mus-11*, appeared to have wild-type levels of homologous integration. However, in each case, some integrants showed anomalous behavior not seen in homologous integrants in wild-type *(140)*. These data suggest that some members of each of the three main epistasis groups may affect processes involved in recombination. Perhaps this is not as surprising as it appears initially, since the *S. cerevisiae* RAD6 group, which is homologous to the *N. crassa* Uvs-2 group, is thought to be involved in directing repair to appropriate points along the chromosome *(5)*; the Uvs-6 group is quite likely to be involved with recombinational repair; and the Uvs-3 group shows the high spontaneous mutation rates coupled with the lack of induced mutation typical of *E. coli recA (170)* and *S. cerevisiae rad6* mutants.

4.6. Photoreactivation

N. crassa has efficient photoreactivation repair *(14,58)*. The product of the photolyase gene has been characterized by isolating the gene using PCR probes to the most conserved amino acid sequences of other photolyase genes *(173)*. Like all known photolyases, the *N. crassa* enzyme contains FAD as the photochemically active chromophore. The secondary chromophores, which act as energy-harvesting antenna pigments, divide the photolyases of microorganisms into two major types: those whose secondary chromophore contains a pterin, 5,10-methylenyltetrahydrofolate, and those that contain 8-hydroxy-5-deazaflavin. Based on structural similarities and on studies of the *N. crassa* photolyase expressed in *E. coli*, the *N. crassa* enzyme more strongly resembles the pterin-type photolyases of *E. coli* and *S. cerevisiae* than the *Anacystis nidulans* deazaflavin-containing photolyase. The *N. crassa* photolyase gene, *phr*, specifies a 615 amino acid polypeptide with a predicted mass of 69.9 kDa. Photolyase from partially purified hyphal extracts showed a very similar molecular weight *(136)*. When the cloned *N. crassa* photolyase gene is expressed in *E. coli*, an absorption maximum of 391 nm is observed, which is higher than the 380 nm absorption maximum observed for other pterin-type photolyases. An unusual absorption shoulder at 465 nm is also seen; this shoulder appears to occur because the FAD chromophore is present primarily in an oxidized form in dark-adapted enzyme, which is rapidly converted to the reduced form on exposure to light *(45,173)*.

In the conidia of three mutants, *upr-1*, *uvs-3*, and *mus-26*, photoreactivation is less effective in increasing survival than in wild-type *(82,130,167)*. However, the photolyase enzyme appears to be normal in both *upr-1* and *uvs-3 (131,165)* as does reversal of dimers in spheroplasts *(172)*, indicating that effects on photoreactivation in these mutant backgrounds may be owing to differences in competition for dimers with other enzymes in the cell. A *phr* mutant has been obtained by using RIP (Section 6.1.) and the mutant shows a defect only in photoreactivation, not dark repair *(75)*.

5. OTHER OBSERVATIONS ON DNA REPAIR MUTANTS

5.1. *Heterokaryon Rescue*

Forced heterokaryons have been used to test the ability of wild-type to rescue mutagen-sensitive strains after mutagen damage. When conidia from a *uvs/uvs*$^+$ heterokaryon are UV-irradiated and examined for survival, the heterokaryotic conidia with wild-type and either *upr-1*, *uvs-2*, *uvs-3*, or *uvs-4* nuclear components show levels of survival similar to homokaryotic wild-type conidia, whereas heterokaryotic conidia containing wild-type and *uvs-6* nuclei are more sensitive than wild-type, although clearly not as sensitive as conidia from a *uvs-6* strain *(145,158,159)*. Therefore, the wild-type products of these genes can diffuse into the nuclei that do not contain the functional gene, and the alleles are fully recessive with the exception of *uvs-6*. In the cases of *uvs-2* and *uvs-3*, the gene products appear likely to be packaged in the conidia, because homokaryotic mutagen-sensitive conidia produced by a heterokaryon, and separated from the other types of conidia before irradiation, show a much greater survival than conidia produced from mutant homokaryotic cultures that are exposed to the same dose. However, homokaryotic *uvs-6* conidia from a *uvs-6*/$^+$ heterokaryon are not rescued *(158)*. Less exacting tests suggest that homokaryotic mutant conidia from a *uvs-4*/$^+$ strain heterokaryon can be rescued, but the homokaryotic *upr-1* conidia from a *upr-1*/$^+$ heterokaryon are not completely rescued *(145)*. Exposure of conidia to γ-irradiation yielded similar effects in experiments analyzing *upr-1*, *uvs-4*, and *uvs-6 (146)*.

5.2. Radiation Sensitivity and Cell-Cycle Control

Coordination of cell-cycle events is very important for normal cell growth. All eukaryotes appear to be able to respond to disruptive events, including DNA damage, by stopping the cell cycle at checkpoints in each of the major phases of the cell cycle (*see* Chapters 17, this vol. and Chapter 21, vol. 2). Mutagen-sensitive mutants in budding and fission yeast and in mammals are known which are unable to stop the cell cycle after suffering DNA damage *(65)*, including the human recessive disease, ataxia telangiectasia, in which cells are unable to repress replicon initiation after damage that causes DNA strand breaks.

Analysis of DNA replication in *N. crassa* indicates that wild-type cells halt DNA synthesis in response to DNA damage by stopping the initiation of new replicons *(94)*. Two of the γ-ray-sensitive mutants of *N. crassa*, *uvs-6* and *mus-9*, show no replicon initiation checkpoint control in response to γ-rays or bleomycin *(94)*. The difficulty of synchronizing Neurospora has prevented examination of the G1/S or G2/M checkpoints in mutagen-sensitive mutants. However, newly developed techniques to induce synchronized DNA synthesis in germinating macroconidia *(126)* should now allow such experiments. There is also evidence that microconidia (Fig. 1) are stopped in G1 *(44)*, which may allow utilization of these cells for cell-cycle studies. The complex effects of DNA damage on cell cycle may also partially account for the variety of effects on protein induction, size, and charge changes observed after exposure of germinating conidia to a wide range of DNA-damaging agents (Section 7.2. *[74]*).

5.3. Effects of Histidine and HU on dNTP Pools

Normal cell function requires the maintenance of a proper balance between dNTP pools. Lowered pool sizes or altered pool ratios lower the fidelity of DNA replication, and both mutation and increase chromosomal instability. Maintenance of the proper pool ratios is very complex, and it seems quite likely that some mutagen-sensitive mutants will have important roles in maintaining dNTP pools. Supporting this hypothesis is the known damage inducibility of a key enzyme, ribonucleotide reductase, which catalyzes the reduction of ribonucleotides to deoxyribonucleotides. This enzyme has a free radical in its active site and is inhibited by the free radical scavenger HU *(46,123)*. Many of the mutagen-sensitive mutants in *N. crassa* are sensitive to HU. In every case, they are also sensitive to the amino acid histidine, which can also act as a free radical scavenger. With one exception, *uvs-4*, they also show increased mitotic chromosomal instability in the duplication test (Section 2.3. *[135]*).

No mutagen-sensitive mutants in other organisms have been reported to be sensitive to histidine, although some are unusually sensitive to HU. Early efforts to understand this sensitivity showed that even wild-type *N. crassa* is slightly sensitive to high concentrations of histidine and shows mitotic chromosomal instability when grown with histidine or HU *(103,106,134)*. One mutant, *uvs-6*, is also very sensitive to ethanol, adenine sulfate, and several other, but not all, amino acids. This mutant normally shows a stop and start growth pattern that is enhanced by histidine. Histidine also induces stop-start growth in the *mei-3* mutant, where it is not normally present, but not in *uvs-3*, which occasionally shows start–stop growth *(103)* without histidine. Histidine has also been shown to cause non-NER-dependent repairable breaks in the DNA of germinating conidia *(73)*. The lack of reports of histidine toxicity in wild-type or mutagen-sensitive mutants in other organisms may be in part the result of the unusual histidine transport in *N. crassa*, since histidine is transported by all three known amino acid transport systems *(112)*, which may allow accumulation of unusually high levels.

To explore the possible connection between dNTP pools and phenotypes of mutagen-sensitive mutants, Srivastava and colleagues examined dNTP pools in wild-type and mutagen-sensitive mutants *(152,153)*. Pools in stationary cultures were about one-half the size of those in rapidly growing cultures. The pool sizes varied from about 4.5 pmol/μg DNA for dGTP in stationary cultures to 29 pmol/μg DNA for dTTP in rapidly growing cultures. This compares well with the smaller values reported for mammalian cells *(123)*. Deoxypyrimidine pools were 1.5–2-fold larger than deoxypurine pools, rather than the three to fivefold difference reported for mammalian cells. This is presumably owing to the inability of *Neurospora* to salvage deoxypyrimidine ribonucleosides by phosphorylation because of the lack of thymidine kinase *(61)*.

When either 11 m*M* histidine or 30 m*M* HU were added to rapidly growing, nonagitated cultures for 2 h, the dGTP pool dropped by approx 40%, whereas the other pools remained the same. In agitated cultures of germinated conidia, DNA synthesis stopped immediately upon the addition of 30 m*M* HU, but resumed slowly after approx 1 h. After the addition of 11 or 44 m*M* histidine to the medium, DNA synthesis also stopped, but after delays of 60 and 30 min, respectively. Since histidine has been reported to cause breaks in DNA *(73)*, this result is consistent with activation of the replicon initiation checkpoint which permits elongation, but blocks initiation. Alternatively, histidine might affect ribonucleotide reductase activity, as does HU, but require more time to reach sufficient levels to inhibit new DNA synthesis by affecting precursor availability. In either case, a series of complex interactions would be initiated that might explain the induction or increase in at least 11 polypeptides seen on exposure of germinating conidia to histidine (Section 7.2.). Interestingly, two of these polypeptides cannot be induced by histidine in the four mutagen-sensitive mutants tested in the Uvs-3 and Uvs-6 epistasis groups *(73)*, although they are induced by other agents in these strains *(74)*.

The dNTP pools were also examined in nine mutagen-sensitive *N. crassa* mutants; eight of those chosen show chromosome instability and sensitivity to histidine and HU, and one, *uvs-2*, does not, although it is sensitive to γ-irradiation. Three patterns were seen. The dNTP pools and ratios were not altered in the *uvs-6* mutant. In *uvs-2*, *mus-11* and *mus-16* mutants, pool ratios were not greatly altered, but pool sizes, especially in rapidly growing cells, were consistently 25-50% larger than seen in wild-type. In the remaining mutants, *mei-2*, *mei-3*, *mus-7*, *mus-9*, and *uvs-3*, pool ratios were altered, with a relative increase in deoxypurine pools, and frequently, a relative decrease in the dCTP pool. All but *uvs-3* also showed an increase in overall pool sizes *(152)*. The amounts of DNA/mg dry wt of hyphae were also altered in many of the mutants. The *mei-2*, *uvs-2*, and *uvs-6* mutants had less DNA than wild-type, while, except for *mus-9*, the remaining mutants had 50% more DNA. There is also some evidence that topoisomerase II activity is greater in *mus-9*, *uvs-2*, and *uvs-6* mutants *(137)*. This may be a further indication of problems with DNA synthesis in these mutants. The complexity of the results makes it difficult to know if these changes are primary or secondary to the defects in the mutants. Again, it is striking that there is no correlation with epistasis grouping. Some, but not all members of each of the three major epistasis groups are affected, and more than one pattern may be seen within a group, suggesting that many of these changes will prove to be secondary effects. Isolation of the genes involved will almost certainly be needed to understand the basis for these effects.

5.4. Endo-Exonuclease: A Possible Repair Enzyme

The major alkaline DNase activity in *N. crassa* is part of a complex endo-exonuclease *(27,51)*. This protein has endonuclease activity on single-stranded DNA and RNA, and exonuclease activity with a strict 5' → 3' polarity on linear double-stranded DNA. It has been suggested that this enzyme is a DNA repair and recombination enzyme on the basis of the similarity of its activities to those of *E. coli* ExoV, the nuclease encoded by the *recBCD* genes (*see* Chapter 8). Although the activities are not identical, the *N. crassa* enzyme can produce single-stranded substrates appropriate for recombination in vitro *(51)*.

The *N. crassa* endo-exonuclease is subject to proteolysis in vitro and probably in vivo. Using antibodies to an active 31-kDa protein purified to near homogeneity *(27)*, an inactive precursor form of 93 kDa, and active forms of 76, 42, 37, and 31 kDa were found. Trypsin digestion activates the 93-kDa form. A GTP-binding protein inhibitor of the active forms is also present. The enzyme is found in extracts of nuclei, mitochondria, and vacuoles. In mitochondria, most is tightly bound to the inner membrane. In nuclei, approx 80% of the enzyme is bound to chromatin, with 20% very tightly bound to the nuclear matrix fraction that remains after 2*M* salt extraction. Similar activities and crossreacting material are found in a number of organisms, including other fungi and humans. Although the activities and molecular weights of the various forms are apparently identical in *N. crassa*, in *S. cerevisiae* the mitochondrial (Nuc) and nuclear (Nud) enzymes are encoded by different genes and the predicted protein of the *NUD1* gene is twice the size of that from the *NUC1* gene *(51)*.

The possibility that the endo-exonuclease is a repair enzyme is further supported by a number of observations. Extracts from both the *N. crassa uvs-3* and the *S. cerevisiae rad52* mutants have substantially reduced amounts of inactive and active enzyme. In wild-type *N. crassa*, exposure to the DNA-damaging agent 4NQO causes a dose-dependent loss of inactive nuclear enzyme and an increase of active enzyme at low doses. In the *uvs-3* mutant, an increase in inactive enzyme, but no change in active enzyme, is seen on exposure to 4NQO *(51,122)*. Loss of the mitochondrial form of the enzyme in *S. cerevisiae* does not affect mutagen sensitivity *(51)*, but overexpression of the nuclear gene, *NUD1*, increases resistance to mutagens and increases mitotic recombination. A *rad52 nud1* double mutant is much more sensitive to γ-irradiation than either single mutant *(26)*.

The data suggest that the *N. crassa* endo-exonuclease may be filling a similar repair role to *E. coli* RecC and be evolutionarily related to it. Antibodies raised to the purified *N. crassa* enzyme cross react with both the *S. cerevisiae* enzymes and the *E. coli* RecC enzyme. Furthermore, the C-terminus of the *S. cerevisiae* Nuc1enzyme shows homology to the C-terminus of the RecC polypeptide *(53)*. In *N. crassa* many nicks appear to be site specific and occur at or near the sequence AGCACT, which although unrelated to the *E. coli* Chi site utilized by RecCD, may act in the same way to direct nuclease activity *(52)*. Isolation of the *N. crassa* genes and further studies of mutants and their interaction with mutagen-sensitive mutants in both *N. crassa* and *S. cerevisiae* will be needed to understand the role of this enzyme in DNA repair and recombination.

6. MUTATION IN *N. CRASSA*

6.1. Repeat-Induced Point Mutation (RIP)

N. crassa has an intriguing spontaneous mutational process. During premeiotic stages, tandem or unlinked duplicated sequences are inactivated by mutations that are frequently associated with methylation. This process is known as RIP *(143)*. In RIP, G-C base pairs are converted to AT base pairs *(19)*. Associated with RIPed regions are asymmetrical cytosine methylations *(148)*. A related process, in which duplicated sequences are reversibly inactivated by methylation, but without mutation, is found in the fungus *Ascobolus immersus (124)*.

Little is known about the genes that are required for RIP. RIP appears to require pairing of chromosomes, since in the presence of two copies of a sequence, neither or both may be RIPed, and in the presence of three copies, either none, two, or all are RIPed, but never just one *(148)*. However, RIP does not require *mei-2 (50)*, although *mei-2* is required for meiotic chromosome pairing *(141)*. Since methylation of cytosines is common in duplications in vegetative cells and in regions that have undergone RIP, it is tempting to speculate that the GC to AT mutations produced by RIP arise by methylation of cytosines followed by deamination. However, a mutant strain with no detectable methylation still undergoes RIP, making this an unlikely mechanism *(148)*.

The process of RIP is important to consider in the context of somatic vs germinal mutation *(155)*. Specifically, unstable duplications that revert frequently during somatic, vegetative growth can potentially be fixed in the genome as stable mutations as a consequence of RIP during the sexual cycle.

6.2. Mutation Analysis

6.2.1. Mutation in Wild-Type Strains

Many of the results from the *ad-3* system have been summarized recently *(32)*. Using this system, it is clear that the proportion of small intragenic lesions compared to large multilocus lesions varies greatly with the mutagen. Approximately 85% of spontaneous mutations are small and within the *ad-3* genes, whereas only 15% consist of larger lesions that include essential material between or outside of these genes (Table 4). Mutagens such as MMS, produce almost all small lesions, whereas others, such as hycanthone methanesulfonate (HYC) produce mainly large deletions. Several mutagens, including ICR-170 and 4NQO, have been found to produce only small deletions or point mutations. The extent of the multilocus deletions also varies with the mutagen. For example, virtually no spontaneous or UV-induced large-deletion mutations include the entire *ad-3A*, *nic-2* region of linkage group I, but the majority of HYC-induced lesions result in large deletions of this region.

As with the *ad-3* system, spontaneous mutations found using the *mtr* system were primarily changes within the gene. Only 9 of 75 spontaneous mutations recovered were the result of large deletions extending outside of the *mtr* gene *(41)*. Since this gene has been sequenced, nucleotide changes have been determined directly *(41,155,169)*. Of 31 somatic mutations analyzed, 15 were base substitutions, 11 were deletions, 4 were

Table 4
Percentages of *ad-3* Mutations in Mutagen Treated and Untreated Cultures Resulting from Intragenic (G) Mutations in the *ad3*A or *ad3B* Genes or Deletions (D) into Neighboring Extragenic Regions[a,b]

	No treatment		UV		MMS		EDB		X-rays		DEO		HYC	
Genotype	No.	%	No.	%	No.	%	No.	%	No.	%	No.	%	No.	%
ad-3	167	100	554	100	530	100	387	100	836	100	162	100	107	100
ad-3[Gc]	141	84.9	525	94.8	501	94.5	327	84.5	642	76.8	94	58	19	17.8
$ad\text{-}3A^{G}$	41	24.7	148	26.7	182	34.3	103	26.6	217	26.0	35	21.6	7	6.5
$ad\text{-}3B^{G}$	104	60.2	377	68.1	319	60.2	224	57.9	425	50.8	59	36.4	12	11.2
$ad\text{-}3^{D}$	22	15.1	29	5.2	29	5.5	60	15.5	194	23.2	68	42	88	82.2
$ad\text{-}3A^{D}$	6	3.6	5	0.9	4	0.8	16	4.1	30	3.6	9	5.9	1	0.9
$ad\text{-}3B^{D}$	12	9.0	11	2.0	16	3.0	8	2.1	71	8.5	26	16.1	6	5.6
$ad\text{-}3A\ ad\text{-}3B^{D}$	4	2.4	11	2.0	8	1.5	32	8.2	79	9.4	22	14.4	3	2.8
$ad\text{-}3B\ nic\text{-}2^{D}$	0	0	1	0.2	0	0	0	0	10	1.2	2	1.3	9	8.4
$ad\text{-}3A\ ad\text{-}3B\ nic\text{-}2^{D}$	0	0	1	0.2	1	0.2	4	1.0	4	0.5	9	5.9	69	64.5

[a]The number of mutations per 10^6 survivors (or range of numbers in several experiments using different doses, e.g., UV): No treatment (spontaneous) = 0.4; UV, 11.7–487; MMS, 13.3–366.7; EDB = ethylene dibromide, 19.3; 250 kVp X-irradiation, 5.0–460.1; DEO = 1,2,7,8-diepoxyoctane, 50.1; HYC = hycanthone methanesulfonate, 35.

[b](F. J. de Serres, personal communication; *31,32,37*); data reprinted by permission of the publisher.

[c]The original publications refer to the intragenic (G) mutations as R (gene/point mutations) because the damage was repairable, i.e., conidia homozygous for these mutations were viable on adenine because only the relevant *ad3* gene was damaged. Mutations that had lost essential sequences in the *ad* region (D) or elsewhere in the genome were designated IR (multilocus) for irreparable because they were inviable on adenine.

frameshifts, and 1 was an insertion *(41)*. A different spectrum of mutational events was observed during meiosis *(169)*. First, hotspots for meiotic mutations were found in which there were often short duplications associated with tandem repeats. Duplications were also much more frequently observed as a consequence of meiotic mutation than mitotic mutation, even outside of mutational hotspots, suggesting that defective recombination may be responsible for some meiotic mutations. Finally, an example of RIP was observed, although there were no duplicated *mtr* sequences in the strains, suggesting either that there was a transient duplication event or that RIP does not always require a duplication.

A mutator strain *(mut-1)* was isolated that showed higher frequencies of spontaneous somatic mutation at several genes. Many of the mutations at *mtr* in the *mut-1* strain were –1 frameshifts that suggested increased slippage during DNA replication *(41)*. This mutation maps to a new gene. The mutant has wild-type sensitivity to mutagens, and although the spontaneous mutation rate is elevated at least 10-fold, UV- and X-ray-induced mutation is similar to that of nonmutator strains. Thus, this gene appears to have no effect on DNA repair.

6.2.2. Mutation in Mutagen-Sensitive Strains

Substantial differences in both mutation induction and the spectrum of mutations induced have been seen between wild-type and mutagen-sensitive strains with a variety of mutagens. In the mutagen-sensitive mutants, four different patterns have been seen. Like many excision repair-defective mutants in other organisms, *mus-18* shows wild-type levels of spontaneous mutation induction, accompanied by high levels of UV- and γ-ray-induced mutation, as do two members of the Uvs-2 epistasis group, *upr-1* and *uvs-2*. The other three members of these groups show no change or slightly lower levels of spontaneous and induced mutation *(135)*. However, it should be noted that *mus-26* was only tested with the *ad-8* reversion assay *(76,82)*. All members of the Uvs-6 group show wild-type levels of mutation. In the Uvs-3 group, *mei-3* shows high levels of both spontaneous and UV-induced mutation *(83)*, whereas the other members of this group show high spontaneous mutation rates, but low or no UV- or γ-ray-induced mutation, similar to *E. coli recA* or *S. cerevisiae rad6* mutants *(76,80,135)*. Results with different test systems have agreed well, with two exceptions. In the *mtr* system *uvs-3* appears to give normal levels of UV induced recessive lethal mutations. This may be because of differences in the type of damage or damage processing system producing recessive lethals compared to those producing forward mutations, which appear not to be induced by UV in *uvs-3 (87)*. This possibility is supported by the appearance of some X-ray-induced mutations in *uvs-3 (76)* and its increased mitotic chromosomal instability *(134)*, which could be the result of deletions. However, the X-ray-induced mutations do not appear to be primarily deletions *(35)*. In an *ad-8* reversion assay, *uvs-2* had increased UV induced mutation as expected, but *upr-1* appeared to have low, rather than high induced mutation. Although this may be caused by the complex dose–response pattern of mutation induction in *upr-1 (30)* as suggested *(82)*, this appears unlikely from the published data.

The mutation spectrum of induced *ad-3B* mutants in several mutagen-sensitive strains has been ascertained by intergenic complementation tests based on earlier reversion studies, which show that nonpolarized, complementing mutants have primarily base pair changes, whereas noncomplementing mutants are usually the result of frameshifts or deletions, and polarized patterns result from both types of changes *(35,93)*. Of the mutants tested (*upr-1*, *uvs-2-6*), only *upr-1* and *uvs-2* showed a change in the spectrum of induced mutations. A significant decrease in nonpolarized, complementing mutations accompanied by a large increase in noncomplementing mutations is seen with UV, MMNG, and 4NQO, and in *uvs-2* with many other mutagens *(32)*. This would occur if loss of either of these gene products converts potential base pair changes into deletions, or if DNA damage, which is usually repaired without errors, is converted into deletions. This conclusion is supported by the observation that mutations induced by the frameshift producing agent ICR 170 did not show this change, whereas γ-ray induced mutations, which should include both base damage and strand breaks, showed a small trend in this direction. Unfortunately, no similar analysis of spontaneous mutation appears to have been undertaken, although this would clearly be of considerable interest for understanding the Uvs-3 epistasis group.

6.2.3. UV Dose Rate Effects

The recessive lethal mutation frequency in response to chronic UV treatment is lower than to acute UV treatment when tested in heterokaryons with the *mtr* system *(154)*. Chronic UV irradiation produced approx 25% as many mutations as the same dose given acutely at 156 times the chronic exposure rate. The same mutation reduction was seen in chronically irradiated conidia whether they were held in water at 22°C or allowed to grow. However, when the chronic exposure was done at 0°C, no reduction in mutation rate was seen. When *uvs-2* mutations were present in both components of the heterokaryon, chronic exposure only reduced the mutation rate by a factor of two. These experiments suggest that chronic treatment allows time for DNA damage repair systems to repair damage before dimers in close proximity can accumulate. There are probably at least two systems involved: *uvs-2*-dependent and *uvs-2*- independent. The spectrum of mutations at the *mtr* locus obtained after acute UV treatment also is different from the spectrum obtained after chronic exposure; chronic UV treatment caused mostly point mutations, whereas acute UV produced deletions *(156)*. Interestingly, although there were large differences in mutagenesis between chronic and acute exposure to UV, there was little difference in survival. In general, chronic doses resulted in similar or slightly lower survival rates than acute UV doses.

7. INDUCED RESPONSES TO DNA DAMAGE

7.1. Repair Induction

There is evidence that *N. crassa* has DNA damage-inducible repair systems. The ability of a second nucleus to rescue an irradiated cell was tested by forcing irradiated germinating conidia, carrying appropriate nutritional markers, to form heterokaryons with germinating conidia of a second strain *(157,158)* The irradiated conidia were given a dose of UV irradiation sufficient to kill at least 95% of the cells. When the irradiated strain carried the *uvs-2* mutation, heterokaryon formation with a wild-type strain rescued about four times as many cells as survived without heterokaryon formation. If the rescuing strain was given a low dose of UV, which was nonlethal to the cells, the number of rescued cells increased two- to fourfold. Wild-type or strains carrying *uvs-3* or *uvs-6* could not be rescued by unirradiated or irradiated wild-type conidia. However, *uvs-2*, *uvs-3*, and *uvs-6* strains showed constitutive rescue of *uvs-2* strains. A low dose of UV to the *uvs-6* strain induced rescue, as in wild-type, but there was little induction of rescue by *uvs-2* strains. The constitutive level of rescue by the *uvs-3* strain was as high as the induced level obtained in rescue experiments with wild-type or *uvs-6*, and could not be increased with an inducing dose of UV. Thus, it appears that the damage remaining in the DNA in *uvs-2* mutant strains can be repaired in a heterokaryon formed after irradiation by a system that does not require the *uvs-2*, *3*, or *6* genes. Further repair can be induced by a low UV or X-ray dose. This inducible repair probably requires the *uvs-2* gene and is already induced in unirradiated *uvs-3* conidia. Since the damage induced in *uvs-3* and *uvs-6* nuclei can be at least partially repaired if the nucleus is part of a heterokaryon that contains *uvs*$^+$ nuclei when irradiated (Section 5.1. *[158]*), it appears that this damage must be altered before the heterokaryon can form in the rescue experiments. Surprisingly, although the mutation rate was increased in rescued *uvs-2* nuclei, it was apparently not affected by induction *(158)*.

Using alkaline gradients and T4 endonuclease to follow loss of dimers, Baker *(7)* has provided further evidence for inducible repair systems in wild-type *N. crassa* that are constitutively induced in *uvs-3*. In germinated conidia during the first hour after irradiation, dimer repair is approximately threefold more rapid in growth medium than in phosphate buffer in wild-type. At least half of this increase in repair is inhibited by the addition of cycloheximide. However, in *uvs-3*, repair in both phosphate buffer and growth medium occurs at the same rate as in wild-type cells incubated in growth medium *(72)*.

7.2. Protein Induction

Howard and Baker have examined the effects of DNA damage-inducing conditions on *N. crassa* by using two-dimensional PAGE analyses of whole-cell extracts to examine changes in cellular polypeptide composition *(72–74)*. Of the many changes found, three are noteworthy. One polypeptide doublet (100:93/94) appears to be induced by almost any exposure to stress-inducing agents, including UV, 4NQO, ionizing radiation, paraquat, heat shock and histidine. Activity of a second, 114:52, is increased on exposure to all agents except heat shock. The third, 75:70, shows a shift to a more basic charge on exposure to any agent, but histidine. Expression of other polypeptides appears to be more specifically modulated.

Examination of the effects of *mei-2*, *mei-3*, *upr-1*, *uvs-2*, *uvs-3*, and *uvs-6* on patterns of polypeptide expression in response to UV revealed relatively minor differences from wild-type, usually in net charge, with the exception of the polypeptides mentioned above. UV was unable to induce the 100:93/94 doublet in *upr-1* and *uvs-2*; in *mei-2*, *mei-3*, and *uvs-6*, the 100:93/94 doublet was induced at lower levels of UV irradiation; in *uvs-3*, this doublet was induced to very high levels by low UV doses. The 100:93/94 doublet and the 114:52 polypeptide were also uninducible by histidine in all the mutants above, except *uvs-2* and *upr-1* mutants. Finally, the 75:70 charge-shift was present in untreated *uvs-3* cells. This protein could be shifted back to its normal charge by exposure of *uvs-3* to histidine. Histidine exposure also prevented the UV-induced charge-shift in wild-type. One possibility is that these proteins are important constituents of the inducible repair system, which appears to be constitutive in *uvs-3* (Section 7.1.), and that this system may also be affected in some part in *mei-2*, *mei-3*, and *uvs-6*.

7.3. *Nature of the Induction Pathway*

In mammals, the AP-1 signaling pathway responds to UV-irradiation *(40)*. In *S. cerevisiae*, the transcriptional activator specified by *GCN4*, which has a central role in responding to changes in amino acid availability *(69)*, has also been implicated in the response to UV mediated by a Ras signaling pathway *(47)*. The *N. crassa* gene *cpc-1* has a similar role to *GCN4* in modulating gene expression in response to amino acid availability *(125)*. The sequences of the AP-1 subunit c-Jun, Gcn4p, and Cpc1p polypeptides are similar. Whether the *cpc-1* product has a role in signaling following UV damage remains to be determined. The *cpc-1* gene certainly has a wider role than in amino acid biosynthesis alone: the laccase and tyrosinase genes are also under its control *(75,161)*. *A. nidulans* amino acid auxotrophs are more sensitive than wild-type to exposure to MMS and other DNA-damaging conditions *(43,88)* and it has been suggested that these auxotrophs are deficient in a Ras signaling pathway *(43)*.

8. FUTURE DIRECTIONS

Two striking observations emerge from the results presented in this chapter: (1) *N. crassa*, and perhaps many filamentous fungi (Chapter 21) and the slime molds, repair UV-induced dimers and other bulky DNA damage, but not via a multiprotein NER–complex (*see* Note Added in Proof); and (2) three epistasis groups appear to be involved in the PRR, mutation-induction, and recombination repair processes handled primarily by *recA* of *E. coli* and the *RAD6* and *RAD52* epistasis groups of yeast. Again, this may be true of many filamentous fungi (Chapter 21). These results may not only be interesting from an evolutionary point of view, but also hold considerable practical importance, since fungi are serious plant and animal pathogens. Differences in repair processes may suggest new methods for controlling fungal infections.

The lack of a multiprotein complex active in NER raises many questions. Will *N. crassa* have the gene and strand-specific repair of transcriptionally active and poised chromatin that appear to be of considerable importance in both *S. cerevisiae* and mammals *(55)*? From both its resistance to UV and the biochemical data *(8,9)*, it is clear that *N. crassa* is very adept at rapidly removing (or possibly altering) both CPDs and (6-4)PDs (Section 4.2.). However, loss of all incision at dimers in the *mus-18* mutant has only a slight effect on survival *(84)*. On the other hand, the *uvs-2* mutant, an apparent homolog of the postreplication defective *rad18* mutant of yeast, is extremely sensitive to UV and many other agents *(164)*. Are dimers actually removed after Mus-18p incision? What is the pathway? Is the *uvs-2* gene involved? If not, what does it do to alleviate the effects of UV and other DNA-damaging agents? Some possibilities have been discussed (Section 4.2.1.). With an isolated gene available, it should be possible to answer many questions. Among them are the nature of the defect in the *uvs-2* mutant; the effect of a null mutant produced by RIP; whether the gene products of *uvs-2* and *mus-8*, the apparent *RAD6* homolog *(150)*, interact, as do the Rad18p and Rad6p proteins of *S. cerevisiae*; and, since there are significant differences between *uvs-2* and *RAD18*, whether there are other homologues to *RAD18* in *N. crassa*.

Understanding the roles of the three epistasis groups that appear to be involved in PRR, recombinational repair, and mutation induction will be a daunting task. However, current efforts to create a physical map of the *N. crassa* genome should enable investigators to take advantage of the ease of *N. crassa* genetics, especially when dealing with mutants with very low transformation *(150)*. The development of strains carrying a thymidine kinase gene *(126)* will also facilitate studies of DNA transactions. Perhaps the most interesting group is Uvs-3, which appears to be involved in an error-prone type of repair. Mutants in the one gene studied extensively, *uvs-3*, show lethal interactions with mutants in genes in the Uvs-6 group. Is this gene involved in some basic aspect of DNA metabolism necessary to several types of repair, or is this the default pathway for repair of some type of common endogenous damage normally handled by the Uvs-6 pathway, possibly recombination repair (Section 4.5.)? Use of the excellent *ad-3* mutation analysis system (Section 2.4.) to analyze the high number of spontaneous mutants found in *uvs-3* may also provide important clues.

Another intriguing area is protein induction after exposure to agents that stress cells. Two of the proteins that are induced by nearly every agent (Section 7.2.) are induced in significant amounts in either wild-type or the *uvs-3* mutant. Microsequencing techniques may make it possible to obtain enough information to carry out reverse genetics to isolate the genes coding for these proteins.

Other areas, such as DSB repair, base excision repair, and mismatch repair remain completely unexplored and a large number of mutants have not had fundamental phenotypic characterization or been placed in epistasis groups (Table 1, footnotes). Considering the surprises *N. crassa* has already provided, exploration of these areas will almost certainly prove to be a fascinating undertaking.

NOTE ADDED IN PROOF

Since many multiprotein nuclear exasion repair proteins also play an important role in transcription and/or recombination, homologs are to be expected in *Neurospora*. Recent evidence from one of our laboratories (H.I.) indicates that homologs of *Saccyharomyces cerevisiae RAD1* and *RAD2* genes are present in *Neurospora*.

ACKNOWLEDGMENTS

We thank Chizu Ishii and Dorothy Newmeyer for critical reading of the manuscript, Fred de Serres and David Perkins for many helpful suggestions, and Terry Chow and Namboori Raju for providing unpublished data. This work was supported in part by a US-Japan Cooperative Research grant from the National Science Foundation.

REFERENCES

1. Aboussekhra, A., R. Chanet, A. Adjiri, and F. Fabre. 1992. Semidominant suppressors of *srs2* helicase mutations of *Saccharomyces cerevisiae* map in the *RAD51* gene, whose sequence predicts a protein with similarities to procaryotic RecA proteins. *Mol. Cell. Biol.* **12:** 3224–3234.
2. Akanmu, D., R. Cecchini, O. I. Aruoma, and B. Halliwell. 1991. The antioxidant action of ergothioneine. *Arch. Biochem. Biophys.* **288:** 10–16.
3. Akins, R. A. and A. M. Lambowitz. 1985. General method for cloning *Neurospora crassa* nuclear genes by complementation of mutants. *Mol. Cell. Biol.* **5:** 2272–2278.
4. Bachman, B. J. and W. N. Strickland. 1965. *Neurospora Bibliography and Index.* Yale University Press, New Haven, CT.

5. Bailly, V., J. Lamb, P. Sung, S. Prakash, and L. Prakash. 1994. Specific complex formation between yeast RAD6 and RAD18 proteins: a potential mechanism for targeting RAD6 ubiquitin-conjugating activity to DNA damage sites. *Genes Dev.* **8:** 811–820.
6. Baker, J. M., J. H. Parish, and J. P. E. Curtis. 1984. DNA-DNA and DNA-protein crosslinking and repair in *Neurospora crassa* following exposure to nitrogen mustard. *Mutat. Res.* **132:** 171–179.
7. Baker, T. I. 1983. Inducible nucleotide excision repair in *Neurospora. Mol. Gen. Genet.* **190:** 295–299.
8. Baker, T. I., C. E. Cords, C. A. Howard, and R. J. Radloff. 1990. The nucleotide excision repair epistasis group in *Neurospora crassa. Curr. Genet.* **18:** 207–209.
9. Baker, T. I., R. J. Radloff, C. E. Cords, S. R. Engel, and D. L. Mitchell. 1991. The induction and repair of (6–4) photoproducts in *Neurospora crassa. Mutat. Res.* **255:** 211–228.
10. Beadle, G. W. and E. L. Tatum. 1941. Genetic control of biochemical reactions in *Neurospora. Proc. Natl. Acad. Sci. USA* **27:** 499–506.
11. Boyce, R. P. and P. Howard-Flanders. 1964. Genetic control of DNA breakdown and repair in *E. coli* K-12. *Z. Vererbungsl.* **95:** 345–350.
12. Boyce, R. P. and P. Howard-Flanders. 1964. Release of ultraviolet light-induced thymine dimers from DNA in *E. coli* K-12. *Proc. Natl. Acad. Sci. USA* **51:** 293–300.
13. Bronner, C. E., D. L. Welker, and R. A. Deering. 1992. Mutations affecting sensitivity of the cellular slime mold *Dictyostelium discoideum* to DNA-damaging agents. *Mutat. Res.* **274:** 187–200.
14. Brown, J. S. 1951. The effect of photoreactivation on mutation frequency in *Neurospora. J. Bacteriol.* **62:** 163–167.
15. Brunelli, J. P. and M. L. Pall. 1993. A series of yeast/*Escherichia coli* λ expression vectors designed for directional cloning of cDNAs and *cre/lox*-mediated plasmid excision. *Yeast* **9:** 1309–1318.
16. Calza, R. E. and A. L. Schroeder. 1982. Postreplication repair in *Neurospora crassa. Mol. Gen. Genet.* **185:** 111–119.
17. Calza, R. E. and A. L. Schroeder. 1982. The role of pyrimidine dimers in postreplication repair in *Neurospora. Mol. Gen. Genet.* **186:** 127–134.
18. Calza, R. E. and A. L. Schroeder. 1983. Release of high molecular weight DNA from *Neurospora crassa* using enzymic digestions. *J. Gen. Microbiol.* **129:** 413–422.
19. Cambareri, E. B., B. C. Jensen, E. Schabtach, and E. U. Selker. 1989. Repeat-induced G-C to A-T mutations in *Neurospora. Science* **244:** 1571–1575.
20. Chang, L.-T. and R. W. Tuveson. 1967. Ultraviolet-sensitive mutants in *Neurospora crassa. Genetics* **56:** 801–810.
21. Chang, L.-T., R. W. Tuveson, and M. H. Munroe. 1968. Non-nuclear inheritance of UV sensitivity in *Neurospora crassa. Can. J. Genet. Cytol.* **10:** 920–927.
22. Chary, P., D. Dillon, A. L. Schroeder, and D. O. Natvig. 1994. Superoxide dismutase *(sod-1)* null mutants of *Neurospora crassa*: oxidative stress sensitivity, spontaneous mutation rate and response to mutagens. *Genetics* **137:** 723–730.
23. Chary, P., R. A. Hallewell, and D. O. Natvig. 1990. Structure, exon pattern, and chromosome mapping of the gene for cytosolic copper-zinc superoxide dismutase *(sod-1)* from *Neurospora crassa. J. Biol. Chem.* **265:** 18,961–18,967.
24. Chary, P. and D. O. Natvig. 1989. Evidence for three differentially regulated catalase genes in *Neurospora crassa*: effects of oxidative stress, heat shock, and development. *J. Bacteriol.* **171:** 2646–2652.
25. Cheng, R., T. I. Baker, C. E. Cords, and R. J. Radloff. 1993. *mei-3*, a recombination and repair gene of *Neurospora crassa*, encodes a RecA-like protein. *Mutat. Res., DNA Repair* **294:** 223–234.
26. Chow, T. Y.-K. Unpublished data.

27. Chow, T. Y.-K. and M. J. Fraser. 1983. Purification and properties of single strand DNA-binding endo-exonuclease of *Neurospora crassa*. *J. Biol. Chem.* **258:** 12,010–12,018.
28. Clark, A. J. and A. D. Margulies. 1965. Isolation and characterization of recombination-deficient mutants of *Escherichia coli* K-12. *Proc. Natl. Acad. Sci. USA* **53:** 451–459.
29. Davis, R. H. 1995. Genetics of *Neurospora*, in *The Mycota: Genetics and Biotechnology*, vol. II (Kück, U., ed.), Springer-Verlag, Berlin, pp. 3–18.
30. de Serres, F. J. 1980. Mutagenesis at the *ad-3A* and *ad-3B* loci in haploid UV-sensitive strains of *Neurospora crassa*. II. Comparison of dose-response curves for inactivation and mutation induced by UV. *Mutat. Res.* **71:** 181–191.
31. de Serres, F. J. 1989. X-ray-induced specific locus mutation in the *ad-3* region of two-component heterokaryons of *Neurospora crassa*. II. More extensive genetic tests reveal an unexpectedly high frequency of multiple-locus mutations. *Mutat. Res.* **210:** 281–290.
32. de Serres, F. J. 1992. Characteristics of spontaneous and induced specific-locus mutation in the *ad-3* region of *Neurospora crassa*: utilization in genetic risk assessment. *Environ. Molec. Mutagen.* **20:** 246–259.
33. de Serres, F. J. 1992. Development of a specific-locus assay in the *ad-3* region of two-component heterokaryons of *Neurospora*: a review. *Environ. Mol. Mutagen.* **20:** 225–245.
34. de Serres, F. J., H. Inoue, and M. E. Schüpbach. 1980. Mutagenesis at the *ad-3A* and *ad-3B* loci in haploid UV-sensitive strains of *Neurospora crassa*. I. Development of isogenic strains and spontaneous mutability. *Mutat. Res.* **71:** 53–65.
35. de Serres, F. J., H. Inoue, and M. E. Schüpbach. 1983. Mutagenesis at the *ad-3A* and *ad-3B* loci in haploid UV-sensitive strains of *Neurospora crassa*. VI. Genetic characterization of *ad-3* mutants provides evidence for qualitative differences in the spectrum of genetic alterations between wild-type and nucleotide excision-repair-deficient strains. *Mutat. Res.* **108:** 93–108.
36. de Serres, F. J., and H. G. Kølmark. 1958. A direct method for determination of forward mutation rates in *Neurospora crassa*. *Nature* **182:** 1249,1250.
37. de Serres, F. J., H. V. Malling, H. E. Brockman, and B. B. Webber. 1995. Quantitative and qualitative aspects of spontaneous specific-locus mutation in the *ad-3* region of heterokaryon 12 of *Neurospora crassa*. *Mutat. Res.* **332:** 45–54.
38. DeLange, A. M., and N. C. Mishra. 1981. The isolation of MMS- and histidine-sensitive mutants in *Neurospora crassa*. *Genetics* **97:** 247–259.
39. DeLange, A. M. and N. C. Mishra. 1982. Characterization of MMS-sensitive mutants of *Neurospora crassa*. *Mut. Res.* **96:** 187–199.
40. Devary, Y., R. A. Gottlieb, T. Smeal, and M. Karin. 1992. The mammalian ultraviolet response is triggered by activation of Src tyrosine kinases. *Cell* **71:** 1081–1091.
41. Dillon, D. and D. Stadler. 1994. Spontaneous mutation at the mtr locus in *Neurospora*: the molecular spectrum in wild-type and a mutator strain. *Genetics* **138:** 61–74.
42. Doetsch, P. W. 1995. What's old is new: an alternative DNA excision repair pathway. *Trends Biochem. Sci.* **20:** 284–286.
43. Donnelly, E., Y. A. Barnett, and W. McCullough. 1994. Germinating conidiospores of Aspergillus amino acid auxotrophs are hypersensitive to heat shock, oxidative stress and DNA damage. *FEBS Lett.* **355:** 201–204.
44. Duran, R. and P. M. Gray. 1989. Nuclear DNA, an adjunct to morphology in fungal taxonomy. *Mycotaxon* **36:** 205–219.
45. Eker, A. P. M. 1994. DNA photolyase from the fungus *Neurospora crassa*. Purification, characterization and comparison with other photolyases. *Photochem. Photobiol.* **60:** 125–133.
46. Elledge, S. J., Z. Zhou, J. B. Allen, and T. A. Navas. 1993. DNA damage and cell cycle regulation of ribonucleotide reductase. *BioEssays* **15:** 333–339.
47. Engelberg, D., C. Klein, H. Martinetto, K. Struhl, and M. Karin. 1994. The UV response involving the Ras signaling pathway and AP-1 transcription factors is conserved between yeast and mammals. *Cell* **77:** 381–390.

48. Fincham, J. R. S. 1989. Transformation in fungi. *Microbiol. Rev.* **53:** 148–170.
49. Forsthoefel, A. M. and N. C. Mishra. 1983. Biochemical genetics of *Neurospora* nuclease I: Isolation and characterization of nuclease *(nuc)* mutants. *Genet. Res.* **41:** 271–286.
50. Foss, H. M. and E. U. Selker. 1991. Efficient DNA pairing in a *Neurospora* mutant defective in chromosome pairing. *Mol. Gen. Genet.* **231:** 49–52.
51. Fraser, M. J. 1994. Endo-exonucleases: Enzymes involved in DNA repair and cell death? *BioEssays* **16:** 761–766.
52. Fraser, M. J., Z. Hataher, and X. Huang. 1989. The actions of *Neurospora* endo-exonuclease on double strand DNAs. *J. Biol. Chem.* **264:** 13,093–13,101.
53. Fraser, M. J., H. Koa, and T. Y. Chow. 1990. *Neurospora* endo-exonuclease is immunochemically related to the recC gene product of *Escherichia coli*. *J. Bacteriol.* **172:** 507–510.
54. Freyer, G. A., S. Davey, J. V. Ferrer, A. M. Martin, D. Beach, and P. W. Doetsch. 1995. An alternative eukaryotic DNA excision repair pathway. *Mol. Cell. Biol.* **15:** 4572–4577.
55. Friedberg, E. C., G. C. Walker, and W. Siede. 1995. *DNA Repair and Mutagenesis.* ASM, Washington, DC.
56. Galloway, A. M., M. Liuzzi, and M. C. Paterson. 1994. Metabolic processing of cyclobutyl pyrimidine dimers and (6–4) photoproducts in UV-treated human cells. *J. Biol. Chem.* **269:** 974–980.
57. Glass, N. L., J. Grotelueschen, and R. L. Metzenberg. 1990. *Neurospora crassa* A mating-type region. *Proc. Natl. Acad. Sci. USA* **87:** 4912–4916.
58. Goodgal, S. H. 1950. The effect of photoreactivation on the frequency of ultraviolet induced morphological mutations in the microconidial strain of *Neurospora crassa*. *Genetics* **35:** 667.
59. Griffiths, A. J. F. 1995. Natural plasmids of filamentous fungi. *Microbiol. Rev.* **59:** 673–685.
60. Griffiths, A. J. F., X. Yang, F. J. Debets, and Y. Wei. 1995. Plasmids in natural populations of *Neurospora*. *Can. J. Botany* **73(Suppl. 1A–D):** S186–S192.
61. Grivell, A. R. and J. F. Jackson. 1968. Thymidine kinase: evidence for its absence from *Neurospora crassa* and some other micro-organisms, and the relevance of this to the specific labelling of deoxyribonucleic acid. *J. Gen. Microbiol.* **54:** 307–317.
62. Hanawalt, P. C., P. K. Cooper, A. K. Ganesan, and C. A. Smith. 1979. DNA repair in bacteria and mammalian cells. *Ann. Rev. Biochem.* **48:** 783–836.
63. Handa, N., P. Ballario, A. Cabibbo, G. Macino, and H. Inoue. 1992. Cloning of *mus-25*, a DNA repair gene of *Neurospora crassa*. *Jpn. J. Genet.* **67:** 574.
64. Hartman, Z. and P. E. Hartman. 1987. Interception of some direct-acting mutagens by ergothioneine. *Environ. Mol. Mutagen.* **10:** 3–15.
65. Hartwell, L. H. and M. B. Kastan. 1994. Cell cycle control and cancer. *Science* **266:** 1821–1828.
66. Hatakeyama, S., C. Ishii, and H. Inoue. 1995. Identification and expression of *Neurospora crassa mei-3* gene which encodes a protein homologous to the Rad51 of *Saccharomyces cerevisiae*. *Mol. Gen. Genet.* **249:** 439–446.
67. Hausner, G., S. Stoltzner, S. K. Hubert, K. A. Nummy, and H. Bertrand. 1995. Abnormal mitochondrial DNA in *uvs-4* and *uvs-5* mutants of *Neurospora crassa*. *Fungal Genet. Newsl.* **42A:** 59.
68. Haynes, R. H. and B. A. Kunz. 1981. DNA repair and mutagenesis in yeast, in *The Molecular Biology of the Yeast* Saccharomyces: *Life Cycle and Inheritance*, vol. 1 (Strathern, J. N., E. W. Jones, and J. R. Broach, eds.), Cold Spring Harbor Laboratory, Cold Spring Harbor, N.Y, pp. 371–414.
69. Hinnebusch, A. G. 1992. General and pathway-specific regulatory mechanisms controlling the synthesis of amino acid biosynthetic enzymes in *Saccharomyces cerevisiae*, in *The Molecular and Cellular Biology of the Yeast* Saccharomyces, vol. 2 (Jones, E. W., J. R. Pringle, and J. R. Broach, eds.), Cold Spring Harbor Laboratory, Cold Spring Harbor, NY, pp. 319–414.

70. Holliday, R. 1965. Radiation sensitive mutants of *Ustilago maydis*. *Mutat. Res.* **2:** 557–559.
71. Holliday, R. 1967. Altered recombination frequencies in radiation sensitive strains of *Ustilago*. *Mutat. Res.* **4:** 275–288.
72. Howard, C. A. and T. I. Baker. 1986. Identification of DNA repair and damage induced proteins from *Neurospora crassa*. *Mol. Gen. Genet.* **203:** 462–467.
73. Howard, C. A. and T. I. Baker. 1988. Relationship of histidine sensitivity to DNA damage and stress induced responses in mutagen sensitive mutants of *Neurospora crassa*. *Curr. Genet.* **13:** 391–399.
74. Howard, C. A. and T. I. Baker. 1989. Inducible responses to DNA damaging or stress inducing agents in *Neurospora crassa*. *Curr. Genet.* **15:** 47–55.
75. Inoue, H. Unpublished data.
76. Inoue, H., R. C. Harvey, D. F. Callen, and F. J. de-Serres. 1981. Mutagenesis at the *ad-3A* and *ad-3B* loci in haploid UV-sensitive strains of *Neurospora crassa*. V. Comparison of dose-response curves of single- and double-mutant strains with wild-type. *Mutat. Res.* **84:** 49–71.
77. Inoue, H. and C. Ishii. 1984. Isolation and characterization of MMS-sensitive mutants of *Neurospora crassa*. *Mutat. Res.* **125:** 185–194.
78. Inoue, H. and C. Ishii. 1985. A new ultraviolet-light sensitive mutant of *Neurospora crassa* with unusual photoreactivation property. *Mutat. Res.* **152:** 161–168.
79. Inoue, H. and C. Ishii. 1990. Designation of newly identified mutagen-sensitive mutations in *Neurospora crassa* and their linkage data. *Fungal Genet. Newsletter* **37:** 20.
80. Inoue, H., T. M. Ong, and F. J. de-Serres. 1981. Mutagenesis at the *ad-3A* and *ad-3B* loci in haploid UV-sensitive strains of *Neurospora crassa*. IV. Comparison of dose-response curves for MNNG, 4NQO and ICR-170 induced inactivation and mutation-induction. *Mutat. Res.* **80:** 27–41.
81. Inoue, H. and A. L. Schroeder. 1988. A new mutagen-sensitive mutant in *Neurospora*, *mus-16*. *Mutat. Res.* **194:** 9–16.
82. Ishii, C. and H. Inoue. 1989. Epistasis, photoreactivation and mutagen sensitivity of DNA repair mutants *upr-1* and *mus-26* in *Neurospora crassa*. *Mutat. Res.* **218:** 95–103.
83. Ishii, C. and H. Inoue. 1994. Mutagenesis and epistatic grouping of the *Neurospora* meiotic mutants, *mei-2* and *mei-3*, which are sensitive to mutagens. *Mutat. Res.* **315:** 249–259.
84. Ishii, C., K. Nakamura, and H. Inoue. 1991. A novel phenotype of an excision-repair mutant in *Neurospora crassa*: mutagen sensitivity of the mutant is specific to UV. *Mol. Gen. Genet.* **228:** 33–39.
85. Kafer, E. 1981. Mutagen sensitivities and mutator effects of MMS-sensitive mutants in *Neurospora*. *Mutat. Res.* **80:** 43–64.
86. Kafer, E. 1983. Epistatic grouping of DNA repair-deficient mutants in *Neurospora*: comparative analysis of two *uvs-3* alleles and *uvs-6*, and their *mus* double mutant strains. *Genetics* **105:** 19–33.
87. Kafer, E. 1984. UV-induced recessive lethals in *uvs* strains of *Neurospora* which are deficient in UV mutagenesis. *Mutat. Res.* **128:** 137–146.
88. Kafer, E. 1987. MMS sensitivity of all amino acid-requiring mutants in *Aspergillus* and its suppression by mutations in a single gene. *Genetics* **115:** 671–676.
89. Kafer, E., and M. Fraser. 1979. Isolation and genetic analysis of nuclease halo *(nuh)* mutants in *Neurospora*. *Mol. Gen. Genet.* **169:** 117–127.
90. Kafer, E. and D. Luk. 1988. Properties and strains of additional DNA repair defective mutants in known and new genes of *Neurospora crassa*. *Fungal Genet. Newsletter* **35:** 11–13.
91. Kafer, E. and D. Luk. 1989. Sensitivity to bleomycin and hydrogen peroxide of DNA repair-defective mutants in *Neurospora crassa*. *Mutat. Res.* **217:** 75–81.
92. Kafer, E. and E. Perlmutter. 1980. Isolation and genetic analysis of *MMS*-sensitive *mus* mutants of *Neurospora*. *Can. J. Genet. Cytol.* **22:** 535–552.

93. Kilbey, B. J., F. J. de Serres, and M. H.V. 1971. Identification of the genetic alteration at the molecular level of ultraviolet light-induced *ad-3B* mutants in *Neurospora crassa*. *Mutat. Res.* **12:** 47–56.
94. Koga, S. J. and A. L. Schroeder. 1987. Gamma-ray-sensitive mutants of *Neurospora crassa* with characteristics analogous to ataxia telangiectasia cell lines. *Mutat. Res.* **183:** 139–148.
95. Kunz, B. A. 1982. Genetic effects of deoxyribonucleotide pool imbalances. *Environ. Mutagen.* **4:** 695–725.
96. Lehmann, A. R., A. M. Carr, F. Z. Watts, and J. M. Murray. 1991. DNA repair in the fission yeast, *Schizosaccharomyces pombe. Mutat. Res.* **250:** 205–210.
97. Leslie, J. F. and N. B. Raju. 1985. Recessive mutations from natural populations of *Neurospora crassa* that are expressed in the sexual diplophase. *Genetics* **111:** 759–777.
98. Macleod, H. and D. Stadler. 1986. Excision of pyrimidine dimers from the DNA of *Neurospora*. *Mol. Gen. Genet.* **202:** 321–326.
99. McClintock, B. 1945. Neurospora. I. Preliminary observations of the chromosomes of *Neurospora crassa*. *Am. J. Bot.* **32:** 671–678.
100. Mehta, R. D. and J. Weijer. 1971. U.V. mutability in gamma-ray-sensitive mutants of *Neurospora crassa*, in *IAEA-SM Symposium on the Use of Radiation and Radioisotopes for Genetic Improvement of Industrial Microorganisms*, vol. 134/13, International Atomic Energy Agency, Vienna, pp. 63–71.
101. Mishra, N. C. and A. M. Forsthoefel. 1983. Biochemical genetics of *Neurospora* nuclease II: Mutagen sensitivity and other characteristics of the nuclease mutants. *Genet. Res.* **41:** 287–297.
102. Navarro, R. E., M. A. Stringer, W. Hansberg, W. E. Timberlake, and J. Aguirre. 1996. *catA*, a new *Aspergillus nidulans* gene encoding a developmentally regulated catalase. *Curr. Genet.* **29:** 352–359.
103. Newmeyer, D. 1984. *Neurospora* mutants sensitive both to mutagens and to histidine. *Curr. Genet.* **9:** 65–74.
104. Newmeyer, D. and D. R. Galeazzi. 1977. The instability of *Neurospora* duplication *Dp(IL* $\rightarrow$ *IR) H4250* and its genetic control. *Genetics* **85:** 461–487.
105. Newmeyer, D. and D. R. Galeazzi. 1978. A meiotic UV-sensitive mutant which causes deletion of duplications in *Neurospora*. *Genetics* **89:** 245–269.
106. Newmeyer, D., A. L. Schroeder, and D. R. Galeazzi. 1978. An apparent connection between histidine, recombination and repair in *Neurospora*. *Genetics* **89:** 271–279.
107. Okaichi, K., T. Mori, and M. Ihara. 1995. Unique DNA repair property of an ultraviolet-sensitive *(radC)* mutant of *Dictyostelium discoideum*. *Photochem. Photobiol.* **61:** 281–284.
108. Orbach, M. 1992. Untitled. *Fungal Genet. Newsletter* **39:** 92.
109. Orbach, M. J. 1994. A cosmid with a HyR marker for fungal library construction and screening. *Gene* **150:** 159–162.
110. Orbach, M. J., D. Vollrath, R. W. Davis, and C. Yanofsky. 1988. An electrophoretic karyotype of *Neurospora crassa*. *Mol. Cell. Biol.* **8:** 1469–1473.
111. Pall, M. L. 1969. Amino acid transport in *Neurospora crassa*. 1. Properties of two amino acid transport systems. *Biochim. Biophys. Acta* **173:** 113–127.
112. Pall, M. L. 1970. Amino acid transport in *Neuropora crassa*. *Biochem. Biophys. Acta* **203:** 139–149.
113. Perkins, D. D. 1992. *Neurospora*: the organism behind the molecular revolution. *Genetics* **130:** 687–701.
114. Perkins, D. D. 1993. Use of a helper strain in *Neurospora crassa* to maintain stocks of *uvs-4* and *uvs-5*, which deteriorate unless sheltered in heterokaryons. *Fungal Genetics Newsletter* **40:** 66.
115. Perkins, D. D. and E. G. Barry. 1977. The cytogenetics of *Neurospora*. *Adv. Genet.* **19:** 133–285.

116. Perkins, D. D., A. Radford, D. Newmeyer, and M. Björkman. 1982. Chromosomal loci of *Neurospora crassa*. *Microbiol. Rev.* **46:** 426–570.
117. Prakash, L. 1981. Characterization of postreplication repair in *Saccharomyces cerevisiae* and effect of *rad6*, *rad18*, *rev3* and *rad52* mutations. *Mol. Gen. Genet.* **184:** 471–478.
118. Raju, N. 1986. Postemeiotic mitoses without chromosome replication in a mutagen-sensitive *Neurospora* mutant. *Exp. Mycology* **10:** 243–251.
119. Raju, N. and D. D. Perkins. 1978. Barren perithecia in *Neurospora crassa*. *Can. J. Genet. Cytol.* **20:** 41–59.
120. Raju, N. B. 1984. Use of enlarged cells and nuclei for studying mitosis in *Neurospora*. *Protoplasma* **121:** 87–98.
121. Raju, N. B. Unpublished data.
122. Ramotar, D., A. H. Auchincloss, and M. J. Fraser. 1987. Nuclear endo-exonuclease of *Neurospora crassa*: Evidence for a role in DNA repair. *J. Biol. Chem.* **262:** 425–431.
123. Reichard, P. 1988. Interactions between deoxyribonucleotide and DNA synthesis. *Ann. Rev. Biochem.* **57:** 349–374.
124. Rossignol, J. L. and G. Faugeron. 1994. Gene inactivation triggered by recognition between DNA repeats. *Experientia* **50:** 307–317.
125. Sachs, M. S. 1996. General and cross-pathway controls of amino acid biosynthesis, in *The Mycota: Biochemistry and Molecular Biology*, vol. III (Brambl, R. and G. A. Marzluf, eds.), Springer-Verlag, Heidelberg, pp. 315–345.
126. Sachs, M. S. Unpublished data.
127. Schiestl, R. H. and T. D. Petes. 1991. Integration of DNA fragments by illegitimate recombination in *Saccharomyces cerevisiae*. *Proc. Natl. Acad. Sci. USA* **88:** 7585–7589.
128. Schroeder, A. L. 1968. Ultraviolet-sensitive mutants in *Neurospora crassa*. *Genetics* **60:** 223 (abstract).
129. Schroeder, A. L. 1970. Ultraviolet-sensitive mutants of *Neurospora*. I. Genetic basis and effect on recombination. *Mol. Gen. Genet.* **107:** 291–304.
130. Schroeder, A. L. 1970. Ultraviolet-sensitive mutants of *Neurospora*. II. Radiation studies. *Mol. Gen. Genet.* **107:** 305–320.
131. Schroeder, A. L. 1972. Photoreactivating enzyme in a UV-sensitive *Neurospora* mutant with abnormal photoreactivation. *Genetics* **71:** s56.
132. Schroeder, A. L. 1974. Properties of a UV-sensitive *Neurospora* strain defective in pyrimidine dimer excision. *Mutat. Res.* **24:** 9–16.
133. Schroeder, A. L. 1975. Genetic control of radiation sensitivity and DNA repair in *Neurospora*, in *Molecular Mechanisms for Repair of DNA*, vol. Part B (Hanawalt, P. C. and R. B. Setlow, eds.), Plenum, New York.
134. Schroeder, A. L. 1986. Chromosome instability in mutagen sensitive mutants of *Neurospora*. *Curr. Genet.* **10:** 381–387.
135. Schroeder, A. L. 1988. Use of *Neurospora* to study DNA repair, in *DNA Repair: A Laboratory Manual of Research Procedures*, vol. 3 (Friedberg, E. C. and P. C. Hanawalt, eds.), Marcel Dekker, Inc., New York, pp. 77–98.
136. Schroeder, A. L. Unpublished data.
137. Schroeder, A. L., M. F. Lavin, and S. Bohnet. 1989. Topoisomerase activity assays in *Neurospora*. *Fungal Genet. Newsletter* **36:** 73,74.
138. Schroeder, A. L. and L. D. Olson. 1980. Mutagen-sensitive mutants in *Neurospora*, in *Conference on DNA Repair and Mutagenesis in Eukaryotes* (de Serres, F. J., W. M. Generoso, and M. D. Shelby, eds.), Plenum, New York, pp. 55–62.
139. Schroeder, A. L. and L. D. Olson. 1983. Mutagen sensitivity of *Neurospora* meiotic mutants. *Can. J. Genet. Cytol.* **25:** 16–25.
140. Schroeder, A. L., M. L. Pall, J. Lotzgesell, and J. Siino. 1995. Homologous recombination following transformation in *Neurospora crassa* wild type and mutagen sensitive strains. *Fungal Genet. Newsletter* **42,** in press.

141. Schroeder, A. L. and N. B. Raju. 1991. *mei-2*, a mutagen-sensitive mutant of *Neurospora* defective in chromosome pairing and meiotic recombination. *Mol. Gen. Genet.* **231:** 41–48.
142. Schüpbach, M. E. and F. J. de Serres. 1981. Mutagenesis at the *ad-3A* and *ad-3B* loci in haploid UV-sensitive strains of *Neurospora crassa*. III. Comparison of dose-response curves for inactivation. *Mutat. Res.* **81:** 49–58.
143. Selker, E. U. 1990. Premeiotic instability of repeated sequences in *Neurospora crassa*. *Ann. Rev. Genet.* **24:** 579–613.
144. Serna, L. and D. Stadler. 1978. Nuclear division cycle in germinating conidia of *Neurospora crassa*. *J. Bacteriol.* **136:** 341–351.
145. Shelby, D. M., F. J. de Serres, and G. J. Stine. 1975. Ultraviolet-inactivation of conidia from heterokaryons of *Neurospora crassa* containing UV-sensitive mutations. *Mutat. Res.* **27:** 45–58.
146. Shelby, M. D., G. J. Stine, and F. J. de Serres. 1975. Gamma-ray inactivation of conidia from heterokaryons of *Neurospora crassa* containing UV-sensitive mutations. *Mutat. Res.* **28:** 147–154.
147. Shinohara, A., H. Ogawa, and T. Ogawa. 1992. Rad51 protein involved in repair and recombination in *S. cerevisiae* is a RecA-like protein. *Cell* **69:** 457–470.
148. Singer, M. J., B. A. Marcotte, and E. U. Selker. 1995. DNA methylation associated with repeat-induced point mutation in *Neurospora crassa*. *Mol. Cell. Biol.* **15:** 5586–5597.
149. Smith, D. A. 1975. A mutant affecting meiosis in *Neurospora*. *Genetics* **80:** 125–133.
150. Soshi, T., E. Käfer, and H. Inoue. 1996. The *mus-8* gene of *Neurospora crassa* encodes a structural and functional homolog of the Rad6 protein of *Saccharomyces cerevisiae*. *Curr. Genet.* **30:** 224–231.
151. Springer, M. L. 1993. Genetic control of fungal differentiation: the three sporulation pathways of *Neurospora crassa*. *Bioessays* **15:** 365–374.
152. Srivastava, B., and A. L. Schroeder. 1989. Deoxyribonucleoside triphosphate pools in mutagen sensitive mutants of *Neurospora crassa*. *Biochem. Biophys. Res. Commun.* **162:** 583–590.
153. Srivastava, V. K., M. L. Pall, and A. L. Schroeder. 1988. Deoxyribonucleoside triphosphate pools in *Neurospora crassa*: effects of histidine and hydroxyurea. *Mutat. Res.* **200:** 45–53.
154. Stadler, D. and H. Macleod. 1984. A dose-rate effect in UV mutagenesis in *Neurospora*. *Mutat. Res.* **127:** 39–47.
155. Stadler, D., H. Macleod, and D. Dillon. 1991. Spontaneous mutation at the mtr locus of *Neurospora*: the spectrum of mutant types. *Genetics* **129:** 39–45.
156. Stadler, D., H. Macleod, and M. Loo. 1987. Repair-resistant mutation in *Neurospora*. *Genetics* **116:** 207–214.
157. Stadler, D. and R. Moyer. 1981. Induced repair of genetic damage in *Neurospora*. *Genetics* **98:** 763–774.
158. Stadler, D. R. 1983. Repair and mutation following UV damage in heterokaryons of *Neurospora*. *Mol. Gen. Genet.* **190:** 227–232.
159. Stadler, D. R. and D. A. Smith. 1968. A new mutation in *Neurospora* for sensitivity to ultraviolet. *Can. J. Genet. Cytol.* **10:** 916–919.
160. Takao, M., R. Yonemasu, K. Yamamoto, and A. Yasui. 1996. Characterization of a UV endonuclease gene from the fission yeast *Schizosaccharomyces pombe* and its bacterial homolog. *Nucleic Acids Res.* **24:** 1267–1271.
161. Tamaru, H., T. Nishida, T. Harashima, and H. Inoue. 1994. Transcriptional activation of a cycloheximide-inducible gene encoding laccase is mediated by *cpc1*, the cross-pathway control gene, in *Neurospora crassa*. *Mol. Gen. Genet.* **243:** 548–554.
162. Tanaka, S., C. Ishii, and H. Inoue. 1989. Effects of heat shock on the induction of mutations by chemical mutagens in *Neurospora crassa*. *Mutat. Res.* **223:** 233–242.

163. Thomas, S. A., M. L. Sargent, and R. W. Tuveson. 1981. Inactivation of normal and mutant *Neurospora crassa* conidia by visible light and near-UV: role of $^1O_2$, carotenoid composition and sensitizer location. *Photochem. Photobiol.* **33:** 349–354.
164. Tomita, H., T. Soshi, and H. Inoue. 1993. The *Neurospora* uvs-2 gene encodes a protein which has homology to yeast Rad18 with unique zinc finger motifs. *Mol. Gen. Genet.* **238:** 225–233.
165. Tuveson, R. W. 1972. Genetic and enzymatic analysis of a gene controlling UV sensitivity in *Neurospora crassa. Mutat. Res.* **15:** 411–424.
166. Tuveson, R. W. 1972. Interaction of genes controlling ultraviolet sensitivity in *Neurospora crassa. J. Bacteriol.* **112:** 632–634.
167. Tuveson, R. W. and J. Mangan. 1970. A UV-sensitive mutant of *Neurospora* defective for photoreactivation. *Mutat. Res.* **9:** 455–466.
168. Vollmer, S. J. and C. Yanofsky. 1986. Efficient cloning of genes of *Neurospora crassa. Proc. Natl. Acad. Sci. USA* **83:** 4869–4873.
169. Watters, M. K. and D. R. Stadler. 1995. Spontaneous mutation during the sexual cycle of *Neurospora crassa. Genetics* **139:** 137–145.
170. Witkin, E. M. 1976. Ultraviolet mutagenesis and inducible DNA repair in *Escherichia coli. Bacteriol. Rev.* **40:** 869–907.
171. Worthy, T. E. and J. L. Epler. 1972. Repair of ultraviolet light-induced damage to the deoxyribonucleic acid of *Neurospora crassa. J. Bacteriol.* **110:** 1010–1016.
172. Worthy, T. E. and J. L. Epler. 1973. Biochemical basis of radiation-sensitivity in mutants of *Neurospora crassa. Mutat. Res.* **19:** 167–173.
173. Yajima, H., H. Inoue, A. Oikawa, and A. Yasui. 1991. Cloning and functional characterization of a eucaryotic DNA photolyase from *Neurospora crassa. Nucleic Acids Res.* **19:** 5359–5362.
174. Yajima, H., M. Takao, S. Yasuhira, J. H. Zhao, C. Ishii, H. Inoue, and A. Yasui. 1995. A eukaryotic gene encoding an endonuclease that specifically repairs DNA damaged by ultraviolet light. *EMBO J.* **14:** 2393–2399.
175. Yanofsky, C. 1946. Gene-enzyme relationships, in *The Bacteria*, vol. 5 (Gunsalus, I. C. and R. Y. Stanier, eds.), Academic, New York, pp. 373–417.
176. Yoon, J. H., B. J. Lee, and H. S. Kang. 1995. The *Aspergillus uvsH* gene encodes a product homologous to yeast RAD18 and *Neurospora* UVS-2. *Mol. Gen. Genet.* **248:** 174–181.

23

Pathways of DNA Repair in *Ustilago maydis*

William K. Holloman, Richard L. Bennett, Allyson Cole-Strauss, David O. Ferguson, Kenan Onel, Mara H. Rendi, Michael L. Rice, Michael P. Thelen, and Eric B. Kmiec

1. INTRODUCTION

1.1. *Networks of DNA Repair and Recombination*

An interrelationship between DNA recombination and repair was hypothesized by Holliday in his thinking about the molecular basis for the non-Mendelian genetic phenomenon of gene conversion *(28)* and was later established by the isolation of mutants simultaneously defective in repair and recombination *(12,29,34)*. Since that time, it has become increasingly apparent that recombination and repair processes intersect at a number of distinct and important junctions in the life of a cell. An example is gene conversion, which for many years was an obscure genetic phenomenon in the province of only a handful of fungal geneticists attending to analysis of aberrant marker segregation during meiosis. Studies showing that gene conversion in fungi results from mismatch repair and discoveries that mutations in human homologs of the fungal mismatch repair genes predispose to an inherited form of colon cancer illustrate the complex interplay between recombination and repair processes, and the cell-cycle machinery.

Studies in a variety of experimental organisms have revealed that numerous genes contribute to maintaining the integrity of the DNA. For example, in the fungus *Saccharomyces cerevisiae*, the number of genes already isolated that correspond to mutants in DNA repair runs to several dozen. To arrange an orderly system for thinking about the function of these genes, investigators have grouped them loosely into broad categories based on phenotype and epistasis analyses of the corresponding mutants. These include the excision repair group of genes, the recombinational repair group, and a post-replicational repair group, also known as the error prone repair pathway *(23)*. In *S. cerevisiae*, the excision repair group functions in removal of adducts from photo-damaged DNA. Mutants defective in these genes tend to be primarily UV-sensitive, yet normal in recombination. Mutants in the recombinational repair group are so named because of the profound deficiency in recombination. These mutants are sensitive to ionizing radiation and are defective in repair of broken DNA molecules. Mutants in the third group are heterogeneous with respect to phenotype, and can display sensitivity to both UV and ionizing radiation. Members of this group also appear to contribute to

From: DNA Damage and Repair, Vol. 1: DNA Repair in Prokaryotes and Lower Eukaryotes
Edited by: J. A. Nickoloff and M. F. Hoekstra ©Humana Press Inc., Totowa, NJ

aspects of mutagenesis. Although this general classification scheme does impart some order to the large number of mutants, there are numerous examples that do not appear to fit easily into one or the other category. For instance, certain mutants in both the excision repair group and the postreplicational repair group exhibit defects in recombination. Furthermore, certain mutants defective in repair, such as the mismatch repair mutants, exhibit no radiation sensitivity. Thus, the three categories represent loose associations that can be useful when comparing genes from different organisms, but do not represent stringent boundaries.

Ustilago maydis was developed as a experimental system for studying DNA repair since it is extremely resistant to radiation and thus was considered to have potential for providing a rich source of information on mechanisms of repair. The first radiation-sensitive mutants isolated in eukaryotes were obtained in *U. maydis (29)*. Three of the mutants originally isolated were characterized in some detail and were revealed to fall into three classes *(31)* that could be categorized as those above. The *uvs3* mutant was found to be defective in excision repair, but proficient in recombination *(67)*. The *rec2* mutant was discovered to be quite sensitive to ionizing radiation and deficient in recombination, in line with the recombinational repair group *(31)*. The *rec1* mutant was found to have an extremely complicated phenotype with certain hallmarks found only in the postreplicational repair group *(31,33)*. No epistasis was observed when the mutants were tested in double- and triple-mutant combinations. Thus, by similar criteria that were used in classifying *S. cerevisiae rad* mutants, the *U. maydis* mutants would appear to fall into place with the paradigm of multiple repair pathways.

2. *REC1*

2.1. *Isolation of the* U. maydis rec1 *Mutant and Cloning of the* REC1 *Gene*

The *rec1* mutant was originally isolated in a screen for mutants sensitive to UV light, but was also found to be sensitive to ionizing radiation and chemical alkylating agents. It has an elevated frequency of mitotic recombination in which normal gene conversion is replaced by an aberrant form of crossing over, which is strongly associated with chromosome breakage and loss. The mutant is also blocked in radiation-induced allelic recombination. Homozygous diploids are unstable during mitotic growth, and exhibit considerable lethal sectoring and striking variation in colony size and morphology. Haploids exhibit considerable lethal sectoring during mitotic growth. Spontaneous mutation is elevated as evidenced by an increase in reversion of particular auxotrophic markers and an increase in forward mutation, but no radiation-enhanced reversion or induction of mutation is observed. Meiosis is aberrant with low spore viability, and with frequent production of diploids and aneuploids. This complicated phenotype was difficult to explain in terms of loss of a single gene function and led to the notion that *REC1* encoded a regulatory factor *(33)*.

To gain insight into the function of *REC1*, the gene was cloned *(26,27,66)*, and the sequence was examined for homology to other genes and for functional motifs *(64)*. The genomic sequence indicated the gene product is encoded by an uninterrupted open reading frame with the predicted product of 522 amino acids and mass of 57 kDa. However, no overall sequence homology was found with any other known protein, and no unambiguous functional motifs were immediately obvious. With no clues from the sequence, biochemical studies were initiated with the expectation that, at the minimum, the Rec1 protein should bind to DNA.

The *REC1* gene was overexpressed in *Escherichia coli*, and purified Rec1 protein was indeed active in DNA binding. The key discovery leading to the realization of the biochemical activity inherent in Rec1 came when DNA binding was assayed in the presence of various cations. When Mg^{2+} was added to reactions, DNA binding activity was greatly diminished. At first, this behavior seemed perplexing, since many proteins that interact with DNA are known to require Mg^{2+}, but when the DNA was examined, it became immediately clear that binding was diminished because the DNA was degraded to mononucleotides *(64)*. This demonstration that Rec1 has exonucleolytic activity was an important advancement in understanding of the biological function of the *REC1* gene, and has provided a framework for beginning to rationalize the complicated phenotype of the *rec1* mutant.

A simple model by which a defective exonuclease might account for the radiation-sensitive phenotype of the *rec1* mutant would hold that inability to excise damage from DNA is directly responsible for the DNA repair defect. Along these same lines, the mutator phenotype might be supposed to arise as a consequence of a dysfunctional exonuclease failing to remove misincorporated bases in DNA. However, other aspects of the phenotype are less easily rationalized by this simple model. Furthermore, the normal removal of pyrimidine dimers from UV-irradiated *rec1* cells would argue against the simple model that Rec1 is an excision repair exonuclease.

2.2. *Structure and Regulation of the* REC1 *mRNA*

The *REC1* gene has an uninterrupted open reading frame predicted to encode a protein of 522 amino acid residues. Nevertheless, an intron is present. This was first discovered during mapping of the 3'-end of the mRNA by S1 analysis and then confirmed by sequence analysis of the cDNA *(50)*. Analysis revealed that an intron of 184 nucleotides was removed and that the splicing event resulted in a change in reading frame in the message. A termination codon different from the one apparent in the genomic open reading frame is brought into frame after splicing, resulting in production of a protein of 463 amino acids with a calculated mass of 51 kDa. Functional activity of the gene in mitotic cells requires removal of the intron. This was demonstrated by constructing an altered form of the *REC1* gene designed to prohibit splicing. The alteration was silent in terms of simply substituting one leucine codon for another, but the mutant gene was not spliced correctly and was unable to complement the radiation sensitivity of a *rec1* mutant after transformation. Wild-type cells transformed with the plasmid expressing the splice-site *rec1* mutant RNA were unchanged in resistance to UV irradiation, indicating that there is no dominant negative phenotype. These results indicate that removal of the intron is necessary and sufficient to confer full biological activity in DNA repair proficiency.

There is no evidence for the presence of the unspliced *REC1* mRNA in mitotic cells. This circumstance of an intron present in an uninterrupted open reading frame is out of the ordinary and raised the question of its biological significance. It is possible that the intron is merely a molecular curiosity that might be predicted to occur with some frequency among genes containing small introns, especially if the location was near the 3'-end of the gene. It would appear that there are no other exact parallels in any other biological system. However, a number of fungal genes are known to be spliced so as to remove in-frame stop codons. The other possibility to consider is whether there are situations where the genomic encoded sequence might be translated. The life cycle of *U. maydis is* complicated and proceeds through several developmental changes that include fusion of haploid cells, hyphal formation, growth of heterokaryons, diploidization, formation of teliospores, and teliospore germination. The analysis described in these studies was performed only on RNA from the mitotically growing yeast-like haploid stage, and no difference was observed in the pattern of splicing regardless of how the cells were grown or stressed. It is possible that there might be as yet undiscovered conditions in mitosis, or a stage in the life cycle where the intron is not removed and the gene product is translated from the transcript of the genomic sequence.

Insight into the biological function of DNA metabolic genes in *S. cerevisiae* and *Schizasaccharomyces pombe* has come from analysis of the regulation of gene expression. In general, DNA metabolic genes that function in replication have been found to be periodically expressed during the cell cycle in G1/S phase, whereas those dedicated to DNA repair tend to be expressed constitutively throughout the cell cycle (e.g., *see* ref. *35*). The level of the *REC1* mRNA was measured in synchronously growing cells, and found to fluctuate periodically with accumulation and depletion in concert with cell division *(51)*.

To define the cell cycle in more detail, the periodicity of histone H4 *(HHF)* gene expression was measured. Histone H4 was chosen as a cell-cycle marker, since histone gene expression is tightly controlled and is restricted to G1 and S phase, and since the level of *HHF* mRNA is known to peak during S phase *(25)*. Therefore, the time of the maxims state of *HHF* mRNA accumulation was taken as an S-phase landmark of the *U. maydis* cell cycle. *REC1* gene expression was found to coincide almost precisely with *HHF* gene expression with the peak in accumulation of *REC1* mRNA slightly preceding *HHF* mRNA *(51)*. These results indicate that *REC1* is expressed periodically, most likely during G1/S phase of the cell cycle. This mode of regulated expression is similar to that seen with genes that function during DNA replication.

2.3. Exonuclease Activity Associated with the Rec1 Protein

Several basic reaction parameters were determined for Recl exonuclease activity. The polarity of exonuclease digestion by Rec1 is 3' → 5'. The enzyme was five times more active on single-stranded DNA compared to duplex DNA. Optimal activity was observed at pH 9.0 in the presence of reducing agent. Mg^{2+} was required as a cofactor. When Mn^{2+} was substituted, the activity was only 5%. An unusual property is that Rec1 has high activity on DNA with a 3'-phosphorothioate ester linkage *(64)*.

In view of the exonuclease activity associated with the *REC1* gene product, a search for sequence similarity with other known exonucleases was carried out. The search revealed no known exonucleases with overall homology. However, the segment of *E. coli* DNA polymerase I (residues 250–460) containing the 3' → 5' exonuclease active site was aligned with 209 residues of the N-terminal portion of *REC1* (residues 40–250) with 20% identity in the overlap. This similarity was then localized more specifically to residues in the three so-called Exo motifs (Fig. 1), which are ordered within the domain of the 3' → 5' exonuclease active site conserved in many other DNA polymerases *(7,9)*. Notable among these is the *E. coli dnaQ* gene product, the 3' → 5' proof-

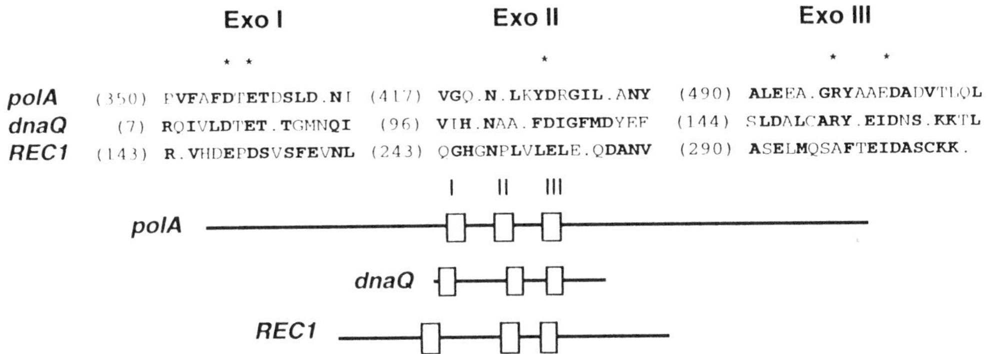

Fig. 1. Comparison of the deduced proteins from the genes encoding *E. coli* DNA polymerase I *(polA)*, ε-subunit of DNA polymerase III holoenzyme *(dnaQ)*, and *U. maydis* Rec1. The positions of the three Exo motif blocks are shown as boxes. The sequences are compared above with bold letters indicating homologies. The numbers indicate the positions in the sequence of the first residue of the block. *Indicates conserved residues that are thought to function in metal coordination based on structural studies in DNA polymerase I.

reading ε subunit of DNA polymerase III holoenzyme (59). The comparison of ε and *REC1* with the DNA pol I 3' → 5' exonuclease active site region indicated the homology in the Exo motifs was higher in the three sequences than with any two of these sequences alone.

The three motifs in DNA polymerases were first identified by crystallographic studies on the Klenow fragment of *E. coli* DNA polymerase I, which revealed amino acid residues essential for metal ion binding, substrate orientation, and catalysis *(6,14,15,22)*. This has been extended from the crystallographic model by sequence alignment to conserved residues in over 30 DNA polymerases *(7)*, although many of these contain little homology in the N-terminal region apart from that observed in the three Exo motifs. That the conserved residues are indeed essential for catalysis has been demonstrated by both genetic and biochemical tests using proteins mutant at specific residues *(14,15)*. Although these three motifs appear to be similar in their sequence composition and spatial arrangement in many DNA polymerases, they appear to be absent in other well-characterized 3' → 5' exonucleases such as *E. coli* exonucleases I and III. It was not possible to make similar homologous alignments of the *REC1* gene product with these latter exonucleases.

2.4. Structure/Function Relationships in the Rec1 Protein

Since it had been established that splicing itself was essential for biological activity, it might have been expected that the exonuclease activity in the spliced gene product would be substantially different from the unspliced gene product. Therefore, the exonuclease activity was measured with various alleles of the gene, including the spliced gene, several deletion mutants constructed by in vitro methods, and strains isolated during mutant hunts (Fig. 2). Deletion mutants ranged in size from 200–457 amino acids in length. Three *rec1* alleles isolated by Holliday had termination codons that generated truncated polypeptides of 400, 157, and 159 amino acids for the *rec1-1*, *rec1-2*,

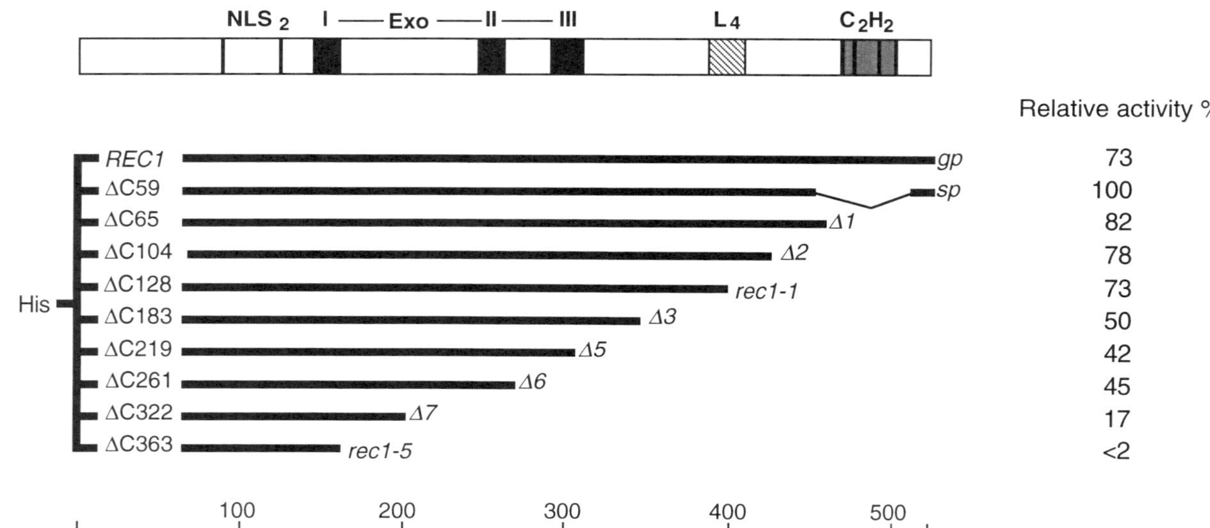

Fig. 2. Schematic representation of the Rec1 protein. Blocks indicate the two nuclear localization signals, the three Exo motifs, the leucine-rich sequence, and the putative Zn finger. Lengths are shown for deletions mutants generated and for *rec1* alleles isolated from mutant hunts. The polypeptides were expressed as fusions with a hexa-histidine leader sequence, and then purified by adsorption onto immobilized Ni^{2+} affinity resin. After specific elusion with imidazole buffer, the exonuclease activity was assayed on 3'-^{35}S-labeled DNA. The relative activity indicates 3'-exonuclease activity. A scale indicating amino acids residues is shown.

and *rec1-5* alleles, respectively. Analysis of exonuclease activity revealed that the specific activities of proteins from the genomic *REC1* ORF, the cDNA *REC1 ORF,* and the *rec1-1* mutant allele, respectively, were all nearly the same. Activity diminished in a graded manner as the C-terminal end point receded through the Exo III sequence motif *(rec1Δ5)* back to the Exo I motif *(rec1Δ7),* and then dropped to an undetectable level as the C-terminal endpoint fell within the Exo I motif *(rec1-5).* Since biological activity is known to be severely perturbed in *rec1-1*, it is apparent that there are conditions in which biological function can be uncoupled from exonuclease activity.

Much of the genetic characterization of *REC1* conducted by Holliday and coworkers derived from studies on the *rec1-1* allele, but with the exception of radiation sensitivity, extensive analysis of other *rec1* alleles has not been reported. In light of the substantial exonuclease activity retained by the *rec1-1* allele, it was of interest to consider that *rec1-1* might not be completely devoid of biological activity as was originally thought by Holliday et al. *(33)*. A property of particular interest was the mutator phenotype, since the role of exonucleases is well documented in the mechanism of mutation induction. Therefore, the mutator phenotype of *rec1-1* was examined and compared to *rec1-5,* the allele encoding a severely truncated polypeptide with no detectable exonuclease activity. When the kinetics of killing of *rec1-1 and rec1-5* following irradiation were measured, identical survival curves were obtained in agreement with the findings of Holliday et al. *(33)* in their study of other *rec1* alleles. Thus, complete loss of function in radiation resistance resulted after removal of the terminal 15% amino acid residues from the carboxy end of the protein, but >70% of exonuclease activity remained in the corresponding mutant protein. On the other hand, the *rec1-1* and *rec1-5* alleles exhibited different rates of spontaneous forward mutation. In comparison with *REC1,* the *rec1-1* allele was elevated about 10-fold in the rate of spontaneous mutation, whereas the *rec1-5* mutant was elevated 100-fold *(50)*. Although the data set is limited, these results suggest that the mutator phenotype becomes more severe with progressive loss of the exonuclease activity.

Two important conclusions have emerged from this analysis. First, the exonuclease activity of the Rec1 protein does not appear to function directly in maintenance of radiation resistance. The *rec1-1* allele, which is extremely radiation-sensitive and which was previously thought to be a null mutant, encodes an protein with a nearly normal level of exonuclease activity. This result strongly suggests that exonuclease activity *per se* is not necessary for radiation resistance. The corollary is that if radiation sensitivity in the *rec1-1* mutant does not result from loss of enzymatic activity inherent in the Rec1 protein, then the radiation sensitivity must be an indirect consequence of the structural change in the mutant protein. Since the Rec1-1 protein differs from the wildtype only by the absence of the last 63 amino acids, which include residues extending from the leucine rich region, it seems likely that this region constitutes a domain that mediates interaction with another component(s) involved in maintaining radiation resistance. The second conclusion is that the mutator phenotype is associated with the exonuclease activity.

2.5. Role of Rec1 in Proofreading or Mismatch Correction?

Proficiency in avoiding mutation has well-known mechanistic precedents in terms of exonuclease function, such as the proofreading by DNA polymerases during DNA synthesis and postsynthesis removal of incorrect nucleotides by mismatch repair systems. In *E. coli* and *S. cerevisiae*, mutation in genes dedicated to either of these systems results in profound increases in mutation rate. The magnitude of the increase in spontaneous mutation rate observed in the case of the *rec1-5* allele compares to that observed in bacterial or yeast mutants defective in proofreading *(24,42)* or mismatch repair *(47)*. Thus, alternative models to account for the mutator phenotype of *rec1-1* are that *REC1* is involved in proofreading or mismatch correction.

There is no associated DNA polymerase activity detectable in preparations of the overexpressed *REC1* gene product, and the size of the protein is well below that of DNA polymerases. Therefore, it seems unlikely that the 3' → 5' exonuclease encoded by *REC1* arises as an activity intrinsically associated with a DNA polymerase. It is possible that *REC1* functions in proofreading during some aspect of DNA synthesis as a separate editing function, as does the *E. coli* DNA polymerase III ε-subunit *(59)*, but this is an unorthodox idea and has little experimental support in eukaryotic systems. Circumstantial evidence in support of the proofreading model is provided by identification of three blocks of amino acid residues in the *REC1* sequence that are similar in sequence and spacing to the three Exo motifs found conserved in over 30 DNA polymerases *(7,9)*. Furthermore, the enzymatic properties of the 3' → 5' exonuclease activity associated with Rec1 and the ε-subunit of DNA polymerase III are quite similar. Nevertheless, evidence against the proofreading model comes from studies on the analysis of exonuclease activity in the *rec1* mutants. The correlation in 3' → 5' exonuclease activity with the extent of deletion in the mutants analyzed is not consistent with the Exo motifs being essential for exonuclease activity in Rec1, as they are in *E. coli* DNA polymerase I *(6)*. Alteration of conserved aspartate residues in the *E. coli* DNA polymerase Exo II or Exo III regions thought to be important in metal binding completely eliminated the exonuclease activity *(15)*. However, when deletions spanning the putative comparable regions in *REC1* were examined, considerable 3' → 5' exonuclease activity remained. Thus, the presence of these signature Exo motifs in *REC1* might only be fortuitous and represent no functional significance in terms of proofreading.

The genetic properties of the *rec1* mutant support a role for *REC1*-encoded exonuclease functions in mismatch repair. In this regard, it is instructive to compare the properties of *rec1 (33)* with *pms1 (41,46,69)* and other mutants of *S. cerevisiae*, such as *mlh1 (53)* and *msh2 (3,55)*, that are defective in homologs of the bacterial *mut* genes, which are components of the *E. coli* methyl-directed mismatch repair system. Like these mutants, *rec1* is a mutator and confers increased rates of forward mutation as well as reversion. Also similar is the meiotic phenotype. *pms1*, *mlh1*, and *msh2* show poor spore viability, which is exacerbated by prolonged vegetative growth of diploids before sporulation, and this compares with *rec1*, in which complete tetrads are extremely rare. Accumulation of deleterious mutations is proposed to be responsible for this poor viability *(41,53,55,69)*. The extreme variation in size and morphology in *rec1* diploids could also reflect the accumulation of mutations. Spontaneous allelic recombination is elevated in all cases. *For pms1* and *msh2*, this has been interpreted to mean that close markers are not coconverted Increased recombination is thought to arise as a result of independent short-tract mismatch repair events. Despite the similarities between *rec1* and the *pms* mutants, the response to radiation differs markedly. *pms1* and *msh2* are not radiation-sensitive nor are they deficient in radiation-induced allelic recombination. This does not necessarily diminish the notion that Rec1 might serve in mismatch repair, especially in light of the evidence that loss of radiation resistance in the *rec1* mutant appears to be mediated through the C-terminal domain of the protein rather than the absence of exonuclease activity.

Identification of the essential components for mismatch repair in eukaryotes is still in the initial stages *(16)*. Based on what is known of the importance of exonucleases in the methyl-directed mismatch repair system of *E. coli (13)*, it is likely that at least one, if not several exonucleases will be found to play essential roles in mismatch repair in eukaryotes. The recent study on the *S. pombe exo1* mutant, which exhibits a mutator phenotype and elevated recombination between close markers, provides strong evidence tying a 5' → 3' exonuclease to a role in correction of mismatched bases *(63)*. It seems very possible that the 3' → 5' exonuclease of Rec1 might also be involved in mismatch repair.

2.6. Checkpoint Defect Caused by Mutation in REC1

In their genetic characterization of *rec1,* Holliday and coworkers *(33)* noticed that there was strong liquid holding recovery following exposure to radiation. This feature describes a phenomenon in which the severity of radiation-induced killing can be partially attenuated by maintaining cells in a nutrient-free medium for several hours before plating *(48)*. The recovery indicates that the mutant retains the ability to repair damage. Thus, the exonuclease deficiency in *rec1* can account for some of the genetic properties, but not the radiation sensitivity. *U. maydis* is extremely resistant to radiation *(43,56)*. Factors that contribute to its survival after irradiation are the predominance of the G2 compared to the G1 phase of the mitotic cell cycle, since G2 cells show greater radiation resistance *(30)*. Given that *U. maydis* cells arrested in growth by starvation are known to re-enter the cell cycle in G2 *(19,30)*, it is possible that cell-cycle dynamics at G2 might also contribute to the repair of DNA damage in *U. maydis.*

One explanation for the pronounced survival of irradiated *rec1* cells following liquid holding is that cell-cycle delay or arrest caused by nutrient deprivation enables DNA repair. This implies that the DNA repair deficiency in *rec1* reflects a cell-cycle defect. Therefore, the artificial induction of cell-cycle arrest following UV irradiation would be predicted to suppress the UV-sensitive phenotype of *rec1* cells. This was tested by irradiating exponentially growing cells with UV light and then incubating in medium containing methyl-1-(butylcarbamoyl)-2-benzimidazole-carbamate (MBC), a microtubule inhibitor that prevents tubulin polymerization *(54)*. This treatment blocks chromosome segregation by inhibiting spindle assembly and holds cells in G2 phase. Killing of the wild-type and the radiation-sensitive mutant *rec2 (31,58)* was unaffected by MBC treatment, but killing of the *rec1* mutant was partially suppressed when cells were held for 8 h in MBC. The partial suppression suggests that the *rec1* mutant is defective in damage-induced cell-cycle arrest and in some other aspect of repair as well.

Flow cytometry of unirradiated cultures of wild-type and *rec1* revealed similar bimodal distributions. This is interpreted as representing approximately equal proportions of cells in G_1 and G_2 phases. Following UV irradiation, there were marked changes in the distributions with significant differences in how the wild-type and *rec1* strains responded. In wild-type cells, it was evident that over the course of 6 h following irradiation, cells accumulated in G_2. However, in *rec1* cells, there was an overall decrease in the steady-state level of cells in G_2 and concomitant accumulation in G_1/S. The results indicate that the dynamics of cell-cycle progression following DNA damage are deranged in *rec1* with an apparent failure in preventing cells from proceeding through the $G_2 \rightarrow$ M checkpoint. The accumulation of cells in G_1/S indicates that another checkpoint at this stage is operational in *rec1*.

Mutants of *S. cerevisiae* and *S. pombe* which trespass illicitly through M phase, can be grouped into categories that define elements of the cell cycle. By studying the response of these mutants to radiation and hydroxyurea, as well as by examining their interaction with genes that encode components of the DNA replication apparatus, investigators have defined *S*-phase, G_1-phase, and G_2-phase pathways for the mitotic checkpoint. These pathways include components that are responsive to DNA damage or DNA replication or both. Mutants of *S. cerevisiae* that are simultaneously defective in the S-phase and DNA damage checkpoint component include *MEC1/ESR1 (36,52,68), RAD53 (4,68)*, and *DUN2/POL2* (*49*; *see* Chapters 17 and 18). In *S. pombe* a number of mutants have been identified as defective in overlapping S- and DNA damage phases, including *rad1*, *rad3*, *rad9*, *rad17*, *hus1*, and *rad26* (*1,2,18,57*; *see* Chapter 20). The cell requires three elements for the effective regulation of ordered progression through the mitotic cycle *(44)*. First, there must be a sensing mechanism by which genomic integrity can be determined and by which DNA damage, replication errors, or incompletely replicated DNA can be detected. Second, there must be a transducer by which this information is conveyed to the cell-cycle machinery. Third, there must be a responder or series of responders that regulate cell-cycle progression allowing for the resolution of the genomic insult. An attractive hypothesis is that the sensor is part of the DNA synthesis machinery. Recent studies on checkpoint control mutants defective for cell-cycle arrest in the presence of unreplicated DNA in *S. cerevisiae* have indicated that the *POL2* gene functions as a checkpoint control *(49)*. This gene encodes DNA polymerase ε, which is essential for DNA replication, but which is also necessary for transcriptional activation of damage-inducible genes and for entry into mitosis. These observations suggest that DNA polymerase ε senses unreplicated DNA as a result of direct physical interaction with DNA, perhaps by sensing DNA synthesis activity or gaps and regions of single-stranded DNA. This is supported by the observation that the *S. pombe* *cdc18*$^+$ gene product serves to couple the onset of S phase to START and to induce cell-cycle delay until the completion of S phase *(37)*. *cdc18*$^+$ is a homolog of the *S. cerevisiae* *CDC6* gene, an essential gene that encodes a protein involved in the initiation of replication *(45)*, and that itself is implicated in delaying entry into mitosis *(10)*.

By analogy with *S. cerevisiae* DNA polymerase ε, which functions directly as a DNA metabolic enzyme, and framed within the context of the function of the *REC1* gene product, which is also a DNA metabolic enzyme, it is possible that the *rec1* mutant is defective for sensing DNA damage and unreplicated DNA. Further support for this notion comes from the periodic mode of expression of the *REC1* gene. This mode of regulation is similar to that of genes in *S. cerevisiae* and *S. pombe* whose function is dedicated to DNA synthesis. One model is that the 3' $\rightarrow$ 5' exonuclease activity encoded by the *REC1* gene functions as part of a surveillance mechanism operating during DNA replication to find and repair stretches of DNA with compromised integrity and to communicate with the cell-cycle machinery.

Interallelic complementation and mutational analyses of the *S. cerevisiae POL2* gene have provided evidence that DNA polmerase ε is comprised of two separable, essential domains, an N-terminal DNA polymerase domain and a C-terminal checkpoint domain *(49)*. The *REC1* gene product also appears to be arranged in a modular fashion. As noted above, there is a substantial difference in the size of the *rec1-1* and *rec1-5* gene products (*see* Section 2.4.). Despite this difference, the two alleles share the same radiation-sensitive phenotype, but are quite different in exonuclease activity. These findings suggest that *REC1* encodes a protein consisting of two domains, an N-terminal exonuclease domain and a C-terminal checkpoint domain. The N-terminal domain may operate directly in repair of mismatched bases, whereas the C-terminal domain may be regulatory and interact with components of the cell-cycle machinery at the G2/M checkpoint.

3. *REC2*

3.1. *Phenotype of the* rec2 *Mutant*

The *rec2* mutant was isolated from the same screen for UV sensitivity as *rec1*. The spectrum of radiation sensitivity observed in *rec2* is complementary to *rec1* (i.e., *rec2* is more sensitive to γ-radiation and less sensitive to UV). The spontaneous level of mitotic allelic recombination is reduced about fivefold, but the usual several hundred-fold increase seen in UV-induced allelic recombination is abolished in *rec2*. Mitotic crossing over in *rec2* is not significantly different from wild-type. A striking effect of *rec2 is* on meiosis, which is completely blocked. Cytological analysis has revealed that the zygotic nucleus fails to divide. No viable haploid meiotic products have been observed. *rec2* haploids appear elongated, a characteristic that is seen in wild-type cells that have been damaged by radiation, and in conditional mutants temperature-sensitive for DNA synthesis after shifting to the restrictive temperature. Mitotic allelic recombination in a *rec1 rec2* double mutant is reduced at least 100-fold.

For exploration of the mechanism of recombination in *U. maydis*, pairs of autonomously replicating plasmids bearing noncomplementing alleles of a selectable marker were used to cotransform the DNA repair-deficient strains mentioned above that were deleted entirely for the genomic copy of the marker *(21)*. Generation of prototrophy proceeded through extrachromosomal recombination. Introduction of double-strand breaks into the plasmid DNA greatly stimulated recombination, but no such stimulation was apparent in the *rec2* mutant, unless the genetic markers on the plasmids were oriented in such a way as to circumvent recombination by a conservative pathway. It is unlikely that the defect in *rec2* is in a mismatch correction step, since artificially formed heteroduplex DNA containing the two allelic markers is highly active in transformation to prototrophy. These results suggest that the *REC2*-dependent pathway is involved in homologous pairing and strand exchange.

3.2. *Isolation and Analysis of the* REC2 *Gene*

The *REC2* gene was isolated by complementing the UV sensitivity of the *rec2-1* mutant and the DNA sequence was determined *(5,58)*. *REC2* encodes a protein of 781 amino acids with a calculated mass of 84 kDa. It features a nuclear localization signal at the N-terminus, an acidic region, an ATP binding loop, and a consensus Cdc2 protein kinase phosphorylation site. No other protein in the database with overall homology was found. However, a 47 amino acid stretch encompassing the nucleotide binding

Walker "A" motif is 39% identical to a region of *E. coli* RecA protein that is highly conserved among all bacterial species known to date. This same region is also conserved among three RecA-related proteins identified in *S. cerevisiae,* namely Dmc1p (42% identical), Rad51p (40% identical), and Rad57p (36% identical). It should be noted that the relationship between Rec2 and RecA as well as the eukaryotic RecA homologs does not result solely from the presence of the Walker "A" type nucleotide binding motif. Other proteins containing this motif were not identified in the homology search. The region of homology has been analyzed extensively in *E. coli* RecA and has been shown to be important for interaction with ATP. Crystallographic analysis indicates the region immediately preceding the Walker "A" type nucleotide binding motif is structurally unusual and consists of a helix and β-sheet separated by a loop *(61,62).* This region has been implicated in mediating a structural transition that is coupled to ATP binding.

To address the biological importance of the region with shared homology to RecA protein, a series of single amino acid substitutions in Rec2 was created *(58).* These included changes at residues within the nucleotide binding motif, at residues highly conserved in the loop structure, and an identical change at a residue that is responsible for the well-known *recA13* allele of *E. coli* that lies within this region. The resulting *rec2* alleles were then introduced into a *rec2* null mutant. Several of the homologous amino acid residues were found to be functionally sensitive to mutation. Mutations in the region that abolish DNA repair and recombination proficiency were found to be coincident with residues essential for the RecA family function in accordance with a model for RecA-ATP interactions.

3.3. Plasmid × Chromosome Recombination in U. maydis

In the absence of *REC2* function, plasmid × chromosome recombination is reduced to <5% of the frequency observed in wild-type *(58).* This reduction indicates that the normal cellular functions operating in recombinational repair and meiosis in *U. maydis* also take part in the more artificial situation of recombination after DNA transformation. However, no improvement in targeting frequency results after increasing the copy number of *REC2.* Thus, the supply of RecA-like recombination function might not be rate-limiting in gene targeting experiments, although the caveat must be raised that no correlation between the level of *REC2* mRNA and Rec2 protein has been established as yet. In contrast, a mutation in the consensus Cdc2 phosphorylation motif (the *rec2-10* allele; Thr697 → Ala) does influence targeting efficiency. No effect on DNA repair proficiency is observed, and no effect on recombination is observed in *rec2-10* cells with this allele, except in the exceptional circumstance in which there is a targeting plasmid construct with a *cis* arrangement of the marker gene and the *rec2-10* allele. When an integrating plasmid with a selectable marker and the *rec2-10* allele is introduced into *rec2* null cells and targeted to a genomic sequence corresponding to the marker, the frequency of targeted integration is six to eight times greater than in the case where the plasmid contained the *REC2* allele. In the *trans* experiment in which the *rec2-10* allele was already integrated in the genome, there is no enhancement in targeting the marker gene. The interpretation of these results is uncertain, but it is possible that the state of phosphorylation has a significant effect on the activity of the *REC2* gene product and that expression of *rec2-10* from a transforming plasmid violates the normal constraints that govern *REC2* regulation. Information from such studies could be helpful in optimizing gene targeting in higher eukaryotes.

Although homologous recombination of plasmid DNA with the genome is strongly decreased in *rec2* strains, it is not completely abolished. This could indicate that an alternative pathway for recombination remains in operation in the absence of *REC2* gene function. Nevertheless, the spectrum of recombination events apparent in the collection of infrequent transformants is similar to wild-type. Given the redundancy of RecA-like functions in yeast, if a second pathway is at work in *U. maydis,* it would seem likely to result from a second RecA-like activity involved in promoting similar types of recombination reactions, rather than through a nonconservative process exemplified by single-strand annealing.

Recombinational repair of double-stranded DNA gaps has been investigated in *U. maydis*. The experimental system was designed for analysis of repair of an autonomously replicating plasmid containing a cloned gene disabled by an internal deletion. Gap repair was dependent on *REC2*. It was discovered that crossing over rarely accompanied gap repair. The strong bias against crossing over was observed in three different genes regardless of gap size. These results indicate that gap repair in *U. maydis* is unlikely to proceed by the mechanism envisioned in the double-strand-break repair model of recombination, which was developed to account for recombination in *S. cerevisiae* (*see* Chapter 16). Experiments aimed at exploring processing of DNA ends were performed to gain understanding of the mechanism responsible for the observed bias. A heterologous insert placed within a gap in the coding sequence of two different marker genes strongly inhibited repair if the DNA was cleaved at the promoter-proximal junction joining the insert and coding sequence, but had little effect on repair if the DNA was cleaved at the promoter-distal junction. Gene conversion of plasmid RFLP markers engineered in sequences flanking both sides of a gap accompanied repair, but was directionally biased. These results are interpreted to mean that the DNA ends flanking a gap are subject to different types of processing. A model featuring a single migrating D-loop has been proposed to explain the bias in gap repair outcome based on the observed asymmetry in processing the DNA ends *(20)*.

3.4. Homologous Pairing Activity of Rec2 Protein

Ongoing parallel investigation of the RecA-like homologous pairing activity purified from extracts of *U. maydis* converged with the *REC2* studies to yield the solution to two problems, namely, the identity of the structural gene encoding the homologous pairing activity and the biochemical function of the *REC2* gene. Detailed studies established that homologous pairing reactions involving a variety of pairs of DNA molecules are catalyzed by the activity in an ATP-dependent manner. The activity is owing to a polypeptide of 70-kDa *(38,39)*. Partial amino acid sequence information of the homologous pairing protein correspond to sequences contained within the *REC2* open reading frame *(40)*. Since the N-terminal sequence is downstream of the presumed initiation codon of *REC2*, it is apparent that the 70-kDa homologous pairing protein is a fragment of the full-length Rec2 polypeptide. The 70-kDa fragment is missing 130–150 N-terminal residues, but retains the region of homology to RecA and likely arises as a proteolytic product of the primary *REC2* translation product.

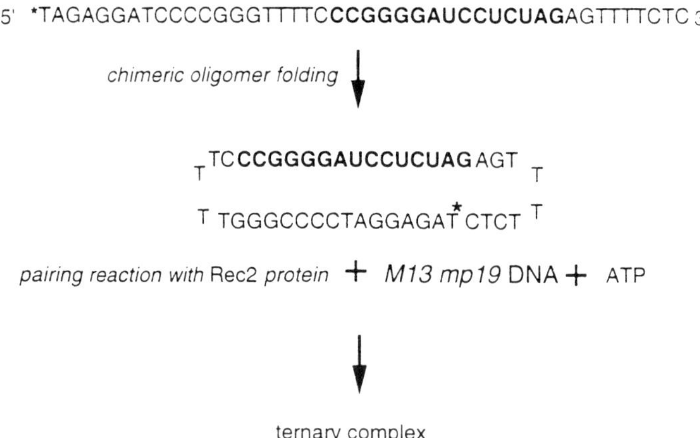

Fig. 3. Homologous pairing of a chimeric hairpin capped oligonucleotide. The oligonucleotide shown was synthesized based on the sequence in the multiple cloning site of M13 mpl9 DNA. The residues in bold are RNA residues, whereas the others are DNA residues. The sequence of RNA residues is antiparallel and complementary to a sequence in the M13 mpl9 multiple cloning site. When folded on itself, the oligonucleotide assumes a duplex structure of RNA/DNA hybrid with hairpin caps. Homologous pairing can be detected in reactions catalyzed by Rec2 protein with ATP as added cofactor. Complex formation is monitored by measuring radiolabeled oligonucleotide trapped on nitrocellulose filters.

The Rec2 protein was purified and shown to be active in catalyzing DNA-dependent ATP hydrolysis and a variety of ATP-dependent homologous pairing reactions. A novel finding was the enhanced activity in pairing RNA/DNA hybrid duplexes with a third DNA strand. Duplex hybrid molecules with hairpin caps on both ends (Fig. 3) were effectively paired with a homologous single strand.

4. *UVS3*

4.1. *Isolation of the* UVS3 *Gene*

The *uvs3* mutant is extremely sensitive to UV light, but exhibits little sensitivity to ionizing radiation. Allelic recombination appears normal, and no effect on crossing over has been observed. Postmeiotic segregation in *uvs3* is no different from wild-type *(32)*.

However, *uvs3* is defective in excision of pyrimidine dimers from photodamaged DNA. By direct measurement, it was observed that all classes of pyrimidine dimers were removed more slowly from the DNA of UV-irradiated *uvs3* compared to wild-type *(67)*. These observations are consistent with a tentative assignment of the *UVS3* gene to the excision repair group.

UVS3 was isolated by complementation of the radiation sensitivity. An 8-kbp complementing fragment was isolated, and this was subcloned to a 4-kbp fragment. DNA sequence analysis revealed an open reading frame with significant homology to the *S. cerevisiae RAD1* gene. The genomic structure has not yet been completely defined because of some ambiguity resulting from the presence of introns; this will be resolved by analysis of a fragment isolated from a cDNA library. While it has not yet been formally established that the cloned DNA fragment encodes the *UVS3* gene, this seems likely given the homology with *RAD1* and the role of *RAD1* in nucleotide excision repair.

The *UVS3* gene provides an additional member to the class of excision repair genes in the *RAD1* class, including the *rad16*$^+$ gene of *S. pombe (11)*, the *mei-9* gene of *Drosophila melanogaster (60)*, and *ERCC4* of *Homo sapiens* (*8*; *see* Chapter 18, vol. 2). These genes have mitotic and meiotic functions in certain instances in addition to the roles in excision repair. An important clue to function came from the discovery in budding yeast that Rad1p protein together with Rad10p protein form a protein complex with single-strand DNA endonuclease activity *(65)*. In the fission yeast and human, there is evidence for concerted interaction of the *RAD1* homologs with other genes as well. The realization in several experimental systems that the excision repair machinery interacts with the transcription apparatus illustrates the global complexity of DNA repair. Investigation of the *UVS3* gene will contribute to the understanding of the excision repair system in *U. maydis* and might also provide unique insights into DNA repair in general.

ACKNOWLEDGMENTS

This chapter is dedicated to Robin Holliday who began the genetic study of DNA repair mutants of *U. maydis* described here. The work summarized was supported in part by grants from the National Institutes of Health.

REFERENCES

1. Al-Khodairy, F. and A. M. Carr. 1992. DNA repair mutants defining G2 checkpoint pathways in *Schizosaccharomyces pombe. EMBO J.* **11:** 1343–1350.
2. Al-Khodairy, F., E. Fotou, K. S. Sheldrick, D. J. F. Griffiths, A. R. Lehmann, and A. M. Carr. 1994. Identification and characterization of new elements involved in checkpoint and feedback controls in fission yeast. *Mol. Biol. Cell.* **5:** 147–160.
3. Alani, E., R. A. G. Reenan, and R. D. Kolodner. 1994. Interaction between mismatch repair and genetic recombination in *Saccharomyces cerevisiae. Genetics* **137:** 1939.
4. Allen, J. B., Z. Zhou, W. Siede, E. C. Friedberg, and S. J. Elledge. 1994. The *SAD1/RAD53* protein kinase controls multiple checkpoints and DNA damage induced transcription in yeast. *Genes Dev.* **8:** 2416–2428.
5. Bauchwitz, R. and W. K. Holloman. 1990. Isolation of the *REC2* gene controlling recombination in *Ustilago maydis. Gene* **96:** 285–288.
6. Beese, L. S. and T. A. Steitz. 1991. Structural basis for the 3' → 5' exonuclease activity of *Escherichia coli* DNA polymerase I: a two metal ion mechanism. *EMBO J.* **10:** 25–33.
7. Bernad, A., L. Blanco, J. M. Lazaro, G. Martin, and M. Salas. 1989. A conserved 3' → 5' exonuclease active site in prokaryotic and eukaryotic DNA polymerases. *Cell* **59:** 219–228.
8. Biggerstaff, M., D. Szymkowski, and R. D. Woods. 1993. Co-correction of the *ERCC1 ERCC4* and xeroderma pigmentosum group F DNA repair defects in vivo. *EMBO J.* **12:** 3685–3692.
9. Blanco, L., A. Bernad, and M. Salas. 1991. Evidence favouring the hypothesis of a conserved 3' → 5' exonuclease active site in DNA-dependent DNA polymerases. *Gene* **112:** 139–144.
10. Bueno, A. and P. Russell. 1992. Dual functions of *CDC6*: A yeast protein required for DNA replication also inhibits nuclear division. *EMBO J.* **11:** 2167–2176.
11. Carr, A. M., H. Schmidt, S. Kirchoff, W. J. Muriel, K. S. Sheldrick, D. J. Griffiths, C. N. Basmacioglu, S. Subramani, M. Clegg, M., A. Nasim, and A. R. Lehmann. 1994. The

rad16 gene of *Schizosaccharomyces pombe:* a homolog of the *RAD1* gene of *Saccharomyces cerevisiae. Mol. Cell. Biol.* **14:** 2029–2040.
12. Clark, A. J. and A. D. Margulies. 1965 Isolation and characterization of recombination deficient mutants of *E. coli* K12. *Proc. Natl. Acad. Sci. USA* **53:** 451–457.
13. Cooper, D. L., R. S. Lahue, and P. Modrich. 1993. Methyl-directed mismatch repair is bidirectional. *J. Biol. Chem.* **268:** 11,823–11,829.
14. Derbyshire, V., P. S. Freemont, M. R. Sanderson, L. Beese, J. M. Friedman, C. M. Joyce, and T. A. Steitz. 1988. Genetic and crystallographic studies of the 3', 5' exonucleolytic site of DNA polymerase I. *Science* **240:** 199–201.
15. Derbyshire, V., N. D. F. Grindley, and C. M. Joyce. 1991 The 3' → 5' exonuclease of DNA polymerase I of *Escherichia coli:* contribution of each amino acid at the active site to the reaction. *EMBO J.* **10:** 17–24.
16. Drummond, J. T., G. M. Li, M. J. Longley, and P. Modrich. 1995. Isolation of an hMSH2–pl60 heterodimer that restores DNA mismatch repair to tumor cells. *Science* **268:** 1909–1912.
17. Enoch, T. and P. Nurse. 1990. Mutation of fission yeast cell cycle control genes abolished dependence of mitosis on DNA replication. *Cell* **60:** 665–673.
18. Enoch, T., A. M. Carr, and P. Nurse. 1992. Fission yeast genes involved in coupling mitosis to completion of DNA replication. *Genes Dev.* **6:** 2035–2046.
19. Esposito, R. E. and R. Holliday. 1964. The effect of 5-fluorodeoxyuridine on genetic replication and mitotic crossing over in synchronized cultures of *Ustilago maydis. Genetics* **50:** 1009–1017.
20. Ferguson, D. O. and W. K. Holloman. 1995. Recombinational repair of gaps in DNA is asymmetric in *Ustilago maydis* and can be explained by a migrating D-loop model. *Proc. Natl. Acad. Sci. USA.* **93:** 5419–5424.
21. Fotheringham, S. and W. K. Holloman. 1991. Extrachromosomal recombination is deranged in the *rec2* mutant of *Ustilago maydis. Genetics* **129:** 1053–1060.
22. Freemont, P. S., J. M. Friedman, L. S. Beese, M. R. Sanderson, and T. A. Steitz. 1988. Cocrystal structure of an editing complex of Klenow fragment with DNA. *Proc. Natl. Acad. Sci. USA* **85:** 8924–8928.
23. Friedberg, E. C., W. Siede, and A. J. Cooper. 1991. Cellular responses to DNA damage in yeast, in *The Molecular Biology of the Yeast Saccharomyces. Genome Dynamics, Protein Synthesis and Energetics* (Broach, J. R., J. R. Pringle, and E. W. Jones, eds.), Cold Spring Harbor Laboratory, Cold Spring Harbor, NY, pp. 147–192.
24. Goodman, M. F., S. Creighton, L. B. Bloom, and J. Petruska. 1993. Biochemical basis of DNA replication fidelity. *Crit. Rev. Biochem. Mol. Biol.* **28:** 83–126.
25. Hereford, L., S. Bromley, and M. A. Osley. 1982. Periodic expression of yeast histone genes. *Cell* **30:** 305–310.
26. Holden, D. W., A. Spanos, and G. R. Banks. 1989. Nucleotide sequence of the REC1 gene of *Ustilago maydis. Nucleic Acids Res.* **17:** 10,489.
27. Holden, D. W., A. Spanos, N. Nauga, and G. R. Banks. 1991 Cloning the *REC1* gene of *Ustilago maydis. Curr. Genet.* **20:** 145–150.
28. Holliday, R. 1964. A mechanism for gene conversion in fungi. *Genet. Res.* **5:** 282–304.
29. Holliday, R. 1965. Radiation sensitive mutants of *Ustilago maydis. Mutat. Res.* **2:** 557–559.
30. Holliday, R. 1965. Induced mitotic crossing-over in relation to genetic replication in synchronously dividing cells of *Ustilago maydis. Genet. Res.* **6:** 104–120.
31. Holliday, R. 1967. Altered recombination frequencies in radiation sensitive strains of *Ustilago maydis. Mutat. Res.* **4:** 275–288.
32. Holliday, R. and J. M. Dickson. 1977. The detection of post-meiotic segregation without tetrad analysis in *Ustilago maydis. Mol. Gen. Genet.* **153:** 331–335.
33. Holliday. R., R. E. Halliwell, M. W. Evans, and V. Rowell. 1976. Genetic characterization of *rec-1,* a mutant of *Ustilago maydis* defective in repair and recombination. *Genet. Res.* **27:** 413–453.

34. Howard-Flanders, P. and L. Theriot. 1966. Mutants of *E. coli* K12 defective in DNA repair and in genetic recombination. *Genetics* **53:** 1137–1150.
35. Johnston, L. H. 1992. Cell cycle control of gene expression in yeast. *Trends Cell Biol.* **2:** 353–357.
36. Kato, R. and H. Ogawa, H. 1994. An essential gene, *ESR1, is* required for mitotic cell growth, DNA repair, and meiotic recombination in *Saccharomyces cerevisiae. Nucleic Acids Res.* **22:** 3104–3112.
37. Kelly, T. J., G. S. Martin, S. L. Forsburg, R. J. Stephen, A. Russo, and P. Nurse. 1993. The fission yeast *cdc18*$^+$ gene product couples S phase to START and mitosis. *Cell* **74:** 371–382.
38. Kmiec, E. and W. K. Holloman. 1982. Homologous pairing of DNA molecules promoted by a protein from Ustilago. *Cell* **29:** 367–374.
39. Kmiec, E. B. and W. K. Holloman. 1994. ATP-dependent DNA renaturation and DNA-dependent ATPase reactions catalyzed by the *Ushilago maydis* homologous pairing protein. *Eur. J. Biochem.* **219:** 865–875.
40. Kmiec, E. B. A. Cole, and W. K. Holloman. 1994. The *REC2* gene encodes the homologous pairing protein of *Ustilago maydis. Mol. Cell. Biol.* **14:** 7163–7172.
41. Kramer, W., B. Kramer, M. S. Williamson, and S. Fogel. 1989. Cloning and nucleotide sequence of DNA mismatch repair gene *PMS1* from *Saccharomyces cerevisiae:* homology of PMS 1 to prokaryotic MutL and HexB. *J. Bacteriol.* **171:** 5339–5346.
42. Kunkel, T. A. 1992. DNA replication fidelity. *J. Biol. Chem.* **267:** 18,251–18,254.
43. Leaper, S., M. A. Resnick, and R. Holliday. 1980. Repair of double-strand breaks and lethal damage in DNA of *Ustilago maydis. Genet. Res.* **35:** 291–307.
44. Li, J. J., and R. J. Deshaies. 1993. Exercising self-restraint: Discouraging illicit acts of S and M in eukaryotes. *Cell* **74:** 223–226.
45. Liang, C., M. Weinreich, and B. Stillman. 1995. ORC and Cdc6p interact and determine the frequency of initiation of DNA replication in the genome. *Cell* **81:** 667–676.
46. Lichten, M., C. Govon, N. P. Schultes, D. Treco, D., J. W. Szostak. 1990. Detection of heteroduplex DNA molecules among the products of *Saccharomyces cerevisiae* meiosis. *Proc. Natl. Acad. Sci. USA* **87:** 7653–7657.
47. Modrich, P 1991. Mechanisms and biological effects of mismatch repair. *Annu. Rev. Genet.* **25:** 229–353.
48. Moustacchi, E. and S. Enteric. 1970. Differential "liquid holding recovery" for the lethal effect and cytoplasmic "petite" induced by UV light in *Saccharomyces cerevisiae. Mol. Gen. Genet.* **109:** 69–83.
49. Navas, T. A., Z. Zhou, and S. J. Elledge. 1995. DNA Polymerase ε links the DNA replication machinery to the S phase checkpoint. *Cell* **80:** 29–39.
50. Onel, K., M. P. Thelen, D. O. Ferguson, R. L. Bennett, and W. K. Holloman. 1995. Mutation avoidance and DNA repair proficiency in *Ustilago maydis* are differentially lost with progressive truncation of the *REC1* gene product. *Mol. Cell. Biol.* **15:** 5329–5338.
51. Onel, K., A. Koff, R. L. Bennett, P. Unrau, P., and W. K. Holloman. 1996. The *REC1* gene of *Ustilago maydis,* which encodes a 3'-5' exonuclease, couples DNA repair and completion of DNA synthesis to a mitotic checkpoint. *Genetics* **143:** 165–174.
52. Paulovich, A. G. and L. H. Hartwell. 1995. A checkpoint regulates the rate of progression through S phase in S. cerevisiae in response to DNA damage. *Cell* **82:** 841–847.
53. Prolla, T. A., D. M. Christie, and R. M. Liskay. 1994. Dual requirement in yeast DNA mismatch repair for *MLH1* and *PMS1,* two homologs of the bacterial mutL gene. *Mol. Cell. Biol.* **14:** 407–415.
54. Quinlan, R. A., C. I. Pogson, and K. Gull. 1981). The influence of the microtubule inhibitor, methyl benzimidazol-2-yl-carbamate (MBC) on nuclear division and the cell cycle in *Saccharomyces cerevisiae. J. Cell. Sci.* **46:** 341–352.

55. Reenan, R. A. G. and R. D. Kolodner. 1992. Characterization of insertion mutations in the *Saccharomyces cerevisiae MSH1* and *MSH2* genes. Evidence for separate mitochondrial and nuclear functions. *Genetics* **132:** 975–985.
56. Resnick, M. A. 1978. Similar responses to ionizing radiation of fungal and vertebrate cells and the importance of DNA double-strand breaks. *J. Theor. Biol.* **71:** 339–346.
57. Rowley, R., S. Subramani, and P. G. Young. 1992. Checkpoint controls in *Schizosaccharomyces pombe radl. EMBO J.* **11:** 1335–1342.
58. Rubin, B. R., D. O. Ferguson, and W. K. Holloman. 1994. Structure of *REC2,* a recombinational repair gene of *Ustilago maydis,* and its function in homologous recombination between plasmid and chromosomal sequences. *Mol. Cell. Biol.* **14:** 6287–6296.
59. Scheuermann, R. H. and H. Echols. 1984. A separate editing exonuclease for DNA replication: the ε subunit of *Escherichia coli* DNA polymerase III holoenzyme. *Proc. Natl. Acad. Sci. USA* **81:** 7747–7751.
60. Sekelsky, J. J., K. S. McKim, G. M. Chin, and R. S. Hawley. 1995. The Drosophila meiotic recombination gene *mei-9* encodes a homologue of the yeast excision repair protein Radl. *Genetics* **141:** 610–627.
61. Story, R. M., I. T. Weber, and T. A. Steitz. 1992. The structure of the *E. coli recA* protein monomer and polymer. *Nature* **355:** 318–325.
62. Story, R. M., D. K. Bishop, N. Kleckner, and T. A. Steitz. 1993. Structural relationship of bacterial recA proteins to recombination proteins from bacteriophage T4 and yeast. *Science* **259:** 1892–1896.
63. Szankasi, P. and G. R. Smith. 1995. A role of exonuclease I from *S. pombe* in mutation avoidance and mismatch correction. *Science* **267:** 1166–1169.
64. Thelen, M. P., K. Onel, and W. K. Holloman. 1994. The *REC1* gene of *Ustilago maydis* involved in the cellular response to DNA damage encodes an exonuclease. *J. Biol. Chem.* **269:** 747–754.
65. Tompkinson, A. E., A. J. Bardwell, L. Bardwell, N. J. Tappe, and E. C. Friedberg. 1993. Yeast DNA repair and recombination proteins Rad1 and Rad10 constitute a single stranded DNA endonuclease. *Nature* **263:** 860–862.
66. Tsukuda, T., R. Bauchwitz, and W. K. Holloman. 1989. Isolation of the *REC1* gene controlling recombination in *Ustilago maydis. Gene* **85:** 335–341.
67. Unrau, P. 1975. The excision of pyrimidine dimers from the DNA of mutant and wild type strains of *Ustilago. Mutat. Res.* **29:** 53–65.
68. Weinert, T. A., G. L. Kiser, and L. H. Hartwell. 1994. Mitotic checkpoint genes in budding yeast and the dependence of mitosis on DNA replication and repair. *Genes Dev.* **8:** 652–665.
69. Williamson, M. S., J. C. Game, and S. Fogel. 1985. Meiotic gene conversion mutants in *Saccharomyces cerevisiae.* I. isolation and characterization of *pms1-1 and pms1-2. Genetics* **110:** 609–646.

24

Processing of DNA Damage in the Nematode *Caenorhabditis elegans*

Phil S. Hartman and Gregory A. Nelson

1. *CAENORHABDITIS ELEGANS* AS A MODEL SYSTEM

1.1. *Overview*

The free-living nematode *Caenorhabditis elegans* has emerged rapidly as an organism with which to study many basic biological phenomena, particularly those related to development. This can be evidenced numerically in many ways; for example, the number of presentations at the biennial *C. elegans* meetings has increased over sevenfold, from 80 in 1979 to 569 in 1995. In addition to numerous review articles, several books are devoted to this nematode, its attributes and various foci of interest *(13,61,66,67)*. The three primary attributes that have rendered *C. elegans* a popular model system are overviewed briefly in the following three sections.

The attributes that have rendered *C. elegans* popular with developmental biologists have also been exploited to examine specific areas in radiation biology, DNA repair, and mutagenesis. Several of the basic DNA repair pathways operative in *C. elegans* have been elucidated. Also, a number of biological end points such as survival and mutagenesis, have been examined so as to address the various mechanisms by which *C. elegans* accommodates DNA damage. Central to these efforts has been the isolation and characterization of radiation-sensitive *(rad)* mutants that modify various biological responses. In particular, these studies provide insights into damage processing, particularly as related to development and aging.

1.2. *Basic Biology of* C. elegans

C. elegans reproduces primarily as a self-fertilizing hermaphrodite, which presents several advantages in mutant isolation and maintenance. Males also exist and allow for genetic manipulations, such as mapping, complementation, and strain construction. Development, from fertilization to adulthood, takes a mere 3.5 d at 20°C, with adults attaining a length slightly in excess of 1 mm (Fig. 1). Its life-span is relatively short as well: 2–3 wk for wild-type under standard laboratory conditions. Growth conditions are quite simple: animals are propagated on agar-filled Petri dishes, where they feed on a lawn of *Escherichia coli*. On one hand, single animals may be readily transferred to single plates using a platinum wire, akin to a bacterial transfer loop. On the other, gram

From: DNA Damage and Repair, Vol. 1: DNA Repair in Prokaryotes and Lower Eukaryotes
Edited by: J. A. Nickoloff and M. F. Hoekstra © Humana Press Inc., Totowa, NJ

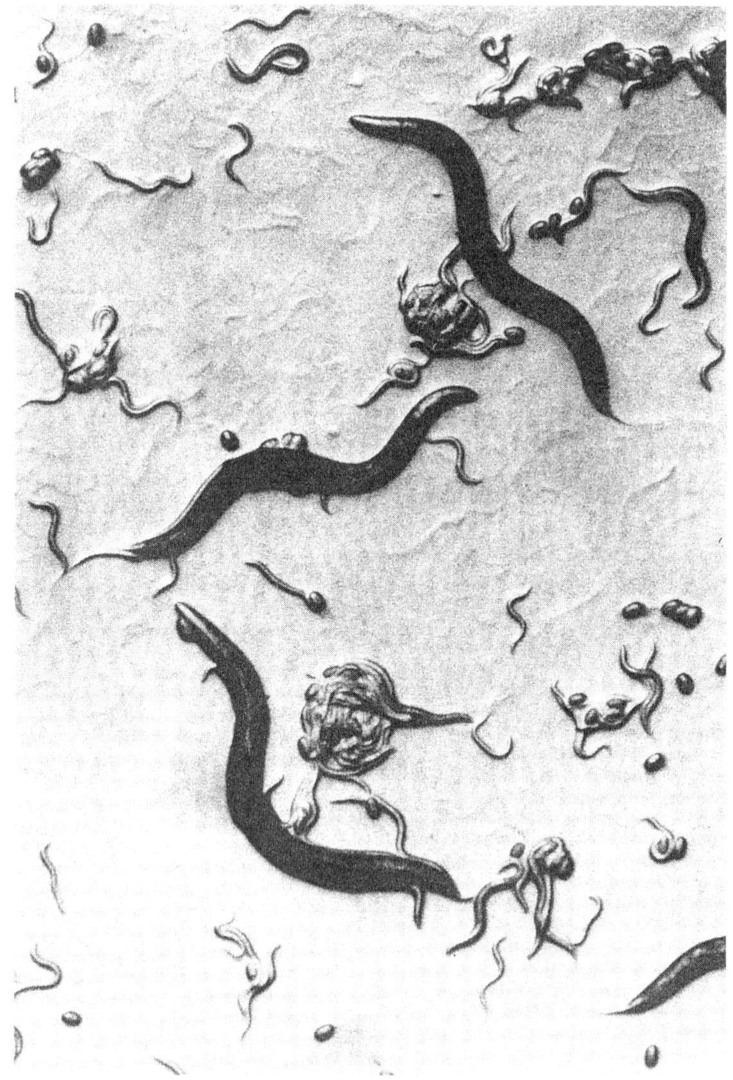

Fig. 1. Photomicrograph of wild-type *C. elegans*. Various developmental stages are present, including three adults, ca. 50 larvae and ca. 50 embryos ("eggs").

quantities of animals can be easily grown and isolated. Finally, stocks are easily maintained, because animals can be resuscitated after storage in liquid nitrogen.

1.3. Genetic Attributes of C. elegans

Following the lead of Brenner *(4)*, whose pioneering efforts were pivotal in establishing and validating *C. elegans* as a model system, tens of thousands of mutants have been isolated in scores of laboratories. These mutations affect a wide variety of processes, including several of particular interest to the readers of this book (e.g., programmed cell death, DNA replication, DNA repair, radiation sensitivity, mutagenesis, aging), and map to one of the five autosomes or the X chromosome (Table 1). In addition, a large number of translocations, duplications, and deficiencies have been isolated and maintained. The *Caenorhabditis* Genetics Center at the University of Minnesota currently maintains over 2150 strains, including at least one allele of the >1200 genetic loci thus far defined by mutation. These strains are distributed widely and without charge, which exemplifies the good will that exists generally in the "worm community."

Through a collaborative project involving The Sanger Centre, Cambridge, and Washington University, substantial efforts have been directed at compiling a physical map of *C. elegans (62)*. A collection of ordered cosmids and YACs provide nearly complete coverage of the genome. They have been partially aligned with the genetic map, allowing investigators quickly to position cloned genes via hybridization to a nylon membrane with ordered YACs. In addition, over 38 Mbp of the estimated 100 Mbp genome has been sequenced, with a 1998 target for completion of the sequencing project. Thus far, approx 50% of the sequences are within genes, which occur at an average frequency of 1 per 5 kbp.

Germ-line transformation can be accomplished in *C. elegans* via microinjection of the gonad. This has facilitated the cloning of many genes via transformation rescue of the mutant phenotype of interest. In addition, the use of reporter constructs has enabled determination of the developmental expression of many genes. Genes have also been cloned in *C. elegans* via transposon tagging. More recently, transposition has been exploited to generate "knockout" mutations. These and several other complementary methodologies are comprehensively detailed by Epstein and Shakes *(13)*.

1.4. Cellular Development of C. elegans

C. elegans executes a largely invariant set of cell divisions, beginning with a single-celled zygote, to a 558 cell first-stage larva, to an adult containing 959 somatic and roughly 2000 germ cells. Embryogenesis is divided into two distinct phases, each occupying approx 6 h. All embryonic DNA synthesis and cell division occur during the initial proliferative phase; a variety of cell interactions and migrations transpire during the second phase, that of morphogenesis. Remarkably, the cell lineage has been completely elucidated using Nomarksi differential interference contrast microscopy *(53,54)* and has thus provided the framework for elegant studies of development.

Experimentally induced perturbations in the lineage can be created in a number of ways, allowing exploitation of the cell lineage. First, there are a number of mutations that alter cell divisions and cell fate within the cell lineage. More often than not, these effects are limited to relatively few cells, suggesting that the mutated genes control developmental decisions (i.e., they act as homeotic genes). Second, a finely tuned laser microbeam has been employed to manipulate the lineage, most commonly by heavily irradiating specific nuclei to ablate specific cells. The general, but not universal, finding is that development of surrounding cells is not influenced by ablation of their neighbors. Third, genetic mosaics can be created via the mitotic loss of a free duplication; these allow determination of the focus of activity of a particular gene *(29)*. Free duplications are chromosomal segments that segregate independently of normal chromosomes; they are thought to be stable owing to the holocentric nature of *C. elegans* chromosomes.

Two interrelated aspects of cellular development in *C. elegans* deserve specific mention given the context of this book. First, the phenomenon of programmed cell death,

Table 1
Genes that Affect the Processing of DNA Damage in *C. elegans*[a]

Gene	Mutant phenotype(s)	Section discussed
age-1	Long-lived	4.7.
	hyperresistant to oxidative stress	
ced-9	Gain-of-function mutation blocks programmed cell death and suppresses supernumerary rounds of intestinal-cell division after ionizing radiation;	4.3.
	bc1-2 homolog	
mev-1	Hypersensitive to oxidative stress;	4.7.
	spontaneous anaphase bridge formation in intestine and enhances radiation-induced anaphase bridge formation;	4.5.
	Cu/Zn superoxide dismutase?	
nuc-1	Major endonuclease (required for DNA degradation during programmed cell death and bacterial digestion);	2.2.
	promotes intestinal anaphase bridge formation after ionizing radiation;	4.5.
rad-1	UV and X-ray hypersensitive for embryogenesis;	
	hypomutable to UV;	4.5.
	hypomutable to ionizing radiation;	4.8.
rad-2	UV and X-ray hypersensitive for embryogenesis;	3.
	potentiates UV and γ-radiation-induced chromosome aberrations in embryos;	4.5.
	promotes radiation-induced anaphase bridge formation in intestine cells;	4.5.
	hypermutable in oocytes for ionizing radiation	
rad-3	MMS and UV hypersensitive at several developmental stages;	3.
	UV hypermutable, but wild-type for ionizing radiation mutagenesis;	4.5.
	excision repair-defective (developmentally regulated);	2.1.2.
	wild-type for radiation-induced anaphase bridge formation in intestine	
rad-4	MMS and UV hypersensitive at embryogenesis;	3.
	suppresses meiotic nondisjunction;	3.
	enhances meiotic recombination;	3.
	wild-type for radiation-induced anaphase bridge formation in intestine	

rad-5	UV hypersensitive at embryogenesis;	3.
	elevated spontaneous mutation frequency (at two loci);	3.
	wild-type for radiation-induced anaphase bridge formation in intestine	
rad-6	UV hypersensitive at embryogenesis;	3.
	reduces meiotic recombination;	3.
	wild-type for radiation-induced anaphase bridge formation in intestine	
rad-7	UV hypersensitive for embryogenesis;	3.
	UV hypomutable;	4.5.
	hypomutable for ionizing radiation; couples to hypersensitivity for gamete viability;	
	promotes intestinal anaphase bridge formation for charged particles, but not γ-rays	
rad-8	UV hypersensitive for embryogenesis;	3.
	hypersensitive to oxidative stress	4.7.
rad-9	UV hypersensitive for embryogenesis;	3.
rec-1	Enhances meiotic recombination;	2.3.
	shares sequence identity with family of helicases, including *rad3* of yeast	

[a]At least one mutant allele exists for each of these genes. In addition, there are other *C. elegans* genes (for which mutants are not available) that likely affect DNA damage processing. These include four genes that encode SODs (Sections 2.2. and 2.3.) and two DNA glycosylases (Section 2.2.). Also, a number of putative genes that likely affect DNA-damage processing have been identified via their homologies to damage-processing genes in other organisms (Section 2.3.).

especially its genetics, has been extensively studied *(10)*. A total of 131 specific cells are destined to die during embryogenesis; additional cells die during larval development. Mutations in a series of *ced (ce*ll *d*eath) genes have been isolated, which modify this apoptotic process. Several of these have been cloned and sequenced. Second, cell-cycle regulation has received some attention. For example, mutants in the *lin-19* and *lin-23* genes display hyperplasia of multiple tissues; both show homology to *cdc* genes in yeast (Kipreos, personal communication).

2. THE BIOCHEMISTRY OF DAMAGE TOLERANCE SYSTEMS

2.1. Repair Systems

2.1.1. Photoreactivation

Three experimental approaches were employed to demonstrate that *C. elegans* does not photoreactivate cyclobutane pyrimidine dimers (CPDs; *25,37)*. First, survival after UV radiation was experimentally identical in animals exposed subsequently to photoreactivating light as compared with those held in the dark. Second, *C. elegans* extracts were incapable of photoreactivating CPDs in *Hemophilus influenza* DNA, as measured by transformation frequencies. Third, using radioimmunoassay (RIA), the number of CPDs in *C. elegans* DNA was not modified by exposure to photoreactivating light. Uniformly negative results were obtained using a variety of synchronized populations, indicating that photoreactivation is universally absent from *C. elegans* as opposed to being limited to specific stages of development.

2.1.2. Excision Repair of UV Radiation-Induced Photoproducts

Removal of CPDs and (6-4)photoproducts ([6-4]PDs), the two most prevalent lesions induced by UV radiation, has been measured in wild-type and *rad* mutant *C. elegans* using RIAs specific to these two photoproducts *(25)*. In wild-type animals, fluences >50 Jm^{-2} saturated repair of both photoproducts; specifically, >70% of these photoproducts were removed after fluences of <50 Jm^{-2} as opposed to <20% removal after 100 Jm^{-2}. The kinetics of repair were measured in four synchronous populations of wild-type animals, ranging from embryos to adults (Fig. 2). *C. elegans* excised both photoproducts much more slowly than did prokaryotes and single-celled eukaryotes. The kinetics of removal were similar to those obtained using cultured mammalian cells; however, unlike mammalian cells, which excised (6-4)PDs much more rapidly than CPDs, both photoproducts were removed at equal rates in the nematode. The initial rates of photoproduct removal were similar at all stages of development tested (Fig. 2). However, the amount of damage unrepaired after 24 h increased throughout development, consistent with the notion that diminution in DNA repair capacity may play a role in aging.

The excision repair capacities of the four most UV-hypersensitive *rad* mutants were also determined using the same RIA *(25)*. Three of the four *(rad-1*, *rad-2*, *rad-7)* had repair kinetics essentially identical to those of the wild-type strain. The fourth, *rad-3*, displayed a slight excision repair defect when embryos were assayed. Conversely, *rad-3* larvae had a much lower capacity to remove both CPDs and (6-4)PDs. This indicates that DNA repair is subject to developmental regulation in *C. elegans*.

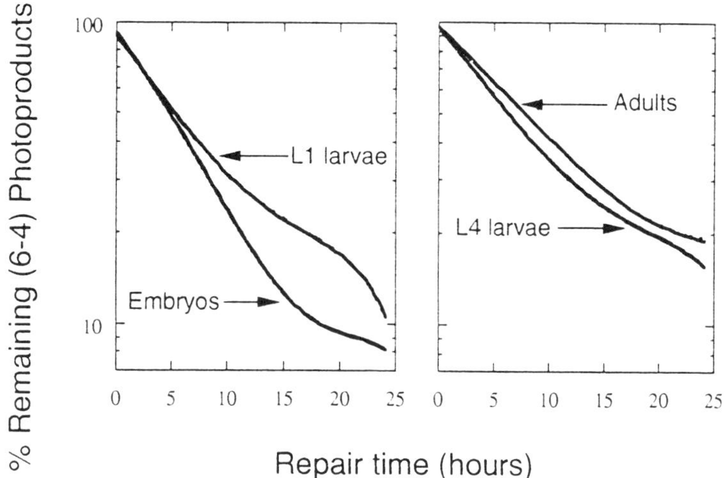

Fig. 2. Repair of UV-induced (6-4)PDs from selected developmental stages of wild-type *C. elegans*. Staged populations were exposed to 50 J/m^2 of 254 nm UV and allowed to repair for the times indicated. Reproduced from ref. *25* with permission.

2.1.3. Strand Break Repair

Using pulsed-field-gel electrophoresis, Nelson *(46)* recently began to explore the repair of ionizing-radiation-induced double-strand breaks (DSBs). The kinetics of repair are apparently more rapid than for UV radiation-induced damage. H. Van Luenen and R. Plasterk (personal communication) have taken a different approach, examining DSB repair at a specific locus. This derives from their investigation of transposition mechanisms (e.g., *7*). Using a construct in which the transposase gene was fused to a heat-shock promoter, they forced expression of the Tc3 transposase to generate many transposition events, largely restricted to the X chromosome. By probing to the region adjacent to the transposon, they were able to detect three fragments using Southern hybridization:

1. The region with the transposon still integrated;
2. The region with the transposon excised and the resultant DSB unrepaired; and
3. The region with the transposon excised and the DSB repaired.

Thirty-nine hours after transposase induction, over 10% of the transposons had excised, each presumably generating a DSB at the same specific locus. Densitometric analysis indicated that repair was efficient; only 2% of the excision events did not culminate in break repair. The kinetics of repair were not affected by mutations in *rad-1*, *rad-2*, or *rad-3*.

2.2. *Enzymology*

Relatively little is known concerning the enzymology of DNA replication and damage processing in *C. elegans*. In fact, in the majority of these cases, enzyme activities have been determined only with crude extracts. The ability to obtain easily gram quantities of synchronous populations makes *C. elegans* attractive for additional, more systematic studies.

The major endonuclease, the apparent product of the X-linked gene *nuc-1*, has been partially purified and characterized *(30)*. The enzyme hydrolyzed double-stranded DNA seven times more rapidly than single-stranded DNA. It is apparently required for repair of radiation-induced damage leading to anaphase bridge formation in certain intestinal cells, but plays no obvious role in replication or meiotic recombination, as evidenced by its wild-type phenotype in tests addressing these functions. The endonuclease is required to digest DNA in cell corpses derived from programmed cell death events and acts to degrade bacterially derived DNA in the lumen of the gut.

Both catalase and superoxide dismutase (SOD) activities have been observed in crude extracts of *C. elegans (1,9,33,41,55,56)*. In one instance, the cytosolic SOD has been purified to homogeneity *(55)*. As discussed in Section 4.7., a number of reports detail correlations between fluctuations in SOD levels and both aging as well as sensitivity to superoxide anion. In addition, genes encoding copper/zinc and manganese SODs from *C. elegans* have been cloned and sequenced (*see* Section 2.3.).

DNA glycosylases acting on uracil- and 3-methyl-adenine-containing DNA were detected in crude extracts of *C. elegans (45)*. The two activities were developmentally regulated; specifically, uracil-DNA glycosylase activity was highest in the embryonic extract, whereas 3-methyladenine-DNA glycosylase was undetectable in embryo extracts and found equally in the three other stages examined.

Substantial efforts have been directed toward eludicating the transposable elements present in *C. elegans* as well as their modes of action. There are at least six different types of transposons (Tc1–Tc6), with Tc1 and Tc3 the best studied. The Tc1 encodes an endonuclease responsible for the excision of the element; its activity has been examined both in vitro and in vivo *(58,59)*. In addition, the DNA binding activities of the Tc3 transposase have been characterized *(7)*.

2.3. Repair Gene Isolation

Several genes with putative functions in DNA damage processing, including four SOD genes, have been isolated and sequenced. One gene, named *sod-1*, encodes a cytosolic copper/zinc SOD; the enzyme was purified to homogeneity by Vanfleteren *(55)*. The genomic sequence contains three introns and totals 608 bp in length *(14)*. The cDNA clone from this gene had been isolated previously *(41)* and shows considerable sequence homology to Cu/Zn SODs from other organisms. The gene resides close to, but is distinct from *age-1*, which influences its expression (*see* Section 4.7.). A second Cu/Zn SOD gene *(sod-4)* was revealed by the genome sequencing project and maps to linkage group III. Since *mev-1* and *sod-4* both affect SOD and are closely linked, N. Ishii (personal communication) is currently exploring the possibility that they are allelic. Two related, but distinct Mn SOD genes *(sod-2* and *sod-3)* have also been cloned and sequenced (Ishii, personal communication; *15*).

A *C. elegans* homolog to the yeast *RAD6* gene (*see* Chapter 15) has been cloned by using PCR degenerate oligonucleotide primers *(65)*. The predicted product is >50% identical to both the yeast and human proteins. In addition to its role in DNA repair and mutagenesis, Rad6p catalyzes conjugation of ubiquitin to proteins, facilitating their degradation. The *C. elegans* genes that participate in this ubiquitin-conjugation system have also been cloned *(65)*; at least one can substitute for its yeast homolog.

Transposon tagging has been employed to clone a gene that confers sensitivity to UV radiation and methyl methanesulfonate *(17)*. Sequencing of the cDNA corresponding to this gene is in progress. In addition, as described in Section 2.3., the Tc1 and Tc3 transposase genes have been cloned, sequenced, and expressed *(7,58,59)*.

A clone with high homology to the 14-3-3 family has been recently identified via its ability to complement functionary the *sme2*-deletion mutant of *Schizosaccharomyces pombe* (Hayashizaki, personal communication). A full-length clone partially rescued the UV-hypersensitivity of *rad24* in *S. pombe,* but paradoxically did not rescue the *sme2* deletion. As with other aspects of DNA repair, *C. elegans* may prove valuable for studies of the developmental regulation of this gene, particularly given that Wang and Shakes *(60)* have isolated another 14-3-3 homolog that is highly expressed in gonads.

The *rec-1* mutation is particularly interesting, because it alters the distribution of crossovers and increases the frequency of recombination across certain genetic intervals *(64)*. The gene has been cloned and sequenced (Rose, personal communication). It shares identity with a family of helicases that includes *RAD3* of yeast and *XP-D* of humans.

The sequencing project has uncovered a number of other *C. elegans* genes with homologies to DNA damage-processing genes in other organisms. These include homologs to several helicases, DNA polymerases, replication factors, topoisomerases, cycling, reverse transcriptases, and yeast *RAD* genes *(62)*. These data have great potential, given the ability to do reverse genetics with *C. elegans* as well as the fact that these putative genes have not only been sequenced, but also positioned on the physical/genetic map (*see* Section 1.3.).

3. RADIATION-SENSITIVE MUTANTS: SPONTANEOUS MUTATION AND RECOMBINATION PHENOTYPES

A series of radiation-sensitive *(rad)* mutants have served to illuminate complexities of DNA damage processing in *C. elegans*. Nine such mutants were isolated and mapped on the basis of their embryonic UV or X-ray hypersensitivities *(23)*. Three of these (*rad-2*, *rad-3*, and *rad-4*) were also hypersensitive to methyl methane sulfonate; *rad-5* showed elevated levels of spontaneous mutability; another *(rad-4)* suppressed the frequency of meiotic nondisjunction by a factor of ten. Goldstein *(16)* demonstrated increased amounts of a specific structure in *rad-4* mutants, observable under the electron microscope, which he terms "disjunction regulator regions" (DRR). A positive correlation exists between meiotic nondisjunction and the number of DRRs in several strains of *C. elegans*, prompting the hypothesis that DRRs either promote disjunction or inhibit nondisjunction. The *rad-1–rad-7* mutants were tested for their effects on recombination between both autosomal and X-chromosomal markers. The *rad-4* mutation caused an enhancement of recombination frequency, whereas the *rad-6* gene led to a reduction of recombination frequency in a large chromosome X interval, but had no significant effect on an autosomal interval. The properties of these and other mutants that affect DNA-damage processing are summarized in Table 1.

4. EFFECTS OF DAMAGE ON VARIOUS BIOLOGICAL END POINTS

As opposed to the more common use of either single-celled organisms or disassociated cells in tissue culture, most studies with *C. elegans* employ living animals composed of a variety of differentiating and differentiated cells. As a consequence of such *in toto* investigations, developmental modulations and influences are necessarily investigated. This is reflected directly in that viability has most often been measured by exposing a staged population of animals to a particular DNA-damaging agent and measuring the fraction of animals that continue development to late larval stage or adulthood.

4.1. Embryonic and Larval Survival

Irradiated embryos were inactivated readily by UV and ionizing radiation, and showed a temporal sensitivity to ionizing radiation that broadly correlates with periods of active mitosis. Using 50-kV X-rays and precisely staged embryos, Ishii and Suzuki *(31)* established the time-course of sensitivity as measured by percent survival to at least the L4 larval stage. Mitotically active embryos had a D_{37} of <10 Gy, roughly comparable to that for mammalian cells. The D_{37} increased dramatically to 184 Gy during the second half of embryogenesis (the morphogenesis phase), during which there were no cell divisions or DNA synthesis. By contrast, irradiation with 254 nm UV showed a different pattern, as wild-type embryos were only ca. threefold more sensitive in the proliferative phase than in the morphogenesis phase *(18)*.

The survival of larvae following UV and X-irradiation has been measured and shown to increase with increasing age of the worms. For X-rays, the D_{37} dose for survival to adult rose from approx 200 Gy for newly hatched L1 larvae to 500 Gy for L2 larvae *(31)*. Hartman *(19)* tested the effects of liquid holding recovery on dauer larvae following 0–200 J/m^2 of 254 nm UV irradiation and found no enhancement of survival to adulthood. UV exposure did, however, impose a substantial developmental delay in recovery from dauer larva to L4 stage (from 1.8 d at 0 dose to 4–6 d at 156 J/m^2).

The importance played by development was dramatically apparent in the striking stage-specific variations in UV hypersensitivity of the *rad* mutants *(18)*. For example, *rad-1* animals were extremely UV hypersensitive if irradiated as embryos, but they displayed wild-type resistance when irradiated as first-stage larvae. Conversely, *rad-3* animals were moderately hypersensitive when irradiated as embryos, but L1s were highly hypersensitive (Fig. 3). In fact, three of the four *rad* mutants examined displayed different hypersensitivity patterns during development. These developmental differences could reflect stage-specific dependence on different DNA repair pathways. For example, rapidly proliferating embryonic cells might be more dependent on a different repair pathway than larval cells, which are mostly postmitotic. Alternatively, DNA repair may be developmentally regulated in *C. elegans*. Direct support for this hypothesis comes from the observation that the excision repair capacity of *rad-3* animals varies substantially throughout development; specifically, the excision repair defect in the mutant is much more pronounced in larvae than in embryos *(25)*. This developmental regulation is biologically relevant, because *rad-3* animals are much more hypersensitive as larvae (38-fold) than as embryos (4.6-fold). Interestingly, none of the four *rad* mutants were hypersensitive to 8-methoxypsoralen photoinactivation *(26)*. Split-dose experiments indicated that DNA–DNA crosslinks were primarily responsible for lethality. Little if any crosslink repair was detected in wild-type, which explains the absence of *rad* mutant hypersensitivity.

Wild-type male embryos and larvae were more sensitive to ionizing (but not UV) radiation than their hermaphrodite counterparts *(24)*. Taking advantage of two muta-

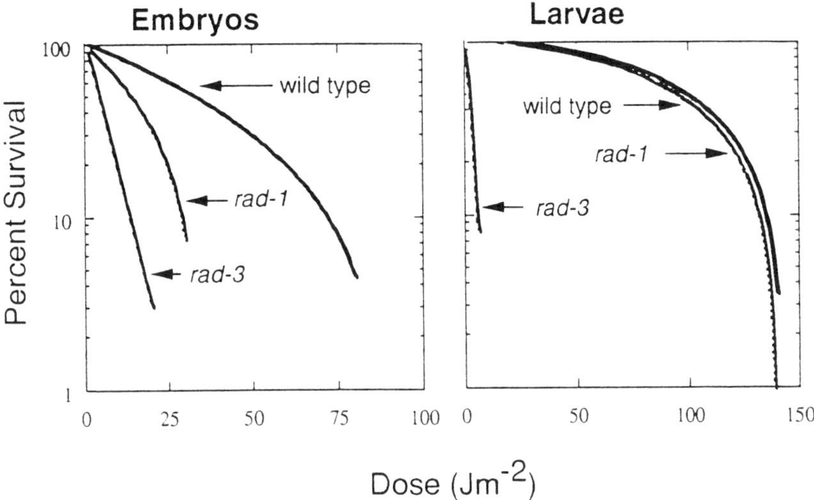

Fig. 3. UV radiation sensitivities of wild-type and two *rad* mutants when irradiated as either embryos or first-stage larvae. Taken from ref. *18* with permission.

tions that uncoupled sexual phenotype from genotype, it was demonstrated that damage to the X chromosome in somatic cells was the primary cause of this sex-specific difference; namely, hermaphrodites possess two X chromosomes whereas males possess only one. This difference in survival likely reflects the fact that recombinational repair is unavailable for the X chromosome of males, but can be utilized to repair X-chromosomal damage in hermaphrodites.

4.2. Damage-Resistant DNA Synthesis

Wild-type *C. elegans* embryos were highly resistant to the effects of UV on DNA synthesis and cell division *(27)*. For example, a fluence of 250 J/m^2 reduced DNA synthesis by half after 3 h, but synthesis returned to normal levels by 12 h. This resistance is much greater than that of mammalian cells, where fluences of >20 J/m^2 resulted in a rapid and permanent cessation of DNA synthesis. This "damage-resistant DNA synthesis" was not a result of a highly efficient excision repair system, because repair was saturated by fluences of greater than 50 J/m^2 (*25*; *see* Section 2.1.2.). Several lines of evidence indicate that it was not owing to poor penetration of germicidal radiation *(27)*.

The nature of this damage-resistant DNA synthesis was examined further using three experimental approaches. First, Hartman and associates *(27)* employed alkaline sucrose centrifugation to demonstrate that the size of nascent fragments exceeded the interdimer distance by up to 19-fold. This indicates that wild-type *C. elegans* embryos readily replicated through noninstructional lesions. Second, Jones and Hartman *(36)* examined DNA from unirradiated and UV-irradiated wild-type embryos using the electron microscope. Large fluences had little effect on either replication bubble size or distances between bubbles, indicating that the damage-resistant DNA synthesis was not grossly aberrant. The small replicon sizes—inferred from the DNA preparations—is consistent with damage-resistant DNA synthesis, because replicon size and recovery of DNA synthesis after irradiation are inversely proportional *(5)*. The third approach involved measurements of lineage-specific cell-cycle delay and is discussed in the next section.

4.3. Specific Embryonic Targets

Several recent studies have focused on the effects of DNA damage on specific cell lineages, thus utilizing the greatest experimental strength of *C. elegans*. Using Nomarski DIC microscopy, Buckles and Hartman *(17)* observed the effects of UV on cellular development of two-celled embryos. They found that the delays imposed by UV radiation were much more profound in cells destined to give rise to the germ line vs somatic progenitors. For example, although exposure to 150 J/m^2 abolished cytokinesis in somatic AB descendants, which give rise to soma exclusively, the next two cell cycles were only 22% longer than in unirradiated controls. Conversely, the equivalent cell cycles, which give rise to the germ-line precursor P_4, averaged 95% longer. This suggests the intriguing possibility that cell-cycle checkpoints are largely inoperative in embryonic progenitors to the soma, but are at least partially functional in the cells which give rise to the germ line. Such an absence in somatic precursors may be a consequence of *C. elegans* essentially invariant pattern of cellular development; i.e., the nematode has evolved efficient mechanisms to replicate through damage in order to ensure the successful and orderly completion of embryogenesis. Alternatively, this difference may reflect that genome integrity is a higher priority in germ-line precursors than in somatic precursors.

Honda and Nelson *(46)* have shown that intestinal nuclei were induced to undergo one or two supernumerary rounds of division in response to embryo exposure to chemically induced oxidative stress. This response occurred in a dose-dependent fashion for embryos of developmental stages corresponding to the middle third of embryogenesis. Specific intestinal cells showed different sensitivities than their neighbors or siblings. Overexpression of the *bcl-2* oncogene analog *ced-9*, which regulates cytoplasmic free radical levels and programmed cell death, suppressed this escape from proliferation regulation. Extra intestinal nuclear divisions can also be induced with densely ionizing particles, such as accelerated iron ions (35% of worms undergo at least one extra division at 32 Gy), but not with up to fourfold higher doses of γ-rays *(46)*. These observations suggest that the regulatory pathway for cell division has a redox-sensitive component, and that densely ionizing particles can produce local oxidative environments similar to those produced by free-radical generating compounds and high O_2 concentrations.

4.4. Gonadogenesis and Fertility

As with the rest of *C. elegans* cellular development, specific cells give rise to the gonad. The developmental program for gonadogenesis and production of a brood of offspring offers a readily quantifiable end point for radiation injury. Successful completion of the program requires multiple rounds of cell division, cell migration, and signal transduction properly sequenced in time and position. When properly integrated, the processes result in a brood of 280 offspring. Defects traceable to inhibition of specific cell functions can also be quantified. Kimble *(39)* has followed the gonad cell lineages and examined results from laser ablation of gonad cells. These studies have shown that germ-line and somatic cells develop mostly autonomously, and that discrete substitutions of cellular identity may occur on loss or damage to individuals.

Brood sizes are readily scored following irradiation of newly hatched larvae containing the discrete cellular target of four gonad primordial cells Z1–Z4. Standard "survival curves" measuring brood size vs dose show a smoothly decreasing fertility with dose using UV *(18)* as well as various ionizing radiation sources *(42)*. Charged particles showed a strong LET dependence. The smoothly varying survival curve was inconsistent with results from laser ablation studies *(39)* if cell death was the primary cause of reduced fertility. Laser ablation of a Z-cell effectively inactivated one-half of the gonad. Losses of cell function(s) without cell death or cell loss followed by repopulation could explain the smooth variation.

The developmental stage-dependence of fertility was examined by exposing synchronized populations of larvae to ionizing radiation at constant dose, but as a function of age *(48)*. Sensitivity increased sharply from 7–18 h posthatching and then decreased sharply again. The period of maximum sensitivity corresponds to an animal of 4–20 germ and 8–12 somatic cells when cell division rate was low. A similar period of sensitivity was evident using near-UV epi-illumination of psoralen-treated worms viewed under a compound microscope *(12)*. These and other data *(12)* suggest that damage to actively transcribing genes is sensitizing relative to the quiescent state. Finally, a variety of experimental evidence indicates that exposure to a single densely ionizing particle was sufficient to inactivate several different cells with specific functions in gonadogenesis *(48)*.

4.5. Gene and Chromosomal Mutations

Several individual loci for which strong selections are available (e.g., *unc-22*, *fem-3*) have been used to generate mutation spectra following radiation and chemical mutagenesis. In addition, the reciprocal translocation eTI (III;V) has been extensively employed as a balancer chromosome to measure mutation frequencies. The molecular spectra of various mutagens has been reviewed recently *(2)*.

UV dose vs response curves were measured for the eT1 region *(8)*. Mutation frequencies rose sharply to 4.7% of surviving F_1 offspring at a fluence of 100 J/m^2, plateauing at higher fluences. The *rad-3* mutation rendered animals hypermutable, whereas *rad-1* and *rad-7* animals were hypomutable. Hartman and coworkers *(22)* measured the mutation frequencies at the *fem-3* locus and found a similar pattern. Stewart and associates *(52)* characterized the spectrum of mutants induced in the eT1 region by 120 J/m^2 of 254 nm UV. Surprisingly, a large fraction of the mutants were not simple intragenic events, but were instead deletions or other chromosomal mutations, a spectrum similar to that recovered after γ-irradiation. By contrast, only 1 of 48 *fem-3* mutations showed a large-scale change with Southern hybridization, suggesting that most were point mutations *(22)*. Thus, the molecular spectrum of mutations recovered with the *fem-3* and *unc-22* systems were quite different, a disparity that has not yet been resolved.

Ionizing radiation is a potent mutagen in *C. elegans* and has been quantified using one specific locus *(unc-22)* and a large autosomal region balanced by the eT1 translocation *(48,49)*. Sparsely ionizing radiation (such as X-rays and γ-rays) produce relatively uniform distributions of ionization in target cells, and particulate radiations produce highly structured tracks (*see* Chapter 5, vol. 2). The dense pattern of ionization in the particle tracks has important biological consequences *(3)*. These include clustering of damage, unresponsiveness to oxygen concentration, complex dose-rate effects, and low repair rates. A measure of the density of ionization is given by the parameter Linear Energy Transfer (LET), which is energy deposited/unit track length.

At the *unc-22* locus, mutation by ionizing radiation increased linearly with doses up to 70 Gy and there was a strong LET dependence as measured with charged particles and neutrons. The spectrum of mutants in *unc-22* was also LET-dependent; specifically, the proportion and size of particle-induced deletions increased with increasing LET. Using the eT1 lethal screen, the linear mutation frequency was confirmed for all radiation species tested. An oxygen enhancement ratio of greater than twofold was observed for γ-rays. This ratio decreased with increasing LET. Mutations in the *rad-1*, *rad-3*, *rad-7*, and *nuc-1* genes had little effect on mutation frequency. Calculations of mutation frequency vs particle fluence resulted in a 0.36 probability of inducing at least one autosomal lethal mutation/incident particle. Mutation frequencies were determined using both *unc-22* and eT1 in dauer larvae, where the targets were mitotic gonadal cells that had not yet differentiated into gametes. Mutation frequencies were linear with dose, but much lower than for mature gametes. The reduced mutation rate in dauer larvae may reflect differences in geometric cross-section, longer available repair times, or regulation of expression via programmed cell death of damaged cells. The mutation frequency versus responses to UV and ionizing radiation in *C. elegans* parallel those in mammalian cells using *HPRT, TK*, and *APRT* loci. relative biological effect (RBE) vs LET relationships for charged particle mutagenesis also parallel the mammalian pattern *(6)*. Thus, the nematode serves well as a higher eukaryotic mutagenesis model, which complements and extends mammalian studies.

C. elegans has been used as a biological dosimeter for assessing risks to astronauts for naturally occurring high LET particles in space *(46,47)*. Nematodes were flown aboard space shuttle Discovery in January 1992, in a configuration that allowed eT1 dauer larvae struck by identified cosmic rays to be assayed for mutation. Fifty-three autosomal mutants were recovered, along with 13 additional *unc-22* mutants, including at least one large deletion. More recently, over *20 fem-3* mutants were recovered from dauer larvae, which were flown aboard space shuttle Atlantis in March 1996 *(17,46)*. Collectively, these data indicate a mutation frequency two to three times higher than the spontaneous frequency of ground controls. Thus, the densely ionizing radiation species present in the space radiation environment are extremely effective mutagens and represent a health risk disproportionate to their absorbed dose. Analysis of space radiation-induced *C. elegans* mutant "spectra" will inform as to the nature of the radiogenic damage and aid in the development of countermeasures.

Sadaie and Sadaie *(50)* detected chromosomal aberrations in embryos irradiated *in utero* with ^{137}Cs γ-rays or 254 nm UV. Aberration induction was approximately linear with doses up to 80 Gy γ-rays. For UV, the dose–response curve rose to 20% at 10 J/m^2 and plateaued. *rad-1* responses were similar to wild-type, but *rad-2* strongly potentiated both responses. Chromosomal aberrations decreased after irradiation, suggesting many were repaired; *rad-2* embryo cells were much less effective than wild-type in repairing the lesions (35 vs 70% restoration at 48 h). When gonads were the target cells, both charged particles and ionizing radiation were efficient means to generate duplications and translocations *(29,46)*.

For a well-defined set of mitotic cells, Nelson and coworkers *(47,48)* have measured stable anaphase bridge formation in binucleate intestinal cells. In a systematic survey of aberration induction by radiations of different quality (e.g., γ-rays, charged particles, neutrons), there was a positive correlation between aberration induction and LET. Aberration induction showed an oxygen enhancement of ca. three, which decreased with increasing LET. Three mutants *(rad-2, mev-1, nuc-1)* were found to potentiate this response. Dose–response curves for individual cells were determined and int-5 cells were substantially more radioresistant than their siblings and neighbors, suggesting cell-specific mechanisms of damage tolerance. The pattern of development in *C. elegans* implicates gonadal regulation of the int-5 response. An additional spatial pattern of sensitivity was revealed in the *mev-1* mutant, where anterior intestinal cells showed a high rate of spontaneous anaphase bridge formation.

4.6. Modulation of Recombination

McKim and Rose *(43)* measured the rate of recombination between markers on chromosome I. They sampled regions of the LGI cluster, presumably associated with the dominant centromere, as well as the relatively gene-deficient left arm. Following irradiation with ^{60}Co γ-rays at 10–40 Gy, recombination in the gene cluster increased linearly with dose, up to a factor of almost three. There was no effect of irradiation on recombination on the left arm. Recombination frequencies were sensitive to the developmental stage of the gametes; specifically, frequencies were highest when animals were irradiated 36 h prior to the initiation of egg laying. Induced sterility of recombinant individuals was also noted (up to 30% vs 0% for nonrecombinants). In the same study, a fivefold increase of meiotic nondisjunction of the X-chromosome was observed following a dose of 20 Gy. The response was developmental stage-dependent and peaked for gametes present 24–36 h prior to egg laying.

Recombination in embryonic intestine cells was demonstrated by Siddiqi and Babu *(51)* in a study of radiation-induced mosaics heterozygous for the recessive *flu-3 II* mutation which causes autofluorescence of homozygous gut cells. With 20 Gy of X-rays, up to 0.06% of intestines had fluorescent patches of cells. As expected, the patch sizes were inversely correlated with the age of the embryo at the time of irradiation. Mosaic patch formation was presumably associated with induced mitotic recombination, but the inconsistency of patch patterns with lineages made their origin somewhat uncertain.

4.7. Aging/Oxidative Stress

Considerable effort has been directed toward understanding the genetic basis of aging and responses to oxidative damage in *C. elegans.* They are treated together here since these investigations have often converged. The relationship between aging and response to oxidative stress in *C. elegans* was first suggested by Ishii's characterization of a *mev-1* mutation, which conferred:

1. Hypersensitivity to oxygen and methyl viologen (a singlet oxygen generator);
2. Reduced levels of SOD; and
3. A reduced life-span, especially under high oxygen concentrations *(33)*.

Ishii and colleagues *(32)* also demonstrated that a mutation in *rad-8*, first isolated in a screen for X-ray-hypersensitive mutants *(23)*, also reduced life-span and caused hypersensitivity to high oxygen concentrations and methyl viologen. Two other *mev* mutants have been isolated and characterized (Ishii, personal communication).

The ability of mutations in several genes to extend life span is particularly intriguing, with some strains living up to five times as long as wild-type animals *(40)*. In the case of *age-1*, the importance of oxidative stress in the aging process was suggested by the fact that *age-1* mutants, which live ca. 60% longer than wild-type, were hyperresistant to oxidative stress *(41,56)*. Moreover, levels of both catalase and SOD increased as *age-1* mutants (but not wild-type animals) grew older. It is curious that life-span can be extended by mutations in disparate genes, such as one affecting sperm production *(spe-26)* and another *(daf-2)* that turns on dauer development (a specialized, facultative larval stage) *(38,57)*. Although a complete explanation for these correlations is not available, Johnson and associates *(42*; Johnson, personal communication) have recently compared these mutants, and observed a positive correlation between life-span and increases in both thermotolerance and resistance to UV radiation. The same group reported recently that deletions in the mitochondrial genome accumulated more slowly in an *age-1* strain than its wild-type equivalent *(44)*.

The polygenes that influence life-span also overlap with those that control sensitivity to free radical damage. These polygenes have been studied using recombinant inbred (RI) strains, generated by crossing two wild-type strains that have life-spans ranging from 10–31 d *(35)*. Both catalase and SOD levels varied in different wild-type strains, which provides an opportunity to map genes that control expression of these two enzymes, as well as control sensitivity to acute hydrogen peroxide exposure *(11)*. Hartman and associates *(21)* observed that there was an inverse relationship between life-span and sensitivity to 95% oxygen and methyl viologen; that is, development of short-lived strains was inhibited more profoundly than was development of long-lived strains. Given that no differences in excision repair were observed between the shortest-lived and the longest-lived RI, and that life-span did not correlate with the sensitivity to UV and ionizing radiation-induced damage *(28)*, it was previously concluded that DNA repair capacity plays, at best, a minor role in the aging process in *C. elegans*. However, it is possible that subtle differences in DNA repair may exist between the RI strains that impact aging.

Johnson and Hartman *(34)* tested the effects of 10–3000 Gy of ^{137}Cs γ-rays on life-span using staged populations of worms and a liquid medium-based method for assaying individual worm survival. They found that 24-h (L3), 48-h (L4) dauer larvae, and 8-d-old adults required in excess of 300–1000 Gy to shorten life-span substantially. These data agree with those of Yeagers *(63)*, who used different culture conditions for assessing survival of dauer larvae. Dauer larvae were most sensitive and a small extension of life-span was sometimes observed. Mutations in a number of genes that confer radiation resistance *(rad-1–rad-7)* had little effect on life-span.

4.8. Developmental Complexities as Revealed by rad *Mutants*

As described in previous sections, nine radiation-sensitive *(rad)* mutants have been isolated and characterized. More extensive work with the four most sensitive *rad* mutants, *rad-1, rad-2, rad-3*, and *rad-7,* initially yielded a consistent picture. For example, double and triple mutants of these were constructed and radiation sensitivities assessed *(20)*. Results suggested that *rad-1* and *rad-2* defined one epistasis group, whereas *rad-3* and *rad-7* defined another. The inference from these data is that, as in yeast, the epistasis groups correspond to different "DNA repair pathways," or perhaps more accurately, different mechanisms for processing DNA damage. Consistent with this model was the observation that *rad-1* and *rad-2* (but not *rad-3* and *rad-7)* were hypersensitive to γ-rays for inhibition of embryogenesis. However, this tidy picture was complicated by additional studies. For example, although *rad-7* is not sensitive to γ-rays for embryogenesis, it is sensitive for gamete survival to both γ-rays and charged particles *(46)*. In addition, *rad-1* and *rad-7* are in different epistatic groups, yet both render animals hypomutable to UV radiation *(8)*. Similarly, *rad-3* and *rad-7* are in the same epistatic group, yet *rad-3* animals are excision-repair-defective, whereas *rad-7* animals are not *(25)*. Also, *rad-3* animals are hypermutable to UV while *rad-7* animals are UV hypomutable. Finally, mutations in *rad-1* and *rad-2* produce a variety of responses, which suggest the genes do not control a single DNA pathway:

1. Whereas *rad-1* is wild-type for radiation-induced anaphase bridge formation, *rad-2* renders animals hypersensitive to this end point *(46)*;
2. *rad-1* animals are hypomutable to γ- and heavy-ion radiation, *rad-2* mutants are hypermutable to high and low LET ionizing radiation but hypomutable to UV *(8,46)*; and
3. *rad-2,* but not *rad-1*, blocks the decrease in embryonic chromosomal aberrations observed with incubation after irradiation *(50)*.

Clearly, further studies are necessary to unravel this bewildering complexity of radiation responses. In addition, the epistatic relationships of *nuc-1*, *mev-1*, and other genes have yet to be determined with respect to the *rad* genes.

5. CLOSING COMMENTS

Given the relative paucity of investigators who employ *C. elegans* to study DNA repair, mutagenesis, and radiation biology, the field will not likely mature to the comprehensive level of, for example, those of yeast, bacterial, or mammalian DNA repair. However, it is equally apparent that the strengths of this nematode can be exploited to provide perspectives and insights that are not as readily available with other organisms. Thus, research with *C. elegans* serves to complement rather than merely duplicate efforts in more established organisms. Perhaps most importantly, and as detailed above, experiments with *C. elegans* measure the effects of radiation on intact organisms, composed of cells with a variety of developmental fates. Conversely, most DNA repair studies involving metazoa employ disassociated cells, usually immortalized, in tissue culture. As a consequence, future research with this nematode should serve to illuminate the developmental modulations of DNA damage processing at the molecular, cellular, and organismal levels.

REFERENCES

1. Anderson, G. L. 1981. Superoxide dismutase activity in dauer larvae of *Caenorhabditis elegans* (Nematode: Rhabditidae). *Can. J. Zool.* **60:** 288–291.
2. Anderson, P. 1995. Mutagenesis, in Caenorhabditis elegans: *Modern Biological Analysis of an Organism*, vol. 48, *Methods in Cell Biology* (Epstein, H. F. and D. C. Shakes, eds.), Academic, New York, pp. 31–58.
3. Blakely, E. A., F. Q. H. Ngo, S. B. Curtis, and C. A. Tobias. 1984. Heavy-ion radiobiology: Cellular studies. *Adv. Radiat. Biol.* **11:** 295–389.

4. Brenner, S. 1974. The genetics of *Caenorhabditis elegans*. *Genetics* **77:** 71–94.
5. Cleaver, J. E., W. K. Kaufmann, L. N. Kapp, and S. D. Park. 1983. Replicon size and excision repair as factors in the inhibition and recovery of DNA synthesis from ultraviolet damage. *Biochim. Biophys. Acta* **739:** 207–215.
6. Chen, D. J., K. Tsuboi, T. Nguyen, and T. C. Yang. 1994. Charged-particle mutagenesis II. Mutagenic effects of high energy charged particles in normal human fibroblasts. *Adv. Space Res.* **14:** 347–354.
7. Colloms, S. D., H. vanLuenen, and R. Plasterk. 1994. DNA binding activities of the *Caenorhabditis elegans* Tc3 transposon. *Nucleic Acid Res.* **22:** 5548–5554.
8. Coohill, T., T. Marshall, W. Schubert, and G. Nelson. 1988. Ultraviolet mutagenesis of radiation-sensitive *(rad)* mutants of the nematode *Caenorhabditis elegans*. *Mutat. Res.* **209:** 99–106.
9. Darr, D. and I. Fridovitch. 1995. Adaptation to oxidative stress in young but not mature or old *Caenorhabditis elegans*. *Free Radical Biol. Med.* **18:** 195–201.
10. Driscoll, M. 1992. Molecular genetics of cell death in the nematode *Caenorhabditis elegans*. *J. Neurosci.* **23:** 1327–1351.
11. Ebert, R. H., V. A. Cherkasova, R. A. Dennis, J. H. Wu, S. Ruggles, T. E. Perrin, and R. J. S. Reis. 1993. Longevity-determining genes in *Caenorhabditis elegans:* chromosomal mapping of multiple noninteractive loci. *Genetics* **135:** 1003–1010.
12. Edgar, L. G. and D. Hirsh. 1988. Use of a psoralen-induced phenocopy to study genes controlling spermatogenesis in *Caenorhabditis elegans*. *Dev. Biol.* **111:** 108–118.
13. Epstein, H. F. and D. C. Shakes (eds.) 1995. Caenorhabditis elegans: *Modern Biological Analysis of an Organism*, vol. 48, *Methods in Cell Biology*, Academic, New York.
14. Giglio, M-P., T. Hunter, J. V. Bannister, W. H. Bannister, and G. J. Hunter. 1994. The manganese *sod* gene of *C. elegans*. *Biochem. Mol. Biol. Int.* **33:** 37–40.
15. Giglio, M.-P., T. Hunter, J. V. Bannister, W. H. Bannister, and G. J. Hunter. 1994. The copper/zinc sod gene of *C. elegans*. *Biochem. Mol. Biol. Int.* **33:** 41–44.
16. Goldstein, P. 1984. The synaptonemal complexes of *Caenorhabditis elegans:* Pachytene karyotype analysis of the *rad-4* radiation-sensitive mutant. *Mutat. Res.* **129:** 337–343.
17. Hartman, P. S. Unpublished results.
18. Hartman, P. S. 1984. UV Irradiation of wild type and radiation-sensitive mutants of the nematode *Caenorhabditis:* fertilities, survival, and parental effects. *Photochem. Photobiol.* **39:** 169–175.
19. Hartman, P. S. 1984. Effects of age and liquid holding on the UV-radiation sensitivities of wild-type and mutant *Caenorhabditis elegans* dauer larvae. *Mutat. Res.* **132:** 95–99.
20. Hartman, P. S. 1985. Epistatic interactions of radiation-sensitive *(rad)* mutants of *Caenorhabditis elegans*. *Genetics* **109:** 81–93.
21. Hartman, P. S., E. Childress, and T. Beyer. 1996. Nematode development is inhibited by methyl viologen and high oxygen concentrations at a rate inversely proportional to life span. *J. Gerontol.* **50A:** B322–B326.
22. Hartman, P. S., D. De Wilde, and V. N. Dwarakanath. 1995. Genetic and molecular analyses of UV radiation-induced mutations in the *fem-3* gene of *Caenorhabditis elegans*. *Photochem. Photobiol.* **61:** 607–614.
23. Hartman, P. S. and R. K. Herman. 1982. Radiation-sensitive mutants of *Caenorhabditis elegans*. *Genetics* **102:** 159–178.
24. Hartman, P. S. and R. K. Herman. 1982. Somatic damage to the X chromosome of the nematode *Caenorhabditis elegans* induced by gamma radiation. *Mol. Gen. Genet.* **187:** 116–119.
25. Hartman, P. S., J. Hevelone, V. Dwarakanath, and D. L. Mitchell. 1989. Excision repair of UV radiation-induced DNA damage in *Caenorhabditis elegans*. *Genetics* **122:** 379–385.

26. Hartman, P. S. and A. Marshall. 1992. Inactivation of wild-type and *rad* mutant *Caenorhabditis elegans* by 8-methoxypsoralen and near ultraviolet radiation. *Photochem. Photobiol.* **55:** 103–111.
27. Hartman, P. S., J. Reddy, and B.-A. Svendsen. 1991. Does trans-lesion synthesis explain the UV-radiation resistance of DNA synthesis in *C. elegans* embryos? *Mutat. Res.* **255:** 163–173.
28. Hartman, P. S., V. J. Simpson, T. Johnson, and D. L. Mitchell. 1988. Radiation sensitivity and DNA repair in *Caenorhabditis elegans* strains with different mean life spans. *Mutat. Res.* **208:** 77–82.
29. Herman, R. K. 1995. Mosaic analysis, in Caenorhabditis elegans: *Modern Biological Analysis of an Organism*, vol. 48, *Methods in Cell Biology* (Epstein, H. F. and D. C. Shakes, eds.), Academic, New York, pp. 123–146.
30. Hevelone, J. and P. S. Hartman. 1988. An endonuclease from *Caenorhabditis elegans:* partial purification and characterization. *Biochem. Gen.* **26:** 447–461.
31. Ishii, N. and K. Suzuki. 1990. X-ray inactivation of *Caenorhabditis elegans* embryos or larvae. *Int. J. Radiat. Biol.* **S8:** 827–833.
32. Ishii, N., N. Suzuki, P. S. Hartman, and K. Suzuki. 1993. The radiation-sensitive mutant *rad-8* of *Caenorhabditis elegans* is hypersensitive to the effects of oxygen on aging and development. *Mech. Ageing Dev.* **68:** 1–10.
33. Ishii, N., K. Takahashi, S. Tomita, T. Keino, S. Honda, K. Yoshino, and K. Suzuki. 1990. A methyl viologen-sensitive mutant of the nematode *Caenorhabditis elegans. Mutat. Res.* **237:** 165–171.
34. Johnson, T. E. and P. S. Hartman. 1988. Radiation effects on life span in *Caenorhabditis elegans. J. Gerontol.* **43:** B137–141.
35. Johnson, T. E. and W. Wood. 1979 Genetic analysis of life-span in *Caenorhabditis elegans. Proc. Natl. Acad. Sci. USA* **90:** 8905–8909.
36. Jones, C. A. and P. S. Hartman. 1996. Replication in UV-irradiated *Caenorhabditis elegans* embryos. *Photochem. Photobiol.* **63:** 187–192.
37. Keller, C. I., J. Calkins, P. S. Hartman, and C. S. Rupert. 1987. UV photobiology of the nematode *Caenorhabditis elegans:* action spectra, absence of photoreactivation and effects of caffeine. *Photochem. Photobiol.* **46:** 483–488.
38. Kenyon, C., J. Chang, E. Gensch, A. Rudner, and R. Tabiang. 1993. A *C. elegans* mutant that lives twice as long as wild type. *Nature* **366:** 451–464.
39. Kimble, J. 1988, Genetic control of sex determination in the germ line in *Caenorhabditis elegans. Phil. Trans. R. Soc. Lond.* **B322:** 11–18.
40. Lakowski, B. and S. Hekimi. 1996. Determination of life-span in *Caenorhabditis elegans* by four clock genes. *Science* **272:** 1010–1013.
41. Larsen, P. L. 1993. Aging and resistance to oxidative damage in *Caenorhabditis elegans. Proc. Natl. Acad. Sci. USA.* **90:** 8905–8909.
42. Lithgow, G. J., T. White, S. Melov, and T. E. Johnson. 1995. Thermotolerance and extended life-span conferred by single-gene mutations and induced by thermal stress. *Proc. Natl. Acad. Sci. USA* **92:** 7540–7544.
43. McKim, K. S. and A. M. Rose. 1994. Spontaneous duplication loss and breakage in *Caenorhabditis elegans. Genome* **37:** 595–606.
44. Melov, S., G. J. Lithgow, D. R. Fisheer, P. M. Tedsco, and T. E. Johnson. 1995. Increased frequency of deletions in the mitochondrial genome with age of *Caenorhabditis elegans. Nucleic Acids Res.* **23:** 1419–1425.
45. Munakata, N. and F. Morohoshi. 1986. DNA glycosylase activities in the nematode *Caenorhabditis elegans. Mutat. Res.* **165:** 101–107.
46. Nelson, G. A. Unpublished results.
47. Nelson, G. A., W. W. Schubert, G. A. Kazarians, G. F. Richards, E. V. Benton, E. R. Benton, and R. Henke. 1994. Radiation effects in nematodes: Results from IML-1 experiments. *Adv. Space Res.* **14:** 87–91.

48. Nelson, G. A., W. W. Schubert, and T. M. Marshall. 1992. Radiobiological Studies with the nematode *Caenorhabditis elegans.* genetic and developmental effects of high LET radiation. *Nucleic Tracks Radial. Meas.* **20:** 227–232.
49. Nelson, G. A., W. W. Schubert, T. M. Marshall, E. R. Benton, and E. V. Benton. 1989. Radiation effects in *Caenorhabditis elegans,* mutagenesis by high and low LET ionizing radiation. *Mutat. Res.* **212:** 181–192.
50. Sadaie, T. and Y. Sadaie. 1989. Rad-2-dependent repair of radiation-induced chromosomal aberrations in *Caenorhabditis elegans. Mutat. Res.* **218:** 25–31.
51. Siddiqi, S. S. and P. Babu. 1980. Genetic mosaics of *Caenorhabditis elegans*: A tissue-specific fluorescent mutant. *Science* **210:** 330–332.
52. Stewart, H. I., R. E. Rosenbluth, and D. L. Baillie. 1991. Most ultraviolet irradiation induced mutations in the nematode *Caenorhabditis elegans* are chromosomal rearrangements. *Mutat. Res.* **249:** 37–54.
53. Sulston, J. E. and H. R. Horvitz. 1977. Post-embryonic cell lineages of the nematode *Caenorhabditis elegans. Dev. Biol.* **56:** 110–156.
54. Sulston, J. E., E. Schierenberg, J. G. White, and J. N. Thomson. 1983. The embryonic cell Lineage of the nematode *Caenorhabditis elegans. Dev. Biol.* **100:** 64–119.
55. Vanfleteren, J. R. 1992. Cu-Zn superoxide dismutase from *Caenorhabdinis elegans:* purification, properties and isoforms. *Comp. Biochem. Physiol.* **B102:** 219–229.
56. Vanfleteren, J. R. 1993. Oxidative stress and ageing in *Caenorhabditis elegans. Biochem. J.* **292:** 605–608.
57. Van Voorhies, W. A. 1992. Production of sperm reduces nematode lifespan. *Nature* **360:** 456–458.
58. Voss, J. C., H. VanLuenen, and R. Plasterk. 1993. Characterization of the *Caenorhabditis elegans* Tcl transposase *in vivo* and *in vitro. Genes Dev.* **7:** 1244–1253.
59. Voss, J. C. and R. Plasterk. 1994. Tc1 transposase of *Caenorhabditis elegans* is an endonuclease with a bipartite DNA binding domain. *EMBO J.* **13:** 6125–6132.
60. Wang, W. and D. C. Shakes. 1994. Isolation and sequence analysis of a *Caenorhabditis elegans* cDNA which encodes a 14-3-3 homologue. *Gene* **147:** 215–218.
61. Wood, W. B. (ed.) 1988. *The Nematode* Caenorhabditis elegans. Cold Spring Harbor Laboratory, Cold Spring Harbor, NY.
62. Wilson, R. 1994. 2.2 Mb of contiguous nucleotide sequence from chromosome III of *C. elegans. Nature* **368:** 32–38.
63. Yeagers, E. 1981. Effect of gamma-radiation on dauer larvae of *Caenorhabditis elegans. J. Nematol.* **13:** 235–237.
64. Zetka, M. and A. Rose. 1995. The genetics of meiosis in *Caenorhabditis elegans. Trends Genet.* **11:** 27–31.
65. Zhen, M., R Heinlein, D. Jones, S. Jentsch, and E. P. M. Candido. 1993. The ubc-2 gene of *Caenorhabditis elegans* encodes a ubiquitin-conjugating enzyme involved in selective protein degradation. *Mol. Cell Biol.* **13:** 1371–1377.
66. Zuckerman, B. M. (ed.) 1980. *Nematodes as Biological Models*, vol. 1, *Behavioral and Developmental Models*. Academic, New York.
67. Zuckerman, B. M. (ed.) 1980. *Nematodes as Biological Models*, vol. 2, *Aging and Other Model Systems*. Academic, New York.

25

DNA Repair in Higher Plants

Anne B. Britt

1. INTRODUCTION: WHY PLANTS?

The study of DNA repair and mutagenesis in humans has direct applications to carcinogenesis and cancer therapy, as well as aging and the induction of developmental abnormalities. Bacterial systems have been useful models for repair processes in humans. Some animal systems (notably *Drosophila melanogaster* and *Caenorhabditis elegans*) will become invaluable as the study of repair processes advances into the area of developmental regulation. We do not know whether the DNA repair strategies of higher plants are similar enough to those of animals that plants can also make a contribution, as a model system, to the study of the carcinogenesis and aging. However, given the radically different developmental strategies of plants vs higher animals, the study of the developmental and environmental regulation of DNA damage repair, damage tolerance, and mutagenesis will certainly benefit from a comparison of the two systems.

DNA repair and damage tolerance is not only relevant to the study of carcinogenesis (although this is certainly the rationale for its high level of support), but also is involved in the both the limitation and production of genetic diversity, through the balance between error-free repair pathways and mutagenic DNA damage tolerance pathways. Genetic diversity is required both for evolution and for the artificial selection of new varieties of plants. Because of fundamental differences in the developmental strategies of plants and animals, plants are far more susceptible than animals to the accumulation of germinal mutations. Although in animals the germ line is effectively set aside from the somatic tissues very early in development, in plants all organs derive from a thin cap of meristematic tissue, which divides and differentiates throughout the life of the plant to produce both somatic and gametic tissue. The continuous growth of the meristem would be expected to result in the accumulation of germ-line mutations. In mangrove trees *(Rhizophora mangle)*, where many cell divisions, as well as many years, separate one generation from the next, the spontaneous mutation rate to albinism has been shown to be 25 times higher than in several annual species *(44)*, supporting the notion that mutations accumulate in the meristems of long-lived plant species.

However, the induction of new, random mutations is probably less damaging to plants than to animals. The flexible meristematic development of plants might enable them to "test" new mutations within an individual, rather than within a population.

From: DNA Damage and Repair, Vol. 1: DNA Repair in Prokaryotes and Lower Eukaryotes
Edited by: J. A. Nickoloff and M. F. Hoekstra © Humana Press Inc., Totowa, NJ

Dominant deleterious mutations in the meristem will soon be outgrown by healthy tissue, whereas an early mutation that provides the meristem with a significant selective advantage over nonmutant tissue may result in enhanced growth and seed production by the mutant clone *(43,93)*. In addition, because plant cells are fixed in place by their walls, tumors generated by somatic mutation cannot metastasize. A second evolutionarily important difference between the two kingdoms is the presence of a haploid (gametophytic) stage of the life cycle in plants. In animals, a germinal mutation is usually selected for or against only in subsequent generations and so contributes to "genetic load." Because higher plant development includes several cell divisions of haploid tissue (as well as selection during the growth of the pollen tube), many cell-lethal mutations, even when recessive, are transmitted at a reduced frequency, or are completely nontransmissible.

As a result of the developmental pattern of plants, the downside of mutagenesis (toxicity and carcinogenicity) is greatly reduced, whereas its beneficial effects (the rare production of novel, useful genotypes) can in some cases be immediately put to use. For this reason, it will be interesting to determine whether the relative levels of expression of error-prone vs error-free repair/tolerance pathways in plants are different from those in animals. It is also possible that the expression of these pathways is regulated by relatively subtle environmental and developmental signals.

A second reason for studying repair processes in plants is the practical understanding of the role of DNA repair in plant resistance to natural and artificial toxins, some of which may act primarily as DNA-damaging agents. Unlike animals, plants are continuously exposed to ultraviolet radiation (UV). Although artificially high doses of UV have an obvious effect on plant growth and fertility, the effects of solar UV are more subtle *(74)*, and we have yet to determine whether they are the result of DNA damage, damage to other cellular components, or a programmed developmental response to UV light. Mutants defective in the repair of UV-induced damage would enable us to distinguish among these possibilities.

Finally, the study of recombination as a damage tolerance pathway has direct applications to the genetic engineering of plants. We currently have little or no understanding of the molecular mechanisms involved in either illegitimate or meiotic recombination in plants. As a result, our ability to transform plants genetically, or to introgress a desirable cluster of genes from one plant species to another, is remarkably clumsy and limited in scope. The study of DNA repair and DNA damage tolerance mechanisms in higher plants will enable us to understand and manipulate the physical processes that are, essentially, the tools of the genetic engineer.

2. REPAIR PATHWAYS IN PLANTS

2.1. *Assaying Repair* In Planta

Assaying the induction and repair of DNA damage in any large, multicellular organism is extremely problematic. Undoubtedly the most serious technical obstacle is the difficulty involved in inducing a homogenous distribution of DNA damage. With the exception of γ- and X-radiation, most biological tissues absorb DNA-damaging agents with a high degree of efficiency, resulting in a high concentration of damage products on the epidermis, and a much lower concentration in the organism's interior. Obviously, it is possible to UV irradiate a plant, grind it up at various times after irradiation, and use a sensitive immunoassay to follow the concentration of antigen per microgram DNA, but the result would be difficult to interpret. A decrease in dimer concentration, though quite real, could represent repair, degradation of overirradiated DNA in the now-dead epidermal cells, or replication of the unirradiated DNA in the interior cells. One solution to this problem is to work with finely divided tissue cultures, although it is not possible to study the developmental regulation of repair processes in such cultures. This method was commonly employed in the past (reviewed in ref. *49*), and has recently been used to study replication and repair in soybean cultures *(12)*. Another method is to work with unicellular plants, pollen grains *(38)*, very thin plants, or epidermal peels of large plants. A third solution might be the use of labeled antibodies for *in situ* detection of damage, as has been applied to the study of the repair of UV-induced damage in the human skin *(21)*. Although this approach can be difficult (though not impossible) to quantify, it can certainly yield qualitative results. *Arabidopsis thaliana*, a popular plant model system owing to its small genome and well-developed genetics, is also well suited to the study of DNA repair because its seedlings are extremely small (the shoot and root are less than a half millimeter in diameter). Because of this, the seedling can be uniformly treated with chemical mutagens, radioactively labeled bases, or longer-wavelength UV radiation.

The actual assay of dimers or other lesions from plant tissues is no more difficult in plants than in animals or microbes. High-mol-wt single-stranded DNA can be isolated from intact plants via standard plant molecular biology techniques *(76)*. These preparations yield DNA suitable for immunoassay, or T4-endonuclease V-related techniques. As in the assay of dimers from other organisms, the sensitivity of T4-endonuclease V-related techniques (Southern hybridization assays for specific sequences *(7)*, alkaline sucrose gradient analysis *(26,58)*, or gel electrophoresis *[64]*) is strictly limited by the frequency of nicks or breaks generated during DNA extraction, purification, and storage. PCR-based assays have similar limitations *(12,29)*. The assay of extremely low (but biologically relevant) dimer frequencies, on the order of 5 dimers/Mb, is possible when plant cells are lysed within agarose plugs *(65)*. Lesion-specific antibodies *(10,81)* have also been successfully applied to the study of damage induction and repair in plants, and this radioimmunoassay is independent of the degree of degradation of the plant DNA.

2.2. Photoreactivation

In some organisms, the biological effects of UV radiation are significantly reduced by subsequent exposure to light in the blue or UV-A range of the spectrum. This phenomenon, known as photoreactivation, is generally caused by the effects of one or more DNA photolyases. Photolyase binds UV-induced pyrimidine dimers and, on absorption of a photon of the appropriate wavelength (in the UV-A to blue spectrum), directly reverses the damage in an error-free manner. Photolyase-related genes can be categorized into several classes based on their function and sequence. Microbial ("type I") cyclobutane dimer-specific photolyase (CPD photolyase) genes have been cloned from a variety of bacteria and fungi, and their sequences display obvious homologies *(95)*. The "metazoan" (or "type II") CPD photolyases, cloned from certain insects, archeabacteria, fish, and marsupial (but not placental) mammals, form an unrelated class of proteins, which, like the microbial enzymes, bind two chromophores and are active when expressed in *E. coli (107)*. A third class of photolyases are the pyrimidine [6–4]pyrimidinone-specific enzymes, which are closely related in sequence to type I

CPD photolyases *(88)*. All of these enzymes are described in detail in Chapters 15, 21, and 22 of this volume and Chapter 2 in vol. 2; we will limit further discussion to what is known about photolyase activities in plants.

Evidence for the biological effects of photoreactivation in plants is complicated by the obvious detrimental effects of growing plants in the dark, but this problem can be alleviated, to some extent, by the use of appropriate controls and filters that absorb the shorter wavelengths required for photoreactivation (≥450 nm), while transmitting photons of longer photosynthetically active wavelengths. Photoreactivation results in the reversal of several UV-induced phenomena in plants, including mutagenesis, chromosome rearrangements *(37)*, inhibition of growth, induction of flavonoid pigments *(4)*, and unscheduled DNA synthesis *(39)*. Light-enhanced repair of dimers from total cellular DNA has been documented in tobacco, *Haplopappus gracilis (90)*, gingko *(91)*, chlamydomonas *(79)*, *Arabidopsis (16,58)*, and wheat *(86)*, and the action spectrum for reversal CPDs by partially purified maize and *Arabidopsis* photolyases has been shown to be similar to that of *E. coli*, a methenyltetrahydrofolate-type photolyase *(36,58)*.

A putative plant photolyase was cloned from wild mustard *(3)* by probing a cDNA library with a degenerate oligonucleotide specific to a conserved region of the microbial CPD photolyases. The clone, labeled SA-*PHR1*, displays significant stretches of similarity to previously cloned microbial photolyases. Moreover, the cDNA hybridizes to an mRNA that is strongly regulated by light; seedlings grown in the dark express low levels of the mRNA, whereas light-grown seedlings express the mRNA at high levels. The protein encoded by this cDNA was overexpressed in *E. coli* and found to bind, like the *E. coli* photolyase, both FAD and methenyltetrahydrofolate *(48)*. The *E. coli*-expressed mustard protein did not, however, display any photolyase activity; it neither enhanced the UV resistance of a photolyase defective host strain, nor repaired thymine dimers in vitro. For this reason, the authors concluded that the SA-*PHR1* clone might represent a blue light photoreceptor, rather than a photolyase. This conclusion was supported by the fact that a constitutively expressed *Arabidopsis* gene known to be involved in the blue light response (*HY4*, also termed *CRY1*, or "cryptochrome") was also found to have a region of substantial homology to the Type I photolyases *(1)*. The *HY4* gene product, when overexpressed in *E. coli*, also binds both FAD and methenyltetrahydrofolate, but fails to exhibit any photoreactivating activity *(48)*.

More recently, a type II CPD photolyase homolog (termed *PHR1*) was cloned from *Arabidopsis (99)*. This sequence is genetically linked to the *UVR2* gene required for the photoreactivation of CPDs, and the independently isolated *uvr2* mutants each carry mutations in this gene *(100,107)*. Thus, this type II CPD photolyase homolog is clearly required for photoreactivation of CPDs in *Arabidopsis*.

Plants, however, possess two distinct, specialized photolyases. In contrast to mammals and some well-characterized microbes, *Arabidopsis* has a light-dependent pathway for the repair of pyrimidine [6–4]pyrimidinone photoproducts [(6–4)PDs] *(16)*. Unlike its CPD-specific photolyase activity, this repair pathway does not require induction by prior exposure to visible light. It also does not require the *UVR1* gene product *(10)*, which is essential for dark repair of 6–4 photoproducts. The ability to photoreactiviate (6–4)PDs and CPDs is encoded by two distinct genes; *Arabidopsis* mutants defective in the repair of CPDs (members of the *uvr2* complementation group) *(100,107)* are proficient in the photoreactivation of (6–4)PDs, whereas a mutant defective in the photoreactivation of (6–4)PDs (the *uvr3* mutant) is proficient in the repair of CPDs *(100)*.

The ability to photoreactivate both of the major UV-induced DNA damage products probably extends to other plants; exposure to visible light greatly enhances the rate of removal of (6–4)PDs from the DNA of wheat seedlings *(86)*. Although photoreactivation of (6–4)PDs has not been observed in microbial or most animal systems, a (6–4)PD-specific photolyase activity has been partially characterized in extracts of *Drosophila* larvae, goldfish *(Carassius auratus) (106)*, frog *(Xenopus laevis)*, and rattlesnake *(Crotalus atrox) (42,89,107)*, and genes encoding this activity on expression in *E. coli* have recently been cloned from *Drosophila* and *Xenopus laevis (88,107)*. A homolog of this (6–4)PD photolyase was recently cloned from *Arabidopsis* (Yamamoto, personal communication). Sequencing of the (6–4)PD photolyase homolog of the (6–4)PD photorepair mutant *uvr3* reveals a nonsense mutation in condon 359 (unpublished results), strongly suggesting that the (6–4)PD photolyase homolog is indeed required for photoreactivation in vivo.

Expression of the CPD photolyase activities of higher plants is known to be regulated by visible light. The CPD photolyase of the common bean is induced twofold by a brief exposure to red light; this effect is partially reversed by subsequent exposure to far red light, suggesting that its induction is phytochrome mediated *(45)*. Similarly, the light-dependent repair of CPDs in *Arabidopsis* requires exposure to visible light prior to as well as after UV irradiation *(16)*. When the cloned *Arabidopsis* CPD-specific photolyase (termed *PHR1/UVR2*) was used as a probe to measure the steady-state levels *UVR2* message, the message was undetectable in the absence of inducing light. The gene was most strongly induced by exposure to UV-A or white light, but not induced by blue or red light *(99)*. This result reinforced the idea that the repair capacity of a plant depends on the quality and timing, as well as quantity, of light in its environment.

The influence of the environment on the steady-state level of pyrimidine dimers, the rate of induction of dimers, and the rate of photoreactivation of dimers has also been illustrated in alfalfa *(84)*. These researchers found that seedlings grown in an essentially UV-free environment had the same steady-state levels of CPDs (approx 6 dimers/megabase) as seedlings grown under unfiltered sunlight. In addition, a given dose of UV was found to induce significantly more dimers in the seedlings grown under artificial light, and these seedlings also had a lower rate of photoreactivation of CPDs than the identical strain grown under natural light. Thus, both the UV transparency and the repair capacity of higher plants is altered substantially in response to the ambient levels of UV and visible radiation. Similar effects have been observed in experiments designed to mimic the effect of ozone depletion that directly measure the effects of enhanced UV-B on crop biomass *(11,24,52)*.

2.3. *Excision Repair*

Excision repair pathways are divided into two classes. The base excision repair pathways are mediated by single-peptide glycosylases and are highly substrate-specific, whereas nucleotide excision repair (NER) pathways are catalyzed by the sequential actions of multisubunit protein complexes with remarkably broad substrate specificities. Generalized NER almost certainly exists in plants, but definitive proof of the existence of this pathway has not yet been established. The lesion-specific glycosylases, on the other hand, are better understood in plants, since they are more amenable to both biochemical and molecular genetic approaches.

2.3.1. Uracil Glycosylase

Uracil accumulates in the genome at a rate approx 100 lesions/cell/d (given a genome size of 3×10^9 bp) owing to hydrolytic deamination of cytosine. Because the production of uracil can lead to mutation (since uracil base pairs with A, resulting in the production of a CG to TA transition), organisms probably produce a uracil glycosylase. Although a gene corresponding to this protein has not yet been identified in plants, the activity has been purified from several plant sources *(8,85)*. Interestingly, there is some evidence that this enzyme is downregulated in fully differentiated plant cells *(30)*.

2.3.2. 3-Methyladenine Glycosylase

3-Methyladenine is a noncoding lesion that, like uracil, occurs spontaneously at a significant rate *(47)*. 3-Methyladenine glycosylases have been identified in bacteria, yeast, mammals, and Arabidopsis, and vary in their substrate specificity. *E. coli* constitutively expresses two 3-methyladenine glycosylases. The first, the product of the *tag* gene, is highly specific for 3-methyladenine, whereas the second, the product of the *alkA* gene, has a broad substrate specificity, cleaving the *N*-glycosylic bond at 7-methylguanine, 3-methylguanine, O^2-methylthymine, and O^2-methylcytosine, as well as 3-methyladenine *(23,41)*. The biological effects of an *alkA* mutation can be suppressed by overexpression of the *tag* gene *(97)*, suggesting that these additional substrates do not play an important role in the lethality induced by methylating agents. Interestingly, the *tag* and *alkA* genes share no significant homology. All of the cloned higher eukaryotic 3-methyladenine glycosylases, including one from Arabidopsis *(71)*, have been isolated via complementation of the MMS sensitivity of the *E. coli* double mutant *(5,14)* (summarized in refs. *22* and *55*). Although the mammalian genes have a high degree of homology to each other, the overall trans-kingdom homology is fairly weak.

2.3.3. UV Endonucleases

Glycosylases and endonucleases specific for CPDs have been observed in bacteria and bacteriophage, and have been useful as diagnostic agents for the assay of UV-induced damage *(25)*. Eukaryotic UV endonucleases that recognize both CPDs and (6–4)PDs and generate an incision immediately 5' to the lesion were recently identified in *Schizosaccharomyces pombe* and *Neurospora crassa (9,96)*. Several groups have described endonucleolytic activities obtained from plant extracts that exhibit some specificity for UV-irradiated DNA *(20,53,92)*. Some of these activities are particularly intriguing in that they do not appear to recognize CPDs, suggesting that the recognition site may be the (6–4)PD. In only one case has a plant UV-specific endonuclease been substantially purified and characterized (the spinach endonuclease SP); this enzyme was suggested to be a single-stranded endonuclease, which apparently recognizes a single-stranded region that is induced by (6–4) PDs, but not by CPDs *(82)*.

2.3.4. Nucleotide Excision Repair

NER differs from base excision repair in two ways: the spectrum of DNA damage products recognized by the repair complex is remarkably wide, and the repair complex initiates removal of the damage by generating nicks on the damaged strand at a specific distance both 5' and 3' of the lesion, which is then excised as an oligonucleotide through the action of a helicase. The excision repair complex will, with varying efficiencies, cleave almost any abnormality in DNA structure—from very small, nondistorting lesions (such as O^6-methylguanine or abasic sites) to very bulky adducts (thymine-psoralen adducts or pyrimidine dimers). Cells cannot produce a specific repair protein for every potential lesion, and NER may exist, in part, to cope with unusual lesions. As discussed above, placental mammals are generally thought to lack photolyase (although this is still a subject of debate), and in mammalian cells, NER may be the sole pathway for the repair of bulky adducts *(70)*.

Light-independent ("dark") repair of CPDs, which might represent either NER or base excision repair, has been observed in several plant species. Early studies, previously reviewed by McLennan *(49)*, involved the use of a germicidal lamp to irradiate cell suspension cultures or protoplasts (for uniformity of UV penetration) producing high concentrations of CPDs. The disappearance of dimers from the nuclear fraction and their reappearance in the cytosol were measured by hydrolyzing the DNA and assaying, via thin-layer chromatography, the fraction of total thymidine bases that were present as dimers. The rate of dark repair of CPDs was found to vary widely from one plant system to another, with high rates of repair demonstrated for carrot suspension cultures *(34)*, and protoplasts of carrot, *E. gracilis*, petunia, and tobacco *(35)*, whereas excision repair of CPDs was undetectable in cultured soybean cells *(68)*. It should be stressed that photoreactivation is generally a more rapid and efficient pathway for the excision of UV-induced dimers and probably provides the bulk of the protection against UV-induced DNA damage. However, excision repair may be essential for the repair of minor, nondimer, UV-induced photoproducts.

Recent technical advances in the assay of UV-induced dimers in plants (discussed above) have enabled investigators to use relatively low doses of UV to study repair in various plant systems. Dark repair rates for CPDs have been assayed in very young Arabidopsis seedlings, where no significant repair of CPDs was detectable in 24 h, although repair of (6–4)PDs was efficient *(10)*. In contrast, rapid dark repair of CPDs was observed in alfalfa *(66)*, and an intermediate level of repair detected in wheat seedlings *(86)* and young Arabidopsis plants *(58)*. Although these plants may actually differ in their inherent capacities for dark repair, these disparities might also result from the differing experimental conditions employed by the investigating groups. It has recently been demonstrated that excision repair in the alfalfa seedling, while efficient and easily detectable at high levels of initial UV damage, is undetectable at lower initial damage levels *(66)*. Extremely high doses of UV can also inhibit repair in plant tissues *(34)*. Thus, although laboratory studies are essential for the determination of the biochemical basis of repair, caution must be used in extrapolating these results to make predictions concerning UV resistance in the field, where growth conditions, the plant tissues employed, and the levels of DNA damage induced by sunlight can radically affect both the extent of damage and the rate of repair.

Several mutants of Arabidopsis displaying UV sensitivity have been isolated, and one of these, termed *uvr1*, has been shown to be defective in the dark repair of (6–4)PDs *(10)*. However, because the dark repair of CPDs was so slow as to be insignificant in the seedlings of the wild-type progenitor strain, it is not yet known whether this mutant represents a defect in a general or specific DNA repair pathway. Similarly, a subclass of the Arabidopsis *uvh* (UV-hypersensitive) mutants have a demonstrable sensitivity to both UV and γ-radiation *(32,40)*, but it is not yet known whether this phenotype reflects a defect in repair, or damage tolerance, or a generalized sensitivity to stress.

2.4. Double-Strand Break Repair

Double-strand breaks (DSBs) are generated in plant DNA through a variety of mechanisms: spontaneous oxidative damage to the genome, treatment with ionizing radiation, the formation and subsequent breakage of a dicentric chromosome, cleavage with artificially-introduced nucleases, and (perhaps) excision of transposable elements. Transforming DNAs are also introduced as linear DNAs or as intact plasmids that are rapidly linearized by the cell. Because the DNA sequences near the ends of chromosomal breaks are rapidly degraded, DSBs generally expand into gaps that cannot simply be relegated together to restore the original sequence. Unlike recombination-proficient yeast cells, which virtually always repair DSBs via homologous recombination, the cells of higher plants behave very much like those of most mammalian tissues; DSBs are simply rejoined, end-to-end, in what appears to be a random fashion. Analysis of repaired DSBs generated by ionizing radiation *(75)*, T-DNA insertion *(27,28,77)*, and transposable element excision *(73)* indicates that DSBs usually, though not always, rejoin at sites with a some sequence identity (2–5 bases). This bias toward rejoining at minor sequence identities is probably simply the result of enhanced stability of the splice joint for ligation, rather than a search for extensive sequence identity associated with homologous recombination (Fig. 1A). In addition, the sites at which the ends are rejoined are often characterized by multiple recombination events, such as an inversion of substantial portions of the target site *(75)*, and the insertion of novel sequences. These may be generated via template switching by a repair polymerase *(69)* or through the transient formation of a covalently closed hairpin loop (Fig. 1B) *(17)*. The natural propensity of mammalian and higher plant cells to incorporate exogenous DNAs into random, rather than homologous sites (the fraction of events owing to homologous recombination among all integration events is approx 10^{-4} *[61,62]*) is particularly vexing to the molecular biologist, since true gene replacement is difficult in these cell types.

Very little is known about the genes required for illegitimate recombination in either plants or animals. X-ray-sensitive mutant animal cell lines exist that are defective in the repair of DSBs, and some of the genes that complement these defects have been cloned (reviewed in ref. *25*; *see* Chapters 16 and 17, vol. 2). *scid* mice, which are severely immunodeficient owing to a defect in V(D)J recombination, are also X-ray-sensitive, and fail to incorporate exogenous DNAs *(33)*. Mutants of Arabidopsis specifically sensitive to the growth-inhibitory effects of ionizing radiation have also been isolated *(19,104)*, and some UV-sensitive Arabidopsis mutants also display sensitivity to ionizing radiation *(40,101)*. It is possible that some of these mutants may be defective in the repair of DSBs, although this remains to be determined.

2.5. Repair of the Organellar Genomes

The plant cell contains three distinct genetic compartments: the nuclear, plastid, and mitochondrial genomes. The organellar DNAs, like the nuclear genome, accumulate UV-induced dimers, alkylated bases, and hydrolytic damage. They are probably subject to an enhanced level of oxidative damage. No DNA repair proteins are known to be encoded by the organellar genomes. Because both the plastid and mitochondrial genomes are isolated from the cytoplasm by a pair of lipid bylayers, any repair proteins imported into these organelles must carry specific targeting sequences, and it is very likely that the three genetic compartments have distinct DNA repair pathways. In this

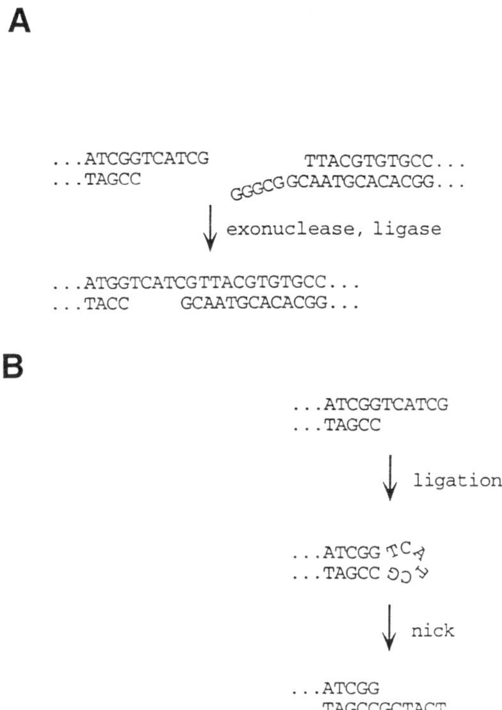

Fig. 1. The repair of double-stranded breaks results in the formation of novel sequences in plants. **(A)** The presence of minor sequence identities helps to stabilize the ligation of two otherwise unrelated sequences. **(B)** The transient formation of a hairpin loop at a double-strand break can generate sequence inversions when the breaks are rejoined. Alternatively, these inverted duplications might be generated via template switching by DNA polymerase.

regard, plants provide us with a unique opportunity to compare and contrast repair strategies in three compartments of a single cell.

Investigation into the repair of organellar damage in higher plants has only recently begun. Young Arabidopsis seedlings were found, surprisingly, to lack any mechanism for the removal of CPDs from their mitochondrial or plastid genomes *(15)*; in contrast, the CPDs in the nuclear genome are efficiently removed via photoreactivation (*see* Section 2.2.). A comparison of the photoreactivation of CPDs in the nuclear vs plastid genomes of soybean suggests that photoreactivation of plastid CPDs does occur in this species, although the nuclear DNA is repaired more efficiently *(12)*. More extensive studies of repair in plastids have been performed using the photosynthetic unicellular alga *Chlamydomonas reinhardtii*, an organism that is particularly amenable to both genetic and biochemical approaches to the repair of plastid DNA (reviewed in ref. *79*). *C. reinhardtii* has been shown to encode distinct photolyases for the repair of CPDs in the plastid vs nuclear genomes.

3. DNA DAMAGE TOLERANCE PATHWAYS

The excision repair pathways described above can all be divided into two steps: first the damaged base is removed, and then the undamaged strand is used as a template to fill the resulting gap. These repair pathways are essentially error-free. If, however, a cell undergoes DNA replication before repair is complete, a noninformational DNA damage product, such as a pyrimidine dimer, will act as a block to DNA replication. DNA polymerase will reinitiate synthesis 3' to the lesion, but a gap remains in the newly synthesized daughter strand at the site opposite the DNA damage product. The resulting single-stranded region can no longer act as a substrate for excision repair, because the sister strand is no longer available as a template. Although one would expect the persistence of such a lesion to be lethal, a variety of organisms have been shown to undergo repeated rounds DNA synthesis and cell division in spite of the continued presence of noninformational lesions. At least two independent pathways that facilitate the replication of damaged chromosomes in spite of the persistence of noninformational lesions are known to exist in various organisms. These damage tolerance pathways include dimer bypass and recombinational "repair."

3.1. Dimer Bypass

Although noninformational lesions normally block DNA replication, some organisms produce a modified polymerase that is capable of performing translesion synthesis. The *E. coli umuC,D* gene products are thought to bind to DNA polymerase and relax its normally stringent requirements for the stable insertion of a new base, thereby enabling it to perform translesion synthesis (*67*; *see* Chapter 12). The altered polymerase generally inserts adenine residues across from UV-induced pyrimidine dimers. As a result, UV-induced thymine dimers are not mutagenic, but cytosine-containing dimers are. Translesion synthesis permits DNA replication (and so enhances survival) at the expense of accuracy. Similarly, the *REV3* gene of *Saccharomyces cerevisiae* produces a nonessential, mutagenic polymerase with a specialized ability to synthesize DNA using damaged templates *(78)*. Humans may produce a modified polymerase with a similar tendency to install As at pyrimidine dimers: sunlight-induced mutations in humans occur mainly at dipyrimidines, and are primarily C → T or CC → TT transversions *(98)*.

Whether mutagenesis in plants occurs as a result of lesion bypass remains to be seen. UV radiation is an excellent source of noninformational DNA damage products, and the spectrum of mutations induced by UV could provide insights into the means by which plants tolerate the persistence of DNA damage. Unfortunately, few UV-induced mutations have been generated, and none have been characterized by sequencing. Because the plant's germ line is shielded from UV during virtually all stages of growth, studies of UV-induced mutations in higher plants have been limited to the mutagenic effects of UV irradiation of pollen. Mutagenesis of pollen has the advantage of enabling the investigator to observe the induction of mutations, such as large deletions, which might otherwise be nontransmissable owing to counterselection during the postmeiotic mitoses and growth of the pollen tube. In fact, UV-induced mutations in maize pollen were generally found to be nontransmissable or to have reduced transmission beyond the first generation, indicating that UV-induced lesions result in large deletions rather than point mutations *(54)*. This finding suggests that translesion synthesis (which induces point mutations) rarely occurs in pollen or during the early stages of embryonic development, and that UV-induced DNA damage results in chromosome breaks and/or recombination. However, one must bear in mind that large chromosomal deletions, which result in the simultaneous loss of many genes, are generally easier to score as mutations than single base changes, since the majority of single base changes fail to affect gene function. It is also possible that dimer bypass is preferentially employed in somatic cell lines (where mutagenesis is relatively inconsequential), but is not expressed during the critical last stage of pollen development, when mutations can no longer be eliminated through diplontic selection *(43)*. Because of its potential role in the creation of genetic diversity (as well as in UV tolerance), more research on trans-lesion synthesis is needed in both plants and animals.

3.2. Recombinational Repair

In contrast to lesion bypass, recombinational "repair" fills the daughter strand gap by transferring a pre-existing complementary strand from a homologous region of DNA to the site opposite the damage. As in the dimer bypass mechanism, the lesion is left unrepaired, but the cell manages to complete another round of replication, and the damaged base is now available as a substrate for excision repair. When the complementary strand is obtained from the newly replicated sister chromatid, the resulting "repair" is error-free. However, if the information is obtained from the homologous chromosome, or perhaps from a similar DNA sequence elsewhere in the genome, there is a possibility that a change will be generated in the gene's sequence either via gene conversion or through the formation of deletions, duplications, and translocations. Although UV irradiation has been shown to induce chromosomal rearrangements in plants *(54)*, including homologous intrachromosomal recombination events *(63)*, it remains to be seen whether the filling of daughter strand gaps via homologous recombination is a significant UV tolerance mechanism in plants. UV light has been shown to induce previously quiescent transposable elements *(94)*; it is possible that this effect is the result of chromosomal rearrangements or other repair-related activities.

A homolog of the *E. coli recA* gene has been cloned from Arabidopsis *(16)*. The predicted amino acid sequence includes a conserved recognition site for the chloroplast stromal processing protease. Southern hybridization analysis using this cDNA as a probe suggests that there is more than one copy of this gene encoded by the Arabidopsis nucleus *(6)*. This chloroplast *recA* homolog may play a role in recombinational "repair." Several other Arabidopsis cDNAs, cloned on the basis of their ability to complement partially the UV-sensitive and recombination-defective phenotype of *E. coli* repair-defective mutants, also appear to possess chloroplast-targeting sequences *(59,60)*. Recently, a strand-transferase activity has been purified to near-homogeneity from the nuclei of broccoli *(87)*. The monomer has a mol wt of approx 95 kDa and, like RecA, requires free 3'-homologous ends for efficient catalysis of strand exchange. Unlike the plastid-directed gene product, this protein almost certainly represents a nuclear, rather than organellar, *recA* homolog.

3.3. Other Damage Tolerance Mechanisms

The two pathways described above permit the cell to replicate in spite of the persistence of dimers, but do not reduce the deleterious effects of DNA damage on transcription. One of the most interesting recent developments in the field of DNA repair is the discovery that the template strand employed for transcription is repaired more rapidly than the nontranscribed strand or nontranscribed regions (*31*; *see* Chapter 9, this vol., and Chapters 10 and 18, vol. 2). In fact, the relationship between repair and transcription is particularly intimate—not only are some repair proteins physically coupled to RNA polymerase, but a subset of those proteins, notably the TFIIH complex, actually act independently both as transcription factors and as repair complexes *(72)*. By selectively removing damage from actively transcribed units, targeted repair substantially reduces the toxic effect of UV. Although preferential repair of transcribed strands has been shown to exist in mammals *(51)*, yeast *(83)*, and *E. coli (50)*, this phenomenon has not yet been investigated in plants.

Lesions opposite a daughter strand gap are particularly problematic, since the damage cannot be repaired via excision repair. If the cell is unfortunate enough not only to replicate its damaged DNA, but also to under go cell division, then the information at the site of the lesion is permanently lost, since no sister chromatid is available to take part in recombinational repair. For this reason, some organisms are capable of detecting genomic damage and will delay cell division until the integrity of the genome is restored. Yeasts (*S. cerevisiae, S. pombe*) damaged in G1 or S phase will delay further DNA synthesis, whereas G2 cells will delay mitosis *(13)*. Cells defective in genes required for the G2 "checkpoint" will proceed with cell division in spite of the presence of gapped DNA, and so exhibit an increase in sensitivity to both the toxic and mutagenic effects of DNA damaging agents. Similar "checkpoint" responses to DNA damage have been observed in other fungi and in mammals (*46*; *see* Chapter 17, this vol., and Chapter 21, vol. 2).

The germination behavior of aged seeds suggests that plants might also delay DNA replication until repair is completed. Although the exact structure and hydration state of DNA in dried seeds are unknown *(57)*, one would expect that seeds, when stored for long periods of time in a desiccated state in which no DNA repair occurs, would progressively accumulate AP sites and other spontaneously generated lesions. Although desiccated seeds accumulate hydrolytic damage at a slower rate than fully hydrated cells *(18)*, long-term seed storage has been correlated with a delay in replicative DNA synthesis, the limited synthesis of low molecular weight, untranslated RNAs, and an increase in unscheduled DNA synthesis *(56)*. These phenomena are consistent with a requirement for a period of genomic repair before cell division can occur in germinating seeds. At least some of the beneficial effects of "osmopriming," a procedure involving partial hydration of seed designed to enhance early and uniform germination, may be owing to DNA repair activities during the priming period *(2)*.

Several labs are currently in the process of isolating Arabidopsis mutants that are hypersensitive to the growth-inhibiting effects of DNA damaging agents *(10,19,32,40)*. Unfortunately, few of these mutants have been characterized in terms of their repair capabilities. Although many of these UV-sensitive mutants will have demonstrable defects in repair, undoubtedly some fraction will display normal rates of repair. This second class of mutants is a particularly interesting one, since it may include mutants defective in damage tolerance. Thus, a screen for UV sensitivity might yield mutants defective in mutagenesis, recombination, transcription, and cell-cycle control. These mutants will be extremely useful in the study of basic biological processes that include, but extend beyond the field of DNA repair.

Table 1
DNA Repair in Plants

	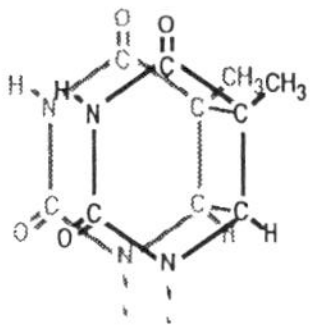 cyclobutane dimers	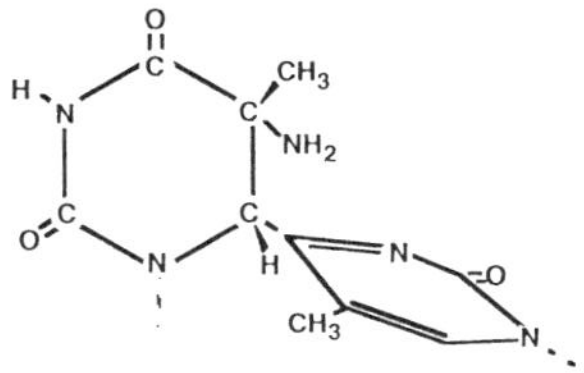 pyrimidine (6-4) pyrimidinone dimers	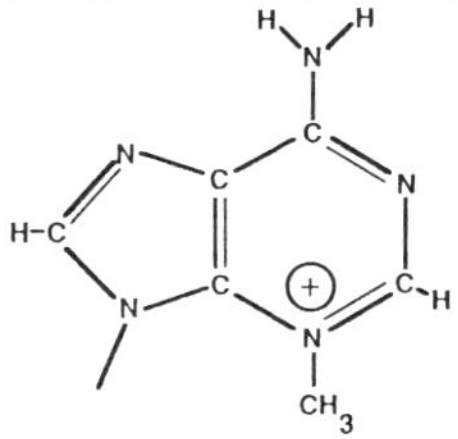 methylated bases	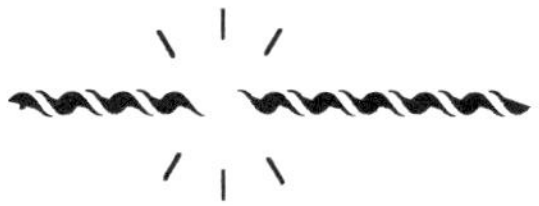 Double strand breaks
Photoreactivation	yes	yes		
Excision repair	yes	yes	yes	
Repair genes cloned	no	no	yes	no
Repair mutants identified	?	yes	no	yes

4. CONCLUSIONS

The study of DNA repair in plants is still in its infancy—many basic repair processes, including NER, strand-specific repair, and the repair of organellar genomes, have not yet been definitively established in plants, much less characterized. Given the powerful investigative tools developed by investigators working in the field of mammalian and microbial repair, there are at present no technical limitations to our abilities to study repair in plants. Similarly, many (though not all) repair-related genes have been shown to be highly conserved between bacteria and mammals; these highly conserved regions should also be useful as probes for plant genes. More interesting, perhaps, is the classical genetic analysis of plant repair (or tolerance) mutants. Some of these mutants may harbor defects in pathways that do not exist in other organisms or that cannot be genetically identified in animals owing to their pleiotropic effects on the rather stringent constraints of animal development. Recent efforts have shown that DNA repair can be assayed in higher plants, that conserved regions of repair genes can be used to isolate plant repair genes, that such genes can also be isolated via complementation of the repair-defective phenotypes of microbes, and that repair-defective mutants of higher plants can be isolated (Table 1, *previous page*). In short, the study of DNA repair and repair-related processes in higher plants is not only interesting and useful, but it is also entirely feasible.

REFERENCES

1. Ahmad, M. and A. R. Cashmore. 1993. The *HY4* gene of *Arabidopsis thaliana* encodes a protein with characteristics of a blue-light receptor. *Nature* **11:** 162–166.
2. Ashraf, M. and C. M. Bray. 1993. DNA synthesis in osmoprimed leek (*Allium porrum* L.) seeds and evidence for repair and replication. *Seed Sci. Res.* **3:** 15–23.
3. Batschauer, A. 1993. A plant gene for photolyase: an enzyme catalyzing the repair of UV-light induced DNA damage. *Plant J.* **4:** 705–709.
4. Beggs, C. J., A. Stolzer-Jehle, and E. Wellmann. 1985. Isoflavonoid formation as an indicator of UV stress in bean (*Phaseolus vulgaris* L.) leaves. *Plant Phys.* **79:** 630–634.
5. Berdal, K. G., M. Bjøras, S. Bjelland, and E. Seeberg. 1990. Cloning and expression in *E. coli* of a gene for an alkylbase DNA glycosylase from *S. cerevisiae*; a homolog to the bacterial *alkA* gene. *EMBO J.* **9:** 4563–4568.
6. Binet, M. N., M. Osman, and A. T. Jagendorf. 1993. Genomic sequence of a gene from *Arabidopsis thaliana* encoding a protein homolog of *E. coli recA*. *Plant Phys.* **103:** 673,674.
7. Bohr, V. A., C. A. Smith, D. S. Okumoto, and P. C. Hanawalt. 1985. DNA repair in an active gene: removal of pyrimidine dimers from the DHFR gene of CHO cells is much more efficient than in the genome overall. *Cell* **40:** 359–369.
8. Bones, A. M. 1993. Expression and occurrence of uracil-DNA glycosylase in higher plants. *Phys. Plant.* **88:** 682–688.
9. Bowman, K. K., K. Sidik, C. A. Smith, J.-S. Taylor, P. W. Doetsch, and G. A. Freyer. 1994. A new ATP-dependent DNA endonuclease from S. pombe that recognizes cyclobutane pyrimidine dimers and 6-4 photoproducts. *Nucleic Acids Res.* **22:** 3026–3032.
10. Britt, A. B., J.-J. Chen, D. Wykoff, and D. Mitchell. 1993. A UV-sensitive mutant of *Arabidopsis* defective in the repair of pyrimidine-pyrimidinone (6-4) dimers. *Science* **261:** 1571–1574.
11. Caldwell, M. M., S. D. Flint, and P. S. Searles. 1994. Spectral balance and UV-B sensitivity of soybean: a field experiment. *Plant Cell Environ.* **17:** 267–276.

12. Cannon, G., L. Hendrick, and S. Heinhorst. 1995. Repair mechanisms of UV induced DNA damage in soybean chloroplasts. *Plant Mol. Biol.* **29:** 1267–1277.
13. Carr, A. M. 1994. Radiation checkpoints in model systems. *Int. J. Rad. Biol.* **66:** S133–S139.
14. Chen, J., B. Derfler, A. Maskati, and L. Samson. 1989. Cloning a eukaryotic DNA glycosylase repair gene by the suppression of a DNA repair defect in *E. coli*. *Proc. Natl. Acad. Sci. USA* **86:** 7961–7965.
15. Chen, J.-J., C.-Z. Jiang, and A. B. Britt. 1996. Little or no repair of cyclobutyl pyrimidine dimers is observed in the organellar genomes of the young Arabidopsis seedling. *Plant Phys.* **111:** 19–25.
16. Chen, J.-J., D. Mitchell, and A. B. Britt. 1994. A light-dependent pathway for the elimination of UV-induced pyrimidine (6-4) pyrimidinone photoproducts in *Arabidopsis thaliana*. *Plant Cell* **6:** 1311–1317.
17. Coen, E. S., T. P. Robbins, J. Almeida, A. Hudson, and R. Carpenter. 1989. Consequences and mechanisms of transposition in *Antirrhinum majus*, in *Mobile DNA* (Berg, D. E. and M. M. Howe, eds.), American Society for Microbiology, Washington, DC, pp. 413–436.
18. Dandoy, E., R. Schyns, R. Deltour, and W. G. Verly. 1987. Appearance and repair of apurinic/apyrimidinic sites in DNA during early germination of *Zea mays*. *Mutat. Res.* **181:** 57–60.
19. Davies, C., D. Howard, G. Tam, and N. Wong. 1994. Isolation of *Arabidopsis thaliana* mutants hypersensitive to gamma radiation. *Mol. Gen. Genet.* **243:** 660–665.
20. Doetsch, P. W., W. H. McCray, and M. R. L. Valenzula. 1989. Partial purification and characterization of an endonuclease from spinach that cleaves ultraviolet light-damaged duplex DNA. *Biochim. Biophys. Acta* **1007:** 309–317.
21. Eggset, G., G. Voldern, and H. Krokan. 1983. UV-induced DNA damage and its repair in human skin *in vivo* studied by sensitive immunohistochemical methods. *Carcinogenesis* **4:** 745–750.
22. Engelward, B. P., M. S. Boosalis, B. J. Chen, Z. Deng, M. J. Siciliano, and L. D. Samson. 1993. Cloning and characterization of a mouse 3-methyladenine 7-methylguanine 3-methylguanine DNA glycosylase whose gene maps to chromosome-11. *Carcinogenesis* **14:** 175–181.
23. Evensen, G. and E. Seeberg. 1982. Adaptation to alkylation resistance involves the induction of a DNA glycosylase. *Nature* **296:** 773–775.
24. Fiscus, E. L. and F. L. Booker. 1995. Is increased UV-B a threat to crop photosynthesis? *Photosyn. Res.* **43:** 81–92.
25. Friedberg, E. C., G. C. Walker, and W. Siede. 1995. *DNA Repair and Mutagenesis*. ASM, Washington, D. C.
26. Ganesan, A. K., C. A. Smith, and A. A. van Zeeland. 1981. Measurement of the pyrimidine dimer content of DNA in permeabilized bacterial or mammalian cells with endonuclease V of bacteriophage T4, in *DNA Repair* (Freidberg, E. C. and P. C. Hanawalt, eds.), Marcel Dekker, New York, pp. 89–98.
27. Gheysen, G., M. Van Monatgu, and P. Zambryski. 1987. Integration of Agrobacterium tumefaciens transfer DNA (T-DNA) involves rearrangements of target plant DNA sequences. *Proc. Natl. Acad. Sci. USA* **84:** 6169–6173.
28. Gheysen, G., R. Villarroel, and M. Van Montagu. 1991. Illegitimate recombination in plants: a model for T-DNA integration. *Genes Dev.* **5:** 287–297.
29. Govan, H. L. I., Y. Valles-Ayoub, J. Braun. 1990. Fine-mapping of DNA damage and repair in specific genomic segments. *Nucleic Acids Res.* **18:** 3823–3830.
30. Gutierrez, C. 1987. Excision repair of uracil in higher plant cells: Uracil-DNA glycosylase and sister-chromatid exchange. *Mutat. Res.* **181:** 111–126.
31. Hanawalt, P. C. 1994. Transcription-coupled repair and human disease. *Science* **266:** 1957,1958.

32. Harlow, G. R., M. E. Jenkins, T. S. Pittalwala, D. W. Mount. 1994. Isolation of *uvh1*, an Arabidopsis mutant hypersensitive to ultraviolet light and ionizing radiation. *Plant Cell* **6:** 227–235.
33. Harrington, J., C.-L. Hsieh, J. Gerton, G. Bosma, M. R. Lieber. 1992. Analysis of the defect in DNA end joining in the murine *scid* mutation. *Mol. Cell. Biol.* **12:** 4758–4768.
34. Howland, G. P. 1975. Dark-repair of ultraviolet-induced pyrimidine dimers in the DNA of wild carrot protoplasts. *Nature* **254:** 160,161.
35. Howland, G. P. and R. W. Hart. 1977. Radiation biology of cultured plant cells, in *Applied Aspects of Plant Cell Tissue and Organ Culture* (Reinert, J. and Y. P. S. Bajaj, eds.), Springer, Berlin, pp. 731–789.
36. Ikenaga, M., S. Kondo, and T. Fuji. 1974. Action spectrum for enzymatic photoreactivation in maize. *Photochem. Photobiol.* **19:** 109–113.
37. Ikenaga, M. and T. Mabuchi. 1966. Photoreactivation of endosperm mutations in maize. *Radiat. Bot.* **6:** 165–169.
38. Jackson, J. F. 1987. DNA repair in pollen. *Mutat. Res.* **181:** 17–30.
39. Jackson, J. F.and H. F. Liskens. 1979. Pollen DNA repair after treatment with the mutagens 4-NQO, ultraviolet and near ultraviolet radiation, and boron dependence of repair. *Mol. Gen. Genet.* **176:** 11–16.
40. Jenkins, M. E., G. R. Harlow, Z. Liu, M. A. Shotwell, J. Ma, and D. W. Mount. 1995. Radiation-sensitive mutants of *Arabidopsis thaliana*. *Genetics* **140:** 725–732.
41. Karran, P., T. Hjelmgren, and T. Lindahl. 1982. Induction of a DNA glycosylase for N-methylated purines is part of the adaptive response to alkylating agents. *Nature* **296:** 770–773.
42. Kim, S.-T., K. Malhotra, C. A. Smith, J.-S. Taylor, and A. Sancar. 1994. Characterization of (6-4) photoproduct DNA photolyase. *J. Biol. Chem.* **269:** 8535–8540.
43. Klekowski, E. J. 1988. *Mutation, Developmental Selection, and Plant Evolution*. Columbia University Press, New York.
44. Klekowski, E. J. and P. Godfrey. 1989. Ageing and mutation in plants. *Nature* **340:** 389–391.
45. Langer, B., E. Wellmann. 1990. Phytochrome induction of photoreactivating enzyme in *Phaseolus vulgaris* L. seedlings. *Photochem. Photobiol.* **52:** 801–803.
46. Li, J. J., R. J. Deshaies. 1993. Excercising self-restraint: discouraging illicit acts of S and M in eukaryotes. *Cell* 74**:** 223–226.
47. Lindahl, T. 1993. Instability and decay of the primary structure of DNA. *Nature* **362:** 709–715.
48. Malhotra, K., S.-T. Kim, A. Batschauer, L. Dawut, and A. Sancar. 1995. Putative blue-light photoreceptors from *Arabidopsis thaliana* and *Sinapus alba* with a high degree of sequence homology to DNA photolyase contain the two photolyase cofactors but lack DNA repair activity. *Biochemistry* **34:** 6892–6899.
49. McLennan, A. G. 1987. The repair of ultraviolet light-induced DNA damage in plant cells. *Mutat. Res.* **181:** 1–7.
50. Mellon, I. and P. Hanawalt. 1989. Induction of the *E. coli* lactose operon selectively increases repair of its transcribed DNA strand. *Nature* **342:** 95–98.
51. Mellon, I., G. Spivak, P. C. Hanawalt. 1987. Selective removal of transcription-blocking DNA damage from the transcribed strand of the mammalian DHFR gene. *Cell* **51:** 241–249.
52. Middleton, E. M. and A. H. Teramura. 1994. Understanding photosynthesis, pigment and growth responses induced by UV-B and UV-A irradiances. *Photochem. Photobiol.* **60:** 38–45.
53. Murphy, T. M., C. P. Martin, and J. Kami. 1993. Endonuclease activity from tobacco nuclei specific for ultraviolet radiation-damaged DNA. *Phys. Plant.* **87:** 417–425.
54. Nuffer, M. G. 1957. Additional evidence on the effect of X-ray and ultraviolet radiation on mutation in maize. *Genetics* **42:** 273–282.
55. O'Connor, T. R. and J. Laval. 1991. Human cDNA expressing a functional DNA glycosylase excising 3-methyladenine and 7-methylguanine. *Biochem. Biophys. Res. Commun.* **176:** 1170–1177.

56. Osborne, D. J. 1983. Biochemical control systems operating in the early hours of germination. *Can. J. Botany* **61:** 3568–3577.
57. Osborne, D. J. 1994. DNA and desiccation tolerance. *Seed Sci. Res.* **4:** 175–185.
58. Pang, Q. and J. B. Hays. 1991. UV-B-inducible and temperature-sensitive photoreactivation of cyclobutane pyrimidine dimers in *Arabidopsis thaliana*. *Plant Physiol.* **95:** 536–543.
59. Pang, Q., J. B. Hays, and I. Rajagopal. 1992. A plant cDNA that partially complements *E. coli recA* mutations predicts a polypeptide not strongly homologous to RecA proteins. *Proc. Natl. Acad. Sci. USA* **89:** 8073–8077.
60. Pang, Q., J. B. Hays, and I. Rajagopal. 1993. Two cDNAs from the plant *Arabidopsis thaliana* that partially restore recombination proficiency and DNA-damage resistance to *E. coli* mutants lacking recombination-intermediate-resolution activities. *Nucleic Acids Res.* **21:** 1647–1653.
61. Paszkowski, J., M. Baur, A. Bogucki, and I. Potrykus. 1988. Gene targeting in plants. *EMBO J.* **7:** 4021–4026.
62. Puchta, H., P. Swoboda, and B. Hohn. 1994. Homologous recombination in plants. *Experientia* **50:** 277–284.
63. Puchta, H., P. Swoboda, and B. Hohn. 1995. Induction of intrachromosomal homologous recombination in whole plants. *Plant J.* **7:** 203–210.
64. Quaite, F. E., B. M. Sutherland, and J. C. Sutherland. 1992. Action spectrum for DNA damage in alfalfa lowers predicted impact of ozone depletion. *Nature* **358:** 576–578.
65. Quaite, F. E., J. C. Sutherland, and B. M. Sutherland. 1994. Isolation of high molecular weight plant DNA for DNA damage quantitation: relative effects of solar 297 nm UVB and 365 nm radiation. *Plant Mol. Biol.* **24:** 475–483.
66. Quaite, F. E., S. Takayanagi, J. Ruffini, J. C. Sutherland, and B. M. Sutherland. 1994. DNA damage levels determine cyclobutyl pyrimidine dimer repair mechanisms in alfalfa seedlings. *Plant Cell* **6:** 1635–1641.
67. Rajagopalan, M., C. Lu, R. Woodgate, M. O'Donnell, M. F. Goodman, and H. Echols. 1992. Activity of the purified mutagenesis proteins UmuC, UmuD', and RecA in replicative bypass of an abasic DNA lesion by DNA polymerase III. *Proc. Natl. Acad. Sci. USA* **89:** 10,777–10,781.
68. Reilly, J. J. and W. L. Klarman. 1980. Thymine dimer and glyceolin accumulation in UV-irradiated soybean suspension cultures. *J. Environ. Exp. Bot.* **20:** 131–133.
69. Saedler, H. and P. Nevers. 1985. Transposition in plants: a molecular model. *EMBO J.* **4:** 585–590.
70. Sancar, A. 1994. Mechanisms of DNA excision repair. *Science* **266:** 1954–1956.
71. Santerre, A. and A. Britt. 1994. Cloning of a 3-methyladenine-DNA glycosylase from *Arabidopsis thaliana*. *Proc. Natl. Acad. Sci. USA* **91:** 2240–2244.
72. Schaeffer, L., R. Roy, S. Humbert, V. Moncollin, W. Vermeulen, J. H. J. Hoeijmakers, P. Chambon, and J.-M. Egly. 1993. DNA repair helicase: A component of BTF2 (TFIIH) basic transcription factor. *Science* **260:** 58–63.
73. Scott, L., D. LaFoe, and C. F. Weil. 1996. Adjacent sequences influence DNA repair accompanying transposon excision in maize. *Genetics* **142:** 237–246.
74. Searles, P., M. Caldwell, and K. Winter. 1995. The response of five tropical dicotyledon species to solar UV-B. *Am. J. Bot.* **82:** 445–453.
75. Shirley, B. W., S. Hanley, and H. Goodman. 1992. Effects of ionizing radiation on a plant genome: analysis of two *Arabidopsis transparent* testa mutations. *Plant Cell* **4:** 333–347.
76. Shure, M., S. Wessler, and N. Fedoroff. 1983. Molecular identification and isolation of the waxy locus in maize. *Cell* **35:** 225–233.
77. Simpson, R. B., P. J. O'Hara, W. Kwok, A. L. Montoya, C. Lichtenstein, et al. 1982. DNA from the A6S/2 crown gall tumor contains scrambled Ti-plasmid sequences near its junctions with plant DNA. *Cell* **29:** 1005–1014.

78. Singhal, R. K., D. C. Hinkle, and C. W. Lawrence. 1992. The *REV3* gene of *S. cerevisiae* is transcriptionally regulated more like a repair gene than one encoding a DNA polymerase. *Mol. Gen. Genet.* **236:** 17–24.
79. Small, G. D. 1987. Repair systems for nuclear and chloroplast DNA in *Chlamydomonas reinhardtii*. *Mutat. Res.* **181:** 31–35.
80. Sonti, R. V., M. Chiurazzi, D. Wong, C. S. Davies, G. R. Harlow, D. W. Mount, and E. R. Signer. (1995) Arabidopsis mutants deficient in T-DNA integration. *Proc. Natl. Acad. Sci. USA* **92:** 11,786–11,790.
81. Stapleton, A. E. and V. Walbot. 1994. Flavonoids can protect maize DNA from the induction of ultraviolet radiation damage. *Plant Phys.* **105:** 881–889.
82. Strickland, J. A., L. G. Marzilli, J. M. Puckett, and P. W. Doetsch. 1991. Purification and properties of nuclease SP. *Biochemistry* **30:** 9749–9756.
83. Sweder, K. and P. Hanawalt. 1992. Preferential repair of cyclobutane pyrimidine dimers in the transcribed strand of a gene in yeast chromosomes and plasmids is dependent on transcription. *Proc. Natl. Acad. Sci.* **89:** 10,696–10,700.
84. Takayanagi, S., J. G. Trunk, J. C. Sutherland, and B. M. Sutherland. 1994. Alfalfa seedlings grown outdoors are more resistant to UV-induced damage than plants grown in a UV-free environmental chamber. *Photochem. Photobiol.* **60:** 363–367.
85. Talpaert-Borlè, M. 1987. Formation, detection, and repair of AP sites. *Mutat. Res.* **181:** 45–56.
86. Taylor, R. M., O. Nikaido, B. R. Jordan, J. Rosamond, C. M. Bray, and A. K. Tobin. 1996. UV-B-induced DNA lesions and their removal in wheat (*Triticum aestivum* L.) leaves. *Plant Cell Environ.* **19:** 171–181.
87. Tissier, A., M. Lopez, and E. Signer. 1995. Purification and characterization of a DNA strand transferase from broccoli. *Plant Phys.* **108:** 379–386.
88. Todo, T., H. Ryo, K. Yamamoto, H. Toh, T. Inui, H. Ayaki, T. Nomura, and M. Ikenaga. 1996. Similarity among the Drosophila (6-4) photolyase, a human photolyase homolog, and the DNA photolyase-blue-light photoreceptor family. *Science* **272:** 109–112.
89. Todo, T., H. Takemori, H. Ryo, M. Ihara, T. Matsunaga, O. Nikaido, K. Sato, and T. Nomura. 1993. A new photoreactivating enzyme that specifically repairs ultraviolet light-induced (6-4) photoproducts. *Nature* **361:** 371–374.
90. Trosko, J. E. and V. H. Mansour. 1968. Response of tobacco and *Haplopappus* cells to ultraviolet radiation after posttreatment with photoreactivating light. *Mutat. Res.* **36:** 333–343.
91. Trosko, J. E. and V. H. Mansour. 1969. Photoreactivation of ultraviolet light-induced pyrimidine dimers in Ginkgo cells grown in vitro. *Mutat. Res.* **7:** 120,121.
92. Velemínsky, J., J. Svachulová, and J. Satava. 1980. Endonucleases for UV-irradiated and depurinated DNA in barley chloroplasts. *Nucleic Acids Res.* **8:** 1373–1381.
93. Walbot, V. 1985. On the life strategies of plants and animals. *Trends Genet.* **1:** 165–169.
94. Walbot, V. 1992. Reactivation of Mutator transposable elements by ultraviolet light. *Mol. Gen. Genet.* **234:** 353–360.
95. Yajima, H., H. Inoue, A. Oikawa, and A. Yasui. 1991. Cloning and functional characterization of a eukaryotic DNA photolyase gene from *Neurospora crassa*. *Nucleic Acids Res.* **19:** 5359–5362.
96. Yajima, H., M. Takao, S. Yasuhira, J. H. Zhao, C. Ishii, et al. 1995. A eukaryotic gene encoding an endonuclease that specifically repairs DNA damaged by ultraviolet light. *EMBO J.* **14:** 2393–2399.
97. Yamamoto, Y. and M. Sekiguchi. 1979. Pathways for repair of DNA damaged by alkylating agent in *E. coli*. *Mol. Gen. Genet.* **171:** 251–256.
98. Ziegler, A., D. J. Lefell, S. Kunala, H. W. Sharma, M. Gailani, J. A. Simon, A. J. Halperin, H. P. Baden, P. E. Shapiro, A. E. Bale, and D. E. Brash. 1994. Mutation hotspots due to sunlight in the p53 gene of nonmelanoma skin cancers. *Proc. Natl. Acad. Sci. USA* **90:** 4216–4220.

99. Ahmad, M., J. Jarilo, L. Klimczak, L. Landry, T. Peng, R. Last, and A. Cashmore. 1997. An enzyme similar to animal type II photolyases mediates photoreactivation in *Arabidopsis*. *Plant Cell* **9:** 199–207.
100. Jiang, C.-Z., J. Yee, D. Mitchell, and A. Britt. 1997. Photorepair mutants of *Arabidopsis*. *Proc. Natl. Acad. Sci.* **94:** 7414–7445.
101. Jiang, C.-Z., C.-N. Yen, K. Cronin, D. Mitchell, and A. Britt. 1997. UV- and gamma-radiation sensitive mutants of *Arabidopsis*. *Genetics*, in press.
102. Kim, S.-T., K. Malhotra, J.-S. Taylor, and A. Sancar, 1996. Purification and partial characterization of (6–4) photolyase from *Xenopus laevis*. *Photochem. Photobio.* **63:** 292–295.
103. Landry, L., A. Stapleton, J. Lim, P. Hoffman, J. Hays, V. Walbot, and R. Last. 1997. An arabidopis photolyase mutant is hypersensitive to UV-B radiation. *Proc. Natl. Acad. Sci.* **94:** 328–332.
104. Masson, J., P. King, and J. Paszkowski. 1997. Mutants of *Arabidopsis thaliana* hypersensitive to DNA damaging treatments. *Genetics* **146:** 401–407.
105. Todo, T., S.-T. Kim, K. Hitomi, E. Otoshi, T. Inui, H. Morioka, H. Kobayashi, E. Ohtsuka, H. Toh, and M. Ikenaga. 1997. Flavin adenine dinucleotide as a chromophore of the Xenopus (6–4)photolyase. *Nucleic Acids Res.* **25:** 764–768.
106. Uchida, N., H. Mitani, T. Todo, M. Ikenaga, and A. Shima. 1997. Photoreactivating enzyme for (6–4) photoproducts in cultured goldfish cells. *Photochem. Photobiol.* **65:** 964–968.
107. Yasui, kA., A. P. M. Eker, S. Yasuhira, H. Yajima, T. Kobayashi, M. Takao, and A. Oikawa. 1994. A new class of DNA photolyases present in various organisms including aplacental mammals. *EMBO J.* **13:** 6143–6151.

26

Modes of DNA Repair in *Xenopus* Oocytes, Eggs, and Extracts

Dana Carroll

1. INTRODUCTION

To judge from the attention paid in the recent popular press, issues of DNA damage and repair in frogs are second in importance only to those involving humans. Precipitous declines in amphibian populations around the world have been attributed partly to deterioration of the atmospheric ozone layer and consequent elevation of radiation-induced DNA damage to hazardous levels *(6)*. An inverse correlation has been found between measured repair capabilities in different species and their rates of population decline *(5)*. Thus, frogs may be sensitive biodetectors of rising radiation levels that endanger all species.

Taking a more microscopic focus, this chapter reviews the uses made of oocytes and eggs of the South African clawed frog, *Xenopus laevis*, in studies of various processes of DNA repair. It begins with a brief introduction to the biological and experimental characteristics of *Xenopus* oocytes and eggs. Next, repair activities are examined on various types of lesions, beginning with double-strand breaks (DSBs), followed by microlesions (e.g., UV damage), and base mismatches. Finally, the advantages this experimental system offers for future studies of DNA repair are anticipated.

2. *XENOPUS* OOCYTES AND EGGS

2.1. *Biological Properties*

Full-grown *Xenopus* oocytes are very large. A stage VI oocyte is a single living cell with a diameter of more than 1 mm and an internal volume of approx 1 μL (Fig. 1). For comparison, this is about 10^5 times the volume of a typical vertebrate somatic cell. Normal cellular components are similarly scaled; for example, the number of ribosomes, content of mitochondrial DNA, and amount of total protein are also equivalent to about 10^5 somatic cells *(24)*. Although the DNA content of an oocyte is only 4*N*, the nucleus (also called the germinal vesicle, or GV) is very large: approx 0.4 mm in diameter, with a volume of about 50 nL (Fig. 1).

The enormous accumulation of cellular components in oocytes reflects the frog's reproductive strategy *(20)*. Fertilization occurs outside the mother's body, and the offspring spend all of their existence exposed to the hazards of their surroundings. In

From: DNA Damage and Repair, Vol. 1: DNA Repair in Prokaryotes and Lower Eukaryotes
Edited by: J. A. Nickoloff and M. F. Hoekstra © Humana Press Inc., Totowa, NJ

Fig. 1. Dimensions of a *Xenopus* oocyte. A full-grown, stage VI oocyte is shown on the left. On the right is an oocyte that has been dissected; its GV lies between the two oocytes. The GV has swollen somewhat and is larger than normal size. The ruler along the bottom of the photograph has millimeter divisions.

order to progress rapidly to a stage with some mobility and independence, the early phases of development are dramatically accelerated. The first 12 cell division cycles occur synchronously and very rapidly; after an initial 90-min delay, cell division occurs every 30 min. The genome is replicated in a compressed S phase, but there is insufficient time to synthesize a full complement of other cellular components *de novo*. Little or no gene transcription occurs during this earliest cleavage stage *(59)*. By mid-blastula, when more normal division cycles resume, each of the several thousand cells in the embryo has acquired most of its contents, except DNA, by partitioning materials that were stored in the oocyte.

Xenopus oocytes have become a popular experimental system because many stored cellular capabilities are accessible by direct injection of substrates. In addition, many thousands of oocytes can be collected surgically from a single female, providing plentiful material for preparation of cell-free extracts. Not only can whole cells be injected very readily, but direct injections into the oocyte nucleus are routine. This has particular importance for studies of DNA repair, since substrates can be delivered to the site of normal DNA metabolism. Because a principal occupation of embryos after fertilization is DNA synthesis, it is not surprising to find that oocytes have a high repair capacity to ensure accurate copying and transmission of the genome in the face of very abbreviated cell division cycles.

It is important to make a distinction between oocytes and eggs of *Xenopus* in both their biological and experimental capabilities. Oocytes are obtained by dissection of ovaries of mature females. They are arrested in the first meiotic prophase and cannot be fertilized. Maturation of oocytes into eggs is induced, naturally or experimentally, by increasing hormone levels. This results in breakdown of the nuclear membrane, progression to metaphase II of meiosis, and deposition of external fertilization layers. A single female may lay 1000 or more eggs that are ready to be fertilized. Only the act of fertilization, or some artificial form of activation, induces the eggs to progress to the first mitotic interphase. Some aspects of the maturation of oocytes can be reproduced in vitro by treatment with progesterone *(85)*, but the resulting "eggs" cannot be reliably fertilized.

2.2. Experimental Features

Circular DNA molecules injected into oocyte nuclei are assembled into chromatin and are transcribed if they carry a functional promoter *(23)*. Simple linear molecules are degraded in oocytes, but circular molecules are stable for days or weeks, depending on the health of the cells *(11,103)*. Oocytes do not support the replication of double-stranded DNA. They are capable, however, of synthesizing a second strand on an injected single-stranded circular molecule *(17)*, suggesting that oocytes have the capacity for repair-type DNA synthesis. In eggs, DNA is efficiently replicated in a process that is not dependent on any particular origin sequences *(25,55)*. Unfertilized eggs do not have an injectable nucleus, but functional nuclear membranes are assembled from stored components around injected DNA *(18)*.

Whole-cell extracts from oocytes or eggs are often prepared simply by disrupting the cells with centrifugal force *(45)*. In the case of eggs, a distinction has been made between low-speed and high-speed extracts: the former supports the assembly of functional nuclear membranes and replication of exogenous DNA, whereas the latter does not, unless it is supplemented with a vesicular fraction *(45)*. Nuclear extracts can be prepared from oocytes by either manual (*see* Fig. 1) *(4,40,62)* or bulk *(41,73,77)* isolation of GVs. These preparations retain many nuclear activities, including chromatin assembly, transcription, RNA processing, recombination, and DNA synthesis.

3. REPAIR OF DOUBLE-STRAND BREAKS

3.1. Overview

In principle there are several ways to fix a broken DNA molecule (Fig. 2).

1. The ends could simply be rejoined without gain or loss of material. This is termed end ligation, and it might occur, for example, by rejoining cohesive ends produced by cleavage with a restriction enzyme.
2. Sequences near the broken ends might be rejoined with some loss of material, but without the need for extensive sequence homology; this is termed nonhomologous end joining.
3. If homologous sequences are present near the ends, they can support restoration of continuous duplex by nonconservative recombination, so named because it results in loss of one of the two repeats and the sequences between them.
4. Finally, homologous sequences elsewhere in the genome can be used to repair the break by a conservative process in which intact strands in the donor serve as templates for synthesis of overlapping sequences across the break.

All but the last of these processes have been documented in *Xenopus*.

Among these alternatives for DSB repair, only end ligation and conservative homologous recombination restore the original DNA structure without loss of information. Conservative homologous recombination proceeds quite efficiently in yeast, both to repair chromosomal damage and to integrate linear exogenous DNAs, and the mechanism is usually referred to as double-strand-break repair (DSBR) (*91*; *see* Chapter 16). In diploid yeast, alignment of the two ends at a chromosomal break with

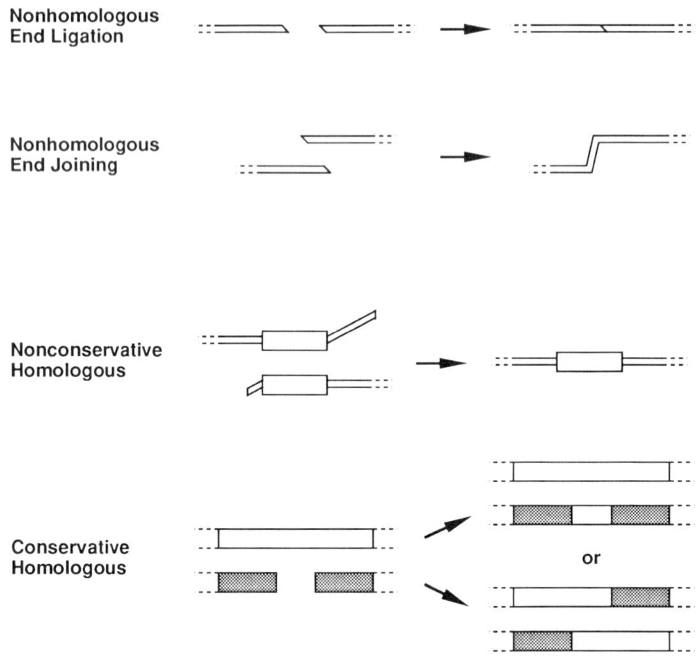

Fig. 2. Alternative modes of repairing DSBs in DNA. In each case, double-stranded DNA is implied, and dashed lines indicate continuous duplex away from the break. The larger boxes in the lower two sections denote homologous DNA sequences; one is shaded in the bottom diagram to help illustrate the crossover and noncrossover options. *See* Section 3.1. for details.

homologous sequences at the allelic site ensures that appropriate rejoining will take place. In organisms with high levels of repeated sequences in their genomes, this mechanism could lead to pairing with, and repair from nonallelic but related sequences. This threatens to undermine the intended repair in two ways. First, if repair of a break in an exon occurs from a distantly related site, e.g., from a pseudogene, the integrity of the coding sequence may not be restored. Second, DSBR is frequently associated with crossing over (*see* Fig. 2) *(63)*, and such exchange between nonallelic sites would cause a potentially deleterious chromosomal translocation.

Nonhomologous end joining and nonconservative homologous recombination both lead to the deletion of some information locally near the break. Although this would be intolerable in the germ line, in somatic cells, it may be preferable to the loss of a whole chromosome or chromosomal segment. Another mechanism for healing DSBs that has been documented is the *de novo* addition of telomeres at the broken ends *(22)*. This is only half a cure, however, since one fragment will necessarily be acentromeric and soon lost from dividing cells.

3.2. Homologous Recombination

As mentioned above, when simple linear DNAs are injected into oocyte nuclei, they are degraded over a period of several hours. However, if a substrate is provided with terminal homologies that can support recombination, recombinant products are formed

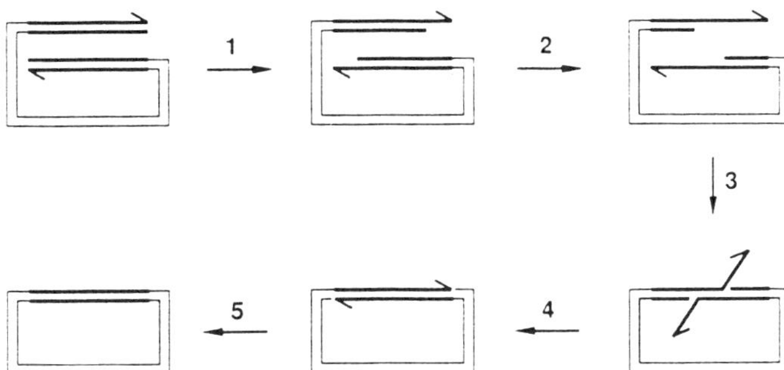

Fig. 3. Diagram of the SSA mechanism. The substrate is a linear molecule with terminal direct repeats (thick lines); both strands are shown, and half arrowheads indicate 3'-ends. Action of a 5' → 3' exonuclease (steps 1 and 2) exposes complementary single strands that anneal (step 3) to form recombination junctions. Further resection by the 5' → 3' exonuclease (step 4) removes the redundant strands, leaving nicks or gaps that can be repaired by DNA polymerase and DNA ligase (step 5) to yield covalently closed products. Although only an intramolecular event is shown, intermolecular recombination proceeds by the same steps. Reproduced from ref. *49* with permission.

efficiently *(11)*. A typical substrate is illustrated in Fig. 3 (upper left). A single oocyte can process more than 10^9 such DNA molecules into intra- and intermolecular recombination products in a few hours with yields ranging from 20% to nearly 100% *(11,80)*. This large capacity and convenient time-course have allowed investigation of the biochemical mechanism of recombination by direct characterization of intermediates. The particular substrate shown in Fig. 3 is designed such that products will necessarily appear to be nonconservative, but injections of other DNAs constructed specifically to monitor conservative recombination have shown that homologous recombination in oocytes is essentially always nonconservative *(33)*.

The accumulated data on homologous recombination in oocytes lead to the conclusion that it proceeds by a mechanism that is usually called single-strand annealing (SSA). As shown in Fig. 3, the key steps in SSA are: resection from the broken ends by a 5' → 3' exonuclease activity *(48)* exposing complementary single strands; pairing to form an annealed junction; and resolution by nuclease, DNA polymerase, and DNA ligase activities. Evidence for this mechanism has come from analysis of recombination intermediates. Using one- and two-dimensional gel electrophoresis, in conjunction with oligonucleotide hybridization, it was shown that intermediates had the strand composition and nuclease sensitivity predicted by the SSA mechanism *(49)*. Perhaps the most compelling evidence came from an electron microscopic study of intermediates, all of which had structures consistent with SSA *(68)*. An example is shown in Fig. 4. Structures predicted by alternative mechanisms based on invasion of duplex regions by homologous single strands were not seen, even though precautions were taken to stabilize such intermediates by psoralen crosslinking prior to isolation *(68)*.

Intermediates predicted by the SSA scheme were constructed and injected into oocytes, where their rapid conversion to completed products confirmed that they are intermediates of the major recombination pathway *(49,50)*. It was shown that the reso-

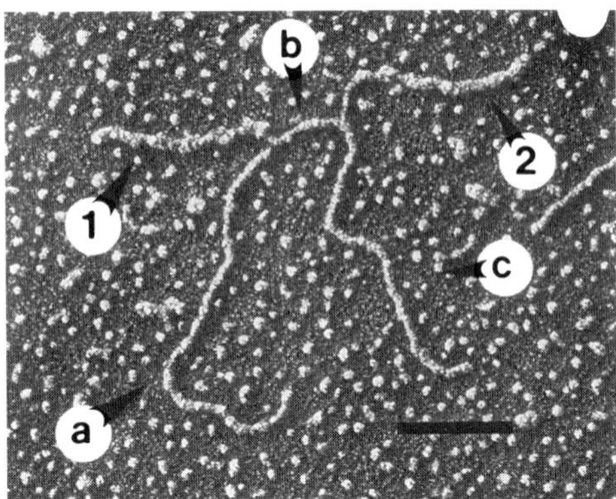

Fig. 4. Electron micrograph of a recombination intermediate. This molecule consists of a linear duplex (a,b,c) with two single-stranded branches (1, 2). It corresponds to an annealed intermediate (Fig. 3, between steps 3 and 4) that has been cleaved with a restriction enzyme outside the homologous overlaps. Single strands were coated with RecA protein, so they appear thicker. The scale bar corresponds to 0.2 μm. Reproduced from ref. *68* with permission.

lution of annealed junctions is accomplished by continued resection by 5' → 3' exonuclease *(49)*. Unlike yeast, and perhaps mammalian cells, oocytes appear to lack activities that can remove single-stranded branches. This conclusion is supported by the observation that terminal nonhomologies interfere with recombination because the 3'-end of the nonhomology is only very slowly removed *(32)*.

To gain some insight into the biological role of SSA recombination, variation of this capacity during oocyte and egg development was investigated *(42)*. Injection of the standard substrate into small oocytes (stages I and II) led to a low level of end ligation, but no homologous recombination. SSA capability was first visible at stage III and continued to increase up to fully grown, stage VI oocytes. Ovulated eggs also support homologous recombination, particularly after activation (which mimics fertilization; *101*). It is not known whether SSA recombination *per se* is a normal cellular process, or whether injection of exogenous DNA reveals activities of enzymes that normally serve other functions. Nonetheless, the pattern of accumulation of activities involved in SSA indicates that they reflect somatic capabilities, stored like much of the rest of oocyte contents for embryogenesis, not utilized for oocyte-specific functions. There is no information to distinguish whether these activities are required along with other DNA metabolic activities during cleavage stages or are utilized only in later phases of development.

Attribution of a somatic function for SSA activities correlates well with observations in other organisms. The primary mechanism of extrachromosomal recombination in cultured mammalian *(46,86)* and plant *(69)* cells has the characteristics of SSA, and yeast *(67,74)* and bacteria *(84,92)* also support this style of recombination. In yeast, at least, the process is mediated by a 5' → 3' exonuclease activity *(87,88)*. In these other organisms, it has not yet been possible to obtain such direct evidence for the details of the SSA model as in oocytes. In some situations, SSA has the consequence of productively repairing DSBs in tandemly repeated sequences *(61)*, and this may be its ultimate function.

3.3. Nonhomologous Recombination

The fact that large oocytes support only homologous recombination greatly simplified the detailed analysis of the SSA mechanism, but nonhomologous recombination has been observed in several preparations from *Xenopus*. Both end ligation and end joining are seen in egg extracts *(66)*, injected eggs *(42)*, and oocyte nuclear (GV) extracts *(40,41,44)*—in the latter two cases coincidentally with SSA.

The most extensive characterization of end-joined products has been done by Pfeiffer and colleagues using extracts of fertilized eggs *(64–66,93)*. They have examined the junction sequences formed by joining many types of noncohesive ends created by restriction enzymes. The initial conclusions were that very little information was lost from broken ends on rejoining, that extensive homologies were not required, but that a match as small as a single base pair would set the register for joining (66). Similar conclusions were reached regarding the junctions formed on injection of linear DNA into activated eggs, although joining was often accompanied by deletion of several tens of base pairs from one or both ends *(44)*.

Combining this information with the observation of SSA recombination mentioned above, it is fair to say that *Xenopus* eggs are very similar to cultured mammalian cells in the ways they deal with exogenous linear DNAs *(70,72)*. In both cases, extensive homologies, if present on the substrate, support nonconservative homologous recombination, but this competes with nonhomologous end joining. The joints formed by the latter mechanism make use of very short matches, in the range of 1–5 bp, but the length of that microhomology does not determine the efficiency of joining.

Continued studies of the junctions formed in egg extracts have added considerable detail to our understanding of how joining proceeds (Fig. 5). When two ends have protruding single strands of the same polarity, partial base matches in the overlaps set the register of joining *(66)*. Internal mismatches are incorporated into the ligated products; terminal mismatches are usually removed exonucleolytically (whether they are on 3'- or 5'-termini), but can sometimes be ligated into the junction *(64,65)*. DNA synthesis fills any gaps left by the alignment or by removal of mismatched nucleotides. Mismatches incorporated by ligation appeared relatively stable, indicating little capability for mismatch repair, but this may have been inactivated in the preparation of these extracts (*see* Section 5.).

When blunt-ended substrates or single-stranded termini of opposite polarities were used, the products were essentially blunt-end joined without loss of sequences from either terminus. Recessed 3'-ends were presumably extended by DNA polymerase, but the case of recessed 5'-ends presents a problem. By ingenious use of dideoxynucleotides, Thode et al. *(93)* showed that the joining of a blunt end to a protruding 3'-end required extension from the blunt 3'-end by DNA polymerase. One possibility is that an untemplated nucleotide is added, as has been observed for other DNA polymerases *(14,15,36)*, and the 25% of termini that now have a match to the last base of the protruding 3'-tail are able to base pair and join with those ends. An alternative explanation *(93)* invokes an end alignment activity that would bring the ends together and allow templated synthesis across the gap and subsequent ligation. This activity would also participate in the alignment of partially matched overlaps and stabilize their association, while DNA polymerase and/or DNA ligase consummates the joining.

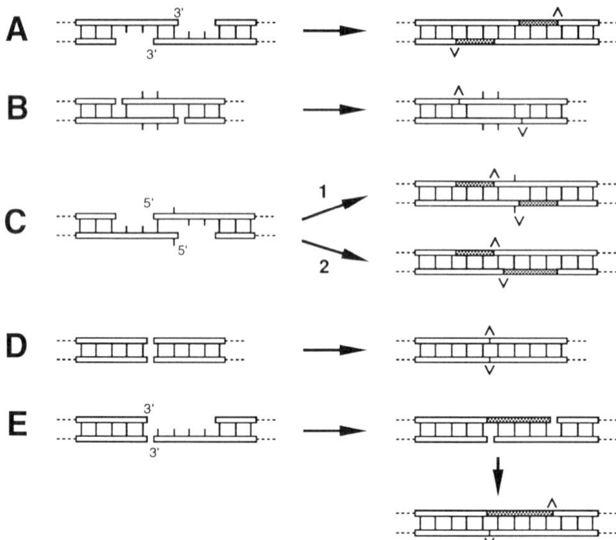

Fig. 5. End-joining events observed in egg extracts. The original strands are represented as open lines; shaded segments result from DNA synthesis. Vertical bars between the two strands denote base pairs; short vertical lines are unpaired bases; and short lines outside the strands represent mispaired bases. Carets indicate sites that are joined by DNA ligase. **(A)** Terminal matches at overlapping 3'-ends are extended by DNA synthesis and joined by ligase. **(B)** Overlaps of either polarity that have matched bases at the junctions can be ligated with retention of the internal mismatches. **(C)** Matches in 5'-overlaps require fill-in from the recessed 3'-ends; terminal mismatches are either incorporated (1) or replaced (2). **(D)** Blunt ends are simply ligated. **(E)** When a blunt end joins a 3'-protruding end, DNA synthesis from the 3'-ending strand at the blunt end is required, and this precedes ligation (ref. *93*; *see* Section 3.3. for additional discussion). Based on refs. *64* and *66*.

Although no direct evidence for this activity has been produced, proteins have been identified in other systems (e.g., Ku) that bind DNA ends and participate in repair (*95,102*; *see* Chapters 16 and 17, vol. 2), and it is quite plausible that one of their functions is to bring ends together for purposes of rejoining.

It is notable that when 5'-protruding ends are joined (as in Fig. 5C), DNA synthesis from the recessed 3'-ends fills the gaps, but this does not happen so rapidly that blunt ends are formed before joining begins. Perhaps end alignment precedes and stimulates this type of repair synthesis. Pfeiffer's group has recently turned to the use of well-defined hairpin substrates that have only one joinable end *(2)*. They showed that efficient joining in egg extracts requires a duplex length of at least 27 bp and that single-stranded tails longer than about 10 bases, whether 3' or 5', are not efficiently joined.

Intrigued by the observation that oocytes do not support end joining, but eggs do, Goedecke et al. *(21)* monitored this capability during in vitro maturation of oocytes. They found that end joining activity appeared very late, many hours after GV breakdown. In fact, efficient end joining may require activation of the egg, which is induced by injection *(21)*, or by treatment with a calcium ionophore *(42)*.

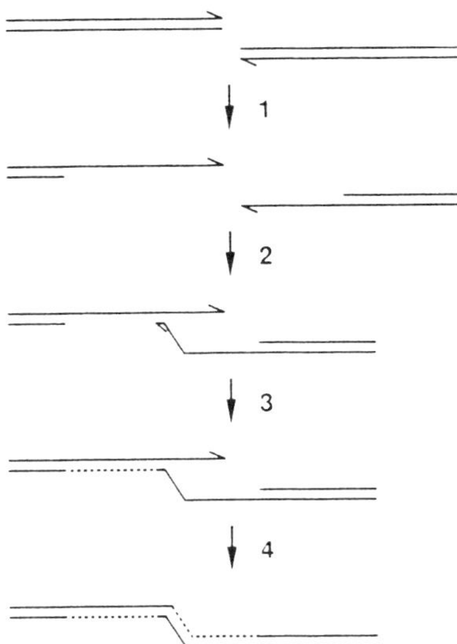

Fig. 6. A possible mechanism for nonhomologous end joining. Resection by 5' → 3' exonuclease activity leaves single-stranded 3'-tails (step 1). One end finds a short complementary region (microhomology) in the other tail (step 2) and is extended by DNA synthesis (step 3). Resolution (step 4) requires removal of the 3'-branch and gap filling by DNA polymerase. The original association between the single strands is not thought to be stable, but the occasional priming of DNA synthesis from the initial contact (step 3) creates a much more stable junction, which allows time for resolution. Symbols as in Fig. 3; dashed lines denote newly synthesized DNA.

Because no evidence of nonhomologous recombination has ever been detected in injected stage VI oocytes, it was surprising to find that GV extracts from those same cells do, under some circumstances, catalyze end joining *(40,41)*. It is possible that during extract preparation some cytoplasmic components were included, that latent activities were inadvertently activated, or that an inhibitory component was lost or inactivated. The junctions were the same as those formed in eggs: sequences were deleted from one or both ends, and microhomologies were recognized at the ends *(44)*. These findings emphasize the fact that oocytes and eggs are very closely related cell types, and suggest that differences in their activities are largely a matter of posttranslational modifications and changes in compartmentalization.

The fact that end joining in GV extracts requires the addition of all four dNTPs was interpreted to signal a critical role for DNA synthesis in this process *(44)*. A plausible model is illustrated in Fig. 6. It envisions 5' → 3' degradation on both participating ends and perhaps slight resection of the 3'-ends; when one 3'-end pairs at a point of microhomology on another 3'-ending single strand, there is a small chance that this structure will be used as a primer–template complex by DNA polymerase; extension by DNA synthesis would greatly stabilize the joint and allow time for ultimate resolution on both sides. This mechanism has also been offered to account for end joining in mammalian cells *(71)*. Even in cases in which unexpected insertions have been found at the joint, these can be explained by synthesis at a primary site initiated by microhomology pairing, withdrawal after synthesis of tens or hundreds of nucleotides, and a microhomology-driven synthesis reaction at a second locus leading to a completed joint *(56)*.

4. REPAIR OF MICROLESIONS

4.1. Overview—Substrate Design

Xenopus oocytes are capable of repairing many types of DNA damage by processes fully consistent with those seen in other organisms, and they offer two distinct advantages. First, substrates for repair can be constructed and characterized thoroughly, then delivered directly to the cell nucleus by microinjection *(9)*. For biochemical studies, this is preferable to treating whole cells with damaging agents and to protocols that involve introduction of the substrate by procedures, such as electroporation or calcium phosphate coprecipitation, that require traversal of the cytoplasm and potential introduction of additional lesions *(100)*. Second, the oocytes have very high capacities for repair of many types of lesions, which facilitates analysis of reaction intermediates.

4.2. UV Damage

A general protocol for examination of repair of UV-induced lesions is to irradiate a well-defined DNA in vitro, then to inject it into the GV *(28,39)*. After incubation for varying periods of time, the DNA is recovered and subjected to an assay that detects remaining damage. By monitoring reactivation of a heavily irradiated bacterial plasmid, Hays et al. *(28)* showed that a single oocyte has the capability of repairing at least 10^{10} UV-induced lesions in just a few hours. Cyclobutane pyrimidine dimers (CPDs) are well repaired in the oocytes. By using prior treatment with photolyase to remove CPDs, the same authors showed that other photoproducts (presumably [6-4] pyrimidine-pyrimidones) are also repaired. Consistent with known oocyte activities, there is no extensive, semiconservative DNA replication associated with this repair, although localized, damage-induced DNA synthesis is readily observed *(28,39,75)*.

Using specific inhibitors, Saxena et al. *(76)* provided evidence that repair of CPDs in oocytes requires DNA polymerase α, and possibly δ. Inhibitors of topoisomerases do not consistently interfere with repair, suggesting that neither topo I nor topo II is required *(39,76)*. Repair appears to be at least partially processive, since multiple lesions in a single DNA molecule are repaired before repair is initiated on other molecules in the same oocyte *(76)*. This processivity could be direct in that the repair machinery scans from one lesion to the next, or indirect, if, for example, nicked or gapped intermediates in the repair of one lesion affect the chromatin structure of the whole molecule and cause linked lesions to be more accessible. Although oocytes contain both a CPD photolyase *(5)* and (probably) a (6-4) photolyase *(35)*, these enzymes do not play a major role, since repair of UV damage proceeds efficiently in the dark *(76)*.

How similar is the mechanism of repair in oocytes to the process of nucleotide excision repair (NER) studied in other systems? Svodboda and colleagues used a substrate with a single defined thymine dimer to show that the incision step is essentially identical to that in a mammalian cell extract *(89)*; the major excised products were 27–29 nt long, and the dimer was located 5 bases from the 3'-end of this fragment. In a high-speed extract of activated eggs that is capable of repairing UV damage, Shivji et al.

(83) showed that repair is dependent on proliferating cell nuclear antigen (PCNA), a processivity factor for DNA polymerase δ, as is true in mammalian cell extracts. A distinction between the roles of PCNA in DNA replication and DNA repair was achieved by showing that the damage-induced Cip1 protein interacts with PCNA to inhibit replicative, but not repair synthesis. A *Xenopus* homolog of the human repair gene XP-G (corresponding to xeroderma pigmentosum complementation group G) has been isolated *(78)*, but not extensively studied. In all these respects, NER in oocytes appears to proceed very much as in other vertebrates.

4.3. Abasic Sites

Abasic sites are intermediates in the repair of many types of lesions that are corrected by base excision repair. Matsumoto and Bogenhagen *(51,52,54)* have made a detailed study of the repair of a tetrahydrofuran residue, which is a close analog of an abasic site, introduced at a specific position in a large DNA molecule. In whole oocyte extracts, this site is incised on its 5'-side by a class II AP endonuclease, which leaves a 3'-OH and 5'-phosphate *(51)*. Removal of the lesion *per se* is performed by a 5' → 3' exonuclease activity and is dependent on the presence of DNA polymerase and dNTPs *(52)*. Excision and resynthesis are very localized, replacing no more than 3–4 bases, including the damaged site *(51)*. Covalently closed products are formed by the action of DNA ligase. Fractionation of the extract has allowed the identification of several specific proteins required for repair. In addition to the AP endonuclease and DNA ligase, one pathway seems to rely on DNA polymerase δ and PCNA, similarly to NER *(83)*. In a potentially separate pathway, DNA polymerase β can replace δ and PCNA *(53,54)*.

4.4. Other Damage

Several other types of DNA damage have also been shown to be repaired in *Xenopus*. Bulky adducts formed by acetylaminofluorene (AAF) were efficiently repaired after injection of damaged DNA into activated eggs *(60)*. In oocytes, AAF-induced repair synthesis was readily detected, but the overall efficiency of repair was quite low *(60)*. In whole oocyte extracts optimized for transcription, AAF adducts were not detectably removed *(12)*. Damage induced by X-irradiation of plasmid DNA was repaired in injected oocytes *(90)*. Some of the lesions were DSBs, and at least a proportion of these were repaired by homologous recombination, but other types of lesions were also apparently remedied.

It is difficult to prepare bacterial plasmid DNA without introducing some single-strand nicks, and several investigators have noted that incubation in oocyte nuclei rapidly reduces the proportion of nicked molecules. Local DNA synthesis is associated with this type of repair, which accounts for a low level of radioactive precursor incorporation in the first 30 min after nuclear injection. Inhibition by aphidicolin suggests the participation of DNA polymerase α and/or δ *(75)*. In high-speed egg extracts, the repair of heat-induced nicks was accompanied by repair synthesis with an average track size of 36 bases *(29)*. Remarkably, the repair in oocytes of simple nicks or UV photoproducts was accompanied by more rapid chromatin assembly than that found with unrepaired DNA in the same oocytes *(75)*. Earlier studies with egg extracts suggested that DNA synthesis, specifically the synthesis of a second strand on a single-stranded substrate, facilitates the assembly of chromatin *(1)*. This may be important for efficient restoration of normal chromatin structure after passage of a replication fork, and also after each episode of DNA repair.

In bacteria *(16)* and in yeast *(31)*, repair of psoralen crosslinks depends on both the normal NER and DSB repair machinery. Since *Xenopus* oocytes have active NER capabilities, it was expected that crosslinks might be processed similarly after injection. In initial experiments, evidence was obtained that DNA molecules carrying multiple random crosslinks are degraded in oocytes, suggesting that DSBs might be introduced *(81)*. The capacity for repairing, or even for initiating repair of crosslinks seems to be much reduced compared to that for UV-induced lesions, indicating that some unique features beyond simple NER may be involved.

5. MISMATCH REPAIR

Brooks et al. *(8)* and Varlet et al. *(97)* prepared covalently closed, double-stranded circular DNA molecules carrying a single mismatched base pair. They examined all 12 possible mismatches, and determined both the efficiency and direction of repair in high-speed (nonreplicating) extracts of activated eggs *(97)*. All of the mismatches were corrected, with yields varying from about 4 to 18%, and each mismatch was corrected in both directions with about equal efficiencies. Mismatch-dependent, repair-associated DNA synthesis was confined to a region of a few hundred bases on either side of the mismatch *(8,97)*. Thus, eggs carry the activities needed for mismatch correction. More recently, a *Xenopus* homolog of the MutS and Msh2 proteins has been isolated *(96)*.

In conjunction with studies of the mechanism of homologous recombination in oocytes *(10)*, heteroduplex DNAs having eight single-base pair mismatches in a region of 1200 bp were injected into oocytes *(43)*. Such heteroduplexes are predicted to be intermediates in the SSA mechanism. Repair was quite efficient, and markers on one strand were essentially always corepaired. This suggests a long-patch mechanism for mismatch removal and resynthesis, which is a feature of the bacterial mismatch repair system *(57)*. Either strand could be replaced and the results suggested that nicks positioned at the ends of the homologous overlap served to initiate repair, as is true of both human cell *(30,94)* and bacterial *(38)* extracts. When covalently closed heteroduplex substrates were injected, repair was still efficient, indicating that some attribute other than a strand interruption can be used in oocytes to initiate mismatch repair *(43)*. At the same time, the pattern of marker recovery was shifted so that one strand became preferred over the other as a repair template *(43)*. These data were tentatively interpreted to reflect recognition of and initiation at specific mismatches, followed by corepair of neighboring sites on the same strand.

6. PERSPECTIVES

6.1. Biochemical Mechanisms

Many types of DNA repair have now been demonstrated in extracts from *Xenopus* oocytes and/or eggs (Table 1). Because of the availability of large quantities of starting material, this opens the door to detailed biochemical characterization of the requirements for each process, to dissection of individual steps along each pathway, and to the purification of participating activities. This approach has been at least partly successful in the case of DNA replication. Evidence for a licensing

Table 1
DNA Repair Activities in *Xenopus*[a]

Lesion(s)	Oocytes	Eggs	Extracts	References
DSB				
Homologous	++	++	++	*11,32,33,40–42,48–50,68,80*
Nonhomologous	–	++	++	*2,21,40–42,44,65–67,93*
UV damage	++		++	*5,28,39,75,76,83,89*
Abasic site			++	*51–54*
Single-strand nick	++	++	++	*29,75*
AAF	+	++	–	*12,60*
X-ray damage	++			*90*
Psoralen crosslinks	+			*81*
Base mismatches	++		++	*8,43,97*

[a]Entries indicate whether repair of the particular type of lesion has been observed by injection of oocytes, by injection of eggs, or by incubation in cell-free extracts of either oocytes or eggs. Double plus (++) denotes robust activity; single plus (+) denotes weak activity; minus (–) indicates activity not detected. No entry means not tested. The absence of activity in one type of extract does not mean that all extract preparations would fail to show activity.

factor that permits the initiation of replication was first obtained in *Xenopus* egg extracts *(7)*, and candidate proteins have recently been isolated from the same source *(13,37,47)*.

As discussed in Chapters 11 and 19, this vol. and Chapter 20, vol. 2, the investigation of mismatch repair mechanisms has been greatly stimulated by the discovery that deficiencies in human mismatch repair genes are associated with a predisposition to certain types of cancer. These studies are more advanced in bacteria, yeast, and mammals than in amphibians, but there is an aspect to which *Xenopus* may be able to contribute. In bacteria, initiation of long-patch mismatch repair requires introduction of a nick by the MutH protein into the strand that will be replaced (*38,57*; *see* Chapter 11). In mammalian cell extracts, repair is similarly nick-dependent and nick-directed *(30,94)*, but no analogue to the MutH function has been identified. In both the egg extract and oocyte injection studies described in Section 5., covalently closed substrates containing mismatches were efficiently repaired. This suggests that the initiation step in mismatch repair, which also selects which strand will be replaced, may be accessible in *Xenopus*. If single mismatched sites are corrected in oocytes, this could be a particularly informative experimental system, since injected DNAs are not replicated and strand interruptions introduced in the course of replication cannot be responsible for initiation of repair.

There is a continuing interplay between studies in other organisms and those in *Xenopus* based on sequence homologies among gene products that catalyze a given reaction. As indicated earlier, *Xenopus* homologs of the XP-G protein involved in NER and the Msh2 mismatch repair protein have been identified. It seems certain that homologs of other repair proteins await identification. Because processes and mechanisms manifested in *Xenopus* oocytes are so similar to those in other organisms and cell types, it is possible to exploit the oocytes as a testing ground for procedures that may be generally applicable. For example, a model gene targeting experiment in oocytes demonstrated the utility of making a DSB in the target to increase recombination efficiency *(80)*. A potentially fruitful area in which little work has apparently been done in *Xenopus* is that of transcription-coupled repair (*82*; *see* Chapters 9 and 15, this vol. and Chapters 10 and 18, vol. 2). It should be possible to insert well-characterized lesions at arbitrary locations with respect to functional promoters in oocytes for RNA polymerases I, II, and III, and to explore subtleties of transcription-coupled repair more readily than in some other systems.

6.2. Are Recombination Processes Really Designed for Repair?

Only during meiosis is homologous recombination required to ensure proper chromosome segregation *(27)*. In somatic or vegetative cells, both prokaryotic and eukaryotic, the function of homologous recombination processes may be primarily for the repair of DSBs. The RecA protein of bacteria potentiates repair by sensing damage and activating a regulon of repair genes, in addition to catalyzing interactions between homologous DNAs that may lead directly to repair (*19*; *see* Chapter 8). Many of the yeast genes required for recombination in mitotic cells were identified originally as genes involved in repair of radiation damage (*19*; *see* Chapter 16). As discussed in Section 3.1., the SSA model of nonconservative homologous recombination, although it can effectively repair DSBs in tandemly repeated sequences *(61)*, has drawbacks as a general repair mechanism, since it is necessarily accompanied by deletion. It should be possible to address the issue of the natural function of SSA by identifying proteins in *Xenopus* that are required for SSA recombination. Although traditional genetic analysis is not currently feasible in frogs, alternative approaches have proven informative *(98)*. Levels of particular gene products can be manipulated by overexpression from introduced RNAs or DNAs, by depletion of mRNAs with antisense oligonucleotides, or by direct inhibition with specific antibodies. Such tools could be used to evaluate the consequences of increasing or decreasing the levels of catalysts of the SSA pathway at specific times in embryonic development. The same considerations apply to nonhomologous end joining. Only one gene that has a pronounced effect on end joining has been identified in yeast *(79)*. If catalysts of this process can be isolated from eggs or egg extracts, they can be used to manipulate activity levels.

It will be important to determine whether there are any enzymes that function uniquely in the homologous or nonhomologous recombination processes. The alternative is that recombination is a side reaction of activities designed for some other purpose. For example, the SSA mechanism can, in principle, be executed by proteins known to be required for DNA replication: 5' $\rightarrow$ 3' exonuclease, DNA polymerase, and DNA ligase *(99)*. If SSA is really a separate, selectively advantageous biological process, it may have some distinguishing features. For example, cells may have activities that catalyze annealing of complementary strands, a unique exonuclease, or a unique class of ligase.

6.3. Developmental Studies

Whether or not it is possible to inhibit individual enzymatic activities effectively in *Xenopus* cells or embryos, the well characterized and accessible stages of amphibian development *(20,34)* make it feasible to inquire when and where particular activities appear during ontogeny. This approach has proved helpful in the cases of homologous and nonhomologous recombination; the timing of their appearance during oocyte *(42)* and egg *(21)* development, respectively, has guided inferences regarding their biological roles (*see* Sections 3.2. and 3.3.).

It seems probable that all modes of DNA repair will be active during the cleavage stages of early embryogenesis, when replication and transmission of the genome are major cellular occupations. One would suppose that accuracy of transmission would be at least as important at this stage as it is earlier or later. On the other hand, cleavage embryos face life without an effective heat-shock response *(3)*, and it is possible that they have jettisoned some other protective functions as well. For example, the checkpoint controls that monitor DNA damage and prevent cell-cycle progression while damage persists may not be operative in such abbreviated cell cycles *(26,58)*. Furthermore, because RNA synthesis is suppressed in cleavage embryos *(59)*, coupling DNA repair to transcription *(82)* is not feasible. Given that some regulatory aspects of normal somatic DNA repair may be missing, it may be even more crucial for *Xenopus* embryos to have high levels of basic repair activities primed to deal with a variety of lesions as they make a figurative run for their lives between fertilization and gastrulation. In addition, it will be interesting to learn how the coordination of repair functions is lost during oocyte or egg development, and then restored later in embryogenesis.

7. CONCLUDING REMARKS

Xenopus oocytes and eggs have shown the capacity to repair essentially all types of DNA damage, from DSBs to UV-induced lesions to base mismatches. To the extent that comparisons have been made, *Xenopus* has repair capabilities and mechanisms that are typical of vertebrate somatic cells. When they have been sought, *Xenopus* homologs of repair proteins identified in other organisms have routinely been found, further underscoring the universality of repair processes. The unusually large size and capacity of the oocytes make possible manipulations of substrates and characterization of repair intermediates that are more difficult in other systems. Extensive use of cell-free extracts of oocytes and eggs has opened the door to biochemical studies of repair mechanisms. Because embryonic stages are well characterized and accessible, detailed investigations of the regulation of DNA repair activities during development are possible. For these reasons—universality, high-capacity, and the availability of in vitro and developmental systems—further studies of DNA repair using *Xenopus* promise to be richly rewarded.

ACKNOWLEDGMENTS

I thank Petra Pfeiffer, Alain Nicolas, and members of my laboratory (past and present) for continuing discussions of *Xenopus* recombination and repair. I am grateful to Petra Pfeiffer, John Wilson, Ray Merrihew, John Hays, and Dan Bogenhagen for their responses to my request for recent information. The manuscript was improved by suggestions from Geneviève Pont-Kingdon, Shawn Christensen, and Dave Segal. The research in my laboratory mentioned in this chapter was supported by grants from the National Institutes of Health and the National Science Foundation.

REFERENCES

1. Almouzni, G. and M. Méchali. 1988. Assembly of spaced chromatin promoted by DNA synthesis in extracts form *Xenopus* eggs. *EMBO J.* **7:** 665–672.
2. Beyert, N., S. Reichenberger, M. Peters, M. Hartung, B. Göttlich, W. Goedecke, W. Vielmetter, and P. Pfeiffer. 1994. Nonhomologous DNA end joining of synthetic hairpin substrates in *Xenopus laevis* egg extracts. *Nucleic Acids Res.* **22:** 1643–1650.
3. Bienz, M. 1984. Developmental control of the heat shock response in *Xenopus*. *Proc. Natl. Acad. Sci. USA* **81:** 3138–3142.
4. Birkenmeier, E. H., D. D. Brown, and E. Jordan. 1978. A nuclear extract of *Xenopus laevis* oocytes that accurately transcribes 5S RNA genes. *Cell* **15:** 1077–1086.
5. Blaustein, A. R., P. D. Hoffman, D. G. Hokit, J. M. Kiesecker, S. C. Walls, and J. B. Hays. 1994. UV repair and resistance to solar UV-B in amphibian eggs: a link to population declines? *Proc. Natl. Acad. Sci. USA* **91:** 1791–1795.
6. Blaustein, A. R. and D. B. Wake. 1995. The puzzle of declining amphibian populations. *Sci. Am.* **272:** 52–57.
7. Blow, J. J. and R. A. Laskey. 1988. A role for the nuclear envelope in controlling DNA replication within the cell cycle. *Nature* **332:** 546–548.
8. Brooks, P., C. Dohet, G. Almouzni, M. Méchali, and M. Radman. 1989. Mismatch repair involving localized DNA synthesis in extracts of *Xenopus* eggs. *Proc. Natl. Acad. Sci. USA* **86:** 4425–4429.
9. Carroll, D. and C. W. Lehman. 1991. DNA recombination and repair in oocytes, eggs, and extracts, in *Methods in Cell Biology,* Xenopus laevis: *Practical Uses in Cell and Molecular Biology*, vol. 36 (Kay, B. K. and H. B. Peng, eds.), Academic, San Diego, pp. 467–486.
10. Carroll, D., C. W. Lehman, S. Jeong-Yu, P. Dohrmann, R. J. Dawson, and J. K. Trautman. 1994. Distribution of exchanges upon homologous recombination of exogenous DNA in *Xenopus laevis* oocytes. *Genetics* **138:** 445–457.
11. Carroll, D., S. H. Wright, R. K. Wolff, E. Grzesiuk, and E. B. Maryon. 1986. Efficient homologous recombination of linear DNA substrates after injection into *Xenopus laevis* oocytes. *Mol. Cell. Biol.* **6:** 2053–2061.
12. Chen, Y.-H., Y. Matsumoto, S. Shibutani, and D. F. Bogenhagen. 1991. Acetylaminofluorene and aminofluorene adducts inhibit *in vitro* transcription of a *Xenopus* 5S RNA gene only when located on the coding strand. *Proc. Natl. Acad. Sci. USA* **88:** 9583–9587.
13. Chong, J. P. J., H. M. Mahbubani, C.-Y. Khoo, and J. J. Blow. 1995. Purification of an MCM-containing complex as a component of the DNA replication licensing system. *Nature* **375:** 418–421.
14. Clark, J. M. 1988. Novel non-templated nucleotide addition reactions catalyzed by procaryotic and eucaryotic DNA polymerases. *Nucleic Acids Res.* **16:** 9677–9686.
15. Clark, J. M. 1991. DNA synthesis on discontinuous templates by DNA polymerase I of *Escherichia coli*. *Gene* **104:** 75–80.
16. Cole, R. S. 1973. Repair of DNA containing interstrand crosslinks in *Escherichia coli*: sequential excision and recombination. *Proc. Natl. Acad. Sci. USA* **70:** 1064–1068.
17. Cortese, R., R. Harland, and D. Melton. 1980. Transcription of tRNA genes *in vivo*. *Proc. Natl. Acad. Sci. USA* **77:** 4147–4151.
18. Forbes, D. J., M. W. Kirschner, and J. W. Newport. 1983. Spontaneous formation of nucleus-like structures around bacteriophage DNA microinjected into *Xenopus* eggs. *Cell* **34:** 13–23.
19. Friedberg, E. C., G. C. Walker, and W. Siede. 1995. *DNA Repair and Mutagenesis.* ASM, Washington, DC.
20. Gerhart, J. C. 1980. Mechanisms regulating pattern formation in amphibian egg and early embryo, in *Biological Regulation and Development*, vol. 2, *Molecular Organization and Cell Function* (Goldberger, R. F., ed.), Plenum, New York, pp. 133–316.

21. Goedecke, W., W. Vielmetter, and P. Pfeiffer. 1992. Activation of a system for the joining of nonhomologous DNA ends during *Xenopus* egg maturation. *Mol. Cell. Biol.* **12:** 811–816.
22. Greider, C. W. 1991. Chromosome first aid. *Cell* **67:** 645–647.
23. Gurdon, J. B. and D. A. Melton. 1981. Gene transfer in amphibian eggs and oocytes. *Ann. Rev. Genet.* **15:** 189–218.
24. Gurdon, J. B. and M. P. Wickens. 1983. The use of *Xenopus* oocytes for the expression of cloned genes. *Methods Enzymol.* **101:** 370–386.
25. Harland, R. M. and R. A. Laskey. 1980. Regulated replication of DNA microinjected into eggs of *X. laevis*. *Cell* **21:** 761–771.
26. Hartwell, L. H. and T. A. Weinert. 1989. Checkpoints: Controls that ensure the order of cell cycle events. *Science* **246:** 629–634.
27. Hawley, R. S. 1988. Exchange and chromosomal segregation in eucaryotes, in *Genetic Recombination* (R. Kucherlapati and G. R. Smith, eds.), American Society for Microbiology, Washington, DC, pp. 497–527.
28. Hays, J. B., E. J. Ackerman, and Q. Pang. 1990. Rapid and apparently error-prone excision repair of nonreplicating UV-irradiated plasmids in *Xenopus laevis* oocytes. *Mol. Cell. Biol.* **10:** 3505–3511.
29. Höfferer, L., K. H. Winterhalter, and F. R. Althaus. 1995. *Xenopus* egg lysates repair heat-generated DNA nicks with an average patch size of 36 nucleotides. *Nucleic Acids Res.* **23:** 1396,1397.
30. Holmes, J. J., S. Clark, and P. Modrich. 1990. Strand-specific mismatch correction in nuclear extracts of human and *Drosophila melanogaster* cell lines. *Proc. Natl. Acad. Sci. USA* **87:** 5837–5841.
31. Jachymczyk, W. J., R. C. von Borstel, M. R. A. Mowat, and P. J. Hastings. 1981. Repair of interstrand crosslinks in DNA of *Saccharomyces cerevisiae* requires two systems for DNA repair: the RAD3 system and the RAD51 system. *Mol. Gen. Genet.* **182:** 196–205.
32. Jeong-Yu, S. and D. Carroll. 1992. Effect of terminal nonhomologies on homologous recombination in *Xenopus laevis* oocytes. *Mol. Cell. Biol.* **12:** 5426–5437.
33. Jeong-Yu, S. and D. Carroll. 1992. Test of the double-strand-break repair model of recombination in *Xenopus laevis* oocytes. *Mol. Cell. Biol.* **12:** 112–119.
34. Kay, B. K. and H. B. Peng, eds. 1991. Xenopus laevis: *Practical Uses in Cell and Molecular Biology, Methods in Cell Biology*, vol. 36, Academic, San Diego.
35. Kim, S.-T., K. Malhotra, J.-S. Taylor, and A. Sancar. 1996. Purification and partial characterization of (6-4) photoproduct DNA photolyase from *Xenopus laevis*. *Photochem. Photobiol.* **63:** 292–295.
36. King, J. S., C. F. Fairley, and W. F. Morgan. 1994. Bridging the gap: the joining of nonhomologous ends by DNA polymerases. *J. Biol. Chem.* **269:** 13,061–13,064.
37. Kubota, Y., S. Mimura, S. Nishimoto, H. Takisawa, and H. Nojima. 1995. Identificaiton of the yeast MCM3-related protein as a component of *Xenopus* DNA replication licensing factor. *Cell* **81:** 601–609.
38. Lahue, R. S., K. G. Au, and P. Modrich. 1989. DNA mismatch correction in a defined system. *Science* **245:** 160–164.
39. Legerski, R. J., J. E. Penkala, C. A. Peterson, and D. A. Wright. 1987. Repair of UV-induced lesions in *Xenopus laevis* oocytes. *Mol. Cell. Biol.* **7:** 4317–4323.
40. Lehman, C. W. and D. Carroll. 1991. Homologous recombination catalyzed by a nuclear extract from *Xenopus* oocytes. *Proc. Natl. Acad. Sci. USA* **88:** 10,840–10,844.
41. Lehman, C. W. and D. Carroll. 1993. Isolation of large quantities of functional, cytoplasm-free *Xenopus laevis* oocyte nuclei. *Anal. Biochem.* **211:** 311–319.
42. Lehman, C. W., M. Clemens, D. Worthylake, J. K. Trautman, and D. Carroll. 1993. Homologous and illegitimate recombination pathways in developing *Xenopus* oocytes and eggs. *Mol. Cell. Biol.* **13:** 6897–6906.

43. Lehman, C. W., S. Jeong-Yu, J. K. Trautman, and D. Carroll. 1994. Repair of heteroduplex DNA in *Xenopus laevis* oocytes. *Genetics* **138:** 459–470.
44. Lehman, C. W., J. K. Trautman, and D. Carroll. 1994. Illegitimate recombination in *Xenopus*: characterization of end-joined junctions. *Nucleic Acids Res.* **22:** 434–442.
45. Leno, G. H. and R. A. Laskey. 1991. DNA replication in cell-free extracts from *Xenopus laevis*, in *Methods in Cell Biology,* Xenopus laevis: *Practical Uses in Cell and Molecular Biology*, vol. 36 (Kay, B. K. and H. B. Peng, eds.), Academic, San Diego, pp. 561–579.
46. Lin, F.-L., K. Sperle, and N. Sternberg. 1984. Model for homologous recombination during transfer of DNA into mouse L cells: role for the ends in the recombination process. *Mol. Cell. Biol.* **4:** 1020–1034.
47. Madine, M. A., C.-Y. Khoo, A. D. Mills, and R. A. Laskey. 1995. MCM3 complex required for cell cycle regulation of DNA replication in vertebrate cells. *Nature* **375:** 421–424.
48. Maryon, E. and D. Carroll. 1989. Degradation of linear DNA by a strand-specific exonuclease activity in *Xenopus laevis* oocytes. *Mol. Cell. Biol.* **9:** 4862–4871.
49. Maryon, E. and D. Carroll. 1991. Characterization of recombination intermediates from DNA injected into *Xenopus laevis* oocytes: evidence for a nonconservative mechanism of homologous recombination. *Mol. Cell. Biol.* **11:** 3278–3287.
50. Maryon, E. and D. Carroll. 1991. Involvement of single-stranded tails in homologous recombination of DNA injected into *Xenopus laevis* oocyte nuclei. *Mol. Cell. Biol.* **11:** 3268–3277.
51. Matsumoto, Y. and D. F. Bogenhagen. 1989. Repair of a synthetic abasic site in DNA in a *Xenopus laevis* oocyte extract. *Mol. Cell. Biol.* **9:** 3750–3757.
52. Matsumoto, Y. and D. F. Bogenhagen. 1991. Repair of a synthetic abasic site involves concerted reactions of DNA synthesis followed by excision and ligation. *Mol. Cell. Biol.* **11:** 4441–4447.
53. Matsumoto, Y. and K. Kim. 1995. Excision of deoxyribose phosphate residues by DNA polymerase β during DNA repair. *Science* **269:** 699–702.
54. Matsumoto, Y., K. Kim, and D. F. Bogenhagen. 1994. Proliferating cell nuclear antigen-dependent abasic site repair in *Xenopus laevis* oocytes: an alternative pathway of base excision DNA repair. *Mol. Cell. Biol.* **14:** 6187–6197.
55. Méchali, M. and S. Kearsey. 1984. Lack of specific sequence requirements for DNA replication in *Xenopus* eggs compared with high sequence specificity in yeast. *Cell* **38:** 55–64.
56. Merrihew, R. V., K. Marburger, S. L. Pennington, D. B. Roth, and J. H. Wilson. 1996. High-frequency illegitimate integration of transfected DNA at preintegrated target sites in a mammalian genome. *Mol. Cell. Biol.* **16:** 10–18.
57. Modrich, P. 1991. Mechanisms and biological effects of mismatch repair. *Annu. Rev. Gen.* **25:** 229–253.
58. Murray, A. W. 1992. Creative blocks: cell-cycle checkpoints and feedback controls. *Nature* **359:** 599–604.
59. Newport, J. and M. Kirschner. 1982. A major developmental transition in early *Xenopus* embryos: I. Characterization and timing of cellular changes at the midblastula stage. *Cell* **30:** 675–686.
60. Orfanoudakis, G., G. Gilson, C. M. Wolff, J. P. Ebel, N. Befort, and P. Remy. 1990. Repair of acetyl-aminofluorene modified pBR322 DNA in *Xenopus laevis* oocytes and eggs; effect of diadenosine tetraphosphate. *Biochimie* **72:** 271–278.
61. Ozenberger, B. A. and G. S. Roeder. 1991. A unique pathway of double-strand break repair operates in tandemly repeated genes. *Mol. Cell. Biol.* **11:** 1222–1231.
62. Paine, P. L., M. E. Johnson, Y.-T. Lau, L. J. M. Tluczek and D. S. Miller. 1992. The oocyte nucleus isolated in oil retains *in vivo* structure and functions. *BioTechniques* **13:** 238–246.
63. Petes, T. D., R. E. Malone, and L. S. Symington. 1991. Recombination in yeast, in *The Molecular and Cellular Biology of the Yeast Saccharomyces: Genome Dynamics, Protein*

Synthesis and Energetics (Broach, J. R., Pringle, C. R. and E. W. Jones, eds.), Cold Spring Harbor Laboratory, Cold Spring Harbor, NY, pp. 407–521.

64. Pfeiffer, P., S. Thode, J. Hancke, P. Keohavong, and W. G. Thilly. 1994. Resolution and conservation of mismatches in DNA end joining. *Mutagenesis* **9:** 527–535.
65. Pfeiffer, P., S. Thode, J. Hancke, and W. Vielmetter. 1994. Mechanisms of overlap formation in nonhomologous DNA end joining. *Mol. Cell. Biol.* **14:** 888–895.
66. Pfeiffer, P. and W. Vielmetter. 1988. Joining of nonhomologous DNA double strand breaks *in vitro. Nucleic Acids Res.* **16:** 907–924.
67. Plessis, A., A. Perrin, J. E. Haber, and B. Dujon. 1992. Site-specific recombination determined by I-*Sce*I, a mitochondrial group I intron-encoded endonuclease expressed in the yeast nucleus. *Genetics* **130:** 451–460.
68. Pont-Kingdon, G., R. J. Dawson, and D. Carroll. 1993. Intermediates in extrachromosomal recombination in *Xenopus laevis* oocytes: characterization by electron microscopy. *EMBO J.* **12:** 23–34.
69. Puchta, H., S. Kocher, and B. Hohn. 1992. Extrachromosomal homologous DNA recombination in plant cells is fast and is not affected by CpG methylation. *Mol. Cell. Biol.* **12:** 3372–3379.
70. Roth, D. and J. Wilson. 1988. Illegitimate recombination in mammalian cells, in *Genetic Recombination* (Kucherlapati, R. and G. R. Smith, eds.), American Society for Microbiology, Washington, DC, pp. 621–653.
71. Roth, D. B., T. N. Porter, and J. H. Wilson. 1985. Mechanisms of nonhomologous recombination in mammalian cells. *Mol. Cell. Biol.* **5:** 2599–2607.
72. Roth, D. B. and J. H. Wilson. 1985. Relative rates of homologous and nonhomologous recombination in transfected DNA. *Proc. Natl. Acad. Sci. USA* **82:** 3355–3359.
73. Ruberti, I., E. Beccari, E. Bianchi, and F. Carnevali. 1989. Large scale isolation of nuclei from oocytes of *Xenopus laevis. Anal. Biochem.* **180:** 177–180.
74. Rudin, N., E. Sugarman, and J. E. Haber. 1989. Genetic and physical analysis of double-strand break repair and recombination in *Saccharomyces cerevisiae. Genetics* **122:** 519–534.
75. Ryoji, M., E. Tominna, and W. Yasui. 1989. Minichromosome assembly accompanying repair-type DNA synthesis in *Xenopus* oocytes. *Nucleic Acids Res.* **17:** 10,243–10,258.
76. Saxena, J. K., J. B. Hays, and E. J. Ackerman. 1990. Excision repair of UV-damaged plasmid DNA in *Xenopus* oocytes is mediated by DNA polymerase α (and/or δ). *Nucleic Acids Res.* **18:** 7425–7432.
77. Scalenghe, F., M. Buscaglia, C. Steinheil, and M. Crippa. 1978. Large scale isolation of nuclei and nucleoli from vitellogenic oocytes of *Xenopus laevis. Chromosoma (Berl.)* **66:** 299–308.
78. Scherly, D., T. Nouspikel, J. Corlet, C. Ucla, A. Bairoch, and S. G. Clarkson. 1993. Complementation of the DNA repair defect in xeroderma pigmentosum group G cells by a human cDNA related to yeast RAD2. *Nature* **363:** 182–185.
79. Schiestl, R. H., J. Zhu, and T. D. Petes. 1994. Effect of mutations in genes affecting homologous recombination on restriction enzyme-mediated and illegitimate recombination in *Saccharomyces cerevisiae. Mol. Cell. Biol.* **14:** 4493–4500.
80. Segal, D. J. and D. Carroll. 1995. Endonuclease-induced, targeted homologous extrachromosomal recombination in *Xenopus* oocytes. *Proc. Natl. Acad. Sci. USA* **92:** 806–810.
81. Segal, D. J. and D. Carroll. In preparation.
82. Selby, C. P. and A. Sancar. 1994. Mechanisms of transcription-repair coupling and mutation frequency decline. *Microbiol. Rev.* **58:** 317–329.
83. Shivji, M. K. K., S. J. Grey, U. P. Strausfeld, R. D. Wood, and J. J. Blow. 1994. Cip1 inhibits DNA replication but not PCNA-dependent nucleotide excision-repair. *Curr. Biol.* **4:** 1062–1068.
84. Silberstein, Z., M. Shalit, and A. Cohen. 1993. Heteroduplex strand-specificity in restriction-stimulated recombination by the RecE pathway of *Escherichia coli. Genetics* **133:** 439–448.

85. Smith, L. D., W. Xu, and R. L. Varnold. 1991. Oogenesis and oocyte isolation, in *Methods in Cell Biology,* Xenopus laevis: *Practical Uses in Cell and Molecular Biology*, vol. 36 (Kay, B. K. and H. B. Peng, eds.), Academic, San Diego, pp. 45–60.
86. Subramani, S. and B. L. Seaton. 1988. Homologous recombination in mitotically dividing mammalian cells, in *Genetic recombination* (Kucherlapati, R. and G. R. Smith, eds.), American Society for Microbiology, Washington, DC, pp. 549–574.
87. Sugawara, N. and J. E. Haber. 1992. Characterization of double-strand break-induced recombination: homology requirements and single-stranded DNA formation. *Mol. Cell. Biol.* **12:** 563–575.
88. Sun, H., D. Treco, and J. W. Szostak. 1991. Extensive 3'-overhanging, single-stranded DNA associated with the meiosis-specific double-strand breaks at the ARG4 recombination initiation site. *Cell* **64:** 1155–1161.
89. Svodboda, D. L., J.-S. Taylor, J. E. Hearst, and A. Sancar. 1993. DNA repair by eukaryotic nucleotide excision nuclease. Removal of thymine dimer and psoralen monoadduct by HeLa cell-free extract and of thymine dimer by *Xenopus laevis* oocytes. *J. Biol. Chem.* **268:** 1931–1936.
90. Sweigert, S. E. and D. Carroll. 1990. Repair and recombination of X-irradiated plasmids in *Xenopus laevis* oocytes. *Mol. Cell. Biol.* **10:** 5849–5856.
91. Szostak, J. W., T. L. Orr-Weaver, R. J. Rothstein, and F. W. Stahl. 1983. The double-strand break repair model for recombination. *Cell* **33:** 25–35.
92. Takahashi, N. K., K. Yamamoto, Y. Kitamura, S.-Q. Luo, H. Yoshikura, and I. Kobayashi. 1992. Nonconservative recombination in *Escherichia coli. Proc. Natl. Acad. Sci. USA* **89:** 5912–5916.
93. Thode, S., A. Schafer, P. Pfeiffer and W. Vielmetter. 1990. A novel pathway of DNA end-to-end joining. *Cell* **60:** 921–928.
94. Thomas, D. C., J. D. Roberts, and T. A. Kunkel. 1991. Heteroduplex repair in extracts of human HeLa cells. *J. Biol. Chem.* **266:** 3744–3751.
95. Troelstra, C. and N. G. J. Jaspers. 1994. Ku starts at the end. *Curr. Biol.* **4:** 1149–1151.
96. Varlet, I., C. Pallard, M. Radman, J. Moreau, and N. de Wind. 1994. Cloning and expression of the *Xenopus* and mouse *Msh2* DNA mismatch repair genes. *Nucleic Acids Res.* **22:** 5723–5728.
97. Varlet, I., M. Radman, and P. Brooks. 1990. DNA mismatch repair in *Xenopus* egg extracts: repair efficiency and DNA repair synthesis for all single base-pair mismatches. *Proc. Natl. Acad. Sci. USA* **87:** 7883–7887.
98. Vize, P. D., D. A. Melton, A. Hemmati-Brivanlou, and R. M. Harland. 1991. Assays for gene function in developing *Xenopus* embryos, in *Methods in Cell Biology,* Xenopus laevis: *Practical Uses in Cell and Molecular Biology*, vol. 36 (Kay, B. K. and H. B. Peng, eds.), Academic, San Diego, pp. 367–387.
99. Waga, S., G. Bauer, and B. Stillman. 1994. Reconstitution of complete SV40 DNA replication with purified replication factors. *J. Biol. Chem.* **269:** 10,923–10,934.
100. Wake, C. T., T. Gudewicz, T. Porter, A. White, and J. H. Wilson. 1984. How damaged is the biologically active subpopulation of transfected DNA? *Mol. Cell. Biol.* **4:** 387–398.
101. Wangh, L. J. 1989. Injection of *Xenopus* eggs before activation, achieved by control of extracellular factors, improves plasmid DNA replication after activation. *J. Cell Sci.* **93:** 1–8.
102. Weaver, D. T. 1995. What to do at an end: DNA double-strand-break repair. *Trends Genet.* **11:** 388–392.
103. Wyllie, A. H., R. A. Laskey, J. Finch, and J. Gurdon. 1978. Selective DNA conservation and chromatin assembly after injection of SV40 DNA into *Xenopus* oocytes. *Dev. Biol.* **64:** 178–188.

Index

Genes and gene products are listed under each species; certain genes are also listed under general topic listings. Typically, genes and gene products are listed by gene names. All plant species except *Arabidopsis* are only listed under "Plants."